Essential Readings in

Wildlife Management & Conservation

Published in Affiliation with The Wildlife Society

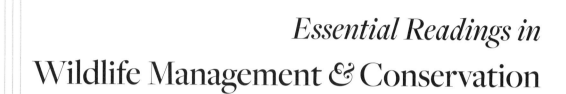

Essential Readings in
Wildlife Management & Conservation

Edited by

Paul R. Krausman *and* Bruce D. Leopold

JOHNS HOPKINS UNIVERSITY PRESS
Baltimore

© 2013 Johns Hopkins University Press
All rights reserved. Published 2013
Printed in the United States of America on acid-free paper

Johns Hopkins Paperback Edition, 2018
9 8 7 6 5 4 3 2 1

Johns Hopkins University Press
2715 North Charles Street
Baltimore, Maryland 21218-4363
www.press.jhu.edu

The Library of Congress has cataloged the hardcover edition of this book as follows:
Essential readings in wildlife management and conservation / edited by Paul R. Krausman and
Bruce D. Leopold.
    p.   cm.
Includes bibliographical references and index.
ISBN 978-1-4214-0818-7 (hdbk. : alk. paper) — ISBN 1-4214-0818-X (hdbk. : alk. paper)
1. Wildlife management. 2. Wildlife conservation. I. Krausman, Paul R., 1946– II. Leopold,
Bruce D. (Bruce David)
SK355.E78 2013
639.9—dc23   2012023929

A catalog record for this book is available from the British Library.

ISBN-13: 978-1-4214-2708-9
ISBN-10: 1-4214-2708-7

Frontispiece: Aldo Leopold writing at the Shack. Dog "Flick" is in the foreground (c. 1940).

Closing photo: Aldo Leopold in his University of Wisconsin office reading. Dog "Gus" is on the floor (c. 1943).

Special discounts are available for bulk purchases of this book. For more information, please contact Special Sales at 410-516-6936 or specialsales@press.jhu.edu.

Johns Hopkins University Press uses environmentally friendly book materials, including recycled text paper that is composed of at least 30 percent post-consumer waste, whenever possible.

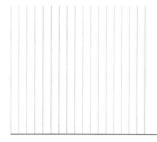

Contents

Habitat

Human Dimensions

Introduction

 It has been over 30 years since The Wildlife Society published *Readings in Wildlife Conservation* (Bailey et al. 1974). That book has been out of print for some time, but it was a staple in undergraduate introductory wildlife management courses during the 15 or so years it was in print. Both of us were required to buy a copy, and the books sit prominently on our shelves.

Every profession has a core literature that is the foundation of the principles on which that vocation depends, and Bailey et al. was part of that core. Wildlife conservation and management is no exception, especially when you consider that the primary value of The Wildlife Society is science-based management. *Wildlife* is any free-ranging vertebrate or invertebrate organism, and *management* is any active or passive action that addresses an animal's reproduction or survival. We define *conservation* as maintaining sustainable natural resources for the well-being of the earth and its peoples. Wildlife management is still a young profession, but our science has come a long way, as have our wildlife population and habitat management techniques and policies.

The trend of summarizing the literature (and data) upon which our management is based actually started with Leopold's (1933) *Game Management*. In this seminal text, Leopold gathered an impressive array of data sets from which he derived his management principles. The field of wildlife management and conservation had come far since Leopold first published *Game Management*, and 40 years later it was appropriate for Bailey et al. to assemble the current literature. Now, almost 40 years after Bailey et al., it's time once again to assemble the key literature that has formed the principles of wildlife management. Our text differs from Bailey et al., a compendium of relevant literature of the time, in that ours is a collection of the classic and fundamental papers that form the core of the science of wildlife management.

Both of us have taught introductory wildlife management classes and advanced undergraduate and graduate courses in the theory and practice of wildlife management. When comparing notes and speaking with other faculty, we found that all were consistently citing a suite of key literature, whose concepts remain relevant today, on how we manage wildlife populations and wildlife habitats, implement policy and administration, and deal with the utilitarian and non-utilitarian user. We also observed that our top students were eager to have access to the core literature and to analyze it in formal and informal discussion groups. Every wildlife management curriculum has some form of undergraduate and graduate seminar course where key concepts and principles are presented and discussed. Thus, a need existed to assemble this literature to facilitate dialogue among professionals.

While we were developing this book, some colleagues expressed concern about why a published collection was necessary, given the level of access students and professionals have to the literature by electronic means. There are many reasons that a printed collection is needed. First, university libraries (at the instigation of publishers) are growing more diligent about stopping unauthorized copying of journal articles for classroom use. With a collection of articles for which permission to reprint has been secured, faculty need not be concerned about violating copyright by providing students with copies of this literature. Second, students, even in this digital age, prefer the written medium for their textbooks. As recently as 12 November 2010, the *Chronicle of Higher Education* reported on a survey by the National Association of College Stores, in which 76% of students surveyed ($n = 627$) said they prefer a printed copy of a book over an e-copy. Third, there are many smaller universities with outstanding wildlife programs that, because of limited budgets, may not have access to all of the key journals (physically or electronically) in our field. Last, a printed collection of core literature is an invaluable resource that is quickly accessed and may be the only means of obtaining it (e.g., when in a remote field location with limited or costly internet access).

This book addresses the needs of wildlife academicians, professionals, students, high school science teachers, and even interested laypersons. We assembled foundational literature to form the basis for a semester-long undergraduate or graduate seminar. For undergraduates, it is an introduction to the literature of the profession; for graduate students, it is the basis for discussing the past and current relevance of the theories developed by the authors and how current research has evolved from these papers. The book may be used throughout an undergraduate curriculum. For example, the section on philosophical roots could be used in an introductory freshman class, while the sections on animals, ecology, and populations and on habitat could be used in an upper-level wildlife management course. We believe that this book also belongs in the library of every wildlife professional.

In addition, this collection of papers would be a useful resource for high school science teachers, who may find the information applicable to their lesson plans. It is consistent with the standards of environmental education established by the National Council for the Accreditation of Teacher Education. The paper on the North American Model of Wildlife Conservation as well as papers by Leopold would be particularly useful for high school classes. Finally, the general public has a greater vested interest in our natural resources than ever before. These papers would be of interest to them and, we hope, to their children, as the essays would expand their understanding of our natural resources and how (and why) they are managed.

The most challenging task in preparing this collection was selecting the 42 articles from the 120 or so we believed were classics in the field. This was an even more daunting task when we considered that since Leopold's 1933 publication, our profession has expanded to include human dimensions, advanced harvest management, and policy and administration. We ultimately decided to collect papers within four major areas: Our Philosophical Roots; Animals, Ecology, and Populations; Habitat; and Human Dimensions. We chose papers that were at the core of the topic at hand and are widely cited. For example, at the time of this writing, Hardin's "The tragedy of the commons" has been cited 15,562 times; Holling's "The components of predation as revealed by a study of small-mammal predation of the European pine sawfly," 1,408 times; and Johnson's "The comparison of usage and availability measurements for evaluating resource preference," 1,697 times.

All papers in this book were obtained from our private collections or libraries and are presented in their original format. Unfortunately, we could not locate pristine copies of all the papers included in this text. Thus, some show more wear than others.

In the first section, the papers address the social and professional basis for the wildlife profession. What makes us a profession? What are the fundamental values of a wildlife biologist? We go from early writings about the scientific method to an essay that remains even more relevant today than when it was first published, "The tragedy of the commons," to the significance and worldwide uniqueness of the North American Model of Wildlife Conservation. The second and third sections deal with animals and their habitats, respectively. The collection also includes a few papers that stirred some controversy when first published. We hope that faculty, and their students, use these papers as launching points to explore current literature on wildlife populations and their habitats and to discover what new theories are being constructed.

Also in the collection are papers on succession, population growth and regulation, interspecific competition, and predation. The processes explored are essential to the dynamics of any animal or plant population. Some of the essays address terms that are often used incorrectly, such as *carrying capacity*, *biodiversity*, and *optimal foraging*. We also included a section on human dimensions, which is likely the most important part of the text. Human intervention, positive and negative, has had and will continue to have a profound impact on the viability of our wildlife populations and their habitats.

The profession of wildlife management and conservation has, for quite some time, been considered an art. There is an art to what we do as wildlife managers, but the profession has evolved as science based. Unlike Leopold, who had to base most of his analyses on state and federal game surveys and on anecdotal data sets, we have amassed a profound collection of science-based management procedures and processes. For each paper selected, we list articles for further reading. We hope that the text will afford both young and seasoned professionals an opportunity to read and appreciate the essential literature in the art and science of wildlife management and conservation.

THIS PROJECT WAS MADE POSSIBLE BY THE SUPPORT of the University of Montana, Mississippi State University, the Boone and Crockett Program in Wildlife Conservation, and the Wildlife Society. Anne-Marie Durey, Laura Andrews, and Melanie Bucci assisted with verification of citations. Laura Andrews procured permissions to publish the papers and Melanie Bucci prepared the index. Three anonymous referees reviewed earlier drafts of this work. We also thank Deborah Stein, Jennifer Malat, Deborah Bors, and Vince Burke of the Johns Hopkins University Press for editorial and technical assistance. Lastly, we thank the many wildlife biologists who, over the past 75 years, shared their landmark research with the scientific community by publishing their work. It is through their science that the profession of wildlife ecology and management is where it is today.

RELATED READING

Bailey, J. A., W. Elder, and T. D. McKinney, editors. 1974. Readings in wildlife conservation. The Wildlife Society, Washington, D.C., USA.

Leopold, A. 1933. Game management. Charles Scribner's Sons, New York, New York, USA.

Real, L. A., and J. H. Brown, editors. 1991. Foundations of ecology: classic papers with commentaries. University of Chicago Press, Chicago, Illinois, USA.

Our Philosophical Roots

Leopold, A. 1925. Wilderness as a form of land use. *The Journal of Land and Public Utility Economics* 1:398–404.

Maintaining wilderness areas has become a goal of many conservation organizations, and controversy exists concerning use of those areas. As early as 1925, Leopold made a strong case for the wilderness as a part of American history that would rapidly disappear unless society recognized its importance. He was one of the first to see that conquering wilderness would result in elimination of wilderness. At the time, the word *wilderness* was not even generally understood, and Leopold begins his essay with a definition: "a wild, roadless area where those who are so inclined may enjoy primitive modes of travel and subsistence, such as exploration trips by pack-train or canoe."

Wilderness is a resource that, in part, is renewable. Its values vary by location: it can be large or small and requires some level of protection. Leopold's objective in the essay was to develop a case for wilderness as part of American culture, unique to much of the world and an important aspect of quality of life in the U.S.

Herein, Leopold addresses the entrance of roads into wilderness and the automobile as the enemy of wilderness, both of which are similar to the continued pressures an increased populace places on our wilderness today. In the 1920s, Leopold was concerned about the elimination of pack trains and the deterioration of outdoor skills. Wilderness in contemporary society is threatened by the energy engines that seek to extract nonrenewable resources, the huge infrastructure associated with energy extraction, and some attempts to prohibit any type of wilderness protection. Americans recognize the culture and importance of wilderness, but the management of wilderness is still controversial, and we have not distanced ourselves from continuing to conquer the wilderness that remains. Because Americans are still debating issues related to wilderness raised by Leopold in the 1920s, this paper indicates how long it takes to make changes that will protect our natural heritage.

RELATED READING

Bleich, V. C. 1999. Wildlife conservation and wilderness management: uncommon objectives and conflicting philosophies. North American Wild Sheep Conference Proceedings 2:195–205.

Bleich, V. C. 2005. In my opinion: politics, promises, and illogical legislation confound wildlife conservation. Wildlife Society Bulletin 33:66–73.

Cole, D. N. 1996. Ecological manipulation in wilderness—an emerging management dilemma. International Journal of Wilderness 2:15–19.

Czech, B., and P. R. Krausman. 1999. Controversial wildlife management issues in southwestern U.S. wilderness. International Journal of Wilderness 5:22–28.

Haufler, J. B., L. G. Adams, J. Bailey, R. G. Brocke, M. J. Conroy, G. Joslin, and K. G. Smith. 1996. Wildlife management in North American wilderness. Wildlife Society Technical Review 96-1. The Wildlife Society, Bethesda, Maryland, USA.

Leopold, A. 1925. A plea for wilderness hunting grounds. Outdoor Life 56:348–350.

WILDERNESS AS A FORM OF LAND USE

By ALDO LEOPOLD

FROM the earliest times one of the principal criteria of civilization has been the ability to conquer the wilderness and convert it to economic use. To deny the validity of this criterion would be to deny history. But because the conquest of wilderness has produced beneficial reactions on social, political, and economic development, we have set up, more or less unconsciously, the converse assumption that the ultimate social, political, and economic development will be produced by conquering the wilderness entirely—that is, by eliminating it from our environment.

My purpose is to challenge the validity of such an assumption and to show how it is inconsistent with certain cultural ideas which we regard as most distinctly American.

Our system of land use is full of phenomena which are sound as tendencies but become unsound as ultimates. It is sound for a city to grow but unsound for it to cover its entire site with buildings. It was sound to cut down our forests but unsound to run out of wood. It was sound to expand our agriculture, but unsound to allow the momentum of that expansion to result in the present overproduction. To multiply examples of an obvious truth would be tedious. The question, in brief, is whether the benefits of wilderness-conquest will extend to ultimate wilderness-elimination.

The question is new because in America the point of elimination has only recently appeared upon the horizon of foreseeable events. During our four centuries of wilderness-conquest the possibility of disappearance has been too remote to register in the national consciousness. Hence we have no mental language in which to discuss the matter. We must first set up some ideas and definitions.

What Is a Wilderness Area?

The term wilderness, as here used, means a wild, roadless area where those who are so inclined may enjoy primitive modes of travel and subsistence, such as exploration trips by pack-train or canoe.

The first idea is that wilderness is a resource, not only in the physical sense of the raw materials it contains, but also in the sense of a distinctive environment which may, if rightly used, yield certain social values. Such a conception ought not to be difficult, because we have lately learned to think of other forms of land use in the same way. We no longer think of a municipal golf links, for instance, as merely soil and grass.

The second idea is that the value of wilderness varies enormously with location. As with other resources, it is impossible to dissociate value from location. There are wilderness areas in Siberia which are probably very similar in character to parts of our Lake states, but their value to us is negligible, compared with what the value of a similar area in the Lake states would be, just as the value of a golf links would be negligible if located so as to be out of reach of golfers.

The third idea is that wilderness, in the sense of an environment as distinguished from a quantity of physical materials, lies somewhere between the class of non-reproducible resources like minerals, and the reproducible resources like forests. It does not disappear proportionately to use, as minerals do, because we can conceive of a wild area which, if properly administered, could be traveled indefinitely and still be as good as ever. On the other hand, wilderness certainly cannot be built at will, like a city park or a tennis court. If we should tear down improvements already made in order to build a wilderness, not only would the cost be prohibitive, but the result would probably be highly dissatisfying. Neither can a wilderness be grown like timber, because it is something more than trees. The practical point is that if we want wilderness, we must foresee our want and preserve the proper areas against the encroachment of inimical uses.

Fourth, wilderness exists in all degrees, from the little accidental wild spot at the head of a ravine in a Corn Belt woodlot to vast expanses of virgin country—

> "Where nameless men by nameless rivers
> wander
> And in strange valleys die strange deaths
> alone."

What degree of wilderness, then, are we discussing? The answer is, *all degrees*. Wilderness is a relative condition. As a form of land use it cannot be a rigid entity of unchanging content, exclusive of all other forms. On the contrary, it must be a flexible thing, accommodating itself to other forms and blending with them in that highly localized give-and-take scheme of land-planning which employs the criterion of "highest use." By skilfully adjusting one use to another, the land planner builds a balanced whole without undue sacrifice of any function, and thus attains a maximum net utility of land.

Just as the application of the park idea in civic planning varies in degree from the provision of a public bench on a street corner to the establishment of a municipal forest playground as large as the city itself, so should the application of the wilderness idea vary in degree from the wild, roadless spot of a few acres left in the rougher parts of public forest devoted to timbergrowing, to wild, roadless regions approaching in size a whole national forest or a whole national park. For it is not to be supposed that a public wilderness area is a new kind of public land reservation, distinct from public forests and public parks. It is rather a new kind of land-dedication within our system of public forests and parks, to be duly correlated with dedications to the other uses which that system is already obligated to accommodate.

Lastly, to round out our definitions, let us exclude from practical consideration any degree of wilderness so absolute as to forbid reasonable protection. It would be idle to discuss wilderness areas if they are to be left subject to destruction by forest fires, or wide open to abuse. Experience has demonstrated, however, that a very modest and unobtrusive framework of trails, telephone line and lookout stations will suffice for protective purposes. Such improvements do not destroy the wild flavor of the area, and are necessary if it is to be kept in usable condition.

Wilderness Areas in a Balanced Land System

What kind of case, then, can be made for wilderness as a form of land use?

To preserve any land in a wild condition is, of course, a reversal of economic tendency, but that fact alone should not condemn the proposal. A study of the history of land utilization shows that good use is largely a matter of good balance—of wise adjustment between opposing tendencies. The modern movements toward diversified crops and live stock on the farm, conservation of eroding soils, forestry, range management, game management, public parks—all these are attempts to balance opposing tendencies that have swung out of counterpoise.

One noteworthy thing about good balance is the nature of the opposing tendencies.   In its more utilitarian aspect, as seen in modern agriculture, the needed adjustment is between economic uses.   But in the public park movement the adjustment is between an economic use, on the one hand, and a purely social use on the other.   Yet, after a century of actual experience, even the most rigid economic determinists have ceased to challenge the wisdom of a reasonable reversal of economic tendency in favor of public parks.

I submit that the wilderness is a parallel case.   The parallelism is not yet generally recognized because we do not yet conceive of the wilderness environment as a resource.   The accessible supply has heretofore been unlimited, like the supply of air-power, or tide-power, or sunsets, and we do not recognize anything as a resource until the demand becomes commensurable with the supply.

Now after three centuries of over-abundance, and before we have even realized that we are dealing with a non-reproducible resource, we have come to the end of our pioneer environment and are about to push its remnants into the Pacific.   For three centuries that environment has determined the character of our development;

it may, in fact, be said that, coupled with the character of our racial stocks, it is the very stuff America is made of. Shall we now exterminate this thing that made us American?

Ouspensky says that, biologically speaking, the determining characteristic of rational beings is that their evolution is self-directed.   John Burroughs cites the opposite example of the potato bug, which, blindly obedient to the law of increase, exterminates the potato and thereby exterminates itself. Which are we?

What the Wilderness Has Contributed to American Culture

Our wilderness environment cannot, of course, be preserved on any considerable scale as an economic fact.   But, like many other receding economic facts, it can be preserved for the ends of sport.   But what is the justification of sport, as the word is here used?

Physical combat between men, for instance, for unnumbered centuries was an economic fact.   When it disappeared as such, a sound instinct led us to preserve it in the form of athletic sports and games.   Physical combat between men and beasts since first the flight of years began was an economic fact, but when it disappeared as such, the instinct of the race led us to hunt and fish for sport.   The transition of these tests of skill from an economic to a social basis has in no way destroyed their efficacy as human experiences—in fact, the change may be regarded in some respects as an improvement.

Football requires the same kind of back-bone as battle but avoids its moral and physical retrogressions.   Hunting for sport in its highest form is an improvement on hunting for food in that there has been added, to the test of skill, an ethical code which the hunter formulates for himself and must often execute without the moral support of bystanders.

In these cases the surviving sport is actually an improvement on the receding economic fact. Public wilderness areas are essentially a means for allowing the more virile and primitive forms of outdoor recreation to survive the receding economic fact of pioneering. These forms should survive because they likewise are an improvement on pioneering itself.

There is little question that many of the attributes most distinctive of America and Americans are the impress of the wilderness and the life that accompanied it. If we have any such thing as an American culture (and I think we have), its distinguishing marks are a certain vigorous individualism combined with ability to organize, a certain intellectual curiosity bent to practical ends, a lack of subservience to stiff social forms, and an intolerance of drones, all of which are the distinctive characteristics of successful pioneers. These, if anything, are the indigenous part of our Americanism, the qualities that set it apart as a new rather than an imitative contribution to civilization. Many observers see these qualities not only bred into our people, but built into our institutions. Is it not a bit beside the point for us to be so solicitous about preserving those institutions without giving so much as a thought to preserving the environment which produced them and which may now be one of our effective means of keeping them alive?

Wilderness Locations

But the proposal to establish wilderness areas is idle unless acted on before the wilderness has disappeared. Just what is the present status of wilderness remnants in the United States?

Large areas of half a million acres and upward are disappearing very rapidly, not so much by reason of economic need, as by extension of motor roads. Smaller areas are still relatively abundant in the mountainous parts of the country, and will so continue for a long time.

The disappearance of large areas is illustrated by the following instance: In 1910 there were six roadless regions in Arizona and New Mexico, ranging in size from half a million to a million acres, where the finest type of mountain wilderness pack trips could be enjoyed. Today roads have eliminated all but one area of about half a million acres.

In California there were seven large areas ten years ago, but today there are only two left unmotorized.

In the Lake states no large unmotorized playgrounds remain. The motor launch, as well as the motor road, is rapidly wiping out the remnants of canoe country.

In the Northwest large roadless areas are still relatively numerous. The land-plans of the Forest Service call for exclusion of roads from several areas of moderate size.

Unless the present attempts to preserve such areas are greatly strengthened and extended, however, it may be predicted with certainty that, except in the Northwest, all of the large areas already in public ownership will be invaded by motors in another decade.

In selecting areas for retention as wilderness, the vital factor of location must be more decisively recognized. A few areas in the national forests of Idaho or Montana are better than none, but, after all, they will be of limited usefulness to the citizen of Chicago or

New Orleans who has a great desire but a small purse and a short vacation. Wild areas in the poor lands of the Ozarks and the Lake states would be within his reach. For the great urban populations concentrated on the Atlantic seaboards, wild areas in both ends of the Appalachians would be especially valuable.

Are the remaining large wilderness areas disappearing so rapidly because they contain agricultural lands suitable for settlement? No; most of them are entirely devoid of either existing or potential agriculture. Is it because they contain timber which should be cut? It is true that some of them do contain valuable timber, and in a few cases this fact is leading to a legitimate extension of logging operations; but in most of the remaining wilderness the timber is either too thin and scattered for exploitation, or else the topography is too difficult for the timber alone to carry the cost of roads or railroads. In view of the general belief that lumber is being overproduced in relation to the growing scarcity of stumpage, and will probably so continue for several decades, the sacrifice of wilderness for timber can hardly be justified on grounds of necessity.

Generally speaking, it is not timber, and certainly not agriculture, which is causing the decimation of wilderness areas, but rather the desire to attract tourists. The accumulated momentum of the good-roads movement constitutes a mighty force, which, skilfully manipulated by every little mountain village possessed of a chamber of commerce and a desire to become a metropolis, is bringing about the extension of motor roads into every remaining bit of wild country, whether or not there is economic justification for the extension.

Our remaining wild lands are wild because they are poor. But this poverty does not deter the booster from building expensive roads through them as bait for motor tourists.

I am not without admiration for this spirit of enterprise in backwoods villages, nor am I attempting a censorious pose toward the subsidization of their ambitions from the public treasuries; nor yet am I asserting that the resulting roads are devoid of any economic utility. I do maintain, (1) that such extensions of our road systems into the wilderness are seldom yielding a return sufficient to amortize the public investment; (2) that even where they do yield such a return, their construction is not necessarily in the public interest, any more than obtaining an economic return from the last vacant lot in a parkless city would be in the public interest. On the contrary, the public interest demands the careful planning of a system of wilderness areas and the permanent reversal of the ordinary economic process within their borders.

To be sure, to the extent that the motor-tourist business is the cause of invasion of these wilderness playgrounds, one kind of recreational use is merely substituted for another. But this substitution is a vitally serious matter from the point of view of good balance. It is just as unwise to devote 100% of the recreational resources of our public parks and forests to motorists as it would be to devote 100% of our city parks to merry-go-rounds. It would be just as unreasonable to ask the aged to indorse a park with only swings and trapezes, or the children a park with only benches, or the motorists a park with only bridle-paths, as to ask the wilderness recreationist to indorse a universal priority for motor roads. Yet that is what our land plans—or rather lack of them—are now doing; and so sacred is our dogma of "development" that there is no effective protest. The inexorable molding of the individual American to a standardized pattern in his economic activities makes all the more undesirable this unnecessary standardization of his recreational tastes.

Practical Aspects of Establishing Wilderness Areas

Public wilderness playgrounds differ from all other public areas in that both their establishment and maintenance would entail very low costs. The wilderness is the one kind of public land that requires no improvements. To be sure, a simple system of fire protection and administrative patrol would be required, but the cost would not exceed two or three cents per acre per year. Even that would not usually be a new cost, since the greater part of the needed areas are already under administration in the rougher parts of the national forests and parks. The action needed is the permanent differentiation of a suitable system of wild areas within our national park and forest system.

In regions such as the Lake states, where the public domain has largely disappeared, lands would have to be purchased; but that will have to be done, in any event, to round out our park and forest system. In such cases a lesser degree of wilderness may have to suffice, the only ordinary utilities practicable to exclude being cottages, hotels, roads, and motor boats.

The retention of certain wild areas in both national forests and national parks will introduce a healthy variety into the wilderness idea itself, the forest areas serving as public hunting grounds, the park areas as public wild-life sanctuaries, and both kinds as public playgrounds in which the wilderness environments and modes of travel may be preserved and enjoyed.

The Cultural Value of Wilderness

Are these things worth preserving? This is the vital question. I cannot give an unbiased answer. I can only picture the day that is almost upon us when canoe travel will consist in paddling in the noisy wake of a motor launch and portaging through the back yard of a summer cottage. When that day comes, canoe travel will be dead, and dead, too, will be a part of our Americanism. Joliet and LaSalle will be words in a book, Champlain will be a blue spot on a map, and canoes will be merely things of wood and canvas, with a connotation of white duck pants and bathing "beauties."

The day is almost upon us when a pack-train must wind its way up a graveled highway and turn out its bell-mare in the pasture of a summer hotel. When that day comes the pack-train will be dead, the diamond hitch will be merely rope, and Kit Carson and Jim Bridger will be names in a history lesson. Rendezvous will be French for "date," and Forty-Nine will be the number preceding fifty. And thenceforth the march of empire will be a matter of gasoline and four-wheel brakes.

European outdoor recreation is largely devoid of the thing that wilderness areas would be the means of preserving in this country. Europeans do not camp, cook, or pack in the woods for pleasure. They hunt and fish when they can afford it, but their hunting and fishing is merely hunting and fishing, staged in a setting of ready-made hunting lodges, elaborate fare, and hired beaters. The whole thing carries the atmosphere of a picnic rather than that of a pack trip. The test of skill is confined almost entirely to the act of killing, itself. Its value as a human experience is reduced accordingly.

There is a strong movement in this country to preserve the distinctive democracy of our field sports by preserving free hunting and fishing, as distinguished from the European condition of commercialized hunting and fishing privileges. Public shooting grounds and organized cooperative relations between sportsmen and landowners are the means proposed for keeping these sports within reach of the American of moderate means. Free hunting and fishing is a most worthy objective, but it deals with only one of the two distinctive characteristics of American sport. The other characteristic is that our test of skill is primarily the act of living in the open, and only secondarily the act of killing game. It is to preserve this primary characteristic that public wilderness playgrounds are necessary.

Herbert Hoover aptly says that there is no point in increasing the average American's leisure by perfecting the organization of industry, if the expansion of industry is allowed to destroy the recreational resources on which leisure may be beneficially employed. Surely the wilderness is one of the most valuable of these resources, and surely the building of unproductive roads in the wrong places at public expense is one of the least valuable of industries. If we are unable to steer the Juggernaut of our own prosperity, then surely there is an impotence in our vaunted Americanism that augurs ill for our future. The self-directed evolution of rational beings does not apply to us until we become collectively, as well as individually, rational and self-directing.

Wilderness as a form of land-use is, of course, premised on a qualitative conception of progress. It is premised on the assumption that enlarging the range of individual experience is as important as enlarging the number of individuals; that the expansion of commerce is a means, not an end; that the environment of the American pioneers had values of its own, and was not merely a punishment which they endured in order that we might ride in motors. It is premised on the assumption that the rocks and rills and templed hills of this America are something more than economic materials, and should not be dedicated exclusively to economic use.

The vanguard of American thought on the use of land has already recognized all this, in theory. Are we too poor in spirit, in pocket, or in idle acres to recognize it likewise in fact?

Leopold, A. 1933. The conservation ethic. *Journal of Forestry* 31:634–643.

The first conservation movement began in the late 1800s when visionaries like Theodore Roosevelt and Gifford Pinchot realized that without definitive action, wildlife in North America would cease to exist. Action was indeed taken, and the greatest conservation experiment the world has seen began. Wildlife management and conservation became wildlife's salvation through the use of science and innovative dedication. Leopold was one of the first individuals to lead the charge for habitat preservation and management, something still being addressed over a half century later. This charge was not just for land but included the ecology of and interrelationships among land, flora, fauna, and humans.

Leopold's objective in this essay was to demonstrate that with an understanding and respect for land, landscapes can be conserved and modified to meet wildlife's habitat requirements. Land health is not just a nice idea; it is necessary for effective conservation. The concept of Leopold's land ethic has been approached from numerous angles worldwide (e.g., ecosystem conservation, landscape ecology) and is as important today—if not more so—as it was in the early 1930s. In this century, we continue to improve management and conservation of flora and fauna and the habitats they depend on for survival.

Contemporary management acknowledges the conservation ethic by recognizing the dynamic nature of ecosystems, collaborating with diverse stakeholders, and establishing partnerships to develop clear objectives so we can monitor, maintain, and enhance native flora and fauna and their habitats. The conservation ethic also embraces the human spirit, and Leopold makes a strong case for society to always consider the aesthetic aspects of land health and management.

RELATED READING

Callicott, J. B., and E. T. Freyfogle, editors. 1999. For the health of the land. Island Press, Washington, D.C., USA.

Cornett, Z. J., and J. W. Thomas. 1988. Integrity as professionalism: ethics and leadership in practice. Pages 71–90 *in* S. Balch, editor. The land ethic: meeting human needs for the land and its resources. Society of American Foresters, Bethesda, Maryland, USA.

Knight, R. L. 1996. Aldo Leopold, the land ethic, and ecosystem management. Journal of Wildlife Management 60:471–474.

Knight, R. L., and S. Riedel. 2002. Aldo Leopold and the ecological conscience. Oxford University Press, New York, New York, USA.

Lannoo, M. J. 2010. Leopold's shack and Ricketts's lab: the emergence of environmentalism. University of California Press, Berkeley, USA.

Leopold, A. 1934. Conservation economics. Journal of Forestry 32:537–544.

Leopold, A. 1966. A Sand County Almanac. Oxford University Press, New York, New York, USA.

Leopold, A. C. 2004. Living with the land ethic. BioScience 54:149–154.

THE CONSERVATION ETHIC[1]

BY ALDO LEOPOLD

University of Wisconsin

The gradual extension of ethical criteria to economic relationships is an historical fact. Economic criteria did not suffice to adjust men to society; they do not now suffice to adjust society to its environment. If our present evolutionary impetus is an upward one, it is ecologically probable that ethics will eventually be extended to land. The present conservation movement may constitute the beginnings of such an extension. If and when it takes place, it may radically modify what now appear as insuperable economic obstacles to better land-use.

WHEN god-like Odysseus returned from the wars in Troy, he hanged all on one rope some dozen slave-girls of his household whom he suspected of misbehavior during his absence.

This hanging involved no question of propriety, much less . of justice. The girls were property. The disposal of property was then, as now, a matter of expediency, not of right and wrong.

Criteria of right and wrong were not lacking from Odysseus' Greece: witness the fidelity of his wife through the long years before at last his black-prowed galleys clove the wine-dark seas for home. The ethical structure of that day covered wives, but had not yet been extended to human chattels. During the three thousand years which have since elapsed, ethical criteria have been extended to many fields of conduct, with corresponding shrinkages in those judged by expediency only.

This extension of ethics, so far studied only by philosophers, is actually a process in ecological evolution. Its sequences may be described in biological as well as philosophical terms. An ethic, biologically, is a limitation on freedom of action in the struggle for existence. An ethic, philosophically, is a differentiation of social from anti-social conduct. These are two definitions of one thing. The thing has its origin in the tendency of interdependent individuals or societies to evolve modes of coöperation. The biologist calls these symbioses. Man elaborated certain advanced symbioses called politics and economics. Like their simpler biological antecedents, they enable individuals or groups to exploit each other in an orderly way. Their first yardstick was expediency.

The complexity. of coöperative mechanisms increased with population density, and with the efficiency of tools. It was simpler, for example, to define the anti-social uses of sticks and stones in the days of the mastodons than of bullets and billboards in the age of motors.

At a certain stage of complexity, the human community found expediency-yardsticks no longer sufficient. One by one it has evolved and superimposed upon them a set of ethical yardsticks. The first ethics dealt with the relationship between individuals. The Mosaic Decalogue is an example. Later accretions dealt with the relationship between the individual and society. Christianity tries to integrate the individual to society, Democracy to integrate social organization to the individual.

---

[1] Fourth Annual John Wesley Powell Lecture, Southwestern Division, American Association for the Advancement of Science, Las Cruces, New Mexico, May 1, 1933.

There is as yet no ethic dealing with man's relationship to land and to the non-human animals and plants which grow upon it. Land, like Odysseus' slave-girls, is still property. The land-relation is still strictly economic, entailing privileges but not obligations.

The extension of ethics to this third element in human environment is, if we read evolution correctly, an ecological possibility. It is the third step in a sequence. The first two have already been taken. Civilized man exhibits in his own mind evidence that the third is needed. For example, his sense of right and wrong may be aroused quite as strongly by the desecration of a nearby woodlot as by a famine in China, a near-program in Germany, or the murder of the slave-girls in ancient Greece. Individual thinkers since the days of Ezekial and Isaiah have asserted that the despoliation of land is not only inexpedient but wrong. Society, however, has not yet affirmed their belief. I regard the present conservation movement as the embryo of such an affirmation. I here discuss why this is, or should be, so.

Some scientists will dismiss this matter forthwith, on the ground that ecology has no relation to right and wrong. To such I reply that science, if not philosophy, should by now have made us cautious about dismissals. An ethic may be regarded as a mode of guidance for meeting ecological situations so new or intricate, or involving such deferred reactions, that the path of social expediency is not discernible to the average individual. Animal instincts are just this. Ethics are possibly a kind of advanced social instinct in-the-making.

Whatever the merits of this analogy, no ecologist can deny that our land-relation involves penalties and rewards which the individual does not see, and needs modes of guidance which do not yet exist. Call these what you will, science cannot escape its part in forming them.

ECOLOGY—ITS RÔLE IN HISTORY

A harmonious relation to land is more intricate, and of more consequence to civilization, than the historians of its progress seem to realize. Civilization is not, as they often assume, the enslavement of a stable and constant earth. It is a state of *mutual and interdependent coöperation* between human animals, other animals, plants, and soils, which may be disrupted at any moment by the failure of any of them. Land-despoliation has evicted nations, and can on occasion do it again. As long as six virgin continents awaited the plow, this was perhaps no tragic matter,—eviction from one piece of soil could be recouped by despoiling another. But there are now wars and rumors of wars which foretell the impending saturation of the earth's best soils and climates. It thus becomes a matter of some importance, at least to ourselves, that our dominion, once gained, be self-perpetuating rather than self-destructive.

This instability of our land-relation calls for example. I will sketch a single aspect of it: the plant succession as a factor in history.

In the years following the Revolution, three groups were contending for control of the Mississippi valley: the native Indians, the French and English traders, and American settlers. Historians wonder what would have happened if the English at Detroit had thrown a little more weight into the Indian side of those tipsy scales which decided the outcome of the Colonial migration into the cane-lands of Kentucky. Yet who ever wondered why the cane-lands, when subjected to the particular mixture of forces represented by the cow, plow, fire, and axe of the pioneer, became bluegrass? What if the plant succession inherent in this "dark and bloody ground" had, under the impact of these forces, given us some worthless sedge, shrub, or weed? Would Boone and Kenton have held out? Would there have been any overflow into Ohio? Any Louisiana Purchase? Any transcontinental union of new states? Any Civil War? Any machine age? Any depression? The subsequent drama of American history, here and elsewhere, hung in large degree on the reaction of particular soils to the impact of particular forces exerted by a particular kind and degree of human occupation. No statesman-biologist selected those forces, nor foresaw their effects. That chain of events which in the Fourth of July we call our National Destiny hung on a "fortuitous concourse of elements," the interplay of which we now dimly decipher *by hindsight only.*

Contrast Kentucky with what hindsight tells us about the Southwest. The impact of occupancy here brought no bluegrass, nor other plant fitted to withstand the bumps and buffetings of misuse. Most of these soils, when grazed, reverted through a successive series of more and more worthless grasses, shrubs, and weeds to a condition of unstable equilibrium. Each recession of plant types bred erosion; each increment to erosion bred a further recession of plants. The result today is a progressive and mutual deterioration, not only of plants and soils, but of the animal community subsisting thereon. The early settlers did not expect this, on the cienegas of central New Mexico some even cut artificial gullies to hasten it. So subtle has been its progress that few people know anything about it. It is not discussed at polite tea-tables or go-getting luncheon clubs, but only in the arid halls of science.

All civilization seem to have been conditioned upon whether the plant succession, under the impact of occupancy, gave a stable and habitable assortment of vegetative types, or an unstable and uninhabitable assortment. The swampy forests of Caesar's Gaul were utterly changed by human use—for the better. Moses' land of milk and honey was utterly changed—for the worse. Both changes are the unpremeditated resultant of the impact between ecological and economic forces. We now decipher these reactions retrospectively. What could possibly be more important than to foresee and control them?

We of the machine age admire ourselves for our mechanical ingenuity; we harness cars to the solar energy impounded in carboniferous forests; we fly in mechanical birds; we make the ether carry our words or even our pictures. But are these not in one sense mere parlor tricks compared with our utter ineptitude in keeping land fit to live upon? Our engineering has attained the pearly gates of a near-millennium, but our applied biology still lives in nomad's tents of the stone age. If our system of land-use happens to be self-perpetuating, we stay. If it happens to be self-destructive we move, like Abraham, to pastures new.

Do I overdraw this paradox? I think not. Consider the transcontinental airmail which plies the skyways of the Southwest—a symbol of its final conquest. What does it see? A score of mountain valleys which were green gems of fertility when first described by Coronado, Espejo, Pattie, Abert, Sitgreaves, and Couzens. What are they now? Sandbars, wastes of cobbles and burroweed, a path for torrents. Rivers which Pattie says were clear, now muddy sewers for the wasting fertility of an empire. A "Public Domain," once a velvet carpet of rich buffalo-grass and grama, now an illimitable waste of rattlesnake-bush and tumbleweed, too impoverished to be accepted as a gift by the states within which it lies. Why? Because the ecology of this Southwest happened to be set on a hair-trigger. Because cows eat brush when the grass is gone, and thus postpone the penalties of over-utilization. Because certain grasses, when grazed too closely to bear seed-stalks, are weakened and give way to inferior grasses, and these to inferior shrubs, and these to weeds, and these to naked earth. Because rain which spatters upon vegetated soil stays clear and sinks, while rain which spatters upon devegetated soil seals its interstices with colloidal mud and hence must run away as floods, cutting the heart out of country as it goes. Are these phenomena any more difficult to foresee than the paths of stars which science deciphers without the error of a single second? Which is the more important to the permanence and welfare of civilization?

I do not here berate the astronomer for his precocity, but rather the ecologist for his lack of it. The days of his cloistered sequestration are over:

"Whether you will or not,
You are a king, Tristram, for you are one
Of the time-tested few that leave the world,
When they are gone, not the same place it was.
Mark what you leave."

Unforseen ecological reactions not only make or break history in a few exceptional enterprises—they condition, circumscribe, delimit, and warp all enterprises, both economic and cultural, that pertain to land. In the cornbelt, after grazing and plowing out all the cover in the interests of "clean farming," we grew tearful about wild-life, and spent several decades passing laws for its restoration. We were like Canute commanding the tide. Only recently has research made it clear that the impliments for restoration lie not in the legislature, but in the former's toolshed. Barbed wire and brains are doing what laws alone failed to do.

In other instances we take credit for shaking down apples which were, in all probability, ecological windfalls. In the Lake States and the Northeast lumbering, pulping, and fire accidentally created some scores of millions of acres of new second-growth. At the proper stage we find these thickets full of deer. For this we naively thank the wisdom of our game laws.

In short, the reaction of land to occupancy determines the nature and duration of civilization. In arid climates the land may be destroyed. In all climates the plant succession determines what economic activities can be supported. Their nature and intensity in turn determine not only the domestic but also the wild plant and animal life, the scenery, and the whole face of nature. We inherit the earth, but within the limits of the soil and the plant succession we also *rebuild* the earth,—without plan, without knowledge of its properties, and without understanding of the increasingly coarse and powerful tools which science has placed at our disposal. We are remodelling the Alhambra with a steam-shovel.

ECOLOGY AND ECONOMICS

The conservation movement is, at the very least, an assertion that these interactions between man and land are too important to be left to chance, even that sacred variety of chance known as economic law.

We have three possible controls: Legislation, self-interest, and ethics. Before we can know where and how they will work, we most first understand the reactions. Such understanding arises only from research. At the present moment research, inadequate as it is, has nevertheless piled up a large store of facts which our land using industries are unwilling, or (they claim) unable, to apply. Why? A review of three sample fields will be attempted.

Soil science has so far relied on self-interest as the motive for conservation. The landholder is told that it pays to conserve his soil and its fertility. On good farms this economic formula has improved land-practice, but on poorer soils vast abuses still proceed unchecked. Public acquisition of submarginal soils is being urged as a remedy for their misuse. It has been applied to some extent, but it often comes too late to check erosion, and can hardly hope more than to ameliorate a phenomenon involving in some degree *every square foot* on the continent. Legislative compulsion might work on the best soils where it is least needed, but it seems hopeless on poor soils where the existing economic set-up hardly permits even uncontrolled private enterprise to make a profit. We must face the fact that, by and large, no defensible relationship between man and the soil of his nativity is as yet in sight.

Forestry exhibits another tragedy—or comedy—of *Homo sapiens*, astride the runaway Juggernaut of his own building, trying to be decent to his environment. A new profession was trained in the confident expectation that the shrinkage in virgin timber would, as a matter of self-interest, bring an expansion of timber-cropping. Foresters are cropping timber on certain parcels of poor land which happen to be public, but on the great bulk of private holdings they have accomplished little. Economics won't let them. Why? He would be bold indeed who claimed to know the whole answer, but these parts of it seem agreed upon: modern transport prevents profitable tree-cropping in cut-out regions until virgin stands in all others are first exhausted; substitutes for lumber have undermined confidence in the future need for it;

carrying charges on stumpage reserves are so high as to force perennial liquidation, overproduction, depressed prices, and an appalling wastage of unmarketable grades which must be cut to get the higher grades; the mind of the forest owner lacks the point-of-view underlying sustained yield; the low wage-standards on which European forestry rests do not obtain in America.

A few tentative gropings toward industrial forestry were visible before 1929, but these have been mostly swept away by the depression, with the net result that forty years of "campaigning" have left us only such actual tree-cropping as is under-written by public treasuries. Only a blind man could see in this the beginnings of an orderly and harmonious use of the forest resource.

There are those who would remedy this failure by legislative compulsion of private owners. Can a landholder be successfully compelled to raise any crop, let alone a complex long-time crop like a forest, on land the private possession of which is, for the moment at least, a liability? Compulsion would merely hasten that avalanche of tax-delinquent land-titles now being dumped into the public lap.

Another and larger group seeks a remedy in more public ownership. Doubtless we need it—we are getting it whether we need it or not—but how far can it go? We cannot dodge the fact that the forest problem, like the soil problem, *is coextensive with the map of the United States.* How far can we tax other lands and industries to maintain forest lands and industries artificially? How confidently can we set out to run a hundred-yard dash with a twenty foot rope tying our ankle to the starting point? Well, we are bravely "getting set," anyhow.

The trend in wild-life conservation is possibly more encouraging than in either soils or forests. It has suddenly become apparent that farmers, out of self-interest, can be induced to crop game. Game crops are in demand, staple crops are not. For farm-species, therefore, the immediate future is relatively bright. Forest game has profited to some extent by the accidental establishment of new habitat following the decline of forest industries. Migratory game, on the other hand, has lost heavily through drainage and over-shooting; its future is black because motives of self-interest do not apply to the private cropping of birds so mobile that they "belong" to everybody, and hence to nobody. Only governments have interests coextensive with their annual movements, and the divided counsels of conservationists give governments ample alibi for doing little. Governments could crop migratory birds because their marshy habitat is cheap and concentrated, but we get only an annual crop of new hearings on how to divide the fast-dwindling remnant.

These three fields of conservation, while but fractions of the whole, suffice to illustrate the welter of conflicting forces, facts, and opinions which so far comprise the result of the effort to harmonize our machine civilization with the land whence comes its sustenance. We have accomplished little, but we should have learned much. What?

I can see clearly only two things:

First, that the economic cards are stacked against some of the most important reforms in land-use.

Second, that the scheme to circumvent this obstacle by public ownership, while highly desirable and good as far as it goes, can never go far enough. Many will take issue on this, but the issue is between two conflicting conceptions of the end towards which we are working.

One regards conservation as a kind of sacrificial offering, made for us vicariously by bureaus, on lands nobody wants for other purposes, in propitiation for the atrocities which still prevail everywhere else. We have made a real start on this kind of conservation, and we can carry it as far as the tax-string on our leg will reach. Obviously, though it conserves our self-respect better than our land. Many excellent people accept it, either because they despair of anything better, or because they fail to see the *universality of the reactions needing control*. That is to say their ecological education is not yet sufficient.

The other concept supports the public program, but regards it as merely extension, teaching, demonstration, an initial nucleus, a means to an end, but not the end itself. The real end is a *universal symbiosis with land,* economic and esthetic, public and private. To this school of thought public ownership is a patch but not a program.

Are we, then, limited to patchwork until such time as Mr. Babbitt has taken his Ph.D. in ecology and esthetics? Or do the new economic formulae offer a short-cut to harmony with our environment?

THE ECONOMIC ISMS

As nearly as I can see, all the new isms—Socialism, Communism, Fascism, and especially the late but not lamented Technocracy—outdo even Capitalism itself in their preoccupation with one thing: The distribution of more machine-made commodities to more people. They all proceed on the theory that if we can all keep warm and full, and all own a Ford and a radio, the good life will follow. Their programs differ only in ways to mobilize machines to this end. Though they despise each other, they are all, in respect of this objective, as identically alike as peas in a pod. They are competitive apostles of a single creed: *salvation by machinery.*

We are here concerned, not with their

proposals for adjusting men and machinery to goods, but rather with their lack of any vital proposal for adjusting men and machines to land. To conservationists they offer only the old familiar palliatives: Public ownership and private compulsion. If these are insufficient now, by what magic are they to become sufficient after we change our collective label?

Let us apply economic reasoning to a sample problem and see where it takes us. As already pointed out, there is a huge area which the economist calls submarginal, because it has a minus value for exploitation. In its once-virgin condition, however, it could be "skinned" at a profit. It has been, and as a result erosion is washing it away. What shall we do about it?

By all the accepted tenets of current economics and science we ought to say "let her wash." Why? Because staple land-crops are overproduced, our population curve is flattening out, science is still raising the yields from better lands, we are spending millions from the public treasury to retire unneeded acreage, and here is nature offering to do the same thing free of charge; why not let her do it? This, I say, is economic reasoning. *Yet no man has so spoken.* I cannot help reading a meaning into this fact. To me it means that the average citizen shares in some degree the intuitive and instantaneous contempt with which the conservationist would regard such an attitude. We can, is seems, stomach the burning or plowing-under of over-produced cotton, coffee, or corn, but the destruction of mother-earth, however "sub-marginal," touches something deeper, some sub-economic stratum of the human intelligence wherein lies that something—perhaps the essence of civilization—which Wilson called "the decent opinion of mankind."

THE CONSERVATION MOVEMENT

We are confronted, then, by a contradiction. To build a better motor we tap the uttermost powers of the human brain; to build a better countryside we throw dice. Political systems take no cognizance of this disparity, offer no sufficient remedy. There is, however, a dormant but widespread consciousness that the destruction of land, and of the living things upon it, is wrong. A new minority have espoused an idea called conservation which tends to assert this as a positive principle. Does it contain seeds which are likely to grow?

Its own devotees, I confess, often give apparent grounds for skepticism. We have, as an extreme example, the cult of the barbless hook, which acquires self-esteem by a self-imposed limitation of armaments in catching fish. The limitation is commendable, but the illusion that it has something to do with salvation is as naive as some of the primitive taboos and mortifications which still adhere to religious sects. Such excrescences seem to indicate the whereabouts of a moral problem, however irrelevant they be in either defining or solving it.

Then there is the conservation-booster, who of late has been rewriting the conservation ticket in terms of "tourist-bait." He exhorts us to "conserve outdoor Wisconsin" because if we don't the motorist-on-vacation will streak through to Michigan, leaving us only a cloud of dust. Is Mr. Babbitt trumping up hard-boiled reasons to serve as a screen for doing what he thinks is right? His tenacity suggests that he is after something more than tourists. Have he and other thousands of "conservation workers" labored through all these barren decades fired by a dream of augmenting the sales of sandwiches and gasoline? I think not. Some of these people have hitched their wagon to a star—and that is something.

Any wagon so hitched offers the discerning politician a quick ride to glory. His agility in hopping up and seizing the reins adds little dignity to the cause, but it does add the testimony of his political nose to an important question: is this conservation something people really want? The political objective, to be sure, is often some trivial tinkering with the laws, some useless appropriation, or some pasting of pretty labels on ugly realities. How often, though, does any political action portray the real depth of the idea behind it? For political consumption a new thought must always be reduced to a posture or a phrase. It has happened before that great ideas were heralded by growing-pains in the body politic, semi-comic to those onlookers not yet infected by them. The insignificance of what we conservationists, in our political capacity, say and do, does not detract from the significance of our persistent desire to do something. To turn this desire into productive channels is the task of time, and ecology.

The recent trend in wild life conservation shows the direction in which ideas are evolving. At the inception of the movement fifty years ago, its underlying thesis was to save species from extermination. The means to this end were a series of restrictive enactments. The duty of the individual was to cherish and extend these enactments, and to see that his neighbor obeyed them. The whole structure was negative and prohibitory. It assumed land to be a constant in the ecological equation. Gun-powder and blood-lust were the variables needing control.

There is now being superimposed on this a positive and affirmatory ideology, the thesis of which is to prevent the deterioration of environment. The means to this end is research. The duty of the individual is to apply its findings to land, and to encourage his neighbor to do likewise. The soil and the plant succession are recognized as the basic variables which determine plant and animal life, both wild and domesticated, and likewise the quality and quantity of human satisfactions to be derived. Gun-powder is relegated to the status of a tool for harvesting one of these satisfactions. Blood-lust is a source of motive-power, like sex in social organization. Only one constant is assumed, and that is common to both equations: the love of nature.

This new idea is so far regarded as merely a new and promising means to better hunting and fishing, but its potential uses are much larger. To explain this, let us go back to the basic thesis—the preservation of fauna and flora.

Why do species become extinct? Because they first become rare. Why do they become rare? Because of shrinkage in the particular environments which their particular adaptations enable them to inhabit. Can such shrinkage be controlled? Yes, once the specifications are known. How known? Through ecological research. How controlled? By modifying the environment with those same tools and skills already used in agriculture and forestry.

Given, then. the knowledge and the desire, this idea of controlled wild culture or "management" can be applied not only to quail and trout, but to *any living thing* from bloodroots to Bell's vireos. Within the limits imposed by the plant succession, the soil, the size of the property, and the gamut of the seasons, the landholder can "raise" any wild plant, fish, bird, or mammal he wants to. A rare bird or flower need remain no rarer than the people willing to venture their skill in *building it a habitat*. Nor need we visualize this as a new diversion for the idle rich. The average dolled-up estate merely proves what we will some day learn to acknowledge: that bread and beauty grow best together. Their harmonious integration can make farming not only a business but an art; the land not only a food-factory but an instrument for self-expression, on which each can play music of his own choosing.

It is well to ponder the sweep of this thing. It offers us nothing less than a renaissance—a new creative stage—in the oldest, and potentially the most universal, of all the fine arts. "Landscaping," for ages dissociated from economic land-use, has suffered that dwarfing and distortion which always attends the relegation of esthetic or spiritual functions to parks and parlors. Hence it is hard for us to visualize a creative art of land-beauty which is the prerogative, not of esthetic priests but of dirt farmers, which deals not with plants but with biota, and which wields not only spade and pruning shears, but also draws rein on those invisible forces which determine the presence or absence of plants and animals. Yet such is this thing which lies to hand, if we want it. In it are the seeds of change, including, perhaps, a rebirth of that social dignity which ought to inhere in land-ownership, but which, for the moment, has passed to inferior professions, and which the current processes of land-skinning hardly deserve. In it, too, are perhaps the seeds of a new fellowship in land, a new solidarity in all men privileged to plow, a realization of Whitman's dream to "*plant companionship as thick as trees along all the rivers of America.*" What bitter parody of such companionship, and trees, and rivers, is offered to this our generation!

I will not belabor the pipe-dream. It is no prediction, but merely an assertion that the idea of controlled environment contains colors and brushes wherewith society may some day paint a new and possibly a better picture of itself. Granted a community in which the combined beauty and utility of land determines the social status of its owner, and we will

see a speedy dissolution of the economic obstacles which now beset conservation. Economic laws may be permanent, but their impact reflects what people want, which in turn reflects what they know and what they are. The economic set-up at any one moment is in some measure the result, as well as the cause, of the then prevailing standard of living. Such standards change. For example: some people discriminate against manufactured goods produced by child-labor or other anti-social processes. They have learned some of the abuses of machinery, and are willing to use their custom as a leverage for betterment. Social pressures have also been exerted to modify ecological processes which happened to be simple enough for people to understand;—witness the very effective boycott of bird-skins for millinery ornament. We need postulate only a little further advance in ecological education to visualize the application of like pressures to other conservation problems.

For example: the lumberman who is now unable to practice forestry because the public is turning to synthetic boards may then be able to sell man-grown lumber "to keep the mountains green." Again: certain wools are produced by gutting the public domain; couldn't their competitors, who lead their sheep in greener pastures, so label their product? Must we view forever the irony of educating our sons with paper, the offal of which pollutes the rivers which they need quite at badly as books? Would not many people pay an extra penny for a "clean" newspaper? Government may some day busy itself with the legitimacy of labels used by land-industries to distinguish conservation products, rather than with the attempt to operate their lands for them.

I neither predict nor advocate these particular pressures—their wisdom or unwisdom is beyond my knowledge. I do assert that these abuses are just as real, and their correction every whit as urgent, as was the killing of egrets for hats. *They differ only in the number of links composing the ecological chain of cause and effect.* In egrets there were one or two links, which the mass-mind saw, believed, and acted upon. In these others there are many links; people do not see them, nor believe us who do. The ultimate issue, in conservation as in other social problems, is whether the mass-mind *wants to* extend its powers of comprehending the world in which it lives, or, granted the desire, *has the capacity to do so.* Ortega, in his "Revolt of the Masses," has pointed the first question with devastating lucidity. The geneticists are gradually, with trepidations, coming to grips with the second. I do not know the answer to either. I simply affirm that a sufficiently enlightened society, by chang-ing its wants and tolerances, can change the economic factors bearing on land. It can be said of nations, as of individuals: "as a man thinketh, so is he."

It may seem idle to project such imaginary elaborations of culture at a time when millions lack even the means of physical existence. Some may feel for it the same honest horror as the Senator from Michigan who lately arraigned Congress for protecting migratory birds at a time when fellow-humans lacked bread. The trouble with such deadly parallels is we can never be sure which is cause and which is effect. It is not inconceivable that the wave phenomena which have lately upset everything from banks to crime-rates might be less troublesome if the human medium in which they run *readjusted its tensions.* The stampede is an attribute of animals interested solely in grass.

Leopold, A. 1943. Wildlife in American culture. *Journal of Wildlife Management* 7:1–6.

It is important that wildlife biologists recognize, and can intelligently convey, the numerous benefits our wildlife resources contribute to society. Many authors have written about the value of wildlife and how society should ethically engage it. Leopold's 1943 article was one of the first to emphasize wildlife's importance in American culture. Leopold wrote about three broad areas: experiences that remind us of our American heritage (i.e., *split-rail value*), experiences that remind us of the conquest of nature by machinery and our dependency on the soil-plant-animal-human food chain. The split-rail value is any experience that reminds us of our American history, our national origins, and evolution. As society moves away from nature and the land, people need to be reminded that their roots lie in the soil and not in an app on an iPod.

The conquest of nature has been and will continue to be a challenge to the preservation of wildlife, especially as habitat is destroyed with various justifications (e.g., energy extraction, housing developments, agriculture). Leopold, however, argues, "We shall achieve conservation when and only when the destructive use of land becomes unethical—punishable by social ostracism." Further, Leopold calls for a new sport: wildlife research, which can be accomplished by anyone interested. This sport will enhance wildlife values in all three areas and is important because it emphasizes the value of society in the conservation and management of wildlife. Leopold concludes this piece with a plea to look closely at animal behavior to answer questions about human behavior and hints at human overpopulation being a concern that cannot be ignored.

RELATED READING

Errington, P. L. 1940. On the social potentialities of wildlife management. Journal of Wildlife Management 4:451–452.

Flader, S. L. 1974. Thinking like a mountain: Aldo Leopold and the evolution of an ethical attitude toward deer, wolves and forests. University of Missouri Press, Columbia, USA.

Kellert, S. R. 1996. The value of life: biological diversity and human society. Island Press, Washington, D.C., USA.

Leopold, A. 1938. Conservation esthetic. Bird Lore 40:101–109.

McHarg, I. L. 1969. Design with nature. Doubleday, New York, New York, USA.

Meine, C. 1988. Aldo Leopold: his life and work. University of Wisconsin Press, Madison, USA.

Shaw, W. W., and E. H. Zube, editors. 1980. Wildlife values. Rocky Mountain Experiment Station, U.S. Forest Service, Ft. Collins, Colorado, USA.

THE JOURNAL OF WILDLIFE MANAGEMENT

| VOLUME 7 | JANUARY 1943 | NUMBER 1 |

WILDLIFE IN AMERICAN CULTURE

Aldo Leopold
University of Wisconsin, Madison, Wis.

The culture of primitive peoples is often based on wildlife. Thus, the plains Indian not only ate buffalo, but buffalo largely determined his architecture, dress, language, arts, and religion.

In civilized peoples the economic base shifts to tame animals and plants, but the culture nevertheless retains part of its wild roots. This paper deals with the value of this wild rootage.

No one can weigh or measure culture, hence I will waste no time trying to do so. Suffice it to say that by common consent of thinking people, there are cultural values in the sports, customs, and experiences which renew contacts with wild things. I venture the opinion that these values are of three kinds.

First, there is value in any experience which reminds us of our distinctive national origins and evolution, i.e., which stimulates awareness of American history. Such awareness is "nationalism" in its best sense. For lack of any other short name, I will call this the "split-rail value." For example: a boy scout has tanned a coonskin cap, and goes Daniel-Booneing in the willow thicket below the tracks. He is re-enacting American history. He is, to that extent, culturally prepared to face the dark and bloody realities of 1943. Again: a farmer boy arrives in the schoolroom reeking of muskrat; he has tended his traps before breakfast. He is

re-enacting the romance of the fur trade. Ontogeny repeats phylogeny in society as well as in the individual.

Second, there is value in any experience which reminds us of our dependency on the soil-plant-animal-man food chain. Civilization has so cluttered this elemental man-earth relation with gadgets and middle-men that awareness of it is growing dim. We fancy that industry supports us, forgetting what supports industry. Time was when education moved toward soil, not away from it. The nursery jingle about bringing home a rabbit skin to wrap the baby bunting in is one of many reminders in folklore that man once hunted to feed and clothe his family.

Third, the conquest of nature by machines has led to much unnecessary destruction of resources. Our tools improve faster than we do. It is unlikely that economic motives alone will ever teach us to use our new tools gently. The only remedy is to extend our system of ethics from the man-man relation to the man-earth relation (1). We shall achieve conservation when and only when the destructive use of land becomes unethical—punishable by social ostracism. Any experience that stimulates this extension of ethics is culturally valuable. Any that has the opposite effect is culturally damaging. For example, we have many bad hunt-

1

ers with good guns. Such a hunter shoots a woodduck, and then tramples the bejeweled carcass into the mud, lest he fall foul of the law. Such an experience is not only devoid of cultural value, it is actually damaging to all concerned. It does physical damage to woodduck, and moral damage to the hunter, and to all fellow-hunters who condone him. No sane person could find anything but minus value in such "sport."

It seems, then, that split-rail and man-earth experiences have zero or plus values, but that ethical experiences may have minus values as well.

This, then, defines roughly three kinds of cultural nutriment available to our outdoor roots. It does not follow that culture is fed. The extraction of value is never automatic; only a healthy culture can feed and grow. Is culture fed by our present forms of outdoor recreation?

The pioneer period gave birth to two ideas which are the very essence of split-rail value in outdoor sports. One is the "go-light" idea, the other the "one-bullet-one-buck" idea. The pioneer went light of necessity. He shot with economy and precision because he lacked the transport, the cash, and the weapons requisite for machine-gun tactics. Let it be clear, then, that in their inception, both of these ideas were forced on us; we made a virtue of necessity.

In their later evolution, however, they became a code of sportsmanship, a self-imposed limitation on sport. On them is based a distinctively American tradition of self-reliance, hardihood, woodcraft, and marksmanship. These are intangibles, but they are not ab-

stractions. Theodore Roosevelt was a great sportsman, not because he hung up many trophies, but because he expressed (2) this intangible American tradition in words any schoolboy could understand. A more subtle and accurate expression is found in the early writings of Stewart Edward White (3). It is not far amiss to say that such men created cultural value by being aware of it, and by creating a pattern for its growth.

Then came the gadgeteer, otherwise known as the sporting-goods dealer. He has draped the American outdoorsman with an infinity of contraptions, all offered as aids to self-reliance, hardihood, woodcraft, or marksmanship, but too often functioning as substitutes for them. Gadgets fill the pockets, they dangle from neck and belt. The overflow fills the auto-trunk, and also the trailer. Each item of outdoor equipment grows lighter and often better, but the aggregate poundage becomes tonnage. The traffic in gadgets adds up to astronomical sums, which are soberly published as representing "the economic value of wildlife." But what of cultural values?

As an end-case consider the duck hunter, sitting in a steel boat behind composition decoys. A put-put has brought him to the blind without exertion. Canned heat stands by to warm him in case of a chilling wind. He talks to the passing flocks on a factory caller, in what he hopes are seductive tones; home lessons from a phonograph record have taught him how. The decoys work, despite the caller; a flock circles in. It must be shot at before it circles twice, for the marsh bristles with other sportsmen, similarly accoutred, who might shoot first. He opens up at 70 yards, for his polychoke is set for infinity, and the ads have told him that Super-Z shells, and plenty of them, have a long reach. The flock flares. A couple of cripples scale off to die elsewhere. Is this sportsman absorbing cultural value? Or is he just feeding minks? The next blind opens up at 75 yards; how else is a fellow to get some shooting? This is duck-shooting, model 1943. It is typical of all public grounds, and of many clubs. Where is the go-light idea, the one-bullet tradition?

The answer is not a simple one. Roosevelt did not disdain the modern rifle; White used freely the aluminum pot, the silk tent, dehydrated foods. Somehow they used mechanical aids, in moderation, without being used by them.

I do not pretend to know what is moderation, or where the line is between legitimate and illegitimate gadgets. It seems clear, though, that the origin of gadgets has much to do with their cultural effects. Homemade aids to sport or outdoor life often enhance, rather than destroy, the man-earth drama; he who kills a trout with his own fly has scored two coups, not one. I use many factory-made gadgets myself. Yet there must be some limit beyond which money-bought aids to sport destroy the cultural value of sport.

Not all sports have degenerated to the same extent as duck hunting. Defenders of the American tradition still exist. Perhaps the bow-and-arrow movement and the revival of falconry mark the beginnings of a reaction. The net trend, however, is clearly toward more and more mechanization, with a corresponding shrinkage in cultural values, especially split-rail values and ethical restraints.

I have the impression that the American sportsman is puzzled; he doesn't understand what is happening to him. Bigger and better gadgets are good for industry, so why not for outdoor recreation? It has not dawned on him that outdoor recreations are essentially primitive, atavistic; that their value is a contrast-value; that excessive mechanization destroys contrasts by moving the factory to the woods or to the marsh.

The sportsman has no leaders to tell him what is wrong. The sporting press no longer represents sport, it has turned billboard for the gadgeteer. Wildlife administrators are too busy producing something to shoot at to worry much about the cultural value of the shooting. Because everybody from Xenophon to Teddy Roosevelt said sport has value, it is assumed that this value must be indestructible.

Among non-gunpowder sports, the impact of mechanization has had diverse effects. The modern field glass, camera, and aluminum bird-band have certainly *not* deteriorated the cultural value of ornithology. Fishing, but for motorized transport, seems less severely mechanized than hunting. On the other hand, motorized transport has nearly destroyed the sport of wilderness travel by leaving only fly-specks of wilderness to travel in.

Fox-hunting with hounds, backwoods style, presents a dramatic instance of partial and perhaps harmless mechanized invasion. This is one of the purest of sports; it has real split-rail flavor; it has man-earth drama of the first water. The fox is deliberately left unshot, hence ethical restraint is also present. But we now follow the chase in Fords!

The voice of Bugle-Anne mingles with the honk of the flivver! However, no one is likely to invent a mechanical foxhound, nor to screw a polychoke on the hound's nose. No one is likely to teach dog-training by phonograph, or by other painless shortcuts. I think the gadgeteer has reached the end of his tether in dogdom.

It is not quite accurate to ascribe all the ills of sport to the inventor of physical aids-to-sport. The advertiser invents ideas, and ideas are seldom as honest as physical objects, even though they may be equally useless. One such deserves special mention: the "where-to-go" department. Knowledge of the whereabouts of good hunting or fishing is a very personal form of property. Perhaps it is like rod, dog, or gun: a thing to be loaned or given as a personal courtesy, or even to be sold man-to-man, as in the guide-sportsman relation. But to hawk it in the marketplace of the sports column as an aid-to-circulation seems to me another matter. To hand it to all and sundry as free public "service" seems to me distinctly another matter. Both tend to depersonalize one of the essentially personal elements in hunting skill. I do not know where the line lies between legitimate and illegitimate practice; I am convinced, though, that "where-to-go" service has broken all bounds of reason.

If the hunting or fishing is good, the where-to-go service suffices to attract the desired excess of sportsmen. But if it is no good, the advertiser must resort to more forcible means. One such is the fishing lottery, in which a few hatchery fish are tagged, and a prize is offered for the fisherman catching the winning number. This curious hybrid between the techniques of science and of the pool hall insures the overfishing of many an already exhausted lake, and brings a glow of civic pride to many a village Chamber of Commerce.

It is idle for the profession of wildlife management to consider itself aloof from these affairs. The production engineer and the salesman belong to the same company; both are tarred with the same stick.

* * * * *

Wildlife management is trying to convert hunting from exploitation to cropping. If the conversion takes place, how will it affect cultural values? It must be admitted that split-rail flavor and free-for-all exploitation are historically associated. Daniel Boone had scant patience with agricultural cropping, let alone wildlife cropping. Perhaps the stubborn reluctance of the one-gallus sportsman to be converted to the cropping idea is an expression of his split-rail inheritance. Probably cropping is resisted because it is incompatible with one component of the split-rail tradition, free hunting.

Mechanization offers no cultural substitute for the split-rail values it destroys; at least none visible to me. Cropping or management does offer a substitute, which to me has at least equal value, wild husbandry (4). The experience of managing land for wildlife crops has the same value as any other form of farming; it is a reminder of the man-earth relation. Moreover ethical restraints are involved; thus managing game without resorting to predator-control calls for ethical restraint of a high order. It may be concluded, then, that game cropping shrinks one value (split-rail) but enhances both others.

* * * * *

If we regard outdoor sports as a field of conflict between an immensely vigorous process of mechanization and a wholly static tradition, then the outlook for cultural values is indeed dark. But why can not our concept of sport grow with the same vigor as our list of gadgets? Perhaps the salvation of cultural value lies in seizing the offensive. I, for one, believe that the time is ripe. Sportsmen can determine for themselves the shape of things to come.

The last decade, for example, has disclosed a totally new form of sport which does not destroy wildlife, which uses gadgets without being used by them, which outflanks the problem of posted land, and which greatly increases the human carrying capacity of a unit area. This sport knows no bag limit, no closed season. It needs teachers, but not wardens. It calls for a new woodcraft of the highest cultural value. The sport I refer to is wildlife research.

Wildlife research started as a professional priestcraft. The more difficult or laborious problems must remain in professional hands, but there are plenty of problems suitable for all grades of amateurs. In the mechanical field, research has long since spread to amateurs. In the biological field the sport-value of amateur work is just beginning to be realized. When amateurs like Margaret Nice outstrip their professional colleagues, a very important new element is added: the element of high stakes open to all comers, the possibility of really outstanding amateur performance.

Ornithology, mammalogy, and botany, as now known to most amateurs, are but kindergarten games compared with researches in these fields. The real game is decoding the messages written on the face of the land. By learning how some small part of the biota ticks, we can guess how the whole mechanism ticks.

Few people can become enthusiastic about research as a sport because the whole structure of biological education is aimed to perpetuate the professional research monopoly. To the amateur is allotted only make-believe voyages of discovery, the chance to verify what professional authority already knows. This is false; the case of Margaret Nice proves what a really enterprising amateur can do. What the youth needs to be told is that a ship is a-building in his own mental dry-dock, a ship with freedom of the seas. If you are a pessimist, you can say this ship is "on order"; if an optimist, you can see the keel.

In my opinion, the promotion of wildlife research sports is the most important job confronting our profession.

* * * * *

Wildlife has still another value, now visible only to a few ecologists, but of potential importance to the whole human enterprise.

We now know that animal populations have behavior patterns of which the individual animal is unaware, but which he nevertheless helps to execute. Thus the rabbit is unaware of cycles, but he is the vehicle for cycles.

We cannot discern these behavior patterns in the individual, or in short periods of time. The most intense scrutiny of an individual rabbit tells us nothing of cycles. The cycle concept springs from a scrutiny of the mass through decades of history.

This raises the disquieting question: do human populations have behavior patterns of which we are unaware, but which we help to execute? Are mobs and wars, unrests and revolutions, cut of such cloth?

Many historians and philosophers persist in interpreting our mass behaviors as the collective result of individual acts of volition. The whole subject-matter of diplomacy assumes that the political group has the properties of an honorable person. On the other hand, some economists (5) see the whole of society as a plaything for processes, our knowledge of which is largely ex-post-facto.

It is reasonable to suppose that our social processes have a higher volitional content than those of the rabbit, but it is also reasonable to suppose that we contain patterns of which nothing is known because circumstance has never evoked them. We may have others the meaning of which we have misread.

This state of doubt about the fundamentals of human population behavior lends exceptional interest, and exceptional value, to the only available analogue: the higher animals. Errington (6), among others, has pointed out the cultural value of these animal analogues. For centuries this rich library of knowledge has been inaccessible to us because we did not know where or how to look for it. Ecology is now teaching us to search in animal populations for analogies to our own problems. The ability to perceive these, and to appraise them critically, is the woodcraft of the future.

To sum up, wildlife once fed us and shaped our culture. It still yields us pleasure for leisure hours, but we try to reap that pleasure by modern machinery and thus destroy part of its value. Reaping it by modern mentality would yield not only pleasure, but wisdom as well.

LITERATURE CITED

1. LEOPOLD, ALDO. 1933. The Conservation Ethic. Journ. Forestry, 31: 634–643.
2. ROOSEVELT, THEODORE, T. S. VAN DYKE, D. G. ELLIOTT, and A. J. STONE. 1902. The Deer Family. New York, Grosset and Dunlap. (See especially the introduction by Roosevelt.)
3. WHITE, STEWART EDWARD. 1903. The Forest. New York, The Outlook Company.
4. LEOPOLD, ALDO. 1938. Conservation Esthetic. Bird Lore, 40: 101–109.
5. BURNHAM, JAMES. 1941. The Managerial Revolution. New York, John Day Publishing Co.
6. ERRINGTON, PAUL L. 1940. On the Social Potentialities of Wildlife Management. Journ. Wildl. Mgt., 4: 451–452.

Chamberlin, T. C. 1965 (reprinted). The method of multiple working hypotheses. *Science* 148:754–759.

Sir Francis Bacon proposed an alternative to deductive reasoning in 1620, which developed into what scientists refer to as the scientific method. The fundamental components of that method have been hypothesis formulation and testing. Thomas C. Chamberlin's paper was a critical step forward for science because he advocated the method of multiple working hypotheses to promote thoroughness. Chamberlin wrote, "The value of a working hypothesis lies largely in its suggestiveness of lines of inquiry that might otherwise be overlooked. . . . In the use of the multiple method, the re-action of one hypothesis upon another tends to amplify the recognized scope of each, and their mutual conflicts whet the discriminative edge of each." His work replaced the method of the ruling theory and the method of the working hypothesis, both of which were influenced by the values of the researcher.

In addition to science, Chamberlin applies the multiple working hypotheses method to social and civic situations. The main objective of his paper was to illustrate that by developing multiple hypotheses, the scientist avoids pet hypotheses, thereby reducing biases and enhancing credibility of experiments. It provides both means for formulating hypotheses that can lead to scientific discovery and an example of how solid science transcends numerous fields.

The original ideas behind this classical work were published by Chamberlin in *Science* in 1890 and read before the Society of Western Naturalists on 25 October 1889. The work was reprinted in the *Journal of Geology* in 1897 (emphasizing geological study) and 1931, in the *Scientific Monthly* in November 1944, and in other outlets. Thomas C. Chamberlin (1843–1928) was a geologist, but his work applies to all scientific fields.

RELATED READING

Anderson, D. R., K. P. Burnham, and W. L. Thompson. 2000. Null hypothesis testing: problems, and an alternative. Journal of Wildlife Management 64:912–923.

Guthery, F. S. 2007. Deductive and inductive methods of accumulating reliable knowledge in wildlife science. Journal of Wildlife Management 71:222–225.

Johnson, D. H. 1999. The insignificance of statistical significance testing. Journal of Wildlife Management 63:763–772.

Matter, W. J., and R. W. Mannan. 1989. More on gaining reliable knowledge: a comment. Journal of Wildlife Management 53:1172–1176.

Medawar, P. B. 1969. Induction and intuition in scientific thought. American Philosophical Society, Philadelphia, Pennsylvania, USA.

Platt, J. R. 1964. Strong inference. Science 146:347–353.

Romesburg, H. C. 1981. Wildlife science: gaining reliable knowledge. Journal of Wildlife Management 45:293–313.

Romesburg, H. C. 1989. More on gaining reliable knowledge: a reply. Journal of Wildlife Management 53:1177–1180.

The Method of Multiple Working Hypotheses

With this method the dangers of parental
affection for a favorite theory can be circumvented.

T. C. Chamberlin

As methods of study constitute the leading theme of our session, I have chosen as a subject in measurable consonance the method of multiple working hypotheses in its application to investigation, instruction, and citizenship.

There are two fundamental classes of study. The one consists in attempting to follow by close imitation the processes of previous thinkers, or to acquire by memorizing the results of their investigations. It is merely secondary, imitative, or acquisitive study. The other class is primary or creative study. In it the effort is to think independently, or at least individually, in the endeavor to discover new truth, or to make new combinations of truth, or at least to develop an individualized aggregation of truth. The endeavor is to think for one's self, whether the thinking lies wholly in the fields of previous thought or not. It is not necessary to this habit of study that the subject-material should be new; but the process of thought and its results must be individual and independent, not the mere following of previous lines of thought ending in predetermined results. The demonstration of a problem

Thomas C. Chamberlin (1843–1928), a geologist, was president of the University of Wisconsin at the time this lecture was written. Later he was professor and director of the Walker Museum of the University of Chicago. In 1893 he founded the *Journal of Geology*, which he edited until his death. In 1908 he was president of the AAAS. The article is reprinted from *Science* (old series), **15**, 92 (1890).

Reprints of this article are available.
Prices *(cash with order)*:

1	50 cents (*or* 25 cents and stamped, self-addressed envelope)
2 to 9	45 cents each
10 to 24	30 cents each
25 or more	20 cents each

Address orders to AAAS, Chamberlin Reprints, 1515 Massachusetts Ave., NW, Washington, D.C. 20005.

in Euclid precisely as laid down is an illustration of the former; the demonstration of the same proposition by a method of one's own or in a manner distinctively individual is an illustration of the latter; both lying entirely within the realm of the known and the old.

Creative study, however, finds its largest application in those subjects in which, while much is known, more remains to be known. Such are the fields which we, as naturalists, cultivate; and we are gathered for the purpose of developing improved methods lying largely in the creative phase of study, though not wholly so.

Intellectual methods have taken three phases in the history of progress thus far. What may be the evolutions of the future it may not be prudent to forecast. Naturally the methods we now urge seem the highest attainable. These three methods may be designated, first, the method of the ruling theory; second, the method of the working hypothesis; and, third, the method of multiple working hypotheses.

In the earlier days of intellectual development the sphere of knowledge was limited, and was more nearly within the compass of a single individual; and those who assumed to be wise men, or aspired to be thought so, felt the need of knowing, or at least seeming to know, all that was known as a justification of their claims. So, also, there grew up an expectancy on the part of the multitude that the wise and the learned would explain whatever new thing presented itself. Thus pride and ambition on the one hand, and expectancy on the other, developed the putative wise man whose knowledge boxed the compass, and whose acumen

found an explanation for every new puzzle which presented itself. This disposition has propagated itself, and has come down to our time as an intellectual predilection, though the compassing of the entire horizon of knowledge has long since been an abandoned affectation. As in the earlier days, so still, it is the habit of some to hastily conjure up an explanation for every new phenomenon that presents itself. Interpretation rushes to the forefront as the chief obligation pressing upon the putative wise man. Laudable as the effort at explanation is in itself, it is to be condemned when it runs before a serious inquiry into the phenomenon itself. A dominant disposition to find out what is, should precede and crowd aside the question, commendable at a later stage, "How came this so?" First full facts, then interpretations.

Premature Theories

The habit of precipitate explanation leads rapidly on to the development of tentative theories. The explanation offered for a given phenomenon is naturally, under the impulse of self-consistency, offered for like phenomena as they present themselves, and there is soon developed a general theory explanatory of a large class of phenomena similar to the original one. This general theory may not be supported by any further considerations than those which were involved in the first hasty inspection. For a time it is likely to be held in a tentative way with a measure of candor. With this tentative spirit and measurable candor, the mind satisfies its moral sense, and deceives itself with the thought that it is proceeding cautiously and impartially toward the goal of ultimate truth. It fails to recognize that no amount of provisional holding of a theory, so long as the view is limited and the investigation partial, justifies an ultimate conviction. It is not the slowness with which conclusions are arrived at that should give satisfaction to the moral sense, but the thoroughness, the completeness, the all-sidedness, the impartiality, of the investigation.

It is in this tentative stage that the affections enter with their blinding influence. Love was long since represented as blind, and what is true in the personal realm is measurably true in the intellectual realm. Important as the intellectual affections are as stimuli and as rewards, they are nevertheless dangerous factors, which menace the integrity of the intellectual processes. The moment one has offered an original explanation for a phenomenon which seems satisfactory, that moment affection for his intellectual child springs into existence; and as the explanation grows into a definite theory, his parental affections cluster about his intellectual offspring, and it grows more and more dear to him, so that, while he holds it seemingly tentative, it is still lovingly tentative, and not impartially tentative. So soon as this parental affection takes possession of the mind, there is a rapid passage to the adoption of the theory. There is an unconscious selection and magnifying of the phenomena that fall into harmony with the theory and support it, and an unconscious neglect of those that fail of coincidence. The mind lingers with pleasure upon the facts that fall happily into the embrace of the theory, and feels a natural coldness toward those that seem refractory. Instinctively there is a special searching-out of phenomena that support it, for the mind is led by its desires. There springs up, also, an unconscious pressing of the theory to make it fit the facts, and a pressing of the facts to make them fit the theory. When these biasing tendencies set in, the mind rapidly degenerates into the partiality of paternalism. The search for facts, the observation of phenomena and their interpretation, are all dominated by affection for the favored theory until it appears to its author or its advocate to have been overwhelmingly established. The theory then rapidly rises to the ruling position, and investigation, observation, and interpretation are controlled and directed by it. From an unduly favored child, it readily becomes master, and leads its author whithersoever it will. The subsequent history of that mind in respect to that theme is but the progressive dominance of a ruling idea.

Briefly summed up, the evolution is this: a premature explanation passes into a tentative theory, then into an adopted theory, and then into a ruling theory.

When the last stage has been reached, unless the theory happens, perchance, to be the true one, all hope of the best results is gone. To be sure, truth may be brought forth by an in-

Thomas Chrowder Chamberlin was noted for his contributions to glaciology and for his part in formulating the Chamberlin-Moulton (planetesimal) hypothesis of the origin of the earth.

vestigator dominated by a false ruling idea. His very errors may indeed stimulate investigation on the part of others. But the condition is an unfortunate one. Dust and chaff are mingled with the grain in what should be a winnowing process.

Ruling Theories Linger

As previously implied, the method of the ruling theory occupied a chief place during the infancy of investigation. It is an expression of the natural infantile tendencies of the mind, though in this case applied to its higher activities, for in the earlier stages of development the feelings are relatively greater than in later stages.

Unfortunately it did not wholly pass away with the infancy of investigation, but has lingered along in individual instances to the present day, and finds illustration in universally learned men and pseudo-scientists of our time.

The defects of the method are obvious, and its errors great. If I were to name the central psychological fault, I should say that it was the admission of intellectual affection to the place that should be dominated by impartial intellectual rectitude.

So long as intellectual interest dealt chiefly with the intangible, so long it was possible for this habit of thought

to survive, and to maintain its dominance, because the phenomena themselves, being largely subjective, were plastic in the hands of the ruling idea; but so soon as investigation turned itself earnestly to an inquiry into natural phenomena, whose manifestations are tangible, whose properties are rigid, whose laws are rigorous, the defects of the method became manifest, and an effort at reformation ensued. The first great endeavor was repressive. The advocates of reform insisted that theorizing should be restrained, and efforts directed to the simple determination of facts. The effort was to make scientific study factitious instead of causal. Because theorizing in narrow lines had led to manifest evils, theorizing was to be condemned. The reformation urged was not the proper control and utilization of theoretical effort, but its suppression. We do not need to go backward more than twenty years to find ourselves in the midst of this attempted reformation. Its weakness lay in its narrowness and its restrictiveness. There is no nobler aspiration of the human intellect than desire to compass the cause of things. The disposition to find explanations and to develop theories is laudable in itself. It is only its ill use that is reprehensible. The vitality of study quickly disappears when the object sought is a mere collocation of dead unmeaning facts.

The inefficiency of this simply repressive reformation becoming apparent, improvement was sought in the method of the working hypothesis. This is affirmed to be *the* scientific method of the day, but to this I take exception. The working hypothesis differs from the ruling theory in that it is used as a means of determining facts, and has for its chief function the suggestion of lines of inquiry; the inquiry being made, not for the sake of the hypothesis, but for the sake of facts. Under the method of the ruling theory, the stimulus was directed to the finding of facts for the support of the theory. Under the working hypothesis, the facts are sought for the purpose of ultimate induction and demonstration, the hypothesis being but a means for the more ready development of facts and of their relations, and the arrangement and preservation of material for the final induction.

It will be observed that the distinc-

Our Philosophical Roots   31

tion is not a sharp one, and that a working hypothesis may with the utmost ease degenerate into a ruling theory. Affection may as easily cling about an hypothesis as about a theory, and the demonstration of the one may become a ruling passion as much as of the other.

A Family of Hypotheses

Conscientiously followed, the method of the working hypothesis is a marked improvement upon the method of the ruling theory; but it has its defects—defects which are perhaps best expressed by the ease with which the hypothesis becomes a controlling idea. To guard against this, the method of multiple working hypotheses is urged. It differs from the former method in the multiple character of its genetic conceptions and of its tentative interpretations. It is directed against the radical defect of the two other methods; namely, the partiality of intellectual parentage. The effort is to bring up into view every rational explanation of new phenomena, and to develop every tenable hypothesis respecting their cause and history. The investigator thus becomes the parent of a family of hypotheses: and, by his parental relation to all, he is forbidden to fasten his affections unduly upon any one. In the nature of the case, the danger that springs from affection is counteracted, and therein is a radical difference between this method and the two preceding. The investigator at the outset puts himself in cordial sympathy and in parental relations (of adoption, if not of authorship) with every hypothesis that is at all applicable to the case under investigation. Having thus neutralized the partialities of his emotional nature, he proceeds with a certain natural and enforced erectness of mental attitude to the investigation, knowing well that some of his intellectual children will die before maturity, yet feeling that several of them may survive the results of final investigation, since it is often the outcome of inquiry that several causes are found to be involved instead of a single one. In following a single hypothesis, the mind is presumably led to a single explanatory conception. But an adequate explanation often involves the co-ordination of several agencies, which enter into the combined result

in varying proportions. The true explanation is therefore necessarily complex. Such complex explanations of phenomena are specially encouraged by the method of multiple hypotheses, and constitute one of its chief merits. We are so prone to attribute a phenomenon to a single cause, that, when we find an agency present, we are liable to rest satisfied therewith, and fail to recognize that it is but one factor, and perchance a minor factor, in the accomplishment of the total result. Take for illustration the mooted question of the origin of the Great Lake basins. We have this, that, and the other hypothesis urged by different students as the cause of these great excavations; and all of these are urged with force and with fact, urged justly to a certain degree. It is practically demonstrable that these basins were river-valleys antecedent to the glacial incursion, and that they owe their origin in part to the pre-existence of those valleys and to the blocking-up of their outlets. And so this view of their origin is urged with a certain truthfulness. So, again, it is demonstrable that they were occupied by great lobes of ice, which excavated them to a marked degree, and therefore the theory of glacial excavation finds support in fact. I think it is furthermore demonstrable that the earth's crust beneath these basins was flexed downward, and that they owe a part of their origin to crust deformation. But to my judgment neither the one nor the other, nor the third, constitutes an adequate explanation of the phenomena. All these must be taken together, and possibly they must be supplemented by other agencies. The problem, therefore, is the determination not only of the participation, but of the measure and the extent, of each of these agencies in the production of the complex result. This is not likely to be accomplished by one whose working hypothesis is pre-glacial erosion, or glacial erosion, or crust deformation, but by one whose staff of working hypotheses embraces all of these and any other agency which can be rationally conceived to have taken part in the phenomena.

A special merit of the method is, that by its very nature it promotes thoroughness. The value of a working hypothesis lies largely in its suggestiveness of lines of inquiry that might otherwise be overlooked. Facts that are

trivial in themselves are brought into significance by their bearings upon the hypothesis, and by their causal indications. As an illustration, it is only necessary to cite the phenomenal influence which the Darwinian hypothesis has exerted upon the investigations of the past two decades. But a single working hypothesis may lead investigation along a given line to the neglect of others equally important; and thus, while inquiry is promoted in certain quarters, the investigation lacks in completeness. But if all rational hypotheses relating to a subject are worked co-equally, thoroughness is the presumptive result, in the very nature of the case.

In the use of the multiple method, the re-action of one hypothesis upon another tends to amplify the recognized scope of each, and their mutual conflicts whet the discriminative edge of each. The analytic process, the development and demonstration of criteria, and the sharpening of discrimination, receive powerful impulse from the co-ordinate working of several hypotheses.

Fertility in processes is also the natural outcome of the method. Each hypothesis suggests its own criteria, its own means of proof, its own methods of developing the truth; and if a group of hypotheses encompass the subject on all sides, the total outcome of means and of methods is full and rich.

The use of the method leads to certain peculiar habits of mind which deserve passing notice, since as a factor of education its disciplinary value is one of importance. When faithfully pursued for a period of years, it develops a habit of thought analogous to the method itself, which may be designated a habit of parallel or complex thought. Instead of a simple succession of thoughts in linear order, the procedure is complex, and the mind appears to become possessed of the power of simultaneous vision from different standpoints. Phenomena appear to become capable of being viewed analytically and synthetically at once. It is not altogether unlike the study of a landscape, from which there comes into the mind myriads of lines of intelligence, which are received and co-ordinated simultaneously, producing a complex impression which is recorded and studied directly in its complexity. My description of this process

T. C. Chamberlin published two papers under the title of "The method of multiple working hypotheses." One of these papers, first published in the *Journal of Geology* in 1897, was quoted by John R. Platt in his recent article "Strong inference" (*Science*, 16 Oct. 1964). Platt wrote: "This charming paper deserves to be reprinted." Several readers, having had difficulty obtaining copies of Chamberlin's paper, expressed agreement with Platt. One wrote that the article had been reprinted in the *Journal of Geology* in 1931 and in the *Scientific Monthly* in November 1944. Another sent us a photocopy. Several months later still another wrote that the Institute for Humane Studies (Stanford, Calif.) had reprinted the article in pamphlet form this year. On consulting the 1897 version, we found a footnote in which Chamberlin had written: "A paper on this subject was read before the Society of Western Naturalists in 1892, and was published in a scientific periodical." Library research revealed that "a scientific periodical" was *Science* itself, for 7 February 1890, and that Chamberlin had actually read the paper before the Society of Western Naturalists on 25 October 1889. The chief difference between the 1890 text and the 1897 text is that, as Chamberlin wrote in 1897: "The article has been freely altered and abbreviated so as to limit it to aspects related to geological study." The 1890 text, which seems to be the first and most general version of "The method of multiple working hypotheses," is reprinted here. Typographical errors have been corrected, and subheadings have been added.

is confessedly inadequate, and the affirmation of it as a fact would doubtless challenge dispute at the hands of psychologists of the old school; but I address myself to naturalists who I think can respond to its verity from their own experience.

Drawbacks of the Method

The method has, however, its disadvantages. No good thing is without its drawbacks; and this very habit of mind, while an invaluable acquisition for purposes of investigation, introduces difficulties in expression. It is obvious, upon consideration, that this method of thought is impossible of verbal expression. We cannot put into words more than a single line of thought at the same time; and even in that the order of expression must be conformed to the idiosyncrasies of the language, and the rate must be relatively slow. When the habit of complex thought is not highly developed, there is usually a leading line to which others are subordinate, and the difficulty of expression does not rise to serious proportions; but when the method of simultaneous vision along different lines is developed so that the thoughts running in different channels are nearly equivalent, there is an obvious embarrassment in selection and a disinclination to make the attempt. Furthermore, the impossibility of expressing the mental operation in words leads to their disuse in the silent process of

thought, and hence words and thoughts lose that close association which they are accustomed to maintain with those whose silent as well as spoken thoughts run in linear verbal courses. There is therefore a certain predisposition on the part of the practitioner of this method to taciturnity.

We encounter an analogous difficulty in the use of the method with young students. It is far easier, and I think in general more interesting, for them to argue a theory or accept a simple interpretation than to recognize and evaluate the several factors which the true elucidation may require. To illustrate: it is more to their taste to be taught that the Great Lake basins were scooped out by glaciers than to be urged to conceive of three or more great agencies working successively or simultaneously, and to estimate how much was accomplished by each of these agencies. The complex and the quantitative do not fascinate the young student as they do the veteran investigator.

**Multiple Hypotheses and
Practical Affairs**

It has not been our custom to think of the method of working hypotheses as applicable to instruction or to the practical affairs of life. We have usually regarded it as but a method of science. But I believe its application to practical affairs has a value coordinate with the importance of the

affairs themselves. I refer especially to those inquiries and inspections that precede the coming-out of an enterprise rather than to its actual execution. The methods that are superior in scientific investigation should likewise be superior in those investigations that are the necessary antecedents to an intelligent conduct of affairs. But I can dwell only briefly on this phase of the subject.

In education, as in investigation, it has been much the practice to work a theory. The search for instructional methods has often proceeded on the presumption that there is a definite patent process through which all students might be put and come out with results of maximum excellence; and hence pedagogical inquiry in the past has very largely concerned itself with the inquiry, "What is the best method?" rather than with the inquiry, "What are the special values of different methods, and what are their several advantageous applicabilities in the varied work of instruction?" The past doctrine has been largely the doctrine of pedagogical uniformitarianism. But the faculties and functions of the mind are almost, if not quite, as varied as the properties and functions of matter: and it is perhaps not less absurd to assume that any specific method of instructional procedure is more effective than all others, under any and all circumstances, than to assume that one principle of interpretation is equally applicable to all the phenomena of nature. As there is an endless variety of mental processes and combinations and an indefinite number of orders of procedure, the advantage of different methods under different conditions is almost axiomatic. This being granted, there is presented to the teacher the problem of selection and of adaptation to meet the needs of any specific issue that may present itself. It is important, therefore, that the teacher shall have in mind a full array of possible conditions and states of mind which may be presented, in order that, when any one of these shall become an actual case, he may recognize it, and be ready for the emergency.

Just as the investigator armed with many working hypotheses is more likely to see the true nature and significance of phenomena when they present themselves, so the instructor equipped with a full panoply of hypotheses ready for application more readily recognizes the actuality of the situation, more accurately measures its significance, and more appropriately applies the methods which the case calls for.

The application of the method of multiple hypotheses to the varied affairs of life is almost as protean as the phases of that life itself, but certain general aspects may be taken as typical of the whole. What I have just said respecting the application of the method to instruction may apply, with a simple change of terms, to almost any other endeavor which we are called upon to undertake. We enter upon an enterprise in most cases without full knowledge of all the factors that will enter into it, or all of the possible phases which it may develop. It is therefore of the utmost importance to be prepared to rightly comprehend the nature, bearings, and influence of such unforeseen elements when they shall definitely present themselves as actualities. If our vision is narrowed by a preconceived theory as to what will happen, we are almost certain to misinterpret the facts and to misjudge the issue. If, on the other hand, we have in mind hypothetical forecasts of the various contingencies that may arise, we shall be the more likely to recognize the true facts when they do present themselves. Instead of being biased by the anticipation of a given phase, the mind is rendered open and alert by the anticipation of any one of many phases, and is free not only, but is predisposed,

to recognize correctly the one which does appear. The method has a further good effect. The mind, having anticipated the possible phases which may arise, has prepared itself for action under any one that may come up, and it is therefore ready-armed, and is predisposed to act in the line appropriate to the event. It has not set itself rigidly in a fixed purpose, which it is predisposed to follow without regard to contingencies. It has not nailed down the helm and predetermined to run a specific course, whether rocks lie in the path or not; but, with the helm in hand, it is ready to veer the ship according as danger or advantage discovers itself.

It is true, there are often advantages in pursuing a fixed predetermined course without regard to obstacles or adverse conditions. Simple dogged resolution is sometimes the salvation of an enterprise; but, while glorious successes have been thus snatched from the very brink of disaster, overwhelming calamity has in other cases followed upon this course, when a reasonable regard for the unanticipated elements would have led to success. So there is to be set over against the great achievements that follow on dogged adherence great disasters which are equally its result.

Danger of Vacillation

The tendency of the mind, accustomed to work through multiple hypotheses, is to sway to one line of policy or another, according as the balance of evidence shall incline. This is the soul and essence of the method. It is in general the true method. Nevertheless there is a danger that this yielding to evidence may degenerate into unwarranted vacillation. It is not always possible for the mind to balance evidence with exact equipoise, and to determine, in the midst of the execution of an enterprise, what is the measure of probability on the one side or the other: and as difficulties present themselves, there is a danger of being biased by them and of swerving from the course that was really the true one. Certain limitations are therefore to be placed upon the application of the method, for it must be remembered that a poorer line of policy consistently adhered to may bring better results than a vacillation between better policies.

There is another and closely allied danger in the application of the method. In its highest development it presumes a mind supremely sensitive to every grain of evidence. Like a pair of delicately poised scales, every added particle on the one side or the other produces its effect in oscillation. But such a pair of scales may be altogether too sensitive to be of practical value in the rough affairs of life. The balances of the exact chemist are too delicate for the weighing-out of coarse commodities. Despatch may be more important than accuracy. So it is possible for the mind to be too much concerned with the nice balancings of evidence, and to oscillate too much and too long in the endeavor to reach exact results. It may be better, in the gross affairs of life, to be less precise and more prompt. Quick decisions, though they may contain a grain of error, are oftentimes better than precise decisions at the expense of time.

The method has a special beneficent application to our social and civic relations. Into these relations there enter, as great factors, our judgment of others, our discernment of the nature of their acts, and our interpretation of their motives and purposes. The method of multiple hypotheses, in its application here, stands in decided contrast to the method of the ruling theory or of the simple working hypothesis. The primitive habit is to interpret the acts of others on the basis of a theory. Childhood's unconscious theory is that the good are good, and the bad are bad. From the good the child expects nothing but good; from the bad, nothing but bad. To expect a good act from the bad, or a bad act from the good, is radically at variance with childhood's mental methods. Unfortunately in our social and civic affairs too many of our fellow-citizens have never outgrown the ruling theory of their childhood.

Many have advanced a step farther, and employ a method analagous to that of the working hypothesis. A certain presumption is made to attach to the acts of their fellow-beings, and that which they see is seen in the light of that presumption, and that which they construe is construed in the light of that presumption. They do not go to the lengths of childhood's method by assuming positively that the good are wholly good, and the bad wholly bad; but there is a strong presumption in their minds that he concerning whom

758

they have an ill opinion will act from corresponding motives. It requires positive evidence to overthrow the influence of the working hypothesis.

The method of multiple hypotheses assumes broadly that the acts of a fellow-being may be diverse in their nature, their moves, their purposes, and hence in their whole moral character; that they may be good though the dominant character be bad; that they may be bad though the dominant character be good; that they may be partly good and partly bad, as is the fact in the greater number of the complex activities of a human being. Under the method of multiple hypotheses, it is the first effort of the mind to see truly what the act is, unbeclouded by the presumption that this or that has been done because it accords with our ruling theory or our working hypothesis. Assuming that acts of similar general aspect may readily take any one of several different phases, the mind is freer to see accurately what has actually been done. So, again, in our interpretations of motives and purposes, the method assumes that these may have been any one of many, and the first duty is to ascertain which of possible motives and purposes actually prompted this individual action. Going with this effort there is a predisposition to balance all evidence fairly, and to accept that interpretation to which the weight of evidence inclines, not that which simply fits our working hypothesis or our dominant theory. The outcome, therefore, is better and truer observation and juster and more righteous interpretation.

Imperfections of Knowledge

There is a third result of great importance. The imperfections of our knowledge are more likely to be detected, for there will be less confidence in its completeness in proportion as there is a broad comprehension of the possibilities of varied action, under similar circumstances and with similar appearances. So, also, the imperfections of evidence as to the motives and purposes inspiring the action will become more discernible in proportion to the fulness of our conception of what the evidence should be to distinguish between action from the one or the other of possible motives. The necessary result will be a less disposition to reach conclusions upon imperfect grounds. So, also, there will be a less inclination to misapply evidence; for, several constructions being definitely in mind, the indices of the one motive are less liable to be mistaken for the indices of another.

The total outcome is greater care in ascertaining the facts, and greater discrimination and caution in drawing conclusions. I am confident, therefore, that the general application of this method to the affairs of social and civic life would go far to remove those misunderstandings, misjudgments, and misrepresentations which constitute so pervasive an evil in our social and our political atmospheres, the source of immeasurable suffering to the best and most sensitive souls. The misobservations, the misstatements, the misinterpretations, of life may cause less gross suffering than some other evils; but they, being more universal and more subtle, pain. The remedy lies, indeed, partly in charity, but more largely in correct intellectual habits, in a predominant, ever-present disposition to see things as they are, and to judge them in the full light of an unbiased weighing of evidence applied to all possible constructions, accompanied by a withholding of judgment when the evidence is insufficient to justify conclusions.

I believe that one of the greatest moral reforms that lies immediately before us consists in the general introduction into social and civic life of that habit of mental procedure which is known in investigation as the method of multiple working hypotheses.

Hardin, G. 1968. The tragedy of the commons. *Science*, New Series 162:1243–1248.

Our natural resources are not infinite; they are limited by environmental and anthropogenic factors. When the colonists first arrived on the shores of North America, natural resources were viewed as property of the people, not kings, queens, or landowners, as was the case in Europe. Numerous court cases in North America supported the concept that wildlife and fisheries resources belonged to the people and were entrusted to state and federal agencies. This "common resource" viewpoint, however, leads to an important concept that affects the heart of sustainability. This concept is as relevant today as it was in 1968, and Hardin's landmark paper on the tragedy of allowing groups of individuals common access to resources brought it to national prominence.

Wildlife and conservation biologists had recognized the problems and concerns of lost habitat and resources for decades. Hardin's objective was to put these concerns in perspective by clearly pointing out the human needs that must be controlled if society is to maintain viable habitats for wildlife in a sustained manner. For example, humans need to consider pollution that is fouling the earth and to develop social arrangements to maintain and improve the environment for all that live there. Society needs to reduce population growth or we will lose important resources, leading to misery. Hardin proposed that the freedom to reproduce be abandoned as a social arrangement.

Natural resource managers are faced with many modern "tragedies of the commons," from use of our National Parks and rare ecosystems to the fundamental concepts of land use to ensure sustainability. Hardin's message is quite simple: if individuals do not have a vested interest in our natural resources, then they will use them until they are gone, or in the case of wildlife, are extinct. The tragedy of the commons represents a fundamental dichotomy; our natural resources belong to everyone, but unless use and access are limited, those resources assigned to everyone will not exist for future generations to enjoy.

RELATED READING

Bromley, D. W. 1992. The commons, property, and common regimes. Pages 3–15 *in* D. W. Bromley, editor. Making the commons work. ICS, San Francisco, California, USA.

Czech, B. 2000. Economic growth as the limiting factor for wildlife conservation. Wildlife Society Bulletin 28:4–15.

Feeny, D., F. Berkes, B. J. McCay, and J. M. Acheson. 1990. The tragedy of the commons: twenty-two years later. Human Ecology 18:1–19.

Hall, C. A. S., P. W. Jones, T. M. Donovan, and J. P. Gibbs. 2000. The implications of mainstream economics for wildlife conservation. Wildlife Society Bulletin 28:16–25.

Hirsch, F. 1995. Social limits to growth. Psychology Press, Routledge & Kegan Paul Limited, UK.

McKean, M. A. 1982. The Japanese experience with scarcity: management of traditional commonlands. Environmental Review 6:63–88.

Ostrom, E., J. Burger, C. B. Field, R. B. Norgaard, and D. Policansky. 1999. Revisiting the commons: local lessons, global challenges. Science 284:278–282.

The Tragedy of the Commons

Garrett Hardin

At the end of a thoughtful article on the future of nuclear war, Wiesner and York (1) concluded that: "Both sides in the arms race are . . . confronted by the dilemma of steadily increasing military power and steadily decreasing national security. *It is our considered professional judgment that this dilemma has no technical solution.* If the great powers continue to look for solutions in the area of science and technology only, the result will be to worsen the situation."

I would like to focus your attention not on the subject of the article (national security in a nuclear world) but on the kind of conclusion they reached, namely that there is no technical solution to the problem. An implicit and almost universal assumption of discussions published in professional and semipopular scientific journals is that the problem under discussion has a technical solution. A technical solution may be defined as one that requires a change only in the techniques of the natural sciences, demanding little or nothing in the way of change in human values or ideas of morality.

In our day (though not in earlier times) technical solutions are always welcome. Because of previous failures in prophecy, it takes courage to assert that a desired technical solution is not possible. Wiesner and York exhibited this courage; publishing in a science journal, they insisted that the solution to the problem was not to be found in the natural sciences. They cautiously qualified their statement with the phrase, "It is our considered professional judgment. . . ." Whether they were right or not is not the concern of the present article. Rather, the concern here is with the important concept of a class of human problems which can be called "no technical solution problems," and, more specifically, with the identification and discussion of one of these.

It is easy to show that the class is not a null class. Recall the game of tick-tack-toe. Consider the problem, "How can I win the game of tick-tack-toe?" It is well known that I cannot, if I assume (in keeping with the conventions of game theory) that my opponent understands the game

perfectly. Put another way, there is no "technical solution" to the problem. I can win only by giving a radical meaning to the word "win." I can hit my opponent over the head; or I can drug him; or I can falsify the records. Every way in which I "win" involves, in some sense, an abandonment of the game, as we intuitively understand it. (I can also, of course, openly abandon the game—refuse to play it. This is what most adults do.)

The class of "No technical solution problems" has members. My thesis is that the "population problem," as conventionally conceived, is a member of this class. How it is conventionally conceived needs some comment. It is fair to say that most people who anguish over the population problem are trying to find a way to avoid the evils of overpopulation without relinquishing any of the privileges they now enjoy. They think that farming the seas or developing new strains of wheat will solve the problem—technologically. I try to show here that the solution they seek cannot be found. The population problem cannot be solved in a technical way, any more than can the problem of winning the game of tick-tack-toe.

What Shall We Maximize?

Population, as Malthus said, naturally tends to grow "geometrically," or, as we would now say, exponentially. In a finite world this means that the per capita share of the world's goods must steadily decrease. Is ours a finite world?

A fair defense can be put forward for the view that the world is infinite; or that we do not know that it is not. But, in terms of the practical problems that we must face in the next few generations with the foreseeable technology, it is clear that we will greatly increase human misery if we do not, during the immediate future, assume that the world available to the terrestrial human population is finite. "Space" is no escape (2).

A finite world can support only a finite population; therefore, population growth must eventually equal zero. (The case of perpetual wide fluctuations above and below zero is a trivial variant that need not be discussed.) When this condition is met, what will be the situation of mankind? Specifically, can Bentham's goal of "the greatest good

for the greatest number" be realized?

No—for two reasons, each sufficient by itself. The first is a theoretical one. It is not mathematically possible to maximize for two (or more) variables at the same time. This was clearly stated by von Neumann and Morgenstern (3), but the principle is implicit in the theory of partial differential equations, dating back at least to D'Alembert (1717–1783).

The second reason springs directly from biological facts. To live, any organism must have a source of energy (for example, food). This energy is utilized for two purposes: mere maintenance and work. For man, maintenance of life requires about 1600 kilocalories a day ("maintenance calories"). Anything that he does over and above merely staying alive will be defined as work, and is supported by "work calories" which he takes in. Work calories are used not only for what we call work in common speech; they are also required for all forms of enjoyment, from swimming and automobile racing to playing music and writing poetry. If our goal is to maximize population it is obvious what we must do: We must make the work calories per person approach as close to zero as possible. No gourmet meals, no vacations, no sports, no music, no literature, no art. . . . I think that everyone will grant, without argument or proof, that maximizing population does not maximize goods. Bentham's goal is impossible.

In reaching this conclusion I have made the usual assumption that it is the acquisition of energy that is the problem. The appearance of atomic energy has led some to question this assumption. However, given an infinite source of energy, population growth still produces an inescapable problem. The problem of the acquisition of energy is replaced by the problem of its dissipation, as J. H. Fremlin has so wittily shown (4). The arithmetic signs in the analysis are, as it were, reversed; but Bentham's goal is still unobtainable.

The optimum population is, then, less than the maximum. The difficulty of defining the optimum is enormous; so far as I know, no one has seriously tackled this problem. Reaching an acceptable and stable solution will surely require more than one generation of hard analytical work—and much persuasion.

We want the maximum good per person; but what is good? To one person it is wilderness, to another it is ski lodges for thousands. To one it is estuaries to nourish ducks for hunters to shoot; to another it is factory land. Comparing one good with another is, we usually say, impossible because goods are incommensurable. Incommensurables cannot be compared.

The author is professor of biology, University of California, Santa Barbara. This article is based on a presidential address presented before the meeting of the Pacific Division of the American Association for the Advancement of Science at Utah State University, Logan, 25 June 1968.

Theoretically this may be true; but in real life incommensurables are commensurable. Only a criterion of judgment and a system of weighting are needed. In nature the criterion is survival. Is it better for a species to be small and hideable, or large and powerful? Natural selection commensurates the incommensurables. The compromise achieved depends on a natural weighting of the values of the variables.

Man must imitate this process. There is no doubt that in fact he already does, but unconsciously. It is when the hidden decisions are made explicit that the arguments begin. The problem for the years ahead is to work out an acceptable theory of weighting. Synergistic effects, nonlinear variation, and difficulties in discounting the future make the intellectual problem difficult, but not (in principle) insoluble.

Has any cultural group solved this practical problem at the present time, even on an intuitive level? One simple fact proves that none has: there is no prosperous population in the world today that has, and has had for some time, a growth rate of zero. Any people that has intuitively identified its optimum point will soon reach it, after which its growth rate becomes and remains zero.

Of course, a positive growth rate might be taken as evidence that a population is below its optimum. However, by any reasonable standards, the most rapidly growing populations on earth today are (in general) the most miserable. This association (which need not be invariable) casts doubt on the optimistic assumption that the positive growth rate of a population is evidence that it has yet to reach its optimum.

We can make little progress in working toward optimum population size until we explicitly exorcize the spirit of Adam Smith in the field of practical demography. In economic affairs, *The Wealth of Nations* (1776) popularized the "invisible hand," the idea that an individual who "intends only his own gain," is, as it were, "led by an invisible hand to promote . . . the public interest" (5). Adam Smith did not assert that this was invariably true, and perhaps neither did any of his followers. But he contributed to a dominant tendency of thought that has ever since interfered with positive action based on rational analysis, namely, the tendency to assume that decisions reached individually will, in fact, be the best decisions for an entire society. If this assumption is correct it justifies the continuance of our present policy of laissez-faire in reproduction. If it is correct we can assume that men will control their individual fecundity so as to produce the optimum population. If the assumption is not correct, we need to reexamine our individual freedoms to see which ones are defensible.

Tragedy of Freedom in a Commons

The rebuttal to the invisible hand in population control is to be found in a scenario first sketched in a little-known pamphlet (6) in 1833 by a mathematical amateur named William Forster Lloyd (1794–1852). We may well call it "the tragedy of the commons," using the word "tragedy" as the philosopher Whitehead used it (7): "The essence of dramatic tragedy is not unhappiness. It resides in the solemnity of the remorseless working of things." He then goes on to say, "This inevitableness of destiny can only be illustrated in terms of human life by incidents which in fact involve unhappiness. For it is only by them that the futility of escape can be made evident in the drama."

The tragedy of the commons develops in this way. Picture a pasture open to all. It is to be expected that each herdsman will try to keep as many cattle as possible on the commons. Such an arrangement may work reasonably satisfactorily for centuries because tribal wars, poaching, and disease keep the numbers of both man and beast well below the carrying capacity of the land. Finally, however, comes the day of reckoning, that is, the day when the long-desired goal of social stability becomes a reality. At this point, the inherent logic of the commons remorselessly generates tragedy.

As a rational being, each herdsman seeks to maximize his gain. Explicitly or implicitly, more or less consciously, he asks, "What is the utility *to me* of adding one more animal to my herd?" This utility has one negative and one positive component.

1) The positive component is a function of the increment of one animal. Since the herdsman receives all the proceeds from the sale of the additional animal, the positive utility is nearly +1.

2) The negative component is a function of the additional overgrazing created by one more animal. Since, however, the effects of overgrazing are shared by all the herdsmen, the negative utility for any particular decision-making herdsman is only a fraction of −1.

Adding together the component partial utilities, the rational herdsman concludes that the only sensible course for him to pursue is to add another animal to his herd. And another; and another. . . . But this is the conclusion reached by each and every rational herdsman sharing a commons. Therein is the tragedy. Each man is locked into a system that compels him to increase his herd without limit—in a world that is limited. Ruin is the destination toward which all men rush, each pursuing his own best interest in a society that believes in the freedom of the commons. Freedom in a commons

brings ruin to all.

Some would say that this is a platitude. Would that it were! In a sense, it was learned thousands of years ago, but natural selection favors the forces of psychological denial (8). The individual benefits as an individual from his ability to deny the truth even though society as a whole, of which he is a part, suffers.

Education can counteract the natural tendency to do the wrong thing, but the inexorable succession of generations requires that the basis for this knowledge be constantly refreshed.

A simple incident that occurred a few years ago in Leominster, Massachusetts, shows bow perishable the knowledge is. During the Christmas shopping season the parking meters downtown were covered with plastic bags that bore tags reading: "Do not open until after Christmas. Free parking courtesy of the mayor and city council." In other words, facing the prospect of an increased demand for already scarce space. the city fathers reinstituted the system of the commons. (Cynically, we suspect that they gained more votes than they lost by this retrogressive act.)

In an approximate way, the logic of the commons has been understood for a long time, perhaps since the discovery of agriculture or the invention of private property in real estate. But it is understood mostly only in special cases which are not sufficiently generalized. Even at this late date, cattlemen leasing national land on the western ranges demonstrate no more than an ambivalent understanding, in constantly pressuring federal authorities to increase the head count to the point where overgrazing produces erosion and weed-dominance. Likewise, the oceans of the world continue to suffer from the survival of the philosophy of the commons. Maritime nations still respond automatically to the shibboleth of the "freedom of the seas." Professing to believe in the "inexhaustible resources of the oceans," they bring species after species of fish and whales closer to extinction (9).

The National Parks present another instance of the working out of the tragedy of the commons. At present, they are open to all, without limit. The parks themselves are limited in extent—there is only one Yosemite Valley—whereas population seems to grow without limit. The values that visitors seek in the parks are steadily eroded. Plainly, we must soon cease to treat the parks as commons or they will be of no value to anyone.

What shall we do? We have several options. We might sell them off as private property. We might keep them as public property, but allocate the right to enter them. The allocation might be on the basis of wealth, by the use of an auction system. It might be on the basis of merit, as defined by some agreed-upon standards. It might be by lottery. Or it might be on a first-come, first-served basis, administered to long queues. These, I think, are all the reasonable possibilities. They are all objectionable. But we must choose—or acquiesce in the destruction of the commons that we call our National Parks.

Pollution

In a reverse way, the tragedy of the commons reappears in problems of pollution. Here it is not a question of taking something out of the commons, but of putting something in—sewage, or chemical, radioactive, and heat wastes into water; noxious and dangerous fumes into the air, and distracting and unpleasant advertising signs into the line of sight. The calculations of utility are much the same as before. The rational man finds that his share of the cost of the wastes he discharges into the commons is less than the cost of purifying his wastes before releasing them. Since this is true for everyone, we are locked into a system of "fouling our own nest," so long as we behave only as independent, rational, free-enterprisers.

The tragedy of the commons as a food basket is averted by private property, or something formally like it. But the air and waters surrounding us cannot readily be fenced, and so the tragedy of the commons as a cesspool must be prevented by different means, by coercive laws or taxing devices that make it cheaper for the polluter to treat his pollutants than to discharge them untreated. We have not progressed as far with the solution of this problem as we have with the first. Indeed, our particular concept of private property, which deters us from exhausting the positive resources of the earth, favors pollution. The owner of a factory on the bank of a stream—whose property extends to the middle of the stream—often has difficulty seeing why it is not his natural right to muddy the waters flowing past his door. The law, always behind the times, requires elaborate stitching and fitting to adapt it to this newly perceived aspect of the commons.

The pollution problem is a consequence of population. It did not much matter how a lonely American frontiersman disposed of his waste. "Flowing water purifies itself every 10 miles," my grandfather used to say, and the myth was near enough to the truth when he was a boy, for there were not too many people. But as population became denser, the natural chemical and biological recycling processes became overloaded, calling for a redefinition of property rights.

How To Legislate Temperance?

Analysis of the pollution problem as a function of population density uncovers a not generally recognized principle of morality, namely: *the morality of an act is a function of the state of the system at the time it is performed* (10). Using the commons as a cesspool does not harm the general public under frontier conditions, because there is no public, the same behavior in a metropolis is unbearable. A hundred and fifty years ago a plainsman could kill an American bison, cut out only the tongue for his dinner, and discard the rest of the animal. He was not in any important sense being wasteful. Today, with only a few thousand bison left, we would be appalled at such behavior.

In passing, it is worth noting that the morality of an act cannot be determined from a photograph. One does not know whether a man killing an elephant or setting fire to the grassland is harming others until one knows the total system in which his act appears. "One picture is worth a thousand words," said an ancient Chinese; but it may take 10,000 words to validate it. It is as tempting to ecologists as it is to reformers in general to try to persuade others by way of the photographic shortcut. But the essense of an argument cannot be photographed: it must be presented rationally—in words.

That morality is system-sensitive escaped the attention of most codifiers of ethics in the past. "Thou shalt not . . ." is the form of traditional ethical directives which make no allowance for particular circumstances. The laws of our society follow the pattern of ancient ethics, and therefore are poorly suited to governing a complex, crowded, changeable world. Our epicyclic solution is to augment statutory law with administrative law. Since it is practically impossible to spell out all the conditions under which it is safe to burn trash in the back yard or to run an automobile without smog-control, by law we delegate the details to bureaus. The result is administrative law, which is rightly feared for an ancient reason—*Quis custodiet ipsos custodes?*—"Who shall watch the watchers themselves?" John Adams said that we must have "a government of laws and not men." Bureau administrators, trying to evaluate the morality of acts in the total system, are singularly liable to corruption, producing a government by men, not laws.

Prohibition is easy to legislate (though not necessarily to enforce); but how do we legislate temperance? Experience indicates that it can be accomplished best through the mediation of administrative law. We limit possibilities unnecessarily if we suppose that the sentiment of *Quis custodiet* denies us the use of administrative law. We should rather retain the phrase as a perpetual reminder of fearful dangers we cannot avoid. The great challenge facing us now is to invent the corrective feedbacks that are needed to keep custodians honest. We must find ways to legitimate the needed authority of both the custodians and the corrective feedbacks.

Freedom To Breed Is Intolerable

The tragedy of the commons is involved in population problems in another way. In a world governed solely by the principle of "dog eat dog"—if indeed there ever was such a world—how many children a family had would not be a matter of public concern. Parents who bred too exuberantly would leave fewer descendants, not more, because they would be unable to care adequately for their children. David Lack and others have found that such a negative feedback demonstrably controls the fecundity of birds (11). But men are not birds, and have not acted like them for millenniums, at least.

If each human family were dependent only on its own resources; if the children of improvident parents starved to death; *if,* thus, overbreeding brought its own "punishment" to the germ line—*then* there would be no public interest in controlling the breeding of families. But our society is deeply committed to the welfare state (12), and hence is confronted with another aspect of the tragedy of the commons.

In a welfare state, how shall we deal with the family, the religion, the race, or the class (or indeed any distinguishable and cohesive group) that adopts overbreeding as a policy to secure its own aggrandizement (13)? To couple the concept of freedom to breed with the belief that everyone born has an equal right to the commons is to lock the world into a tragic course of action.

Unfortunately this is just the course of action that is being pursued by the United Nations. In late 1967, some 30 nations agreed to the following (14):

The Universal Declaration of Human Rights describes the family as the natural and fundamental unit of society. It follows that any choice and decision with regard to the size of the family must irrevocably rest with the family itself, and cannot be made by anyone else.

It is painful to have to deny categorically the validity of this right; denying it, one feels as uncomfortable as a resident of Salem, Massachusetts, who denied the reality of witches in the 17th century. At the present time, in liberal quarters, something like a taboo acts to inhibit criticism of the United Nations. There is a feeling that the United Nations is "our last and best hope," that we shouldn't find fault with it; we shouldn't play into the hands of the archconservatives. However, let us not forget what Robert Louis Stevenson said: "The truth that is suppressed by friends is the readiest weapon of the enemy." If we love the truth we must openly deny the validity of the Universal Declaration of Human Rights, even though it is promoted by the United Nations. We should also join with Kingsley Davis (15) in attempting to get Planned Parenthood-World Population to see the error of its ways in embracing the same tragic ideal.

Conscience Is Self-Eliminating

It is a mistake to think that we can control the breeding of mankind in the long run by an appeal to conscience. Charles Galton Darwin made this point when he spoke on the centennial of the publication of his grandfather's great book. The argument is straightforward and Darwinian.

People vary. Confronted with appeals to limit breeding, some people will undoubtedly respond to the plea more than others. Those who have more children will produce a larger fraction of the next generation than those with more susceptible consciences. The difference will be accentuated, generation by generation.

In C. G. Darwin's words: "It may well be that it would take hundreds of generations for the progenitive instinct to develop in this way, but if it should do so, nature would have taken her revenge, and the variety *Homo contracipiens* would become extinct and would be replaced by the variety *Homo progenitivus*" (16).

The argument assumes that conscience or the desire for children (no matter which) is hereditary—but hereditary only in the most general formal sense. The result will be the same whether the attitude is transmitted through germ cells, or exosomatically, to use A. J. Lotka's term. (If one denies the latter possibility as well as the former, then what's the point of education?) The argument has here been stated in the context of the population problem, but it applies equally well to any instance in which society appeals to an individual exploiting a commons to restrain himself for the general good—by means of his conscience. To make such an appeal is to set up a selective system that works toward the elimination of conscience from the race.

Pathogenic Effects of Conscience

The long-term disadvantage of an appeal to conscience should be enough to condemn it; but has serious short-term disadvantages as well. If we ask a man who is exploiting a commons to desist "in the name of con-science," what are we saying to him? What does he hear? —not only at the moment but also in the wee small hours of the night when, half asleep, he remembers not merely the words we used but also the nonverbal communication cues we gave him unawares? Sooner or later, consciously or subconsciously, he senses that he has received two communications, and that they are contradictory: (i) (intended communication) "If you don't do as we ask, we will openly condemn you for not acting like a responsible citizen"; (ii) (the unintended communication) "If you do behave as we ask, we will secretly condemn you for a simpleton who can be shamed into standing aside while the rest of us exploit the commons."

Everyman then is caught in what Bateson has called a "double bind." Bateson and his co-workers have made a plausible case for viewing the double bind as an important causative factor in the genesis of schizophrenia (17). The double bind may not always be so damaging, but it always endangers the mental health of anyone to whom it is applied. "A bad conscience," said Nietzsche, "is a kind of illness."

To conjure up a conscience in others is tempting to anyone who wishes to extend his control beyond the legal limits. Leaders at the highest level succumb to this temptation. Has any President during the past generation failed to call on labor unions to moderate voluntarily their demands for higher wages, or to steel companies to honor voluntary guidelines on prices? I can recall none. The rhetoric used on such occasions is designed to produce feelings of guilt in noncooperators.

For centuries it was assumed without proof that guilt was a valuable, perhaps even an indispensable, ingredient of the civilized life. Now, in this post-Freudian world, we doubt it.

Paul Goodman speaks from the modern point of view when he says: "No good has ever come from feeling guilty, neither intelligence, policy, nor compassion. The guilty do not pay attention to the object but only to themselves, and not even to their own interests, which might make sense, but to their anxieties" (18).

One does not have to be a professional psychiatrist to see the consequences of anxiety. We in the Western world are just emerging from a dreadful two-centuries-long Dark Ages of Eros that was sustained partly by prohibition laws, but perhaps more effectively by the anxiety-generating mechanism of education. Alex Comfort has told the story well in *The Anxiety Makers* (19); it is not a pretty one.

Since proof is difficult, we may even concede that the results of anxiety may sometimes, from certain points of view, be desirable. The larger question we should ask is whether, as a matter of policy, we should ever encourage the use of a technique the tendency (if not the intention) of which is psychologically pathogenic. We hear much talk these days of responsible parenthood; the coupled words are incorporated into the titles of some organizations devoted to birth control. Some people have proposed massive propaganda campaigns to instill responsibility into the nation's (or the world's) breeders. But what is the meaning of the word responsibility in this context? Is it not merely a synonym for the word conscience? When we use the word responsibility in the absence of substantial sanctions are we not trying to browbeat a free man in a commons into acting against his own interest? Responsibility is a verbal counterfeit for a substantial *quid pro quo*. It is an attempt to get something for nothing.

If the word responsibility is to be used at all, I suggest that it be in the sense Charles Frankel uses it (20). "Responsibility," says this philosopher, "is the product of definite social arrangements." Notice that Frankel calls for social arrangements—not propaganda.

Mutual Coercion
Mutually Agreed upon

The social arrangements that produce responsibility are arrangements that create coercion, of some sort. Consider bank-robbing. The man who takes money from a bank acts as if the bank were a commons. How do we prevent such action? Certainly not by trying to control his behavior solely by a verbal appeal to his sense of responsibility. Rather than rely on propaganda we follow Frankel's lead and insist that a bank is not a commons; we seek the definite social arrangements that will keep it from becoming a commons. That we thereby infringe on the freedom of would-be robbers we neither deny nor regret.

The morality of bank-robbing is particularly easy to understand because we accept complete prohibition of this activity. We are willing to say "Thou shalt not rob banks," without providing for exceptions. But temperance also can be created by coercion. Taxing is a good coercive device. To keep downtown shoppers temperate in their use of parking space we introduce parking meters for short periods, and traffic fines for longer ones. We need not actually forbid a citizen to park as long as he wants to; we need merely make it increasingly expensive for him to do so. Not prohibition, but carefully biased options are what we offer him. A Madison Avenue man might call this persuasion; I prefer the greater candor of the word coercion.

Coercion is a dirty word to most liberals now, but it need not forever be so. As with the four-letter words, its dirtiness can be cleansed away by exposure to the light, by saying it over and over without apology or embarrassment. To many, the word coercion implies arbitrary decisions of distant and irresponsible bureaucrats; but this is not a necessary part of its meaning. The only kind of coercion I recommend is mutual coercion, mutually agreed upon by the majority of the people affected.

To say that we mutually agree to coercion is not to say that we are required to enjoy it, or even to pretend we enjoy it. Who enjoys taxes? We all grumble about them. But we accept compulsory taxes because we recognize that voluntary taxes would favor the conscienceless. We institute and (grumblingly) support taxes and other coercive devices to escape the horror of the commons.

An alternative to the commons need not be perfectly just to be preferable. With real estate and other material goods, the alternative we have chosen is the institution of private property coupled with legal inheritance. Is this system perfectly just? As a genetically trained biologist I deny that it is. It seems to me that, if there are to be differences in individual inheritance, legal possession should be perfectly correlated with biological inheritance—that those who are biologically more fit to be the custodians of property and power should legally inherit more. But genetic recombination continually makes a mockery of the doctrine of "like father, like son" implicit in our laws of legal inheritance. An idiot can inherit millions, and a trust fund can keep his estate intact. We must admit that our legal system of private property plus inheritance is unjust—but we put up with it because we are not convinced, at the moment, that anyone has invented a better system. The alternative of the commons is too horrifying to contemplate. Injustice is preferable to total ruin.

It is one of the peculiarities of the warfare between reform and the status quo that it is thoughtlessly governed by a double standard. Whenever a reform measure is proposed it is often defeated when its opponents triumphantly discover a flaw in it. As Kingsley Davis has pointed out (21), worshippers of the status quo sometimes imply that no reform is possible without unanimous agreement, an implication contrary to historical fact. As nearly as I can make out, automatic rejection of proposed reforms is based on one of two unconscious assumptions: (i) that the status quo is perfect; or (ii) that the choice we face is between reform and no action; if the proposed reform is imperfect, we presumably should take no action at all, while we wait for a perfect proposal.

But we can never do nothing. That which we have done for thousands of years is also action. It also produces evils. Once we are aware that the status quo is action, we can then compare its discoverable advantages and disadvantages with the predicted advantages and disadvantages of the proposed reform, discounting as best we can for our lack of experience. On the basis of such a comparison, we can make a rational decision which will not involve the unworkable assumption that only perfect systems are tolerable.

Recognition of Necessity

Perhaps the simplest summary of this analysis of man's population problems is this: the commons, if justifiable at all, is justifiable only under conditions of low-population density. As the human population has increased, the commons has had to be abandoned in one aspect after another.

First we abandoned the commons in food gathering, enclosing farm land and restricting pastures and hunting and fishing areas. These restrictions are still not complete throughout the world.

Somewhat later we saw that the commons as a place for waste disposal would also have to be abandoned. Restrictions on the disposal of domestic sewage are widely accepted in the Western world; we are still struggling to close the commons to pollution by automobiles, factories, insecticide sprayers, fertilizing operations, and atomic energy installations.

In a still more embryonic state is our recognition of the evils of the commons in matters of pleasure. There is almost no restriction on the propagation of sound waves in the public medium. The shopping public is assaulted with mindless music, without its consent. Our government is paying out billions of dollars to create supersonic transport which will disturb 50,000 people for every one person who is whisked from coast to coast 3 hours faster. Advertisers muddy the airwaves of radio and television and pollute the view of travelers. We are a long way from outlawing the commons in matters of pleasure. Is this because our Puritan inheritance makes us view pleasure as something of a sin, and pain (that is, the pollution of advertising) as the sign of virtue?

Every new enclosure of the commons involves the infringement of somebody's personal liberty. Infringements made in the distant past are accepted because no contemporary complains of a loss. It is the newly proposed infringements that we vigorously oppose; cries of "rights" and "freedom" fill the air. But what does "freedom" mean? When men mutually agreed to pass laws against robbing, mankind became more free, not less so. Individuals locked into the logic of the commons are free only to bring on universal ruin once they see the necessity of mutual coercion, they become free to pursue other goals. I believe it was Hegel who said, "Freedom is the recognition of necessity."

The most important aspect of necessity that we must now recognize, is the necessity of abandoning the commons in breeding. No technical solution can rescue us from the misery of overpopulation. Freedom to breed will bring ruin to all. At the moment, to avoid hard decisions many of us are tempted to propagandize for conscience and responsible parenthood. The temptation must be resisted, because an appeal to independently acting consciences selects for the disappearance of all conscience in the long run, and an increase in anxiety in the short.

The only way we can preserve and nurture other and more precious freedoms is by relinquishing the freedom to breed, and that very soon. "Freedom is the recognition of necessity"—and it is the role of education to reveal to all the necessity of abandoning the freedom to breed. Only so, can we put an end to this aspect of the tragedy of the commons.

REFERENCES

1. J. B. Wiesner and H. F. York, *Sci. Amer.* **211** (No. 4), 27 (1964).
2. G. Hardin, *J. Hered.* **50**, 68 (1959); S. von Hoernor, *Science* **137**, 18 (1962).
3. J. von Neumann and O. Morgenstern, *Theory of Games and Economic Behavior* (Princeton Univ. Press, Princeton, N.J., 1947), p.11.
4. J. H. Fremlin, *New Sci.*, No. 415 (1964), p. 285.
5. A. Smith, *The Wealth of Nations* (Modern Library, New York, 1937), p. 423.
6. W. F. Lloyd, *Two Lectures on the Checks to Population* (Oxford Univ. Press, Oxford, England, 1833), reprinted (in part) in *Population, Evolution, and Birth Control*, G. Hardin, Ed. (Freeman, San Francisco, 1964), p. 37.
7. A. N. Whitehead, *Science and the Modern World* (Mentor, New York, 1948), p. 17.
8. G. Hardin, Ed. *Population, Evolution, and Birth Control* (Freeman, San Francisco, 1964), p. 56.
9. S. McVay, *Sci. Amer.* **216** (No. 8), 13 (1966).
10. J. Fletcher, *Situation Ethics* (Westminster, Philadelphia, 1966).
11. D. Lack, *The Natural Regulation of Animal Numbers* (Clarendon Press, Oxford, 1954).
12. H. Girvetz, *From Wealth to Welfare* (Stanford Univ. Press, Stanford, Calif., 1950).
13. G. Hardin, *Perspec. Biol. Med.* **6**, 366 (1963).
14. U. Thant, *Int. Planned Parenthood News*, No. 168 (February 1968), p. 3.
15. K. Davis, *Science* **158**, 730 (1967).
16. S. Tax, Ed., *Evolution after Darwin* (Univ. of Chicago Press, Chicago, 1960), vol. 2, p. 469.
17. G. Bateson, D. D. Jackson, J. Haley, J. Weakland, *Behav. Sci.* **1**, 251 (1956).
18. P. Goodman, *New York Rev. Books* **10**(8), 22 (23 May 1968).
19. A. Comfort, *The Anxiety Makers* (Nelson, London, 1967).
20. C. Frankel, *The Case for Modern Man* (Harper, New York, 1955), p. 203.
21. J. D. Roslansky, *Genetics and the Future of Man* (Appleton-Century-Crofts, New York, 1966), p. 177.

Thomas, J. W. 1986. Effectiveness—the hallmark of the natural resource management professional. *Transactions of the North American Wildlife and Natural Resources Conference* 51:27–38.

Wildlife management has come a long way since Genghis Khan established some of its first known principles, such as creating food plots for game animals. The role of the wildlife biologist has certainly evolved from one of studying and managing wild animals. Because of the array of issues wildlife biologists have to address (e.g., all aspects of animal biology, habitats, vegetation, and human dimensions including politics), the task of describing a wildlife biologist can be daunting. Indeed, the complexity of the profession makes it difficult to find a single description of the wildlife professional that applies to everyone in the field. Jack Ward Thomas' objective was to identify critical traits of successful wildlife and natural resource professionals.

First, their occupations should be more than a way to pay the bills; their work is a vocation, not a mere job. "For a vocation, the driving mechanism was not the boss but the will; the goal became not money but mission. A true vocation is the work one most needs to do, and it is the work the world most needs to have done." Besides biology, wildlife biologists need to understand and be able to appropriately use biopolitics, economics, and communication to be "effective for wildlife and the sound, holistic management of all renewable natural resources." In addition, adhering to professional ethics is critical. Other important aspects of being an effective wildlife professional include dressing appropriately, working with professional societies, and never giving up your dedication. This essay by Thomas provides an anchor for all wildlife biologists, as he discusses what it means to be a natural resource management professional and how "we few, we fortunate few, we band of brothers and sisters, are privileged indeed to stand this watch. What more could we ask but to be here in this time and place with a chance to make a difference?"

RELATED READING

Allen, D. L. 1962. Our wildlife legacy. Funk and Wagnalls, New York, New York, USA.

Benson, W. H. 1995. Better science makes for better decisions. Environmental Toxicology and Chemistry 14:1811–1812.

Fraidenburg, M. E. 2007. Intelligent courage: natural resource careers that make a difference. Krieger Publishing, Malabar, Florida, USA.

Gross, L. 2000. Education for a biocomplex future. Science 288:807.

Kessler, W. B. 1995. Wanted: a new generation of environmental problem solvers. Wildlife Society Bulletin 23:594–599.

Kroll, A. J. 2007. Integrating professional skills in wildlife student education. Journal of Wildlife Management 71:226–230.

Effectiveness—The Hallmark of the Natural Resource Management Professional[1]

Jack Ward Thomas
USDA Forest Service
Forestry and Range Sciences Laboratory
La Grande, Oregon

Prologue

What I have to say here is, I believe, germane to all the natural resource management professions. It is, however, written from the perspective of a wildlife biologist talking to other wildlife biologists.

Wildlife Biologists—Where Are We Coming From? ·

Some 25 years ago, I was working for the Texas Game and Fish Commission when a young biologist with the ink drying on his diploma came to work under my supervision. At the end of his first day, spent clearing brush from deer census lines, he exuberantly remarked, "Four years ago I couldn't even spell wildlife biologist and here I are one." We laughed at his joke and shared his joy.

Later, as we sat by the fire and talked late into the night, it became clear to me that he did not have a vocation. He exhibited an overwhelming concentration on *his* dreams, *his* needs, *his* desires. The new job was merely a means to those ends. He had not recognized that this new job was work the world needed to have done.

Maturity brought that recognition. A job evolved into a vocation. With vocation came self–imposed obligations: to grow, to improve, to strive, to serve, to be his best. A job was easy compared to a vocation. For a vocation, the driving mechanism was not the boss but the will; the goal became not money but mission.

Vocation, from the Latin verb *vocare,* is work to which one is called to by the gods (Morris 1976). Frederick Buechner (1973) proposed two criteria by which a true vocation could be judged: (1) it is work the individual most needs to do; and (2) it is work the world most needs to have done.

I believe that many (and certainly the most effective) resource management professionals have a vocation. Those who have a vocation have a precious and rare possession.

The Chinese have a blessing (sometimes described as a curse): "May you live in interesting times." If it is a blessing, we are doubly blessed. We have a vocation and, considering the critical importance of enlightened management of natural resources at this juncture in history, we live in the most interesting of times, the most critical of times, the most challenging of times.

How do we define ourselves? As usual, Leopold (1949:vii) probably said it best:

---

[1] This is a revised version of a paper first given at the Western Section of the Wildlife Society meeting in 1985 and printed in the *Cal–Neva Wildlife* Transactions (1985), under the title "Professionalism—Commitment Beyond Employment."

Effectiveness: Hallmark of Professionalism ◆ 27

"There are some who can live without wild things and some who cannot." We cannot or, at least, we choose not to.

Commitment beyond employment is required to produce wildlife biologists who continuously become all that they can be. I call such people "professionals." That word carries meaning for me beyond the dictionary definition of a profession as "an occupation or vocation requiring training in the liberal arts or the sciences and advanced study in a specialized field" (Morris 1976:1045). That's not nearly enough.

A sense of professionalism lies solely with the individual. Professionalism does not depend on professional societies and organizations, nor on employers. Professionalism is a reflection, through behavior, of vocation with its inherent commitment, and sharply focused will. Those who have those attributes will find or make a way to express their sense of professionalism. Once the individual has defined "professional" in his or her own mind and seared those standards into the soul, a standard for the conduct of a career has been established. This process of definition is personal and individual. Such people never allow their vision of their professionalism to rest in the hands of another.

Philosophical Positions—Contrasts and Conflicts

Many (if not most of us) have something of a split mind about who wildlife biologists are and what they do. The dilemma is manifest in the name given our profession—wildlife management. We have lived with the name so long that we fail to see that the words can be perceived as diametrically opposed in meaning. This is typified by the wildlife biologist's struggle to ensure that wildlife exists in habitats that are more and more controlled by human activity.

I see signs that this does not quite make sense to many wildlife biologists and it shows up in job dissatisfaction and in emotional distress (Kennedy and Mincolla 1985b). For example, consider the role of biologists in the management of our national forests. First and foremost, the wildlife biologist is quite likely dedicated to the welfare of wildlife on wild land. Second, the wildlife biologist is likely to be charged with helping convert wilderness into managed forests that produce wildlife.

The managed forest is comparatively tame and controlled compared to wilderness. The wildlife in the managed forest is as wild as its ancestors, but it is now a product of an environment more and more controlled by humans. Therefore, many wildlife biologists, philosophically dedicated to the preservation of wildness, participate in the purposeful dilution of wildness in order to preserve or produce wildlife in a managed environment. The wildlife is as wild as ever, but the environment is increasingly tame. Being a participant in this process forces many wildlife biologists to face an unanticipated paradox that leaves some confused and unsettled. The wildlife probably does not perceive a difference in the evolving habitat, but some biologists do and have moments of doubt. Those who doubt, perhaps, sense Leopold's (1949) observation that "Man always kills the thing he loves, and so we the pioneers have killed our wilderness. Some say we had to. . . ." Probably so, but for many biologists, it still hurts.

I suspect that there are very different philosophies among natural resource management professionals concerning how man relates to the natural world. Remember, there are no inherent rights or wrongs in these philosophical positions—they merely are. Some groups tend to be anthropocentric in philosophical position and take a utilitarian view of land—i.e., land exists for and is to be managed to satisfy people's needs (Devall and Sessions 1984, Leopold 1949).

Many wildlife biologists, I dare say, are mainly biocentric in philosophy (Kennedy and Mincolla 1985b) and view humans as part of nature (Leopold 1949), and subscribe to the admonition voiced by Sessions (1977:450) to be concerned with ". . . organic wholeness, [and to] love that, not man apart from that. . . ."

When biocentrists are employed by management agencies that, by law and tradition, are essentially anthropocentric in outlook and mission, there is apt to be friction (Devall and Sessions 1984). It is unlikely that most of those involved will recognize the problem for what it is—a basic difference in philosophy.

I am bemused by wildlife biologists who have an anthropocentric view of handling populations of game animals and predators and a biocentric view of forest and range management. Self-examination of basic philosophies and prejudices can be revealing and productive for us all.

The system for dealing with the management of public lands that has evolved in the United States has, in far too many cases, produced an adversarial relationship between wildlife biologists and foresters. In the formalized system that now exists, the land–use planning and allocation procedure can be referred to as *advocacy planning*. In advocacy planning, each interest group is expected to strive for satisfaction of its own welfare. Because compromise is inevitable as the culmination of such a process, each interest group feels that it has lost—a little or a lot. Relationships are apt to become a bit strained. Managers are given "targets" for various products from the forest. The best–defined and driving mechanism for the overall process tends to be timber harvest, followed by stand regeneration. Wildlife goals and objectives are much more difficult to define and quantify. As a result, objectives for wildlife have usually entered into the equation as constraints. Consider the definition of constraint: "A constraining agency or force; a repression of one's own feelings, behavior or action" (Morris 1976:286). So long as wildlife considerations are operative in the management arena as constraints, there will be intensifying conflict. Wildlife must be considered as a desired product—not as a constraint—to receive adequate attention (Thomas 1985).

Wildlife biologists that have a biocentric philosophy should recognize that, if they work for a land management agency or state game and fish department, they are facing an inherently anthropocentric orientation in the work place. Merely recognizing the situation can help biocentrists to be more effective. At least it can help them understand and deal with flashes of schizophrenia that come in the night.

Effectiveness—the Measure of Success

In the end, the measure of success for a professional is demonstrated effectiveness in achieving objectives. The following are considerations in enhancing effectiveness.

Biopolitics—Achieving Results in the Real World

Wildlife biologists are trained to be concerned with the art and science of managing wildlife, habitats and the users of wildlife. Another facet of natural resource management—biopolitics—is not as well understood nor as skillfully practiced as it should be. In fact, biology and politics personify opposing views, in the purist's

Effectiveness: Hallmark of Professionalism ◆ 29

mind, of wildlife management. Biology implies the gathering, analysis, interpretation and application of data in a methodical and peer–approved process to achieve goals dispassionately derived. Politics, on the other hand, is defined as "the methods or tactics involved in managing a state or government" or the "partisan or factional intrigue within a given group" (Morris 1976:1015). In the sense of natural resource management, biology is never pure and politics are not necessarily corrupting. All data are collected, all analyses conducted and all conclusions drawn by individuals with a point of view established through education and experience (Livingston 1981). And decisions are made within the context of laws, courtrooms, policies and financial resources.

Biopolitics is concerned with interactions between biological facts and theory and the reconciliation of the desires of individuals and organizations within the constraints of law (Peek et al. 1982). It is "the art of resolving biological . . . management problems in a biologically sound and politically acceptable manner" (Greenley 1971:505).

We need to remember that, in most areas of natural resource management, the body politic sets the goals for most endeavors. Appropriately qualified people are then employed to achieve those goals. There is no guarantee that all such goals are well–stated, appropriate, needed or even achievable. Yet the professional must strive to achieve the goals, change the goals or, if the conflict with conscience is too great, to refuse to participate or to resign.

There is nothing inherently wrong with biopolitics. In fact, it is the essence of natural resource management in government agencies and in those agencies' relationships with constituencies (Allen 1962, Poole 1980). Unfortunately, most natural resource managers did not learn about biopolitics in school—not that it exists or how to practice the art.

No natural resource manager can be truly effective, over the long term, without mastery of biopolitics. So far as I know, there are few formal training programs and no degrees in biopolitics. Neophyte natural resource managers learn biopolitics from apprenticeship to a master practitioner if they are lucky and from experience if they are not. They remain perpetually naive and ineffective if they remain ignorant of the art. Perhaps it could be said that natural resource managers are not properly and fully educated until they understand and demonstrate mastery of biopolitics.

The effective wildlife manager is expert in biology *and* is a politician. Biologists know something about what makes elk or deer or ducks or woodpeckers tick. And they know that laws, land–use planning processes, agencies, governing boards and landowners largely determine the goals and objectives for management. A good biopolitician combines biological and political skills to achieve goals and objectives in an acceptable manner while considering prevailing circumstances, and legal and ethical constraints. The fate of wildlife in America depends and will continue to depend largely on effective application of biopolitics (Peek et al. 1982).

Economics—Does It Make The World Go Around?

Money does *not* make the world go around. But biologists just may be a part of the tiny minority of the American population that believes that. Most of the rest of society operates on the premise that money does, indeed, make the world spin on its axis. Biologists are not required to change their views on this matter. But we must recognize that economics dominate biopolitical decisions and the exploitation by and the allocation of natural resources among user groups. More and more, the fate of wildlife is being determined primarily by consideration of cost–benefit ratios when various alternatives of land management are considered. That probably always has been true, but the law now requires consideration of cost–benefit ratios in the management of federal lands.

When wildlife biologists were forcibly thrust into the arena of formalized cost–benefit analysis, they quickly found that, with the exception of game species in some states, wildlife does not have market value. That means that wildlife's value must be indirectly estimated. Value estimates so derived are, in practice, easily distinguished from real dollars and have been, in my view, notoriously ineffective in influencing resource allocation and management decisions to favor wildlife. Craig Rupp, speaking as a Regional Forester for the U.S. Forest Service, summed it up perfectly: "The times are changing. Today it's a matter of dollars and cents. That makes it tough on uses that don't produce much income . . ." (Findley 1982:313). This remains true despite the fact that overemphasis on cost–benefit analysis can lead to ecologically and socially inappropriate decisions.

That observation is difficult to dispute, particularly as it relates to the production of game species for recreational hunting. If wildlife doesn't produce income at least equal to costs the landowner incurs in producing wildlife, there is apt to be a continuing loss of wildlife habitat and wildlife. Purposeful provision of wildlife needs on evermore intensively managed lands will, almost inevitably, exact significant opportunity costs (Thomas 1984). Costs that exceed benefits produce cost–benefit ratios that are unfavorable to maintenance or enhancement of wildlife and their habitats.

This does not mean that all natural resource management issues will, or should be, settled primarily on the basis of economics—but, considering the track record, that's probably the way to bet. Fortunately, there are exceptions. Sometimes there is support from an aroused public to accomplish something because it is so obviously "right" that the cost–benefit ratios are put aside. We chose, at least for now, to have grizzly bears, whooping cranes, peregrine falcons and spotted owls. We chose, at least for now, to clean up Chesapeake Bay though it might be better economics to let it turn into another Houston ship channel. But such decisions are fragile and each will be subject, over and over again, to new arguments based on cost–benefit ratios. This will remain so until the science and art of economics are sophisticated and mature enough to encompass the full measure of costs and benefits. The effective natural resource manager, then, understands: (1) economics; (2) the role of economic considerations in decision making; (3) the capitalistic nature of the economy; and (4) increasing expectations that government assets will produce revenue. In the meantime, those interested in sound natural resource management must continue to strive to ensure that decisions concerning natural resources are made considering factors beyond cost–benefit ratios. These include ecological, social, aesthetic and ethical considerations.

We should remain cognizant of Leopold's (1949:225) exhortation that "The fallacy the economic determinists have tied around our collective neck, and which we now need to cast off, is the belief that economics determines all land-use. . . ." But, while knowing and believing that, biologists must be prepared to live, work and be effective in an atmosphere increasingly permeated by economic determinism.

Effectiveness: Hallmark of Professionalism ◆ 31

Communication Skills

Biologists cannot be completely effective without possessing and exercising good communication skills. That includes being able to write in both technical and popular style, converse intelligibly and speak persuasively to groups.

In my youth, I had a vision of what biologists were like. They bore close resemblance to the cartoon character naturalist Mark Trail, who with pipe clenched in teeth, paddled their canoes (with a big shaggy dog in the front) into the glowing sunset. Such paragons communed with nature, avoided people and their works, and were unhurried and at peace. I grew up, became a wildlife manager, and found out that dreams are not necessarily harbingers of the future.

I suspect that the last thing most wildlife managers ever wanted to be in their youth was a salesperson. I gradually discovered that wildlife managers, the truly effective ones, were also salespersons—for wildlife, for programs, for proposals that benefit wildlife and for good stewardship of natural resources.

Today's wildlife biologists stand their watch during a critical time for wildlife in our country and the world. How wildlife fares in the long run probably does not depend on a census perfectly done or a new piece of information on elk behavior or whether a hunting season runs for 7 or 10 days. It does depend, however, on effective communication among biologists, others interested in wildlife and natural resources, and the general public.

We have an obligation, as professionals, to be effective. To be effective, we must communicate well and often. There is no dearth of information on or training in how to improve one's communication skills. The key is to try—over and over. Given the stresses and strains of today's climate as influenced by conflicts over spending, allocation of natural resources and the general economic climate, it is tempting to hunker down and hide from the tempest. That temptation must be resisted. If there has ever been a time for speaking up for proper natural resource management, it is now.

Getting Your Head Straight

The effective professional is, by definition, a winner. By "winner," I don't mean (necessarily) a quick climb up some bureaucratic ladder, making more money or receiving accolades. After all, our profession is not a competitive sport. I mean being effective for wildlife and the sound, holistic management of all renewable natural resources.

There are very few total victories for those interested in wildlife and absolutely none that are final. We have to win for wildlife and appropriate natural resource management what we can, where we can, how we can and be proud, rejuvenated and encouraged by each success.

I watched a situation where several biologists helped consider the fate of a pristine watershed on a national forest. They looked at the fish and wildlife situation carefully and professionally, mustered the available information, and concluded and recommended that the area be included in an adjoining wilderness area. After considering additional pertinent information, the decision makers decided otherwise.

The watershed was allocated to be managed forest and alternatives were considered. The biologist's first recommendation was for "backcountry" status. Again, the decision was otherwise. An alternative was selected, however, in which fish and wildlife received high emphasis. The biologists were ready with recommendations as how to accomplish the goals. Most important, perhaps, they learned something at every step about how to play the game and to be effective, and they came away determined to do a better job next time.

Winners or losers? these biologists played hard, fair, truthfully, ethically and effectively in the only game in town. In the end, wildlife and holistic forest management were well-served by their efforts. I say they were winners.

Only when we do less than our best, are less than truthful or are less effective than we can be are we losers in the professional sense. So, we must think of ourselves as winners. We must always focus on next time—always next time. Yesterday's victories and defeats are, indeed, yesterday. Next time—always next time. We must believe we serve a good and necessary cause. For, I think, people become what they believe in their hearts. Attitude is crucial to effectiveness, and the professional is obligated to be effective.

Doing the Best You Can With What You've Got

Perhaps the greatest challenge that faces wildlife management professionals is the organization and synthesis of information on wildlife into a form that can be applied in management and evaluation. To say "we don't know enough" is to take refuge behind a half-truth and ignore the fact that decisions will be made regardless of the information available. In my opinion, it is far better to examine available knowledge, synthesize it, and combine it with expert opinion on how the system operates, and make predictions about the consequences of alternative management actions. What results are working hypotheses—places to start, ways to derive tentative responses to questions to which there are no certain answers (Thomas 1979). Ecology is made up of successive approximations—there is no final truth (Franklin 1985).

Yet those who produce and certainly those who apply models and other approximations in natural resource management need to hear a whisper saying over and over, "You are dealing only with the essence of what is—nature seen through a glass darkly. it is not real—it is but the shimmering image of the moment that will change as the viewer's perspective and need changes" (Thomas in press).

Do the best you can with what you've got. But remember to tell the truth, all the truth, all the time, about where the information came from, about the assumptions involved and about the level of confidence that you have in the product. Credibility requires that, and credibility is a prerequisite to effectiveness.

Continuing Education—Staying Sharp

The professional is always in the process of education. University diplomas are not proof of education or of competence. Such training is and has always been inadequate. It always will be. A university degree is merely a ticket to board, a license to learn, a platform on which new learning, new skills and experience can be structured and from which wisdom can emerge (Cutler 1982). University degrees signify the beginning of real learning not its terminus. Yet, of all our failings in striving for professionalism, we fail most grievously in continuing to learn and grow as we should (Krausman 1979, Nelson 1980). There is no excuse for that failure.

To my dismay, there are those who are satisfied with an appropriate degree(s), a certificate of blessing from this trade union or that, and a lifetime of going through

Effectiveness: Hallmark of Professionalism ♦ 33

the prescribed drills. To my mind, such are functionaries—not professionals. To those with vocation, there can be no cessation from learning, no respite from desire to improve, no relief from the demand to serve.

Education cannot make a professional, but a professional cannot exist without appropriate education. And, for the professional, the need for more training and new and better skills never ends.

Universities, professional societies and agencies are paying more and more attention to the needs of professionals in terms of continuing education. Approaches run from short courses, seminars and video tapes to more and better publications. Some employers are unable or unwilling to provide employees such training. That's no excuse. Pay your own way for training sessions. Step up your reading. There is more and better literature than ever before in wildlife biology. But we can't stop there. We must learn more about economics, forestry, range management, fisheries, land-use planning, politics, sociology, philosophy and history. Biologists operate in an increasingly complex world, and if we are to be effective agents for the overall good management of natural resources, we must be conversant in other fields (Cutler 1982, Kennedy and Mincolla 1982 and 1985a).

Yet, we often hear the refrain, "I'm so busy I don't have time to read, to study, to learn." I don't agree. We wouldn't and shouldn't accept such a statement from the lawyers, airplane pilots, financial advisors, physicians and other professionals we employ. We do not and should not accept such statements from any person who aspires to be a professional. The professional strives for and aspires to excellence.

Appearance—Seeing is Critical to Believing

Some time ago, during his anti-establishment period, a colleague had occasion to deliver what could have been a very important briefing to some agency heads. After the briefing, one of them quietly said, "I suspect that what you said was important. But, frankly, I had a hard time hearing you because of the way you look."

The colleague grumbled and rationalized, but came to the inescapable conclusion that his appearance had detracted from his effectiveness. A too-rare chance to really do something for wildlife had been lost. He never lost another chance for that reason.

Too often, we let the dress code of our particular subculture get in the way of our effectiveness to do something for wildlife and for society. Too often people can't hear us because of how we dress or act or talk.

Dress and behavior should be suitable to the occasion. There is a time for field clothes and a time for suits—not because of anything so mundane as an appropriate professional image but because of necessity to enhance effectiveness. Professionals have the obligation to be effective.

Ethics and the Professional Society

As I implied earlier, the definition of professionalism that determines the actions and attitudes of individuals is self-defined and largely self-imposed. Ethics, on the other hand, has to do with a standard of behavior within a group or profession. Most organizations that feature themselves as the standard bearer of any group of people aspiring to professional status, sooner or later comes up with a code of ethics. Those organizations relating to the natural resource management professions are much the same.

The Wildlife Society has a code of ethics and standards for professional conduct and standards of behavior for Certified Wildlife Biologists (The Wildlife Society 1978). They are flowery but good words for professional wildlife biologists to live by. In simple words, they say:

1. Tell folks that your prime responsibility is to the public interest, the wildlife resource and the environment.
2. Don't perform professional services for anybody whose sole or primary intent is to damage the wildlife resource.
3. Work hard.
4. Don't agree to perform tasks for which you aren't qualified.
5. Don't reveal confidential information about your employer's business.
6. Don't brag about your abilities.
7. Don't take or offer bribes.
8. Uphold the dignity and integrity of your profession.
9. Respect the competence, judgment and authority of other professionals.

Implied but not specifically mentioned is the requirement simply to tell the truth. More and more lately, I seem to find myself advising troubled colleagues to tell the truth. It seems so simple. Yet, it can be so liberating. We live in an age of euphemisms, half–truths, obfuscations, double–talk and double–think. This atmosphere has closed in on us so gradually, so cloaked in the camouflage of the committee or team report, so justified by the need to get the job done, that we've come to consider such things the norm. Tell the truth, all the truth, all the time. It's the right thing, the healthy thing, the professional thing to do.

The Professional Society—the Professional's Prop

Some definitions of a profession indicate that the members are organized into an association that is responsible for maintaining and improving the quality of the service. Other definitions say that a profession is defined by the existence of a body of knowledge or literature.

The Wildlife Society serves that role for wildlife biologists. Many of us belong to other professional societies as well. There is no conflict and much benefit in that. The Wildlife Society gives voice and definition to our profession. I cannot imagine our profession existing without it.

Yet, probably more wildlife biologists do not belong to The Wildlife Society than do. But that's the norm for other societies in the natural resource management professions. To paraphrase John Kennedy, it has never occurred to me to ask "What does The Wildlife Society do for me?" The opposite tack seems more appropriate—"What can we do for The Wildlife Society, for the profession, for wildlife?" And truly, service is its own reward, yielding benefits far in excess of the individual's contributions in time and money.

There are those who need The Wildlife Society, who believe in its goals and who are willing to support it with money and service. There are those who don't. Just maybe, the problem does not lie primarily with The Wildlife Society.

That doesn't mean I always agree with the Society's decisions. But I have little

Effectiveness: Hallmark of Professionalism ◆ 35

respect for those who, upon losing an argument, withdraw support from the Society. We should be bigger than that—the stakes are too high and we are too few to make such action laudable. In short, professional involvement is a required commitment beyond employment.

Summary

Those are my ideas of what commitments beyond employment are required for wildlife biologists to be professionals. I started with an observation about vocation, about how precious and how rare it is to have a vocation instead of a job. More and more of our colleagues—disappointed by disparagement, discouraged by de-emphasis on environmental concerns, beaten down by budget cuts on top of budget cuts—are saying things like, "That's it, I'm putting in the hours I have to and no more" or, basically, "To hell with it."

I've felt the temptation—but it is wrong. If you have a vocation, don't let other people or circumstances make that vocation into a mere job. All that you have can be stripped away—wealth, possessions, status, job, loved ones. The only thing that belongs to you forever unless *you* give it up, is what's in your head and in your heart. Hang on tight. A sense of vocation is a truly rare and precious possession. It is what, down deep, spells the difference between professionals and functionaries.

Cervante's character Don Quixote, in his perceived madness, saw things differently and, strangely enough, more clearly than other men. He recognized that the quest, the striving, was everything. In a musical version of the story, he dared "to dream the impossible dream." We pursue what some say is an impossible dream of maintaining wildlife as a continuing part of our nation's and the world's fabric. Impossible? It is we who bear much of the responsibility for the answer to that question.

We can't afford dropouts nor the insidious slow poisoning of cynicism. Those so afflicted need to cure themselves or move aside. The stakes are too high and there are too many young people who want to try, who don't know they can't succeed, to make such indulgence acceptable. Those with battle scars and gray in their hair must be steadfast and optimistic. We owe that to our successors.

For those interested in wildlife management, indeed in the management of natural resources, these are confusing and often discouraging times. Natural resource management professionals have great responsibilities to keep the faith and serve steadfastly as advocates and agents of good stewardship and management. These are indeed interesting times—times of testing. It is useless to look back for the good old days—they are gone. It is pointless to look around for others to lead—they aren't there. For better or worse, we're it. Whether we recognize it or not, we are agents of change in how natural resources are treated, considered and used. If we succeed, there will be accolades from historians. If we fail, historians will, doubtless, take little note—but history will be much different. In my opinion, we stand at one of those moments in history that is a watershed for our nation in terms of how we treat our natural resources. We need to be fully aware of *who* we are, *what* we are and *where* we are in history.

These are, indeed, interesting times, exciting times, critical times. When the history of conservation in the United States in the 20th century is written, this period will loom as large, for good or ill, I believe, as the did times of Pinchot and Roosevelt.

36 ◆ *Trans. 51ˢᵗ N. A. Wildl. & Nat. Res. Conf.*

When I consider the role that natural resource management professionals can play in human affairs at this juncture in history, the words of Shakespeare's Henry V as he contemplated victory over the superior French force at Agincourt come to mind:

> [This day] . . . shall ne'er go by,
> from this day to the ending of the world,
> But we in it shall be remembered—
> We few, we happy few, we band of brothers. . . .

We few, we fortunate few, we band of brothers and sisters, are privileged indeed to stand this watch. What more could we ask but to be here in this time and place with a chance to make a difference? We will be remembered kindly by history if we fulfill our charge of continuing on course to achieving a maturity and skill wherein we can provide the needs of mankind while preserving ". . . the integrity, stability, and beauty of the biotic community" (Leopold 1949:225). It is, indeed, a noble quest and a worthy vocation.

References Cited

Allen, D. L. 1962. Our wildlife legacy. Funk and Wagnalls Co., New York. pp. 293–308.

Buechner, F. 1973. Wishful thinking. Harper and Row, New York. 100 pp.

Cutler, M. R. 1982. What kind of wildlifers will be needed in the 1980's? Wildl. Soc. Bull. 10(1):75–79.

Devall, B., and G. Sessions. 1984. The development of natural resources and the integrity of nature. Environ. Ethics 6(4):293–322.

Findley, R. 1982. Our national forests: problems in paradise. Nat. Geogr. 162:313.

Franklin, J. J. 1985. Observations from outside. Park Sci. 5(2):19.

Greenley, J. C. 1971. The effects of biopolitics on proper game management. Proc. West. Assoc. State Game and Fish Comm. 51:505–509.

Kennedy, J. J., and J. A. Mincolla. 1982. Career evolution of young 400–series U.S. Forest Service professionals. Career Development Project Report 1, College of Natural Resources, Utah State Univ., Logan. 26 pp.

Kennedy, J. J., and J. A. Mincolla. 1985a. Career development and training needs of entry-level wildlife/fisheries managers in the USDA Forest Service. Career Development Project Report 4. College of Natural Resources, Utah State Univ., Logan. 69 pp.

Kennedy, J. J., and J. A. Mincolla. 1985b. Early career development of fisheries and wildlife biologists in two Forest Service Regions. Trans. North Am. Wildl. and Natural Resour. Conf. 50:425–435.

Krausman, P. R. 1979. Continuing (?) education for wildlife administrators. Wildl. Soc. Bull. 7:57–58.

Leopold, A. 1949. A Sand County almanac and sketches here and there. Oxford Univ. Press., New York. 226 pp.

Livingston, J. A. 1981. The fallacy of wildlife conservation. McClellan and Stewart, Ltd. Toronto, Ontario. 117 pp.

Morris, W., ed. 1976. The American heritage dictionary. Houghton Mifflin Co., Boston, Mass. 1,550 pp.

Nelson, L., Jr. 1980. Final report on the pilot program of continuing education in wildlife ecology and management. The Wildlifer 182:43–44.

Peek, J. M., R. J. Pedersen, and J. W. Thomas. 1982. The future of elk and elk hunting. Pages 599–626 in Elk of North America—ecology and management, edited by J. W. Thomas and D. E. Toweill. Stackpole Books, Harrisburg, Pa.

Poole, D. A. 1980. The legislative process and wildlife. Pages 489–494 in Wildlife management techniques manual, edited by S. D. Schemnitz. The Wildlife Society, Washington, D.C.

Sessions, G. 1977. Spinoza and Jeffers on man in nature. Inquiry 20:481–528.

Effectiveness: Hallmark of Professionalism ◆ 37

Thomas, J. W., ed. 1979. Preface. Pages 6–7 *in* Wildlife habitats in managed forests—the Blue Mountains of Oregon and Washington. U.S. Dept. Agric., Agric. Handb. No. 553. U.S. Gov. Print. Off., Washington, D.C.

———. 1984. Fee–hunting on the public's lands?—an appraisal. Trans. North Am. Wildl. and Nat. Resour. Conf. 49:455–468.

———. 1985. Toward the managed forest—going places we've never been. For. Chron. 61(2):168–172.

———. In press. Wildlife habitat modeling—cheers, fears, and introspection. *In* Proceedings modeling habitat relationships of terrestrial vertebrates. Univ. Wisconsin Press, Madison.

Wildlife Society (The). 1978. The Wildlife Society program for certification of professional wildlife biologists. The Wildl. Soc., Washington, D.C. 7 pp.

Decker, D. J., C. C. Krueger, R. A. Baer, Jr., B. A. Knuth, and M. E. Richmond. 1996. From clients to stakeholders: a philosophical shift for fish and wildlife management. *Human Dimensions of Wildlife* 1:70–82.

Most of the research and, thus, science concerning wildlife management prior to the 1990s dealt with the animal and its habitat. As human dimensions grew to be a vital component of wildlife management, papers were published that laid the foundation for key concepts and ideologies concerning how people fit into the management and conservation equation. This is such a paper, and it is the first that used the term *stakeholder* (i.e., "any citizen potentially affected by or having a vested interest [a stake] in an issue, program, action or decision leading to an action"). The term has been used widely since then and represents a philosophical shift in wildlife management as the public became more interested and, hence, involved in wildlife management, conservation, and policy. During this period, state and federal agencies expanded their management to include stakeholder input by holding public forums or scoping sessions.

The objective of the authors was to demonstrate that the older model of clients (i.e., hunters, trappers, and fishermen) was outdated and needed to be replaced for effective management. Decker et al. make their case by outlining the manifold public attitudes and interests in wildlife issues, the increasing expectation of public participation in wildlife management issues (both from the public and agency personnel), the broadening of the traditional client model that had dominated wildlife management, and ways to develop diverse views. The authors also propose communication strategies to encourage understanding between stakeholders and managers while keeping management of wildlife resources effective and a top priority.

This paper is included in the section Our Philosophical Roots because the stakeholder concept had been mentioned as early as 1933 when Leopold (1933:422) wrote: "Herein lies the social significance of game management. It promulgates no doctrine, it simply asks for land and the chance to show that farm, forest, and wildlife products can be grown on it, to the mutual advantage of each other, of the landowner, and of the public." It took decades, however, before the concept was embedded in contemporary wildlife management.

RELATED READING

Baker, S. V., and J. A. Fritsch. 1997. New territory for deer management: human conflicts on the suburban frontier. Wildlife Society Bulletin 25:404–407.

Carpenter, L. H., D. J. Decker, and J. F. Lipscomb. 2000. Stakeholder acceptance capacity in wildlife management. Human Dimensions of Wildlife 5:5–19.

Chase, L. C., T. M. Schusler, and D. J. Decker. 2000. Innovations in stakeholder involvement: what's the next step? Wildlife Society Bulletin 28:208–217.

Curtis, P. D., and J. R. Hauber. 1997. Public involvement in deer management decisions: consensus versus consent. Wildlife Society Bulletin 25:399–403.

Leopold, A. 1933. Game management. Charles Scribner's Sons, New York, New York, USA.

Weiss, C. H. 1983. Toward the future of stakeholder approaches in evaluation. Pages 83–96 *in* A. S. Bryk, editor. Stakeholder-based evaluation. Jossey-Bass, San Francisco, California, USA.

Human Dimensions of Wildlife
Spring 1996

Volume 1 Number 1
pp. 70-82

From Clients to Stakeholders: A Philosophical Shift for Fish and Wildlife Management

Daniel J. Decker
Human Dimensions Research Unit
Department of Natural Resources
Cornell University

Charles C. Krueger
Richard A. Baer, Jr.
Department of Natural Resources
Cornell University

Barbara A. Knuth
Human Dimensions Research Unit
Department of Natural Resources
Cornell University

Milo E. Richmond
New York Cooperative Fish and Wildlife Research Unit
Department of Natural Resources
Cornell University

Abstract: Fish and wildlife management in North America has been experiencing a fundamental philosophical shift among professional managers and policy makers about who are the beneficiaries of management. This has been reflected in broadening notions of who should be considered in decision making; not just traditional clients who pay for and receive services of managers, but all stakeholders in fish and wildlife management. The term "stakeholder" has emerged to represent any citizen potentially affected by or having a vested interest (a stake) in an issue, program, action or decision leading to an action. The stakeholder approach in management decision making recognizes a larger set of beneficiaries of management (including, in concept, future generations) than the traditional concepts of constituencies and clients, or customers, a term currently popular among fish and wildlife agencies. The stakeholder approach requires: (1) identification of important stakeholders, (2) flexibility in selection of methods for incorporating stakeholder input in decision making to account for specific contexts, (3) development of a professional management philosophy strong enough to resist powerful special interests when broader public interests are in the balance, (4) development of ways to weigh stakeholder views on issues in management decision making, and (5) establishment of effective strategies for communication between managers and stakeholders and among stakeholders to encourage understanding and compromise.

Key words: stakeholder, management, decision making, constituents, interest groups, publics, clients, users.

Fish and wildlife in North America belong to the people as shared resources. The management and decision-making rules that govern their use and sustainability are similar to those found in the management of any common property. Although patterns of consumption and management rules vary across cultures, regions, and resources, the pervasive problem for professionals responsible for managing "common" natural resources (Bromley, 1991, 1992; Hardin, 1968; McKean, 1982) is the same: recognizing the interests of numerous individuals and coordinating resource use by them to optimize value while simultaneously sustaining the resource. This problem is made more challenging in the context of managing fish and wildlife as a common property resource because of: (1) the diverse range of public interests, concerns and uses of fish and wildlife; (2) the increasing public expectation for citizen participation in management decision making; and (3) the broadening view among managers about who are the beneficiaries of fish and wildlife management. Increasingly, people who have interests in fish and wildlife but are not anglers, hunters, and trappers have communicated to policy makers and managers their desires to have their interests addressed. Representing a philosophical shift for many managers, they have sought to understand the fish and wildlife interests and concerns of such people and to consider these in management decision making.

Adoption of this new, broadened perspective about whose interests and concerns should be considered and who should have input in fish and wildlife management decisions is a vital step in keeping the profession in a viable, central role in conservation. Failure to recognize and consider the breadth of public interests in fish and wildlife can diminish management credibility and effectiveness. Despite the importance of this broader perspective, the transition has been subtle, slow, and resisted by some fish and wildlife professionals.

This paper shares our perception of this broader view of the beneficiaries of fish and wildlife management, and argues the importance to the profession of adopting it. We begin with a review of the evolution of thinking vis-a-vis beneficiaries of fish and wildlife management over the last few decades. We then discuss the stakeholder approach to management, and note some challenges for the profession as it widely adopts this new approach.

Stakeholder: A New and Critical Concept for Fish and Wildlife Management

We encourage the fish and wildlife management profession to adopt and use the term "stakeholders" to refer to the beneficiaries of fish and wildlife management. The concept of stakeholder originated in the field of program evaluation, where it was used in the early 1970s (Bryk, 1983) to represent someone with a vested interest (a stake) in an issue or program (Gold, 1983). Included as stakeholders are all those who may be affected by a program, as well as those who make decisions about how the program is managed (Weiss, 1983). Environmental dispute resolution practitioners and theorists embraced the concept of including all stakeholders in dispute resolution efforts (Crowfoot & Wondolleck, 1990; Susskind & Cruikshank, 1987). Definition of the term "stakeholder" related to environmental disputes varies, but typically includes individuals and groups who have: (1) legal standing, (2) great political influence, (3) power to block implementation of a decision, or (4) sufficient moral claims (Susskind & Cruikshank, 1987).

The term stakeholder is already being used in the fish and wildlife management literature (e.g., Decker & Krueger, 1993, p. 55), but has not been well defined nor consistently applied. We propose that stakeholders in fish and wildlife management be defined simply as those individuals and groups who may be affected by or can affect fish and wildlife management decisions and programs. Applying this definition liberally, people with many kinds and degrees of stakes may be stakeholders in a management decision, which can be a cumbersome notion to operationalize. Realistically, fish and wildlife managers must use judgment about which stakes and stakeholders to emphasize in decision-making processes. The discussion provided here is intended to help professionals reflect on that responsibility as they engage in this new way of doing business.

Clients and Constituents: Original Foci of Fish and Wildlife Management

Prior to about 1970, the people of primary concern to the fish and wildlife management profession were anglers, hunters, and trappers. These resource users fit the traditional definition of a "constituency;" a group of people (constituents) who authorize or support the efforts of others (professionals) to act on their behalf. Those professionals, the fish and wildlife managers, attended to the user groups' interests through their decisions and actions. The conventional vocabulary of the day also referred to anglers, hunters, and trappers as "clients;" people who pay (e.g., through license fees or earmarked taxation) for professional or expert services.

The terms "constituents" and "clients" reflected a special relationship between the professional fish and wildlife manager and the recognized direct beneficiaries of their work. The professional was the expert, paid to make decisions and carry out actions for the benefit of the client. In this relationship, the professional typically was assumed to possess superior knowledge about all aspects of the management process, including knowing what the "right" management goals and objectives were as well as the best ways to achieve them. Because the professional manager had few constituencies to serve, an accurate understanding of their wants and needs was easy to maintain. Thus, most decisions could be made about management objectives without systematic studies of users or comprehensive citizen participation processes.

The client-manager system functioned well for many years because it embraced a mutually shared, narrow set of values regarding fish and wildlife. This close-knit and functional relationship was the working model in most states until the early 1970s when some managers began to consider other groups and vested interests, especially those focused on wildlife. Increasingly, managers found that the concept "constituent" did not work because growing numbers of people interested in resource management were not supporters of the *status quo* (Decker & Krueger, 1993). The concept of "client," someone who receives a service for a fee, was obsolete because not all those interested in management paid for it, nor did all those who paid for management receive a service (Decker & Krueger, 1993). The concept of "user" did not apply, because not all those interested in fish and wildlife management personally used the resource (Decker & Krueger, 1993). To the contrary, among those with growing, visible interest in fish and wildlife management were landowners who posted land against hunting and fishing (Brown & Thompson, 1976) and farmers who expressed their displeasure about crop damage from wildlife (Brown & Decker, 1979; Brown, Decker, & Dawson, 1978). In addition, growing numbers of nonconsumers sought to have fish and wildlife resources protected for their own recreational, environmental, ecological, or humane/animal rights interests. Landowners, farmers, and people with these other interests, however, were for the most part not regarded as potential beneficiaries of fish and wildlife management. Rather, their problems and concerns typically were viewed as impediments to be overcome to achieve management objectives for the "real" clients—anglers, hunters, and trappers.

New Interest Groups and Publics: Change During the 1970s and '80s

During the late 1960s and early 1970s, growing public interest in the environment, punctuated by Earth Day and evidenced in such legislation as the Multiple Use-Sustained Yield Act, National Environmental Policy Act, and the Endangered Species Act, markedly influenced the fish and wildlife profession. Managers began to realize that people other than their traditional clientele were expressing keen interest in all fish and wildlife (not just those consumed or causing nuisance). This period of heightened interest in environmental issues was accompanied by the creation of many local, state, and national environmental groups that vigorously sought consideration in management and policy decisions. Their views were expressed in powerful ways, including state and federal legislation, agency requirements for citizen participation and, with increasing frequency, through the courts, all of which have had major impacts on policy.

The fish and wildlife management literature of the time began to carry many references to "interest groups" and "publics." These two terms helped focus managers' attention on the fact that groups of people other than traditional consumptive users had interests in wildlife and reason to be considered beneficiaries of management. National surveys sponsored every five years by the U.S. Fish and Wildlife Service confirmed that the number of nonconsumptive wildlife users was increasing. Early research in the human dimensions area documented the broad sets of beliefs and attitudes people held about wildlife in addition to valuing wildlife for recreational use (Kellert, 1980). Expressions of people's beliefs, attitudes and interests ranged from concern about problems wildlife cause for people (e.g., car collisions, Lyme disease) to advocacy for the existence of rare and endangered species (e.g., desert pupfish, piping plovers). Furthermore, the animal rights movement received considerable media exposure during the 1980s. These groups typically profess the view that animals should not be managed by people, that animals should not be purposefully harmed by humans, and therefore that rights of animals should supersede human desires for their use for purposes such as hunting, fishing and trapping. These ideas are far removed from the philosophical foundation of traditional fish and wildlife management.

Another trend emerged during the 1970s—many people sought a greater part in governmental decisions at all levels, local to national. The traditional model for management of wild animals was becoming cumbersome and unacceptable (Richmond, 1973). A new era of citizen participation in government was launched, and many people with interests in wildlife expressed a desire to participate in decision making. By the late 1980s, citizen participation became a common activity for some fish and wildlife agencies, and increased their credibility among key publics (Stout, Decker, & Knuth, 1992).

One outcome of the increase in citizen participation in decision making for fish and wildlife management was a reduction of managers' control over the decisions. Increased accountability to the public was unsettling for a profession that considered nearly all decision making for management as its prerogative alone. Managers were faced with the new problem of responding to a greater variety of interests and concerns expressed in the public forum while still meeting legislated natural resource management mandates of stewardship and retaining the authority to apply professional judgment when appropriate.

Complicating this further, a new type of fish and wildlife professional has graduated from our colleges and universities and has been employed by our management agencies since the mid-1970s. Like their predecessors, these aspiring new professionals are dedicated and well educated but do not necessarily share the traditional values that, for example, place hunting as one of the highest priority uses of wildlife; nor do they uniformly regard hunters as the primary beneficiaries of management. Not every new fish and wildlife graduate hunts or fishes! Though these new professionals may support or be neutral toward such traditional uses, many are interested in managing fish and wildlife for a variety of values, sometimes different from the values that motivated their predecessors. The infusion of these different perspectives within the profession has contributed to the changing view of the range of beneficiaries of wildlife management.

The Stakeholder Approach: Considering All Beneficiaries of Management

Since the late 1980s, many professional managers have become active in ensuring that a broader range of interests and concerns (i.e., stakes) in the management of fish and wildlife are considered in decision making. A variety of methods are employed routinely to gain input and frequently to involve the range of stakeholders in management decisions. In this new stakeholder approach, managers seek to include all people who may be affected by a management decision (whether or not they recognize it themselves), not just those who make their views known to managers. Representation by an organized group is not a requirement for having one's interests considered in a management decision. For example, managers recognize that deer cause millions of dollars of damage to motor vehicles and to homeowners' landscape plantings, yet such interests are seldom represented by special interest groups. Deer managers themselves, therefore, have increasingly sought to ensure that these stakeholders' concerns are given fair consideration in management decisions. In addition, fish and wildlife professionals, especially governmental agency employees, have special "trusteeship" responsibility for the future and therefore responsibility for ensuring that tomorrow's citizens (not just future anglers and hunters) are considered stakeholders in today's management decisions (i.e., need to consider opportunity costs or options for future generations, sometimes referred to as existence and bequest values [Bishop, 1987; Steinhoff, Walsh, Peterle, & Petulla, 1987]). Thus, the manager's responsibility in the stakeholder approach (i.e., ensuring that interests and concerns of all significant stakeholders are considered in management decisions) is substantially greater than in the client-manager or constituency-manager systems described earlier. The stakeholder approach is similar to the type of thinking that currently prevails in understanding how the concept of the commons actually works (e.g., Bromley, 1991, 1992; Hardin, 1968; Oakerson, 1992).

The broadening perspectives about the beneficiaries of management and about the interests and concerns that should be considered in management decisions reflect some important characteristics of the fish and wildlife profession that need to be reinforced. First, adoption of a stakeholder approach indicates that fish and wildlife management has the capacity to be adaptive and dynamic in recognizing new needs and changing to improve effectiveness, reflecting a capacity to deal with diverse current and future needs. Second, apparently many people drawn to fish and wildlife management careers understand the difference between reacting to pressures from special interest groups to yield certain decisions, and responding to the full spectrum of current and future societal needs for management of fish and wildlife on a sustainable basis; these professionals try to understand and address broad societal values, rather than limit their focus only to interests that are consistent with their own personal desires and priorities.

The changes that have occurred during the last 25 years reflect a maturation process in the fish and wildlife management profession as it has responded to broader societal changes. During this period, the profession's perspective about who has a legitimate interest or stake in management has expanded. People with traditional interests have not been culled or ignored, but they now have to share the attention of fish and wildlife managers with other stakeholders. Tensions within the profession and between the profession and its traditional "clients" have arisen because of this evolution, but we believe the outcome will be management that serves a broader cross-section of society.

The future of fish and wildlife management depends on managers' responsiveness to the full spectrum of society's values without falling victim to the special interest politics of one or a few stakeholder groups. Identifying and considering the values of all significant stakeholders are essential steps in decision processes that will sustain public support. We do not, however, advocate adopting a populist approach, in which managers try in vain to implement every action desired by each stakeholder group (to attempt this would be chaotic given the contradictory goals of some groups, or even to follow the wishes of a majority could be inappropriate if it is ill-advised for biological reasons). In considering the interests of diverse stakeholders, farsighted and successful public resource managers should not abrogate their responsibility for stewardship and public trust and likewise should not become brokers (Nielsen, 1985), simply doling out resources to highly vocal or single-minded interests. Rather, they should work hard to identify the range of interests that exists pertaining to fish and wildlife management. Given these interests and biological information, managers need to create a vision for the future and develop long-range goals that keep them on track when special interest politics attempts to derail their efforts. (We argue later in this article that much of this "visioning" activity occurs because of an abrogation of responsibility of legislators in providing clear and precise legislation.)

Ultimately, due to the expanded notion of whose values ought to be considered in fish and wildlife management decisions, managers are faced with confronting ethical questions in decision making (Decker, Shanks, Nielsen, & Parsons, 1991). This responsibility is perhaps more evident under the stakeholder approach to management than in previous client-centered approaches. Values such as fairness, justice, and long-term concern for the sustainability of resources are morally and professionally defensible in our society and should be used, along with legislated mandates, as guides in decision-making. In the stakeholder approach one responsibility of the manager is to remind stakeholders of the importance of these overarching issues, help identify the consequences of alternative management decisions vis-a-vis these ethical concerns, and rely on professional judgment when these fundamental values are being encroached. When doing so, however, managers must distinguish between what is scientifically defensible and what is morally or legally defensible (Decker, et al., 1992).

The stakeholder approach suggests that a partnership of professional managers and a diverse body of stakeholders work together to identify management goals and solutions to problems. Fish and wildlife professionals must help people, in an unbiased manner, recognize the full range of short- and long-term consequences of management actions being considered. Communication and trust between managers and stakeholders and among different groups of stakeholders will be critically important. In summary, managers must take responsibility for ensuring: (a) that decision-making processes take into consideration the breadth of relevant stakeholder needs and interests, even those not advocated by special-interest organizations, (b) that those needs and interests are given weight in decisions (Decker & Lipscomb, in review), and (c) that such decisions reflect the overall public interest.

Challenges of the Stakeholder Approach

We realize that adopting the stakeholder approach in fish and wildlife management entails challenges that require further consideration. Here we can do little more than list some of them, and encourage debate and discussion throughout the profession.

One challenge is that more public interests must be brought to the decision-making process and weighed by fish and wildlife managers. This may mean refusing to cave in to the most powerful lobbies or resisting making seemingly "safe" decisions based solely on public opinion surveys (Decker, 1994). In fulfilling these responsibilities, fish and wildlife managers must develop a management philosophy that enables them, when necessary, to resist persistent pressure from particular stakeholders. Because political pressures can be so strong, managers should not be expected to do this on their own. The fundamental responsibility for framing basic guidelines for fish and wildlife management ought to rest with legislatures, where dialogue and debate can be fully public and open to all contenders. Given broad legislative mandates as a foundation, professional fish and wildlife managers can build the framework for effective management that involves stakeholder input processes that, ideally, are insulated from politics. The problem with this ideal, however, was identified earlier—the short-term goals of many politically powerful interest groups have great influence on public officials, a situation that works against the stakeholder approach. Legislators often do not want to bear the political consequences that may be generated by framing clear and precise legislation, preferring instead to give wide latitude to agency personnel in carrying out a particular policy (Lowi, 1979, p. 301). In raising this issue, we are not advocating that legislators micromanage our fish and wildlife resources, but rather that they clearly define overall mandate and societal values that fish and wildlife managers should strive to fulfill.

A second challenge stems from the fact that judgments about which stakes and stakeholders to consider in a particular situation are not always clear. If interpreted too broadly, stakeholder becomes synonymous with "citizen" and thus becomes either useless or redundant. If interpreted too narrowly, the larger public rightly can object that it is simply a pretext for empowering particular special interest groups. What stakes and which stakeholders to include in management decisions needs to vary with the circumstances, regions of the country, and management problems that are present. It is unlikely that every possible stakeholder will have sufficient weight to be included in a management decision. This leads to a related problem, professionals' use of the "responsibility for future generations" perspective to assume veto authority over stakeholder decisions. It is essential to avoid having expression of this perspective construed by participating stakeholders as contempt for them, as if they had no concern for future generations themselves.

Finally, we note that debates about fish and wildlife are likely to become even more controversial over the next few decades, which will make application of the stakeholder approach more difficult and more important. To take a specific case, a good deal of recent thinking about how we should treat animals challenges customary beliefs at a fundamental level. Traditional liberalism holds that it is wrong to restrict an individual's freedom unless it can be shown that significant harm to other persons will likely result from not doing so. But animals are not persons and thus are not taken as morally considerable under this worldview. Today, however, some advocates of animal rights/welfare want to enlarge the umbrella of protection offered by the state. Peter Singer (1990, pp. 18-19), for instance, believes that animals should receive the same moral consideration as that given humans, claiming that "adult chimpanzees, dogs, pigs, and members of many other species far surpass the brain-damaged infant in their ability to relate to others, act independently, be self-aware, and any other capacity that could reasonably be said to give value to life." He concludes that "there will surely be some nonhuman animals whose lives, by any standards, will be more valuable than the lives of some humans." Depending on the extent to which this judgment is shared by citizens, it has potentially radical implications for fish and wildlife management.

What the animal rights/welfare issue makes clear is that arguments for the moral considerability of animals typically rest on either religious beliefs or on secular metaphysical beliefs that function very much like traditional religious beliefs. Such beliefs typically are part of competing worldviews, which, among other things, means that communication and the achievement of agreement on fish and wildlife management policies is likely to be difficult in a stakeholder approach, as it will be in any other approach we can identify.

Summary and Conclusion

Overall, the stakeholder approach is desirable in modern fish and wildlife management for at least three reasons. First, the approach considers the broad range of interests that exists now and is open to others that may be expressed in the future. Second, the inclusive nature of the approach can result in more segments of the public understanding and supporting management decisions. Third, fish and wildlife resources controlled and managed as common property ought to reflect the reasonable views of as many segments of the public as possible without greatly favoring any particular "client," "constituency," or "special interest" group. The keys to improving implementation will include: (1) expanding the manager's view of who is substantially affected by fish and wildlife management and therefore is a stakeholder in management decisions and actions, (2) identifying and understanding stakeholder views, (3) seeking compromise between competing demands (stakes) when appropriate (i.e., without risking the long term integrity or sustainability of fish and wildlife resources), and (4) improving communication between managers and stakeholders.

A broad range of stakeholder values ought to be considered in decision making for fish and wildlife management. However, as we noted earlier, no simple answer exists to the question of "How?" The advantages and disadvantages of various forums for public involvement to encourage exchange among stakeholders and between them and managers must still be assessed. Methodology for weighing the needs and interests of certain stakeholders compared to others needs further development. We do not have a specific, concrete definition of "the public interest" to guide managers; ideally legislative bodies have provided such guidance, but we know that is not typically the case. Defining appropriate solutions to fish and wildlife management issues must occur with the help of clear-thinking individuals perceptive of the needs and interests of others. With widespread adoption of the stakeholder approach will come the experience and evaluation needed to yield practical answers to these questions. It will be a learning process for the fish and wildlife management profession.

Evidence of the success that can be realized by fish and wildlife management agencies that take a stakeholder approach is starting to accumulate. For example, the decision-making process for determining

white-tailed deer population objectives in New York State has incorporated input from diverse stakeholders through an approach where a
citizen task force is established for each deer management unit on a five-
year cycle (Stout, Decker, & Knuth, 1992). Fisheries management also has
had successes to reinforce application of the stakeholder approach, such
as the public consultation activities associated with the international issue
of lake trout restoration in Lake Ontario (Lange & Smith, in press).

We would have preferred to conclude with recommendations that
were more specific and more immediately applicable "on the ground" for
the manager. But to attempt to do so would require oversimplification and
distortion of a very complex situation. We believe that to remain
successful, fish and wildlife professionals of the 1990s and beyond will
find the stakeholder approach useful, but they will have to modify and
apply it to local and regional issues in imaginative ways. Management
decisions were relatively simple when there were just a few well-defined
client groups. But those days are gone, and the future outlook is not clear.
Like explorers of old, fish and wildlife managers will face both the dangers
and the thrills of navigating uncharted waters. To be successful, fish and
wildlife professionals of the future will have to develop a performance
record and seek a widely-recognized image of giving unprejudiced
consideration to all significant stakeholder interests in management
decisions. Just as management of other common property resources
provides occasional basis for dispute, professional fish and wildlife
managers will continue to struggle with how best to handle irreconcilable
differences.

Not every fish and wildlife professional will agree with our perspective on the stakeholder approach, but we believe that it is an essential
element of responsive, adaptive management. In our opinion the future
of the profession will be inherited by those managers who adopt, refine
and practice the evolving stakeholder approach for fish and wildlife
management.

*This paper is a contribution of New York Federal Aid in Wildlife Restoration
Grant WE-173-G and Cornell University Agricultural Experiment Station Hatch
Project NY 147403. We thank J.W. Enck, C.A. Loker, G.R. Parsons, J.C. Proud, W.F.
Siemer, R.J. Stout and the journal reviewers for their helpful suggestions.*

References

Bishop, R.C. (1987). Economic values defined. In D.J. Decker & G.R. Goff
(Eds.), *Valuing wildlife: Economic and social perspectives* (pp. 24-33). Boulder,
CO: Westview Press.

Bromley, D.W. (1991). *Environment and economy: Property rights and
public policy*. Oxford: Blackwell.

Bromley, D.W. (1992). The commons, property, and common property regimes. In D.W. Bromley (Ed.), *Making the commons work* (pp. 3-15). San Francisco: ICS Press.

Brown, T.L., & Decker, D.J. (1979). Incorporating farmers' attitudes into management of white-tailed deer in New York. *Journal of Wildlife Management 43(1)*, 236-239.

Brown, T.L., Decker, D.J., & Dawson, C.P. (1978). Willingness of New York farmers to incur white-tailed deer damage. *Wildlife Society Bulletin 6(4)* 235-239.

Brown, T.L., & Thompson, D.Q. (1976). Changes in posting and landowner attitudes in New York State, 1963-1973. *New York Fish and Game Journal 23*, 101-137.

Bryk, A.S. (Ed.). (1983). Stakeholder-based evaluation. *New directions for program evaluation. No. 17*. San Francisco: Jossey-Bass.

Crowfoot, J.E., & Wondolleck, J.M. (1990). *Environmental disputes: Community involvement in conflict resolution*. Washington, D.C.: Island Press.

Decker, D.J. (1994). What are we learning from human dimensions studies in controversial wildlife situations? Some observations and comments. *Proceedings of the Association of Midwest Fish and Wildlife Agencies Annual Meeting 61*, 25-27.

Decker, D.J., Brown, T.L., Connelly, N.A., Enck, J.W., Pomerantz, G.A., Purdy, K.G., & Siemer, W.F. (1992). Toward a comprehensive paradigm of wildlife management: Integrating the human and biological dimensions. In W. R. Mangun (Ed.), *American fish and wildlife policy: The human dimension* (pp. 33-54). Carbondale, IL: Southern Illinois University Press.

Decker, D.J., & Krueger, C.C. (1993). Communication: Catalyst for effective fishery management. In C.C. Kohler & W.A. Hubert (Eds.), *Inland fisheries management in North America* (pp. 55-75). Bethesda, MD: American Fisheries Society.

Decker, D.J., & Lipscomb, J.F. (in review). Giving weight to stakeholders in wildlife management decision making: A conceptual foundation. *Human Dimensions of Wildlife*.

Decker, D.J., Shanks, R.E., Nielsen, L.A., & Parsons, G.R. (1991). Ethical and scientific judgements in management: Beware of blurred distinctions. *Wildlife Society Bulletin 19*, 523-527.

Gold, N. (1983). Stakeholder and program evaluation: Characterizations and reflections. In A.S. Bryk (Ed.), *New directions for program evaluation. No. 17* (pp. 63-72). San Francisco: Jossey-Bass.

Hardin, G. (1968). The tragedy of the commons. In G. Hardin & J. Baden (Eds.), *Managing in the commons* (pp. 16-30). San Francisco: W.H. Freeman.

Kellert, S.R. (1980). Public attitudes, knowledge and behaviors toward wildlife and natural habitats. *Transactions of the North American Wildlife & Natural Resources Conference 45*, 111-124.

Lange, R.E., & Smith, P.A. (in press). Lake Ontario fishery management: The lake trout restoration issue. *Journal of Great Lakes Research*.

Lowi, T.J. (1979). *The end of liberalism: The second republic of the United States*. New York: W.W. Norton & Company.

McKean, M.A. (1982). The Japanese experience with scarcity: Management of traditional commonlands. *Environmental Review 6(2)*, 63-88.

Nielsen, L.A. (1985). Philosophies for managing competitive fishing. *Fisheries 10(3)*, 5-7.

Oakerson, R.J. (1992). Analyzing the commons: A framework. In D.W. Bromley (Ed.), *Making the commons work,* (pp. 41-59). San Francisco: ICS Press.

Richmond, M.E. (1973). Land animals. *Annals of the New York Academy of Sciences 216,* 121-127.

Singer. P. (1990). *Animal liberation.* (2nd ed.) New York: New York Review of Books.

Steinhoff, H.W., Walsh, R.G., Peterle, T.J., & Petulla, J.M. (1987). Evolution of the valuation of wildlife. In D.J. Decker & G.R. Goff (Eds.), *Valuing wildlife: Economic and social perspectives* (pp. 34-48). Boulder, CO: Westview Press.

Stout, R.J., Decker, D.J. & Knuth, B.A. (1992). Evaluating citizen participation: Creating a communication partnership that works. *Transactions of the North American Wildlife & Natural Resources Conference 57,* 135-140.

Susskind, L., & Cruikshank, J. (1987). *Breaking the impasse: Consensual approaches to resolving public disputes.* USA: Basic Books, Inc.

Weiss, C.H. (1983). Toward the future of stakeholder approaches in evaluation. In A.S. Bryk (Ed.), *New directions for program evaluation. No. 17* (pp. 83-96). San Francisco: Jossey-Bass.

Organ, J. F., S. P. Mahoney, and V. Geist. 2010. Born in the hands of hunters: the North American Model of Wildlife Conservation. *The Wildlife Professional* 4:22–27.

Wildlife conservation and management in the U.S. and Canada has evolved as a unique and very successful experiment over the past century and a half. The North American Model of Wildlife Conservation (the Model) remains the envy of nations around the world. The Model was developed in response to the drastic drop in wildlife populations and a lack of management in the U.S. and Canada. John Organ and his coauthors' objectives were to outline the origins of the Model and describe the principal components that have established its foundation, also known as the Model's seven sisters. The seven sisters are wildlife as a public trust resource, elimination of markets for game, wildlife allocation by law, killing for legitimate purposes only, wildlife as an international resource, decisions based on science, and the democracy of hunting.

This Model has directed management and conservation of wildlife resources in North America and helped sustain the countries' biodiversity. Like all models, it is imperfect, and there are certainly aspects that can be improved. This is especially true when considering human attitudes in North America. Only a small percentage of the population has a passion to conserve wildlife, and many of these are sportsmen and sportswomen who have led the conservation movement.

RELATED READING

Allen, D. L., and the Committee on North American Wildlife Policy. 1973. Report of the committee on North American wildlife policy. Transactions of the North American Wildlife and Natural Resources Conference 38:153–181.

Geist, V., S. P. Mahoney, and J. F. Organ. 2001. Why hunting has defined the North American Model of Wildlife Conservation. Transactions of the North American Wildlife and Natural Resources Conference 66:175–185.

Leopold, A. 1930. Report of the committee on American wild life policy. American Game Conference 16:196–210.

Mahoney, S. P., and D. Cobb. 2010. Future challenges to the Model. The Wildlife Professional 4:83–85.

Organ, J. F., R. M. Muth, J. E. Dizard, S. J. Williamson, and T. A. Decker. 1998. Fair chase and humane treatment: balancing the ethics of hunting and trapping. Transactions of the North American Wildlife and Natural Resources Conference 63:528–543.

Prukop, J., and R. J. Regan. 2005. In my opinion: the value of the North American model of wildlife conservation—an International Association of Fish and Wildlife Agencies position. Wildlife Society Bulletin 33:374–377.

Born in the Hands of Hunters

The North American Model of Wildlife Conservation

Wildlife conservation in the United States and Canada has evolved over the last century and a half to acquire a form distinct from that of any other nation in the world. It's a conservation approach with irony at its core—sparked by the over-exploitation of wildlife, then crafted by hunters and anglers striving to save the resources their predecessors had nearly destroyed. Now a series of principles collectively known as the North American Model of Wildlife Conservation (Geist 1995, Geist et al. 2001), it helps sustain not only traditional game species but all wildlife and their habitats across the continent. The key to its future lies in understanding its origins.

By John F. Organ, Ph.D., Shane P. Mahoney, and Valerius Geist, Ph.D.

Historical Context

The North American Model (the Model) has deep social and ecological roots. In the early days of North American exploration, English and French settlers came from cultures where wildlife at various times in their histories was the private property of an elite landed gentry (Manning 1993). The explorations of these settlers were driven by the incredible wealth of North America's renewable natural resources—and by an unfettered opportunity to exploit it. Today, wildlife conservation in Canada and the United States reflects this historic citizen access to the land and its resources. Indeed, the idea that natural resources belong to the citizenry drives democratic engagement in conservation and forms the heart of North America's unique approach (Krausman 2009).

After resource exploitation fueled the expansion of people across the continent, the Industrial Revolution brought social changes that indelibly marked the land and its wildlife. In 1820, 5 percent of Americans lived in cities, but by 1860, 20 percent were urban dwellers, marking the greatest demographic shift ever to occur in America (Riess 1995). Markets for wildlife arose to feed these urban masses and to festoon a new class of wealthy elites with feathers and furs. Market hunters plied their trade first along coastal waters and interior forests. With the advent of railways, hunters exploited the West, shipping products from bison, elk, and other big game back to eastern cities. The march of the market hunter left once abundant species teetering on the brink of extinction.

By August 1886—when Captain Moses Harris led cavalry troops into Yellowstone National Park to take over its administration and stop rampant poaching—bison, moose, and elk had ceased to exist in the U.S. as a viable natural resource (U.S. Dept. Interior 1987). The Army takeover of Yellowstone is symbolic of the desperate actions taken to protect the remnants of American wildlife from total extinction. Ironically, the sheer scale of the slaughter was to have some influence in engendering a remarkable new phenomenon: the conservation ethic (Mahoney 2007).

Northern shovelers (*Anas clypeata*) take to the air over Laguna Atascosa National Wildlife Refuge in Texas.
Credit: Steve Hillebrand/USFWS

Credit: René Monsalve

John F. Organ, Ph.D., CWB, is Chief of Wildlife and Sport Fish Restoration for the U.S. Fish and Wildlife Service Northeast Region and Adjunct Associate Professor of Wildlife Conservation at the University of Massachusetts, Amherst.

Coauthors

Shane P. Mahoney is Executive Director for Sustainable Development and Strategic Science in the Department of Environment and Conservation, Government of Newfoundland and Labrador and Founder and Executive Director of the Institute for Biodiversity, Ecosystem Science, and Sustainability at the Memorial University of Newfoundland and Labrador.

Valerius Geist, Ph.D., is Professor Emeritus of Environmental Science at the University of Calgary in Alberta, Canada.

Credit: Jim Peaco/NPS

Credit: National Archives

Some 40,000 bison pelts in Dodge City, Kansas (right) await shipment to the East Coast in 1878—evidence of the rampant exploitation of the species. The end of market hunting and the continuing conservation efforts have given bison a new foothold across parts of their historic range, including Yellowstone National Park (above).

criticized Roosevelt for his limited experience in the West and for presenting hunting myths as fact. Roosevelt went to talk with Grinnell, and upon comparing experiences the two realized that big game had declined drastically. Their discussion inspired them to found the Boone and Crockett Club in 1887, an organization whose purpose would be to "take charge of all matters pertaining to the enactment and carrying out of game and fish laws" (Reiger 1975).

Roosevelt and Grinnell agreed that America was strong because, like Canada, its people had carved the country from a wilderness frontier with self-reliance and pioneer skills. With the demise of the frontier and a growing urban populace, however, they feared that America would lose this edge. They believed that citizens could cultivate traditional outdoor skills and a sense of fair play through sport hunting, thereby maintaining the character of the nation (Brands 1997).

The increasing urban population found itself with something that farmers did not have: leisure time. The challenges of fair-chase hunting became a favored pastime of many, particularly those of means. Conflicts soon arose between market hunters, who gained fortune on dead wildlife, and the new breed of hunters who placed value on live wildlife and the sporting pursuit of it.

These "sport" hunters organized and developed the first wildlife hunting clubs (such as the Carroll's Island Club, founded in Maryland in 1832) where hunters protected game from market hunters. Recreational hunters also pushed for laws and regulations to curtail market hunting and overexploitation. The New York Sportsmen's Club, for example, drafted laws recommending closed seasons on deer, quail, woodcock, and trout—laws which passed in 1848 (Trefethen 1975).

Pioneers in Conservation

An early advocate of game protection, Yale-educated naturalist George Bird Grinnell acquired the sporting journal *Forest and Stream* in 1879 and turned it into a clarion call for wildlife conservation. Grinnell had accompanied George Armstrong Custer on his first western expedition in 1874, where he saw herds of bison and elk. A decade later, in 1885, Grinnell reviewed *Hunting Trips of a Ranchman* by fellow New Yorker Theodore Roosevelt. In that review, Grinnell

Endorsing these ideals, influential members of the Boone and Crockett Club used their status to great advantage, helping to create some of North America's most important and enduring conservation legacies. In 1900, for example, Congressman John Lacey of Iowa drafted the Lacey Act, making it a federal offense to transport illegally hunted wildlife across state borders. Canadian Charles Gordon Hewitt wrote the Migratory Bird Treaty of 1916 to protect migratory birds from egg and nest collectors and unregulated hunting. And during his presidency from 1901 to 1909, Theodore Roosevelt protected more than 230 million acres of American lands and waters, doing more to conserve wildlife than any individual in U.S. history.

The Canadian effort revolved around the Commission on Conservation, founded in 1909 under the guidance of Prime Minister Wilfrid Laurier and noted conservationist Clifford Sifton, who served as the Commission's chairman and was eventually knighted for his efforts. Established to combat resource exploitation, the Commission—and its prestigious panel of scientists, academics, and policymakers—sought to provide scientific guidance on the conservation of natural resources. Working committees conducted research on agricultural lands, water, energy, fisheries, forests, wildlife, and other natural-resource issues, eventually publishing the first comprehensive survey of Canadian resources and the challenges to their conservation.

Emergence of a Profession

By the early 20th century, much of the infrastructure of wildlife conservation was already in place. In the 1920s, however, leading conservationists recognized that restrictive game laws alone were insufficient to stem wildlife's decline. To help address such concerns, ecologist Aldo Leopold and other conservationists published *American Game Policy* in 1930, which proposed a program of restoration to augment existing conservation law. "For the first time," writes Leopold biographer Curt Meine, "a coherent national strategy directed the previously disparate activities of sportsmen, administrators, researchers, and ... landowners" (Meine 1991).

Leopold and others also promoted wildlife management as a *profession*, advocating for trained biologists, stable funding for their work, and university programs to educate future professionals. Within 10 years many of these goals had been realized. Among them:

- *Wildlife curriculum.* In 1933, the University of Wisconsin launched the first wildlife management curriculum, a program that taught wildlife science, setting a standard for other universities.
- *Cooperative Wildlife Research Units.* Federal legislation in 1935 established a nationwide network of what are now known as Cooperative Research Units, where federal and state agencies and universities cooperate in fish and wildlife research and training.
- *Professional societies.* In 1937, W. L. McAtee, Aldo Leopold, and others founded The Wildlife Society, the first professional scientific society for those working in wildlife management and conservation. Said McAtee, "The time is ripe for inaugurating a professional

society" to promote discourse on issues facing wildlife conservation.
- *Funding legislation.* Congress passed the Duck Stamp Act of 1934 and the Federal Aid in Wildlife Restoration Act of 1937 (or the Pittman-Robertson Act) to provide reliable funding sources for federal and state wildlife conservation. (See article on page 35.)

Though initially launched in the U.S., these initiatives were endorsed and mirrored by Canadian policies and programs. In both nations, subsequent decades have brought expanded conservation legislation—such as the U.S. Endangered Species Act and Canadian Species at Risk Act—as well as partnership programs to promote and fund wildlife conservation, including the U.S. Migratory Bird Joint Ventures and the Teaming with Wildlife coalition.

The Model's Seven Pillars

Such key conservation laws and programs were built upon a firm foundation—the seven underlying principles of the North American Model (Geist *et al.* 2001). Those principles have stood the test of time, proving resilient to sweeping social and ecological changes (Mahoney and Jackson 2009). Will they stand the test of the future? That question can't be answered without a strong understanding of the principles themselves.

1. Wildlife as a Public Trust Resource. The heart of the Model is the concept that wildlife is

A Colorado hunter fires a Hawken muzzle-loading rifle, a primitive firearm first used on the American frontier in the 1820s. Sportsmen today carry on the tradition begun by early pioneers and trappers, tempered by the understanding that wildlife is a public trust resource to be killed only for legitimate purposes.

Credit: Dennis McKinney/Colorado Division of Wildlife

owned by no one and is held by government in trust for the benefit of present and future generations. In the U.S., the common-law basis for this principle is the Public Trust Doctrine, an 1841 Supreme Court Decision declaring that wildlife, fish, and other natural resources cannot be privately owned (*Martin v. Waddell*). In drafting the Public Trust Doctrine, Supreme Court Chief Justice Roger Taney drew upon the *Magna Carta*, which in turn was rooted in ancient Greek and Roman law. A subsequent Supreme Court Decision in 1896 regarding illegal transport of hunted ducks across a state border firmly made wildlife a trust resource (Geer v. Connecticut). Today, however, each state or province has its own laws regarding wildlife as a public trust. Those laws face potential erosion from multiple threats—such as claims of private ownership of wildlife, commercial sale of live wildlife, limits to public access, and animal-rights philosophy—

Credit: John Gilbert

Jennifer Vashon, a biologist for the Maine Department of Inland Fisheries and Wildlife, retrieves Canada lynx kittens for study. Her research team will measure the cats, determine their sex, collect DNA, and tag them for monitoring. Such work—funded in part by hunting license fees—informs the management of this rare species.

which are prompting moves for model language to strengthen existing laws (Batcheller *et al.* 2010).

2. Elimination of Markets for Game. Historically, the unregulated and unsustainable exploitation of game animals and migratory birds for the market led to federal, provincial, and state laws that greatly restricted the sale of meat and parts from these animals. Those restrictions proved so successful that today there is an overabundance of some game species—such as snow geese (*Chen caerulescens*) and white-tailed deer (*Odocoileus virginianus*) in suburban areas—which may warrant allowing hunting and the sale of meat under a highly

regulated regime. Such regulated hunting and trade could enhance public appreciation of hunting as a management tool by reducing human-wildlife conflicts with overabundant species. In addition, trapping of certain mammal species in North America and commerce in their furs are permitted, but are managed sustainably through strict regulation such that the impacts on populations lie within natural ranges (Prescott-Allen 1996). Unfortunately, trade in certain species of amphibians and reptiles still persists with little oversight, and should be curtailed through tighter restrictions.

3. Allocation of Wildlife by Law. As a trustee, government manages wildlife in the interest of the beneficiaries—present and future generations of the public. Access and use of wildlife is therefore regulated through the public law or rule-making process. Laws and regulations, such as the Migratory Bird Treaty Act, establish the framework under which decisions can be made as to what species can be hunted, what species cannot be harmed due to their imperiled status, and other considerations relative to public use of or impact on wildlife.

4. Kill Only for Legitimate Purpose. Killing wildlife for frivolous reasons has long been deemed unacceptable. The U.S. Congress passed a bill against "useless" slaughter of bison in 1874 (Geist 1995), and the "Code of the Sportsman" as articulated by Grinnell mandated that hunters use without waste any game they killed (Organ *et al.* 1998). Today, 13 states and provinces have "wanton waste" laws requiring hunters to salvage as much meat from legally killed game as possible. In Canada, the Royal Commission on Seals and Sealing recognizes that harvest of wildlife must have a practical purpose if it is to remain acceptable in society (Hamilton *et al.* 1998). Food, fur, self-defense, and property protection are generally considered legitimate purposes for the taking of wildlife. Other practices that conflict with this principle—such as prairie dog shoots or rattlesnake roundups—are under increasing scrutiny (see page 58).

5. Wildlife as an International Resource. One of the greatest milestones in the history of wildlife conservation was the signing of the Migratory Bird Treaty in 1916. Noted Canadian entomologist C. Gordon Hewitt, who masterminded the treaty, saw the protection of migratory songbirds as essential to the protection of agricultural crops against insect pests. Affecting far more than hunted wildlife, this was the first significant treaty that provided for international management of terrestrial wildlife resources. The impetus, of course, was that because some wildlife species migrate across borders, a nation's management policies—or lack thereof—can have consequences for wildlife living in neighboring countries. International commerce in wildlife, for example, has significant potential effects on a species' status. To address this issue, in 1973, 80 countries signed the first Convention on International Trade in Endangered Species of Wild Fauna and Flora (CITES). Today there are 175 parties to the treaty.

6. Science-based Wildlife Policy. Science as a basis for informed decision-making in wildlife management has been recognized as critical to wildlife conservation since the founding days of North American conservation (Leopold 1933). The subsequent application of this principle has led to many advances in management of diverse species, often under highly complex circumstances such as adaptive management of waterfowl harvest (Williams and Johnson 1995). Unfortunately, funding has been largely inadequate to meet the research needs of management agencies. In addition, a trend towards greater influence in conservation decision making by political appointees versus career managers profoundly threatens the goal of science-based management (Wildlife Management Institute 1987, 1997). So, too, do the divisions within the wildlife science community itself, which often splits along a human-versus-animal divide. The integration of biological and social sciences, which Leopold hoped would be one of the great advances of the 20th century, is necessary to meet the conservation challenges of the 21st century.

7. Democracy of Hunting. Theodore Roosevelt believed that society would benefit if all people had an access to hunting opportunities (Roosevelt *et al.* 1902). Leopold termed this idea the "democracy of sport" (Meine 1988)—a concept that sets Canada and the U.S. apart from many other nations, where the opportunity to hunt is restricted to those who have special status such as land ownership, wealth, or other privileges. Yet some note that the greatest historical meaning of the public trust is that certain interests—such as access to natural resources—are so intrinsically important that their free availability marks a society as one of citizens rather than serfs (Sax 1970).

Moving Beyond the Model
Bedrock principles of the North American Model of Wildlife Conservation evolved during a time when game species were imperiled and ultimately led to

Credit: Parks Canada

Elk in Canada's Waterton Lakes National Park are part of the "international herd," which regularly crosses the U.S.-Canada border. The North American Model holds that wildlife is an international resource and should be protected as such.

a continent-wide resurgence of wildlife at a scale unparalleled in the world, as evidenced by the restoration of deer, elk, waterfowl, bear, and many other species. It is clear that these principles have served wildlife conservation well beyond hunted species and helped sustain the continent's biodiversity, especially through the millions of acres of lands purchased with hunter dollars for habitat protection and improvement. Indeed, the structure of modern endangered species legislation harkens back to the old game laws, where the focus was on prevention of take.

As wildlife conservation advances into the 21st century, these founding principles should be safeguarded and improved, and new approaches to biodiversity conservation should be developed that go beyond what the Model currently provides. A U.S.-Canadian treaty securing the Model and improvements in wildlife law would be the most powerful form of protection. As we seek solutions to new challenges, we should remember that only a minority of our citizens have a passion for the perpetuation of wildlife, and among those, the people who call themselves sportsmen and sportswomen have been answering this call for well over one hundred years. Wildlife can ill afford to lose them in a future that is anything but secure. ■

This article has been reviewed by subject-matter experts.

For a full bibliography, go to *www.wildlife.org*.

Animals, Ecology & Populations

Grinnell, J. 1922. The role of the "accidental." *The Auk* 39:373–380.

Grinnell's early work conveyed much about the role of the accidental individual, or individuals, that influence(s) the distribution of wildlife and their rate of spread. Who would have thought in 1922 how important the topic of accidental colonization of plants and animals would be today? Every continent on Earth is faced with addressing the impacts of nonnative, invasive species. Grinnell's objective in this short but insightful essay was to outline the conditions related to distribution (e.g., land, water, climate, humidity, forage, successful breeding places, refuges) that change the role of a one-time occurrence of a species to a potential survival mechanism for the species.

The paper is important because it provides the fundamental basis of species range expansions. The reasons for species distribution are complicated and Grinnell was instrumental in outlining many of them. Although based on avian species numbers in California, Grinnell's essay applies to all dispersers and colonizers. While Grinnell's essay refers to natural expansion of species distributions, it also has relevance to introduced invasive and exotic plant and animal species, which are an increasingly important management issue in terms of preserving the ecological integrity of our ecosystems.

RELATED READING

Brown, J. H. 1984. On the relationship between abundance and distribution of species. The American Naturalist 124:255–279.

Chornesky, E. A., and J. M. Randall. 2003. The threat of invasive species to biological diversity: setting a future course. Annals of the Missouri Botanical Garden 90:67–76.

Elton, C. S. 1958. The ecology of invasions by animals and plants. Methusen, London, UK.

Mack, R. N., D. Simberloff, W. M. Lonsdale, H. Evans, M. Clout, and F. A. Bazzaz. 2000. Biotic invasions: causes, epidemiology, global consequences, and control. Ecological Applications 10:689–710.

Sax, D. F., and J. H. Brown. 2000. The paradox of invasions. Global Ecology and Biogeography 9:363–371.

Vila, M., E. Weber, and C. M. D'Antonio. 2000. Conservation implications of invasion by plant hybridization. Biological Invasions 2:207–217.

Williamson, M., and A. Fitter. 1996. The varying success of invaders. Ecology 77:1661–1666.

THE ROLE OF THE "ACCIDENTAL."*

BY JOSEPH GRINNELL.

The total number of species and subspecies of birds recorded upon definite basis from California amounts at the present moment to 576. Examination of the status of each species, and classification of the whole list according to frequency of observation, show that in 32 cases out of the 576 there is but one occurrence known. In 10 cases the presence of the species has been ascertained twice, in 6 cases three times, and for all the rest there are 4 or more

* Contribution from the Museum of Vertebrate Zoology of the University of California.

records of occurrence. Some 500 species can be called regularly migrant or resident.

Examination of the records for the past 35 years shows that the proportion of one-occurrence cases is continually increasing. In other words, the state is so well known ornithologically that regular migrants and residents have all or nearly all of them been discovered, and *their* number now remains practically constant, while more and more non-regulars are coming to notice. This might be explained on the ground that there are continually more and better-trained observers on the lookout for unusual birds. This is probably correct, partially. But also, I believe, there is indicated a continual appearance, within the confines of the state, through time, of additional species of extra-limital source.

In published bird lists generally, species which have been entered upon the basis of one occurrence only, are called "accidentals." This is true of county lists, of state lists, and, quite patently, of the American Ornithologists' Union's 'Check-list of North American Birds'. The idea in the adoption of the word "accidental" seems to have been that such an occurrence is wholly fortuitous, due to some unnatural agency (unnatural as regards the behavior of the bird itself) such as a storm of extraordinary violence, and that it is not likely to be repeated. This understanding of the word "accidental" is borne out by the explicit meaning given it in the 'Century Dictionary,' for instance, which is "taking place not according to the usual course of things," "happening by chance or accident, or unexpectedly." Now the way in which the word is used by ornithologists is really a misapplication of the term; for, as I propose to show, the occurrence of individual birds a greater or less distance beyond the bounds of the plentiful existence of the species to which they belong is the *regular thing, to be expected.* There is nothing really "accidental" about it; the process is part of the ordinary evolutionary program.

However, as I have intimated, the word is firmly fixed in distributional literature. We had better continue to use it; but let us do so with the understanding that it simply means that any species so designated has occurred in the locality specified on but one known occasion. No special significance need to be implied.

Accidentals are recruited mostly from those kinds of birds which are strong fliers. It is true that the majority belong to species of distinctly migratory habit. But some of our accidentals exemplify the most sedentary of species. Examples of one-instance occurrences, in other words "accidentals," are as follows: the Western Tanager in Wisconsin, the Louisiana Water-Thrush in southern California, Townsend's Solitaire in New York, the Catbird on the Farallon Islands, the Tennessee Warbler in southern California, and Wilson's Petrel on Monterey Bay. In the North American list some of the accidentals come from South America, some even from Asia and from Europe.

I would like to emphasize the point now that there is no species on the entire North American list, of some 1250 entries, that is not just as likely to appear in California sooner or later as some of those which are known to *have* occurred. Expressing it in another way, it is only a matter of time theoretically until the list of California birds will be identical with that for North America as a whole. On the basis of the rate for the last 35 years, $1\frac{3}{5}$ additions to the California list per year, this will happen in 410 years, namely in the year 2331, if the same intensity of observation now exercised be maintained. If observers become still more numerous and alert, the time will be shortened.

It will be observed that there are now many more one-occurrence, "accidental," cases than there are two-occurrence cases, and that there are more of the two-occurrence group than there are three-occurrence, and so forth, there being a regular reduction in the intervals so that, if we just had enough observations, a smooth curve would probably result. If the one-instance occurrences should continue to accumulate without any modification of the process, in the course of about 300 years there would be more of these "accidentals" in California than of regularly resident species, and the other groups would grade down in a steeper curve. I attempted to carry out the figures, which seem to behave according to some mathematical formula; but when I came to deal with $\frac{3}{5}$ of an occurrence I decided it was profitless to go on!

It is evident, however, that another process takes place, of quite opposite effect. With the lapse of time second-occurrence cases replace one-occurrence cases, to be followed by a third order of accretion, and this by a fourth; and the process might continue on ad infinitum, until theoretically, sometime after the full number of the North American list had been reached, our state list would no longer contain any accidentals at all. To cite an example, the eastern White-throated Sparrow was recorded first from California, and then as an accidental, on December 23, 1888; in 1889 a second specimen was taken; in 1891, a third was taken; and so on until in 1921, 19 occurrences have been recorded. This species is considered now simply as rare, certainly not accidental, casual, or even specially noteworthy save from a very local standpoint.

It comes to the mind here that if observations could be carried on so comprehensively as to bring scrutiny each year of every one of the 200,000,000 birds in California, this being the estimated minimum population maintained within the state from year to year, a great many more accidentals would be detected than are now known, and in addition some birds now known from but a few records, or even as accidentals, would come to be considered of frequent, though not necessarily regular, occurrence. With the White-throated Sparrow it is not impossible that a thousand of the birds have wintered in California in certain years.

Some of the considerations in the preceding paragraphs, while of interest in themselves perhaps, have confessedly been rather beside the issue. For the definite question which I wished to ask and which I will now briefly discuss is as to the function or role played by accidentals. Are they a mere by-product of species activity or do they in themselves constitute part of a mechanism of distinct use to the species?

The rate of reproduction in all birds, as with other animals, is so great that the population rapidly tends toward serious congestion except as relieved by death of individuals from various causes or else by expansion of the area occupied. The individuals making up a given bird species and occupying a restricted habitat may be likened to the molecules of a gas in a container which are continually beating against one another and against the confining walls, with resulting pressure outwards. But there is an essential difference in the case of the bird in that the number of individual units is being augmented 50 per cent, 100 per cent, in some cases even 500 per cent, at each annual period of reproduction, with correspondingly reinforced outward pressure.

The force of impingement of the species against the barriers which operate to hem it in geographically, results in the more than normally rapid death of those individuals which find themselves under frontier conditions. There follows, through time and space both, a continual flow of the units of population from the center or centers toward the frontiers.

The common barriers which delimit bird distribution are as follows: Land to aquatic species and bodies or streams of water to terrestrial species, the climatic barriers of temperature, up or down beyond the limits to which the species may be accustomed, and of atmospheric humidity beyond critical limits of percentage; the limits of occurrence of food as regards amount and kind with respect to the inherent food-getting and food-using equipment of the species concerned; and the limits of occurrence of breeding places and safety refuges of a kind prescribed by the structural characters of the species requiring them.

An enormous death rate results from the process of trial and error where individuals are exposed wholesale to adverse conditions. This can be no less, on an average, than the annual rate of increase, if we grant that populations are, on an average, maintaining their numbers from year to year in statu quo. But before the individuals within the metropolis of a species succumb directly or indirectly to the results of severe competition, or those at the periphery succumb to the extreme vicissitudes of unfavorable conditions of climate, food or whatnot obtaining there, the latter have served the species invaluably in *testing out* the adjoining areas for possibly new territory to occupy. These *pioneers* are of exceeding importance to the species in that they are continually being centrifuged off on scouting expeditions (to mix the metaphor), to seek new country which may prove fit for occupancy. The vast majority of such individuals, 99 out of every hundred perhaps, are foredoomed to early destruction without any opportunity of breeding. Some few individuals may get back to the metropolis of the species. In the relatively rare case two birds comprising a pair, of greater hardihood, possibly, than the average, will find themselves a little beyond the confines of the metropolis of the species, where they will rear a brood successfully and thus establish a new outpost. Or, having gone farther yet, such a pair may even stumble upon a combination of conditions in a new locality the same as in its parent metropolis, and there start a new detached colony of the species.

It is this rare instance of success that goes to justify the prodigal expenditure of individuals by the species. Such instances, repeated, result in the gradual extension of habitat limits on the part especially of species in which the frontier populations are in some degree adaptable—in which they can acquire modifications which make them fit for still farther peripheral invasion against forbidding conditions.

Incidentally, the great majority of these pioneers are, I believe, birds-of-the-year, in the first full vigor of maturity; such birds are innately prone to wander; and furthermore it is the autumnal season when the movement is most in evidence, a period of food-lessening when competitive pressure is being brought to bear upon the congested populations within their normal habitats. The impetus to go forth is derived from several sources.

The "accidentals" are the exceptional individuals that go farthest away from the metropolis of the species; they do not belong to the ordinary mob that surges against the barrier, but are among those individuals that cross through or over the barrier, by reason of extraordinary complement of energy, in part by reason of hardihood with respect to the particular factors comprising the barrier, and in part of course, sometimes, through merely fortuitous circumstances of a favoring sort.

Geologists tell us that barriers of climate are continually moving about over the earth's surface, due to uplift and depression, changes in atmospheric currents, and a variety of other causes. Animal populations are by them being herded about, as it were, though that is too weak a word. The encroaching barrier on the one side impinges against the population on that side; the strain may be relieved on the opposite side, *if* the barrier on that side undergoes parallel shifting, with the result that the species as a whole may, through time, flow in a set direction. If anything should happen that a barrier on one side impinged on a species without corresponding retreat of the barrier on the other side, the habitat of the species would be reduced like the space between the jaws of a pair of pliers, and finally disappear: the species would be extinct.

But, in the case of persistence, it is the rule for the population, by means of those individuals and descent lines on the periphery of the metropolis of the species, to keep up with the receding barrier and not only that but to press the advance. I might picture the behavior of the population of a given bird as like the behavior of an active amoeba. This classic animal advances by means of outpushings here and there in reaction to the environment or along lines of least resistance. The whole mass advances as well. The particles of protoplasm comprising the amoeba may be likened to the individuals comprising the entire population of the animal in question, the mass of the amoeba to the aggregate of the population.

It is obvious that the interests of the individual are sacrificed in the interests of the species. The species will not succeed in maintaining itself except by virtue of the continual activity of pioneers, the function of which is to seek out new places for establishment. Only by the service of the scouts is the army as a whole able to advance or to prevent itself being engulfed: in the vernacular, crowded off the map—its career ended.

The same general ideas that I have set forth with regard to birds, who happen to be endowed with means of easy locomotion, hold, I believe, also for mammals, and probably in greater or less degree for most other animals. I can conceive of a snail in the role of an "accidental," an individual which has wandered a few feet or a few rods beyond the usual confines of the habitat of its species. Given the element of time (and geologists are granting this element in greater and greater measure of late), the same processes will hold for the slower moving creatures as they seem to do for those gifted with extreme mobility.

Migration, by the way, looks to me to be just a phase of distribution, wherein more or less regular seasonal shifting of populations takes place in response to precisely the same factors as hem in the ranges of sedentary species.

The continual wide dissemination of so-called accidentals, has, then, provided the mechanism by which each species as a whole spreads, or by which it travels from place to place when this is necessitated by shifting barriers. They constitute sort of sensitive tentacles, by which the species keeps aware of the possibilities of areal expansion. In a world of changing conditions it is necessary that close touch be maintained between a species and its geographical limits, else it will be cut off directly from persistence, or a rival species, an associational analogue, will get there first, and the same fate overtake it through unsuccessful competition—supplantation.

Museum Vert. Zool., Univ. of California, Berkeley, Calif.
   (September 7, 1921.)

Elton, C. S. 1924. Periodic fluctuations in the numbers of animals: their causes and effects. *British Journal of Experimental Biology* 2:119–163.

Understanding temporal patterns of wildlife population numbers is fundamental for effective wildlife management. Wildlife cycles have received considerable attention and remain an important ecological process requiring further study. Theories abound regarding the cause of wildlife cycles, including moon phases, sunspots, influence of predators, food quality and quantity, and complex environmental linkages. Lotka (1925) and Volterra (1926) independently raised awareness about wildlife cycles with their predator-prey models. Based on predictions from mathematical models, they concluded that the predator and prey populations would cycle. Using these theoretical works biologists began a rigorous examination of existing data, conducting experiments to examine wildlife cycles and, if they existed, to then understand their causes and ecological importance.

Elton's objective was to provide a comprehensive overview of literature on fluctuations in animal populations attributed to climatic variations. This original paper reviewed data from studies of lemmings, hares and rabbits, mice and small mammals, and birds, among others. Elton concluded that fluctuations in number are widespread in many birds and mammals and are caused by climatic change and natural selection. He also discussed how fluctuations in these animals influenced other species in the community. This was the first paper to seriously examine the empirical data regarding the biological process of animal cycles. It remains a fundamental reference for research studies examining wildlife cycles and their relevance to ecosystem function and stability.

RELATED READING

Bulmer, M. G. 1974. A statistical analysis of the 10-year cycle in Canada. Journal of Animal Ecology 43:701–718.

Cole, L. C. 1951. Population cycles and random oscillations. Journal of Wildlife Management 15:233–252.

Griffin, P. C., and L. S. Mills. 2009. Sinks without borders: snowshoe hare dynamics in a complex landscape. Oikos 118:1487–1498.

Keith, L. B., and L. A. Windberg. 1978. A demographic analysis of the snowshoe hare cycle. Wildlife Monographs 58:1–70.

Krebs, C. J., M. S. Gaines, B. L. Keller, J. H. Meyers, and R. H. Tamarin. 1973. Population cycles in small rodents. Science 179:34–41.

Lotka, A. J. 1925. Elements of physical biology. Williams and Wilkins, Baltimore, Maryland, USA.

Volterra, V. 1926. Variations and fluctuations of the number of individuals in animal species living together. Appendix, pages 409–448 *in* R. Chapman. Animal Ecology. McGraw-Hill, New York, New York, USA.

Reproduced with permission.

PERIODIC FLUCTUATIONS IN THE NUMBERS OF ANIMALS: THEIR CAUSES AND EFFECTS

By C. S. ELTON.

Dept. of Zoology and Comparative Anatomy, The University Museum, Oxford.

CONTENTS

1. Introduction.

THIS paper has involved the study of facts from many subjects both inside and outside the domain of biology, and I am much indebted to the following gentlemen, who have generously helped me in various ways: Mr G. M. B. Dobson for much valuable advice on meteorology and on methods of analysing curves; Mr D. Brunt for allowing me to consult his unpublished work on periodicities in climate; Professor H. H. Turner for advice on astrophysical questions; Dr T. G. Longstaff for notes on a mouse plague in England; Dr S. Ekman for a list of Swedish lemming years; Dr J. Grieg for notes on a lemming year in Norway; Mr P. Uvarov for help with literature on locusts. I wish to thank all these gentlemen, and also in particular Mr J. S. Huxley who has given me much invaluable advice and help during the preparation of the paper.

2. Climatic Cycles.

It will be shown in the body of this paper that the periodic fluctuations in the numbers of certain animals there dealt with, must be due to climatic variations. This conclusion follows from the biological evidence alone. Therefore, in order that the facts may be appreciated in their proper bearing, a short

119

C. S. Elton

summary is given below of what is known on the subject of short-period pulsations of climate, apart from those deduced from animals. (Full treatment is given in the works of Brooks, 1922; Huntington, 1914, 1922, 1923; Humphreys, 1920; Shaw, 1923; and others.)

It is only within the last few decades that we have begun to reach some understanding on the subject of variations or pulsations of climate. Although there is still a great deal of controversy over many questions, on certain points agreement has been reached. Firstly, climate never remains constant; there are the daily and monthly variations we call weather, and there are recurrent seasonal changes; there are variations from year to year and over several years, and finally there are big climatic pulsations such as ice-ages. There is still much disagreement as to the causes of the larger pulsations, but for the present purpose it is only necessary to discuss the minor pulsations of climate of the order of two to twenty years, since it is the effects of these which are so clearly shown by animals and plants. For some time it was thought that these climatic pulsations were simply the chance result of an enormous number of interacting factors, and therefore impossible to analyse. But recent work has shown that there are certain pulsations of climate going on, which are definitely periodic. Much, however, still remains obscure, and the following account owing to its brevity will necessarily give a rather more clear-cut picture of climatic cycles than is really justifiable.

The ultimate source of nearly all the income of energy which the earth receives is the sun. From it we get a continual stream of radiant energy, part of which is intercepted and circulates in the form of our weather and climatic movements, and through animals and plants, before passing off again into space. Plants actually utilise about 2 per cent. of the sunlight that reaches the surface of the earth.

Now it is clear that variations in climate might be caused either by changes in the sun, or by terrestrial events, or by some outside agency besides the sun. The sun varies about 5 per cent. in its output of energy, and this variation has now been definitely proved to affect the earth's climate. The

120

Periodic Fluctuations in Numbers of Animals

number of sunspots has been observed regularly for over a hundred and fifty years, and these have a well-marked period whose average is 11.2 years, but which actually varies from nine to thirteen years in length. Also, although the minima all approximate to zero, the maxima vary in size, so that there is a well-marked major sunspot periodicity shown by the line joining the maxima of the 11-year periods (fig. 6).

Increase in the number of sunspots is accompanied by increased output of energy from the sun. The evidence for the influence of the sunspot periodicity on the earth's climate may be summarised as follows, and unless otherwise indicated is given in full in the works of the writers mentioned above.

(1) The average annual temperature of the whole earth appears to be, with certain exceptions dealt with later, lower at sunspot maxima, and higher at minima (see fig. 1). At the same time, the number of years for which suitable records exist is not large enough to make this correlation completely convincing. This paradox of a cool earth when there is a hot sun is possibly to be explained by the action of increased radiation of very short wave-length upon the upper atmosphere, rendering it less penetrable to the longer rays (light, infra-red).

(2) The pressures and rainfall of various parts of the world vary in relation to the 11-year sunspot periodicity. The careful work of Walker (1923), who uses the method of correlation coefficients, has proved this clearly. But whereas some regions have more rain at sunspot maxima, others have more at minima, so that we can divide the world into regions according to which way their climate varies.

The effect of the solar variation is probably in the nature of an interference with the normal balance of the atmosphere causing increased or decreased oscillation on a large scale.

(3) Kullmer has shown that the tracks of storms in North America shift about periodically, and that so far as the short series of observations goes, this shift is correlated with the sunspot period. This really means a periodic slight shift of the great climatic zones, and would be a fact of great importance, even if it were shown not to be caused by the solar variation.

(4) Huntington and Douglass have shown that the red-wood tree (Sequoia) bears a very clear record of past climate

121

in its annual rings, and they have been able to carry this record back three or four thousand years. The trees respond to changes in outer conditions of climate by varying the amount of growth of the wood, and if a large number of trees are measured and the rings of individual years compared, very accurate curves can be constructed. Here again there is a well-marked 11-year period, coinciding with that of the sunspots. There are also longer periods. Similarly an 11-year is shown in certain trees from Germany. (This has been attributed to periodic thinning of the forest, but the shape and period of the curve do not support this view.)

(5) Changes in the level of Lake Victoria (Brooks, 1923) and some other lakes are correlated in a very marked way with the sunspot period (*e.g.* a correlation coefficient of +0.87 between Lake Victoria level and sunspots). Here the level is the resultant of rainfall and evaporation, and there is thus an approach to the degree of integration of climatic factors which a living organism is capable of attaining. These are the main lines of evidence showing the importance of solar variation in the earth's climate. It has recently been shown (Hale, 1924) that the sunspots change their magnetic polarity every eleven years, so that there is apparently a 22-year period going on as well.

The amount of variation in temperature on the earth between sunspot maxima and minima is about 0.6° C. in the tropics, and less as one goes further away from the equator, and this is quite enough to be important. It represents from 1/10 to 1/20 of an ice-age, so to speak. A change of 0.5° C. will shift the isothermal line about eighty miles horizontally (Humphreys, 1920), and this is obviously enough to have great effects on animals and plants. It is important to note that a variation in the meteorological observations which may appear quite small, may have enormous effects on living organisms.

The only other factor outside the earth which has been suggested as a cause of fluctuations in climate of the short-period type in is the tidal effect of the moon. The effect on the atmosphere has been shown to be negligible; but Pettersson (1912) has proved clearly that the movements of

122

Periodic Fluctuations in Numbers of Animals

the herring in the Skagerak and Kattegat are determined by periodic movements of the lower layers of the sea, which in turn are caused by the moon. This short period is one of 18.6 years. He therefore suggests that these hydrographical changes affect the climate of those regions.

Now with regard to terrestrial causes, the only big factor which acts at short intervals and is known to affect climate,

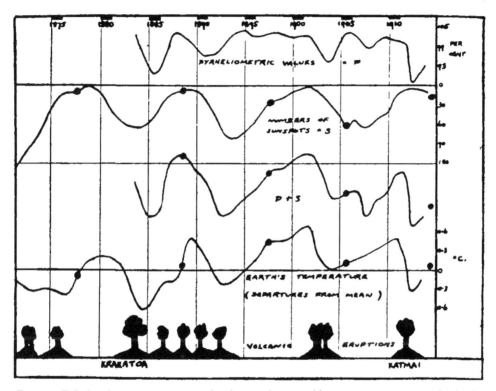

FIG. 1.—Relation between sunspots, volcanic eruptions, earth's average temperature (surface); and maxima in the numbers of the Canadian rabbit (varying hare). (Former from Humphreys, 1920; latter from Hewitt, 1921, see fig. 5). Curve A: Pyrheliometer measurements showing amount of sunlight cut off by volcanic dust. Curve B: Sunspot curve (inverted). Curve C: A and B combined. Curve D: Earth's temperature. (The black dots are rabbit maxima; see p. 135.)

is the eruption of volcanoes. This is, of course, irregular in occurrence. If an eruption is of the kind that sends out quantities of dust, part of the sun's radiation may be prevented from reaching the earth's surface, and a general cooling may result. Such a case of cooling is well seen in the results of the eruption of Katmai in 1912 (fig. 1). If the sunspot curve be combined with the curve (obtained by means of the pyrheliometer) representing the radiation cut

123

off by the volcanic dust, the result resembles that of the earth's temperature much more closely than the sunspot curve alone. Thus the effects of other factors may always be upset temporarily by volcanic eruptions; but the effects of the latter do not last for more than a year or two, and are sometimes confined to one hemisphere only.

It is highly probable that there are other terrestrial factors, hitherto undiscovered, which are influencing our climate. In any case the effects of even a few factors, such as changes in the sun and volcanic eruptions, become very complicated and involved. It has to be remembered that while some of the effects of solar variation might be felt at once, in the case of others there may be a considerable lag. Thus it has been shown (Wiese, 1922) that a cold year in the north of Siberia causes quantities of ice to appear round Spitsbergen about four and a half years later, owing to the time taken for the ice to drift that distance.

There are cycles of climate which cannot be accounted for in any simple way by known causes, *e.g.* the well-known Brückner cycle of about thirty-five years, clearly shown in rainfall and in the changes in level of the Caspian Sea. Separate meteorological factors for any one place show considerable and usually rather irregular variations. Mr D. Brunt has generously allowed me to consult his unpublished work on periodicities in pressure, rainfall, and temperature for various places in Europe (obtained by careful harmonic analysis). These show that the variations from year to year can be analysed into a good many periods whose length varies from $1\frac{1}{2}$ to 35 years. Among these is one of $3\frac{1}{2}$ years which occurs in the temperatures of six stations in Europe. It is not, however, very marked, but is of interest in connection with the period in lemming numbers described later, which is of the same length.

Chree (1924) has recently shown that the meteorological data for Kew show no marked correlation with the 11-year sunspot cycle, and this is also true of Mr Brunt's results, except in the case of Edinburgh. It is the writer's opinion that these facts do not disprove the idea that the solar variation may have important effects on animals and plants through climatic changes, since living organisms integrate the separate meteorological factors. That is to say, the original energy from the sun is split up into numerous separate manifestations (those studied by the meteorologist) which are reintegrated by the organism. A good climate, like a good dinner, is more than the mere sum of its separate parts. It is therefore important to note that the best evidence for the existence of 11-year cycles of climate is from lake-levels, which to some extent integrate climatic factors like an organism, from tree-rings, and as will be shown, from animals. This is what we should expect.

3. Periodic Fluctuations in Animal Numbers.

Lemmings.—The lemmings are a group of rodents, resembling in appearance small guinea-pigs, which live in arctic and sub-arctic countries, where they occupy the ecological niche of the mice and rabbits of lower latitudes. Evidence is given below that the numbers of lemmings fluctuate periodically all over the arctic regions.

1. *Southern Norway.*—The Norwegian Lemming (*Lemmus lemmus*) inhabits the arctic regions of Norway down to sea-level, and occurs further south in the mountains. It follows the zone of arctic-alpine tundra, above the tree limit, which in the mountains of South Norway is about 3000 to 4000 feet above sea-level. The data about the lemming are taken from Collett (1911).

For many years the lemmings have periodically forced themselves upon public attention in Southern Norway by migrating down in swarms into the lowland in autumn, and in many cases marching with great speed and determination into the sea, in attempting to swim across which they perish. The details of the fate of the migrants do not concern us here and are fully described by Collett. The main point is that all these lemmings die, and none return to the mountains, the chief cause of death being an epidemic bacterial disease. It seems pretty certain that the immediate cause of the migration is overpopulation in the usual habitat. Lemming pairs usually have their own territory, nearly all the migrants

125

are young animals of that year, and while large numbers are concerned in migrating, each individual still remains solitary and pursues its own independent course. These facts, taken with the stupendous numbers of migrants, and the fact that a certain number of lemmings stay behind and do not migrate, show that the phenomenon is analogous to infanticide among human beings as a method of preventing overpopulation. As a matter of fact the epidemic also attacks the stay-at-home animals, and thus the population is still further reduced. The lemming-years are such conspicuous phenomena that it is safe to assume that all which have occurred (since about 1860) have been recorded. Lemming-years in Norway have the status of great floods or terrible winters.

It should be remembered that lemming migration is only an indicator of overpopulation which passes a certain point, and therefore lemmings might have a small maximum which did not lead to migration. A study of the lemming-years of South Norway leads to the following results :—

(*a*) Lemming overproduction occurs periodically every few years and culminates in an autumn migration. It is usually found either that some of the migrants of that year survive the first winter and carry on the invasion of the lowland in the next summer, or else overproduction in one district is followed in the next year by overproduction in a neighbouring one; so a lemming maximum may cover one or two years. For this reason, owing to the difficulty of knowing when the maximum has ended, in the following treatment, the first year of onset of migration is taken as the maximum, except in 1890 and 1894, when the migration was very small compared to that of the next year.

(*b*) The area of great over-increase varies in size and position, but often includes several distinct, and for lemmings isolated, mountain blocks.

(*c*) With one exception, a lemming maximum has occurred every three or four years (occasionally two or five). See diagram C, fig. 2.

(*d*) The exception noted above is the year 1898. It is probable, in view of the evidence to follow (p. 132), that there was a maximum of lemmings in the mountains, but not

126

sufficient to cause a migration, especially as there was one in North Finland in 1897.

(*e*) If we assume a maximum in 1898, the mean period between maxima is 3.6 years (1862-1909).

I have not been able to obtain full information of lemming-years since 1910, but Dr Grieg informs me that there was one in the Hardanger region in 1922-1923. It is highly probable that records will be found for 1914 and 1918.

The Wood Lemming (*Lemmus* (*Myopus*) *schisticolor*) which lives lower down on the mountains than its ally, is also subject to "good years" followed by migration. So far as the data go, over-increase is found to take place in the same years as the common lemming (1883, 1888, 1891-92, 1894-95, 1902, 1906).

2. *Southern Sweden.*—The lemming-years in Sweden have not been recorded so regularly as those in Norway, possibly because the cultivated land is not so often invaded. The records of some of the years are probably based on abundance and not on migration, so that they do not afford such an accurate index of the actual maximum as those of South Norway. Dr Sven Ekman has very kindly allowed me to use his list of the known records for Sweden, Lapland, and Finland, compiled from his own observations and from the literature. The years for Sweden south of lat. 66° N. are as follows: 1862-64, 1868-69, 1871-72, 1875-77, 1880, 1884, 1890, 1895, 1901.

These, except that recorded for 1880, all agree with the South Norway lemming-years within a year's variation either way. The exception may be due to the record not being made at the actual maximum, or it may be a real exception.

3. *Canada.*—The Barren Grounds of Canada are inhabited by lemmings of several species, chiefly the Hudson Bay Lemming (*Lemmus* (*Dicrostonyx*) *richardsoni*), and the Tawny Lemming (*L. trimucronatus*). There is practically no direct evidence as to the periodicity of their fluctuations, or details about them. But they are well known to fluctuate, and according to Rae (quoted by Barrett-Hamilton, 1910), the North American species migrate in certain years after the manner of the Norwegian species. There was a big migration

C. S. Elton

of Lemmus at Point Barrow, Alaska, in 1888 (Allen, 1903). However, we can attack the question from another angle.

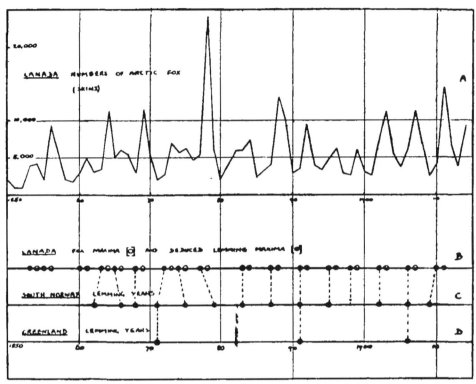

FIG. 2.—Years of lemming maxima in Canada, the mountains of South Norway, and Greenland. The years for Canada are deduced from the curve (A) of arctic fox skins taken annually by the Hudson Bay Company. (Data for Canada from Hewitt, 1921 ; for Norway from Collett, 1911 ; for Greenland from Winge, 1902, and Manniche, 1910).

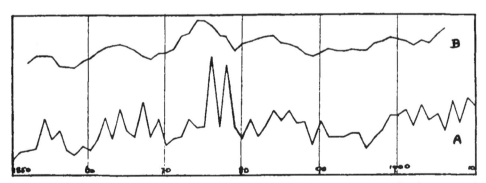

FIG. 3.—The curve A shows the effect of eliminating a period of 3.6 years in the curve of arctic fox skins (fig. 2 A). Curve B shows the effect of smoothing the lower curve (A) by taking 5-year averages for each year, *i.e.* two years before and two years after.

The lemming is the main source of food of the arctic fox; there is not the smallest doubt that the latter depends very largely on the lemming, just as its allies further south depend

128

Periodic Fluctuations in Numbers of Animals

on hares, rabbits, and mice (Hewitt, 1921 ; Stefanssen, 1921). As will be shown later, the number of red foxes fluctuates with the number of rabbits (and probably mice), so we should expect the number of arctic fox to give a good indication of the number of lemmings. This is in fact so. The curve A in fig. 2 shows the variation from year to year of the number of arctic fox skins taken by the Hudson Bay Company (from Hewitt, 1921). If we plot the year before each fox maximum as in diagram B, we obtain the years when there was presumably a maximum of lemmings. The agreement with lemming-years in Norway is seen to be remarkable. In three cases the maximum in Canada occurs a year before, and in three cases a year after that in South Norway, so that the variations cancel out and leave us with the same mean period : 3.6 years. The reason why the maximum of fox skins occurs in the year after that of the lemmings is clearly shown by Manniche's study of a lemming-year in Greenland (Manniche, 1910). In the autumn of 1906 there were colossal numbers of lemmings, but these mostly disappeared before the winter was far advanced. He attributed this death to the cold early winter with little snow to protect the animals, but it is highly probable that disease was an important cause. The foxes swarmed in autumn, partly owing to immigration, but also presumably more young grew up successfully. They fed entirely on lemmings. But in the winter the lemmings had died or were inaccessible under the snow. Foxes make caches of lemmings for the winter (Hewitt, 1921), but by the spring these must have been used up, for Manniche was able to trap large numbers of foxes owing to their hunger. Now the Hudson Bay Company fur returns for any year include the catch of the winter before ; in fact this is the main item, *i.e.* a maximum of skins in 1907 means many foxes caught in the winter 1906-1907, and in the spring. Such abundance will, of course, be the result of the year 1906, since arctic foxes mate in the spring and the young do not grow up until the autumn. So any abundance in fur returns for one year necessarily implies abundant food supply in the preceding year.

4. *Greenland.*—There are only incomplete data for Greenland, but the three certain years of lemming maxima, in every case accompanied by abundance of foxes, fit in remarkably well with the rest (diagram D, fig. 2).

5. *The Pribiloff Islands.*—Preble and MacAtee (1923) state: "As in other regions the lemmings of Saint George suffer considerable fluctuations in numbers." The data about years of abundance are not sufficiently definite to warrant any conclusions about the periodicity, but Elliot (1884) records them as very abundant in 1874, which agrees with Canada. He also states that there is a very marked variation in the severity of the climate with a period of about four or five years. This is only a rough estimate based on a year's stay and the accounts of the natives, and it is probably the 3.6-year period which is referred to. The results are very marked; the ice-pack may or may not stay round the islands in winter, and this influences the movements of the walrus, which in 1879-80 were driven south, with the result that most of the inhabitants starved. On the other hand some plants, *e.g.* Elymus, only ripens seed in the occasional warm summers.

6. *North Scandinavia.*—The foregoing account (section 1) only dealt with the mountains of southern Norway. It remains to consider the fluctuations of lemmings in the arctic regions of Scandinavia. Now the recorded lemming-years imply in every case that a migration of lemmings has taken place. It is obvious that the phenomenon of migration is far more striking than a mere increase in the numbers. The spectacle of processions of lemmings ecstatically throwing themselves over the ends of railway bridges, and falling to an apparently useless death below; the sea strewn with dead lemmings like leaves on the ground after a storm; lemmings making a bee-line across crowded traffic oblivious to danger; all these things are bound to make people talk. But in order that migration may be employed as a means of relieving congestion in an area, it is clearly necessary that there shall be somewhere to migrate to. In the mountains of southern Norway the lemming invades the lowlands. But up in the arctic regions the lemming-habitat runs down to

130

Periodic Fluctuations in Numbers of Animals

sea-level, so that there is no new land which it can occupy—only the snowfields on the mountains. It is probably for this reason that there are not much data about lemming-years in northern Norway and Sweden, and it seems likely that only in occasional years of great production does a migration take place. The absence of any very striking migration phenomena in Canada and Greenland confirms this idea. Rasmussen (1921), Winge (1902), Manniche (1910), and others record cases of lemmings migrating singly out over the sea-ice in definite directions; but the phenomenon is not a big one and never attains the same dimensions as in southern Norway. There is a possible explanation of this. In the arctic regions the enemies of lemmings are able to work day and night owing to the continuous summer daylight, whereas further south the lemming is protected by darkness to a large extent, and in fact comes out chiefly at night. This means that in arctic countries the enemies of lemmings will exercise a more efficient control over the numbers of the latter, and thus very excessive overproduction will be less likely.

So lemming-years in northern Norway are irregular, but when they do occur, always coincide with the southern ones within a year (1878, 1883, 1894, 1902, 1906).

The years which Dr Ekman has given me for northern Sweden are 1862-63, 1871-72, 1883, 1902-4, 1907, 1911, which also agree with the others within a year, either way, except that the one of 1902 apparently lasted three years.

7. *Finland.*—Dr Ekman's data also include some records of lemming-years for northern Finland. These are 1862-64, 1871-72, 1880, 1890-91, 1893 and 1895, 1902-4. These are not quite so regular but agree, except for 1880.

8. *Siberia.*—Allen (1903) says, of the lemmings (Lemmus and Dicrostonyx) of north-eastern Siberia, "the statements made by other writers concerning their comparative scarcity and abundance in different years are verified by the natives and whites here." No details of the years are given.

Effect of Lemming Fluctuations.—The lemming fluctuations have very powerful effects on the animals associated with them. This has already been shown for the arctic fox. In Norwegian lemming-years vast crowds of birds (owls, hawks,

131

Animals, Ecology & Populations   101

ravens, etc.), and mammals (stoats, foxes, etc.), are attracted to the mountains, and their numbers increase not only by their immigration but probably by their larger and more successful broods, due to the abundance of food. Similarly the arctic carnivores are profoundly affected. It is probable that many skuas and snowy owls (Manniche, 1910, etc.) only breed at all in lemming-years. There is not space here to follow out the readjustments of the food-cycle which result from the lemming-years. Two examples must suffice. The Short-eared Owl (*Asio flammeus*) collects in numbers and battens upon lemmings in South Norway lemming-years, and the Peregrine Falcon (*Falco peregrinus*), which hardly ever visits that country in normal years, comes and eats the owls (Berg, 1913). Again, in Greenland, an abundance of lemmings causes the arctic foxes to neglect other sources of food such as Ptarmigan and other birds. The latter accordingly increase. But in the next spring and summer, when there are few lemmings, these birds fare badly (Manniche, 1910).

The Causes of Lemming Fluctuations.—It is clear that the causes of these fluctuations might lie either with the lemmings themselves or with their environment. It is possible to conceive that there might be some rough natural period in the increase of lemmings' numbers (in the sense of having an increase in their "bank balance" of numbers every year), which was terminated after a few years by migration and disease following upon overpopulation, and that the population was thus reduced and the process started all over again. A little consideration will show that such an explanation of lemming periodicity is quite untenable. It is inconceivable that such a process could cause synchronised maxima on the various mountain blocks of southern Norway, which as far as lemmings are concerned, are isolated from one another, or again in the different districts of Scandinavia. When we find further that the lemming maxima are practically synchronous all over the arctic regions and the mountains of southern Scandinavia, any such "natural rhythm" becomes out of the question. Of course the natural rate of increase is a very fundamental factor in determining the size of periodicity into which the fluctuations will fit. The cause

132

Periodic Fluctuations in Numbers of Animals

of the periodicity must therefore lie with the environment, and here the only possible factor which is acting in a similar way all over these regions is climate. We do not know how this climatic factor acts, whether directly, or indirectly through plants, or other animals, but there can hardly be any doubt that we have here to look for a periodic climatic effect whose period is about 3.6 years on the average, and which acts over the whole of the arctic regions and in the Norwegian mountains. It will be shown later that it probably occurs in temperate regions also.

The curve of the arctic fox skins (fig. 2) shows on casual inspection no distinct traces of any other periodicity besides the short 3.6-year one. But if the effects of this periodicity are eliminated interesting results are obtained. I am much indebted to Mr G. M. B. Dobson for showing me the method of carrying out this elimination (due to the late Prof. Chrystal), and for actually demonstrating the existence of other periodicities underlying the 3.6-year one in the fox curve. The method is graphic and consists essentially in shifting the curve half a period (in this case 1.8 years) to one side, and making a new curve based on the means of these two superimposed curves. In this way the 3.6-year period is largely eliminated, but the method is only absolutely accurate in the case of a regular symmetrical curve. It will be seen in fig. 3 that there is left a curve which is irregular in detail, but which shows a well-marked 10 to 11-year period. This comes out more clearly if 5-year averages are taken (upper curve in fig. 3). This agrees with the period found among so many of the southern animals (Canadian rabbit, etc.), which are discussed later.

I shall attempt to show that the 10 to 11-year period of the rabbit may be due primarily to the 11-year period in the sun. If this is so, the fox curve indicates that the solar period is having a slight effect in the arctic regions. This agrees with the conclusion reached from other evidence, that the effect of the solar variation is greatest in the tropics and gets fainter farther from the equator. It will be seen that there is no known cause for this well-marked $3\frac{1}{2}$-year period in the arctic climate. It might be caused by some

C. S. Elton

short period in the sun, or by some unknown terrestrial factor, or the complicated interaction of several such factors.

Since the curve for the arctic fox shows a well-marked 3.6-year periodicity with a less distinct 10 or 11-year period underlying it, it might be thought probable that the curve for the numbers of the southern red fox, which shows a definite periodicity of about ten years, would have a concealed short

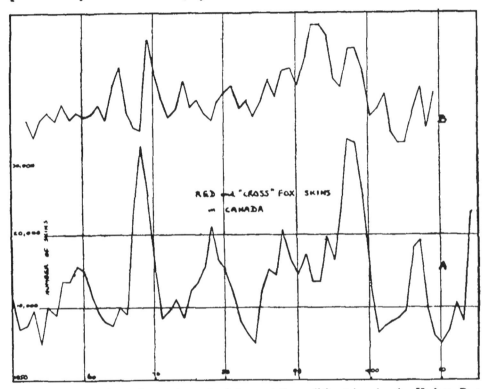

FIG. 4.—Curve A shows the numbers of red fox and "cross" fox taken by the Hudson Bay Company in Canada each year. Curve B shows the same curve after a period of 10 years has been eliminated from it. The period remaining is one of about 3½ years.

period. This is found to be the case. Curve A, fig. 4, shows the number of red fox, and the variety known as the cross fox, taken from year to year in Canada (Hewitt, 1921). Curve B shows the effect of eliminating a period of ten years from the curve. There is left a well-marked short period of about 3½ years which resembles the arctic fox curve (fig. 1A). The comparatively rough method of analysis does not allow of an accurate comparison being made. Now the southern fox feeds both on rabbits and mice. The rabbits cause the 10-year period in the fox, and it is shown later that there is a

134

Periodic Fluctuations in Numbers of Animals

short period of $3\frac{1}{2}$ years in Norwegian mice. It is very probable that the short period in the red fox curve is caused by variations in the supply of mice in Canada. It might also be caused by climate acting directly; but the absence of any such period in the lynx curve goes against this idea, since the lynx eats rabbits only. In any case the ultimate cause must be climate.

It appears then that there are two periods in the numbers of foxes and their food, and that the effects of the shorter one are more marked in the arctic, while those of the longer one are more marked in the temperate regions. This difference must be to a large extent due to the difference in prey, *i.e.*

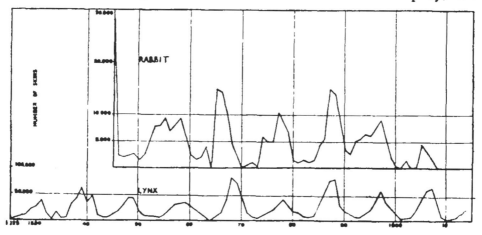

FIG. 5.—Curves show the number of skins of Canadian rabbit (varying hare) and lynx taken annually by the Hudson Bay Company (from Hewitt, 1921).

lemmings in the one place and mice and rabbits in the other, but the climatic difference may also be of the same nature.

Finally, it should be noted that the lemming differs from most of the other fluctuating rodents in that it does not have more young in a brood in good years; it probably has more broods in a given time, and the young grow up more successfully under the good conditions.

The Varying Hare.—The violent fluctuations in the numbers of the Canadian rabbit or varying hare (*Lepus americanus*) are well known through the writings of Seton (1920), Hewitt (1921), Preble (1908), and others. About every ten years the numbers of the rabbits increase to a maximum, just as in the shorter cycle of the lemmings, and then almost the entire

135

C. S. Elton

population is killed off by an epidemic disease. The fur returns of the Hudson Bay Company given in Hewitt's book, form a good record of these fluctuations since the year 1845 (fig. 5).

As will be seen there is an average periodicity of about ten years between 1845 and 1905. The next maximum was in 1914 (Hewitt, 1921). The returns include partly the year before, and it is not possible to say in which year each maximum really came; but as the method of collecting is the same throughout, we can compare the different parts of the curve with each other. The curve of course represents the total for the whole country; but the maxima do not occur in the same year in all parts of the country (although there is only a variation of a few years, otherwise there would not be these regular maxima shown by the totals). The curve of numbers for a single district, *e.g.* the Mackenzie River region, would be far less symmetrical, showing a very steep drop after the year of maximum. It seems well established that the number of young in a brood and the number of broods in a season vary according to the year in the cycle. In years of maximum abundance there are eight to ten in a brood and two or three broods in a year. In bad years there are three in a brood, and only one brood in the season. This fact at once marks off the whole phenomenon as something more than a cycle due to the time taken by the rabbits to increase after decimation by an epidemic. The process must be extra-ordinarily regular, for the Indians of the Athabasca-Mackenzie region are in the habit of counting the numbers of embryos in the rabbits which they catch, in order to get an idea of what the rabbit "crop" is going to be like in the next year (Preble, 1908). This cycle of reproductive activity has been treated so far as a mysterious thing, usually attributed to some obscure "physiological rhythm." There is nothing in physiology which offers the remotest suggestion that any such cycle (considered as an independent rhythm going on in the animals) could possibly exist. When we reflect that the rhythm would have to act in a mixed population of animals of all ages, and through several generations, and more or less simultaneously (within a few years) all over Canada, such an explanation becomes increasingly unlikely. And what is more it is quite

136

Periodic Fluctuations in Numbers of Animals

unnecessary. The facts could quite easily be accounted for by variations in the conditions under which the animals live. Now we know that there are periodic climatic changes going on in North America with about the same period as that of the rabbit-fluctuations (*cf.* the Sequoia evidence, and Küllmer's shift of the storm tracks). So it seems impossible to escape the conclusion that the cycle in reproductive capacity and numbers is caused by some climatic cycle which varies the conditions. Increased young and more broods might be due to better food supply (climate acting through plants). The evidence about mice suggests that this is an important factor. Better conditions in the physical environment (temperature, evaporation, etc.) might produce the same result. The fact is, we know very little about the physiology of reproduction in this species, and not much about that of any mammals. All that can be said, and that can be said with some certainty, is that the variations in reproductive capacity, and consequently in numbers of this rabbit from year to year, are due directly or indirectly to variations in climate. The period of ten or eleven years at once suggests that of the sun, since it is fairly certain that the latter has some effect on the climate of North America. The chief objection raised by meteorologists against the existence of such climatic periods dependent on the sun is that the effects would be too small to be of importance. But in addition to the fact already pointed out, that what is small meteorologically is often of very great importance biologically ; it should be remembered that in the discussions on the subject, the *direct* effect of sunlight is usually left out, chiefly because the technique for measuring it is still in an early stage. Sunlight is of immense importance to plant-life (probably more as to number of hours than as to intensity). In any solar variation the light must vary with the other things, and it is perfectly possible, indeed likely, that the 11-year variation in the sun produces effects by this means, on plants and therefore indirectly upon animals.

Fig. 6 shows all the maxima since 1845, marked on the sunspot curve. It will be seen that there is a rabbit maximum just before or on, or just after each sunspot minimum, except in 1905, when there was a small maximum near the sunspot

C. S. Elton

maximum. In other words, in one of the sunspot periods during this time there were two rabbit maxima instead of one. It will also be noticed that the rabbit maximum comes nearer the sunspot minimum after an unusually high sunspot maximum preceding it. This may indicate that worse weather resulting from a high maximum causes the recovery in numbers of rabbits to take a year or two longer.

Now though the rabbit maximum of 1905 does not agree with the sunspot curve, fig. 1 shows that it follows the curve of average earth-temperature, just in the same way as former maxima. In other words, the curve of earth-temperature falls abruptly two or three years before its usual time, and this appears to be explained by the volcanic eruptions of 1902 and 1903. This idea is confirmed by the marked drop in the

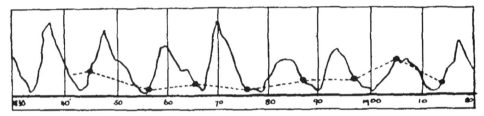

FIG 6.—Years of maxima of the Canadian rabbit (varying hare), from fig. 5, marked on the sunspot curve. (Latter from Huntington, 1923.)

pyrheliometric curve in 1903. If this is so, it is clear that the discrepancy in the rabbit curve would be explained by the volcanic eruptions. Similarly in 1912, when another rabbit maximum might have been expected, the eruption of Katmai occurred and caused another sudden drop in earth-temperature, which soon recovered, so that the maximum of rabbits came in 1914. This hypothesis, for of course it is at present nothing more, does seem to cover the facts, and its truth will be tested by the relations of later rabbit maxima to the sunspot and earth-temperature curves. It is worth noting that the 1905 rabbit maximum for the whole of Canada was a very small one; locally, however, the rabbits reached their customary vast numbers, so the area in which it took place must have been smaller than usual. If we allow for the irregularity in 1905, the rabbit period agrees very well with that of the sunspots: *i.e.* omitting the one in 1905, the

138

Periodic Fluctuations in Numbers of Animals

average period for rabbits between 1845 and 1914 is 11.5. But the 1914 one should have been in 1912, and this brings the average to 11.1, which is about that of the sunspots (= 11.2). These fluctuations in the numbers of rabbits have an immense influence on the numbers of the carnivorous animals which live on them. This is well shown by the curve for lynx numbers (fig. 5). Owing to the fact that the lynx migrates about over great distances in search of food, its curve is much smoother than that of the rabbits. There is a delay of from one to three years in the effects of rabbits on lynxes, caused by factors in the trapping which are fully explained by Hewitt (1921).

A similar fluctuation occurs among other animals that eat rabbits: the red fox and its varieties the black and cross fox, the martin, etc. The fox curve has already been discussed, and it will be remembered that it showed a short period of $3\frac{1}{2}$ years as well as that due to the rabbits. It is important that the rabbit curve itself when analysed shows no regular underlying short period, but only quite irregular variations. So the short period of the fox must have some other cause. This might be either the fluctuations of mice which form part of its food, or some direct effect of climate, *e.g.* hard winters.

Other Rabbits and Hares.—Among all the species of rabbits, jack-rabbits, and hares, there exists the same phenomenon of periodic increase followed by epidemics which kill off most of the population. Palmer (1896) states that the jack-rabbits of the U.S.A. (really hares; known as "narrow-gauge mules" in California) have comparatively few natural enemies, and that their checks are lack of food, climate, and disease. In some cases migration takes place; *e.g.* Anthony has recently noted that the Little Chief hare (Ochotona) has periodic emigrations in America. Other references to such periodicities are contained in the works of Howell (1923), Nelson (1909), Bailey (1905).

The data about the actual years of maxima are too scanty and casual to allow any safe conclusions to be drawn about the length of the period in these species.

Mice.—Plagues of mice have often been recorded in ancient times, in various parts of the world, and these are due to

139

C. S. Elton

abnormal increase in numbers similar to that found among the rabbit and lemmings. A vivid description is given by Holinshed of a mouse plague which took place in 1581 in Essex (quoted by Maxwell, 1893). Similar bad plagues occurred in Scotland in 1892-93, when the mice destroyed the grass on wide areas of hill pasture (Harting, 1892; Maxwell, 1893). The abundance of mice attracted large crowds of kestrels and short-eared owls. These are always looked on as saviours of the farmer, though as a matter of fact they only prolong the plague by keeping the numbers of mice down below the density necessary to cause an epidemic.

Plagues of this type of course cause enough damage to attract attention, and so they are recorded. But where the increase is less marked we have not usually any record. Even where the increase has been reported, the species of mouse has often not been identified. In addition mice are not so conspicuous as their larger allies; there are no commercial records of skins, and no regular migration-phenomena as in the lemmings.

It follows from this that, although the periodic increase of mice has attracted general attention throughout history, there is not the same favourable material for analysing the periodicity as in some of the other rodents. But the data, though scattered, are sufficient to make it clear that a definite periodicity exists, and that the fluctuations are due to climate. The following are the main points:—

1. Mouse plagues in Great Britain. The species of mouse which is usually responsible for plagues in England and Scotland is the short-tailed field mouse or common grass mouse (*Microtus hirtus*). But other species are often abundant at the same time as it, e.g. the long-tailed field mouse (*Apodemus sylvaticus*) and the bank vole (*Euotomys glareolus*). But the first is the species which is usually in question when "vole plagues" are spoken of. As has been mentioned, the records of plagues are only indicators of unusually big increase, and therefore provide only incomplete data for working out the periodicities. They are to be compared with the lemming-years of northern Scandinavia, which come only at irregular intervals, while there is almost certainly a regular periodicity

140

Periodic Fluctuations in Numbers of Animals

in numbers which, however, does not always become sufficiently marked to be noticed. With this reservation in mind we may examine the data of mouse-plagues in this country. Fig. 7 shows the years in which any mouse plague has been recorded in any part of Great Britain. The full details of localities are to be found in Barrett-Hamilton (1910), except that of 1923 which I have on the information of Dr Longstaff. The latter will be referred to again later, since it furnishes an important clue to the way in which the plagues come about. In every case except 1900 the species mainly or wholly responsible was the short-tailed field mouse. In 1900 the records are of the other two species. It is plain from the diagram that mouse plagues have occurred about every eleven

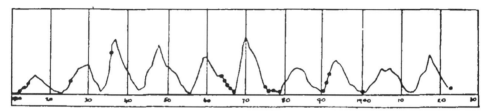

FIG. 7.—Recorded years of mouse plagues in Great Britain, marked on the sunspot curve. (Mouse data up to 1900 from Barrett-Hamilton, 1910; the one in 1923 from Dr Longstaff; sunspot curve from Huntington, 1923.) Allowance has to be made for the delayed effect of the climatic factor, as in the case of beech-mast (see p. 22).

years or multiple of eleven years, and that they seem to come round about the sunspot minimum. This is, of course, only an indication and does not prove much. But in view of other evidence we may say that there is usually a marked maximum in the numbers of mice somewhere every eleven years or so.

2. It is important to find out in what way climate affects mice, and how it can control their numbers. The year may be divided into two sections as far as mice (or indeed almost any animals) are concerned. Firstly, the breeding season, when the animals are increasing in numbers and are also being reduced by checks. Secondly, the non-breeding season (autumn and winter), when there is no increase to counterbalance the mortality. There will be a certain number of mice in any one district in autumn, and this number is the resultant of rate of increase as balanced against mortality through checks. In spring there are a certain number left,

141

after the checks of the winter. Now the essential thing in determining whether there is a fluctuation over a series of years is increase or decrease in the latter number. Better conditions in summer with slightly worse winter conditions would have the same result as worse conditions in summer and much better ones in winter. But if both improve simultaneously there is bound to be an increase in numbers. Now in the case of the mice there are four main factors which are thought to affect them in an important way. Firstly, mild winters favour them, and leave a larger stock to start the next season's population. Secondly, better physical conditions during the breeding season. Thirdly, abundant food supply at any time, but especially in the breeding season. This would be the effect of climate favouring plant life in some way, *e.g.* big crops of beech-mast or acorns. Fourthly, a favourable growing season leaves plenty of plant cover in the winter, which gives the mice protection from their enemies. There must, of course, be other factors as well. But the point is that not only are all these factors connected with climate, but the last three are all the result of the same kind of climatic complex during the breeding season.

Dr T. G. Longstaff has kindly given me some notes on a mouse plague in the New Forest, and the conditions accompanying it. In 1921 there was a very dry and hot year, which evidently caused the beech flowers to form very successfully in autumn, so that in the autumn of 1922 there was a colossal crop of beech-mast. In the spring of 1923 the woods were carpeted with sheets of beech-seedlings, and in May, June, and July there was a plague of mice in the district near by (long- and short-tailed field mice). They were found dead in large numbers at the end of this time, evidently killed by an epidemic. Now the winters over this period were all mild, and this apparently combined with the delayed effect of the 1921 summer in making a good food supply, to cause a great increase in the numbers of mice, which eat beech-mast among other things. The chances are that other plants prospered in the same way as the beech, and so the other foods may have been more abundant too.

142

Periodic Fluctuations in Numbers of Animals

The scattered notes on the subject in the phenological reports of the *Quarterly Journal of the Meteorological Society* show that there has been an unusually heavy beech-mast crop in Britain every eleven years.

Enormous crops: 1900, 1902, 1911, 1912, 1922. (Crops larger than usual, but not nearly as big as these, occurred in 1896, 1905, in places.)

The records do not go back with any completeness beyond this point.

This periodicity in beech crops is suggestive, but there are not sufficient records to prove the hypothesis that the crops are correlated with the sunspot cycle.

The great mouse plague in southern Scotland in 1892-93 was associated with mild winters and "growthy" summers; the grass crops were unusually fine until the mice ate them up.

3. In several cases these big mouse plagues have happened simultaneously in different parts of the world. This is clear if we take the regions for which there are good records :—

> 1875 : The Athabasca-Mackenzie ("A-M") region in northern America (Preble, 1908); South Scotland (Harting, 1892; Maxwell, 1893); Galicia and Hungary (Maxwell, 1893); Norway (Collett, 1911).
> 1891-93 : The A-M region; South Scotland; Norway; Thessaly (Maxwell, 1893).
> 1900 : The A-M region; England (Barrett-Hamilton, 1910).

Such facts would be explained by the hypothesis that there was some climatic factor acting widely, such as that caused by solar changes.

4. If these mouse fluctuations depend to any large extent on the solar variations, then they should conform to the map given by Walker (1923) which divides the world into areas according to the way in which they act with reference to the sunspots. Though the data are very scanty, the mouse years do in fact agree with this map. (In Norway, however, a shorter period occurs, of $3\frac{1}{2}$ years. See next section.)

All the places mentioned above have more rain and storms

143

C. S. Elton

at sunspot maxima. On the other hand, the only mouse plague which I can find that occurred at a sunspot maximum is that recorded by Hudson (1892) in La Plata in 1872-73. Now this area is one of those which have more rain and storm at minima. So the facts, as far as they go, do not conflict with the idea that the mouse periodicity is under the control of the solar variation.

5. The curves given on p. 17 show that there is a short period in the numbers of foxes in temperate North America, of about $3\frac{1}{2}$ years. It was suggested that this was due to fluctuations in the numbers of mice, which might have a period of that length. There is evidence in support of this, from Norway. There, various species of mice occur in unusually large numbers in the lowlands and also in the mountains in the same year as, or within a year of the lemming maxima, in one place or another. The following years are given by Collett (1911):—

Euotomys glareolus (bank vole): 1887, 1903, 1906.
E. rutilus: 1876, 1880, 1883, 1903, 1906-7, 1910.
E. rufocanus: 1872, 1876, 1880, 1887, 1904, 1907.
Microtus agrestis (short-tailed field mouse): 1876, 1880, 1882-83, 1887-88, 1891, 1894-95, 1897, 1906.
M. ratticeps: 1863, 1872, 1876, 1891, 1894-95, 1897, 1906.

Thus in or within a year of every lemming maximum there has been an abundance of from two to four species of mice. This suggests the probability that there is a $3\frac{1}{2}$-year period in the numbers of mice, but that increase in some places, *e.g.* England, never reaches the dimensions of a "plague," unless the additional force of some longer period is added. Whereas the 11-year period acts only faintly on the lemmings in the arctic, it is much more strongly marked farther south, and may well affect the mouse numbers in this way. Perhaps such a hypothesis sounds rather too rigid, but then no one would ever have believed that rabbits and lemmings could behave with such clockwork regularity.

6. We have not got direct evidence that the number of young in a brood varies according to the year, as in rabbits,

144

but there is indirect evidence. It is noticeable that in all the text-books (Barrett-Hamilton, 1910, etc.) the mice are said to have about twice as many mammæ as they have young in brood. Thus the bank vole, short-tailed field mouse, and dormouse have eight mammæ, while the usual number of young is four, five, four respectively. The long-tailed field mouse has six mammæ and four or five young. This is explained if we assume that like the rabbit their number in a brood goes up to double in good seasons.

Results of Mouse Fluctuations.—One of the most striking and universal results is that the short-eared owl (*Asio flammeus*) very soon turns up and gorges on the mice. The owls often stop on migration at places which they do not visit in ordinary times at all. The number of eggs in a clutch in these years is nine to fourteen instead of four to eight, and they have two broods instead of one in the season. This gathering of the owls has been recorded from South America (Hudson, 1892), Texas (Bailey, 1905), Canada (Hewitt, 1921), Britain (Barrett-Hamilton, 1910; Harting, 1892), Norway (Collett, 1911; Berg, 1913).

It seems likely that the fluctuations in the numbers of mice must have very important effects on the regeneration of forest trees. Watt (1923) has shown by experiment that in a normal year, when the beech-mast crop is small, *e.g.* 1921, about 98 per cent. of it is eaten up by mice before it has a chance to germinate at all. That which does germinate succumbs to the attacks of mice plus a great many other enemies. In a year like 1922 when there was an enormous crop, only about 40 per cent. was found to be eaten by mice before germination had time to start. If the numbers of mice remained constant, the thing would be comparatively simple. But as we have seen, they do not remain constant. What happens seems to be this. There is a huge supply of food in the beech-woods in autumn, and this attracts a big crowd of mice from round about, as well as the original population. They are able to survive the winter better, owing to its mildness, and to the abundant food supply, but do not succeed in eating all the beech-nuts by the spring. The tree has thus got ahead up to this point, and as a result we have the

C. S. Elton

sheets of seedlings in the woods. The mice now increase to such an extent that an epidemic results, and most of them are killed off. From the point of view of the tree, the sacrifice of half its seed crop has caused the mice to increase too much and die off, thus removing the danger to its seedlings to a large degree. Whether this has actually evolved as an adaptation, or occurred ready-made, there is no doubt that this certainly acts as an adaptation for the tree.

Rodents are evidently a big factor in affecting the natural regeneration of forest trees. Similar relations between good years and heavy crops are found for the oak. In this case the big crop often comes in the year before that of the beech, but also sometimes in the same year. Here also the mice must be an important factor, since they feed extensively on acorns. The hornbeam also has periodic heavy crops in England, after the manner of beech (Christy, 1924). Pearson (1923) states of the Western Yellow Pine in America that in ordinary years rodents eat up the entire seed-crop, but in good years some is left over. In all such cases the tree has to have a much bigger excess of seed than might be supposed, owing to increase in the mice as well. Show (1924) says also that rodents almost defeated attempts at artificial sowing of pines in California. A further complication exists in some cases, for according to Hofman (1920) rodents aid the regeneration of the Western Yellow Pine and the Douglas Fir in North America by making many caches of seeds in the ground, which get forgotten. When the forest is cut or burnt, and regeneration occurs, these caches play a large part in causing the Douglas Fir to prevail over other trees owing to the abundance of "planted" seeds.

Other Mice.—There is evidence of periodic fluctuations in the numbers of a good many species of mice besides those already mentioned.

Deer-mice (Peromyscus), and jumping mice in North America (Howell, 1923).

The Texas cotton rat (*Sigmodon hispidus*), which had tremendous invasions of neighbouring regions in 1854 and 1889-90, and attracted vast crowds of

146

Periodic Fluctuations in Numbers of Animals

hawks, owls, skunks, weasels, snakes, etc. (Bailey, 1905).

The Baird wood-rat (*Neotoma micropus*) of Texas and Mexico (Preble, 1908).

The bushy-tailed wood-rat (*N. cinerea occidentalis*) in Northern America (Anthony, 1923).

The water-rat in Britain and Norway (Barrett-Hamilton, 1910; Collett, 1911).

It is pretty clear from what has been said that mice do fluctuate in numbers periodically, and that this is due to climatic variations. Further work is required before we can say definitely what the exact periodicities are in each case.

Other Rodents.—Periodic fluctuations occur among some other species. The musk-rat curve of skins (Canada) shows a short 3 or 4-year period mainly. Seton says that these fluctuations are probably related in the main to the amount of water "which, as is well known, is cyclic in the North-West" (Seton, 1920). The difference between a good and a bad year must be enormous, for he mentions that they used to catch twenty musk-rats in an hour after sundown in 1900 in the Mackenzie region; in 1907 only seventeen were seen in six months. He also states that red squirrels fluctuate like the musk-rat. Anthony (1923) says that the Yellow-bellied Wood-chick (*Marmota flaviventer*) of the U.S.A. has emigrations at certain times. No doubt many other species will be found to fluctuate in numbers, when naturalists have got rid of the obsession that numbers always stay constant.

The beaver forms an interesting exception to the other rodents. The Hudson Bay Company fur records show that the beaver does not have any regular short periodicity in its numbers, although there is a general "secular" trend in the curve depending on the gradual exploitation of the country followed by exhaustion of the stock of beavers. Now the beaver does not depend on the yearly income of plant-food which is rescued by plants out of the sun's radiation. It eats the bark of trees almost entirely, and when the trees are used up in the neighbourhood of its colony, it moves on somewhere else. Thus the beaver leads an ideal existence,

C. S. Elton

living entirely on capital. What is more, it regulates its
water-supply by building dams, and so is to a large extent
independent of variations in rainfall, unlike the musk-rat.
Also its aquatic life in stream conditions causes it to live
to a large extent in water with a constant temperature
Bailey states that there is never any overcrowding in beaver
colonies, and that disease seems to be unknown among them
(Bailey, 1922). Further, there are four mammæ in the female,
and the normal number of young is four. There are certain
enemies which act as checks, but a very important factor in
regulating the numbers is rivalry among the males. The

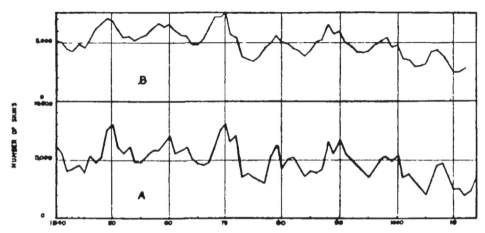

FIG. 8.—Curve A shows number of skins of the fisher taken annually by the Hudson Bay
Company in Canada (from Hewitt, 1921). Curve B shows the effect of eliminating a
3.6-year period in Curve A. Note the almost smooth 10 to 11-year period.

beaver is polygamous and there are severe fights among the
males for possession of the females. We see, then, in the
beaver an animal which has so regulated its existence that
it is independent of short-period climatic variations. Whether
this is an advantage in the long run is another matter, which
will be gone into later on.

Other Mammals.—The fisher, which is a near relation of
the martin, but has a much more general diet of all kinds
of other animals, has a well-marked periodicity, shown by
the Canadian fur returns (Hewitt, 1921). Fig. 8 shows that
the curve is rather irregular, but has a general 10-year
periodicity corresponding to that of the rabbit. But as the
rabbit is not the main source of its food, it is plain that the

148

Periodic Fluctuations in Numbers of Animals

same climatic influence must be affecting the other animals on which the fisher preys. Fig. 8 also shows the effect of assuming a periodicity of 3.6 years and eliminating it. The 10-year curve comes out quite smooth, except in the early part of the nineteenth century, when the fur returns are not so regular. This proves that there is a short period present, but as to its cause we are in ignorance.

The mink and wolverine have curves which resemble that of the fisher, showing well-marked periods of about $3\frac{1}{2}$ and 10 to 11 years. That of the wolverine is not, however, so clear as the others, since the numbers taken are not very large.

The skunk has a conspicuous 10 to 11-year period, but no clear short period. Like the fisher, the skunk has a wide range of food, so that the fluctuations in its numbers must be caused by fluctuations among many other animals, or else by the direct effects of climatic variations, or by both. Periodic abundance of shrews (*Sorex araneus*) is also recorded by Collett, for Norway (Collett, 1911).

The mouse marsupial, *Sminthopsis crassicaudata*, which inhabits the deserts of Australia, has ten young in a brood in good years (i.e. wet ones), and only four or five in bad (i.e. dry) ones (Buxton, 1923).

We have seen that the numbers of many different orders of mammals fluctuate in a periodic way, and the phenomenon is obviously of wide occurrence in mammals. The rodents show it most clearly, but it occurs as well in insectivores, carnivores, and marsupials. Further work will, no doubt, disclose its existence in other groups of mammals.

Birds.—Pallas's Sandgrouse (*Syrrhaptes paradoxus*).—This species lives in the sandy deserts of Central Asia, e.g. the Gobi Desert, where it depends for food on the salt-soil plants. Periodically the sandgrouse leaves its usual home and migrates in large numbers into China on the east, and Europe on the west. The bird is a very strong flyer, and is recorded periodically from the British Isles. These records give a delicate index of the years of maximum exodus from Asia. There has been a very big invasion every 22.5 years on the average, the years being 1863, 1888, 1908 (Witherby,

149

1920). There have been minor invasions as well, and these were in (1872), 1876 (1889, 1890, 1891), 1899 (1904, 1906, 1909). Of the two in 1872 and 1876, the latter was the greater, so that 1876 and 1899 are to be considered as minor maxima, while the others (in brackets) are clearly associated with the big years, *i.e.* form part of the same maximum. There was apparently no visit in 1918 or 1919.

It seems that there is an invasion usually every eleven years or so, while every alternate one is very big. We should expect another big visit about 1930. The cause of the invasion appears to be the onset of glaze-frosts or heavy snow in early spring, which prevent the birds from getting at their food (Dresser, 1871). This of course constitutes relative overpopulation. The net result of the whole phenomenon is that thousands of birds perish, and this is precisely the same result as when rabbits increase in numbers and die from disease, or when lemmings trek out of their usual habitat and are destroyed.

The average period of the sandgrouse maxima is 11.2 years, which is exactly that of the sunspots. Whether the 22.5-year period is anything to do with the change in magnetic polarity of the sunspots it is difficult to say!

Other Birds.—The bird-census work recently started in the U.S.A. seems to show that the numbers of birds fluctuate in a regular way. Both in Europe and America the bad winter of 1917-18 killed enormous numbers of small birds, and it was not until 1920-22 that they recovered their former numbers. This is proved by actual census figures for the U.S.A., 1916-20 (Cooke, 1923). There are numerous other birds which have big invasions of neighbouring countries at intervals. Among others may be mentioned the Waxwing, Mealy Redpoll, Crossbill (Witherby, 1920); and the Nutcracker (Simroth, 1908; Blasius, 1886). The Siberian race of the Nutcracker invades Europe in particularly large numbers every ten to eleven years, or every twenty to twenty-two years, like the Sandgrouse, *e.g.* 1844, 1864, 1885, 1896, 1907. Simroth has pointed out that this is correlated with the sunspot cycle. The periodic increase of owls, hawks, and skuas has already been pointed out. De Lary (1923) states

150

Periodic Fluctuations in Numbers of Animals

that the cuckoo and some other birds vary in their dates of arrival in France in spring according to the sunspot cycle.

Insects.—Periodic variations in the numbers of insects are well known, and may either be caused by over-increase of individuals, leading to migration and subsequent destruction, or else under-production of food leading to the same thing. In both cases the migrants usually perish, and the net result is the same as if they had been wiped out by disease. The best-known examples are found among butterflies (Poulton and Walker, 1921; Walker, 1914), and in the locust family (Uvarov, 1923). Unfortunately, although in most cases the migration must be due to some climatic cause, and can sometimes be proved to be so, the periodicities are not well known. In the case of the locust the only records of such things are the invasions of other countries. Now insects always depend on the direction of the wind in their migrations, flying either with, against, or obliquely to it. The result is that a regular periodic migration from one centre appears as irregular invasion when considered from the point of view of the country invaded, since the wind may be different at different migration times, and the countries visited there-fore different (cf. the confused results of trying to analyse locust visitations on this method, by Kulagin, 1921). But it is also significant that mammals and birds do show much more clearly than other animals the effects of 3 to 10-year climatic variations. This may be because they have tempera-ture regulation, and therefore are not affected so much by chance weather variations, such as sudden hard frost for a day. Insects and other cold-blooded animals are much more at the mercy of these kind of variations. Mammals and birds "smooth" these minor variations from day to day and month to month, and therefore give an average result for the whole year.

But whether insects and other cold-blooded animals really have regular fluctuations or not, they must be strongly affected, firstly by these short-period climatic variations, and secondly, by the direct and indirect results of the fluctuations of the other animals, which must have profound effects ecologically. Finally, as examples of the same kind of thing in widely different groups, one may mention the periodic invasion of freshwater regions by the shipworm (*Teredo navalis*) during very dry summers in Holland, in 1730-32, 1770, 1827, 1858-59 (Kofoid, 1921); also the correlation between the moon and herrings' movements (see p. 122), and between periodic snowfall variation in Norway and fishery results (Sund, 1924). There is also to be considered the yearly fluctuation of most animals, which may have effects similar to the larger ones. For instance, *Daphnia pulex* forms large numbers of winter eggs in the autumn, only a few of which succeed in hatching in spring to start the next year's stock. These, and irregular fluctuations, are just as important as the others, biologically, though less so from the point of view of meteorology.

4. Discussion.

We have already shown some of the economic consequences of these fluctuations in the numbers of animals. It is of vital interest to the forester to know the relations between rodents and the germination and survival of tree-seedlings; the possibility of predicting the years in which plagues of mice will occur may affect him enormously. Again, in the fur industry, efficient prediction of the number of skins which will be obtained is of importance. In particular, however, the meteorologist and the geophysicist are provided with a new method of getting at the facts of climatic cycles. Considering what comparatively good results can be obtained from the correlation of facts which have usually been recorded on account of their intrinsic interest, or because they touched the pocket of the observer, it should be possible to attain much more complete knowledge of these things if systematic observations were started. The idea of using animals and plants as indicators of climatic events is of course an old one. Since about 1870 the Royal Meteorological Society has carried out systematic phenological observations, published each year in the journal of that society. But the notes have been almost confined to records of first flowering and first arrivals of migratory birds, etc. Many of these do not sum up climate

152

Periodic Fluctuations in Numbers of Animals

any better than a meteorologist can. Many of the occurrences such as the opening of flowers depend on single environmental factors, e.g. the snowdrop (*Galanthus nivalis*) opens when the temperature rises above 10° C. and the hive-bee sallies forth at the same temperature (Church, 1908). Thermometer readings would tell one just as much as these observations, and more accurately. Now the only point in using animals and plants in meteorology is because they are capable of integrating and accumulating the factors of climate to a large extent ; these integrations are not, however, shown by the kind of phenomena which are usually recorded, for the reason given above, and also because of the exceptional cases which give a false idea even of the simple weather events. But when we consider the rate of growth and reproduction of animals it is clear that we are dealing with quite a different kind of weather indicator. The rate of metabolism is altered by all kinds of different factors, and the end-result may be the same, e.g. dry winds causing evaporation might have the same effect as lower temperature. So the first point is that we are dealing with a rate of something happening, instead of with a single event, and that the rate depends on the combined integrated effects of a great many environmental factors. The second point is that we are dealing with a geometrical rate of increase or varying acceleration of a rate. The result is that the climatic variations are shown in a highly magnified way. Variations in meteorological factors (e.g. comparisons between average rainfall at sunspot maxima and minima) are always very small when measured directly ; the climatic effects are the result of these small differences accumulating.

The fluctuations in the numbers of animals may be compared to the beam of light which is used to magnify the small angular variations of a fine galvanometer ; the original variation is shown on an exaggerated scale.

Plants do not give such good records because they are mostly longer lived and therefore the regulation of their numbers is quite different. We do get variations in the rate of growth shown well in the annual rings of trees, but not in the immensely magnified form found in the variations in number of animals.

153

C. S. Elton

But among the most important consequences of these fluctuation phenomena are their effects on evolution and other biological questions, and it is with these that we shall now deal. The conclusions that follow do not depend for their validity on the truth or otherwise of the hypotheses that have been advanced in this paper to account for the fluctuations in the numbers of animals. They follow from the fact of fluctuation, and about this there is not the slightest doubt.

The amount of fluctuation varies widely in different animals, and may in some cases be practically nil, as in the beaver. But among a great number of species, variation in numbers plays a big part in their lives, and it is with these species alone that we shall be concerned, in what follows. The climatic theory of the cause of animal fluctuations has this significance, if it be accepted : it means that these short-period variations of climate have been causing similar phenomena during the whole time of the evolution of life on the earth. For the causes of these variations have almost certainly been in existence during the whole of this time (the Sequoia record, and Chinese observations carry the sunspot variation certainly back four thousand years, and there is no reason to believe that they were ever absent before that time. Also the astronomers are agreed on this point, on general grounds.)

A knowledge of the animal periodicity alone, puts us in a position to study natural selection in action in the field, since if we know within a year or two when the critical period for the animal will be ; it is possible to arrange intensive field-work. This subject could be profitably taken up by ecologists, since the study of the regulation of animal numbers forms about half the subject of ecology, although it has hitherto been almost untouched. If the climatic cause of the fluctuation could be predicted, the position would be better still.

The following suggestions as to the effect of periodic fluctuations in the numbers of animals upon the method of evolution are speculative, but at the same time are mostly self-evident when once thought of; that is, they evidently must be factors affecting the animals. The question of whether they are important factors will have to be decided by further work.

154

Periodic Fluctuations in Numbers of Animals

(1) I am assuming here that natural selection must be sometimes effective in changing the race, and the only question is when, how, and how much. (In face of the enormous number of facts that are explained by natural selection as a hypothesis, it is for those who do not believe in it to disprove its truth.) In all the rodents whose periodic overpopulation is ended by epidemics, there must be strong selection of individuals for resistance to the disease, and this does not happen continuously, or even every year, but once every three or four or every eleven years. In the intervals the parasite which is the cause of the disease must also be undergoing severe individual selection for dispersal powers, since the numbers of its host are much reduced. So there is alternate selection of the host and the parasite. We do not know to what extent this selection really changes the race, i.e. whether animals survive on their genotypic merits, but at any rate it seems very probable. Again, there is periodic severe climate, such as the bad winters which occur in England at intervals. There is some evidence that selection really does happen in these years in such a way as to alter the race. The Dormouse (*Muscardinus avellanarius*) hibernates in winter in a nest. After the severe winter of 1860-61 many were found dead in their nests in Essex. The continental race habitually survives the more severe winters there, apparently owing to the fact that it makes a thicker nest and is thus more efficiently insulated (Barrett-Hamilton, 1910). If this is so, we have a case where selection of an instinct has differentiated two races without any accompanying obvious change in structure. In all these cases the factor causing the selection acts at intervals; so that if the climate of the country is becoming progressively worse, the mice in it are not subjected to a minute change in conditions every year, but have the eleven years' accumulated change acting suddenly at the end of that time. Thus the animals are tested by selection for some characters at intervals of eleven years, or of three or four years in the case of the lemming. The subject of natural selection for adaptation to climatic conditions must be very complicated, since there are not only these short periodic pulsations of climate but also the longer ones of hundreds or thousands of

155

C. S. Elton

years (cf. Pettersson, 1912; Huntington, 1914), and then the still larger ones of the size of ice-ages.

A further factor in this periodic selection is that while in maximum years there is selection of individuals, e.g. for resistance to disease, ability to escape enemies, ability of the males to secure females in competition with other males, etc., in the minimum years there is selection for resistance to bad climatic conditions, and if the animals are few and scattered, for ability of the males to find their mates, although there will be little competition for them when once found. There will thus be different types of selection at the maximum and the minimum in numbers. If such selection is effective in changing the genotypic constitution of the species, it is obvious that the alternate selection in different directions is an important phenomenon.

In the case of many factors, they are acting on animals more or less all the time, but increase in severity periodically e.g. the attacks of enemies (cf. the arctic fox and ptarmigan). Now it is well known that mass selection does not necessarily select out animals on their genotypic merits, especially when the character on which it is acting is highly modifiable. Thus selection of normal intensity might produce no effect on the hereditary make-up of the species. But if the intensity increases periodically and becomes very great, then selection may begin to discriminate by its severity between the genotypic constitution of individuals.

(2) It is obvious that if natural selection for a particular character were only effective say, once every ten years, there would be no selection in the intervals, and therefore if any new mutation affecting that character were to arise and establish itself in the intervals it would be temporarily indifferent, i.e. neither adaptive or harmful. But, say, five years later, it would be acted on by the periodic selection and might then turn out to be either good or bad, or might remain indifferent. Now it usually is supposed that a new character could hardly ever establish itself or spread in a species without the aid of natural selection. On this basis, characters which were only selected at all every ten years would therefore only be able to arise and spread at all in

156

one year out of ten. But if we could show a way in which an indifferent mutation, i.e. one which at the moment was neither adaptive or harmful, could arise and spread in the species, without the aid of natural selection, there would be far more material for the periodic selection to act on. There can be little doubt that there are many cases of indifferent characters existing in a species, and what is more forming the characters which distinguish closely allied species. The great problem has always been to explain how such indifferent characters could become established in a population at all often. The only adequate explanation put forward so far is that these characters are a by-product of hereditary factors concerned with adaptive characters; but this does not seem to cover the facts at all completely, although it undoubtedly explains some cases. At first sight it seems very unlikely that any new character could spread or even survive, unless selection were acting in its favour. For instance, in *Gammarus pulex*, the freshwater amphipod, the chances of any one young animal reaching the adult state and breeding are about one in a thousand. It is clear that if the rate of mutation is slow, a new mutation arising in a single animal apparently has an extraordinarily small chance of establishing itself and breeding, much less of spreading in the population, unless selection is acting in its favour. But this difficulty vanishes when we realise that animal populations do not remain stable. The case of Gammarus is a good one. There is a stream near Liverpool, in which *G. pulex* lives, but which dries up periodically, all except some deep ponds near the source. The Gammarus stock in the stream of course is wiped out periodically, and the bare area of several miles of stream is recolonised by the animals from the ponds. Now while this recolonisation is going on, there is practically nothing checking the numbers of the shrimps, and consequently the chance of survival of any individual may be one in two, or even a dead certainty. In other words, the struggle for existence ceases temporarily. If several animals are responsible for starting the new stock, practically all their families will grow up, and any mutation that happened to occur or be contained as a recessive at such a time of spread with no checks, would

157

automatically spread unless it was actually very harmful. This idea does seem to be confirmed by the excessive number of varieties of *G. pulex* in this stream (which are at the moment under experimental analysis). The fluctuation among rabbits presents an exactly similar appearance. There is great abundance followed by almost complete destruction of the rabbits, and then increase to fill up the places that have just been emptied. Any new variation arising at this time would inevitably spread as in the case of Gammarus, and if it was useful would be selected favourably at the next periodic "examination"; if it was absolutely indifferent from the point of view of adaptation, it would remain as a character, and its further history would depend on other considerations; cf. Haldane's striking mathematical analysis of the spread of mutations under the action of natural selection (1924). The difficulty of imagining how a mutation gets a start in a population is removed when the full significance of fluctuations in numbers is realised. The chief question which arises is whether the chance of a new variation cropping up in the small nucleus of individuals that start the new stock is not very small. If the idea put forward here is correct, the method of regulating its numbers is a vital factor in deciding the evolutionary path of a species. It is interesting to note that the beaver, which we have seen to lead a very regular sort of life, with no fluctuation in its numbers, has practically no geographical races, and in this is unlike the mice. This may be a coincidence, or it may be that the beaver has obtained temporary security at the expense of ultimate progress, owing to having now no mechanism for accumulating the "capital" of variations on which natural selection could act.

(3) When rabbits, lemmings, mice, etc., die off through epidemics, there are remarkably few animals left in the district. The example of the musk-rat is not an unusual one. Observers describe the emptiness of the mountain tundras in Norway after a lemming-year has occurred. This has an important bearing in the way mentioned above, in that the new young are able to spread into what is

158

Periodic Fluctuations in Numbers of Animals

practically an empty habitat, in a way entirely comparable to the colonising of bare ground by plants; but it must also have an important influence in keeping the specific characters uniform in the population. If the whole population of thousands of individuals arises afresh from a few animals every eleven years, then the genotypic constitution of the species will tend to become comparatively pure. Or, to put the matter another way, every time a minimum in numbers occurs, there is the chance of the less common genes becoming extinct, so that the total number of genes for any one character would tend to become reduced. Hagedoorn (1921) has already pointed out the importance of any fluctuation in the numbers of animals in causing uniformity, but he definitely states that the numbers of all common species remain constant from year to year. The fluctuations to which he refers are the irregular ones such as the colonisation of new empty habitats by a few individuals, or irregular climatic and biological factors. The facts about periodic fluctuations in animals, therefore, merely make his conclusions of wider application. He also suggests that although the numbers of a species remain constant usually (which is, however, not the case), only a certain number of individuals breed each year out of the population which has survived. This would have the same effect as fluctuation in the actual numbers. There is practically no evidence of this, however, but it may sometimes be important. Hagedoorn also tried to prove that genotypic purity was bound to occur in a stable population; but this has been criticised on mathematical grounds, and is probably not the case.

The degree to which a species does remain uniform in characters will depend therefore not only on the factors usually quoted, such as natural selection and crossing, but also on the extent of its periodic fluctuations. This does not contradict the other idea about establishment of new variations, since the latter process may not happen very often. In fact there is in many animals a mechanism by which variation in the population are alternately acquired and weeded out, or kept as indifferent characters, which in turn may become useful at some crisis in the history of the animal at a later date.

159

C. S. Elton

(4) These ideas, which follow from the realisation of the existence of fluctuations in the numbers of animals, do help to explain some of the difficulties which arise in studying the method of evolution. The fluctuation in numbers explains (at any rate theoretically and in the animals which do fluctuate) several things; firstly, how by a temporary cessation of the struggle for existence, new mutations which are indifferent (i.e. neither adaptive nor harmful) can occasionally get a footing and spread in the population; secondly, how, in all probability, some characters are selected periodically; thirdly, how the species remains on the whole uniform in composition. A fourth important conclusion follows from the recognition of factors other than selection and crossing causing uniformity. One of the great problems in adaptation is how an animal can have, as it often does, a particular structure or habit, which has only a general adaptive significance. Thus, in the case of certain colour patterns, such as ruptive coloration for protection, there often seems no point in having that one particular pattern rather than another, although the general survival significance is clear enough. It is difficult to see how natural selection could discriminate to such a fine point, so as to eliminate other equally good patterns. The difficulty is explained when we realise that the uniformity caused by fluctuating numbers acts without reference to adaptation. There is, so to speak, an entrance examination by natural selection, which weeds out the worst candidates, but the final selection is by lot.

5. Summary.

1. Four main points are dealt with :—

(a) The widespread existence of fluctuations in the numbers of animals.

(b) The existence, in many birds and mammals, of periodic fluctuations (p.f.).

(c) The cause of the latter, which must be some periodic climatic change acting over wide areas.

(d) The effects of fluctuations in general, and in particular of the p.f., on the method of evolution and other biological phenomena.

160

Periodic Fluctuations in Numbers of Animals

2. A short sketch is given of what is known about short-period climatic cycles (2 to 20 years), and their causes.

3. P.f. of lemmings have an average period of about $3\frac{1}{2}$ years. The maxima in numbers occur synchronously in North America and Europe, and probably all round the arctic regions.

The varying hare in Canada has a period of 10 to 11 years.

5. The only regular periods shown by the animals dealt with are the short one of $3\frac{1}{2}$ years and the longer one of 10 to 11 years. The former is probably more marked in the arctic and the other further south.

6. The sandgrouse p.f. point to the existence of an 11-year climatic cycle in the deserts of Central Asia.

7. The effects of these p.f. on evolution must be very great, although at present problematical; but the following suggestions are made :—

(*a*) Natural selection of some characters must be periodic.

(*b*) There will be different types of natural selection at the maxima and minima of numbers.

(*c*) The struggle for existence, and therefore natural selection, tend to cease temporarily during the rapid expansion in numbers from a minimum, and new mutations have then a chance to get established and spread, i.e. without the aid of natural selection. This might happen only rarely.

(*d*) This would explain the origin and survival of non-adaptive characters in a species.

(*e*) On the other hand periodic reduction in numbers will act as an important factor causing uniformity in the species.

(*f*) The opposing factors (*c*) and (*e*) will vary much in different species, and the problem will require the combined attentions of mathematicians, and of ecologists working on the methods of regulation of the numbers of animals.

(*g*) This mechanical uniformity factor, since it acts independently of natural selection, explains how a particular structure or habit may evolve, when it only has a general adaptive significance.

161

C. S. Elton

6. References.

Allen, J. A. (1903), "Report on the Mammals collected in North-Eastern Siberia by the Jesup North Pacific Expedition . . ." *Bull. Amer. Mus.*, 19, 101.

Anthony, A. W. (1923), "Periodical Emigrations of Mammals," *Journ. Mammal*, 4, 60.

Bailey, V. (1905), "A Biological Survey of Texas," *N. Amer. Fauna*, 25.

Bailey, V. (1922), "Beaver Habits, Beaver Control, and . . . Beaver Farming," *U.S. Dept. Agric. Bull.*, 1078.

Barrett-Hamilton, G. E. H., and Hinton, M. A. C. (1910), *A History of British Mammals*, London.

Berg, B. (1913), "Der Wanderfalke und die Lemmingzuge," *Novit. Zool.* (Tring), 20.

Blasius, R. (1886), *Der Wanderzug der Tannenhener durch Europa im Herbst*, 1885, *und Winter*, 1885-86, Vienna.

Brooks, C. E. P. (1922), *The Evolution of Climate*, London.

Brooks, C. E. P. (1923), "Variations in the Levels of the Central African Lakes, Victoria and Albert," *Geophys. Mem. Meteor. Office* (London), 20.

Buxton, P. A. (1923), *Animal Life in Deserts*, London.

Chree, C. (1924), "Periodicities, Solar and Meteorological," *Quart. Journ. Royal Meteor. Soc.*, 50, No. 210 p. 87.

Christy, M. (1924), "The Hornbeam (*Carpinus*, L.) in Britain," *Journ. Ecology*, 12, No. 1, p. 39.

Collett, R. (1911-12), *Norges Pattedyr*, Christiania.

Cooke, M. T. (1923), "Report on Bird Censuses in the United States, 1916-20," *U.S. Dept. Agric. Bull.*, 1165.

Dresser, H. E. (1871-81), *A History of the Birds of Europe*, London.

Elliot, H. W. (1884), *Report on the Seal Islands of Alaska*, Washington.

Hagedoorn, A. L., and A. C. (1921), *The Relative Value of the Processes causing Evolution*, The Hague.

Haldane, J. B. S. (1924), "A Mathematical Theory of Natural Selection," *Trans. Cambr. Phil. Soc.*, 23, No. 2, p. 19.

Hale, G. E. (1924), "Sunspots as Magnets and the Periodic Reversal of their Polarity," *Nature*, 113, No. 2829, p. 105.

Harting, J. E. (1892), "The Plague of Field Voles in the South of Scotland," *Zoologist*, 16, 161.

Hewitt, C. G. (1921), *The Conservation of the Wild Life of Canada*, New York.

Hofman, J. V. (1920), "The Establishment of a Douglas Fir Forest," *Ecology*, 1, No. 1, p. 49.

Howell, A. B. (1923), "Periodic Fluctuations in the Numbers of Small Mammals," *Journ. Mammal*, 4, 149.

Hudson, W. H. (1892), *The Naturalist in La Plata*, London.

Humphreys, W. J. (1920), *Physics of the Air*, Philadelphia.

Huntington, E. (1914), *The Climatic Factor as illustrated in Arid America*, Washington.

Huntington, E. (1922), *Climatic Changes : their Nature and Causes.* New Haven.

Huntington, E (1923), *Earth and Sun.* New Haven.

Kofoid, C. A. (1921), *Report of the San Francisco Bay Marine Piling Survey; the Biological Phase*, San Francisco.

Kulagin, N. M. (1921), "On Appearances of Locusts in the 18th and 19th Centuries in Europe," *Proc.*, 2nd All-Russian Entomo-Phytopathological Conference in Petrograd in 1920. (In Russian.)

Lary, R. E. de (1923), "Arrival of Birds in Relation to Sunspots," *Auk*, 1923, 414.

Periodic Fluctuations in Numbers of Animals

Manniche, A. L. V. (1910), "The Terrestrial Mammals and Birds of North-East Greenland," *Medd. om Grønland*, Copenhagen, 45, No. 1.

Maxwell, H. E., and others (1893), "The Plague of Field Voles in Scotland," *Zoologist*, 17, 121.

Nelson, E. W. (1909), "The Rabbits of North America," *N. Amer. Fauna*, 29.

Palmer, T. S. (1896), "The Jack-rabbits of the United States," *U.S. Dept. Agric. Divn. Ornith. and Mammal. Bull.*, 8.

Pearson, G. A. (1923), "Natural Reproduction of Western Yellow Pine in the South-West," *U.S. Dept. Agric. Bull.*, 1105.

Pettersson, O. (1912), "The Connection between Hydrographical and Meteorological Phenomena," *Quart. Journ. Roy. Meteor. Soc.*, 38, No. 163, p. 173.

Poulton, E. B., and Walker, J. J. (1921), "Notes on the Migration of Lepidoptera," *Proc. Ent. Soc.*, London, p. 12.

Preble, E. A. (1908), "Biological Survey of the Athabasca-Mackenzie Region," *N. Amer. Fauna*, 27.

Preble, E. A., and MacAtee, W. L. (1923), "A Biological Survey of the Pribiloff Islands, Alaska ; (1) Birds and Mammals," *N. Amer. Fauna*, No. 46.

Rasmussen, K. (1921), *Greenland by the Polar Sea*, London.

Seton, E. T. (1920), *The Arctic Prairies*, London.

Shaw, N. (1923), *Forecasting Weather*, London.

Show, S. B. (1924), "Some Results of Experimental Forest Planting in Northern California," *Ecology*, 5, No. 1, p. 83.

Simroth, H. (1908), "Der Einfluss der letzten Sonnenfleckenperiod auf die Tierwelt," *Kosmos*, 9.

Stefansson, V. (1921), *The Friendly Arctic*, London.

Sund, O. (1924), "Snow and the Survival of Cod Fry," *Nature*, 118, No. 2821, p. 163.

Uvarov, B. P. (1923), "Quelques problemes de la biologie des sauterelles," *Ann. des Epiphyties*, 9, No. 2, p. 84.

Walker, G. T. (1923), "Correlation in the Seasonal Variations of Weather: A Preliminary Study of World Weather," *Mem. Indian Meteor. Dept.*, Calcutta, 24, Part 4.

Walker, J. J. (1914), "The Geographical Distribution of *Danaida plexippus*, L. . . . with Especial Reference to its Recent Migrations," *Ent. Mon. Mag.* (2), 25, 181.

Watt, A. S. (1923), "On the Ecology of British Beechwoods, with Special Reference to their Regeneration," *Journ. Ecol.*, 11, No. 1, p. 1.

Wiese, W. (1922), "Ice in the Greenland Sea and North Atlantic Weather," *Ann. Hydrog. und Marit. Meteor.*, 10.

Winge, H. (1902), "Grønlands Pattedyr," *Medd. om Grønland*, Copenhagen, 21, Part 2.

Witherby, H. F. (1920), *A Practical Handbook of British Birds*, London.

Nicholson, A. J. 1933. The balance of animal populations. *Journal of Animal Ecology* 2:132–178.

The most critical concept in population ecology that relates directly to effective wildlife population management is *population regulation,* defined as maintaining numbers consistent with available resources. Much of what wildlife biologists do in the field centers on obtaining information on animal numbers so that wildlife populations can be managed in a sustained manner. It is not surprising, therefore, that the literature abounds in articles reporting research on the mechanisms of wildlife population regulation. Nicholson and his colleagues, in the 1930s and 1940s, entered into one of the greatest debates on regulation, centered on whether it was caused by abiotic, biotic, or intrinsic mechanisms, or some combination of them.

Nicholson's 1933 paper is one of the most important published as he (and others) searched for the single cause that regulated populations. Nicholson argues that there is a balance of nature tied to the density of the controlled population. Furthermore, he makes a case that competition is the only factor that can regulate populations, which gives him credit for leading that biological school of thought. Nicholson's publication of this paper led to much further debate in population dynamics. Others claimed different factors as the main cause of population regulation (e.g., climate, predators).

During this time, it was interesting to understand what regulated populations, but the findings for an individual species were, and are, difficult to generalize to all species. This single issue was instrumental in creating the controversy about the different theories presented. As with many scientific arguments, Nicholson's ideas have been challenged and improved on as biologists continue to seek causal effects and further develop theories of population control and regulation. Nicholson's concepts pertaining to the biological influence on regulation of wildlife numbers remain as relevant today as they were 80 years ago.

RELATED READING

Bakker, K. 1964. Backgrounds of controversies about population theories and their terminologies. Journal of Applied Etymology 53:187–208.
Burk, C. J. 1973. The Kaibab deer incident: a long-persisting myth. BioScience 23:113–114.
Chitty, D. 1967. The natural selection of self-regulatory behavior in animal populations. Proceedings of the Ecological Society of Australia 2:51–78.
Christian, J. J., and D. E. Davis. 1964. Endocrines, behavior, and population. Science 146:1550–1560.
Pimentel, D. 1961. Animal population regulation by the genetic feedback mechanism. American Naturalist 95:65–79.
Rosenzweig, M. L., and R. H. MacArthur. 1963. Graphical representation and stability conditions of predator-prey interactions. American Naturalist 97:209–223.
Wynne-Edwards, V. C. 1965. Self-regulating systems in populations of animals. Science 147:1543–1548.

Reprinted by permission of John Wiley & Sons Ltd., publisher.

THE BALANCE OF ANIMAL POPULATIONS

By A. J. NICHOLSON, D.Sc.

*(Division of Economic Entomology, Commonwealth Council
for Scientific and Industrial Research, Canberra, Australia.)*

(With eleven Figures in the Text.)

FOREWORD.

For some years the writer, with the generous collaboration of Prof. V. A. Bailey[1], has been investigating the problem of competition in animal populations. An endeavour has been made to investigate the influence of competition on animal populations in all the principal kinds of situations known to exist between animals, and between animals and their environments, and to examine the major types of other factors known to influence populations, in order to see whether these are capable of nullifying, or otherwise influencing, the effects of competition. The scope of the work is thus very wide, and a full presentation of the conclusions, and of the arguments upon which they are based, cannot be given here.

The present paper forms an introduction to later and more detailed publications. Its object is first, to show why it is believed that the competition existing between animals when searching is of fundamental importance in the limitation of animal populations, and secondly, to give an outline of the major conclusions of biological interest that have been reached. Nothing more is attempted at present than to show the reasonableness of these conclusions, rigorous proof being postponed to future publications. It is thought that the "bird's-eye view" of the subject thus given will not only prepare the way for the more detailed work to be presented later, but will also provide biologists with useful hypotheses, the proof or disproof of which by experiment and observation should materially advance our knowledge of animal populations and their natural control.

PART I. THE NATURAL LIMITATION OF POPULATIONS.

THE EXISTENCE OF BALANCE.

It appears to be usual at present to deny the existence of the "balance of nature," about which there has been so much vague talk. Therefore, before considering the question of the balance of animal populations, we must examine the arguments for and against the existence of such balance.

There is an outstanding feature of animal populations with which all must be familiar. In any given place the population densities of animals are ob-

---

[1] Associate Professor of Physics, University of Sydney.

served to change in close association with the changing seasons. Similarly, for any given species of animal, the population density differs in different places, and usually there is a fairly evident relation between the differences in the population densities and the differences in climate, or other environmental conditions. These observations clearly indicate that animal populations must exist in a state of balance, for they are otherwise inexplicable.

Let us take a simple analogy. A balloon rises until the weight of air displaced exactly balances the weight of the balloon, but if ballast be then discarded the balloon again rises until it reaches a new position of balance. Because a balloon in the air is a balanced system, there is a relation between the weight it carries and the height it reaches; without balance, the height reached would be indeterminate.

The balance of animal populations is similar to that of a balloon acted upon by the changing temperatures of night and day. Such a balloon rises and falls in relation to the change in temperature, for this varies the volume of the balloon and the density of the surrounding air. The balloon is continually in a state of tending towards a position of stationary balance, but continues to rise and fall because the position of stationary balance is changing all the time. It may be remarked that even when moving uniformly the balloon is in a state of balance, friction with the air making up the deficiency either in the weight of the balloon or in the weight of air displaced, so governing the speed with which the balloon rises or falls.

The observed fact that there is a relation between the population densities of animals and environmental conditions can be explained only in terms of balance, just as the relation between the weight carried and the height reached by a balloon can be explained only in this way. Without balance the population densities of animals would be indeterminate, and so could not bear a relation to anything.

The evidence for the existence of balance is not confined to logical deduction from the known facts about animal populations existing in a state of nature. Experiments dealing specifically with populations prove that the latter do reach a state of stationary balance under constant environmental conditions. Thus Pearl (**18**, ch. 2), working with *Drosophila melanogaster* bred on banana-agar in milk bottles, showed that the density progressively approaches an asymptotic population, which is clearly a population in a state of balance. However, owing to difficulties of manipulation, his experiments mostly ended when the asymptotic population was closely approached. It is of particular interest that different races of *Drosophila*, bred under identical conditions, have different asymptotic populations.

Chapman (**3**), working with the flour-beetle, *Tribolium confusum*, showed that populations of this beetle grow until a particular density is reached, beyond which there is no further growth. This density is independent of the absolute quantity of flour available. In other words, the number of beetles is directly proportional to the quantity of flour available, when a state of balance is reached.

MacLagan (**15**, p. 126), working with *Smynthurus viridis*, has shown that increasing density progressively decreases the rate of population growth, and this is the essential mechanism for the production of balance in populations.

The most satisfactory experimental demonstration of the production of balance in animal populations is given by Holdaway (**9**). He shows that under given conditions populations of *Tribolium confusum* reach and maintain a particular density; and also that when the conditions are varied (he varied humidity) the density reached is also varied. He brings out clearly the important point that it is the interaction of the insects themselves that produces balance and so limits density, while physical factors, by modifying this interaction, influence the position at which the insects themselves limit their population.

ARGUMENTS USED TO DISPROVE THE EXISTENCE OF BALANCE IN NATURE.

Having very briefly considered the evidence provided by observation and experiment for the existence of balance in animal populations, let us now examine the arguments that have been used to support the contention that there is no such balance.

The claim that animal populations are not in a state of balance is usually based upon the observation that animals do not maintain constant population densities. Clearly this argument is illogical, for, if a population is in a state of balance with the environment, its density must necessarily change in relation to any changes of the environment. A population density that does not change with the environment is evidently not in a state of balance, but is fixed independently of the environment.

Another argument, which has received much prominence in recent years, is that, because a close association between variations in the population densities of animals and changes in climate has been demonstrated, therefore climate is almost wholly responsible for the determination of the densities of animals—while other factors, such as the availability of food and space, and the presence of natural enemies, are of negligible importance (cf. Bodenheimer (**2**) and Uvarov (**23**)). The fallacy of this argument is made evident by the following analogy: If we examine the surface of the ocean, we observe that it rises and falls in relation to the position of the moon. From this we do not conclude that the depth of the ocean is determined by the position of the moon, but merely that change in depth is associated with the position of the moon. The depth of the ocean is determined by the shape of the ocean bottom, by the quantity of water present, and by the balance of the surface, governed by gravity. Similarly, the observed association between changes in the population densities of animals and changes in climate does not show that the densities are determined by climate, but merely that climate may vary the densities. These, it will be shown, are determined by other factors besides climate, and are limited and held in a state of balance by competition.

Climate, by itself, cannot determine the population densities of animals, for it does not possess the property necessary for the production of balance; and it has already been shown that, unless there is balance, densities are indeterminate and cannot be related to anything. For the production of balance, it is essential that a controlling factor should act more severely against an average individual when the density of animals is high, and less severely when the density is low. In other words, *the action of the controlling factor must be governed by the density of the population controlled.* Clearly no variation in the density of a population of animals can modify the intensity of the sun, or the severity of frost, or of any other climatic factor. The necessity for this interdependence of population densities and the action of the controlling factors has been pointed out by several writers (cf. Howard and Fiske (**11**, p. 107), Nicholson (**17**, p. 83), Fisher (**7**, p. 42), Martini (**16**)), but its paramount importance in the study of the population problem has generally been lost to sight by biologists.

A moment's reflection will show that any factor having the necessary property for the control of populations must be some form of competition. If the severity of its action against an average individual increases as the density of animals increases, the decreased chance of survival, or of producing offspring, is clearly brought about by the presence of more individuals of the same species in the vicinity. This can only mean that the decreased chance of survival is due to increased competition of some kind.

In recent years the importance of the influence of climate on the activities of animals has been demonstrated by many workers. They have shown that changes in population densities are associated with changes in climate; but, for the reasons just given, this does not support the hypothesis many of them stress so strongly, namely, that the climate is mainly, if not wholly, responsible for the determination of the population densities of animals. The point of view of these workers is well expressed by Uvarov in the following passages: "...the evidence collected in this section, as well as in the whole of this paper, should go far towards proving that the key to the problem of balance in nature is to be looked for in the influence of climatic factors on living organisms" (**23**, p. 161). "It is a balance resembling that of a cork floating on the surface of a running stream, with its whirlpools, eddies and back currents, while the wind blowing with varying force now ripples the surface gently, now causes it to rise and fall in great waves. The cork rises and falls with them, comes into collision with other floating objects, now rides on a calm surface and is dried by the sun, now is again rolled over and over by a storm and is beaten by rain, but always continues its journey with ever-changing velocity and along a fantastically tortuous course" (**23**, p. 162).

This simile is apt, and forcibly illustrates the vicissitudes of an animal population. But stress is placed on the agitation of the cork, whereas its balance on the surface of the water is fundamentally important. Were it not for this balance, the cork would simply fall out of the system, and its interesting career would be brought to an untimely end. So it is with animal populations. Were it not for the fact that competition holds populations in a state of balance with their environments, there would be no limited populations for physical factors to vary. The populations would either disappear altogether or increase without limit.

In support of the contention that climate is mainly, if not wholly, responsible for the limitation of animal populations, much stress is commonly placed upon the fact that climate is known to destroy large numbers of animals. Quantitative observations in the field have proved that climate sometimes destroys far more individuals of a species than all other factors put together (cf. Bodenheimer (2)). However, the belief that this proves the important part played by climate in determining the densities of animals is based upon the confusion of two distinct processes, namely, destruction and control.

Let us take an example to illustrate the distinction between these processes. We will suppose that the animals in a certain population would increase one hundredfold in each generation if unchecked, and also that, on the average, climate destroys 98 per cent. of the animals. It is clear that the number of animals would be doubled in each successive generation if no other factors operated. Climate could never check this progressive increase, for it would continue to destroy only 98 per cent., its action being uninfluenced by the density of the animals. If, however, there is some other factor, such as a natural enemy, the action of which is governed by the density of animals, the destruction of the remaining 1 per cent. necessary to check increase would soon be accomplished. If this example were observed in nature, one would be tempted to conclude that, because climate destroys 98 per cent. of the animals while the natural enemy destroys only 1 per cent., the limitation of the population is mainly due to the influence of climate. However, it is clear that the natural enemy is wholly responsible for control, because climate, by itself, would permit the density of the population to become indefinitely great. It is not even necessarily true that the destruction caused by climate reduces the density of animals. Considerations will be given later which show that in such a situation the effect of climate is almost as likely to increase the number of survivors in an average generation as to decrease this number!

If an attempt be made to assess the relative importance of the various factors known to influence a population, no reliance whatever must be placed upon the proportion of animals destroyed by each. Instead, we must find which of the factors are influenced, and how readily they are influenced, by changes in the density of animals.

Investigations on the influence of climate on the life and activity of animals are in themselves of great interest. Also, they are of great economic importance, for they enable us to predict the possible limits of distribution of pests, and under what conditions increases or decreases in abundance may be expected. But, until such time as they take balance into account, they cannot give us any information about the values at which the densities of animals will be limited under given conditions, and this is the most important problem of all in economic biology.

Views on the Mechanism of Balance.

Having shown that the evidence for the existence of balance in animal populations is overwhelmingly great, and that the arguments commonly used when denying balance are unsatisfactory, we will now examine the views generally held about the mechanism of balance.

In almost all discussions of the "balance of nature" there is a strong implication that evolution is responsible for this. It seems generally to be considered that natural selection is the mechanism of evolution that produces and maintains balance, though how it can do so is never clearly stated. This common belief that natural selection has two distinct functions, namely, selection and the production of balance, appears to be responsible for most of the criticism to which the theory of natural selection has been subjected. Elton clearly recognises the difficulty, and remarks (5, pp. 24–5): "We have had to abandon the simple idea of the balance of nature, which was supposed to be produced by the natural selection of more or less fixed instincts, physiological reactions and structures....I do not wish to cast doubt here on the general ability of natural selection to produce adaptations. I am merely raising the question whether it is able to account for the regulation of numbers in an animal community."

The function of natural selection is to select—not to produce balance. Its mechanism is very simple. All animals produce a surplus of offspring, which is somehow destroyed. Natural selection merely causes a greater proportion of the more perfectly adapted individuals than of the less perfect ones to be amongst the survivors. This tends progressively to improve adaptation, but it has nothing whatever to do with the production of balance or the limitation of populations.

The idea that natural selection produces balance appears to be due to a misconception about the meaning of "adaptation." It seems generally to be thought that adaptation is the precise adjustment of the properties of species so that they balance the inherent resistance of the environment. This idea is expressed by Fisher (7, p. 38) in the following passage: "An organism is regarded as adapted to a particular situation, or to the totality of situations which constitute its environment, only in so far as we can imagine an assemblage of slightly different situations, or environments, to which the animal would on the whole be less well adapted; and equally only in so far as we can imagine an assemblage of slightly different organic forms, which would be less well adapted to that environment. This I take to be the meaning which the word is intended to convey, apart altogether from the question whether organisms really are adapted to their environments."

Observation and experiment show that such precise adjustment of the properties of animals to those of their environments does not exist. Thus Uvarov (**23**, p. 152) points out that "The common, though never proved, assumption that every organism is perfectly adjusted to its environment, requires a complete agreement between the environmental relations of the insect and its natural enemies, particularly parasites," and investigation has shown that this complete agreement is often lacking. Similarly, MacLagan (**15**, p. 144) has recently shown that no one set of conditions provides the optima of all the physiological processes of *Smynthurus viridis*, so that clearly this insect can never be precisely adjusted to its environment. Also, we know that species range over environments differing markedly from one another. Therefore, if a species be assumed to be perfectly adjusted to one kind of environment, it clearly cannot be so adjusted to the other kinds of environment in which it also occurs.

Adaptation is not the exact balancing of the properties of the environment by those of a species. It is simply the possession by animals of properties that enable them to exist in a given environment. If individuals with improved properties appear, this type ultimately displaces the original normal type of the species, so improving adaptation; but this adaptation has nothing to do with perfect adjustment or balance.

Even if we assume that adaptation is the qualitative balance of the properties of animals and those of their environments, and that this balance can be produced by natural selection, this does not help to explain the limitation of populations—for qualitative balance would be maintained whatever the density of a species might be. That is to say, there would be a neutral equilibrium, leaving the population densities indeterminate.

However, natural selection cannot produce even this neutral equilibrium. A number of arguments in support of this contention might be given, but it is sufficient at present to mention just one. We will assume that a population is in a state of qualitative balance with its environment, and that within the population there are some individuals having a greater survival value than the average. Clearly natural selection should tend to preserve such individuals at the expense of the others. But, if it does so, the properties of the population are improved, so over-balancing those of the environment and permitting a progressive and indefinite increase of the population. If natural selection is to produce and maintain balance, in such a situation it must preserve the normal individuals at the expense of those that have the greatest survival value! It is thus evident that natural selection is not only incapable of producing and preserving qualitative balance, but actually, by altering the properties of populations, must continually tend to destroy any such balance that may already exist.

Lotka (**12** and **13**), and later Volterra (**24–28**, and see Chapman (**4**, pp. 409–48)) and Bailey (**1**), give mathematical theories which conclude that the interaction of animals may itself lead to a condition of balance, and that the position of stationary balance depends upon the properties of the animals, the nature of their interaction and the properties of the environment. They also conclude that this interaction causes an oscillation about the steady density. However, consideration of the mathematical papers of these and other authors must be postponed to subsequent publications, though it may be mentioned that a brief account of the relation of Volterra's work to the present investigation has already been given (Bailey (**1**, p. 76)).

OTHER VIEWS ON THE LIMITATION OF POPULATIONS.

Another mechanism of balance is sometimes described, though it is generally put forward as an alternative to what is considered to be the untenable hypothesis of balance in nature. This mechanism is made clear in the following passages: "The organism considered may increase in numbers for a considerable time but it obviously cannot go on increasing indefinitely. As it increases in numbers it necessarily spreads, both in space and time, and during its spread it moves to points outside its optimum environment, when its rate of multiplication immediately diminishes" (Thompson (**22**, p. 57)). "When the population increases seriously far above its optimum, there is a shifting of the animals to other places, so that the abnormal density is relieved....If the population is less than it might be, the gaps soon fill in from the outside by a similar process of adjustment. Migration in response to an innate sense of harmony with the habitat makes possible a solution of the animal population problem" (Elton (**5**, p. 61)).

It will be observed that the migration referred to here is not necessarily a concerted movement of a large section of a population to a distant place, but is more likely to be a diffusion of individuals within, or beyond, the normal area of distribution of the species. The migrating animals may reach new and incompletely stocked places suitable for their existence, but in general they are likely to spend their time in unsuitable places, so that their chance of survival is reduced. In short, pressure of population induces migration, and this in turn decreases the chance of survival of an average individual. Thus this mechanism possesses the property essential for the limitation of populations, and automatically causes a population always to tend towards a position of balance.

It is of interest to note that with this mechanism it is possible that the whole of the surplus animals may be destroyed by climatic factors, and yet climate is not responsible for control. If increasing pressure of population in favourable areas leads to migration, the degree of migration is governed by the severity of competition. The proportion of individuals destroyed by climate is dependent on the proportion competition forces out of the favourable parts of the environment. Therefore it is competition, not climate, that regulates the fraction of animals destroyed, and so limits the density and holds the population in a state of balance with the environment.

Mention needs to be made of Chapman's concept of "biotic potential," which he defines as "the inherent property of an organism to reproduce and to survive" (**4**, p. 182). His idea appears to be that biotic potential, when pitted against environmental resistance, determines the density of a species; and that, as biotic potential and density can be measured, environmental resistance can be expressed quantitatively in terms of these factors. Space does not permit of detailed discussion of his arguments, so attention will be confined to his main thesis, namely, that the resistance of the environment to organisms is similar to the resistance of a conductor to an electric current. The reader is expected to see a striking resemblance between electrical potential (i.e. work done on a unit electric charge) and biotic potential (i.e. *possible* rate of increase). The whole argument seems to depend upon this double use of the word "potential."

Chapman continually reiterates the importance of his hypothesis in enabling us to express biotic factors in a quantitative way. However, he makes only one attempt to give an example of such quantitative expression. Having experimentally determined the density of *Tribolium confusum* under certain conditions, he proceeds to calculate environmental resistance by means of his formula. A glance at this example (**3**, p. 120) shows that the only use made of the experimentally determined density is to multiply *both* sides of the equation by it. Clearly the result is independent of density, and Chapman's hypothesis completely fails to help us to deal quantitatively with animal populations.

COMPETITION.

It has already been shown that competition is capable of producing balance, and so of limiting animal populations. Indeed, any factor that produces balance is almost necessarily some form of competition, for balance can be produced only if increasing density decreases the chance of survival of an average individual. There is, of course, nothing new in the idea that competition is the major factor limiting populations, for this idea is fundamental to the epoch-making work of Malthus (**14**) on the population problem.

Many organisms alter the chemical and physical nature of their environment, and, generally, the increase in density of such organisms causes the character of the environment to become progressively less suitable for themselves. This clearly is a form of competition and provides the mechanism necessary for balance. It appears to be of major importance in the control of plants and micro-organisms, and not only limits the densities of species but also plays an important part in determining what other species may exist in the same environment. Competition of this type is sometimes also of importance to animals. For example, by destroying vegetation grazing animals alter the light, humidity and other physical factors near the ground, so making the environment unsuitable for many animals and plants that live on or close to the ground, while making it suitable for other animals and plants that would have been unable to live in the shade of the vegetation. Thus the alteration of the physical qualities of the environment by the activities of animals may be of the utmost importance in determining what kinds of animals can live together in any given environment.

Though a species may sometimes limit its own density by altering the chemical and physical qualities of its environment, such limitation seems to be unusual amongst animals. Generally speaking, animals appear to be limited in density, either directly or indirectly, by the difficulty they experience in finding the things they require for existence, or by the ease with which they are found by natural enemies. We are thus faced with the problem of the competition that exists amongst animals when searching, and the whole of the investigation that follows is concerned with this important problem.

Examination of the problem of searching shows that it is fundamentally very simple, provided the searching within a population is random. Before further investigation, therefore, we must find if it may safely be claimed that the searching within animal populations is actually random.

It is important to realise that we are not concerned with the searching of individuals, but with that of whole populations. Many individual animals follow a definite plan when searching. For example, a fox follows the scent of a rabbit, and a bee systematically moves from flower to flower without returning on its course. However, there is nothing to prevent an area that has been searched by an individual from again being searched systematically by another, or even the same, individual. If individuals, or groups of individuals, search independently of one another, the searching within the *population* is unorganised, and therefore random. Systematic searching by individuals improves the efficiency of the individuals, but otherwise the character of the searching within a population remains unaltered. Therefore, when investigating the problem of competition, we may safely assume that the searching is random.

We will now examine this question of random searching. The area searched by animals may be measured in two distinct ways. We may, as it were, follow the animals throughout the whole of their wanderings and measure the area they search, without reference to whether any portions have already been searched, and so measured, or not. This we will call the area *traversed* by the animals. On the other hand, we may measure only the previously unsearched area the animals search. This we will call the area *covered* by the animals. Thus the area *traversed* represents the total amount of searching carried out by the animals, while the area *covered* represents their successful searching, i.e. it is the area within which the objects sought have been found.

Suppose we now take a unit of area, say a square mile, and consider what happens at each step when animals traverse a further tenth of that area. When the animals begin to traverse the first tenth of the area, no part of the area has already been searched, so in traversing one-tenth the animals also cover one-tenth of the area. At the beginning of the next step only nine-tenths of the area remains unsearched, so, as the animals search at random, only nine-tenths of the second tenth of the area they search is previously unsearched area. Consequently after traversing two-tenths of the area the animals have covered only 1·9-tenths. Similarly, after traversing three-tenths of the area the animals have covered only 2·71-tenths. The calculation may be continued in this way indefinitely, and it is clear from the character of the problem that, at each step of one-tenth of area traversed, the animals cover a smaller fraction of area than in the preceding step. Also, because at each step the animals cover only one-tenth of the previously unsearched area, the whole area can never be completely searched. This, however, is true only if the total area occupied by the animals (not the unit of area considered) is very large. When the results of such a pro-

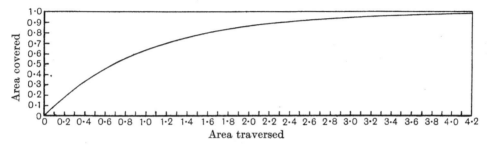

Fig. 1. The "competition curve."

gressive calculation are plotted, they are found to approximate to the curve shown in Fig. 1.

Such a calculation gives the general character of the effects produced by progressively increasing competition, and the original work on the present investigation was based on a curve calculated in this way. However, such a curve only approximates to the true form. Thus, when the animals have nearly completed their search of the first tenth of the area, only slightly more than nine-tenths of the area remains unsearched. Consequently even while traversing the first tenth of the area the animals spend some small part of the time searching over areas that have already been searched, and the same type of error runs through the remainder of the calculation. The calculation could give the correct result only if it were based upon indefinitely small steps. For this reason I approached Prof. V. A. Bailey, who gave me the formula (see Bailey (**1**, p. 69)) upon which the "competition curve" shown in Fig. 1 is based. This curve was then used throughout the investigation.

It may be mentioned that Fiske (**8**) used a curve, based on arithmetical calculations, which was essentially similar to the "competition curve," and that

Thompson (**21**) made use of a formula (given him by Prof. Deltheil of Toulouse) virtually the same as that on which the "competition curve" is based. However, both these authors were concerned only with the problem of superparasitism, a very specialised form of competition, and neither attempted to use the curve or formula for the investigation of the general problem of competition.

Examination of Fig. 1 shows that as the area traversed increases there is a progressive diminution in the rate of increase of the area covered. That is to say, the searching animals have progressively increasing difficulty in finding the things they seek. This is shown more directly in Fig. 2, which is easily derived from Fig. 1. Thus there is a definite relation (expressed in Fig. 2) between

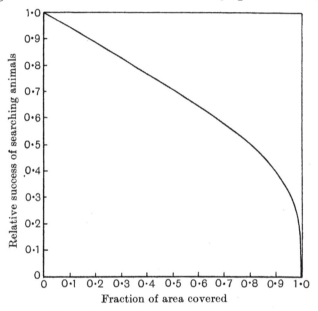

Fig. 2. Curve showing how increasing intensity of competition decreases the success of an average individual in finding the things it seeks. The success of an individual when free from competition is represented by unity.

the reduction of the success of the searching animals and the increase of the intensity of competition; and, with random searching, this relation is perfectly general, and so is independent of the properties of animals and those of their environments. Consequently, the "competition curve" has a general application to all problems of random searching.

It should be noticed, however, that the "competition curve" represents a probability. If small numbers of animals and small areas are taken, it is likely that the relation between the area traversed and that covered will not be found to be exactly as shown in Fig. 1. This does not indicate that there is anything wrong with the "competition curve," but merely that small samples of a statistical population are not necessarily completely representative of the large population from which they are taken. As in the present investigation we are always concerned with large populations, it is safe to conclude that the probability represented by the "competition curve" corresponds closely to what actually takes place in nature.

Before making use of the "competition curve" to investigate the limitation of the densities of animals it is necessary to define certain properties.

The number of times a population of animals would be multiplied in each generation, if unchecked, will be referred to as the *power of increase* of the species under the given conditions. The value of this property determines the fraction of the animals that needs to be destroyed in each generation in order to prevent increase in density.

The area *effectively* traversed by an average individual during its lifetime will be referred to as the *area of discovery* of the species under the given conditions. For example, if an average individual fails to capture, say, half the objects of the required kind that it meets, then the area of discovery is half the area traversed. The value of the area of discovery is determined partly by the properties of the searching animals, and partly by the properties of the objects sought. Thus it is dependent upon the movement, the keenness of the senses, and the efficiency of capture of an average individual when searching, and also upon the movement, size, appearance, smell, etc., and the efficiency in resisting capture of an average object of the kind sought. Consequently, under given conditions, a species has a different area of discovery for each kind of object it seeks. The value of the area of discovery defines the efficiency of a species in discovering and utilising objects of a given kind under given conditions, and determines the density of animals necessary in order to cause any given degree of intraspecific competition.

It is particularly important to recognise that these two properties respectively embrace all those things that influence the possible rate of increase of the animals, and all those that influence the efficiency of the animals in searching. Thus they are not merely properties of species, but properties of species when living under given conditions—the same species may have different properties in different places, or in the same place at different times. Also, it is important to notice that climatic conditions and other environmental factors play their part in determining the values of these properties, for they influence the vitality and activity of animals. Consequently, though such factors may not be specifically mentioned, they appear implicitly in all investigations in which values are given to the powers of increase and areas of discovery of animals.

Further increase of a population is prevented when all the surplus animals are destroyed, or when the animals are prevented from producing any surplus. When this happens the animals are clearly in a state of stationary balance with their environments, and maintain their population densities unchanged from generation to generation under constant conditions. This will be referred to as the *steady state*, and the densities of the animals when at this position of balance as their *steady densities* under the given conditions.

It is a relatively simple matter to determine the steady densities of animals when we know the values of their areas of discovery and powers of increase. Let us take a simple example. We will suppose that an entomophagous parasite attacks a certain species of host, that one host individual provides sufficient food for the full development of one parasite, that the area of discovery of the parasite is 0·04, that the power of increase of the host is 50, and that no factors operate other than those specifically mentioned. Clearly the steady state will be reached when the parasites are sufficiently numerous to destroy 49 out of every 50 hosts, and when there are sufficient hosts to maintain this density of parasites.

The parasites, then, are required to destroy 98 per cent. of the hosts, and so to *cover* 0·98 of the area occupied by the animals. To do this, it is necessary for the parasites to *traverse* an area of 3·91, as will be seen from examination of the "competition curve" (Fig. 1).

The required density of parasites, therefore, is 3·91/0·04, i.e. 98 (approx.).

But, in order that the density of the parasites may be maintained exactly, each parasite is required to find on the average one host[1].

Hence, the parasites are required to find 98 hosts in the area of 0·98 they cover, so that the steady density of hosts is 98/0·98, i.e. 100.

The steady densities calculated, of course, are the number of animals per unit of area. It is always convenient to choose a large unit for the measurement of area, so that the areas of discovery of the animals are represented by fractions, for the densities of animals can then be given in whole numbers. If small units of measurement are used the character of the results obtained is actually unaffected, but the densities calculated have to be expressed as small fractions of an animal per unit of area, which is undesirable.

It should also be noticed that the densities calculated are those within the areas in which the animals interact, and not necessarily within the whole countryside. Thus, if the animals can live only in areas containing a certain kind of vegetation, then the calculated densities are those within such areas, while the intervening areas in which the vegetation is unsuitable for the animals are ignored. Other things being equal, then, the density of a species within the whole countryside varies directly with the fraction of the countryside that provides suitable conditions for the species. It will be shown later that this is only approximately true.

The example just given is an exceptionally simple one, but it illustrates how the problem of the competition of animals may be investigated arithmetically. By assuming different properties for the animals, and different kinds of association between animals, or between animals and their environments, it is possible in this way to investigate how competition influences the

[1] Though only the female parasites search for hosts, it is convenient to consider the searching of an average individual, both males and females being included when making the average, and so establishing the value of the area of discovery.

densities of animals in all the principal types of situation known to exist in nature. Originally this arithmetical method was used for the investigation of almost all the problems that are here presented, but it is cumbersome and makes an adequate presentation of the subject very difficult. I therefore placed the problems before Prof. V. A. Bailey, who undertook the onerous task of investigating them mathematically[1]. The results obtained arithmetically were confirmed by the mathematical investigation, and the work was extended to certain problems that were not amenable to the arithmetical method.

In order to make any progress it was clearly necessary first to investigate ideally simple situations, even though these may have no counterpart in nature. Later, new factors were introduced, one at a time, in order to see how these influence the results already obtained. In this way the major types of factors known to influence animals were investigated. Unless some important factor capable of nullifying the effects of competition has been overlooked, which seems unlikely, the results should correspond to the facts of nature, for we know that animals do search for the things they require, and that competition does exist. It will later be seen that in their general features the results obtained in this investigation correspond to the observed facts, but more detailed observation and experiment is necessary in order to show whether there is a similar correspondence in detail.

The more important conclusions that have been obtained by a study of the problem of the competition that exists between animals when searching are given in Part II of this paper. These conclusions inevitably follow from the formulation of the problem already given, but for their rigorous proof much use needs to be made of mathematics, and it is thought desirable to postpone this to subsequent publications. However, when possible, brief general arguments are given to show that the conclusions are entirely reasonable.

PART II. THE EFFECT OF COMPETITION ON THE DENSITIES OF ANIMAL POPULATIONS.

In this part we shall investigate the way in which competition may determine the densities of animal populations in the types of situation that are believed to be of greatest importance under natural conditions. It must be realised, however, that though natural enemies and the availability of food and of suitable locations are factors that *may* control the densities of populations, they are not *always* control factors. For example, a species that is controlled by the availability of suitable locations is unaffected by the attack of a natural enemy if the latter merely destroys some of the surplus individuals that otherwise would die for lack of suitable places in which to live.

---

[1] Though arithmetic is a branch of mathematics, it is convenient in this paper to speak of *arithmetical* work when the arguments are conducted on definite numerical examples, and of *mathematical* work when the reasoning is by means of mathematical symbols.

(1) Two conclusions of fundamental importance.

It has been shown (pp. 135 and 142) that intraspecific competition acts to populations in the way necessary to produce balance, and it is clear that the position of stationary balance is that at which the surplus animals are destroyed, or at which no surplus of animals is produced. This leads inevitably to the following conclusion:

C 1. Intraspecific competition automatically regulates the severity of its action to the requirements of each population, provided the inherent resistance of the environment is sufficiently low to permit the species to exist.

This means that, if animals find an environment suitable for their existence, they will increase in density until such time as the severity of competition, together with the inherent resistance of the environment, just counterbalances the innate tendency of the animals to increase in density. For example, let us suppose that conditions are suitable for the existence of a certain species of natural enemy. In this situation competition influences the density of prey available to the natural enemies. Starting with few natural enemies, their density (and consequently competition) increases until the density of prey is such that the surplus natural enemies fail to find sufficient prey for survival. Clearly, then, if individual natural enemies require large numbers of prey, decrease in the density of prey is arrested when there is a just sufficiently high density of prey for the requirements of the natural enemies. Similarly, when individual natural enemies require very small numbers of prey, balance is reached when the density of prey is very low. This is only one of a number of apparently paradoxical conclusions that follow immediately from *C* 1.

C 2. For the steady state to exist, each species must possess some advantage over all other species with respect to some one, or group, of the control factors to which it is subject.

A *control factor* is one which responds to increase in the density of a given species either (1) by increasing the severity of its action against the species (as do natural enemies), or (2) by causing intraspecific competition to decrease the chance of survival of individuals (as do limited supplies of food or of suitable places in which to live). It should be noticed that (1) is actually a special case of (2).

When the conditions of *C* 2 are not fulfilled, other species maintain the action of all the control factors to which a given species is subject at a greater severity than this species can withstand. Moreover, even when these conditions are fulfilled the given species may not be able to exist in the steady state, for its special advantage with respect to certain of its control factors may be more than counteracted by special disadvantages with respect to other control factors. However, if the species is not so affected by disadvantageous characters, its density tends to reach such a value that intraspecific competition, together with environmental resistance (which includes that due to the

10–2

presence of other species), just balances its natural tendency to increase in density.

(2) The interaction of parasites and hosts.

Entomophagous parasites (which are really predators of a special kind) possess certain properties that make the investigation of their competition comparatively simple. They can produce offspring only when they find hosts. Consequently the number of offspring they produce is directly dependent on the number of hosts they find, and the number of hosts they destroy is also directly dependent on the number they find. This is strictly true only on the assumption that parasites possess an unlimited supply of eggs, but investigation has shown that unless the number of eggs produced is very small, the limitation of the egg supply affects the steady densities of the animals so little that it may be neglected. It is of interest to note that one of the fundamental hypotheses of Volterra's work (**24–28**) is that the number of offspring produced by predators is proportional to the number of encounters with prey. Consequently, though he speaks of the "eating species" (*specie mangiante*) and the "eaten species" (*specie mangiata*), he actually deals with parasites possessing an unlimited egg supply. Predators (using the term in its generally accepted sense) do not eat prey in proportion to the number found, for hunger and satiety commonly determine whether the prey met are eaten or not. Similarly, the number of offspring produced, though it may be influenced, is clearly not proportional to the number of prey eaten by a predator.

Because of this direct dependence of the number of hosts destroyed, and of the number of offspring produced by parasites, on the success of the parasites when searching, the arithmetical and mathematical investigations of the interactions of animals have been mainly devoted to the interaction of hosts and parasites. So far it has not been found possible to formulate the problem of the interaction of true predators and their prey in a completely satisfactory way for mathematical treatment. However, the investigation of the special kind of predators we here call parasites (i.e. entomophagous parasites), makes it possible for us to deduce the kind of effects that should be produced under given circumstances when true predators and their prey interact, as will shortly be shown.

For arithmetical convenience it was found necessary to confine attention to what may be referred to as the problem of "discontinuous generation." That is to say, a definite succession of generations was assumed; so, for example, it was considered that all parasite eggs are laid at the same time, and that all the hosts reach the vulnerable stage simultaneously. The mathematical treatment was also mainly devoted to this problem, but later Prof. Bailey extended the investigation to the problem of "continuous generation" (**1**), that is to say, the problem of interacting animals that may reach any stage of their life cycles at any time. Actually the "discontinuous problem" is a special case of the "continuous problem." However, it should be noticed that "discontinuous genera-tion" approximates closely to what we know to happen commonly amongst animals in nature, and notably amongst insects, to which this portion of the investigation particularly applies. The "continuous problem" will not be pre-sented in this paper, for it can be investigated only by the use of mathematics, but it may be remarked that the conclusions reached are in agreement with those obtained by the investigations of the "discontinuous problem."

Before proceeding to examine the question of the interaction of hosts and parasites, it is necessary to define several convenient terms.

A *specific parasite* is one that attacks only a single species of host; a *general parasite* is one that attacks several species of hosts; and a species of parasite that attacks several given species of hosts is said to be a *common parasite* of the given hosts. Similarly we speak of specific, general and common hosts, accord-ing as they are attacked by one species, by several species, or by several given species of parasites. It should be particularly noticed that these relations refer only to the given environment. Thus a parasite known to attack many species of hosts is a specific parasite in a given environment when it attacks only one species of host there.

The steady density of a species at the beginning of a generation is referred to as the *initial steady density*, and that at the end of a generation as the *final steady density* of the species under the given conditions.

We will first confine attention to the steady state, leaving till later the question of whether animals do actually maintain their respective steady densities, and also postponing the investigation of the oscillations in density that occur when hosts and parasites interact.

A. *The steady state.*

(a) *The interaction of a specific parasite and a specific host.*

The steady state exists when there is a sufficient density of parasites to destroy exactly the surplus of hosts produced, and when there is a sufficient density of hosts exactly to maintain the density of parasites necessary to de-stroy this surplus. Thus there is a mutual adjustment of the steady densities of interacting hosts and parasites; the parasites destroy the surplus hosts, and the hosts maintain the parasites while preventing any surplus of parasites from being produced.

The power of increase (see p. 144) of the host species determines what fraction of the hosts the parasites must destroy in the steady state, and so also determines what fraction of the environment the parasites must cover (see p. 142). It is clear that there is only one density at which parasites with a given area of discovery can exactly cover a given fraction of the environment, so it follows at once that:

C 3. The steady density of a parasite species depends alone upon its area of discovery and the fraction of hosts that is required to be destroyed in the steady state.

From this conclusion a number of others may readily be derived.

C 4. The steady density of a host species varies inversely as the number of parasites that develop at the expense of a single host individual.

Thus, if twice as many parasites can develop at the expense of a single host individual, only half as many hosts are necessary to support the steady density of parasites. So, because at this density the parasites destroy the surplus fraction of hosts, it is clear that the steady density of the host species is halved.

C 5. If some parasites are destroyed during development, the steady density of the parasite species is unaltered, but that of the host species is increased, and varies inversely as the fraction of parasites that escape destruction before maturity.

No matter how many parasites are destroyed during development, the same density of mature parasites is required in order to find and destroy the surplus of hosts. Consequently there must be a sufficient density of hosts to support the parasite offspring that are destroyed, in addition to those that ultimately survive.

C 6. The steady density of the host species and that of the parasite species vary inversely as the area of discovery of the parasite species.

If the area of discovery of the parasite species is doubled, then only half the density of parasites is necessary in order to find the surplus hosts, and so half the density of hosts supports the required density of parasites.

C 7. Reducing the power of increase of a host species decreases the initial steady density and increases the final steady density of the host; it also decreases the steady density of the parasite species.

It is easy to see that the steady density of parasites should be reduced, because, for balance, they are required to destroy a smaller fraction of hosts, and they can do this only if their density is reduced. Also, it is clear that the initial steady density of hosts should be reduced, for a lower density of hosts supports the required density of parasites. However, the parasites leave a larger fraction of a reduced density of hosts, and it is impossible to show by means of simple argument whether this means that the final steady density of hosts is increased or decreased. A special numerical example, illustrated in Fig. 3, is therefore used in order to show that the final steady density of hosts is increased. The curves in this figure may be derived easily from the "competition curve" (Fig. 1) by means of simple arithmetical calculations. Thus great fecundity itself *tends* to decrease the density of survivors.

C 8. The steady density of a parasite species is lowered when some of the hosts are destroyed by factors other than the parasite.

Clearly the destruction of some of the hosts by some other factor reduces the fraction of hosts to be destroyed by the given parasite. In other words, the *effective power of increase* of the host, with reference to the given parasite species, is reduced. The arguments supporting *C* 7 therefore also apply here.

C 9. According to whether a destructive environmental factor operates before, after, or at the same time as a given parasite species, both the initial and final steady densities of the host species are respectively lower than, higher than, or the same as they would be if the parasite were alone responsible for destroying the surplus hosts.

The destruction of some of the hosts by another factor clearly reduces the effective power of increase of the host with reference to the given parasite. Consequently the parasites are required to commence to operate in a lower density and to leave a higher density of hosts than they otherwise would (*C* 7). If, then, the parasite operates after the other destructive factor, the density of hosts reaching the stage attacked by the parasite is decreased, the final steady density of hosts is increased and the initial steady density of hosts is also

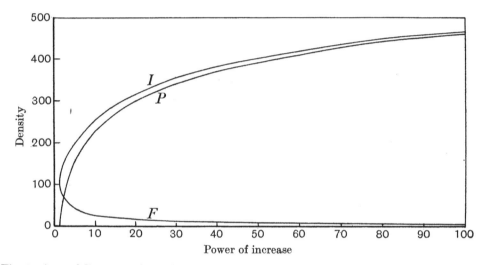

Fig. 3. Area of discovery of parasite species 0·01. *I*, the initial, and *F*, the final steady density curve of the host species. *P*, the steady density curve of the parasite species.

increased, for the true power of increase of the host is unaltered and the initial steady density is equal to the final steady density multiplied by the power of increase. Similarly, if the parasite operates before the other destructive factor, the initial steady density of hosts is clearly decreased, and the final steady density of hosts, left after the operation of the other destructive factor, is also decreased. When the other destructive factor operates at the same time as the given parasite, the situation is evidently intermediate between the two just examined, so one might expect that there would be little, if any, change in the density of the host. However, to prove that the density of the host is actually unaltered does not appear to be possible without the use of mathematics, or of numerical examples.

C 10. The steady density of a host species is but little increased owing to the limitation of the egg supply of its parasites, unless the maximum number of eggs a parasite may lay is very small.

In the steady state on the average (including males and females) a parasite is required to lay only one egg. It is easy to show that, with random distribution, the number of parasites required to lay more than five eggs (say, 10 eggs per female) is extremely small. The egg supplies of parasites are much greater than this, so unless there is an extremely high mortality before the parasites reach maturity, it is unlikely that the limitation of the egg supply will make any appreciable difference in the number of hosts destroyed. However, it should be noticed that during violent fluctuations the abundance of hosts may so greatly exceed that of their parasites that the parasites are unable to attack all the hosts they find. If so, the rate at which the parasite population catches up to that of the hosts may be somewhat retarded, but even so the steady densities of the animals are unaltered.

(b) Several specific parasites attack a common host.

C 11. When several species of specific parasites simultaneously attack a common host species, the steady state is practically impossible.

It is inconceivable that the properties of two or more species should be identical. Consequently, when several species of specific parasites simultaneously attack a common host species, that parasite which is capable of maintaining the steady density of the host at the lowest value must inevitably and completely displace the other species of parasites, however slight its advantage may be. This is because individuals of the most potent species compete for the same hosts as do individuals of the other parasite species, and so at all times have a greater chance of producing offspring. The consequent reduction in the densities of the less potent species does not reduce the severity of the competition to which they are exposed, for this is maintained by the most potent species.

C 12. If several species of specific parasites attack a common host species in succession, the steady state may exist, provided each species of parasite has a greater area of discovery than that of the parasite preceding it in the series.

A parasite species that attacks a common host species at a later stage in its life cycle than do other parasites has the special disadvantage of being able to attack only such hosts as are left by the preceding parasites. Each species of parasite operates in a lower density of hosts than does the preceding parasite, so it is conceivable that the special disadvantage of operating in a lower density of hosts might be counterbalanced by the possession of a greater area of discovery than that of the preceding parasite. Actually a precise counterbalancing of the special disadvantage by the value of the area of discovery is not essential, for competition provides a mechanism of adjustment. Each species of parasite operates at a time when there are no other parasites, so an increase or decrease in its density varies the success of an average individual in finding hosts. Consequently an adjustment of the density of each of the species of parasites may permit a state of balance to exist. Clearly, however, there must be a limit to the possibility of such adjustment, for any one of the parasite species may have such an area of discovery that even in the absence of intraspecific competition its efficiency is less than sufficient to counterbalance that of the other parasite species.

Consequently:

C 13. For the steady state to exist, the relative values of the areas of discovery of successive parasite species must lie between certain limits.

The limits of the permissible areas of discovery of parasites in a special numerical example involving two parasite species are shown in Fig. 4. The curves in this figure may be derived fairly easily from the "competition curve"

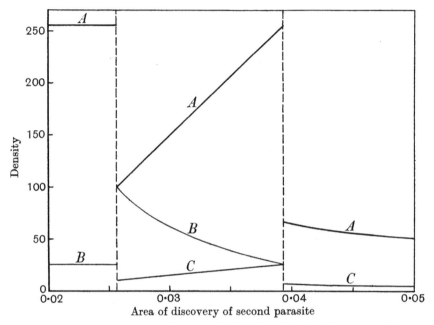

Fig. 4. Power of increase of host species 10. *A*, initial steady density curves of host species. *B* and *C*, steady density curves of hosts left respectively by the first and second parasite species. The region in which balance is possible between the two parasite species is delimited by dotted lines.

(Fig. 1) by means of arithmetical calculation. It will be observed that the second parasite can operate in association with the given first parasite only when its area of discovery lies between 0·0256 and 0·0392. When the area of discovery of the second parasite is less than 0·0256 the first parasite is relatively too powerful, and so operates alone, while when its area of discovery is more than 0·0392 the second parasite is so powerful that it displaces the first parasite.

C 14. As the area of discovery of any one of several successive species of specific parasites is increased, the density, and the fraction, of hosts attacked by this parasite is decreased.

This is clearly illustrated by Fig. 4 and by other numerical examples that have been examined, but its general truth can be proved only by the use of mathematics. Though this conclusion at first seems paradoxical, it is actually only an extension of *C* 6.

C 15. As the area of discovery of any one of several successive species of specific parasites is increased, the steady density of this parasite decreases.

This follows at once from *C* 14, and is an extension of *C* 6.

C 16. As the area of discovery of any one of several successive species of specific parasites is increased, the initial and final steady densities of the host are equally as likely to be increased as decreased.

For the rigorous proof of this conclusion it is necessary to use mathematics. However, the middle part of Fig. 4 illustrates the more paradoxical part of the conclusion, namely, that the initial and final steady densities of a host may be increased as a result of an increase in the area of discovery of one of its parasites. Whether an increase or a decrease in the initial and final steady densities of a host is produced is dependent on whether the given parasite occupies an even or an odd position in the series of successive parasites. More detailed investigation shows that the steady density of the host at any other stage of its life cycle may be either increased or decreased, according to the circumstances, by the increase in the area of discovery of one of the parasite species.

C 17. When a new specific parasite species is introduced to attack a host species, that is already attacked by specific parasites, the density of the host at the stage attacked by the introduced parasite is increased.

Fig. 4 shows that when the area of discovery of a parasite has the highest permissible value, the parasite destroys no hosts and so is present in indefinitely small numbers. The general truth of this has been proved mathematically, and it may be inferred from *C* 15. We may reasonably assume that the situation before the introduction of a new parasite is essentially the same as if the new parasite were present in indefinitely small numbers, its area of discovery having the highest permissible value. The reduction of the area of discovery of this parasite to such a value that the parasite can destroy some of the hosts (see *C* 14) is equivalent to the introduction of the new parasite. As the parasite originally found no hosts, when its area of discovery had the highest permissible value, but is caused to find an appreciable number when its area of discovery is reduced (i.e. when the new parasite is introduced), it is clear that a higher density of hosts must reach the stage it attacks. This is illustrated by Fig. 4, for it shows that, in the presence of the second parasite, the first parasite always leaves a higher density of hosts than it does when it operates alone.

C 18. When a new specific parasite species is introduced to attack a host species already attacked by specific parasites, the initial and final steady densities of the host may be either increased or decreased.

This follows from *C* 16, for it has been shown (*C* 17) that the introduction of a parasite is equivalent to lowering the area of discovery of a parasite from its highest permissible value. More detailed investigation shows that the introduction of a new parasite may either increase or decrease the density of the host species at any stage of its life cycle, other than that attacked by the introduced parasite.

C 19. As the power of increase of a host species is raised, the densities of hosts left by alternate species of specific parasites, including the last in the series, are decreased, while the densities of hosts left by the remaining parasites are increased.

Mathematical or arithmetical investigation is necessary for the proof of this conclusion. For a particular numerical example, it is illustrated by Fig. 5, which has been derived arithmetically from the "competition curve" (Fig. 1). It

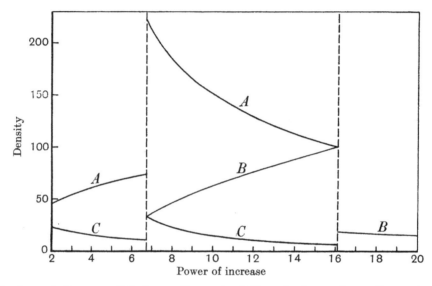

Fig. 5. Area of discovery of first parasite species 0·01, and of second parasite species 0·03. A, initial steady density curves of host species. B and C, steady density curves of hosts left respectively by the first and second parasites. The region in which balance is possible between the two parasite species is delimited by dotted lines.

will be observed that the steady state for the two given parasites and the given host is possible only between certain limits, that is, when the power of increase of the host lies between 6·7 and 16·1. It follows that:

C 20. When the power of increase of a host species is modified, the steady state between species of specific parasites that attack it may be rendered impossible, though this was previously possible.

A general conclusion of importance to subsequent investigations is that:

C 21. As the area of discovery of any one of several species of specific parasites that attack a common host species is increased, the number of hosts attacked by the given parasite in the steady state is decreased, whatever may be the relation in time between the given parasite and the other parasites, provided that the steady state is maintained by all the interacting animals.

This has already been shown to be true when several species of specific parasites attack a common host species in succession (C 14). More detailed investigation shows that this is also true when the various species of parasites attacking a common host species have overlapping periods of attack, and when the parasites attack the common host simultaneously. In spite of C 11, the latter situation is of some importance to subsequent investigations.

(c) Several specific hosts are attacked by a common parasite.

C 22. If two or more species of specific hosts are attacked by a common parasite species, one of the hosts must completely displace the others, unless the properties of the hosts are identical.

This follows at once from C 2, and clearly that species of host capable of maintaining the density of the common parasite at the highest value must completely displace the other hosts, however slight its special advantage may be.

C 23. If the properties of the host species are identical, the steady density of the total of host species is equal to the steady density any one of them would maintain if it were alone attacked by the given species of parasite.

This is evident, for, when the individual hosts all have the same properties, it is immaterial whether they be considered to belong to one or to several species. The balance between the host species is clearly neutral, that is to say, any given fraction of the total hosts may consist of any one of the host species, and this fraction tends to be maintained from generation to generation. This situation is evidently of no more than theoretical interest, but its examination will assist us in subsequent investigations.

C 24. The effectiveness of a species of parasite in controlling the density of a particular host species at a low value is increased if the parasite also attacks other species of hosts.

This increased effectiveness may be represented by the area of discovery a *specific* parasite would require to possess in order to control the density of the host at the same value. This is referred to as the *equivalent area of discovery* of the given general parasite for the given host under the given conditions. Thus, let us suppose that equal numbers of two species of hosts, having identical properties, are attacked by a common parasite species. It follows from C 23 that the steady density of each of these hosts is half what it would be if the host were alone attacked by the given parasite. From C 6 it is clear that this is the effect that would be produced by a specific parasite with an area of discovery twice as great as that of the given common parasite, so the presence of the other host species has doubled the equivalent area of discovery of the common parasite for each of the hosts. Similarly, if the common parasite destroys twice as many individuals of one of the species (A) as of the other (B), then the steady density of A is two-thirds, and that of B one-third, of what it would be if the host (A or B) were alone attacked by the given parasite. Consequently the equivalent area of discovery of the given parasite for A is one and a half times, and for B three times, its actual area of discovery. These and other similar numerical examples show that:

C 25. The equivalent area of discovery of a parasite species for a given host species varies inversely as that fraction of the total of attacked hosts which consists of hosts of the given species.

It should also be noticed that:

C 26. If a general parasite is caused to attack more hosts of one species, then fewer individuals of its other host species are available for attack, in the steady state.

This conclusion follows immediately from C 23, C 24 and C 25.

(d) Several hosts and several parasites interact.

This is the general problem of interacting hosts and parasites; any of the hosts and parasites may be specific or general, and they may interact in any conceivable combination.

C 27. For the steady state to exist, there must be at least as many parasite species as host species.

This follows from C 2, for the only control factors in the present situation are parasites, and so each host must possess an advantage over all other hosts with respect to at least one parasite.

C 28. When several species of general hosts are attacked by a common parasite species, balance may be established between them, even though their properties differ greatly.

When several species of specific hosts are attacked by a common parasite, one of them inevitably displaces the others (C 22), but when each species of host is also attacked by other parasite species, any tendency for one of the hosts to be displaced tends to cause a reduction in the severity of the attack of one or more of its other parasite species. Consequently, if the situation complies with C 2 and C 27, it should in general be possible for all the species of hosts to exist in the steady state, for each host may regulate the density of one or more of its parasites to its own requirements. However, the steady state is not inevitable, for it may happen that no possible reduction in the severity of the attack of those parasites that are regulated by the given species of host is sufficient to counterbalance the severity of the attack of a common parasite that is maintained by some other host species.

C 29. Balance between general parasite species that attack a common host species may be established even though the properties of the parasites differ greatly.

In the present situation the only control factors of the parasites are hosts, so (by C 2) each parasite must possess an advantage over all the other parasites with respect to some one, or group, of hosts. When a host is wholly controlled by a given parasite, this parasite clearly cannot be displaced, for the density of the given host may always be adjusted to its requirements (see C 1). However, when a parasite is controlled by hosts that are partly under the control of other parasites, balance is likely but not inevitable. Thus, when the severity of the action of other parasites tends to displace the given parasite from a common host, other species of hosts that are partly under the control of the given parasite tend to increase (cf. C 26), and so tend to counteract the displacement of the given parasite. But the degree to which the given parasite may regulate the density of its hosts to its own requirements is here limited, and so the reaction of the hosts it partially controls may not be sufficient to prevent the displacement of the parasite.

C 30. The equivalent areas of discovery of general parasite species that attack a common host species automatically reach the values necessary for balance.

This gives the mechanism of C 29. Let us suppose that two general parasite species attack a common host species, and that one of them (A) is in process of displacing the other (B). This is the situation that would exist if the parasites were specific and the area of discovery of B were relatively too low to permit of balance with A, so the equivalent area of discovery (see discussion of C 24) of B is temporarily too low to permit of balance with A. But the discussion of C 29 shows that as A reduces the number of individuals of the given host species available to B, other host species available to B tend to increase. Consequently the equivalent area of discovery of B for the given host progressively increases (C 25) and would become indefinitely large if B were caused to attack a very small number of hosts of the given species. Before this extreme could be reached the equivalent area of discovery of B is bound to attain a sufficient value to permit of balance with A, when the tendency for the displacement of B would be arrested. However, if B does not wholly control the density of any one of its hosts the reaction of the hosts other than the given one may not be as just described, and so the equivalent area of discovery of B for the given host may never reach the value required for balance. If so, B ultimately vanishes. This can happen only with species that do not comply with the conditions of C 1.

C 31. An increase in the area of discovery of a parasite species for a given host species may decrease the steady density of the parasite or leave it unchanged.

The steady density of the parasite may be (1) independent of, (2) wholly dependent on, or (3) partly dependent on the density of the host for which its area of discovery is increased. Clearly in situation (1) the density of the parasite is unchanged, and in situation (2) the density is decreased (C 6 and C 15). Situation (3) is intermediate between (1) and (2), so the density of the parasite might reasonably be expected to be generally decreased, but there is reason to believe that under certain circumstances it may be increased.

C 32. An increase in the area of discovery of a parasite species for a given host species may either increase or decrease the steady density of the host at any given stage of its life cycle.

It has already been shown that increase in the area of discovery of any one of several specific parasites that attack a common host decreases the density of hosts at the stage attacked by the given parasite (*C* 14), while the initial and final steady densities are equally as likely to be increased as decreased.

Now let us examine a situation identical with that just considered except that the density of the given parasite is uninfluenced by the given host (e.g. because it also attacks a specific host for which its area of discovery remains unaltered). Clearly, when the area of discovery of the given parasite is increased for the given host, this parasite will attack a larger fraction of the surplus hosts of the given species than before. Consequently it will behave like a *specific* parasite with a *decreased* area of discovery (*C* 14), and its equivalent area of discovery for the given host will therefore be decreased (*C* 30). So, in the present situation the increased area of discovery of the given parasite causes a decrease in the host density wherever it caused an increase in the preceding situation, and *vice versa*.

These examples show that whether the density of the given host at any particular stage is increased or decreased, by a given change in the properties of one of the animals, is determined by the nature of the interrelations of the animals. The same result is obtained when more complex situations (e.g. the interaction of general hosts and general parasites) are examined.

C 33. Raising the power of increase of a host species in general tends to increase the steady density of the host, but it may decrease the density at any stage in the life cycle, this being most probable at the final stage.

As a larger fraction of the host offspring is surplus, for the steady state to exist the parasites are required collectively to increase in density. If all the species of parasites attacking the given host species also attack hosts of other species, then any increase in the available density of the given host automatically tends to cause the parasites to reduce the density of their other hosts (cf. *C* 26). Consequently, the density of the given host must increase sufficiently, not only to support the required increased density of the parasites, but also to take the place of the reduction in the densities of the other hosts of the parasites. On the other hand, if the parasites of the given host are specific, the final steady density of the host is decreased, while the density of the host at any other stage may be either increased or decreased, according to the number of parasites in the series (*C* 19). Consequently, if general parasites are mainly, under the control of the given species of host (i.e. almost specific), it is possible that the density of the host may be decreased at any stage, and particularly at the final stage.

C 34. General parasites and general hosts with widely different properties may interact and maintain balance, and this balance is unlikely to be destroyed even by great modifications of the properties of any of the animals, or of those of the environment.

This follows from *C* 28–*C* 33, but the limitation of *C* 27 should be noticed.

C 35. The introduction of a new parasite species in general tends to reduce the steady density of the host species it attacks, but it may increase the density of the host, and, if it is a general parasite, it may exterminate the host.

This follows from *C* 33, for the successful introduction of a new parasite causes a smaller fraction of the surplus hosts to be destroyed by the original parasites, i.e. the effective power of increase of the host, with reference to the original parasites, is reduced. If the introduced parasite is a general one, attacking other species of hosts in the new country besides the one particularly considered, these other hosts may maintain its density at such a value that it destroys more than the surplus of the given host. If so, the given host is exterminated.

(e) The influence of hyperparasites.

C 36. In general, hyperparasites should produce little effect on the densities of hosts, though they tend to increase these densities.

Recognising that the fraction of parasites destroyed by true hyperparasites is generally small, this conclusion follows from *C* 6, so far as it applies to specific parasites and hosts. If, however, the host is also attacked by other species of parasites, the increase in its density, which tends to be produced as a result of the attack of the given species by hyperparasites, is likely to be modified by the action of these other parasite species. The attack of hyperparasites, as it reduces efficiency, is equivalent to reducing the area of discovery of the attacked parasite species; so it is seen from *C* 32 that the density of a host may even be decreased, when one of the species of parasites that attack it is in turn attacked by hyperparasites.

However, if a hyperparasite attacks a certain species of parasite more efficiently than this parasite attacks its host, the hyperparasite destroys the whole of the surplus offspring produced by the parasite, and maintains the density of the parasite below that necessary to destroy the surplus of hosts. Consequently the hyperparasite completely frees the host from the control of this parasite. This situation seems very improbable.

B. *Interspecific oscillation.*

So far we have confined attention to the steady densities of interacting hosts and parasites. It is now necessary to consider whether there is anything to cause the interacting animals to reach, and to maintain, these densities.

A little thought shows that any departure from the steady density sets up a reaction that *tends* to cause a return to this density. Thus an increase in the density of the parasites causes them to destroy more than the surplus of hosts, so reducing the density of hosts in the next generation, and this in its turn reduces the density of parasites. Similarly, an increase in the density of hosts causes the parasites to produce more offspring, and this increase in the density of parasites causes them to destroy more than the surplus of hosts. So any departure from the steady density sets up a reaction that tends to cause a return to this density, but there is always a delay in the appearance of the reaction.

Now let us suppose that the density of a host is above its steady value and is in process of being reduced by some species of parasite. When the host reaches its steady density, the density of the parasite is a result of the host density in the *preceding* generation, when it was above the steady value. Consequently there are more than sufficient parasites to destroy the surplus hosts, so the host density is still further reduced in the following generation. This is a new displacement from the steady density of the host, in the opposite direction to the original one. Consequently it sets up new reactions, of opposite sign to the original ones. Clearly, then, the densities of the interacting animals should oscillate about their steady values.

In order to investigate the character of such interspecific oscillation it is necessary either to use mathematics or to examine special numerical examples. For the present, attention is confined to the latter method.

C 37. When a specific parasite and a specific host interact, an oscillation in the densities of the animals about their steady values is produced, and this increases in amplitude with time.

This conclusion is illustrated by Fig. 6 for a particular numerical example. Further investigation and other numerical examples show that, when a specific parasite and a specific host interact:

C 38. Interspecific oscillation is independent of the area of discovery of the parasite.

C 39. The greater the power of increase of the host, the greater is the violence of interspecific oscillation, and the shorter is the period of oscillation, the minimum period being four generations.

Comparison of Figs. 6 and 7 demonstrates this for particular numerical examples.

It is also found that when a species of host is attacked by several successive species of specific parasites, and when it is attacked by a specific parasite that is also attacked by a hyperparasite, the interspecific oscillation is of the same increasing type.

C 40. When interacting animals are subject to increasing oscillation, they tend to be maintained permanently at densities much below their steady values, the animals being irregularly distributed in small groups, the positions of which are continually changing.

In outline, the argument on which this conclusion is based is as follows. With increasing oscillation the densities of the interacting animals must sooner or later be reduced to very low values. In the next generation it is clear that

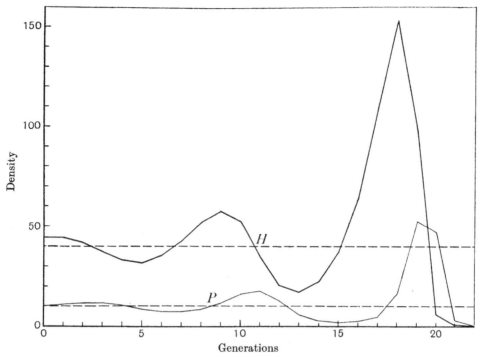

Fig. 6. Interaction of a specific host species (*H*) and a specific parasite species (*P*). Power of increase of host 2. Area of discovery of parasite 0·035. Arbitrary initial displacement of host density from 40 to 44. *H*, host curve; *P*, parasite curve. The steady densities are represented by dotted lines. Parasite curve drawn to half the vertical scale of host curve.

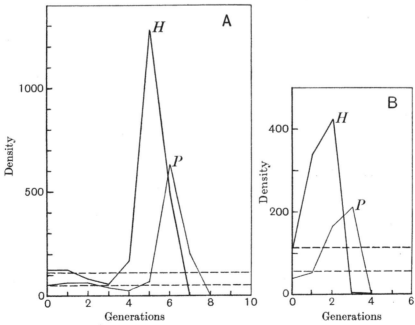

Fig. 7. Interaction of a specific host species (*H*) and a specific parasite species (*P*). Power of increase of host 50. Area of discovery of parasite 0·035. Arbitrary initial displacement of host density from 114 to 126 in A. Arbitrary initial displacement of parasite density from 111·7 to 80 in B. Parasite curve drawn to half the vertical scale of host curve.

the animals must be scattered in the environment in more or less widely separated groups, for each female forms a centre from which its offspring diffuse. Some of these groups are found by parasites, and it is likely that these groups will be exterminated[1] during the next few generations, the course of events being similar to that shown in Fig. 7 B. The other groups of hosts increase geometrically for a few generations, but are sooner or later found by parasites and are ultimately exterminated; in the meantime there has been a migration of hosts, some of which have established new groups. Thus, instead of there being a simultaneous oscillation of the animals throughout the whole environment, there are large numbers of independent local systems of oscillation, all phases of oscillation being represented in the environment at any given time. The fact that parasites can develop only in areas in which hosts occur means that such areas are searched more intensively than the rest of the environment. Consequently, the parasites do not have to cover a fraction of the whole environment equal to the fraction of hosts that is surplus, but only need to cover a much smaller fraction in order to find the surplus hosts. Therefore the density of parasites, and consequently the density of hosts, necessary for balance under the given conditions is much below the calculated values of the steady densities of the animals.

C 41. When there is an upper limit to the density an animal may reach, interspecific oscillation may be maintained indefinitely.

When a species increases in density it impairs its environment, for example by reducing the amount of available food. Therefore it is possible that a host may be prevented from reaching a very high density which otherwise is permitted by its parasite. If so, the parasite cannot reduce the host to a very low density as a result of interspecific oscillation, for very low densities can be produced only as an effect of preceding very high densities. Therefore the scattering of animals in small groups that is produced at very low densities (see C 40) does not occur, and interspecific oscillation persists indefinitely, its amplitude remaining constant under constant conditions.

There is reason to suspect that the fluctuations in the numbers of soil organisms observed by Cutler, and referred to by Russell in the following passages, were interspecific oscillations of the present type. "One of the most striking properties of the soil population shown by all the groups for which examination is possible, is that the numbers do not remain constant, even under apparently constant conditions, but fluctuate continuously in number.... Bacteria did not fluctuate in numbers when grown by themselves in sterilised soil; they rose to high numbers and remained at approximately a constant level. Their numbers fell, however, as soon as the soil amoebae were introduced, but no constant level was reached; instead there were continuous fluctuations as in normal soils" (**19**, p. 331). "There is a sharp inverse relationship between

---

[1] Compared with a species population the number of individuals in a group is small, so at the minima of violent oscillations the groups of individuals are generally exterminated (see p. 143).

11-2

the numbers of bacteria and those of active amoebae; when the numbers of amoebae fall, those of the bacteria rise, and, conversely, when the numbers of amoebae rise, those of the bacteria fall" (**19**, p. 317).

C 42. When a parasite attacks other host species in addition to the one it controls, interspecific oscillation is reduced in violence and its amplitude is likely to decrease with time.

The violence of interspecific oscillation when specific hosts and parasites interact is clearly due to the powerful reaction of any one of the species to any departure of the density of the other from its steady value. However, when the controlled species is only one of several hosts a given species of parasite attacks, if the density of the controlled host is, say, doubled, this does not

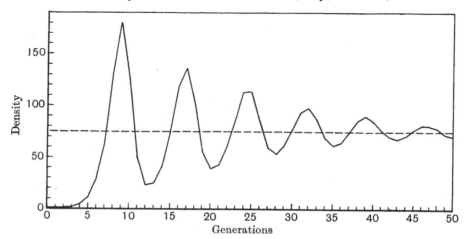

Fig. 8. Interaction of a specific host species and a general parasite species. Power of increase of host 10. Area of discovery of parasite 0·01. Steady density of specific host 75·5, and that of some other host of the given parasite 180. The density of the latter is independent of the activity of the given parasite and remains indefinitely at 180. Initially the given parasite is in equilibrium with this host and the specific host is introduced at a density of 0·1. Host density only is shown.

double the density of the parasite in the next generation; it only increases the density to some lesser degree, for the density of the parasite is also in part determined by its other hosts, the densities of which are controlled by other factors. Consequently, the violence of the reaction of the given parasite is reduced, and if this reduction is great it is evident that interspecific oscillation may decrease with time, rather than increase. Figs. 8 and 9 show that this is so. In Fig. 9 the controlled host has a smaller influence on the density of the given parasite than it has in Fig. 8. These figures, and the foregoing argument, show that:

C 43. The smaller the influence of the controlled host on the density of the parasite, the smaller is the oscillation.

It is of interest to note that the curve in Fig. 9 resembles the well-known form of a growth of population curve (see Pearl (**18**)). When the influence of the controlled host on the given parasite is even less than in Fig. 9 the resemblance is very close.

The considerations given in this section show that:

C 44. Only when the parasites are not specific do the calculated steady densities correspond with those which animals actually tend to maintain.

However, there is reason to believe that the relative values of the densities specific parasites and their hosts tend to maintain should correspond to the relative values of their respective steady densities, but this correspondence may be modified if the various species have different powers of diffusion. It should also be noticed that under certain circumstances the densities of specific hosts and parasites are caused continually to oscillate about their respective steady values (*C* 41).

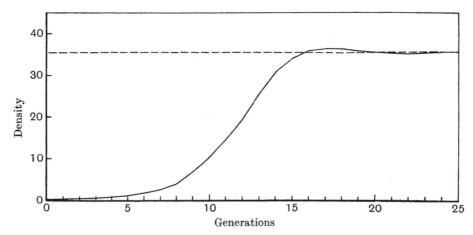

Fig. 9. Interaction of a specific host species and a general parasite species. Data as in Fig. 8, except that at equilibrium the specific host has a density of 35·5 and the other host a density of 220.

(3) THE INTERACTION OF PREDATORS AND PREY.

A. *The steady state.*

Parasites of the kind so far considered, i.e. entomophagous parasites, are actually predators of a special type, and their hosts are prey, for, unlike true parasites, these parasites destroy and consume their hosts. The mechanism of balance between true predators and their prey is essentially similar to that between such parasites and their hosts, for the predators destroy the surplus of hosts and the hosts support a density of predators necessary to do this. As a result, it is found that the conclusions reached by the investigation of the interaction of parasites and hosts apply in a general way to interacting predators and their prey, but need to be modified in detail. For the present it is sufficient to confine attention to certain features in which true predators differ markedly from entomophagous parasites. The most important difference

11–3

between parasites and predators is that, while the number of offspring produced by parasites is proportional to the number of hosts they find, there is no simple relation between the number of prey found by predators and the number of offspring they produce. All mature predators tend to produce a surplus of offspring, so it is clear that in the steady state this surplus must in some way be destroyed. Consequently:

C 45. The observed destruction of surplus predators does not indicate that the efficiency of a predator species is impaired.

It is generally recognised that in nature few animals die as a direct result of starvation. It is reasonable to believe then, that surplus predators are destroyed as an indirect effect of malnutrition, for predators can be controlled by their prey only if their chance of survival is influenced by their success in finding prey. Clearly, a half-starved predator should be more easily attacked than a fully vigorous one, and the observed fact that predators are commonly heavily attacked by other predators shows that this is a likely method of control. Also, it appears to be the only one that can cause the densities of predators to bear any relation to the densities of their prey. So:

C 46. The surplus of predators is destroyed by other predators as an indirect effect of difficulty in finding sufficient prey.

C 47. Cannibalism increases the efficiency of predators and enables them to maintain their prey at a lower density than would otherwise be possible.

When the surplus predators, instead of dying of starvation, are eaten by other predators of the same species, the survivors obtain at second-hand some of the nutriment provided by the prey that were eaten by the surplus predators. In other words, to the normal area of discovery of a surviving predator is added some fraction of the area of discovery of each of the surplus predators it eats. Thus the efficiency of the average surviving predator is increased by cannibalism, so permitting the steady density of predators to be maintained in a lower density of prey than would otherwise be possible.

C 48. The attack of primary predators by other species of primary predators tends to increase the efficiency of the predators in maintaining their own densities, and in controlling the densities of their prey at low values.

The surviving predators are collectively supported by the prey eaten by the attacked predators, as well as by the prey they themselves have eaten. Consequently the density of predators necessary to destroy the surplus of hosts can be supported in a lower density of prey than would otherwise be possible. The predator species share this advantage between them.

C 49. The attack of predators by special natural enemies may either increase or decrease the efficiency of the predators, but in general produces little effect.

If special natural enemies of primary predators attack these when very young, they decrease the intensity of competition between the survivors, so increasing the efficiency of the primary predators in controlling their prey. If the special natural enemies attack primary predators when almost mature, the steady density of primary prey must be increased, for the prey are required to support to maturity those predators that are ultimately destroyed by natural enemies as well as those that ultimately reproduce. In general, the effect of special natural enemies is to destroy some of the surplus primary predators that in any case must be destroyed, so they only reduce the efficiency of primary predators in that they prevent these predators from enjoying to the full the benefits of cannibalism (C 47 and C 48).

C 50. It is possible that special natural enemies may prevent primary predators from controlling the densities of their prey.

If special natural enemies can control predators at a lower density than that necessary for the destruction of the surplus primary prey, the prey are completely freed from the control of the given predators and their density progressively increases until arrested by some new factor.

B. *Interspecific oscillation.*

C 51. The interspecific oscillation of predators and prey should in general be less violent than that of parasites and hosts.

In contrast to parasites, predators present the following features. (1) Their offspring commence to search shortly after they are born, so a variation in the number of offspring produced by predators causes an almost immediate variation in the intensity of attack delivered against the prey. (2) Predators deliver an increased severity of attack in the same generation in which an increase in the density of prey takes place, for fewer predators then die as a result of difficulty in finding prey. (3) The attack of predators by other primary predators permits more predators to survive when prey are scarce than would otherwise be possible (C 47 and C 48). (4) When the density of prey increases above normal, this does not cause a corresponding increase in the number of offspring produced by predator individuals (see p. 166). (5) An increase in the density of prey increases the density of surviving predators, so, though individually these may produce no more offspring than usual, the total number of offspring produced in the predator population may be considerably increased.

It will be observed that, in comparison with parasites and hosts, (1) and (2) reduce the delay in the production of reactions, (3) and (4) reduce the intensity of the reactions of predator densities to variations in the densities of prey, while (5) shows that an increase in the density of prey causes a subsequent increase in the density of predators. Thus the conditions necessary for the production of interspecific oscillation exist, but the lesser delay and intensity of the reactions should reduce the violence of oscillation (see C 42). It may therefore be concluded that:

C 52. In general, predators and their prey should tend to be maintained at their steady densities, for increasing oscillation is unlikely to be produced between predators and prey.

(4) COMPETITION BETWEEN ANIMALS FOR SUITABLE LOCATION AND FOOD.

The precise places in which the habitats of animals exist, i.e. where all conditions are suitable for the existence of animals, are here referred to as *suitable locations*. Also, the word *food* is here used in a restricted sense that does not include hosts and prey, for we are now concerned with the competition of animals when searching for objects which, unlike hosts and prey, are *produced* in quantities that are entirely independent of the activities of the searching animals.

C 53. The minimum density of suitable locations necessary for the existence of a species varies inversely as the area of discovery, and also tends to vary inversely as the power of increase, of the species.

If inability to find suitable locations causes exactly the surplus of animals to be destroyed when there is no competition, i.e. when the density of the given species is indefinitely small, it is clear that the density of suitable locations has then the minimum permissible value for the existence of the given species. If the area of discovery is, say, doubled, the minimum permissible density of suitable locations is halved, for only thus can exactly the surplus of animals be destroyed without competition. If the power of increase is, say, doubled, the fraction of animals required to survive in the steady state is halved, and this should be supported if the minimum permissible density of suitable locations is also halved. However, more detailed investigation shows that this is only approximately true, for the fraction of its area of discovery traversed by an average animal before coming to rest in a suitable location varies with the fraction of animals that succeed in finding suitable locations.

C 54. Before all suitable locations are occupied, the densities of animals are limited by difficulty in finding unoccupied ones.

Balance is reached when difficulty in finding suitable locations, plus the intensity of competition, causes the destruction of exactly the surplus of animals. In Fig. 10 is shown the fraction of suitable locations occupied in the steady state when the density of suitable locations has any given value (measured in terms of the lowest permissible value). It is clear from this that animals should occupy almost the whole of the available suitable locations unless the power of increase is very low or the density of suitable locations is less than about four times the lowest permissible value. However, as the density of suitable locations is independent of the activity of the searching animals, it is likely at times to approach the lowest permissible value. Consequently, even when animals are controlled by the availability of suitable locations, they must be expected sometimes to leave many suitable locations unoccupied.

C 55. When food is continually produced, the density of animals is independent of, and the density of food always present is inversely proportional to, the area of discovery of the animals.

In the steady state there can be only one density of animals, namely, that at which food is consumed at the rate at which it is produced. If, then, the area of discovery is, say, doubled, the density of animals is unaltered, while the amount of food existing at any moment in the environment is halved, for the animals hunt for food twice as quickly as before.

C 56. The density of food always present is independent of the rate of production of food.

If the rate of production of food is, say, doubled, the density of animals is also doubled (see *C* 55) and so the food is found twice as quickly as before. Consequently, the amount of food existing in the environment at any moment is unaltered.

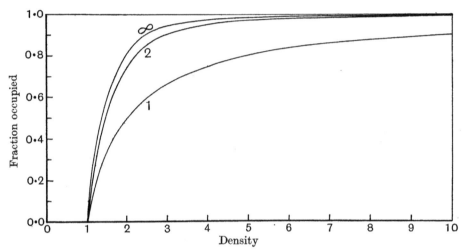

Fig. 10. Curves showing the fraction of suitable locations occupied in the steady state when the density of suitable locations has any given value up to 10, the lowest permissible density being the unit of measurement. The curves are for animals respectively having powers of increase of infinity, 2 and 1.

C 57. The density of animals is proportional to the amount of food produced in the lifetime of an animal.

For the steady state, food must be consumed at the rate it is produced. Assuming that the amount of food consumed per animal is constant, if its lifetime is, say, doubled, an animal consumes food at half the rate, and so double the density of animals is required to consume food at the rate it is produced.

C 53–*C* 57 are strictly true only on the following assumptions: (1) there is a continuous and constant generation of animals; (2) the surplus animals do not find any food; (3) the food always remains suitable for consumption by the animals. If these assumptions are not made the conclusions need to be modified, but in general character they remain true.

C 58. If the food ceases to be edible after it has been exposed for a certain period, there is a lower limit to the rate of production of food for any given species of animal, at less than which the species cannot exist.

If no animals operate there is clearly a limited density of food remaining edible at all times. If this density is such that, in the absence of competition, more than the surplus of animals fail to find food, the species clearly cannot exist.

C 59. Even though the density of a species is wholly controlled by the availability of food, in the steady state there must always be some food remaining unattacked in the environment.

Only if no time elapsed between the production of food and its discovery and consumption by animals could this be untrue. So animals must not be expected to increase in density until all food is consumed.

C 60. The control of the densities of animals by the availability of food is consistent with the observed fact that few animals die as a direct result of starvation.

As an indirect effect of difficulty in finding food the surplus animals are likely to be destroyed by natural enemies (cp. *C* 46) or adverse weather conditions, while the discovery of insufficient food may lower the birthrate of animals.

C 61. If there are fluctuations in the rate of production of food, the densities of animals fluctuate in relation to these, with some delay; and the quantity of food existing in the environment is not maintained by the animals at an absolutely constant value, but fluctuates in fairly close relation to the fluctuations in the rate of food production.

The density of animals continually tends to reach that necessary for balance with the rate of food production (*C* 57) and so fluctuates, but the nature of the reactions of the animals does not permit of precise correspondence between the fluctuations of food and of animals, so here the relation necessary for *C* 56 does not exist.

C 62. When animals are controlled by the availability of food there is little tendency to produce oscillations in density, and the densities should strongly tend to remain at their steady values.

If the density of animals is above the steady value there is an almost immediate reduction in the density of food and, in turn, an almost immediate decrease in the density of animals. Similarly, if the density of animals is too low, there is an almost immediate increase in the density of food, and so of surviving animals. As interspecific oscillation is primarily dependent on the delay in the production of reactions (see p. 161), there is little likelihood that animals controlled by the availability of food should oscillate in density.

(5) The influence of animal behaviour on populations.

One type of behaviour has already been considered and it has been shown (p. 141) that, so long as animals, or groups of animals, search independently, the searching within a population must be random. This remains true, however systematically the individual animals (or groups of animals) may search.

Systematic searching by individuals increases the areas of discovery of animals to some extent, but does not otherwise influence populations.

C 63. The territory habit produces a stable equilibrium of the population density.

When animals lay claim to territories (see Howard (**10**), and Elton (**6**)), it is clear that in any given area there is room for only a limited number of territories. Consequently the surplus individuals are continually harried by their more fortunate brethren, or are forced into unsuitable environments, and so their chance of survival, and of producing offspring, is greatly reduced. The system of balance resembles that of water in an overflowing reservoir; and there is no tendency to produce oscillations in density. If the animals having territories are natural enemies, they clearly tend to destroy a constant absolute number of prey. Consequently the fraction of prey they destroy tends to vary inversely as the density of prey, so the territory habit tends to make natural enemies behave in the opposite way to control factors (see Strickland (**20**)).

C 64. Certain predators (e.g. birds) tend to concentrate attention upon those parts of the environment in which prey are abundant, and so tend to damp fluctuations in the densities of their prey.

Many predators have two modes of locomotion (e.g. flight and hopping, or running and creeping), one of which is used in a cursory random survey of the environment for aggregations of prey or suitable places in which prey might occur, while the other is used for a slower and more detailed random survey of places in which prey are likely to occur. As different species of prey have different habitats they commonly occur in different places, so at any time the more abundant species are likely to be particularly heavily attacked by predators of the present type. This tends to damp any fluctuations in the densities of the prey, and also provides a mechanism by which a single species of predator may simultaneously control the densities of a number of different species of prey, for each species of prey tends to reach such a density that this causes the predators to destroy its surplus numbers.

C 65. When natural enemies in times of scarcity will attack prey other than the preferred species, this tends to damp the oscillation of the density of the preferred species.

Clearly a reduction in the density of the preferred species does not cause a corresponding reduction in the density of the natural enemies. The intensity of the reaction of the natural enemies being reduced, the violence of interspecific oscillation should also be reduced.

C 66. The polyphagous habit appears to be an almost essential adaptation of large natural enemies in order that the disadvantage of great size may be counterbalanced.

On the average, in traversing a given distance an animal must find sufficient food to provide the energy necessary for traversing that distance. Also, a large animal expends more energy, and so must find more food, than a small one when traversing a given distance. Small natural enemies tend to control the densities of their prey at the low values that are sufficient for their requirements, so, if large natural enemies are to exist in the same environments as small ones, they must be able to utilise more sources of food than the small natural enemies. However, this is not necessarily true if the large natural enemies are much more efficient in finding and attacking prey than the small ones, or if the large natural enemies can attack species of large prey that are unavailable to the small natural enemies.

(6) THE INFLUENCE OF THE PHYSICAL ENVIRONMENT ON POPULATIONS.

C 67. The distribution and densities of all animals are ultimately dependent upon the physical environment.

The physical environment ultimately limits the distribution of all plants. Natural enemies are dependent upon their prey and these, in their turn, upon their food plants. Consequently, the organisms with which given species of animals interact, and so the densities of the species, are ultimately determined by the physical environment.

C 68. Physical factors that are uninfluenced by the densities of animals cannot directly determine these densities.

Arguments supporting this conclusion have already been given (pp. 134–37). However, there are certain exceptional situations in which animals, like plants and micro-organisms, modify the physical nature of their environments. When this happens the densities of animals may be controlled by the reactions of physical factors, but even these are effects of competition (p. 140).

C 69. Physical factors may indirectly influence the densities of animals by modifying the positions at which other factors produce balance between animals and their environments.

Climate, in particular, may influence the vitality and activity of animals, so modifying the values of the powers of increase and areas of discovery, and consequently the steady densities, of animals.

C 70. The destructive action of physical factors may either increase or decrease the steady densities of animals.

This follows at once from *C* 9, *C* 19 and *C* 33, for when some fraction of the surplus hosts is destroyed by physical factors, the effective power of increase of the host with reference to its parasites is reduced. This conclusion applies when the physical factors are normal ones that operate continuously or periodically. An abnormal destructive factor inevitably reduces the density of a species as a direct and immediate effect, but if such a factor continues to operate for a sufficiently long period its indirect effects may counteract or reverse the direct effects it tends to produce.

C 71. Migration, provided it is dependent upon the severity of competition, may control the densities of animals against which physical factors alone operate.

Competition forces the surplus animals into unsuitable portions of the environment, where they are destroyed by the physical factors (see p. 139).

C 72. Climatic oscillation causes interspecific oscillation of the decreasing type to be maintained indefinitely.

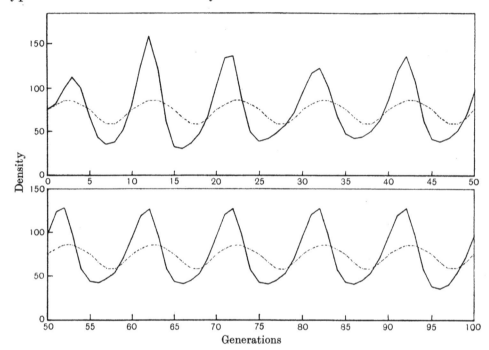

Fig. 11. Interaction of a specific host species and a general parasite species when under the influence of fluctuating climatic conditions. Data as in Fig. 8, except that the animals commence
to operate in a state of equilibrium and, as an effect of climate, the power of increase of the
specific host species changes in the following regular sequence: 10, 11, 12, 12, 11, 10, 9, 8, 8,
9, 10, and so on. Dotted line shows the steady density, and solid line the actual density of
specific host species.

Fig. 11 and Figs. 8 and 9 are based on similar data, except that for Fig. 11 it is assumed that the value of the power of increase of the host periodically changes in relation to some periodic climatic change. In Fig. 11 the interacting animals start in a state of equilibrium, but subsequent variations of the power of increase of the given host cause a departure from the steady state. For about the first fifty generations the oscillations of the host density vary in amplitude but subsequently reach a particular amplitude which is maintained indefinitely under the influence of the given constant climatic oscillation. Thus the inter-specific oscillation, instead of dying out as it does in Fig. 8, is indefinitely maintained, and this is due to the fact that fluctuating climatic conditions continually apply a small external force to the oscillatory system.

C 73. Interspecific oscillation tends to take the period of climatic oscillation that governs it.

This is shown by Fig. 11 and other similar examples that have been examined. However, when the periods of interspecific and of climatic oscillation are widely different this conclusion is no longer true.

C 74. Seasonal fluctuations in the densities of animals are more violent than one would expect from the known influence of climatic conditions on the properties of the interacting animals, and this violence is wholly due to the influence of interspecific oscillation.

The effects produced on the steady density of the host by climatic oscillation alone are shown by the dotted line in Fig. 11.

C 75. Irregular climatic fluctuations tend to produce closely associated irregular fluctuations in the densities of animals, but delayed effects cause further irregular fluctuations to be produced that have no apparent relation to climatic conditions.

The first six oscillations in Fig. 11 vary in amplitude, though they are all under the influence of precisely the same seasonal oscillation. Consequently, the amplitude is governed, not alone by the prevailing conditions, but also by the character of the preceding fluctuations. So a given abnormal climatic fluctuation not only influences the densities of animals subject to it, but also influences the densities of animals in subsequent generations. In addition, the resultant variation in the density of any given species causes subsequent variations in the densities of the species with which it interacts, and through these the impulses pass to all other species in the environment, with varying degrees of delay. This clearly should cause irregular fluctuations in the densities of animals that have no evident relation to variations in the physical environment.

C 76. Competition regulates the densities of animals in erratically and progressively changing environments just as it does in undisturbed environments.

Just as a cork moves in relation to the waves of a storm-tossed ocean only because it is in a state of balance on the surface of the water, so the densities of animals can vary in relation to variations in their environments only because competition is always tending to cause them to reach the position of balance corresponding to the prevailing conditions (see p. 135).

(7) IMPLICATIONS OF THE CONCLUSIONS REACHED.

A. *Economic biological control.*

The significance of many of the conclusions already reached in the question of the utilisation of natural enemies to control pests is evident, and at present it is desired merely to mention certain points that are of particular importance to this question.

All species that are not progressively increasing in density are necessarily controlled; so the function of economic control is not to bring under control a pest that was previously uncontrolled, but to modify the control so that the density of the pest is reduced to a point at which the economic damage caused is negligible.

The importance of a factor in controlling a pest is not even suggested by the fraction of pests it destroys (p. 136). When searching for the control factors it is necessary to find which are influenced, and how readily they are influenced, by changes in the density of the pest (pp. 135 and 137).

Specific parasites, provided their life cycles conform closely with those of their hosts, should control pests at very low densities, the pests occurring in widely separated groups, the positions of which frequently change (C 40). General parasites should commonly control their hosts at higher densities than specific ones, the distribution of their hosts being fairly uniform (C 42), but, unlike specific parasites, a general parasite, when introduced, may exterminate a particular host (C 35). The effect produced on the density of a pest by the introduction of a new natural enemy depends not only upon the properties of this natural enemy, but also upon the nature of the complex of animals with which the pest originally interacted (C 18 and C 35).

It seems probable that the adverse influence of hyperparasites on the control of pests has been over-emphasised (C 36 and C 45–49), and that lack of success which has been attributed to the presence of hyperparasites was probably due to some other cause. However, it must not be forgotten that a hyperparasite may completely free a pest from the control of a particular natural enemy (C 36 and C 50).

B. *Natural selection.*

It has already been shown (p. 137) that natural selection not only cannot produce and preserve the "balance of nature," but must inevitably destroy the qualitative balance it is supposed to produce. Balance is produced and maintained by competition, not by natural selection.

When individuals having advantageous properties appear in a population, their chance of being amongst the survivors is greater than that of the normal individuals. In the course of time, then, the fraction of the population consisting of individuals with advantageous properties increases, so improving the average properties of the animals, and increasing the degree to which they overbalance the inherent resistance of the environment. But the severity of the action of intraspecific competition is automatically regulated to the requirements of each population (C 1). Thus the progressively increasing proportion of favoured to normal individuals causes a progressive increase in the severity of intraspecific competition. As the original degree of competition was sufficient to destroy the surplus normal individuals, the increased severity of competition must cause the destruction of more than the surplus of normal individuals. Ultimately, therefore, the favoured type must completely displace the original normal type, though this would have remained successful indefinitely had the more favoured type not appeared.

Thus competition maintains balance in spite of the disturbing influence of natural selection, and adaptation that leads to the preservation of a type has nothing to do with a precise qualitative adjustment of its properties to those of the environment. Also, from C 31, C 32 and C 33, it is clear that the selection of a type with advantageous properties does not necessarily increase the density of the species. This and other arguments (see Nicholson (**17**)) show that natural selection operates without reference to the success of species. It simply favours those individuals that possess some advantage over their fellows, while leaving competition to determine the densities of the species and to maintain them in a state of balance with the environment.

SUMMARY.

1. The whole of this communication is essentially the summary of a much larger work, so it has been impossible here to do more than give an outline of its main features.

2. The densities of animal populations are known to bear a relation to the environmental conditions to which they are subject, and the existence of this relation shows that populations must be in a state of balance with their environments.

3. For balance, it is essential that the action of a controlling factor should be governed by the density of the population controlled, and competition seems to be the only factor that can be governed in this way.

4. Examination of the competition to which animals are subject shows that it is generally competition between animals when seeking the things they require for existence, or competition between natural enemies that hunt for them.

5. Detailed investigation of the question of competition between searching animals shows that, whether the animals are controlled by competition for food, or for suitable places in which to live, or by the competition of their natural enemies, there is for each species a particular density, referred to as the "steady density," at which balance exists. The value of this steady density depends upon the properties of the species, the properties of the species with which it interacts, the nature of this interaction, and the properties of the environment.

6. Competition always *tends* to cause animals to reach, and to maintain, their steady densities. Factors, such as climate and most kinds of animal behaviour, whose action is uninfluenced by the densities of animals, cannot themselves determine population densities, but they may have an important influence on the values at which competition maintains these densities.

7. When the densities of animals are controlled by natural enemies, this interaction itself produces oscillation, even in a constant environment. With

very simple types of interaction, such as that of a specific entomophagous parasite and its host, the oscillation of density increased in amplitude with time.

8. According to the circumstances, the final result of such oscillation is either the perpetual maintenance of the oscillation with constant amplitude under constant conditions, or the population is broken into widely separated small groups of individuals, the positions of which continually change with time. In the first situation the densities of the animals fluctuate about their steady values, but in the second the average densities are permanently maintained much below their steady values.

9. With more complex types of interaction the oscillation decreases in amplitude with time, so that in a constant environment the densities are caused eventually to reach, and to remain at, their steady values.

10. However, the environment does not remain constant, and it is found that periodic environmental changes, such as those of the seasons, tend to impress their period upon interspecific oscillation. Consequently the oscillation that is produced by the interaction of animals should generally be found to correspond in time with seasonal and other periodic environmental changes, but its violence is greater than that of the oscillation which these environmental changes could themselves produce.

11. Also, environmental oscillation applies an external force which prevents interspecific oscillation of the decreasing type from dying out. Irregular environmental fluctuations tend to cause corresponding irregular fluctuations in the densities of animals, but, owing to the delayed effects produced by interspecific oscillation, further irregularities in the densities of animals are subsequently produced that have no evident relation to environmental conditions.

ACKNOWLEDGMENTS.

I wish to express my great indebtedness to Prof. V. A. Bailey for the time and energy he has devoted to the mathematical investigation of the problems here presented. He has not only greatly increased the value of the work by giving rigorous proofs of the conclusions I had obtained by less precise methods, but has also extended the investigation to certain problems which could not be adequately investigated by the rather crude methods I had used.

The work on the present investigation has been lightened during the past few years by the interest and encouragement of my colleagues, originally at the Department of Zoology of the University of Sydney, and later at the Division of Economic Entomology of the Commonwealth Council for Scientific and Industrial Research, at Canberra, and I particularly wish to mention my indebtedness to Dr R. J. Tillyard, Mr G. A. Currie, Dr I. M. Mackerras and Dr P. D. F. Murray. I also wish to thank Prof. W. E. Agar and Prof. A. M. Carr-Saunders for their generous help and for the useful suggestions they made.

REFERENCES.

(1) **Bailey, V. A. (1931)**. "The interaction between hosts and parasites." Q. J. Math. 2, 68–77.

(2) **Bodenheimer, F. S. (1928)**. "Welche Faktoren regulieren die Individuenzahl einer Insektenart in der Natur?" Biol. Zbl. 48, 714–39.

(3) **Chapman, R. N. (1928)**. "The quantitative analysis of environmental factors." Ecology, 9, 111–22.

(4) **Chapman, R. N. (1931)**. "Animal ecology." New York and London.

(5) **Elton, C. (1930)**. "Animal ecology and evolution." Oxford.

(6) **Elton, C. (1932)**. "Territory among wood ants (*Formica rufa* L.) at Picket Hill." J. Animal Ecology, 1, 69–76.

(7) **Fisher, R. A. (1930)**. "The genetical theory of natural selection." Oxford.

(8) **Fiske, W. F. (1910)**. "Superparasitism: an important factor in the natural control of insects." J. Econ. Entom. 3, 88–97.

(9) **Holdaway, F. G. (1932)**. "An experimental study of the growth of populations of the Flour Beetle, *Tribolium confusum* Duval, as affected by atmospheric moisture." Ecol. Monographs, 2, 262–304.

(10) **Howard, H. E. (1920)**. "Territory in bird life." London.

(11) **Howard, L. O.** and **Fiske, W. F. (1912)**. "The importation into the United States of the parasites of the gipsy-moth and the brown-tail moth." U.S. Dept. of Agric., Bureau of Entom., Bull. 91.

(12) **Lotka, J. A. (1920)**. "Analytical note on certain rhythmic reactions in organic systems." Proc. Nat. Acad. Sci. 6, 410–15.

(13) **Lotka, J. A. (1925)**. "Elements of physical biology." Baltimore.

(14) **Malthus, T. R.** "On the principle of population." London.

(15) **MacLagan, D. S. (1932)**. "An ecological study of the 'Lucerne Flea' (*Smynthurus viridis* Linn.)." Bull. Ent. Res. 23, 101–45.

(16) **Martini, E. (1931)**. "Zur Gradationslehre." Verh. Deuts. Ges. Angew. Ent., Mitglieder-Versamml. Rostock, Aug. 1930, 19–26. Berlin.

(17) **Nicholson, A. J. (1927)**. "A new theory of mimicry in insects." Australian Zoologist, 5, Part 1. Sydney.

(18) **Pearl, R. (1926)**. "The biology of population growth." London.

(19) **Russell, E. J. (1927)**. "Soil conditions and plant growth." Rothamsted Monographs on Agricultural Science, London.

(20) **Strickland, E. H. (1928)**. "Can birds hold injurious insects in check?" Scientific Monthly, 26, 48–56.

(21) **Thompson, W. R. (1924)**. "La théorie mathématique de l'action des parasites entomophages et le facteur du hasard." Ann. Fac. Sci. Marseille, Ser. 2, 2, 69–89.

(22) **Thompson, W. R. (1930)**. "The biological control of insect and plant pests." Publ. Empire Marketing Board 29, H.M. Stationery Office, London.

(23) **Uvarov, B. P. (1931)**. "Insects and climate." Trans. Ent. Soc. London, 79, 1–247.

(24) **Volterra, V. (1926)**. "Variazioni e fluttuazioni del numero d'individui in specie animali conviventi." Mem. Accad. Naz. Lincei (Sci. Fis. Mat. e Nat.), Ser. 6, 2, No. 3.

(25) **Volterra, V. (1927)**. "Sulle fluttuazioni biologiche." Rend. R.A. dei Lincei, Ser. 6, 6.

(26) **Volterra, V. (1927)**. "Leggi delle fluttuazioni biologiche." Rend. R.A. dei Lincei, Ser. 6, 6.

(27) **Volterra, V. (1927)**. "Sulla periodicità delle fluttuazioni biologiche." Rend. R.A. dei Lincei, Ser. 6, 6.

(28) **Volterra, V. (1931)**. "Leçons sur la théorie mathématique de la lutte pour la vie." Cahiers scientifiques, No. 7. Gauthier-Villars et Cie, Paris.

Dice, L. R. 1938. Some census methods for mammals. *Journal of Wildlife Management* 2:119–130.

Obtaining information on population density has always been an essential aspect of wildlife management and conservation. However, wildlife biologists had to start from scratch when trying to census various game species and non-game wildlife. Many techniques to estimate numbers based on harvest were borrowed from the fisheries profession. Yet one of the most perplexing tasks for wildlife biologists in the 1930s and even now is determining animal numbers.

Fortunately, many of the problems encountered in wildlife management (e.g., certain genetic questions, zoogeography, behavior) do not require estimates of population size. But many other issues do (e.g., those related to habitat use, dispersal, rate of increase, responses to management actions), especially as we study threatened and endangered species. The level of accuracy required also varies with the question being asked. Some studies will require accurate population estimates; for others, trend data can suffice.

Lee R. Dice was one of the first wildlife biologists to tackle the census issue and to assimilate the collective knowledge about animal census into one source. He discusses census methods used for mammals in the late 1930s and identifies numerous challenges to obtaining accurate and precise data: animal behavior, study area size, sample size, stability of the population during sampling, disturbance caused by the researcher, marking, and overlapping home ranges. Many of these challenges remain important eight decades later as wildlife biologists continue to strive either to find more accurate and precise censusing techniques for studies that require them or to find—without accurate population estimates—ways to measure how populations respond to various influences.

RELATED READING

Burnham, K. P., and W. S. Overton. 1979. Robust estimation of population size when capture probabilities vary among animals. Ecology 60:927–936.

Caughley, G., and J. Goddard. 1972. Improving the estimates from inaccurate censuses. Journal of Wildlife Management 36:135–140.

Darroch, J. N. 1958. The multiple-recapture census: I. Estimation of a closed population. Biometrika 45:343–359.

Lancia, R. A., W. L. Kendall, K. Pollock, and J. D. Nichols. 2005. Estimating the number of animals in wildlife populations. Pages 106–153 *in* C. E. Braun, editor. Techniques for wildlife investigations and management. The Wildlife Society, Bethesda, Maryland, USA.

Neff, D. J. 1968. The pellet-group count technique for big game trend, census, and distribution: a review. Journal of Wildlife Management 32:597–614.

Otis, D. L., K. P. Burnham, G. C. White, and D. R. Anderson. 1978. Statistical inference from capture data on closed animal populations. Wildlife Monographs 62:1–135.

Ralph, C. J., and J. M. Scott, editors. 1981. Estimating numbers of terrestrial birds. Studies in Avian Biology 6:1–630.

SOME CENSUS METHODS FOR MAMMALS

Lee R. Dice

Accurate information about animal populations is desirable in almost every ecological investigation. It is often important to know the population density of a species at a certain place, the total of its individuals living in a given area, or the fluctuations in its numbers from year to year. Information of this kind is especially needed by the wild-life manager and the conservationist. In practically every study of natural history the taking of one or more censuses would be desirable, if only they could be taken accurately and without too great effort.

Unfortunately, it is very difficult to obtain an accurate census of any kind of mammal even on a small area. Many species are shy and avoid the presence of man. In addition, many forms are nocturnal, and nearly all the smaller species make use of burrows in the ground or of holes in trees, in which they spend most or all the daylight hours.

In spite of these difficulties technics are being developed for making counts of the numbers of certain species on restricted areas. Jack rabbits, deer, and some other kinds of large mammals can be counted more or less accurately by means of drives. Counts of some of the smaller mammals can be secured by intensive trapping. Indications of abundance are given by counts of tracks and of other signs. Although we have yet much to learn about how to secure a satisfactory enumeration of any species even for a small area, good progress is

being made in developing useful census methods.

The taking of a complete census of any mammalian species over a large area is usually impracticable and calculations of the size of populations must in nearly all cases be based on counts or estimates of the animals present on relatively small sample plots. In selecting these sample plots all the precautions necessary in taking samples for any statistical purpose must be observed. It is especially important to avoid any bias, conscious or unconscious, in choosing the areas. Either an especially good habitat or an especially bad one might contain quite a different concentration of the species under consideration than the average of the whole area.

No single statistical sample, unless it be very large, can give more than a rough approximation of the size of the population from which it was taken. If possible, therefore, several sample plots should always be counted and the mean of these samples used as the basis of calculation. With figures for a number of samples at hand there are simple statistical methods available for determining the probable accuracy of their means. It is obvious that the greater the number of samples and also the greater the size of the individual samples the more accurate the calculation of the population will be.

The greatest difficulty in making an accurate count of the animals on any area is to prevent the animals which are

119

on the area leaving before they are counted, and to prevent other individuals from outside the area coming in. It is not at all safe to assume that the outwanderings and inwanderings will balance each other. Timid animals, such as deer, may be disturbed by the placing of the men who are to make a drive, and some of the animals may escape from the area to be driven before the lines are established. On the other hand, if the count is made by trapping over a period of several days, some animals from outside the area trapped may in the course of their wanderings come inside the area and be caught (Chitty, 1937: 42–52). Furthermore, if the individuals living on an area be removed, other individuals may move in to take their place. Some of the unreasonably high counts per acre reported for small mammals are probably due to this cause.

One way of preventing the movements of animals on or off a census area is to select an isolated tract which is surrounded by habitats unsuitable for the species being counted (Hamilton, 1937: 780–781). The isolated area can either be a woodlot surrounded by fields, if a forest inhabiting form is being studied, or a meadow surrounded by woods, if the form to be studied is limited to meadows. It is not certain however, that an isolated area will usually have the same density of population as a larger area of the same habitat type. The boundaries of the individual home ranges of the mammals at the edges of an isolated area would necessarily be somewhat more restricted than on a larger area. Further, the habitat conditions are not the same at the edge, as in the interior of a

woodlot or meadow. Some species are probably more abundant at the edge of the forest than they are in its interior, other species are most abundant away from the forest border. These sources of error might lead to erroneous conclusions if the population density in a small area is used as a basis for calculating the population of a larger area. Nevertheless, the determination of the populations of small isolated areas should give much information about total populations. Some species of small mammals occur mostly in restricted local habitats and these species must properly be studied in such situations.

Theoretically, it should be possible to build a fence around a small selected area within a large uniform habitat and then to count the animals inside. The labor involved would be considerable, however, and great care would have to be taken that the disturbance of building the fence would not drive some of the more timid forms off the area. Furthermore, the fence would interfere with the normal movements of the animals onto and off the area and this unnatural barrier would, therefore, prevent the securing of information about normal populations.

The disturbance caused by the investigator walking over, and setting traps on an area may drive off some of the animals. Intensive trapping in a small area may even damage the vegetation, nest sites, and runways so greatly as seriously to alter the suitability of the habitat for small mammals.

Not all the individual mammals which occur on any given area at a particular time are necessarily resident there. At certain seasons of the year a considerable proportion of the members of the species may be wanderers with no fixed home. Due attention must be given in any computation of animal populations to the possible presence of such unattached individuals. Young animals of the year are especially likely to be wanderers, and the trapping records, therefore, should distinguish the numbers of young and of adults taken. Some young rodents mate and become established at fixed home sites before they acquire the adult pelage, and these sexually mature individuals must be distinguished from the juvenile non-breeding animals.

The males of some species wander more widely than the females, and the numbers of each sex, therefore, must be recorded separately. Though young animals and males may be more inclined to wander than adult females it must never be assumed without positive information that all adult females have fixed home ranges nor that all young animals and all males are wanderers.

One of the most accurate methods for taking a census of the small mammals on a given plot is to trap and mark each animal on the area, liberating it where it was captured. Care must be taken not to attract an unusually large number of individuals to the area by the food provided as bait. If the trapping is efficiently carried out and is continued for a sufficiently long time it should be possible to determine the home range of each individual living on the area. Then by plotting these individual ranges on a map, the resident population of each species on any part or on all of the area can be computed and the number of wandering individ-

uals present at any given time can be counted with considerable accuracy. This method is being used by W. H. Burt with great success in studies of the woodmouse on the George Reserve of the University of Michigan.

The greatest objection to this method of determining populations is the laborious trapping required. Further the method is not applicable to those animals which are trap shy and which therefore cannot be enticed a second time into a trap.

The population of a diurnal species of mammal on a specified area could be determined by a rapid count of all the individuals present, if these had previously been trapped, given distinctive marks that could be read at a distance, and released. Some individual mammals can be recognized by reason of deformities, or other individual peculiarities such as a bobbed tail, crippled leg, or white spot. However, these natural marks are seldom sufficient to distinguish all the individuals of a species living in a given area. We have made some attempts to apply recognition marks to trapped mammals of a few species by clipping off the hair of certain parts of the tail or of the upper parts of the body, and also by applying spots of dye to the pelage. So far these attempts have not been very successful, both because the marks are hard to read at a distance, even with field glasses, and because both the dyed and the clipped areas soon disappear. Further experiments with individual recognition marks on deer, squirrels, chipmunks, and other diurnal animals are desirable.

Some attempts have been made to mark individual animals so that they may be identified by their tracks. A coyote sometimes has a deformed foot which leaves a distinguishing mark. Adolph Stebler, of the Michigan Game Division, has removed toes from the feet of coyotes, hoping that the marked animals could be recognized by their tracks. R. L. Trippensee, while at the University of Michigan, tried the application of soluble stains to the feet of cottontail rabbits. However, these technics are still in the experimental stage.

In the summers of 1935 and 1937, during field work sponsored by the Carnegie Institution of Washington, I made an attempt to work out methods of counting the numbers of small mammals on measured plots. Some of these studies were made in the sand hills of Nebraska, some in the Black Hills of South Dakota, and a few in the mountains of southern New Mexico. Metal box traps taking the mammals alive were used exclusively, and all the individuals captured were removed from the area. Most of the traps were of mouse size, but a few were of chipmunk size.

The traps were set at first in a checkerboard pattern, with traps 1 rod apart in each direction, and in plots made up of multiples of a square chain. A checkerboard setting is undoubtedly best for long-continued studies which aim to determine the home ranges of all the individual animals on a plot. It was soon discovered that to set and to examine traps distributed in a checkerboard pattern, however, requires a very large amount of walking, with numerous trips back and forth across the plots. Furthermore, if the traps must be set at particular points they

will not always be placed in the most suitable spots for capturing small mammals. The setting of traps in the checkerboard pattern was, therefore, abandoned as inefficient in securing a rapid count of the mammals present on a plot.

A simpler method of setting traps for small mammals over measured areas is to mark out plots measuring 4 or more chains on a side (putting wire stakes with small flags of red cloth at the corners of each square chain) and setting the traps in line, 4 or 5 traps to the chain, completely across the plot. The return trap line is similarly set along the next surveyed transect, which is just 1 chain length away. The area is, therefore, covered by traps set 1 rod or less apart in lines 1 chain apart.

The plots used in these studies were laid out in fractions of acres rather than in metric divisions because in the United States large areas are always measured in acres and square miles. A surveyor's chain is 66 feet long, and a square chain is one-tenth of an acre, so the sizes of the plots are easily calculated.

The quadrats covered by these traps were usually entirely trapped out in 3 to 5 nights. On some plots in yellow pine habitat I set new lines of traps in the opposite direction across the quadrat after the other traps had been set for 5 days, but failed to secure any additional mammals, except at the edges of the quadrat. In certain other types of habitat a longer or a shorter trapping period might be required completely to denude an area of its small mammal population.

With any method of trapping on an area which is divided into equal parts it is possible and desirable to keep the records according to the subunits. It will then be possible to compare the catch in the outside trap lines around the area with the catch in the inner lines, and perhaps determine the amount of inwandering during the period of trapping. It should also be possible to learn by comparison of the numbers trapped in the different subunits whether there are variations in abundance correlated, perhaps, with differences in habitat types.

In thick brush or timber and on very steep slopes it is impracticable to run traps in straight lines. In such situations the traps can be set in irregular lines following trails or any other feasible routes across the area. If the lines are close enough together to cover all the home ranges of the resident small mammals, the entire population should be removed from the area within a few days. Measurements with a steel tape and compass along the boundary lines will afford sufficiently accurate data for the computation of the area covered by the traps. This method of setting the traps avoids the labor of surveying the exact location of the trap lines. It may be employed in any situation where it is desired to know the area in the plot which is trapped, but where the exact spacing of the trap lines is not considered important.

The area which will be denuded of small mammals by any setting of traps extends of course beyond the area actually covered by the traps, unless this area is surrounded by unfavorable habitats. If the individuals of the species concerned are not established in definite home ranges, but are moving about at random, it will be impossible to calculate exactly the area trapped out by a trap line or trap quadrat. But if the mammals are living in fixed homes and have definite home ranges, which they travel over at frequent intervals, it should be possible to estimate the average distance from which the animals come to the traps. There is good evidence that many species of small mammals do have more or less definite home ranges at least during certain parts of the year.

The average shape of an individual cruising range should theoretically approach a circle in form. If adjacent territories are defended by their occupants, the home ranges should tend toward a hexagonal shape. However, it is difficult to calculate the number of circular or hexagonal individual ranges included in any given quadrat. For statistical purposes, therefore, the individual ranges may be treated as though they were square. The mean width of an individual range will accordingly be considered to be the square root of the average area ranged over by an individual of the sex and species under consideration.

An efficient line of traps should capture eventually all the individual mammals having home ranges which overlap the line at any place where there is a trap. If the traps are placed closely enough together in the line so as to be practically continuous the influence of the trapline should extend on each side to an average distance of half the mean width of an individual cruising range for the sex and species concerned. Different species undoubtedly vary considerably in the mean widths of their cruising ranges and the distance on each side of a trap line or plot trapped out varies, therefore, with the species concerned.

It is obvious then that in order to calculate the area trapped out by a given setting of traps it is necessary to know the average area covered by the cruising range of a single individual of the species concerned. This information can best be secured by intensive live-trapping and marking as already described. If the mean width of a home territory for the sex and age class considered is already known, then the total area trapped out by a given setting of traps is the area covered by the traps plus $\frac{1}{2}$ the mean width of a home territory on each side of the trap quadrat.

Because the cruising ranges of the males may be larger than those of the females it follows that the area denuded of males by any given setting of traps may be larger than the area denuded of females. The calculation of the size of the male population per unit area must, therefore, be made independently of that of the female population.

Furthermore, if the trapping is carried out at a season when some of the animals do not have fixed home ranges, it will be necessary to determine from how great distances these wanderers have come to the traps. In some cases it will be impossible to make such a determination. If the traps are only left set for a few days, however, it may be permissible to assume an arbitrary distance beyond the trap lines from which these invaders have come. Possibly the arbitrary assumption of the same area as is calculated to have been denuded of adult males will sometimes be sufficient for a rough calculation of the population of wandering animals.

A calculation that should give a rough indication of the average size of the individual home ranges on any given quadrat is here suggested. This method assumes that all the individuals of the species, sex, and age concerned are living in more or less fixed home ranges which adjoin but do not overlap each other. The formula to be used is: $Nx^2 = (A+x)(B+x)$, in which, N is the number of adults of the sex or age group being considered known to be living on the quadrat; x the mean width of an individual home range; A the width of the quadrat from outside trap to outside trap; and B the length of the quadrat. If a complete record is secured of all the individuals living on the area, the value of N is the number of individuals of that sex and age class trapped. If the plot is not rectangular in form a slight modification of the formula will be necessary, the requirement being that to the area covered by the traps is added a surrounding strip $\frac{1}{2}x$ wide.

As an illustration of the use of this formula let us assume a hypothetical quadrat on which traps have been set over a rectangular plot 100 by 200 yards in outside dimensions, and, further, let us assume that on this plot there have been trapped a total of 10 adult females. Then, interpolating in the formula we have:

$$10x^2 = (100+x)\ (200+x)$$
$$10x^2 = 20,000+300x+x^2$$
$$0 = 20,000+300x-9x^2.$$

This is obviously a quadratic equation, which when solved algebraically, gives 66.6 yards for the value of x. Therefore 66.6 yards is the mean width of the home range of an adult female for the data and assumptions given. The area from which adult females were secured by the traps extends on the average one-half of 66.6, which is 33.3 yards, on each side of the plot beyond the setting of traps. The average area of the home range of an adult female is x^2, which is 4,444 square yards.

The average size of the home range of an adult male can be calculated by the same formula. The method cannot be used for the calculation of the number of individuals per unit area if these individuals do not have fixed home ranges.

One of the most important sources of error in calculating the size of individual home ranges by this method lies in the probability that the individual ranges of the species and sex concerned may not completely fill the area where the trapping is done. The resulting error would give too large a figure for the mean width of a home range and would, therefore, indicate that the area denuded by the trapping was larger than actually the case. This in turn would give too small a figure for the total population per unit area calculated from this data.

Other possibilities of error are that some of the adults may not have fixed home ranges; that the ranges of certain of the adults of the same sex overlap; that some individuals escape capture; and that some of the immature animals also have fixed ranges. An inspection of the data may determine which of these sources of error is likely to be present and a correction of the estimated size of the average home range may then be possible.

For the calculation of the numbers of wandering animals per unit area the most accurate basic data will be secured from intensive trapping conducted on a quadrat for only one, or at most a few nights. If all the individuals present on the trapping area could be secured in one night this would give the most accurate record of the number usually present on the area at one time. If the trapping is continued on later nights, wandering individuals may come in and thereby increase the apparent concentration on the area. It is not known, however, how much error is produced by including in the calculations all the individuals taken over a trapping period of say 3 days, which is the usual duration of trapping on an area.

Traps set for only one night cannot be expected to catch all the animals with fixed home ranges impinging on the trap line, for some of these animals may not visit every part of their ranges every night. Ordinarily, trapping must be continued until no more adults are being taken. It should theoretically be possible by setting the traps very closely together to secure all the small mammals on an area in one night, but this would require an impracticably large number of traps. The intensive trapping desirable for securing an accurate count of the number of wandering animals present on a quadrat is, therefore, not the type of trapping best suited for determining the number of individuals having fixed ranges on the area.

The outside lines of traps on a quadrat should catch a greater number of wandering animals than the traps toward the middle of the quadrat. By comparing the numbers taken in the two situations it should theoretically be possible to estimate the proportion of wanderers taken. The difference in the numbers taken in the inner and outer traps should be greatest when the traps are kept set for a long period of time. Little useful information will be given by such a comparison, however, unless the size of the quadrat is many times greater than that of an individual home range.

A buffer line or zone of traps partly or completely surrounding the quadrat being studied has been used by some investigators (Elton, Davis and Findlay, 1935: 283–284) to prevent the in-wandering of non-resident animals. Such a line of traps cannot ever be completely successful, however, in preventing wandering individuals from reaching the area, for some will certainly pass between the traps in the buffer zone. Furthermore, some of the individuals living on the quadrat will likely be caught in the outside line of traps, and it will be impossible to distinguish these from animals coming in from outside. It takes a large number of traps to surround a quadrat and it seems wasteful of time and energy not to make use of the information secured from these traps. A surrounding line of traps in my opinion does no more than increase the size of the quadrat, and it is still necessary to determine the distance beyond the traps from which the animals have come.

The larger the area covered by trapping operations of uniform intensity, the more accurate can be the determination of the populations of the included mammalian species. All the calculations are based upon averages and no statistical average can be very reliable when based only on a few data. Computations from any number of individuals less than 10 cannot give a very reliable average for the size of a home range.

On large quadrats, also, there is proportionally less opportunity for movements of the inhabitants on or off the plot. This is true because the area of a square plot varies as the square of the length of the side, while the length of the boundary of the plot varies directly with the length of the side. Furthermore, on large plots any error in calculating the width of the surrounding strip from which animals are trapped is minimized. In the case of small quadrats the area included in the surrounding strip may form a considerable proportion of the whole area from which animals are taken. With large plots, on the contrary, the surrounding strip forms only a small part of the total area from which the mammals are collected, and errors introduced in the calculations by failure to estimate correctly the width of the surrounding strip are relatively unimportant. The size of quadrat required in order that these errors will be negligible depends of course upon the size of the average home range and will have to be calculated separately for each species.

It is not necessary to kill or remove the animals taken on a quadrat in order to secure an accurate count. If taken in live traps the animals can be released without harm. If the trapping is to be continued longer than one night the animals should be marked before releasing, so that they can be recognized when subsequently trapped. A small notch cut in the ear, or a tuft of hair clipped from some part of the body will be sufficient unless it is desired to make a long-continued study of individual home ranges.

An expanding quadrat can be used to cover a large area with a limited number of traps. The traps are first set in the middle of the area. As soon as these traps have denuded their vicinity they are removed and reset in lines surrounding the first settings. The amount of expansion possible will of course be limited by the number of traps available and the distance they are set apart.

A moving quadrat (Townsend, 1935: 16) differs from an expanding quadrat in that the expansion is on only one side. Wandering animals can be expected to come into a moving quadrat not only from the front and from both sides, but some may catch up with it from behind.

While traps set in large quadrats will undoubtedly give the most accurate information about the populations of small mammals, there are many occasions when it is impracticable to set traps in this manner. At such times traps set in straight or nearly straight lines of measured length will give much information about relative populations and under favorable circumstances it may be possible to estimate actual concentrations per unit areas from the data so obtained.

A line of thoroughly efficient traps should denude an area extending for one-half the mean width of a cruising range on each side of the traps, and also for one-half the mean width of a cruising range beyond each end of the line. This is on the assumption that all the individuals of the species being trapped are living in relatively fixed home ranges. If the mean width of a cruising range for the sex and species concerned is known from previous studies it will be possible to calculate the area trapped out by the trap line and from this to determine the population of the species per unit area.

If the average size of a home range for the species concerned is unknown it is still possible to make an estimate of the mean width of a home range by a simple computation. Divide the length of the line of traps by 1 less than the number of adults of the species and sex concerned, captured during the period of trapping. The quotient should be the mean width of a home range. It is assumed in making this computation that all the individuals of the sex and species considered have fixed ranges which adjoin but do not overlap, and that all whose ranges overlap the trap line are captured. The computation will not give a very accurate figure for the mean width of an individual range unless the line is long enough to cover a considerable number of ranges.

If the trap line extends completely across a habitat which is surrounded by other types of habitat unfavorable to the species being studied, the calculation should be corrected as follows: To find the mean width of a cruising range divide the length of the trap line by the exact number of individuals of the species and sex captured. In any other situation the traps on the end of the line will capture individuals whose ranges extend beyond the trap line. As usually set, the trap line should capture a number of individuals 1 greater than the number of individual ranges included in the length of the line.

If several trap lines have been run in the same season through uniform habitat in a given district, it is statistically proper to combine the several means of cruising range width obtained and thus get a more exact combined mean. The precaution must be observed not to use figures from trap lines set so closely together that some individual ranges might be touched by two lines.

The length of the trap line should be measured with a steel tape if possible. Counts of paces should be resorted to only as an emergency measure.

Moderate curves in the trap line should not cause any important error in the computations, but very sharp curves in the line should be avoided. The sharpest curve allowable would have a radius equal to half the mean width of the largest home range of any of the species considered.

The single trap line of measured length cannot be expected to give as reliable data on mammal populations as the trap quadrat. Only rarely will it be possible to calculate accurately the average size of the home range of any species and of the population per unit area from information secured entirely from single trap lines. Nevertheless, if trap lines are run in a standardized manner in different places or at different times they should give a reliable indication of the comparative abundance of species taken on all of them.

It might be supposed that counts of the tracks of any selected species of mammal crossing a given length of trail could be used according to the above suggested method to calculate the average size of the home range and the population of the species. Before the method can be used in that way, however, there will need to be demonstrated that the individuals of the species concerned have mutually exclusive ranges. Further, it would have to be made certain that no individual is counted twice. It will be difficult to determine from tracks alone whether one or several individuals have made tracks which cross the trail. Only rarely will there be distinctive features or sizes which positively identify the maker of a certain track. The number of tracks of any species crossing a unit length of trail does give an index of relative abundance, and it is possible that this index of abundance may later be correlated with actual population densities determined by other methods.

The number of animals caught per "trap-night" has occasionally been used as an index of abundance, following an early suggestion by Grinnell. From the preceding discussion it will be evident, however, that the number of animals secured on a trap line depends largely on the sizes of the individual cruising ranges of the several species present in the region. If all the individuals with cruising ranges crossing the trap line are ultimately caught the number taken is not directly related to the number of traps nor to the length of time they are set, but is directly related to the length of the trap line. I believe, therefore, that the statement of the numbers of individuals of each sex and age taken per unit length of line is a better method of stating abundance than is the catch per trap-night.

To supplement actual census methods, which often are expensive or impracticable, simple methods of obtaining indices of abundance are greatly to be desired. Attempts have been made to use for this purpose counts of tracks, of homes, and of other evidences of the presence of certain species. MacLulich (1937: 25–39), in a study of the cycles of abundance of the snowshoe hare, has employed an interesting statistical method to show that counts of droppings on a series of small quadrats give a reliable index of the relative abundance of the animals.

It is not always necessary that indices of abundance be directly convertible into census figures, for the knowledge that certain species are increasing or decreasing or that they are most abundant in certain types of habitat will often be of great value to wild-life administrators. As yet no standardized methods have come into general use for securing indices of the abundance for mammalian species. This is, however, a most promising field for experiment.

The use of the frequency index for comparing the abundance of mammals at different times and places was suggested several years ago (Dice, 1931: 378). It is true that the frequency index shows primarily the uniformity of distribution of a species rather than its comparative numbers (Cain, 1932: 478). However, it has been shown by McGinnies (1934: 281) for plants that there is a correlation between the frequency index and abundance. Any one kind of mammal probably has everywhere the same general type of local distribution, and therefore its frequency index for different locations or for different years should give a good indication of its comparative abundance. Where signs of occurrence rather than actual counts of individuals must be depended upon for comparing the abundance of animals, the frequency index is of special value. This is shown in its use (under the name of "trace census") by Elton, Davis, and Findlay (1935: 280–281) in a study of fluctuations of the numbers of voles.

No one census method can be used for all kinds of mammals. The several species differ so much in size and in habits that many different methods will have to be devised, perhaps a distinct method for each species. It is obvious that a very different technic must be used for counting the numbers of mice and of grizzly bears. However, some methods may, with slight modification, serve for the enumeration of several different species.

In any study of population the determination of the size of the individual home range is of great importance. Except for the rare instances in which the whole population of a circumscribed area can be counted, estimates of populations will usually involve a determination of the average size of the individual range. Studies of home ranges, to be most useful, should include a detailed determination for each species of the mean and extreme areas included in the cruising ranges of males and of females, and also of old and of young individuals. Due consideration must be given to possible variation in the sizes of the cruising ranges at the various seasons and in the various types of habitats occupied. The studies should determine also whether the individual ranges are mutually exclusive, or if several individuals may under certain conditions occupy parts of the same range at the same time. Of importance in this connection is the determination of the location of the nest in relation to the home range and of the relationship between the abundance of suitable nest sites and the spacing of the individual ranges. It is to be hoped that in the future field naturalists will give increasing attention to detailed studies of the individual ranges and of the populations of all kinds of mammals.

LITERATURE CITED

CAIN, STANLEY A.
   1932. Concerning certain phytosociological concepts. Ecol. Monog. 2: 475–508, 10 figs.
CHITTY, DENNIS
   1937. A ringing technique for small mammals. Jour. Anim. Ecol. 6: 36–53, 5 figs.
DICE, LEE R.
   1931. Methods of indicating the abundance of mammals. Jour. Mammalogy 12: 376–381.
ELTON, CHARLES, D. H. S. DAVIS, and G. M. FINDLAY
   1935. An epidemic among voles (*Microtus agrestis*) on the Scottish border in the spring of 1934. Jour. Anim. Ecol. 4: 277–288, 1 photo.
HAMILTON, WILLIAM J., JR.
   1937. The biology of microtine cycles. Jour. Agric. Res. 54: 779–790, 2 figs.
MCGINNIES, W. G.
   1934. The relation between frequency index and abundance as applied to plant populations in a semi-arid region. Ecology 15: 263–282, 4 figs.
MACLULICH, D. A.
   1937. Fluctuations in the numbers of the varying hare (*Lepus americanus*). Univ. Toronto Studies, Biol. Ser. no. 43, 1–136, 18 figs.
TOWNSEND, M. T.
   1935. Studies on some of the small mammals of central New York. Roosevelt Wild Life Annals 4: 6–120, 4 maps, 22 figs. 8 pls.

Lee R. Dice
University of Michigan
Ann Arbor, Michigan

Errington, P. L. 1942. On the analysis of productivity in populations of higher vertebrates. *Journal of Wildlife Management* 6:165–181.

In Errington's time, the studies of wildlife ecology and management were at their infancy. Basic data were being collected on wildlife populations, but little was known about how to compute unbiased estimates of population parameters, which parameters reflected the dynamics of wildlife populations, or which biotic and abiotic factors were critically important in determining the distribution and abundance of animal populations. Errington was the first to examine life history characteristics of selected birds and mammals so that biologists could understand how species exist and maintain abundance, despite drastic annual losses of young and adult animals.

This paper was written at a time when many incorrect assumptions were being made (e.g., that every loss of an egg would result in a loss to the population). Pioneers in the study of life history characteristics—including Errington—cautioned biologists that poor assumptions could lead to erroneous interpretations of data. In this article, Errington makes a case for better comprehension of compensatory trends in reproduction and loss rates. This led him to examine how populations compensate for all types of losses, which is important to understand in wildlife management and conservation. In his work, Errington outlined the numerous population dynamics that influence wildlife and the concept of compensatory reproduction. The concepts of compensatory and additive mortality factors remain important aspects of the wildlife profession still under debate today.

RELATED READING

Bartmann, R. M., G. C. White, and L. H. Carpenter. 1992. Compensatory mortality in a Colorado mule deer population. Wildlife Monographs 121:1–39.

Bishop, C. J., G. C. White, D. J. Freddy, B. E. Watkins, and T. R. Stephenson. 2009. Effect of enhanced nutrition on mule deer population rate of change. Wildlife Monographs 172:1–28.

Nichols, J. D., M. J. Conray, D. R. Anderson, and K. P. Burnham. 1984. Compensatory mortality in waterfowl populations: a review of evidence and implications for research and management. Transactions of the North American Wildlife and Natural Resources Conference 49:535–554.

Roseberry, J. L. 1979. Bobwhite population responses to exploitation: real and simulated. Journal of Wildlife Management 43:285–305.

Unsworth, J. W., D. F. Pac, G. C. White, and R. M. Bartmann. 1999. Mule deer survival in Colorado, Idaho, and Montana. Journal of Wildlife Management 63:315–326.

White, G. C., and R. M. Bartmann. 1998. Effect of density reduction on overwinter survival of free-ranging mule deer fawns. Journal of Wildlife Management 62:214–225.

ON THE ANALYSIS OF PRODUCTIVITY IN POPULATIONS OF HIGHER VERTEBRATES[1]

Paul L. Errington

The natural productivity of mammals and birds, especially of game species, has received a great deal of investigative attention during the past decade.

Valuable information has resulted from this work, but investigators, by assuming (or by leaving the reader to infer) that the loss of a clutch of eggs or a litter of young *necessarily* means that much net loss in productivity, may introduce errors of such magnitude that critics could, in many instances, appropriately ask whether the conclusions were not as misleading as instructive.

It may be conceded that various species for reasons of latitude, climate, rarity, inherent shortness and lack of synchrony of sex rhythm (Allen, 1934) physiology, psychology, etc., may have distinctly limited opportunities to reproduce. Late young—even of species that readily renest throughout a long breeding season—may be, moreover, few in number, subject to pronounced if not lethal handicaps, or undesirable from special human viewpoints.

Nevertheless, calculations of so-called life equations that ignore, for example, renesting when its compensatory value is substantial can lead pretty wide of the truth; and the "common sense" reasoning of administrators and sports-men to the effect that every game bird egg that fails to hatch is the equivalent of one less hunting season pheasant (or quail or duck or something else) should hardly be surprising as long as comparable assumptions continue to be made by those whose studies are accepted as technical background. We have only to consider standard practices in poultry management to see how immense numbers of eggs would never be laid in the first place were it not for the stimulus resulting from removal of eggs from laying hens. Among wild-living birds, the flicker (*Colaptes auratus*) has long been known for its propensity for continuing to lay after egg losses (Bent, 1939: 272). Recognition of some of the "fallacies of misplaced concreteness" (Whitehead, 1925; Henderson, 1941) that are so prevalent in life history analyses should be possible without unduly minimizing loss types of real importance.

My own endeavors to evaluate renestings and related phenomena were inspired by Stoddard's (1931) classic investigation of the bobwhite (*Colinus virginianus*) in Georgia and Florida, which clearly distinguishes between percentage and significance of nesting losses. On pages 224 and 225 of his book, we may not only read that "from 60 to 80 per cent of unsuccessful nesting attempts may normally be expected at large over the Southeastern quail territory" but also that, in the course of a season, "the majority of the pairs sooner or later are successful in hatching

[1] Journal Paper No. J-674 of the Iowa Agricultural Experiment Station, Ames, Iowa, Project No. 498. Fish and Wildlife Service (U. S. Department of the Interior), Iowa State College, Iowa State Conservation Commission, and the American Wildlife Institute cooperating.

165

off a brood." In a Wisconsin study, over 80% of the hens may, on fair evidence, be judged to have produced young, despite a loss of 51% of the nests (Errington, 1933). Later, a study of the ring-necked pheasant (*Phasianus colchicus torquatus*) in northwestern Iowa demonstrated some degree of success for only 23.1% of 445 nests under regular observation (Hamerstrom, 1936), but analysis of the data suggested that from 70% to 80% of the hens hatched out the one brood per season that typically satisfies them (Errington and Hamerstrom, 1937).

How widespread such compensatory reproduction may be is imperfectly indicated by current literature. The Hungarian partridge (*Perdix perdix*) has some renesting ability (Yeatter, 1934: 81; Middleton, 1936: 804; Errington and Hamerstrom, 1938). Prairie chickens (*Tympanuchus cupido americanus*) probably re-nest although to a lesser extent than bobwhites, Hungarian partridges, and pheasants (Hamerstrom, 1939: 115); and this statement may be applicable to other grouse (Tetraonidae). One would not expect waterfowl to do much if any renesting during the short summer of the Arctic regions, but, farther south, where the active breeding season extends over a period of months, renesting by Anatidae, Rallidae, and perhaps other groups surely offsets some of the early failures.

It has long been known that many passerine birds respond to nest failures by laying another clutch of eggs, usually some days after the failure. This is mentioned in recent monographic writings by Miller (1931) for shrikes (*Lanius* spp.), Nice (1937) for the song sparrow (*Melospiza melodia*), Kendeigh and Baldwin (1937) for the house wren (*Troglodytes aedon*), Hann (1937) for the oven-bird (*Seiurus aurocapillus*), Erickson (1938) for the wren-tit (*Chamaea fasciata*), Wallace (1939) for Bicknell's thrust (*Hylocichla minima bicknelli*), and Blanchard (1941) for the white-crowned sparrow (*Zonotrichia leucophrys*).

Of the wren-tit, Erickson (1938: 282) writes that "Pairs which successfully fledged young have been watched, but no evidence of repetition of nesting cycle, once it is completed, has been found" and she has "records of as many as four and five attempts to raise young on the part of one pair, and of the initiation of a new cycle when the brood was lost when about ten days old." For multiple-brooded species, the difficulty of distinguishing between failure-induced layings and the normal complement of clutches naturally complicates analyses; and, after studying Nice's (1937) voluminous data on the song sparrow and Emlen's (1941) experimental analysis of the breeding cycle in the tricolored red-wing (*Agelaius tricolor*), I must confess to a doubt that the complexities of the breeding equation in such forms would permit an accurate measurement of compensatory reproduction.

Among mammals, the equivalents of these compensating phenomena are doubtless less pronounced, yet they could very possibly be of frequent occurrence in species that characteristically give birth to litters at short intervals during a long breeding season. The muskrat (*Ondatra zibethicus*) is the only species in the latter category with which I am intimately familiar, and the findings from eight years of work on population recovery in Iowa (largely unpublished) suggest that poor success early in the breeding season is a major stimulus to exceptionally long-continued breeding. One of the clearest examples came out of the work at Round Lake, near Ruthven, Iowa, in 1935, a year of notable overpopulation. On roughly one-half of the marsh, losses from causes other than intraspecific strife were not great, but breeding stopped by mid-summer, apparently in consequence of population tensions. On the other half, losses through predation and disease predominated, intraspecific antagonism was far less intense, and nearly all of the late summer litters—usually the third and fourth of the season for individual females (Errington, 1937b and unpublished)—were born on sections of the marsh where known early-summer mortality of young had been highest.

I have been unable to find reference to compensatory reproduction (under that or any other name) in the literature devoted to mice of the genera *Microtus* and *Peromyscus*, but some of the data would seem to suggest the operation of such a mechanism. One may ask, for instance, how much Bailey's (1924: 528) record of 17 consecutive litters born within a year to a captive *Microtus pennsylvanicus* was conditioned by human removal of young from the cage.

On the whole, if an investigator wonders that the bird or mammal he is studying can exist at all in the face of the losses of eggs or young that take place—and the species not only exists but maintains abundance, often in spite of drastic annual losses of adults—he may well try to inquire into the matter of possible compensations.

This paper is prepared in the hope that it may be helpful to workers who are conscientiously striving to make their calculations of productivity as nearly accurate as they can. Suggestions are intended for broad, rather than highly specific, guidance, for each species patently has problems that are more or less peculiar to itself and which, furthermore, vary with locality and season. The procedures outlined have their crudities and their limitations of scope, and I do not claim more for them than that they are the product of direct and indirect experience with several life history investigations of higher vertebrates.

GROUNDWORK FOR AN ANALYSIS OF COMPENSATORY REPRODUCTION

"Renesting" of birds is a term that applies to reproductive efforts resulting primarily from previous failures. In view of the technical obstacles to tracing renestings in multiple-brood small birds and the equivalent in multiple-litter mammals, this exposition will treat with the simpler sort of problem that an investigator can reasonably hope to work out.

Let us then consider, as representative of a number of familiar species, a hypothetical bird that is satisfied with the hatching of one brood a year and that renests after ordinary failures at least during the early part of an April to July general nesting season. We may remain in touch with actuality by following the trends of Hamerstrom's (1936) data on the ringnecked pheasant, with sufficient changes made in numerical values to aid the reader in keeping track of the sources of the figures in the simple algebraic equations to follow.

Table 1 presents the kinds of data that can be obtained for almost any avian species through employment of usual techniques (Errington, 1932). Although the breeding season for our hypothetical species is divided into half-month periods, an investigator whose data are exceptionally complete might find shorter divisions advantageous. Establishing the date of be-

gins just after the laying of the last egg of a complete clutch), and about 20 days for the laying of the clutch (our species lays an egg a day on most but not all days of this period), we get a time interval of about 46 days and trace the laying of the first egg to the second half of April.

TABLE 1

HYPOTHETICAL SAMPLES OF CORRECTED[1] DATA ON SUCCESS AND FAILURE OF CLUTCHES FOR WHICH TIME OF LAYING OF FIRST EGG IS KNOWN

Periods in which clutches were begun	Number of clutches	Successful[2] clutches	Failed clutches
Periods of initial nesting attempts:			
First half of April	19	2	17
Second half of April	35	12	23
First half of May	52	22	30
Second half of May	54	26	28
Periods following initial nesting attempts:			
First half of June	45	21	24
Second half of June	20	10	10
First half of July	8	3	5
Second half of July	1	0	1

[1] Correction for analytical purposes made by excluding from the table clutches barely begun before failure and the failure of which obviously had little or no significance from the standpoint of productivity. In practice, these excluded clutches are mainly single abandoned eggs, either in nests or "dropped" at random.

[2] A clutch is listed as successful if any young are hatched from it.

ginning of clutches is largely done by calculating backward from facts known about each clutch, the rate of egg-laying and the length of the incubation period for the species. Undetected losses of individual eggs are an occasional source of error—usually insufficient to result in assignment of a clutch to the wrong half-month period.

Hypothetical examples of dating:

Clutch "A," 18 eggs, was known to have hatched a few days before the last observation on June 10. Allowing 3 days for time between dates of hatching and the visit at which the fact of hatching was noted, 23 days for incubation (which for our hypothetical species be-

Clutch "B," 17 eggs, was destroyed by mowing on July 20, shortly before hatching. Although we may not be able to determine the exact date of laying the first egg, we can surely place it in the first half of June.

Clutch "C," 21 eggs, was destroyed by predators about May 30, after having previously lost 2 eggs through unknown cause. Four eggs recovered for examination contained embryos of about 10 days' development. Allowing 26 days for laying of the clutch and 10 for incubation, we assign the beginning to the second half of April.

Clutch "D," 10 eggs, was laid in the same nest as Clutch "E," 19 eggs; from

the end of a temporarily double rate of egg deposition (2 eggs a day for 3 consecutive days), the nest was judged to have been deserted by hen "E" soon after hen "D" apparently began laying (about May 24) and deserted in turn by hen "D" about June 4. All eggs were unincubated when examined June 9, after it had been made sure that hens were no longer using the nest, so the time of laying of first egg for clutch "E" was determined as the first half of May.

Clutch "F," 3 eggs, was abandoned after disturbance, May 7, long before completed. It obviously was begun early in the first half of May.

Clutch "G," 2 eggs laid "parasitically" in the nest of another species[2] between June 4 and 8, was considered in the same category as a "dropped" egg or a single-egg abandoned clutch, hence omitted from analytical calculations (see first footnote to Table 1) even though date of laying could be assigned to a half-month period. Parenthetically, it may be explained that this sort of "parasitism" typically has no reproductive value, in contrast with the "parasitism" of the cowbird (*Molothrus ater*) (Friedmann, 1929; Nice, 1937: 152–165).

The fundamental question of what proportion of clutches constitute renesting attempts is not one to which re-

[2] Such "parasitism" was frequently observed among ground-nesting game birds of northwestern Iowa (Hamerstrom, 1936: 180; Bennett, 1936; Errington and Hamerstrom, 1938). Low (1941: 510–511), studying mutual "parasitism" by redhead (*Nyroca americana*) and ruddy (*Erismatura jamaicensis rubida*) ducks, found that "the amount of promiscuous egg laying was directly proportional to the amount of unsuccessful nesting during each season of study."

liable answers may always be forthcoming. Bennett (1938: 57–59), in his monograph on the blue-winged teal (*Querquedula discors*), designated as renestings the late, poorly-constructed nests concerning which there could be little doubt, but, as I see it, these may strictly be considered only as *late* renestings. Incomplete or unincubated clutches broken up or abandoned early in the nesting season are surely followed in a large number of instances by renestings that cannot on basis of dating or appearance be distinguished from initial attempts; and, in these instances a renesting often stands as good chance to produce a full-sized, early-season brood as many clutches that were actually the first laid by individual females. Indeed, if an investigator is working with a species, the hens of which are habitually careless about where they lay their first few eggs, and gives "dropped" eggs and freely deserted single-egg nests the status of lost clutches in his notes, he may well consider whether most of his "regular" nests should not be counted as renestings.

Calculations of possible time of renesting of hens after failures are beset by two analytical hazards in particular. One relates to the proportion of nest failures due to mortality or incapacitation of the hen, and the investigator who finds this factor to be either immaterial or easily measurable is fortunate. Let us say that, for our hypothetical species, the losses of 11 of the clutches begun up to the middle of July and dated as in Table 2 were known to have followed or accompanied losses of the hens.

The second hazard arises from uncertainties in judging when to expect a hen to resume laying after a failure. In many common species, if failure occurs early or in mid-season before the clutch is complete, renesting may follow much as a continuation of the failed effort, though perhaps with some "egg-dropping." When failure occurs during incubation, the probable time of renesting

and, following Riddle (1911: 482), we read that the "time interval between the beginning of rapid growth of the 6 mm. egg, and the breaking of the egg from the ovarian follicle (ovulation) is normally between five and eight days." We also learn from the experiences of poultrymen that prolonged incubation of a failed clutch may deplete a bird's

TABLE 2

HYPOTHETICAL SAMPLES OF CALCULATIONS ON POSSIBLE TIME OF RENESTING OF HENS AFTER FAILURES[1]

Periods in which failed clutches were begun	Number of clutches failing	Number of failures positively terminating breeding efforts, as through death of hens	Number of failed clutches for which renesting attempts may be expected in specific half-month periods						
			Periods of initial nesting attempts			Periods following initial nesting attempts			
			Second half of April	First half of May	Second half of May	First half of June	Second half of June	First half of July	Second half of July
Periods of initial nesting attempts:									
First half of April	17 of 19		1	7	6	2			
Second half of April	23 of 35			11	5	4			
First half of May	30 of 52	1			13	9	3	4	
Second half of May	28 of 54	5				15	3	4	1
Periods following initial nesting attempts:									
First half of June	24 of 45	2					12	4	3
Second half of June	10 of 20	1						3	2
First half of July	5 of 8	2							1

[1] See text for explanation of methods.

should be calculated with reference to records from artificial propagation, all available sources of information including the literature, experimentation with captives, and observations of marked individuals living in the wild.

To avoid making our hypothetical problem too complex, let us say that reproduction in our species is in many ways comparable to that of the domestic fowl. It has either a long sex rhythm or one that recurs so frequently as to be virtually continuous for the duration of the laying season. The most rapid resorption of ova after cessation of laying may be expected, as for the fowl (Stieve, 1918), in the first 5 days;

physical reserve to the point that she may require a rest before renewing her reproductive efforts—in a closely related wild bird possibly a week or two, judging from the rapidity with which pheasants recover weight lost from hunger (Errington, 1939a: 35).

Table 2 shows the renesting periods assigned to hens after failures, as estimated from the status of clutches at time of failure and from careful guesses as to physical condition of hens that lost clutches during incubation. Previous elimination from the tabular data of single-egg, and some other very small, lost clutches (first footnote, Table 1) reduces the error due to certain hens losing one clutch and starting another in the same half-month period; at any rate, it should be less hazardous to assume, for convenience of analysis, that renesting doesn't occur within a short period than that it doesn't occur within the nesting season. The use of shorter than half-month periods would reduce possible errors from this source still further, provided that data were available concerning sufficient nests (several hundred or more) to permit subdividing without getting down to samples too small to be representative. Ample, good-quality, field data on broods are about the best safeguard against gross imperfections in appraisals of renesting activities.

Taking the hypothetical examples of clutches "A" to "G," we may attempt to predict periods of renesting:

Clutch "A," through success, terminated the season's breeding for the hen.[3]

Clutch "B," destroyed near its hatching point on July 20, was begun in the first half of June; but, considering the lateness of the season and the breeding status of the hen at the time of failure, it is unlikely that renesting followed.

The loss of clutch "C," begun in the second half of April and destroyed after about 10 days of incubation about May 30, was probably followed by a delayed nesting. After 10 days of incubation, the hen's ova would be pretty well resorbed, and the bird would require something like a week or perhaps longer to resume laying; irrespective of uncertainties as to the exact date of re-

nesting, the time could reasonably be set as within the first half of June.

Renesting almost certainly followed the abandonment of clutches "D" and "E," within the first half of June and the second half of May, respectively.

The hen deserting the small, very incomplete clutch "F" on May 7 may be expected to continue her laying scarcely without interruption and, even if she "dropped" an egg or two about the countryside, to have another clutch started in the second week of May. In calculations using half-month periods, clutch "F," being followed by renesting in the period of its origin, would be ignored, though it could be handled in calculations having a weekly basis. On the other hand, clutches in the class of "G" (including the "dropped" and the single abandoned eggs in nests) could best be treated as if they had never existed, unless the tabulation periods could be shortened much more than now looks feasible.[4]

Completeness of clutches is hard to recognize before incubation, and even then error may have been introduced by undetected disappearances of eggs and contributions to a clutch by more than one hen. This granted, we may still find computations of the average number of eggs in complete clutches according to half-month (or other length) periods useful, not only for what help they may later be in dating the origin of broods but also because of the clues often afforded for comparison with other

[3] The reader should not be misled by the fact that typically one-brood-per-season species may have second broods as a result of manipulation (Bissonnette and Csech, 1941) or because of other special circumstances.

[4] In practice, a limit to the shortening of the tabulation periods for the nesting data may be imposed by the relatively greater difficulty of assigning brood data—which are highly important for purposes of correlations—to short periods.

species. The number of successful nests yielding data on number of young hatched is likely to be less than the number of nests known to be successful,

Let us now say that our hypothetical species can be studied during the post-nesting months and that data on broods proved valuable in the guidance and

TABLE 3

HYPOTHETICAL SAMPLES OF DATA ON SIZE OF COMPLETE CLUTCHES AND NUMBER OF YOUNG HATCHED PER SUCCESSFUL CLUTCH

Periods in which clutches were begun	Number of complete clutches	Average number of eggs laid per complete clutch	Number of successful clutches	Average number of young hatched per successful clutch
Periods of initial nesting attempts:				
First half of April	9	19.4	2	13.0
Second half of April	9	13.7	10	10.6
First half of May	32	13.3	21	10.0
Second half of May	28	10.6	25	9.1
Periods following initial nesting attempts:				
First half of June	23	9.2	19	8.0
Second half of June	16	9.0	10	7.7
First half of July	7	7.9	3	6.5
Second half of July	1	7.0	0	

for sometimes so many of the shells of hatched clutches become scattered or destroyed that an investigator has all he can do to make sure of the fact of hatching; and, in addition to the losses of individual eggs, there are the young that die soon after hatching and are for various reasons undiscovered at the time the nesting site is examined.

The seasonal decrease in size of hypothetical clutches and in number of young hatched is more uniform in Table 3 than would always be found in actuality; otherwise, the tabulations follow the pattern of the northwestern Iowa pheasant data. Considerable individual variation in size of completed clutches and of newly hatched broods may also be expected throughout the nesting season, so it doesn't follow that a clutch has to be early if large, or late if small.

verification of calculations having to do with the nesting, as well as serving as indicators of early juvenile mortality. Table 4 summarizes certain data obtained up to about the middle of August, after which identity as specific

TABLE 4

HYPOTHETICAL SAMPLES OF BROODS OBSERVED IN THE FIELD AND CALCULATED TIME OF BEGINNING OF THE CLUTCHES FROM WHICH THEY WERE HATCHED

Periods in which clutches were begun	Broods observed
Periods of initial nesting attempts:	
First half of April	8
Second half of April	65
First half of May	110
Second half of May	80
Periods following initial nesting attempts:	
First half of June	20
Second half of June	5
First half of July	2

broods tended to become obscured through separation and interchange of members.

Ages of broods observed in the field were estimated from the size of the young. The approximate dates of laying of the first eggs in clutches from which the broods hatched were computed from the hatching dates with the assistance of the data given for each half-month period of Table 3.

For an example, we may trace a brood of 8 young of an estimated age of 7 weeks on August 10. These young would have hatched from a clutch for which incubation would have begun about the last of May. From Table 3, we may see that the average number of eggs in complete clutches begun in the first half of May was 13.3 and that the average number of young hatching was 10; for the second half of May, the averages were 10.6 eggs and 9.1 young. Refinement of calculations would be spurious because of variables—especially since the beginning date for the clutch would fall close to the middle of the month—but assignment to the first half of May would seem most in keeping with probabilities.

Obtaining requisite quantitative data on the young that a species has to show for a breeding season may be as important as anything an investigator can do in a study of productivity, yet data of this type are not presented in the literature to the extent that one might think they could be. They are often among the most difficult data to amass, however, particularly for species having secretive young that live in habitats where visibility is poor. To my inability to obtain more data on broods in dense summer vegetation may be ascribed some of the main shortcomings of my early attempts to analyze productivity of bobwhites in southern Wisconsin (Errington, 1933). The faculty that muskrats have for staying out of sight between their weaning and "kit" stages (Errington, 1939c: 472–474) would not be such a handicap to detailed correlation of breeding data and population recovery in Iowa if a method could be discovered for closely determining ages of the general run of subadults taken by trappers; growth and developmental rates have been shown by tagging and other data to be so variable (Errington, 1939c; 1941a: 85–86) that the prospects of determining ages of unmarked winter specimens to within a half-month, or a month, of their actual time of birth seem discouraging.

DERIVATION OF A PROCEDURE FOR CALCULATING PERIODS AT WHICH A REPRODUCTIVE STAGE BEGINS

This section could logically relate to groups other than single-brood birds, but its scope will be restricted to our hypothetical species and only to periods at which the hens begin to lay.

While the reproductive activities of a limited number of individuals may under favorable conditions be studied to excellent advantage through marking for field recognition (Nice, 1937; Edminster, 1938a; Emlen, 1939; 1940; Wright, 1939; Trippensee, 1941), the algebraic treatment of mass field data may hold even greater promise for some types of studies intended primarily to give a general picture.

The basic procedure that seems most adapted for use with our hypothetical species is to handle the data for *each* of the half-month periods by means of the formula: number of hens laying multiplied by fraction of clutches succeeding equals number of broods. The number of hens laying is synonymous with the number laying for the first time only during the early part of the laying season; thereafter, values for the number of hens beginning to lay in a given period must be indirectly derived. It is to be emphasized that calculations as to periods of first-laying for the season must be governed by the data given in Tables 1, 2, and 4.

The fewness of observed broods hatching from clutches begun either before the middle of April or after the middle of June defines the span of the season for effective laying, and, of the broods observed, by far the greater number were produced from clutches begun before the first of June (Table 4). As the laying season is distinctly on the wane by June, there should be scant risk in assuming that few if any of the hens lay their first eggs later than the second half of May—an assumption which we may say is borne out by gonadal studies of hens collected at random for specimens or carcasses of victims of accident and predation. Let us also say that the proportion of non-breeding hens shown by the specimen postmortems was negligible.

Before setting up and solving the equations, let us designate the number of hens beginning to lay in the first half of April as "a"; the number beginning to lay in the second half of April as "b"; in the first half of May, "c"; and in the second half of May, "d." Hens "a," "b," "c," and "d" comprise the total number of breeding hens of which the combined efforts (first nestings plus renestings) are necessary to produce the total number of broods hatched from clutches begun in April and May. The number of hens and the number of broods should be regarded as samples—each as valid in relation to the other as is consistent with the tabulated data—not as absolute densities unless the investigator actually does know the total number of hens and broods living on a unit of land.

From Table 1, we see that 2 of 19 clutches begun in the first half of April were successful and, from Table 4, that 8 broods were credited to clutches so begun. The number of hens laying is near enough to the number beginning to lay in the period, or "a," so we may set up the equation, $\frac{2a}{19}=8$, with expectation only of the minimal inaccuracies inherent in the data as subdivided into half-month periods. The value of "a" is then 76 hens.

The number of breeding hens for the second half of April includes both those laying for the first time in the period ("b") and the "a" hens now renesting. The fraction of the "a" hens renesting in the second half of April is $\frac{1}{19}$ according to Table 2, or $\frac{76}{19}$ or 4 hens. The 4 hens plus "b," multiplied by the fraction of success for the period ($\frac{12}{35}$, as in Table 1) equals the number of broods (65, Table 4), and we have the equation $\frac{12}{35}(b+\frac{a}{19})=65$. Substituting the known value for "a," we have $\frac{12}{35}(b+4)=65$, in which the value of "b" proves to be 186 hens.

In the next equation, we have, in addition to the hens laying their first eggs in the first half of May, the "a" and "b" hens that are renesting in the period. In Table 2, the fraction of first half of April clutches followed by renestings in the first half of May is $\frac{7}{19}$ and the number of "a" hens thus renesting is $\frac{7a}{19}$; we also see that $\frac{11}{35}$ of the second half of April clutches are followed by renestings and that the number of "b" hens renesting is $\frac{11b}{35}$. The fraction of nests succeeding for the period is $\frac{22}{52}$ (Table 1) and the number of broods is 110 (Table 4). In the equation, $\frac{22}{52}(c+\frac{11b}{35}+\frac{7a}{19})=110$, we may substitute the known values for "a" and "b" and get $\frac{22}{52}(c+58.5+28)=110$, or a value of 174 hens for "c."

For the second half of May, the nesting effort consists of all period renestings and the initial laying of the "d" hens. By applying the data given for the period in Tables 1, 2, and 4 and following the procedure outlined, we may set up the equation, $\frac{22}{44}(d+\frac{13c}{52}+\frac{5b}{35}+\frac{6a}{19})$ $=80$, which, after substitution of known values, gives us $\frac{22}{44}(d+43.5+26.6+24)=80$, or 72 hens as the value for "d."

Our calculations therefore show that a population of about 508 hens—with the fortunes indicated by the tabular samples—would be required to produce the 263 broods credited to April and May; 51.8% of the hens have thus far succeeded, and some success for clutches begun after May is shown by data on later broods (Table 4).

As the breeding season runs into June, inaccuracies in data and calculations may be expected apart from those resulting from diverse origin of data and the statistical weaknesses of the smaller samples. The Table 4 data, because of difficulties in seeing newly hatched birds during the last days that general brood counts retain any reliability, are suspected of showing a lower than actual proportion of late broods; but the greatest doubts pertain to the re-

nesting schedule of Table 2 for June and July.

Balancing the data of Tables 1 and 2 against the corresponding brood data of Table 4 for the first half of June, we may set up a test equation, $\frac{2}{5}(\frac{15d}{54}$ $+\frac{9c}{52}+\frac{4b}{35}+\frac{2a}{19})=20$, in which we do not have to solve for any new values but have only to substitute available figures for the hens. Since 1,667.4 does not equal 900, the nesting data and renesting calculations are not even in approximate agreement with the brood data, and we may feel that our suspicions are founded.

The above discrepancy may of course cast some question upon the accuracy of the April–May equations, in which event about all that can be said is that the calculations for these months followed fact-guided sequences and that the collective error should be much less before the breeding slackens and during times when fewer technical obstacles are imposed by field conditions and behavior of the birds.

CALCULATION OF PRODUCTIVITY FOR A BREEDING SEASON

Continuing with our hypothetical species but using 100 hens as a sample, we may attempt a calculation of nesting productivity on a percentage basis.

In the first half of April, 76 or 15% of the 508 "a," "b," "c," and "d" hens, or 15 of the 100-hen sample, are judged to be laying for the first time. From Table 1, we may see that $\frac{2}{19}$ of the clutches begun in the period are successful, so from the 15 hens we get 1.6 successful clutches and 1.6 hens retiring for the season from further breeding because of their success. This would leave 13.4 unsuccessful "a" hens, of which, according to Table 2, $\frac{1}{17}$ or 0.8 are scheduled to renest in the second half of April; $\frac{7}{17}$ or 5.5 in the first half of May; $\frac{8}{17}$ or 4.7 in the second half of May; and $\frac{2}{17}$ or 1.6 in the first half of June. Table 3 shows that an average of 13 young or a total of 21 hatched from the 1.6 successful clutches begun in the period.

In the second half of April, we find that 185 or 36.4% of the "a," "b," "c," and "d" hens, or 36.4 of the 100-hen sample, are laying for the first time, and that by adding to the 36.4 the 0.8 due to renest in this period, we get a laying population of 37.2 hens, of which the current efforts of $\frac{12}{31}$ are successful.

Procedures for using the tabulated data should by now be sufficiently clear as to make unnecessary further step-by-step exposition.[5]

No data show success for clutches begun in the second half of July, so the possible nesting attempts indicated for this period in Table 2 may be ignored. Possibly only a few of the renestings scheduled for the first half of July and even the second half of June are valid. Be that as it may, since fewer than 9% of the hens by the most generous of calculations begin successful clutches after the middle of June, our figures for the entire nesting season should not be thrown off very far by any likely errors; and we may calculate that about 70 of

the 100 hens were successful and hatched out in the vicinity of 650 young, or averages of 6.5 young per original hen and over 9 per successful hen. Over 8 of the original 100 hens were listed as lost during the nesting season, which would probably leave a post-nesting population of about 90 hens and about 20 hens unsuccessful but alive.

The calculations that between 70% and 80% of the pheasant hens studied in northwestern Iowa produced young were verified by hen:brood ratios obtained in late July and early August. "During this period of approximately two weeks, between the virtual termination of the nesting season and the main moulting season . . . 106 hen pheasants and 83 broods totalling 441 young were seen. . . . The average size of the broods was 5.31 young at an average age of a little less than 6 weeks, or an average of 4.16 young for all hens seen at this time of the season" (Errington and Hamerstrom, 1937: 13). The same authors (page 16) showed an average decline in size of broods from 8.7 young at hatching to 5.26 at six to seven weeks.

It should not be necessary to list hen:brood ratios or data on seasonal shrinkages in size of broods for our hypothetical species to emphasize the fact that without such data neither confirmation nor disproof of the nesting season calculations may be very convincing.

As indicated earlier in this writing, brood data, in addition to being valuable for checking against nesting data, may also be some of the best indices of juvenile mortality at times when this factor may otherwise be difficult if not

---

[5] Readers may also consult Errington and Hamerstrom (1937) if they wish to see application to a bona fide problem, especially pages 10 and 11 (Table 5). As two confusing errors were missed in proof-reading the paper for the original publication, "52" should be substituted for "42" in the last equation on page 8 and "Table 4" for "Table 6" in the first line of the second column on page 12.

impossible to measure or even to detect qualitatively. All work on broods should be guided and appraised with reference to social habits of the species, reactions to observers, and the many other pertinent variables.

Another index of seasonal productivity that may be useful is the ratio of adults to grown young shown by the general population at nonbreeding periods of the year. Except when migration or segregation is differential, or sampling methods are selective rather than random, many investigators may profitably inspect the bags of hunters (or the catches of trappers, when furbearers are the objects of study) to procure sex and age ratios. With some species, as the quails, the season's grown young may be sorted from older specimens on the basis of external characters (Stoddard, 1931: 75; Sumner, 1935: 225–226; Starker Leopold, 1939); with others, postmortem inspection of gonads is necessary.

Finally, if the species worked with is, like the northern bob-white, essentially sedentary, localized in ecological "islands," and easily counted in winter, comparison of population densities from the end of one winter to the beginning of the next may be a good measure of recovery from breeding stock. The method must be used with discretion, however, for pronounced variables may have to be considered even for the same areas (Errington and Hamerstrom, 1936: 312–333; Errington, 1939b; 1941b).

Let us now suppose that we have managed to work out in a satisfactory manner the productivity of our hypothetical species for a year or, better, for a term of years. We have pieced together a "life-equation" in which we can account for gains and losses, item by item. Yet, this balance sheet alone tells only part of what we need to know for an understanding of productivity. We should not overlook differences associated with geography, climate, human practices in land use, availability and juxtaposition of food and cover, and the presence of specific diseases and enemies; but, aside from these, there may be population phenomena of sufficient importance that their neglect in an analysis may nullify many of the broader conclusions arrived at. On the principal areas where recovery data on bobwhite populations are available in north-central states, the rate of annual increase has tended to be in inverse ratio to the breeding densities, almost irrespective of the variables of weather, agriculture, and predation (Errington and Hamerstrom, 1936: 421–423; 1937: 17; Errington, 1937c; 1941b; Aldo Leopold and Errington, MS);[6] and analytical dissertations on good or bad breeding seasons fall very short of their potentialities if made without reference to density factors when these are the most influential of all. On the other hand, in southeastern United States, the bobwhite appears to be much more sensitive to local influences other than density factors (Stoddard, 1931; Komarek, 1937; Errington and Stoddard, 1938; Stoddard and Komarek, 1941).

An investigator should consider as

[6] Recovery rates of bobwhite populations may, however, show atypical depression as an apparent result of space-competition with pheasants and some other wild gallinaceous birds or at the "trough" of periodic cycles of other species (Errington, 1941b: 94–100; Aldo Leopold and Errington, MS).

well as he can the possibility of many types of losses being intercompensatory in their net significance—losses that remain about the same in their aggregate despite changes in intensity of depredations by specific enemies and other local and seasonal variables (Errington, Hamerstrom, and Hamerstrom, 1940). Sometimes this intercompensation appears to be a function of population density *per se* (McAtee, 1932); sometimes of population density in relation to the capacity for accommodation of a habitat (Errington, 1937c; Errington, Hamerstrom, and Hamerstrom, 1940). Sometimes, the intercompensation looks very nearly complete; sometimes, partial; sometimes, inconsequential; furthermore, it may sometimes be spectacular at one season of the year, and unrecognizable at another.

Edminster (1939) found that predator control reduced nesting losses of the ruffed grouse (*Bonasa umbellus*) on New York experimental areas but that little or no reduction of brood losses was accomplished either by "complete" or selective control. King (1937: 525) states that juvenile mortality of the ruffed grouse in Minnesota " . . . is normally at least 75 per cent, some years even larger," but that " . . . the several hundred nests for which we have records show nest destruction from all causes to be slightly less than three per cent." During the strongly upgrade and peak years of the periodic grouse cycle, the data from good range presented by King (1937: 529) and Edminster (1938b: 826–828[7]), as well as Errington's (1937a) figures for very low ruffed grouse densities in marginal en-

[7] In form corrected for error (Errington, Hamerstrom, and Hamerstrom, 1940: 814).

vironment, show broadly inverse rates of increase from breeding stock similar to those noted for northern bobwhite populations (Aldo Leopold and Errington, MS).

Furniss (1938: 24) reports practically the same percentage of Saskatchewan duck nests hatched in each of two years, although the percentages of nests lost from different causes varied materially. Comparing the results of a nesting study of waterfowl in Canada with the first year's work on the Lower Souris Refuge in North Dakota, Kalmbach (1938: 614) shows that losses of duck nests from the principal predators were about the same on both areas but occurred chiefly through the medium of crows (*Corvus brachyrhynchos*) in the former case and of skunks (*Mephitis*) in the latter; artificial reduction of skunks on the Lower Souris was followed by a greatly lowered loss rate from these animals and a higher total of hatched nests for the second year, but the percentage of loss from factors other than skunks increased in a way that may suggest partial compensation.

Losses of young muskrats from predators (particularly minks, *Mustela vison*), disease, and miscellaneous agencies may be spectacularly heavy, and losses from intraspecific attack may at times equal or exceed the combined mortality from other causes. The Iowa research findings, however, bear out the concept that some of the heaviest losses follow, instead of govern, the directions taken by population curves. Lessening of juvenile mortality from predation and disease tended to be counterbalanced by increased killing of young by older muskrats. Conversely, when predation or disease losses were heavy, muskrat intraspecific attacks tended to diminish in proportion.

The aforementioned examples are not listed with intent of introducing confusion into phenomena from which meaning is at best hard enough to extract. Rather, they should warn against placing too much confidence in what may seem obvious or expected.

Correlation of data in a study of productivity admittedly entails tedious work, and the temptation to oversimplify is understandable. Even when there may not be serious hiatuses in the data, it is not always easy to keep in mind just what one may be figuring and why, to be sure just how much one should even attempt to dissociate and appraise the inevitable field variables.

SUMMARY

This paper was prepared to assist students of productivity in avian and mammalian populations to guard against errors arising from neglect of compensatory trends in reproductive and loss rates. Some degree of increased breeding as a result of losses of eggs or of young can be expected for a considerable variety of higher vertebrates, and an analysis of productivity of a hypothetical species (similar in behavior to the ring-necked pheasant) is outlined as an illustration of technique.

LITERATURE CITED

ALLEN, A. A.
1934. Sex rhythm in the ruffed grouse (Bonasa umbellus Linn.) and other birds. Auk, 51: 180–199.

BAILEY, VERNON
1924. Breeding, feeding, and other life habits of meadow mice (Microtus). Journ. Agricultural Research, 27: 523–536.

BENNETT, LOGAN J.
1936. The ring-necked pheasant as a nesting parasite of other game birds. Iowa State College Journ. Science, 10: 373–375.
1938. The blue-winged teal. xiv + 144 pp. Collegiate Press, Ames, Iowa.

BENT, ARTHUR CLEVELAND
1939. Life histories of North American woodpeckers. U. S. National Museum, Bul. 174, viii + 334 pp.

BISSONNETTE, THOMAS HUME, and ALBERT GEORGE CSECH
1941. Light-induced egg-production in large pens followed by normal nesting in pheasants. JOURN. WILDLIFE MANAGEMENT, 5: 383–389.

BLANCHARD, BARBARA D.
1941. The white-crowned sparrows (Zonotrichia leucophrys) of the Pacific seaboard: Environment and annual cycle. University of California Publications in Zoology, 46: 1–178.

EDMINSTER, FRANK C.
1938a. The marking of ruffed grouse for field identification. JOURN. WILDLIFE MANAGEMENT, 2: 55–57.
1938b. Productivity of the ruffed grouse in New York. Trans. North American Wildlife Conference, American Wildlife Institute, 3: 825–833.
1939. The effect of predator control on ruffed grouse populations in New York. JOURN. WILDLIFE MANAGEMENT, 3: 345–352.

EMLEN, JOHN T., JR.
1939. Seasonal movements of a low-density valley quail population. JOURN. WILDLIFE MANAGEMENT, 3: 118–130.
1940. Sex and age ratios in survival of the California quail. JOURN. WILDLIFE MANAGEMENT, 4: 92–99.
1941. An experimental analysis of the breeding cycle of the tricolored red-wing. Condor, 43: 209–219.

Erickson, Mary M.
　　1938. Territory, annual cycle, and numbers in a population of wrentits (Chamaea fasciata). University of California Publications in Zoology, 42: 247–334.

Errington, Paul L.
　　1932. Suggestions as to nesting studies on Iowa game birds. Iowa Bird Life, 2: 46–48.
　　1933. The nesting and the life-equation of the Wisconsin bob-white. Wilson Bulletin, 45: 122–132.
　　1937a. Winter carrying capacity of marginal ruffed grouse environment in north-central United States. Canadian Field-Naturalist, 51: 31–34.
　　1937b. The breeding season of the muskrat in northwest Iowa. Journ. Mammalogy, 18: 333–337.
　　1937c. What is the meaning of predation? Annual Report Smithsonian Institution, 1936: 243–252.
　　1939a. The comparative ability of the bob-white and the ring-necked pheasant to withstand cold and hunger. Wilson Bulletin, 51: 22–37.
　　1939b. Suggestions for appraising effects of predation on local areas managed for bobwhite. Trans. North American Wildlife Conference, American Wildlife Institute, 4: 422–425.
　　1939c. Observations on young muskrats in Iowa. Journ. Mammalogy, 20: 465–478.
　　1941a. Versatility in feeding and population maintenance of the muskrat. Journ. Wildlife Management, 5: 68–69.
　　1941b. An eight-winter study of central Iowa bob-whites. Wilson Bulletin, 53: 85–102.

Errington, Paul L., and F. N. Hamerstrom, Jr.
　　1936. The northern bob-white's winter territory. Iowa Agricultural Experiment Station Research Bulletin 201: 301–443.

　　1937. The evaluation of nesting losses and juvenile mortality of the ring-necked pheasant. Journ. Wildlife Management, 1: 3–20.
　　1938. Observations on the effect of a spring drought on reproduction in the Hungarian partridge. Condor, 40: 71–73.

Errington, Paul L., Frances Hamerstrom, and F. N. Hamerstrom, Jr.
　　1940. The great horned owl and its prey in north-central United States. Iowa Agricultural Experiment Station Research Bulletin 277: 757–850.

Errington, Paul L., and H. L. Stoddard
　　1938. Modifications in predation theory suggested by ecological studies of the bob-white quail. Trans. North American Wildlife Conference, American Wildlife Institute, 3: 736–740.

Friedmann, Herbert
　　1929. The cowbirds; a study in the biology of social parasitism. xvii +421 pp. Charles C. Thomas Springfield, Ill., and Baltimore, Md.

Furniss, O. C.
　　1938. The 1937 waterfowl season in the Prince Albert District, Central Saskatchewan. Wilson Bulletin, 50: 17–27.

Hamerstrom, F. N., Jr.
　　1936. A study of the nesting habits of the ring-necked pheasant in northwest Iowa. Iowa State College Journ. Science, 10: 173–203.
　　1939. A study of Wisconsin prairie chicken and sharp-tailed grouse. Wilson Bulletin, 51: 105–120.

Hann, Harry W.
　　1937. Life history of the oven-bird in southern Michigan. Wilson Bulletin, 49: 145–237.

Henderson, L. J.
　　1941. The study of man. Science, 94: 1–10.

Kalmbach, E. R.
　　1938. A comparative study of nesting waterfowl on the Lower Souris Refuge. Trans. North

American Wildlife Conference, American Wildlife Institute, 3: 610–623.

KENDEIGH, S. CHARLES, and S. PRENTISS BALDWIN
1937. Factors affecting yearly abundance of birds. Ecological Monographs, 7: 91–124.

KING, RALPH T.
1937. Ruffed grouse management. Journ. Forestry, 35: 523–532.

KOMAREK, E. V.
1937. Mammal relationships to upland game and other wildlife. Trans. North American Wildlife Conference, American Wildlife Institute, 2: 561–569.

LEOPOLD, A. STARKER
1939. Age determination in quail. JOURN. WILDLIFE MANAGEMENT, 3: 261–265.

LOW, JESSOP B.
1941. Nesting of the ruddy duck in Iowa. Auk, 58: 506–517.

MCATEE, W. L.
1932. Effectiveness in nature of the so-called protective adaptations in the animal kingdom, chiefly as illustrated by the food habits of Nearctic birds. Smithsonian Miscellaneous Collections, 85 (7) (Publ. 3125): 1–201.

MIDDLETON, A. D.
1936. Factors controlling the population of the partridge (*Perdix perdix*) in Great Britain. Proc. Zoological Society of London, 1935: 795–815.

MILLER, ALDEN H.
1931. Systematic revision and natural history of the American shrikes (Lanius). University of California Publications in Zoology, 38: 11–242.

NICE, MARGARET MORSE
1937. Studies in the life history of the song sparrow I. Trans. Linnaean Society of New York, 4: vi +247 pp.

RIDDLE, OSCAR
1911. On the formation, significance and chemistry of the white and yellow yolk of ova. Journ. Morphology, 22: 455–491.

STIEVE, H.
1918. Über experimentell, durch veränderte äussere Bedingungen hervorgerufene Rückbildungsvorgänge am Eierstock des Haushuhnes (Gallus domesticus). Archiv f. Entwickelungsmechanik der Organismen, 44: 530–588.

STODDARD, HERBERT L., and others
1931. The bobwhite quail; its habits, preservation and increase. xxix +559 pp. Scribner's, N. Y.

STODDARD, HERBERT L., and ED. V. KOMAREK
1941. Predator control in southeastern quail management. Trans. North American Wildlife Conference, American Wildlife Institute, 6: 288–293.

SUMNER, E. LOWELL, JR.
1935. A life history study of the California valley quail, with recommendations for conservation and management. California Fish and Game, 21(3–4): 165–256 and 277–342.

TRIPPENSEE, R. E.
1941. A new type of bird and mammal marker. JOURN. WILDLIFE MANAGEMENT, 5: 120–124.

WALLACE, GEORGE J.
1939. Bicknell's thrush, its taxonomy, distribution, and life history. Proc. Boston Society of Natural History, 41: 211–402.

WHITEHEAD, ALFRED NORTH
1925. Science and the modern world. xi +296 pp. Macmillan, N. Y.

WRIGHT, EARL G.
1939. Marking birds by imping feathers. JOURN. WILDLIFE MANAGEMENT, 3: 238–239.

YEATTER, RALPH E.
1934. The Hungarian partridge in the Great Lakes Region. University of Michigan School of Forestry and Conservation Bulletin 5: 1–92.

Paul L. Errington
Iowa State College
Ames, Iowa

Lindeman, R. L. 1942. The trophic-dynamic aspect of ecology. *Ecology* 23:399–418.

Lindeman, Odum (1957), and Tansley (1935) were among a group of scientists who felt that energy, not biomass, was the key to understanding how ecosystems function. Lindeman's paper became the foundation for future studies and concepts related to the dynamic flow of energy in plant and animal communities. His objective was to explain the advances in ecology that shed new light on ecological succession. Thus, he changed how scientists viewed ecosystems: from numbers or biomass of individuals to energy flow.

This article was written at the crossroads between descriptive and quantitative ecology and offered several key insights: the major role of trophic function (especially quantitative relations) in determining community patterns through succession; the validity of theory in ecology; and the integration of trophic relations into community change. These insights could easily have been suppressed, because the paper was initially rejected by the journal *Ecology.* Cook (1977) outlines interesting aspects of this article's publication and provides an inside view of the scientific review process. Lindeman died when only 27, after the paper was finally accepted but before it was published. Science was well served that the paper did not perish. Although Lindeman is sometimes credited with coining the term "ecosystem," it was Tansley who first used the term in 1935.

RELATED READING

Augustine, D. J., and S. J. McNaughton. 1998. Ungulate effects on the functional species composition of plant communities: herbivore selectivity and plant tolerance. Journal of Wildlife Management 62:1165–1183.

Brown, J. H., J. F. Gillooly, A. P. Allen, V. M. Savage, and G. B. West. 2004. Toward a metabolic theory of ecology. Ecology 85:1771–1789.

Cook, R. E. 1977. Raymond Lindeman and the trophic-dynamic concept in ecology. Science 198:22–26.

Hobbs, N. T. 1996. Modification of ecosystems by ungulates. Journal of Wildlife Management 60:695–713.

Leibold, M. A. 1996. A graphical model of keystone predators in food webs: trophic regulation of abundance, incidence, and diversity patterns in communities. The American Naturalist 147:784–812.

Odum, H. T. 1957. Trophic structure and productivity of Silver Springs, Florida. Ecological Monographs 27:55–112.

Polis, G. A., and D. R. Strong. 1996. Food web complexity and community dynamics. The American Naturalist 147:813–846.

Tansley, A. G. 1935. The use and abuse of vegetational concepts and terms. Ecology 16:284–307.

Wiens, J. A., and J. T. Rotenberry. 1981. Habitat associations and community structure of birds in shrubsteppe environments. Ecological Monographs 51:21–42.

Used by permission of the Ecological Society of America.

THE TROPHIC-DYNAMIC ASPECT OF ECOLOGY

RAYMOND L. LINDEMAN

Osborn Zoological Laboratory, Yale University

Recent progress in the study of aquatic food-cycle relationships invites a re-appraisal of certain ecological tenets. Quantitative productivity data provide a basis for enunciating certain trophic principles, which, when applied to a series of successional stages, shed new light on the dynamics of ecological succession.

"COMMUNITY" CONCEPTS

A chronological review of the major viewpoints guiding synecological thought indicates the following stages: (1) the static species-distributional viewpoint; (2) the dynamic species-distributional viewpoint, with emphasis on successional phenomena; and (3) the trophic-dynamic viewpoint. From either species-distributional viewpoint, a lake, for example, might be considered by a botanist as containing several distinct plant aggregations, such as marginal emergent, floating-leafed, submerged, benthic, or phytoplankton communities, some of which might even be considered as "climax" (cf. Tutin, '41). The associated animals would be "biotic factors" of the plant environment, tending to limit or modify the development of the aquatic plant communities. To a strict zoologist, on the other hand, a lake would seem to contain animal communities roughly coincident with the plant communities, although the "associated vegetation" would be considered merely as a part of the environment[1] of the animal community. A more "bio-ecological" species-distributional approach would recognize both the plants and animals as co-constituents of restricted "biotic" communities, such as "plankton communities," "benthic communities," etc., in which members of the living community "co-act" with each other and "re-act" with the non-living environment (Clements and Shelford, '39; Carpenter, '39, '40; T. Park, '41). Coactions and reactions are considered by bio-ecologists to be the dynamic effectors of succession.

The trophic-dynamic viewpoint, as adopted in this paper, emphasizes the relationship of trophic or "energy-availing" relationships within the community-unit to the process of succession. From this viewpoint, which is closely allied to Vernadsky's "biogeochemical" approach (cf. Hutchinson and Wollack, '40) and to the "oekologische Sicht" of Friederichs ('30), a lake is considered as a primary ecological unit in its own right, since all the lesser "communities" mentioned above are dependent upon other components of the lacustrine food cycle (cf. figure 1) for their very existence. Upon further consideration of the trophic cycle, the discrimination between living organisms as parts of the "biotic community" and dead organisms and inorganic nutritives as parts of the "environment" seems arbitrary and unnatural. The difficulty of drawing clear-cut lines between the living *community* and the non-living *environment* is illustrated by the difficulty of determining the status of a slowly dying pondweed covered with periphytes, some of which are also continually dying. As indicated in figure 1, much of the non-living nascent ooze is rapidly reincorporated through "dis-

---

[1] The term *habitat* is used by certain ecologists (Clements and Shelford, '39; Haskell, '40; T. Park, '41) as a synonym for *environment* in the usual sense and as here used, although Park points out that most biologists understand "habitat" to mean "simply the place or niche that an animal or plant occupies in nature" in a species-distributional sense. On the other hand, Haskell, and apparently also Park, use "environment" as synonymous with the *cosmos*. It is to be hoped that ecologists will shortly be able to reach some sort of agreement on the meanings of these basic terms.

399

solved nutrients" back into the living "biotic community." This constant organic-inorganic cycle of nutritive substance is so completely integrated that to consider even such a unit as a lake primarily as a biotic community appears to force a "biological" emphasis upon a more basic functional organization.

This concept was perhaps first expressed by Thienemann ('18), as a result of his extensive limnological studies on the lakes of North Germany. Allee ('34) expressed a similar view, stating: "The picture than finally emerges . . . is of a sort of superorganismic unity not alone between the plants and animals to form biotic communities, but also between the biota and the environment." Such a concept is inherent in the term *ecosystem*, proposed by Tansley ('35) for the fundamental ecological unit.[2] Rejecting the terms "complex organism" and "biotic community," Tansley writes, "But the more fundamental conception is, as it seems to me, the whole *system* (in the sense of physics), including not only the organism-complex, but also the whole complex of physical factors forming what we call the environment of the biome. . . . It is the systems so formed which, from the point of view of the ecologist, are the basic units of nature on the face of the earth. . . . These *ecosystems*, as we may call them, are of the most various kinds and sizes. They form one category of the multitudinous physical systems of the universe, which range from the universe as a whole down to the atom." Tansley goes on to discuss the ecosystem as a category of rank equal to the "biome" (Clements, '16), but points out that the term can also be used in a general sense, as is the word "community." The *ecosystem* may be formally defined as the system composed of physical-chemical-biological processes active within a space-time unit of any

magnitude, i.e., the biotic community *plus* its abiotic environment. The concept of the ecosystem is believed by the writer to be of fundamental importance in interpreting the data of dynamic ecology.

TROPHIC DYNAMICS

Qualitative food-cycle relationships

Although certain aspects of food relations have been known for centuries, many processes within ecosystems are still very incompletely understood. The basic process in trophic dynamics is the transfer of energy from one part of the ecosystem to another. All function, and indeed all life, within an ecosystem depends upon the utilization of an external source of energy, solar radiation. A portion of this incident energy is transformed by the process of photosynthesis into the structure of living organisms. In the language of community economics introduced by Thienemann ('26), autotrophic plants are *producer* organisms, employing the energy obtained by photosynthesis to synthesize complex organic substances from simple inorganic substances. Although plants again release a portion of this potential energy in catabolic processes, a great surplus of organic substance is accumulated. Animals and heterotrophic plants, as *consumer* organisms, feed upon this surplus of potential energy, oxidizing a considerable portion of the consumed substance to release kinetic energy for metabolism, but transforming the remainder into the complex chemical substances of their own bodies. Following death, every organism is a potential source of energy for saprophagous organisms (feeding directly on dead tissues), which again may act as energy sources for successive categories of consumers. Heterotrophic bacteria and fungi, representing the most important saprophagous consumption of energy, may be conveniently differentiated from animal consumers as special-

---

[2] The ecological system composed of the "biocoenosis + biotop" has been termed the *holocoen* by Friederichs ('30) and the *biosystem* by Thienemann ('39).

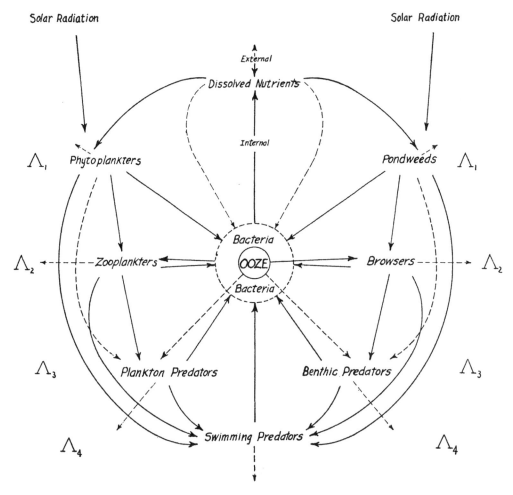

FIG. 1.   Generalized lacustrine food-cycle relationships (after Lindeman, '41b).

ized *decomposers*[3] of organic substance. Waksman ('41) has suggested that certain of these bacteria be further differentiated as *transformers* of organic and inorganic compounds. The combined action of animal consumers and bacterial decomposers tends to dissipate the potential energy of organic substances, again transforming them to the inorganic state. From this inorganic state the autotrophic plants may utilize the dis-

---

[3] Thienemann ('26) proposed the term *reducers* for the heterotrophic bacteria and fungi, but this term suggests that decomposition is produced solely by chemical reduction rather than oxidation, which is certainly not the case. The term *decomposers* is suggested as being more appropriate.

solved nutrients once more in resynthesizing complex organic substance, thus completing the food cycle.

The careful study of food cycles reveals an intricate pattern of trophic predilections and substitutions underlain by certain basic dependencies; food-cycle diagrams, such as figure 1, attempt to portray these underlying relationships. In general, predators are less specialized in food habits than are their prey. The ecological importance of this statement seems to have been first recognized by Elton ('27), who discussed its effect on the survival of prey species when predators are numerous and its effect in enabling predators to survive when their usual prey are only periodically abundant. This ability on the part of predators, which tends to make the higher trophic levels of a food cycle less discrete than the lower, increases the difficulties of analyzing the energy relationships in this portion of the food cycle, and may also tend to "shorten the food-chain."

Fundamental food-cycle variations in different ecosystems may be observed by comparing lacustrine and terrestrial cycles. Although dissolved nutrients in the lake water and in the ooze correspond directly with those in the soil, the autotrophic producers differ considerably in form. Lacustrine producers include macrophytic pondweeds, in which massive supporting tissues are at a minimum, and microphytic phytoplankters, which in larger lakes definitely dominate the production of organic substance. Terrestrial producers are predominantly multicellular plants containing much cellulose and lignin in various types of supporting tissues. Terrestrial herbivores, belonging to a great number of specialized food groups, act as *primary consumers* (sensu Jacot, '40) of organic substance; these groups correspond to the "browsers" of aquatic ecosystems. Terrestrial predators may be classified as more remote (secondary, tertiary, quaternary, etc.) consumers, according to whether they prey upon herbivores or upon other predators; these correspond roughly to the benthic predators and swimming predators, respectively, of a lake. Bacterial and fungal decomposers in terrestrial systems are concentrated in the humus layer of the soil; in lakes, where the "soil" is overlain by water, decomposition takes place both in the water, as organic particles slowly settle, and in the benthic "soil." Nutrient salts are thus freed to be reutilized by the autotrophic plants of both ecosystems.

The striking absence of terrestrial "life-forms" analogous to plankters[4] (cf.

figure 1) indicates that the terrestrial food cycle is essentially "mono-cyclic" with macrophytic producers, while the lacustrine cycle, with two "life-forms" of producers, may be considered as "bi-cyclic." The marine cycle, in which plankters are the only producers of any consequence, may be considered as "mono-cyclic" with microphytic producers. The relative absence of massive supporting tissues in plankters and the very rapid completion of their life cycle exert a great influence on the differential productivities of terrestrial and aquatic systems. The general convexity of terrestrial systems as contrasted with the concavity of aquatic substrata results in striking trophic and successional differences, which will be discussed in a later section.

Productivity

Definitions.—The quantitative aspects of trophic ecology have been commonly expressed in terms of the productivity of the food groups concerned. Productivity has been rather broadly defined as the general rate of production (Riley, '40, and others), a term which may be applied to any or every food group in a given ecosystem. The problem of productivity as related to biotic dynamics has been critically analyzed by G. E. Hutchinson ('42) in his recent book on limnological principles. The two following paragraphs are quoted from Hutchinson's chapter on "The Dynamics of Lake Biota":

The dynamics of lake biota is here treated as primarily a problem of energy transfer . . . the biotic utilization of solar energy entering the lake surface. Some of this energy is transformed by photosynthesis into the structure of phytoplankton organisms, representing an energy content which may be expressed as Λ_1 (first level). Some of the phytoplankters will be eaten by

---

[4] Francé ('13) developed the concept of the *edaphon*, in which the soil microbiota was represented as the terrestrial equivalent of aquatic plankton. This concept appears to have a number of adherents in this country. The author feels that this analogy is misleading, as the edaphon, which has almost no producers, represents only a dependent side-chain of the terrestrial cycle, and is much more comparable to the lacustrine microbenthos than to the plankton.

zooplankters (energy content Λ_2), which again will be eaten by plankton predators (energy content Λ_3). The various successive levels (i.e., stages[5]) of the food cycle are thus seen to have successively different energy contents (Λ_1, Λ_2, Λ_3, etc.).

Considering any food-cycle level Λ_n, energy is entering the level and is leaving it. The rate of change of the energy content Λ_n therefore may be divided into a positive and a negative part:

$$\frac{d\Lambda_n}{dt} = \lambda_n + \lambda_n',$$

where λ_n is by definition positive and represents the rate of contribution of energy from Λ_{n-1} (the previous level) to Λ_n, while λ_n' is negative and represents the sum of the rate of energy dissipated from Λ_n and the rate of energy content handed on to the following level Λ_{n+1}. The more interesting quantity is λ_n which is defined as the true *productivity* of level Λ_n. In practice, it is necessary to use mean rates over finite periods of time as approximations to the mean rates λ_0, λ_1, λ_2. . . .

In the following pages we shall consider the quantitative relationships of the following productivities: λ_0 (rate of incident solar radiation), λ_1 (rate of photosynthetic production), λ_2 (rate of primary or herbivorous consumption), λ_3 (rate of secondary consumption or primary predation), and λ_4 (rate of tertiary consumption). The total amount of organic structure formed per year for any level Λ_n, which is commonly expressed as the annual "yield," actually represents a value uncorrected for dissipation of energy by (1) respiration, (2) predation, and (3) post-mortem decomposition. Let us now consider the quantitative aspects of these losses.

Respiratory corrections.—The amount of energy lost from food levels by catabolic processes (respiration) varies considerably for the different stages in the life histories of individuals, for different levels in the food cycle and for different seasonal temperatures. In terms of annual production, however, individual deviates cancel out and respiratory differences between food groups may be observed.

[5] The term *stage*, in some respects preferable to the term *level*, cannot be used in this trophic sense because of its long-established usage as a successional term (cf. p. 23).

Numerous estimates of average respiration for photosynthetic producers may be obtained from the literature. For terrestrial plants, estimates range from 15 per cent (Pütter, re Spoehr, '26) to 43 per cent (Lundegårdh, '24) under various types of natural conditions. For aquatic producers, Hicks ('34) reported a coefficient of about 15 per cent for Lemna under certain conditions. Wimpenny ('41) has indicated that the respiratory coefficient of marine producers in polar regions (diatoms) is probably much less than that of the more "animal-like" producers (peridinians and coccolithophorids) of warmer seas, and that temperature may be an important factor in determining respiratory coefficients in general. Juday ('40), after conducting numerous experiments with Manning and others on the respiration of phytoplankters in Trout Lake, Wisconsin, concluded that under average conditions these producers respire about $\frac{1}{3}$ of the organic matter which they synthesize. This latter value, 33 per cent, is probably the best available respiratory coefficient for lacustrine producers.

Information on the respiration of aquatic primary consumers is obtained from an illuminating study by Ivlev ('39a) on the energy relationships of *Tubifex*. By means of ingenious techniques, he determined calorific values for assimilation and growth in eleven series of experiments. Using the averages of his calorific values, we can make the following simple calculations: *assimilation* (16.77 cal.) − *growth* (10.33 cal.) = *respiration* (6.44 cal.), so that respiration in terms of growth = $\frac{6.44}{10.33} = 62.30$ per cent. As a check on the growth stage of these worms, we find that $\frac{growth}{assimilation} = 61.7$ per cent, a value in good agreement with the classical conclusions of Needham ('31, III, p. 1655) with respect to embryos: the efficiency of all developing embryos is numerically similar, between 60 and 70 per cent, and independent of temperature within the range of biological tolerance. We may therefore conclude that the worms were growing at nearly maximal efficiency, so that the above respiratory coefficient is nearly minimal. In the absence of further data, we shall tentatively accept 62 per cent as the best available respiratory coefficient for aquatic herbivores.

The respiratory coefficient for aquatic predators can be approximated from data of another important study by Ivlev ('39b), on the energy transformations in predatory yearling carp. Treating his calorific values as in the preceding paragraph, we find that *ingestion* (1829 cal.) − *defecation* (454 cal.) = *assimilation* (1375 cal.), and *assimilation* − *growth* (573 cal.) = *respiration* (802 cal.), so that respiration in terms of growth $= \frac{802}{573} = 140$ per cent, a much higher coefficient than that found for the primary consumer, *Tubifex*. A rough check on this coefficient was obtained by calorific analysis of data on the growth of yearling green sunfishes (*Lepomis cyanellus*) published by W. G. Moore ('41), which indicate a respiratory coefficient of 120 per cent with respect to growth, suggesting that these fishes were growing more efficiently than those studied by Ivlev. Since Moore's fishes were fed on a highly concentrated food (liver), this greater growth efficiency is not surprising. If the maximum growth efficiency would occur when $\frac{growth}{assimilation}$ = 60–70 per cent (AEE of Needham, '31), the AEE of Moore's data (about 50 per cent) indicates that the minimum respiratory coefficient with respect to growth might be as low as 100 per cent for certain fishes. Food-conversion data from Thompson ('41) indicate a minimum respiratory coefficient of less than 150 per cent for young black bass (*Huro salmoides*) at 70° F., the exact percentage depending upon how much of the ingested food (minnows) was assimilated. Krogh ('41) showed that predatory fishes

have a higher rate of respiration than the more sluggish herbivorous species; the respiratory rate of *Esox* under resting conditions at 20° C. was $3\frac{1}{2}$ times that of *Cyprinus*. The form of piscian growth curves (cf. Hile, '41) suggests that the respiratory coefficient is much higher for fishes towards the end of their normal life-span. Since the value obtained from Ivlev (above) is based on more extensive data than those of Moore, we shall tentatively accept 140 per cent as an average respiratory coefficient for aquatic predators.

Considering that predators are usually more active than their herbivorous prey, which are in turn more active than the plants upon which they feed, it is not surprising to find that respiration with respect to growth in producers (33 per cent), in primary consumers (62 per cent) and in secondary consumers (>100 per cent) increases progressively. These differences probably reflect a trophic principle of wide application: the percentage loss of energy due to respiration is progressively greater for higher levels in the food cycle.

Predation corrections.—In considering the predation losses from each level, it is most convenient to begin with the highest level, Λ_n. In a mechanically perfect food cycle composed of organically discrete levels, this loss by predation obviously would be zero. Since no natural food cycle is so mechanically constituted, some "cannibalism" within such an arbitrary level can be expected, so that the actual value for predation loss from Λ_n probably will be somewhat above zero. The predation loss from level Λ_{n-1} will represent the total amount of assimilable energy passed on into the higher level (i.e., the true productivity, λ_n), plus a quantity representing the average content of substance killed but not assimilated by the predator, as will be discussed in the following section. The predation loss from level Λ_{n-2} will likewise represent the total amount of assimilable energy passed on to the next level (i.e., λ_{n-1}), plus a similar factor for unassimilated material, as illustrated by the data of tables II and III. The various categories of parasites are somewhat comparable to those of predators, but the details of their energy relationships have not yet been clarified, and cannot be included in this preliminary account.

Decomposition corrections. — In conformity with the principle of Le Chatelier, the energy of no food level can be completely extracted by the organisms which feed upon it. In addition to the energy represented by organisms which survive to be included in the "annual yield," much energy is contained in "killed" tissues which the predators are unable to digest and assimilate. Average coefficients of indigestible tissues, based largely of the calorific equivalents of the "crude fiber" fractions in the chemical analyses of Birge and Juday ('22), are as follows:

Nannoplankters.....................ca. 5%
Algal mesoplankters................ 5–35%
Mature pondweeds.................. ca. 20%
Primary consumers................. ca. 10%
Secondary consumers.............. ca. 8%
Predatory fishes................... ca. 5%

Corrections for terrestrial producers would certainly be much higher. Although the data are insufficient to warrant a generalization, these values suggest increasing digestibility of the higher food levels, particularly for the benthic components of aquatic cycles.

The loss of energy due to premature death from non-predatory causes usually must be neglected, since such losses are exceedingly difficult to evaluate and under normal conditions probably represent relatively small components of the annual production. However, considering that these losses may assume considerable proportions at any time, the above "decomposition coefficients" must be regarded as correspondingly minimal.

Following non-predated death, every organism is a potential source of energy for myriads of bacterial and fungal

saprophages, whose metabolic products provide simple inorganic and organic solutes reavailable to photosynthetic producers. These saprophages may also serve as energy sources for successive levels of consumers, often considerably supplementing the normal diet of herbivores (ZoBell and Feltham, '38). Jacot ('40) considered saprophage-feeding or coprophagous animals as "low" primary consumers, but the writer believes that in the present state of our knowledge a quantitative subdivision of primary consumers is unwarranted.

Application.—The value of these theoretical energy relationships can be illustrated by analyzing data of the three ecosystems for which relatively comprehensive productivity values have been published (table I). The summary ac-

TABLE I. *Productivities of food-groups in three aquatic ecosystems, as g-cal/cm²/year, uncorrected for losses due to respiration, predation and decomposition. Data from Brujewicz ('39), Juday ('40) and Lindeman ('41b).*

	Caspian Sea	Lake Mendota	Cedar Bog Lake
Phytoplankters: Λ_1...........	59.5	299	25.8
Phytobenthos: Λ_1.............	0.3	22	44.6
Zooplankters: Λ_2............	20.0	22	6.1
Benthic browsers: Λ_2.........		1.8*	0.8
Benthic predators: Λ_3.........	20.6		0.2
Plankton predators: Λ_3........		0.9*	0.8
"Forage" fishes: $\Lambda_3(+\Lambda_2?)$......	0.6	?	0.3
Carp: $\Lambda_3(+\Lambda_2?)$...............	0.0	0.2	0.0
"Game" fishes: $\Lambda_4(+\Lambda_3?)$......	0.6	0.1	0.0
Seals: Λ_5...................	0.01	0.0	0.0

* Roughly assuming that $\frac{2}{3}$ of the bottom fauna is herbivorous (cf. Juday, '22).

count of Brujewicz ('39) on "the dynamics of living matter in the Caspian Sea" leaves much to be desired, as bottom animals are not differentiated into their relative food levels, and the basis for determining the annual production of phytoplankters (which on theoretical grounds appears to be much too low) is not clearly explained. Furthermore, his values are stated in terms of thousands of tons of dry weight for the Caspian Sea as a whole, and must be roughly transformed to calories per square centimeter of surface area. The data for Lake Mendota, Wisconsin, are taken directly from a general summary (Juday, '40) of the many productivity studies made on that eutrophic lake. The data for Cedar Bog Lake, Minnesota, are taken from the author's four-year analysis (Lindeman, '41b) of its food-cycle dynamics. The calorific values in table I, representing annual production of organic matter, are uncorrected for energy losses.

TABLE II. *Productivity values for the Cedar Bog Lake food cycle, in g-cal/cm²/year, as corrected by using the coefficients derived in the preceding sections.*

Trophic level	Uncorrected productivity	Res-pira-tion	Pre-da-tion	De-com-posi-tion	Cor-rected pro-duc-tivity
Producers: Λ_1	70.4 ± 10.14	23.4	14.8	2.8	111.3
Primary consumers: Λ_2	7.0 ± 1.07	4.4	3.1	0.3	14.8
Secondary consumers: Λ_3	1.3 ± 0.43*	1.8	0.0	0.0	3.1

* This value includes the productivity of the small cyprinoid fishes found in the lake.

Correcting for the energy losses due to respiration, predation and decomposition, as discussed in the preceding sections, casts a very different light on the relative productivities of food levels. The calculation of corrections for the Cedar Bog Lake values for producers, primary consumers and secondary consumers are given in table II. The application of similar corrections to the energy values for the food levels of the Lake Mendota food cycle given by Juday ('40), as shown in table III, indicates that Lake Mendota is much more productive of producers and primary consumers than is Cedar Bog Lake, while the production of secondary consumers is of the same order of magnitude in the two lakes.

In calculating total productivity for Lake Mendota, Juday ('40) used a blanket correction of 500 per cent of the annual production of all consumer levels for "metabolism," which presumably includes both respiration and predation. Thompson ('41) found that the "carry-

TABLE III. *Productivity values for the Lake Mendota food cycle, in g-cal/cm²/year, as corrected by using coefficients derived in the preceding sections, and as given by Juday ('40).*

Trophic Level	Uncor-rected pro-duc-tivity	Res-pira-tion	Pre-da-tion	De-com-posi-tion	Cor-rected pro-duc-tivity	Juday's cor-rected pro-duc-tivity
Producers: Λ_1	321*	107	42	10	480	428
Primary con-sumers: Λ_2	24	15	2.3	0.3	41.6	144
Secondary consumers: Λ_3	1†	1	0.3	0.0	2.3	6
Tertiary con-sumers: Λ_4	0.12	0.2	0.0	0.0	0.3	0.7

* Hutchinson ('42) gives evidence that this value is probably too high and may actually be as low as 250.

† Apparently such organisms as small "forage" fishes are not included in any part of Juday's balance sheet. The inclusion of these forms might be expected to increase considerably the productivity of secondary consumption.

ing-capacity" of lakes containing mostly carp and other "coarse" fishes (primarily Λ_3), was about 500 per cent that of lakes containing mostly "game" fishes (primarily Λ_4), and concluded that "this difference must be about one complete link in the food chain, since it usually requires about five pounds of food to produce one pound of fish." While such high "metabolic losses" may hold for tertiary and quaternary predators under certain field conditions, the physiological experiments previously cited indicate much lower respiratory coefficients. Even when predation and decomposition corrections are included, the resultant productivity values are less than half those obtained by using Juday's coefficient. Since we have shown that the necessary corrections vary progressively with the different food levels, it seems probable that Juday's "coefficient of metabolism" is much too high for primary and secondary consumers.

Biological efficiency

The quantitative relationships of any food-cycle level may be expressed in terms of its efficiency with respect to lower levels. Quoting Hutchinson's ('42) definition, "the efficiency of the productivity of any level (Λ_n) relative to the productivity of any previous level (Λ_m) is defined as $\dfrac{\lambda_n}{\lambda_m}$ 100. If the rate of solar energy entering the ecosystem is denoted as λ_0, the efficiencies of all levels may be referred back to this quantity λ_0." In general, however, the most interesting efficiencies are those referred to the previous level's productivity (λ_{n-1}), or those expressed as $\dfrac{\lambda_n}{\lambda_{n-1}}$ 100. These latter may be termed the *progressive efficiencies* of the various food-cycle levels, indicating for each level the degree of utilization of its potential food supply or energy source. All efficiencies discussed in the following pages are progressive efficiencies, expressed in terms of relative productivities $\left(\dfrac{\lambda_n}{\lambda_{n-1}} 100\right)$. It is important to remember that efficiency and productivity are not synonymous. Productivity is a rate (i.e., in the units here used, cal/cm²/year), while efficiency, being a ratio, is a dimensionless number. The points of reference for any efficiency value should always be clearly stated.

The progressive efficiencies $\left(\dfrac{\lambda_n}{\lambda_{n-1}} 100\right)$ for the trophic levels of Cedar Bog Lake and Lake Mendota, as obtained from the productivities derived in tables II and III, are presented in table IV. In view of the uncertainties concerning some of the Lake Mendota productivities, no definite conclusions can be drawn from their relative efficiencies. The Cedar Bog Lake ratios, however, indicate that the progressive efficiencies increase from about 0.10 per cent for production, to 13.3 per cent for primary consumption, and to 22.3 per cent for secondary consumption. An uncorrected efficiency of tertiary consumption of 37.5 per cent ± 3.0 per cent (for the weight ratios of "carnivorous" to "forage" fishes in Alabama ponds) is indicated in data published by Swingle and Smith ('40). These progressively increasing efficiencies may well represent a fundamental trophic principle, namely, that the consumers at progressively higher levels in the food cycle are progressively more efficient in the use of their food supply.

At first sight, this generalization of increasing efficiency in higher consumer groups would appear to contradict the previous generalization that the loss of energy due to respiration is progressively greater for higher levels in the food cycle. These can be reconciled by remembering that increased activity of predators considerably increases the chances of encountering suitable prey. The ultimate effect of such antagonistic principles would present a picture of a predator completely wearing itself out in the process of completely exterminating its prey, a very improbable situation. However, Elton ('27) pointed out that food-cycles rarely have more than five trophic levels. Among the several factors involved, increasing respiration of successive levels of predators contrasted with their successively increasing efficiency of predation appears to be important in restricting the number of trophic levels in a food cycle.

The effect of increasing temperature is alleged by Wimpenny ('41) to cause a decreasing consumer/producer ratio, presumably because he believes that the "acceleration of vital velocities" of consumers at increasing temperatures is more rapid than that of producers. He

TABLE IV. *Productivities and progressive efficiencies in the Cedar Bog Lake and Lake Mendota food cycles, as g-cal/cm²/year*

	Cedar Bog Lake		Lake Mendota	
	Productivity	Efficiency	Productivity	Efficiency
Radiation.........	≦118,872		118,872	
Producers: Λ_1......	111.3	0.10%	480*	0.40%
Primary consumers:				
Λ_2............	14.8	13.3%	41.6	8.7%
Secondary consumers: Λ_3....	3.1	22.3%	2.3†	5.5%
Tertiary consumers: Λ_4....	—	—	0.3	13.0%

* Probably too high; see footnote of table III.
† Probably too low; see footnote of table III.

cites as evidence Lohmann's ('12) data for relative *numbers* (not biomass) of Protophyta, Protozoa and Metazoa in the centrifuge plankton of "cool" seas (741:73:1) as contrasted with tropical areas (458:24:1). Since Wimpenny himself emphasizes that many metazoan plankters are larger in size toward the poles, these data do not furnish convincing proof of the allegation. The data given in table IV, since Cedar Bog Lake has a much higher mean annual water temperature than Lake Mendota, appear to contradict Wimpenny's generalization that consumer/producer ratios fall as the temperature increases.

The Eltonian pyramid

The general relationships of higher food-cycle levels to one another and to community structure were greatly clarified following recognition (Elton, '27) of the importance of size and of numbers in the animals of an ecosystem. Beginning with primary consumers of various sizes, there are as a rule a number of food-chains radiating outwards in which the probability is that predators will become successively larger, while parasites and hyper-parasites will be progressively smaller than their hosts. Since small primary consumers can increase faster than larger secondary consumers and are so able to support the latter, the animals

at the base of a food-chain are relatively abundant while those toward the end are progressively fewer in number. The resulting arrangement of sizes and numbers of animals, termed the pyramid of Numbers by Elton, is now commonly known as the Eltonian Pyramid. Williams ('41), reporting on the "floor fauna" of the Panama rain forest, has recently published an interesting example of such a pyramid, which is reproduced in figure 2.

The Eltonian Pyramid may also be expressed in terms of biomass. The weight of all predators must always be much lower than that of all food animals, and the total weight of the latter much lower than the plant production (Bodenheimer, '38). To the human ecologist, it is noteworthy that the population density of the essentially vegetarian Chinese, for example, is much greater than that of the more carnivorous English.

The principle of the Eltonian Pyramid has been redefined in terms of productivity by Hutchinson (unpublished) in the following formalized terms: the rate of production cannot be less and will almost certainly be greater than the rate of primary consumption, which in turn cannot be less and will almost certainly be greater than the rate of secondary consumption, which in turn . . . , etc. The energy-relationships of this principle may be epitomized by means

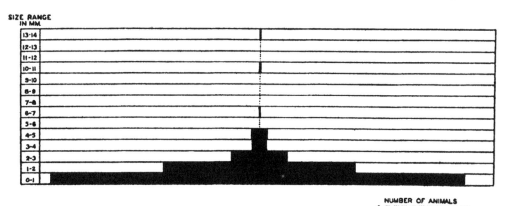

Fig. 2. Eltonian pyramid of numbers, for floor-fauna invertebrates of the Panama rain forest (from Williams, '41).

of the productivity symbol λ, as follows:

$$\lambda_0 > \lambda_1 > \lambda_2 \ldots > \lambda_n.$$

This rather obvious generalization is confirmed by the data of all ecosystems analyzed to date.

TROPHIC-DYNAMICS IN SUCCESSION

Dynamic processes within an ecosystem, over a period of time, tend to produce certain obvious changes in its species-composition, soil characteristics and productivity. Change, according to Cooper ('26), is the essential criterion of succession. From the trophic-dynamic viewpoint, succession is the process of development in an ecosystem, brought about primarily by the effects of the organisms on the environment and upon each other, towards a relatively stable condition of equilibrium.

It is well known that in the initial phases of hydrarch succession (oligotrophy → eutrophy) productivity increases rapidly; it is equally apparent that the colonization of a bare terrestrial area represents a similar acceleration in productivity. In the later phases of succession, productivity increases much more slowly. As Hutchinson and Wollack ('40) pointed out, these generalized changes in the rate of production may be expressed as a sigmoid curve showing a rough resemblance to the growth curve of an organism or of a homogeneous population.

Such smooth logistic growth, of course, is seldom found in natural succession, except possibly in such cases as bare areas developing directly to the climax vegetation type in the wake of a retreating glacier. Most successional seres consist of a number of *stages* ("recognizable, clearly-marked subdivisions of a given sere"—W. S. Cooper), so that their productivity growth-curves will contain undulations corresponding in distinctness to the distinctness of the stages. The presence of stages in a successional sere apparently represents the persistent influence of some combination of limiting factors, which, until they are overcome by species-substitution, etc., tend to decrease the acceleration of productivity and maintain it at a more constant rate. This tendency towards *stage-equilibrium* of productivity will be discussed in the following pages.

Productivity in hydrarch succession

The descriptive dynamics of hydrarch succession is well known. Due to the essentially concave nature of the substratum, lake succession is internally complicated by a rather considerable influx of nutritive substances from the drainage basin surrounding the lake. The basins of lakes are gradually filled with sediments, largely organogenic, upon which a series of vascular plant stages successively replace one another until a more or less stable (climax) stage is attained. We are concerned here, however, primarily with the productivity aspects of the successional process.

Eutrophication. — Thienemann ('26) presented a comprehensive theoretical discussion of the relation between lake succession and productivity, as follows: In oligotrophy, the pioneer phase, productivity is rather low, limited by the amount of dissolved nutrients in the lake water. Oxygen is abundant at all times, almost all of the synthesized organic matter is available for animal food; bacteria release dissolved nutrients from the remainder. Oligotrophy thus has a very "thrifty" food cycle, with a relatively high "efficiency" of the consumer populations. With increasing influx of nutritives from the surrounding drainage basin and increasing primary productivity (λ_1), oligotrophy is gradually changed through mesotrophy to eutrophy, in which condition the production of organic matter (λ_1) exceeds that which can be oxidized (λ_1') by respiration, predation and bacterial decomposition. The oxygen supply of the hypolimnion becomes depleted, with disastrous effects on the oligotrophconditioned bottom fauna. Organisms especially adapted to endure temporary anaerobiosis replace the oligotrophic species, accompanied by anaerobic bacteria which during the stagnation period cause reduction rather than oxidation of the organic detritus. As a result of this process, semi-reduced organic ooze, or *gyttja*, accumulates on the bottom. As oxygen supply thus becomes a limiting factor of productivity, relative efficiency of the consumer groups in utilizing the synthesized organic substance becomes correspondingly lower.

The validity of Thienemann's interpretation, particularly regarding the trophic mechanisms, has recently been challenged by Hutchinson ('41, '42), who states that three distinct factors are involved: (1) the edaphic factor, representing the potential nutrient supply (primarily phosphorus) in the surrounding drainage basin; (2) the age of the lake at any stage, indicating the degree of utilization of the nutrient supply; and (3) the morphometric character at any stage, dependent on both the original morphometry of the lake basin and the age of the lake, and presumably influencing the oxygen concentration in the hypolimnion. He holds that true eutrophication takes place only in regions well supplied with nutrients, lakes in other regions developing into "ideotrophic types." The influx of phosphorus is probably very great in the earliest phases, much greater than the supply of nitrogen, as indicated by very low N/P ratios in the earliest sediments (Hutchinson and Wollack, '40). A large portion of this phosphorus is believed to be insoluble, as a component of such mineral particles as apatite, etc., although certainly some of it is soluble. The supply of available nitrogen increases somewhat more slowly, being largely dependent upon the fixation of nitrogen by microorganisms either in the lake or in the surrounding soils. The photosynthetic productivity (λ_1) of lakes thus increases rather rapidly in the early phases, its

quantitative value for lakes with comparable edaphic nutrient supplies being dependent on the morphometry (mean depth). Since deep lakes have a greater depth range for plankton photosynthesis, abundant oxygen and more chance for decomposition of the plankton detritus before reaching the bottom, such deep lakes may be potentially as productive as shallower lakes, in terms of unit surface area. Factors tending to lessen the comparative productivity of deep lakes are (1) lower temperature for the lake as a whole, and (2) greater dilution of nutrients in terms of volume of the illuminated "trophogenic zone" of the lake. During eutrophication in a deep lake, the phosphorus content of the sediment falls and nitrogen rises, until a N/P ratio of about 40/1 is established. "The decomposition of organic matter presumably is always liberating some of this phosphorus and nitrogen. Within limits, the more organic matter present the easier will be such regeneration. It is probable that benthic animals and anion exchange play a part in such processes" (Hutchinson, '42). The progressive filling of the lake basin with sediments makes the lake shallower, so that the oxygen supply of the hypolimnion is increasingly, and finally completely, exhausted during summer stagnation. Oxidation of the sediments is less complete, but sufficient phosphorus is believed to be regenerated from the ooze surface so that productivity in terms of surface area remains relatively constant. The nascent ooze acts as a trophic buffer, in the chemical sense, tending to maintain the productivity of a lake in stage-equilibrium (*typological equilibrium* of Hutchinson) during the eutrophic stage of its succession.

The concept of eutrophic stage-equilibrium seems to be partially confused (cf. Thienemann, '26; Hutchinson and Wollack, '40) with the theoretically ideal condition of complete *trophic equilibrium*, which may be roughly defined as the dynamic state of continuous, complete utilization and regeneration of chemical nutrients in an ecosystem, without loss or gain from the outside, under a periodically constant energy source—such as might be found in a perfectly balanced aquarium or terrarium. Natural ecosystems may tend to approach a state of trophic equilibrium under certain conditions, but it is doubtful if any are sufficiently autochthonous to attain, or maintain, true trophic equilibrium for any length of time. The biosphere as a whole, however, as Vernadsky ('29, '39) so vigorously asserts, may exhibit a high degree of true trophic equilibrium.

The existence of prolonged eutrophic stage-equilibrium was first suggested as a result of a study on the sediments of Grosser Plöner See in Germany (Groschopf, '36). Its significance was recognized by Hutchinson and Wollack ('40), who based their critical discussion on chemical analyses (ibid.) and pollen analyses (Deevey, '39) of the sediments of Linsley Pond, Connecticut. They reported a gradual transition from oligotrophy to eutrophy (first attained in the oak-hemlock pollen period), in which stage the lake has remained for a very long time, perhaps more than 4000 years. They report indications of a comparable eutrophic stage-equilibrium in the sediments of nearby Upper Linsley Pond (Hutchinson and Wollack, unpublished). Similar attainment of stage-equilibrium is indicated in a preliminary report on the sediments of Lake Windermere in England (Jenkin, Mortimer and Pennington, '41). Every stage of a sere is believed to possess a similar stage-equilibrium of variable duration, although terrestrial stages have not yet been defined in terms of productivity.

The trophic aspects of eutrophication cannot be determined easily from the sediments. In general, however, the ratio of organic matter to the silt washed into the lake from the margins affords an approximation of the photosynthetic productivity. This undecomposed organic matter, representing the amount of

energy which is lost from the food cycle, is derived largely from level Λ_1, as plant structures in general are decomposed less easily than animal structures. The quantity of energy passed on into consumer levels can only be surmised from undecomposed fragments of organisms which are believed to occupy those levels. Several types of animal "microfossils" occur rather consistently in lake sediments, such as the carapaces and postabdomens of certain cladocerans, chironomid head-capsules, fragments of the phantom-midge larva *Chaoborus*, snail shells, polyzoan statoblasts, sponge spicules and rhizopod shells. Deevey ('42), after making comprehensive microfossil and chemical analyses of the sediments of Linsley Pond, suggested that the abundant half-carapaces of the planktonic browser *Bosmina* afford "a reasonable estimate of the quantity of zooplankton produced" and that "the total organic matter of the sediment is a reasonable estimate of the organic matter produced by phytoplankton and littoral vegetation." He found a striking similarity in the shape of the curves representing *Bosmina* content and total organic matter plotted against depth, which, when plotted logarithmically against each other, showed a linear relationship expressed by an empirical power equation. Citing Hutchinson and Wollack ('40) to the effect that the developmental curve for organic matter was analogous to that for the development of an organism, he pressed the analogy further by suggesting that the increase of zooplankton (*Bosmina*) with reference to the increase of organic matter (λ_1) fitted the formula $y = bx^k$ for allometric growth (Huxley, '32), "where $y = Bosmina$, $x =$ total organic matter, $b =$ a constant giving the value of y when $x = 1$, and $k =$ the 'allometry constant,' or the slope of the line when a double log plot is made." If we represent the organic matter produced as λ_1 and further assume that *Bosmina* represents the primary consumers (λ_2), neglecting benthic browsers, the formula becomes $\lambda_2 = b\lambda_1{}^k$. Whether this formula would express the relationship found in other levels of the food cycle, the development of other stages, or other ecosystems, remains to be demonstrated.[6] Stratigraphic analyses in Cedar Bog Lake (Lindeman and Lindeman, unpublished) suggest a roughly similar increase of both organic matter and *Bosmina* carapaces in the earliest sediments. In the modern senescent lake, however, double logarithmic plottings of the calorific values for λ_1 against λ_2, and λ_2 against λ_3, for the four years studied, show no semblance of linear relationship, i.e., do not fit any power equation. If Deevey is correct in his interpretation of the Linsley Pond microfossils, allometric growth would appear to characterize the phases of pre-equilibrium succession as the term "growth" indeed implies.

The relative duration of eutrophic stage-equilibrium is not yet completely understood. As exemplified by Linsley Pond, the relation of stage-equilibrium to succession is intimately concerned with the trophic processes of (1) external influx and efflux (partly controlled by climate), (2) photosynthetic productivity, (3) sedimentation (partly by physiographic silting) and (4) regeneration of nutritives from the sediments. These processes apparently maintain a relatively constant ratio to each other during the extended equilibrium period. Yet the food cycle is not in true trophic equilibrium, and continues to fill the lake with organic sediments. *Succession* is

---

[6] It should be mentioned in this connection that Meschkat ('37) found that the relationship of population density of tubificids to organic matter in the bottom of a polluted "Buhnenfeld" could be expressed by the formula $y = a^x$, where y represents the population density, x is the "determining environmental factor," and a is a constant. He pointed out that for such an expression to hold the population density must be maximal. Hentschel ('36), on less secure grounds, suggested applying a similar expression to the relationship between populations of marine plankton and the "controlling factor" of their environment.

continuing, at a rate corresponding to the rate of sediment accumulation. In the words of Hutchinson and Wollack ('40), "this means that during the equilibrium period the lake, through the internal activities of its biocoenosis, is continually approaching a condition when it ceases to be a lake."

Senescence.—As a result of long-continued sedimentation, eutrophic lakes attain senescence, first manifested in bays and wind-protected areas. Senescence is usually characterized by such pond-like conditions as (1) tremendous increase in shallow littoral area populated with pondweeds and (2) increased marginal invasion of terrestrial stages. Cedar Bog Lake, which the author has studied for several years, is in late senescence, rapidly changing to the terrestrial stages of its succession. On casual inspection, the massed verdure of pondweeds and epiphytes, together with sporadic algal blooms, appears to indicate great photosynthetic productivity. As pointed out by Wesenberg-Lund ('12), littoral areas of lakes are virtual hothouses, absorbing more radiant energy per unit volume than deeper areas. At the present time the entire aquatic area of Cedar Bog Lake is essentially littoral in nature, and its productivity per cubic meter of water is probably greater than at any time in its history. However, since radiant energy (λ_0) enters a lake only from the surface, productivity must be defined in terms of surface area. In these terms, the present photosynthetic productivity pales into insignificance when compared with less advanced lakes in similar edaphic regions; for instance, λ_1 is less than $\frac{1}{3}$ that of Lake Mendota, Wisconsin (cf. table IV). These facts attest the essential accuracy of Welch's ('35) generalization that productivity declines greatly during senescence. An interesting principle demonstrated in Cedar Bog Lake (Lindeman, '41b) is that during late lake senescence general productivity (λ_n) is increasingly influenced by climatic factors, acting through water level changes, drainage, duration of winter ice, snow cover, etc., to affect the presence and abundance of practically all food groups in the lake.

Terrestrial stages.—As an aquatic ecosystem passes into terrestrial phases, fluctuations in atmospheric factors increasingly affect its productivity. As succession proceeds, both the species-composition and the productivity of an ecosystem increasingly reflect the effects of the regional climate. Qualitatively, these climatic effects are known for soil morphology (Joffe, '36), autotrophic vegetation (Clements, '16), fauna (Clements and Shelford, '39) and soil microbiota (Braun-Blanquet, '32), in fact for every important component of the food cycle. Quantitatively, these effects have been so little studied that generalizations are most hazardous. It seems probable, however, that productivity tends to increase until the system approaches maturity. Clements and Shelford ('39, p. 116) assert that both plant and animal productivity is generally greatest in the subclimax, except possibly in the case of grasslands. Terrestrial ecosystems are primarily convex topographically and

thus subject to a certain nutrient loss by erosion, which may or may not be made up by increased availability of such nutrients as can be extracted from the "C" soil horizon.

Successional productivity curves.—In recapitulating the probable photosynthetic productivity relationships in hydrarch succession, we shall venture to diagram (figure 3) a hypothetical hydrosere, developing from a moderately deep lake in a fertile cold temperate region under relatively constant climatic conditions. The initial period of oligotrophy is believed to be relatively short (Hutchinson and Wollack, '40; Lindeman '41a), with productivity rapidly increasing until eutrophic stage-equilibrium is attained. The duration of high eutrophic productivity depends upon the mean depth of the basin and upon the rate of sedimentation, and productivity fluctuates about a high eutrophic mean until the lake becomes too shallow for maximum growth of phytoplankton or regeneration of nutrients from the ooze. As the lake becomes shallower and more senescent, productivity is increasingly influenced by climatic fluctuations and

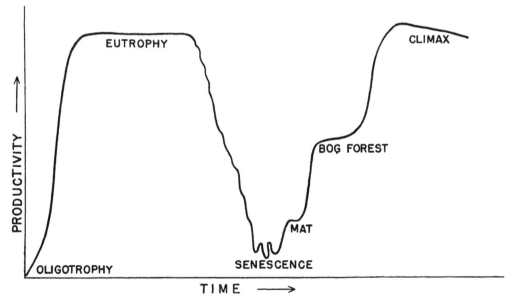

FIG. 3. Hypothetical productivity growth-curve of a hydrosere, developing from a deep lake to climax in a fertile, cold-temperate region.

gradually declines to a minimum as the lake is completely filled with sediments.

The terrestrial aspects of hydrarch succession in cold temperate regions usually follow sharply defined, distinctive stages. In lake basins which are poorly drained, the first stage consists of a mat, often partly floating, made up primarily of sedges and grasses or (in more coastal regions) such heaths as *Chamaedaphne* and *Kalmia* with certain species of sphagnum moss (cf. Rigg, '40). The mat stage is usually followed by a bog forest stage, in which the dominant species is *Larix laricina*, *Picea mariana* or *Thuja occidentalis*. The bog forest stage may be relatively permanent ("edaphic" climax) or succeeded to a greater or lesser degree by the regional climax vegetation. The stage-productivities indicated in figure 3 represent only crude relative estimates, as practically no quantitative data are available.

Efficiency relationships in succession

The successional changes of photosynthetic efficiency in natural areas (with respect to solar radiation, i.e., $\frac{\lambda_1}{\lambda_0} 100$) have not been intensively studied. In lake succession, photosynthetic efficiency would be expected to follow the same course deduced for productivity, rising to a more or less constant value during eutrophic stage-equilibrium, and declining during senescence, as suggested by a photosynthetic efficiency of at least 0.27 per cent for eutrophic Lake Mendota (Juday, '40) and of 0.10 per cent for senescent Cedar Bog Lake. For the terrestrial hydrosere, efficiency would likewise follow a curve similar to that postulated for productivity.

Rough estimates of photosynthetic efficiency for various climatic regions of the earth have been summarized from the literature by Hutchinson (unpublished). These estimates, corrected for respiration, do not appear to be very reliable because of imperfections in the original observations, but are probably of the correct order of magnitude. The mean photosynthetic efficiency for the sea is given as 0.31 per cent (after Riley, '41). The mean photosynthetic efficiency for terrestrial areas of the earth is given as 0.09 per cent $\pm$ 0.02 per cent (after Noddack, '37), for forests as 0.16 per cent, for cultivated lands as 0.13 per cent, for steppes as 0.05 per cent, and for deserts as 0.004 per cent. The mean photosynthetic efficiency for the earth as a whole is given as 0.25 per cent. Hutchinson has suggested (cf. Hutchinson and Lindeman, '41) that numerical efficiency values may provide "the most fundamental possible classification of biological formations and of their developmental stages."

Almost nothing is known concerning the efficiencies of consumer groups in succession. The general chronological increase in numbers of *Bosmina* carapaces with respect to organic matter and of *Chaoborus* fragments with respect to *Bosmina* carapaces in the sediments of Linsley Pond (Deevey, '42) suggests progressively increasing efficiencies of zooplankters and plankton predators. On the other hand, Hutchinson ('42) concludes from a comparison of the P : Z (phytoplankton : zooplankton) biomass ratios of several oligotrophic alpine lakes, ca 1 : 2 (Ruttner, '37), as compared with the ratios for Linsley Pond, 1 : 0.22 (Riley, '40) and three eutrophic Bavarian lakes, 1 : 0.25 (Heinrich, '34), that "as the phytoplankton crop is increased the zooplankton by no means keeps pace with the increase." Data compiled by Deevey ('41) for lakes in both mesotrophic (Connecticut) and eutrophic regions (southern Wisconsin), indicate that the deeper or morphometrically "younger" lakes have a lower ratio of bottom fauna to the standing crop of plankton (10–15 per cent) than the shallower lakes which have attained eutrophic equilibrium (22–27 per cent). The ratios for senescent Cedar Bog Lake, while not directly comparable because of its essentially littoral nature, are even higher. These meager data suggest that the efficiencies of consumer groups may increase throughout the aquatic phases of succession.

For terrestrial stages, no consumer efficiency data are available. A suggestive series of species-frequencies in mesarch succession was published by Vera Smith-Davidson ('32), which indicated greatly increasing numbers of arthropods in successive stages approaching the regional climax. Since the photosynthetic productivity of the stages probably also increased, it is impossible to determine progressive efficiency relationships. The problems of biological efficiencies present a practically virgin field, which appears to offer abundant rewards for studies guided by a trophic-dynamic viewpoint.

In conclusion, it should be emphasized that the trophic-dynamic principles indicated in the following summary cannot be expected to hold for every single case, in accord with the known facts of biological variability. *à priori*, however, these principles appear to be valid for the vast majority of cases, and may be expected to possess a statistically significant probability of validity for any case selected at random. Since the available data summarized in this paper are far too meager to establish such generalizations on a statistical basis, it is highly important that further studies be initiated to test the validity of these and other trophic-dynamic principles.

Summary

1. Analyses of food-cycle relationships indicate that a biotic community cannot be clearly differentiated from its abiotic environment; the *ecosystem* is hence regarded as the more fundamental ecological unit.

2. The organisms within an ecosystem may be grouped into a series of more or less discrete trophic levels (Λ_1, Λ_2, Λ_3, . . . Λ_n) as producers, primary consumers, secondary consumers, etc., each successively dependent upon the preceding level as a source of energy, with the producers (Λ_1) directly dependent upon the rate of incident solar radiation (productivity λ_0) as a source of energy.

3. The more remote an organism is from the initial source of energy (solar radiation), the less probable that it will be dependent solely upon the preceding trophic level as a source of energy.

4. The progressive energy relationships of the food levels of an "Eltonian Pyramid" may be epitomized in terms of the productivity symbol λ, as follows:

$$\lambda_0 > \lambda_1 > \lambda_2 . . . > \lambda_n.$$

5. The percentage loss of energy due to respiration is progressively greater for higher levels in the food cycle. Respiration with respect to growth is about 33 per cent for producers, 62 per cent for primary consumers, and more than 100 per cent for secondary consumers.

6. The consumers at progressively higher levels in the food cycle appear to be progressively more efficient in the use of their food supply. This generalization can be reconciled with the preceding one by remembering that increased activity of predators considerably increases the chances of encountering suitable prey.

7. Productivity and efficiency increase during the early phases of successional development. In lake succession, productivity and photosynthetic efficiency increase from oligotrophy to a prolonged eutrophic stage-equilibrium and decline with lake senescence, rising again in the terrestrial stages of hydrarch succession.

8. The progressive efficiencies of consumer levels, on the basis of very meager data, apparently tend to increase throughout the aquatic phases of succession.

Acknowledgments

The author is deeply indebted to Professor G. E. Hutchinson of Yale University, who has stimulated many of the trophic concepts developed here, generously placed at the author's disposal several unpublished manuscripts, given valuable counsel, and aided the final development of this paper in every way possible. Many of the concepts embodied in the successional sections of this paper were developed independently by Professor Hutchinson at Yale and by the author as a graduate student at the University of Minnesota. Subsequent to an exchange of notes, a joint preliminary abstract was published (Hutchinson and Lindeman, '41). The author wishes to express gratitude to the mentors and friends at the University of Minnesota who encouraged and helpfully criticized the initial development of these concepts, particularly Drs. W. S. Cooper, Samuel Eddy, A. C. Hodson, D. B. Lawrence and J. B. Moyle, as well as other members of the local Ecological Discussion Group. The author is also indebted to Drs. J. R. Carpenter, E. S. Deevey, H. J. Lutz, A. E. Parr, G. A. Riley and V. E. Shelford, as well as the persons mentioned above, for critical reading of preliminary manuscripts. Grateful acknowledgment is made to the Graduate School, Yale University, for the award of a Sterling Fellowship in Biology during 1941–1942.

LITERATURE CITED

Allee, W. C. 1934. Concerning the organization of marine coastal communities. Ecol. Monogr., **4**: 541–554.

Birge, E. A., and C. Juday. 1922. The inland lakes of Wisconsin. The plankton. Part I. Its quantity and chemical composition. Bull. Wisconsin Geol. Nat. Hist. Surv., **64**: 1–222.

Bodenheimer, F. S. 1938. Problems of Animal Ecology. London. Oxford University Press.

Braun-Blanquet, J. 1932. Plant Sociology. N. Y. McGraw-Hill Co.

Brujewicz, S. W. 1939. Distribution and dynamics of living matter in the Caspian Sea. Compt. Rend. Acad. Sci. URSS, **25**: 138–141.

Carpenter, J. R. 1939. The biome. Amer. Midl. Nat., **21**: 75–91.

——. 1940. The grassland biome. Ecol. Monogr., **10**: 617–687.

Clements, F. E. 1916. Plant Succession. Carnegie Inst. Washington Publ., No. 242.

—— and V. E. Shelford. 1939. Bio-Ecology. N. Y. John Wiley & Co.

Cooper, W. S. 1926. The fundamentals of vegetational change. Ecology, **7**: 391–413.

Cowles, H. C. 1899. The ecological relations of the vegetation of the sand dunes of Lake Michigan. Bot. Gaz., **27**: 95–391.

Davidson, V. S. 1932. The effect of seasonal variability upon animal species in a deciduous forest succession. Ecol. Monogr., **2**: 305–334.

Deevey, E. S. 1939. Studies on Connecticut lake sediments: I. A postglacial climatic chronology for southern New England. Amer. Jour. Sci., **237**: 691–724.

——. 1941. Limnological studies in Connecticut: VI. The quantity and composition of the bottom fauna. Ecol. Monogr., **11**: 413–455.

——. 1942. Studies on Connecticut lake sediments: III. The biostratonomy of Linsley Pond. Amer. Jour. Sci., **240**: 233–264, 313–338.

Elton, C. 1927. Animal Ecology. N. Y. Macmillan Co.

Francé, R. H. 1913. Das Edaphon, Untersuchungen zur Oekologie der bodenbewohnenden Mikroorganismen. Deutsch. Mikrolog. Gesellsch., Arbeit. aus d. Biol. Inst., No. 2. Munich.

Friederichs, K. 1930. Die Grundfragen und Gesetzmässigkeiten der land- und forstwirtschaftlichen Zoologie. 2 vols. Berlin. Verschlag. Paul Parey.

Groschopf, P. 1936. Die postglaziale Entwicklung des Grosser Plöner Sees in Ostholstein auf Grund pollenanalytischer Sedimentuntersuchungen. Arch. Hydrobiol., **30**: 1–84.

Heinrich, K. 1934. Atmung und Assimilation im freien Wasser. Internat. Rev. ges. Hydrobiol. u. Hydrogr., **30**: 387–410.

Hentschel, E. 1933–1936. Allgemeine Biologie des Südatlantischen Ozeans. Wiss. Ergebn. Deutsch. Atlant. Exped. a. d. Forschungs- u. Vermessungsschiff "Meteor" 1925–1927. Bd. XI.

Hicks, P. A. 1934. Interaction of factors in the growth of *Lemna:* V. Some preliminary observations upon the interaction of temperature and light on the growth of *Lemna*. Ann. Bot., **48**: 515–523.

Hile, R. 1941. Age and growth of the rock bass *Ambloplites rupestris* (Rafinesque) in Nebish Lake, Wisconsin. Trans. Wisconsin Acad. Sci., Arts, Lett., **33**: 189–337.

Hutchinson, G. E. 1941. Limnological studies in Connecticut: IV. Mechanism of intermediary metabolism in stratified lakes. Ecol. Monogr., **11**: 21–60.

——. 1942. Recent Advances in Limnology (*in manuscript*).

—— and R. L. Lindeman. 1941. Biological efficiency in succession (Abstract). Bull. Ecol. Soc. Amer., **22**: 44.

—— and Anne Wollack. 1940. Studies on Connecticut lake sediments: II. Chemical analyses of a core from Linsley Pond, North Branford. Amer. Jour. Sci., **238**: 493–517.

Huxley, J. S. 1932. Problems of Relative Growth. N. Y. Dial Press.

Ivlev, V. S. 1939a. Transformation of energy by aquatic animals. Internat. Rev. ges. Hydrobiol. u. Hydrogr., **38**: 449–458.

——. 1939b. Balance of energy in carps. Zool. Zhurn. Moscow, **18**: 303–316.

Jacot, A. P. 1940. The fauna of the soil. Quart. Rev. Biol., **15**: 28–58.

Jenkin, B. M., C. H. Mortimer, and W. Pennington. 1941. The study of lake deposits.
Nature, 147: 496–500.

Joffe, J. S. 1936. Pedology. New Brunswick,
New Jersey. Rutgers Univ. Press.

Juday, C. 1922. Quantitative studies of the
bottom fauna in the deeper waters of Lake
Mendota. Trans. Wisconsin Acad. Sci.,
Arts, Lett., 20: 461–493.

——. 1940. .The annual energy budget of an
inland lake. Ecology, 21: 438–450.

Krogh, A. 1941. The Comparative Physiology
of Respiratory Mechanisms. Philadelphia.
Univ. Pennsylvania Press.

Lindeman, R. L. 1941a. The developmental
history of Cedar Creek Bog, Minnesota.
Amer. Midl. Nat., 25: 101–112.

——. 1941b. Seasonal food-cycle dynamics in
a senescent lake. Amer. Midl. Nat., 26:
636–673.

Lohmann, H. 1912. Untersuchungen über das
Pflanzen- und Tierleben der Hochsee, zu-
gleich ein Bericht über die biologischen Ar-
beiten auf der Fahrt der "Deutschland" von
Bremerhaven nach Buenos Aires. Veröffentl.
d. Inst. f. Meereskunde, N.F., A. Geogr.-
naturwissen. Reihe, Heft 1, 92 pp.

Lundegårdh, H. 1924. Kreislauf der Kohlen-
säure in der Natur. Jena. G. Fischer.

Meschkat, A. 1937. Abwasserbiologische Untersuchungen in einem Buhnenfeld unterhalb
Hamburgs. Arch. Hydrobiol., 31: 399–432.

Moore, W. G. 1941. Studies on the feeding
habits of fishes. Ecology, 22: 91–95.

Needham, J. 1931. Chemical Embryology.
3 vols. N. Y. Cambridge University Press.

Noddack, W. 1937. Der Kohlenstoff im Haus-
halt der Natur. Zeitschr. angew. Chemie,
50: 505–510.

Park, Thomas. 1941. The laboratory population as a test of a comprehensive ecological
system. Quart. Rev. Biol., 16: 274–293,
440–461.

Rigg, G. B. 1940. Comparisons of the development of some Sphagnum bogs of the Atlantic coast, the interior, and the Pacific
coast. Amer. Jour. Bot., 27: 1–14.

Riley, G. A. 1940. Limnological studies in
Connecticut. III. The plankton of Linsley
Pond. Ecol. Monogr., 10: 279–306.

——. 1941. Plankton studies. III. Long Island Sound. Bull. Bingham Oceanogr. Coll.
7 (3): 1–93.

Ruttner, F. 1937. Limnologische Studien an
einigen Seen der Ostalpen. Arch. Hydrobiol., 32: 167–319.

Smith-Davidson, Vera. 1932. The effect of
seasonal variability upon animal species in a
deciduous forest succession. Ecol. Monogr.,
2: 305–334.

Spoehr, H. A. 1926. Photosynthesis. N. Y.
Chemical Catalogue Co.

Swingle, H. S., and E. V. Smith. 1940. Experiments on the stocking of fish ponds.
Trans. North Amer. Wildlife Conf., 5: 267–
276.

Tansley, A. G. 1935. The use and abuse of
vegetational concepts and terms. Ecology,
16: 284–307.

Thienemann, A. 1918. Lebensgemeinschaft
und Lebensraum. Naturw. Wochenschrift,
N.F., 17: 282–290, 297–303.

——. 1926. Der Nahrungskreislauf im Wasser. Verh. deutsch. Zool. Ges., 31: 29–79.
(or) Zool. Anz. Suppl., 2: 29–79.

——. 1939. Grundzüge einen allgemeinen
Oekologie. Arch. Hydrobiol., 35: 267–285.

Thompson, D. H. 1941. The fish production
of inland lakes and streams. Symposium
on Hydrobiology, pp. 206–217. Madison.
Univ. Wisconsin Press.

Tutin, T. G. 1941. The hydrosere and current
concepts of the climax. Jour. Ecol. 29: 268–
279.

Vernadsky, V. I. 1929. La biosphere. Paris.
Librairie Felix Alcan.

——. 1939. On some fundamental problems
of biogeochemistry. Trav. Lab. Biogeochem.
Acad. Sci. URSS, 5: 5–17.

Waksman, S. A. 1941. Aquatic bacteria in
relation to the cycle of organic matter in
lakes. Symposium on Hydrobiology, pp.
86–105. Madison. Univ. Wisconsin Press.

Welch, P. S. 1935. Limnology. N. Y. Mc-
Graw-Hill Co.

Wesenberg-Lund, C. 1912. Über einige eigen-
tümliche Temperaturverhältnisse in der Lito-
ralregion. . . . Internat. Rev. ges. Hydrobiol. u. Hydrogr., 5: 287–316.

Williams, E. C. 1941. An ecological study of
the floor fauna of the Panama rain forest.
Bull. Chicago Acad. Sci., 6: 63–124.

Wimpenny, R. S. 1941. Organic polarity:
some ecological and physiological aspects.
Quart. Rev. Biol., 16: 389–425.

ZoBell, C. E., and C. B. Feltham. 1938. Bacteria as food for certain marine invertebrates.
Jour. Marine Research, 1: 312–327.

ADDENDUM

While this, his sixth completed paper, was in
the press, Raymond Lindeman died after a long
illness on 29 June, 1942, in his twenty-seventh
year. While his loss is grievous to all who knew
him, it is more fitting here to dwell on the
achievements of his brief working life. The
present paper represents a synthesis of Lindeman's work on the modern ecology and past
history of a small senescent lake in Minnesota.
In studying this locality he came to realize, as
others before him had done, that the most profitable method of analysis lay in reduction of all
the interrelated biological events to energetic
terms. The attempt to do this led him far

beyond the immediate problem in hand, and in stating his conclusions he felt that he was providing a program for further studies. Knowing that one man's life at best is too short for intensive studies of more than a few localities, and before the manuscript was completed, that he might never return again to the field, he wanted others to think in the same terms as he had found so stimulating, and for them to collect material that would confirm, extend, or correct his theoretical conclusions. The present contribution does far more than this, as here for the first time, we have the interrelated dynamics of a biocoenosis presented in a form that is amenable to a productive abstract analysis. The question, for instance, arises, "What determines the length of a food chain?"; the answer given is admittedly imperfect, but it is far more important to have

seen that there is a real problem of this kind to be solved. That the final statement of the structure of a biocoenosis consists of pairs of numbers, one an integer determining the level, one a fraction determining the efficiency, may even give some hint of an undiscovered type of mathematical treatment of biological communities. Though Lindeman's work on the ecology and history of Cedar Bog Lake is of more than local interest, and will, it is hoped, appear of even greater significance when the notes made in the last few months of his life can be coordinated and published, it is to the present paper that we must turn as the major contribution of one of the most creative and generous minds yet to devote itself to ecological science.

G. Evelyn Hutchinson.

Yale University.

Burt, W. H. 1943. Territoriality and home range concepts as applied to mammals. *Journal of Mammalogy* 24:346–352.

Territoriality and home ranges of animals have been recognized and have been important to biologists and naturalists for years. Regrettably, these concepts were not addressed in much detail until the publication of Burt's paper on territoriality and home range brought them to the forefront of wildlife studies. Burt reviews the concepts applied to territoriality and home range and distinguishes between them. His ideas were formulated from field observations and studies. He concludes the paper by stating, "I think it will be evident that more critical studies in the behavior of wild animals are needed. We are now spending thousands of dollars each year in an attempt to manage some of our wild creatures, especially game species. How can we manage any species until we know its fundamental behavior patterns?"

As society strives to maintain and enhance habitats for wild animals, biologists and politicians should rely on a sound grasp of the biology of species. Burt's work initiated half a century's worth of studies on territoriality and home ranges, which have provided a much deeper understanding of wildlife habitats and home ranges. As more knowledge was gained, biologists further refined their techniques to better comprehend these important concepts. Radio- and satellite telemetry have certainly enabled wildlife biologists to more fully understand the concepts that Burt so eloquently discussed in 1943.

RELATED READING

Don, B. A., and K. Rennolls. 1983. A home range model incorporating biological attraction points. Journal of Animal Ecology 52:69–81.

Dunn, J. E., and P. S. Gipson. 1977. Analysis of radio telemetry data in studies of home range. Biometrics 33:85–101.

Hornocker, M. G. 1969. Winter territoriality in mountain lions. Journal of Wildlife Management 33:457–464.

Howard, H. E. 1920. Territory in bird life. John Murray, London, UK.

Nice, M. M. 1941. The role of territory in bird life. American Midland Naturalist 26:441–487.

Samuel, M. D., and E. O. Garton. 1985. Home range: a weighted normal estimate and tests of underlying assumptions. Journal of Wildlife Management 49:513–519.

Swihart, R. K., and N. A. Slade. 1985. Testing for independence of observations in animal movements. Ecology 66:1176–1184.

Tinbergen, N. 1957. The functions of territory. Bird Study 4:14–27.

Worton, B. J. 1989. Kernel methods for estimating the utilization distribution in home-range studies. Ecology 70:164–168.

Reprinted by Permission of Allen Press, Inc.

TERRITORIALITY AND HOME RANGE CONCEPTS AS APPLIED TO MAMMALS

By William Henry Burt

TERRITORIALITY

The behavioristic trait manifested by a display of property ownership—a defense of certain positions or things—reaches its highest development in the human species.   Man considers it his inherent right to own property either as an individual or as a member of a society or both.   Further, he is ever ready to protect that property against aggressors, even to the extent at times of sacrificing his own life if necessary.   That this behavioristic pattern is not peculiar to man, but is a fundamental characteristic of animals in general, has been shown for diverse animal groups.   (For an excellent historical account and summary on territoriality, with fairly complete bibliography, the reader is referred to a paper by Mrs. Nice, 1941).   It does not necessarily follow that this trait is found in all animals, nor that it is developed to the same degree in those that are known to possess it, but its wide distribution among the vertebrates (see Evans, L. T., 1938, for reptiles), and even in some of the invertebrates, lends support to the theory that it is a basic characteristic of animals and that the potentialities are there whether the particular animal in question displays the characteristic.   Heape (1931, p. 74) went so far as to say:

"Thus, although the matter is often an intricate one, and the rights of territory somewhat involved, there can, I think, be no question that territorial rights are established rights amongst the majority of species of animals.   There can be no doubt that the desire for acquisition of a definite territorial area, the determination to hold it by fighting if necessary, and the recognition of individual as well as tribal territorial rights by others, are dominant characteristics in all animals.   In fact, it may be held that the recognition of territorial rights, one of the most significant attributes of civilization, was not evolved by man, but has ever been an inherent factor in the life history of all animals."

Undoubtedly significant is the fact that the more we study the detailed behavior of animals, the larger is the list of kinds known to display some sort of territoriality.   There have been many definitions to describe the territory of different animals under varying circumstances.   The best and simplest of these, in my mind, is by Noble (1939); "territory is any defended area."   Noble's definition may be modified to fit any special case, yet it is all-inclusive and to the point.   Territory should not be confused with "home range"—an entirely different concept that will be treated more fully later.

The territoriality concept is not a new one (see Nice, 1941).   It has been only in the last twenty years, however, that it has been developed and brought to the front as an important biological phenomenon in the lower animals. Howard's book "Territory in Bird Life" (1920) stimulated a large group of workers, chiefly in the field of ornithology, and there has hardly been a bird life-history study since that has not touched on this phase of their behavior.

In the field of mammals, much less critical work has been done, but many of the older naturalists certainly were aware of this behavior pattern even though they did not speak of it in modern terms. Hearne (1795) apparently was thinking of property rights (territoriality) when he wrote about the beaver as

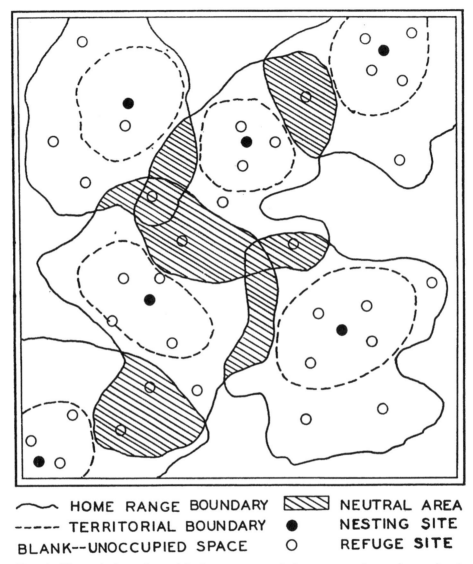

```
⌒⌒ HOME RANGE BOUNDARY   ▨▨ NEUTRAL AREA
----- TERRITORIAL BOUNDARY   ● NESTING SITE
BLANK--UNOCCUPIED SPACE   ○ REFUGE SITE
```

Fig. 1. Theoretical quadrat with six occupants of the same species and sex, showing territory and home range concepts as presented in text.

follows: "I have seen a large beaver house built in a small island, that had near a dozen houses under one roof; and, two or three of these only excepted, none of them had any communication with each other but by water. As there were beavers enough to inhabit each apartment, it is more than probable that each family knew its own, and always entered at their own door without having any further connection with their neighbors than a friendly intercourse" (in Morgan, 1868, pp. 308–309). Morgan (*op. cit.*, pp. 134–135), also writing of the beaver, made the following observation; "a beaver family consists of a male and female, and their offspring of the first and second years, or, more properly, under two years old. . . . When the first litter attains the age of two years, and in the third summer after their birth, they are sent out from the parent lodge." Morgan's observation was later confirmed by Bradt (1938). The works of Seton are replete with instances in the lives of different animals that indicate territorial behavior. In the introduction to his "Lives" Seton (1909) states "In the idea of a home region is the germ of territorial rights." Heape (1931) devotes an entire chapter to "territory." Although he uses the term more loosely than I propose to, (he includes home ranges of individuals and feeding ranges of tribes or colonies of animals), he carries through his work the idea of defense of an area either by an individual or a group of individuals. Not only this, but he draws heavily on the literature in various fields to support his thesis. Although the evidence set forth by Seton, Heape, and other early naturalists is of a general nature, mostly garnered from reports by others, it cannot be brushed aside in a casual manner. The old time naturalists were good observers, and, even though their techniques were not as refined as those of present day biologists, there is much truth in what they wrote.

A few fairly recent published observations on specific mammals serve to strengthen many of the general statements made by earlier workers. In speaking of the red squirrel (*Tamiasciurus*), Klugh (1927, p. 28) writes; "The sense of ownership seems to be well developed. Both of the squirrels which have made the maple in my garden their headquarters apparently regarded this tree as their private property, and drove away other squirrels which came into it. It is quite likely that in this case it was not the tree, but the stores that were arranged about it, which they were defending." Clarke (1939) made similar observations on the same species. In raising wild mice of the genus *Peromyscus* in the laboratory, Dice (1929, p. 124) found that "when mice are placed together for mating or to conserve cage space it sometimes happens that fighting takes place, especially at first, and sometimes a mouse is killed. . . . Nearly always the mouse at home in the cage will attack the presumed intruder." Further on he states, "However, when the young are first born, the male, or any other female in the same cage, is driven out of the nest by the mother, who fiercely protects her young." Similarly, Grange (1932, pp. 4–5) noted that snowshoe hares (*Lepus americanus*) in captivity "showed a definite partiality for certain spots and corners to which they became accustomed" and that "the female would not allow the male in her territory (cage) during late pregnancy and the males themselves were quarrelsome during the breeding season."

Errington (1939) has found what he terms "intraspecific strife" in wild muskrats (*Ondatra*). Much fighting takes place when marshes become overcrowded, especially in fall and winter during readjustment of populations. "But when invader meets resident in the tunnel system of one of [the] last lodges to be used in a dry marsh, conflict may be indeed savage." Gordon (1936) observed definite territories in the western red squirrels (*Tamiasciurus fremonti* and *T. douglasii*) during their food gathering activities. He also performed a neat experiment with marked golden mantled squirrels (*Citellus lateralis chysodeirus*) by placing an abundance of food at the home of a female. This food supply attracted others of the same species. To quote Gordon: "she did her best to drive away the others. Some of her sallies were only short, but others were long and tortuous. There were rather definite limits, usually not more than 100 feet from the pile, beyond which she would not extend her pursuit. In spite of the vigor and the number of her chases (one day she made nearly 60 in about 6 hours) she never succeeded in keeping the other animals away." This individual was overpowered by numbers, but, nevertheless, she was using all her strength to defend her own log pile. To my knowledge, this is the best observation to have been published on territorial behavior in mammals. I have observed a similar situation (Burt, 1940, p. 45) in the eastern chipmunk (*Tamias*). An old female was watched fairly closely during two summers. Having marked her, I was certain of her identity. "Although other chipmunks often invaded her territory, she invariably drove them away [if she happened to be present at the time]. Her protected area was about fifty yards in radius; beyond this fifty-yard limit around her nesting site she was not concerned. Her foraging range (*i.e.*, home range) was considerably greater than the protected area (territory) and occasionally extended 100 or more yards from her nest site." From live trapping experiments, plotting the positions of capture of individuals on a map of the area covered, I interpreted (*op. cit.*, p. 28) the results to mean that there was territorial behavior in the white-footed mouse (*Peromyscus leucopus*), a nocturnal form. When the ranges of the various individuals were plotted on a map, I found that "the area of each of the breeding females is separate—that although areas sometimes adjoin one another, they seldom overlap." Carpenter (1942) writes thus: "The organized groups of every type of monkey or ape which has been adequately observed in its native habitat, have been found to possess territories and to defend these ranges from all other groups of the same species." In reporting on his work on the meadow vole (*Microtus pennsylvanicus*), Blair (1940, pp. 154–155) made the statement "It seems evident that there is some factor that tends to make the females occupy ranges that are in part exclusive; . . . . Possibly there is an antagonism between the females, particularly during the breeding season, but the available evidence does not indicate to me that they have definite territories which they defend against all trespassers. It seems highly probable that most mammalian females attempt to drive away intruders from the close vicinity of their nests containing young, *but this does not constitute territoriality in the sense that the term has been used* by Howard (1920), Nice (1937), and others *in reference to the breeding territories of birds.*" (Ital. mine.) To quote Howard (1920, pp. 192–193): "But the Guillemot is generally surrounded by other Guillemots, and the birds are often so densely packed along the ledges that there is scarcely standing room, so it seems, for all of them. Nevertheless the isolation of the individual is, in a sense, just as complete as that of the individual Bunting, for each one is just as vigilant in resisting intrusion upon its few square feet as the Bunting is in guarding its many square yards, so that the evidence seems to show that that part of the inherited nature which is the basis of the territory is much the same in both species." Blair, in a later paper (1942, p. 31), writing of *Peromyscus maniculatus gracilis*, states: "The calculated home ranges of all sex and age classes broadly overlapped one another. Thus there was no occupation of exclusive home ranges by breeding females. . . . That individual woodland deer-mice are highly tolerant of one another is indicated by the foregoing discussion of overlapping home ranges of all sex and age classes." Reporting on an extensive field study of the opossum, Lay (1942, p. 149) states that "The ranges of individual opossums overlapped so frequently that no discernible tendency towards establishment of individual territories could be detected. On the contrary, tracks rarely showed that two or more opossums traveled together." It seems quite evident that both Blair and Lay are considering the home range as synonymous with the territory ·when in fact they are two quite distinct concepts. Further, there is no concrete evidence in either of the above papers for or against territoriality in the species they studied. It is to be expected that the territory of each and every individual will be trespassed sooner or later regardless of how vigilant the occupant of that territory might be.

It is not intended here to give a complete list of works on territorial behavior. The bibliographies in the works cited above lead to a great mass of literature on the subject. The point I wish to emphasice is that nearly all who have critically studied the behavior of wild mammals have found this behavioristic trait inherent in the species with which they worked. Also, it should be stressed, there are two fundamental types of territoriality in mammals—one concerns breeding and rearing of young, the other food and shelter. These two may be further subdivided to fit special cases. Mrs. Nice (1941) gives six major types of territories for birds. Our knowledge of territoriality in mammals is yet too limited, it seems to me, to build an elaborate classification of types. Some day we may catch up with the ornithologists.

HOME RANGE

The home range concept is, in my opinion, entirely different from, although associated with, the territoriality concept. The two terms have been used so loosely, as synonyms in many instances, that I propose to dwell briefly on them here. My latest Webster's dictionary (published in 1938), although satisfactory in most respects, does not list "home range," so I find no help there. Seton (1909) used the term extensively in his "Lives" where he explains it as follows: "No wild animal roams at random over the country: each has a home region, even if it has not an actual home. The size of this home region corresponds somewhat with the size of the animal. Flesh-eaters as a class have a larger home region than herb-eaters." I believe Seton was thinking of the adult animal when he wrote the above. We know that young adolescent animals often do a bit of wandering in search of a home region. During this time they do not have a home, nor, as I consider it, a home range. It is only after they establish themselves, normally for the remainder of their lives, unless disturbed, that one can rightfully speak of the home range. Even then I would restrict the home range to that area traversed by the individual in its normal activities of food gathering, mating, and caring for young. Occasional sallies outside the area, perhaps exploratory in nature, should not be considered as in part of the home range. The home range need not cover the same area during the life of the individual. Often animals will move from one area to another, thereby abandoning the old home range and setting up a new one. Migratory animals have different home ranges in summer and winter—the migratory route is not considered part of the home range of the animal. The size of the home range may vary with sex, possibly age, and season. Population density also may influence the size of the home range and cause it to coincide more closely with the size of the territory. Home ranges of different individuals may, and do, overlap. This area of overlap is neutral range and does not constitute part of the more restricted territory of animals possessing this attribute. Home ranges are rarely, if ever, in convenient geometric designs. Many home ranges probably are somewhat ameboid in outline, and to connect the outlying points gives a false impression of the actual area covered. Not only that, it may indicate a larger range than really exists. A calculated home range based on trapping records, therefore, is no more than a convenient index to size. Overlapping of home ranges, based on these calculated areas, thus may at times be exaggerated. From trapping records alone, territory may be indicated, if concentrations of points of capture segregate out, but it cannot be demonstrated without question. If the occupant of an area is in a trap, it is not in a position to defend that area. It is only by direct observation that one can be absolutely certain of territoriality.

Home range then is the area, usually around a home site, over which the animal normally travels in search of food. Territory is the protected part of the home range, be it the entire home range or only the nest. Every kind of mammal may be said to have a home range, stationary or shifting. Only those that protect some part of the home range, by fighting or agressive gestures, from others of their kind, during some phase of their lives, may be said to have territories.

SIGNIFICANCE OF BEHAVIORISTIC STUDIES

I think it will be evident that more critical studies in the behavior of wild animals are needed. We are now spending thousands of dollars each year in an attempt to manage some of our wild creatures, especially game species. How can we manage any species until we know its fundamental behavior pattern? What good is there in releasing a thousand animals in an area large enough to support but fifty? Each animal must have so much living room in addition to other essentials of life. The amount of living room may vary somewhat, but for a given species it probably is within certain definable limits. This has all been said before by eminent students of wildlife, but many of us learn only by repetition. May this serve to drive the point home once more.

LITERATURE CITED

BLAIR, W. F. 1940. Home ranges and populations of the meadow vole in southern Michigan. Jour. Wildlife Management, vol. 4, pp. 149–161, 1 fig.

———— 1942. Size of home range and notes on the life history of the woodland deermouse and eastern chipmunk in northern Michigan. Jour. Mamm., vol. 23, pp. 27–36, 1 fig.

BRADT, G. W. 1938. A study of beaver colonies in Michigan. Jour. Mamm., vol. 19, pp. 139–162.

BURT, W. H. 1940. Territorial behavior and populations of some small mammals in southern Michigan. Miscl. Publ. Mus. Zool. Univ. Michigan, no. 45, pp. 1–58, 2 pls., 8 figs., 2 maps.

CARPENTER, C. R. 1942. Societies of monkeys and apes. Biological Symposia, Lancaster: The Jaques Cattell Press, vol. 8, pp. 177–204.

CLARKE, C. H. D. 1939. Some notes on hoarding and territorial behavior of the red squirrel *Sciurus hudsonicus* (Erxleben). Canadian Field Nat., vol. 53, no. 3, pp. 42–43.

DICE, L. R. 1929. A new laboratory cage for small mammals, with notes on methods of rearing Peromyscus. Jour. Mamm., vol. 10, pp. 116–124, 2 figs.

ERRINGTON, P. L. 1939. Reactions of muskrat populations to drought. Ecology, vol. 20, pp. 168–186.

EVANS, L. T. 1938. Cuban field studies on territoriality of the lizard, Anolis sagrei. Jour. Comp. Psych., vol. 25, pp. 97–125, 10 figs.

GORDON, K. 1936. Territorial behavior and social dominance among Sciuridae. Jour. Mamm., vol. 17, pp. 171–172.

GRANGE, W. B. 1932. Observations on the snowshoe hare, Lepus americanus phaeonotus Allen. Jour. Mamm., vol. 13, pp. 1–19, 2 pls.

HEAPE, W. 1931. Emigration, migration and nomadism. Cambridge: W. Heffer and Son Ltd., pp. xii + 369.

HEARNE, S. 1795. A journey from Prince of Wale's fort in Hudson's Bay, to the Northern Ocean. London: A. Strahan and T. Cadell, pp. xliv + 458, illustr.

HOWARD, H. E. 1920. Territory in bird life. London: John Murray, pp. xii + 308, illustr.

KLUGH, A. B. 1927. Ecology of the red squirrel. Jour. Mamm., vol. 8, pp. 1–32, 5 pls.

LAY, D. W. 1942. Ecology of the opossum in eastern Texas. Jour. Mamm., vol. 23, pp. 147–159, 3 figs.

MORGAN, L. H. 1868. The American beaver and his works. Philadelphia: J. B. Lippincott and Co., pp. xv + 330, illustr.

NICE, M. M. 1941. The role of territory in bird life. Amer. Midl. Nat., vol. 26, pp. 441–487.

NOBLE, G. K. 1939. The role of dominance in the life of birds. Auk, vol. 56, pp. 263–273.

SETON, E. T. 1909. Life-histories of northern animals. An account of the mammals of Manitoba. New York City: Charles Scribner's Sons, vol. 1, pp. xxx + 673, illustr., vol. 2, pp. xii + 677–1267, illustr.

———— 1929. Lives of game animals, Doubleday, Doran and Co., Inc., 4 vols., illustr.

Museum of Zoology, Ann Arbor, Michigan.

Cole, L. C. 1954. The population consequences of life history phenomena. *The Quarterly Review of Biology* 29:103–137

In the past, human demography often served as the basis for understanding population dynamics of non-humans. The work of human demographers, mostly life insurance actuaries, created a collection of mathematical relationships and analytical tools used to evaluate dynamics of animal populations, including their survival, mortality, and fecundity rates; age structure relationships; and sex ratio. From these calculations, it became possible to establish rates of increase. Lamont C. Cole was the first to illustrate how natural history writers often ignored quantifying their data on life history characteristics to such an extent that comparisons between species and populations were difficult.

Cole's objective was to definitively evaluate and demonstrate the interrelation of wildlife population parameters and how they relate to species selection of specific strategies to deal with the vagaries of environments. Herein, Cole examines critical life history characteristics (e.g., reproduction, survival, population growth) and demonstrates the importance of expressing population growth numerically by algebraic and geometric expressions. "Having established this fact, it is shown that the exact computational methods lead to identical conclusions when condensed over the long time scale which is of interest in adaptational and evolutionary considerations." Thus, Cole examines and compares some life history patterns with simple equations and reports that the results have bearing on the possible adaptive value of genetically induced changes of life history features. For example, the age at which reproduction begins is one of the most influential characteristics of a species, but, regrettably, these data are often ignored or not recorded. This is one of the first papers encouraging field biologists to accurately collect life history data to better understand the dynamics and population growth of wild populations. Most of Cole's concepts are applicable to wildlife populations today.

RELATED READING

Ar, A., and Yom-Tov, Y. 1978. The evolution of parental care in birds. Evolution 32:655–669.
Cole, L. C. 1957. Sketches of general and comparative demography. Cold Spring Harbor Symposia on Quantitative Biology 22:1–15.
McCullough, D. R. 1990. Detecting density dependence: filtering the baby from the bathwater. Transactions of the North American Wildlife and Natural Resources Conference 55:534–543.
Partridge, L., and P. H. Harvey. 1988. The ecological context of life history phenomena. Science 241:1449–1455.
Stearns, S. C. 1976. Life-history tactics: a review of the ideas. The Quarterly Review of Biology 51:3–47.
Stearns, S. C. 1977. The evolution of life history traits: a critique of the theory and a review of the data. Annual Review of Ecology and Systematics 8:145–171.
White, G. C. 2000. Population viability analysis: data requirements and essential analyses. Pages 288–331 *in* L. Boitani, and T. K. Fuller, editors. Research techniques in animal ecology: controversies and consequences. Columbia University Press, New York, New York, USA.

VOL. 29, NO. 2   June, 1954

THE QUARTERLY REVIEW
of BIOLOGY

THE POPULATION CONSEQUENCES OF LIFE HISTORY PHENOMENA

By LAMONT C. COLE

Department of Zoology, Cornell University

PREFACE

FEW branches of biology have attracted more analytical mathematical treatment than has the study of populations. Despite this, one may read in the most complete treatise of ecology yet published (Allee, et al., 1949, p. 271) that "theoretical population ecology has not advanced to a great degree in terms of its impact on ecological thinking." This unfortunate gap between the biologists and the mathematicians has elicited comments which need not be repeated in detail here (Allee, 1934; Gause, 1934; Allee et al., 1949, p. 386). The neglect of the analytical methods by biologists may be attributed in part to the tendency of writers in this field to concentrate on the analysis of human populations and in part to skepticism about the mathematical methods of analysis. Early analyses of population growth (Verhulst, 1838, 1845; Pearl and Reed, 1920) employed human populations as examples, although it is clear from other publications (e.g., Pearl and Miner, 1935; Pearl, 1937) that comparative and general population studies were the principal interest of some of these students. Similarly, the pioneer works of Lotka (1907b, 1910, 1925) were very general in conception but made their greatest impact in the field of demography (Dublin and Lotka, 1925; Dublin, Lotka, and Spiegelman, 1949). The skepticism expressed by biologists toward theoretical studies has ranged from antagonism (Salt, 1936) to ap-

proval given with the warning that "... for the sake of brevity and to avoid cumbersome expressions, variables are omitted and assumptions made in the mathematical analyses which are not justified by the biological data" (Allee, 1934). It may be unfortunate that warnings about mathematical oversimplification are especially pertinent in connection with the study of interactions between species (Ross, 1911; Lotka, 1920, 1925; Volterra, 1927, 1931; Nicholson and Bailey, 1935; Thompson, 1939), which is just that portion of the subject which has remained most closely associated with general ecology. Hence we have a situation in which the analytical theories which are recognized by ecologists deal with complex phenomena and are susceptible to cogent criticisms (e.g., Smith, 1952) while the simpler analysis of the ways in which differences between the life histories of species may result in different characteristics of their populations has remained relatively unexplored. It is the purpose of the present paper to consider some parts of this neglected branch of ecology which has been called "biodemography" by Hutchinson (1948).

It is possible, but often impracticable, to compute exactly the characteristics of the hypothetical future population obtained by assuming an unvarying pattern of the pertinent life history features which govern natality and mortality. It is often more practicable to employ approximate methods of computation of the type which have

103

aroused skepticism among biologists. It will be shown that the two approaches can be reconciled and that for many cases of ecological interest they lead to identical conclusions. Some of these conclusions reached by the writer have appeared surprising when first encountered, and they seem to give a new perspective to life history studies. They also suggest that pertinent bits of information are frequently ignored in life history studies simply because their importance is not generally recognized.

The total life history pattern of a species has meaning in terms of its ability to survive, and ecologists should attempt to interpret these meanings. The following sections are intended primarily to indicate some of the possibilities in this direction. The writer wishes to express his gratitude to Professor Howard B. Adelmann for a critical reading of the manuscript of this paper, for suggesting numerous ways of clarifying the text and improving terminology, and for translating from the Latin parts of the text from Fibonacci (1202). Thanks are also due to Professors Robert J. Walker and Mark Kac who have been consulted about technical mathematical questions raised by the writer while considering various phases of this subject.

INTRODUCTION

If it is to survive, every species must possess reproductive capacities sufficient to replace the existing species population by the time this population has disappeared. It is obvious that the ability of the ancestors of existing species to replace themselves has been sufficient to overcome all environmental exigencies which have been encountered and, therefore, that the physiological, morphological, and behavioral adaptations that enable offspring to be produced and to survive in sufficient numbers to insure the persistence of a species are of fundamental ecological interest.

On the other hand, it is conceivable that reproductive capacity might become so great as to be detrimental to a species. The many deleterious effects of overcrowding are well known. It also seems obvious that a species which diverts too large a proportion of its available energies into unnecessary, and therefore wasteful, reproduction would be at a disadvantage in competition with other species.

In this paper it will be regarded as axiomatic that the reproductive potentials of existing species are related to their requirements for survival; that any life history features affecting reproductive potential are subject to natural selection; and that such features observed in existing species should be considered adaptations, just as purely morphological or behavioral patterns are commonly so considered.

Some of the more striking life history phenomena have long been recognized as adaptations to special requirements. The great fecundity rather generally found in parasites and in many marine organisms is commonly regarded as an adaptation insuring the maintenance of a population under conditions where the probability is low that any particular individual will establish itself and reproduce successfully. Again, parthenogenesis obviously favors the rapid growth of a population because every member of a population reproducing in this fashion can be a reproductive female. In turning seasonally to parthenogenesis, organisms like cladocerans and aphids are responding in a highly adaptive way during a limited period of time when the environmental resources are sufficient to support a large population. Parthenogenesis, hermaphroditism, and purely asexual reproduction may clearly offer some advantages under conditions that restrict the probability of contacts between the sexes. Protandry, as exhibited, for example, by some marine molluscs, and various related phenomena where population density affects the sex ratio (Allee et al., 1949, p. 409) may be considered as compromise devices providing the advantages of biparental inheritance while maintaining an unbalanced sex ratio which makes most of the environmental resources available to reproductive females.

Reproductive potentialities may be related to the success of a species in still other ways. It was an essential part of Darwin's thesis that the production of excess offspring provided a field of heritable variations upon which environmental conditions could operate to select the most favorable combinations. A high degree of fecundity may also aid the dispersal of species. An extreme example of this is afforded by the ground pine, *Lycopodium* (Humphreys, 1929), whose light windborne spores may be scattered literally over the whole face of the earth and so make it likely that all favorable habitats will come to be occupied. Another adaptational interpretation of the overproduction of offspring postulates that the excessive production of young fish which are frequently cannibalistic is a form of maternal provisioning, the majority of the young serving merely as food for the few that ultimately mature.

Many additional examples of life history phenomena that have been regarded as adaptive could be cited. Here, however, we wish rather to call attention to the striking fact that in modern ecological literature there have been relatively few attempts to evaluate quantitatively the importance of specific features of life histories. The apparent mathematical complexity of the general problem is undoubtedly partly responsible for this. When the biologist attempts to compute from observed life history data the numbers of organisms of a particular type that can be produced in a given interval of time he may find it necessary to make assumptions which biologists in general would hesitate to accept. And even with these simplifying assumptions the computations may become so tedious as to make the labor involved seem unjustifiable in view of the seemingly academic interest of the result. In particular, such computations involve biological parameters which are not necessarily fixed characteristics of the species and which are not ordinarily expressible in convenient mathematical form. It is necessary to know the way in which the chance of dying (or of surviving) and the reproductive activities vary during the life span of an individual. These quantities are nicely summed up by the familiar life-table function, survivorship (l_x), which is defined as the probability of surviving from birth to some age x, and by the age-specific birth rate (b_x), which is defined as the mean number of offspring produced during the interval of age from age x to age $x + 1$. The biologist immediately recognizes that these quantities vary with environmental conditions and that he cannot expect to obtain a realistic result if he must assume, for example, that the probability of surviving a day, a week, or a month, is the same for individuals born in the autumn as for those born in the spring. He also recognizes that the population consists of discrete units and that offspring are produced in batches (here called litters whether in plants or animals) rather than continuously; hence he necessarily regards with suspicion any formulation of the problem in terms of differential equations where these considerations are apparently ignored.

Actually a tremendous variability is observed in life history phenomena which could affect the growth of populations. Some organisms are semelparous, that is to say, they reproduce only once

in a lifetime and in these semelparous forms reproduction may occur at the age of only 20 minutes in certain bacteria (Molisch, 1938), of a few hours in many protozoa, or of a few weeks or months in many insects. Many semelparous plants and animals are annuals; in other semelparous organisms reproduction may occur only after a number of years of maturation, for example, two or more years in dobson flies and Pacific salmon, and many years in "century plants" (*Agave*) and the periodic cicada or "17-year locust" (*Magicicada septendecim*). The number of potential offspring produced by semelparous individuals varies from two in the case of binary fission to the literally trillions (2×10^{13}) of spores produced by a large puffball (*Calvatia gigantea*).

In iteroparous forms, that is to say, those which reproduce more than once in a lifetime, the period of maturation preceding the first production of prospective offspring may vary from as little as a few days in small crustaceans to over a century in the giant sequoia (U. S. Forest Service, 1948), and practically any intermediate value may be encountered. After the first reproduction has occurred in iteroparous organisms it may be repeated at various intervals—for example, daily (as in some tapeworms), semiannually, annually, biennially, or irregularly (as in man). As in semelparous organisms, the litter size of iteroparous forms may also vary greatly; here it may vary from one (as is usual, for example, in man, whales, bovines, and horses) to many thousands (as in various fishes, tapeworms, or trees). The litter size may be constant in a species, vary about some average, or change systematically with the age of the parent, in which case it may increase to some maximum (as in tapeworms) or climb to a maximum and then decline as in some cladocerans (Banta et al., 1939; Frank, 1952). Furthermore, individuals may live on after their reproduction has ceased completely, and this post-reproductive period may amount to more than one-half of the normal life span (Allee et al., 1949, p. 285).

There is similar variability in the potential longevity of individual organisms. Man, various turtles, and trees may survive more than a century, while, on the other hand, the life span of many other species is concluded in hours or days. Innumerable intermediate values of course occur.

Additional sources of variation (such as biased sex ratios and the occurrence of asexual reproduction in developmental stages so as to result in the production of many offspring from one egg or spore) force the conclusion that the number of theoretical combinations of observed life history phenomena must greatly exceed the number of known species of organisms. And if all these phenomena have potential adaptive importance the interpretation of the possible merits of the particular combination of features exhibited by a species presents a problem of apparent great complexity.

The usual mathematical approach to the problem of potential population growth is straightforward. It is assumed that the growth of a population at any instant of time is proportional to the size of the population at that instant. If r is the factor of proportionality and P_x represents the population size at any x time this leads to the differential equation

$$\frac{dP}{dx} = rP \qquad (1)$$

which upon integration gives:

$$P_x = Ae^{rx} \qquad (1')$$

where A is a constant. This is an equation of continuous compound interest at the rate r or of a geometric progression where the ratio between the sizes of the populations in two consecutive time intervals, say years, is e^r.

While formulas (1) and (1') represent only the usual starting point for mathematical discussions of population growth, they already exhibit points about which there has been, and still is, a great deal of controversy. Explicit statements to the effect that human populations potentially increase by geometric progression can be traced back at least to Capt. John Graunt (1662), who estimated that a human population tends to double itself every 64 years (which would correspond to $r = .0108$ in formula 1). This belief in geometric progression as the form of potential population increase was endorsed by numerous students prior to the great controversy initiated by Malthus in 1798 (see review by Stangeland, 1904). Among these early writers we may here note only Linnaeus (1743), who considered the problem of geometric increase in the progeny of an annual plant, and Benjamin Franklin (1751), who estimated that the population of "America" could double at least every 20 years (corresponding to $r = .035$), and who clearly regarded the geometric nature of potential population increase as a general organic phenomenon.

The great controversy over growth in human populations which was initiated by the publication in 1798 of Malthus' *Essay on Population* engendered numerous arguments regarding geometric progression as the potential form of population growth. This controversy is still alive and in much its original form, with the "Neo-Malthusian" position maintaining that potential population growth is indeed in the form of a geometric progression, whereas the capacity of the environment to absorb population is necessarily limited, and with their opponents denying both the geometric progression and the finite capacity of the environment. Essentially the modern arguments against the Malthusian thesis, although not presented in modern concise form, are to be found in the treatise by Sadler (1830) which, whatever its shortcomings from the modern point of view, contains in places (especially in the appendix to Book IV) a very remarkable pre-Darwinian statement of such ecological phenomena as food chains, species interactions, and the balance of numbers between predators and prey.

The entire problem of potential population growth and its relationship to the resources of the environment is clearly one of the fundamental problems of ecology, but one which has never been adequately summarized in a way to reconcile the mathematical approaches, such as those of Lotka (1925), Volterra (1927), Kuczynski (1932, 1935), Kostitzin (1939), and Rhodes (1940), and the purely biological approaches which have concentrated on life history features such as longevity, fecundity, fertility, and sex ratios. In the present paper we will consider the mathematical form of potential population growth and certain subsidiary phenomena and the way in which these are related to particular life history phenomena. It is hoped that this will bring to attention some of the possible adaptive values of observed life history phenomena and will lead ecologists to a greater consideration of population problems which are essentially ecological. Life history features do in fact control potential population growth, as Sadler recognized, but the quantitative relationships have still been so insufficiently elucidated that even today ecologists generally do not attempt to answer queries such as the following, written by Sadler in 1830 (Vol. 2, p. 318):

"For instance, how would those who have the folly to suppose that population in this country advances too fast by one per cent., so operate, had they even their wish, as to diminish the number of marriages by one in one hundred, or otherwise contract the fecundity of the existing number by about one twenty-fifth part of a birth each, or calculate, upon their own erroneous suppositions, the term of that postponement of marriage on which they insist so much, so as to produce this exact effect? The very idea is, in each instance, absurd to the last degree."

FUNDAMENTAL CONSIDERATIONS

Sadler (1830) makes clear in numerous places his belief that ". . . the geometrical ratio of human increase is, nevertheless, in itself, an impossibility . . ." (Vol. 2, p. 68). However, when one examines his argument it is apparent that he is not actually opposing the principle that with fixed life history features populations would grow at compound interest, but rather is proposing the thesis that life history features change with population density, e.g., his fundamental thesis: "The prolificness of human beings, otherwise similarly circumstanced, varies inversely as their numbers" (Vol. 2, p. 252). Some of Sadler's computations assuming fixed ages at marriage and fecundity rates, in fact, lead to geometric progressions.

The modern conception of population growth regards the *potential* rate of increase as a more or less fixed species characteristic (cf. Chapman, 1935) governed by life history features; but it considers that this potential rate is ordinarily only partially realized, the "partial potential" characteristic of a particular situation being dependent on environmental conditions. Ecologists commonly associate this concept of "biotic potential" with the name of Chapman (1928, 1935), but actually the concept of populations as systems balanced between a potential ability to grow and an "environmental resistance" dates back at least to the Belgian statistician Quetelet (1835), who considered (p. 277) that potential population growth is a geometric progression, while the resistance to population growth (by analogy with a body falling through a viscous medium) varies as the square of the rate of growth. Only three years later Quetelet's student and colleague Verhulst (1838) set forth the thoroughly modern concept that potential population growth is a geometric progression corresponding to our formula (1'), and that the environmental resistance varies inversely with the unexploited opportunities for growth. By this conception, if K represents the capacity of the environment or the ultimate size which the population can attain, the resistance to population growth increases as $K - P$, the amount of space

remaining to be occupied, decreases. As the simplest case Verhulst considered that the resistance is related in a linear manner to the remaining opportunities for growth and thus derived the familiar logistic function as a representation of population growth (for discussion see Allee et al., 1949).

The modern mathematical formulation of population growth, as given, for example, by Rhodes (1940), proceeds by expressing the environmental resistance as some function of population size, $f(P)$, and writing a differential equation of the type

$$\frac{dP}{dx} = rPf(P). \tag{2}$$

By employing different functions for $f(P)$, any number of population growth laws may be derived and the mathematical connection between P and x determined, providing equation (2) can be integrated. Rhodes gives several examples of the procedure.

Formula (1'), the equation of the geometric progression representing population growth in an unlimited environment, represents the special case of formula (2) where the factor $f(P)$ is replaced by a constant, most conveniently by the constant value unity. By the foregoing interpretation it is clear that the constant r must be regarded as a quantity of fundamental ecological significance. It is to be interpreted as the rate of true compound interest at which a population would grow if nothing impeded its growth and if the age-specific birth and death rates were to remain constant.

Quite recently a number of ecologists have recognized the importance of a knowledge of the value of r for non-human populations and have computed its value for various species by employing empirical values of age-specific birth rates and survivorship (Leslie and Ranson, 1940; Birch, 1948; Leslie and Park, 1949; Mendes, 1949; Evans and Smith, 1952). While Chapman's term "biotic potential" would seem to have ecological merit as the name for this parameter r it has been variously called by Lotka the "true," the "incipient," the "inherent," and the "intrinsic" rate of increase, and by Fisher (1930) the "Malthusian parameter" of population increase. Probably for the sake of stabilizing nomenclature it is advisable to follow the majority of recent writers and refer to r as "the intrinsic rate of natural increase."

In the works of Dublin and Lotka (1925),

Kuczynski (1932), and Rhodes (1940) on human populations and in the papers mentioned above dealing with other species, the value of r has typically been determined by some application of three fundamental equations developed by Lotka (1907a, b; Sharpe and Lotka, 1911). He showed that if the age-specific fecundity (b_x) and survivorship (l_x) remained constant, the population would in time assume a fixed or "stable" age distribution such that in any interval of age from x to $x + dx$ there would be a fixed proportion (c_x) of the population. Once this stable age distribution is established the population would grow exponentially according to our formula (1') and with a birth rate per head, β. Then the following equations relate these quantities:

$$\int_0^\infty e^{-rx} l_x b_x \, dx = 1 \qquad (3)$$

$$\int_0^\infty e^{-rx} l_x \, dx = 1/\beta \qquad (4)$$

and

$$\beta e^{-rx} l_x = c_x. \qquad (5)$$

While the use of formulas (3), (4), and (5) to compute the value of r often presents practical difficulties owing to the difficulty of approximating the functions l_x and b_x by a mathematical function, and also because the equations usually must be solved by iterative methods, it may fairly be stated that Lotka's pioneer work establishing these relationships provided the methods for interpreting the relationships between life history features and their population consequences.

However, the exceedingly important ecological questions of what potential advantages might be realized if a species were to alter its life history features have remained largely unexplored. Doubtless, as already noted, this is largely to be explained by a certain suspicion felt by biologists toward analyses such as those of Lotka, which seem to involve assumptions very remote from the realities of life histories as observed in the field and laboratories. A particularly pertinent statement of this point of view is that of Thompson (1931), who recognized the great practical need for methods of computing the rate of increase of natural populations of insects adhering to particular life history patterns but who insisted that the reproductive process must be dealt with as a discontinuous phenomenon rather than as a compound

interest phenomenon such as that of formula (1'). His methods of computation were designed to give the exact number of individuals living in any particular time period and, while he recognized that the population growth can be expressed in an exponential form such as (1'), he rejected its use on these grounds:

"In the first place, the constant (r) cannot be determined until the growth of the population under certain definite conditions has been studied during a considerable period; in the second place, no intelligible significance can be attached to the constant after its value has been determined; in the third place, the growth of the population is considered in this formula to be at every moment proportional to the size of the population, which is not true except with large numbers and over long periods and cannot be safely taken as a basis for the examination of experimental data."

In the following sections of the present paper an effort will be made to reconcile these two divergent points of view and to show under what conditions Thompson's "discontinuous" approach and the continuous methods lead to identical results. Practical methods of computation can be founded on either scheme, and there are circumstances where one or the other offers distinct advantages. It is hoped that a theoretical approach to population phenomena proceeding from exact computational methods will clarify the meaning of some of the approximations made in deriving equations such as (3), (4), and (5) by continuous methods, and will stimulate students of ecology to a greater interest in the population consequences of life history phenomena.

Before proceeding to a discussion of potential population growth, one point which has sometimes caused confusion should be mentioned. This concerns the sex ratio and the relative proportions of different age classes in the growing population. Once stated, it is obvious that if a population is always growing, as are the populations in the models used for determining potential population growth, then each age and sex class must ultimately come to grow at exactly the same rate as every other class. If this were not the case the disproportion between any two classes would come to exceed all bounds; the fastest growing class would continue indefinitely to make up a larger and larger proportion of the total population. It is thus intuitively recognizable that with fixed life history features there must ultimately be a fixed sex ratio and a stable age distribution. In discussing potential population growth it is often convenient to confine our attention to females or even to a restricted age class, such as the annual births, while recognizing that the ultimate growth rate for such a restricted population segment must be identical to the rate for the entire population.

SIMPLEST CASES OF POPULATION GROWTH

Non-overlapping generations

The simplest possible cases of population growth from the mathematical point of view are those in which reproduction takes place once in a lifetime and the parent organisms disappear by the time the new generation comes on the scene, so that there is no overlapping of generations. This situation occurs in the many plants and animals which are annuals, in those bacteria, unicellular algae, and protozoa where reproduction takes place by fission of one individual to form two or more daughter individuals, and in certain other forms. Thus in the century plants (*Agave*) the plant dies upon producing seeds at an age of four years or more, the Pacific salmon (*Oncorhynchus*) dies after spawning, which occurs at an age of two to eight years (two years in the pink salmon *O. gorbuscha*), and cicadas breed at the end of a long developmental period which lasts from two years (*Tibicen*) to 17 years in *Magicicada*. For many other insects with prolonged developmental stages such as neuropterans and mayflies potential population growth may be considered on the assumption that generations do not overlap.

In these cases, perhaps most typically illustrated in the case of annuals, the population living in any year or other time interval is simply the number of births which occurred at the beginning of that interval. Starting with one individual which is replaced by b offspring each of which repeats the life history pattern of the parent, the population will grow in successive time intervals according to the series: $1, b, b^2, b^3, b^4, \cdots b^x$. Hence the number of "births," say B_x, at the beginning of any time interval, T_x, is simply b^x which is identical with the population, P_x, in that interval of time. If the population starts from an initial number P_0 we have:

$$P_x = P_0 b^x \qquad (6)$$

which is obviously identical with the exponential formula (1'), $P_x = Ae^{rx}$, where the constant A is precisely P_0, the initial population size, and $r = \ln b$; the intrinsic rate of increase is equal to the natural logarithm of the litter size.

If litter size varies among the reproductive individuals, with each litter size being characteristic of a fixed proportion of each generation, it is precisely correct to use the average litter size, say $\bar{b}$, in the computations, so that we have $r = \ln \bar{b}$. Furthermore, if not all of the offspring are viable, but only some proportion, say l_1, survive to reproduce, we shall have exactly $r = \ln \bar{b}^{l_1}$. Thus, mortality and variations in litter size do not complicate the interpretation of population growth in cases where the generations do not overlap. On the other hand, even in species which reproduce only once, if the generation length is not the same for all individuals, this will lead to overlapping generations, and the simple considerations which led to formula (6) will no longer apply. In other words, we can use an average figure for the litter size b but not for the generation length x. It will be shown in the next section, however, that the more general formula (1') is still applicable.

In these simplest cases the assumption of a geometric progression as the potential form of population growth is obviously correct, and numerous authors have computed the fantastic numbers of offspring which could potentially result from such reproduction. For example, according to Thompson (1942), Linnaeus (1740?) pointed out that if only two seeds of an annual plant grew to maturity per year, a single individual could give rise to a million offspring in 20 years. (In all editions available to the present writer this interesting essay of Linnaeus' is dated 1743, and the number of offspring at the end of twenty years is stated by the curious and erroneous figure 91,296.) That is, $P_{20} = 2^{20} = e^{20 \ln 2} = 1,048,576$. Additional examples are given by Chapman (1935, p. 148).

Formulas (1') or (6) may, of course, also be used in an inverse manner to obtain the rate of multiplication when the rate of population growth is known. For the example given by Molisch (1938, p. 25), referring to diatoms reproducing by binary fission where the average population was observed to increase by a factor of 1.2 per day, we have $1.2 = e^{x \ln 2}$, where x is the number of generations per day. Solving for $1/x$, the length of a generation, we obtain $1/x = \dfrac{\ln 2}{\ln 1.2} = \dfrac{.69315}{.18232} = 3.8$ days.

Overlapping generations

Interest in computing the number of offspring which would be produced by a species adhering to a constant reproductive schedule dates back at least to Leonardo Pisano (=Fibonacci) who, in the year 1202, attempted to reintroduce into Europe the study of algebra, which had been neglected since the fall of Rome. One of the problems in his *Liber Abbaci* (pp. 283–84 of the 1857 edition) concerns a man who placed a pair of rabbits in an enclosure in order to discover how many pairs of rabbits would be produced in a year. Assuming that each pair of rabbits produces another pair in the first month and then reproduces once more, giving rise to a second pair of offspring in the second month, and assuming no mortality, Fibonacci showed that the number of pairs in each month would correspond to the series

$$1, 2, 3, 5, 8, 13, 21, 34, 55, \text{etc.,}$$

where each number is the sum of the two preceding numbers. These "Fibonacci numbers" have a rather celebrated history in mathematics, biology, and art (Archibold, 1918; Thompson, 1942; Pierce, 1951) but our present concern with them is merely as a very early attempt to compute potential population growth.

Fibonacci derived his series simply by following through in words all of the population changes occurring from month to month. One with sufficient patience could, of course, apply the same procedure to more complicated cases and could introduce additional variables such as deductions for mortality. In fact, Sadler (1830, Book III) did make such computations for human populations. He was interested in discovering at what ages persons would have to marry and how often they would have to reproduce to give some of the rates of population doubling which had been postulated by Malthus (1798). To accomplish this, Sadler apparently employed the amazingly tedious procedure of constructing numerous tables corresponding to different assumptions until he found one which approximated the desired rate of doubling.

Although we must admire Sadler's diligence, anyone who undertakes such computations will find that it is not difficult to devise various ways of systematizing the procedure which will greatly reduce the labor of computation. By far the best of these methods known to the present writer is that of Thompson (1931), which was originally suggested to him by H. E. Soper.

In the Soper-Thompson approach a "generation law" (G) is written embodying the fixed life history features which it is desired to consider. The symbol

T^x stands for the xth interval of time, and a generation law such as $G = 2T^1 + 2T^2$ would be read as "two offspring produced in the first time interval and two offspring produced in the second time interval." This particular generation law might, for example, be roughly applicable to some bird such as a cliff swallow, where a female produces about four eggs per year. Concentrating our attention on the female part of the population, we might wish to compute the rate of population growth which would result if each female had two female offspring upon attaining the age of one year and had two more female offspring at the age of two years. The fundamental feature of the Thompson method is the fact that the expression:

$$\frac{1}{1 - G} \qquad (7)$$

is a generating function which gives the series of births occurring in successive time intervals. In the algebraic division the indices of the terms T^1, T^2, etc., are treated as ordinary exponents and the number of births occurring in any time interval T^x is simply the coefficient of T^x in the expansion of expression (7). Thus, for our example where $G = 2T^1 + 2T^2$ we obtain:

$$\frac{1}{1 - 2T^1 - 2T^2} = 1 + 2T^1 + 6T^2 +$$

$$16T^3 + 44T^4 + 120T^5 + 328T^6 + \cdots,$$

showing that one original female birth gives rise to 328 female offspring in the sixth year. The series could be continued indefinitely to obtain the number of births any number of years hence. However, in practice it is not necessary to continue the division. In the above series the coefficient of each term is simply twice the sum of the coefficients of the two preceding terms; hence the generation law gives us the rule for extending the series. $G = 2T^1 + 2T^2$ instructs us to obtain each new term of the series by taking twice the preceding term plus twice the second term back. In the case of the Fibonacci numbers we would have $G = T^1 + T^2$, telling us at once that each new term is the sum of the two preceding it.

From the birth series we can easily obtain the series enumerating the total population. If each individual lives for λ years, the total population in T^x will be the sum of λ consecutive terms in the expansion of the generating function. Multiplying formula (7) by the length of life expressed in the form $1 + T^1 + T^2 + T^3 + \cdots + T^{\lambda-1}$ will give the population series. In our above example if we assume that each individual lives for three years, although, as before, it only reproduces in the first two, we obtain for the population

$$\frac{1 + T^1 + T^2}{1 - 2T^1 - 2T^2} = 1 + 3T^1 + 9T^2 +$$

$$24T^3 + 66T^4 + 180T^5 + 492T^6 + \cdots,$$

a series which still obeys the rule $G = 2T^1 + 2T^2$.

Thompson's method for obtaining the exact number of births and members of the population in successive time intervals is very general. As in the case of non-overlapping generations, the coefficients in the generation law may refer to average values for the age-specific fecundity. Also the length of the time intervals upon which the computations are based can be made arbitrarily short, so that it is easy to take into account variations in the age at which reproduction occurs. For the above example, time could have been measured in six-month periods rather than years so that the generation law would become $G = 2T^2 + 2T^4$, with the same results already obtained.

Furthermore, the factor of mortality can easily be included in the computations. For example, suppose that we wish to determine the rate of population growth for a species where the females have two female offspring when they reach the age of one, two more when they reach the age of two, and two more when they reach the age of three. Neglecting mortality, this would give us the generation law $G = 2T^1 + 2T^2 + 2T^3$. If we were further interested in the case where not all of the offspring survive for three years, the coefficients in the generation law need only be multiplied by the corresponding survivorship values. For example, if one-half of the individuals die between the ages of one and two, and one half of the remainder die before reaching the age of three we would have $l_1 = 1$, $l_2 = \frac{1}{2}$, $l_3 = \frac{1}{4}$, and the above generation law would be revised to $G = 2T^1 + T^2 + \frac{1}{2}T^3$. The future births per original individual would then be

$$\frac{1}{1 - G} = 1 + 2T^1 + 5T^2 + 25/2T^3 + 31T^4 +$$

$$151/2T^5 + \cdots.$$

Very generally, if the first reproduction for a species occurs at some age α and the last reproduction occurs at some age ω, and letting b_x and l_x represent respectively the age-specific fecundity

and survivorship, we may write the generation law as:

$$G = l_\alpha b_\alpha T^\alpha + l_{\alpha+1} b_{\alpha+1} T^{\alpha+1} + \cdots$$

$$+ l_\omega b_\omega T^\omega = \sum_{x=\alpha}^{\omega} l_x b_x T^x. \tag{8}$$

Therefore, in the Thompson method we have a compact system of computation for obtaining the exact number of births and the exact population size at any future time, assuming that the significant life history features (α, ω, l_x, and b_x) do not change.

Not all of the possible applications of Thompson's method have been indicated above. For example, formula (7) may be used in an inverse manner so that it is theoretically possible to work back from a tabulation of births or population counts made in successive time intervals and discover the underlying generation law. Formulas (7) and (8), together with the procedure of multiplying the birth series by the length of life expressed as a sum of T^x values, provide the nucleus of the system and offer the possibility of analyzing the potential population consequences of essentially any life-history phenomena. The system has the merit of treating the biological units and events as discontinuous variates, which, in fact, they almost always are. The members of populations are typically discrete units, and an event such as reproduction typically occurs at a point in time with no spreading out or overlapping between successive litters. While survivorship, l_x, as a population quantity, is most realistically regarded as continuously changing in time, the product $l_x b_x$ which enters our computations by way of formula (8) is typically discontinuous because of the discontinuous nature of b_x.

It is quite obvious that equations of continuous variation such as (1') are often much more convenient for purposes of computation than the series of values obtained by expanding (7). This is especially true in dealing with the life histories of species which have long reproductive lives. In writing a generation law for man by (8) we should have to take α at least as small as 15 years and ω at least as great as 40 years, since for the population as a whole reproduction occurs well outside of these extremes and it would certainly be unrealistic to regard b_x as negligibly small anywhere between these limits. Thus there would be at least 25 terms in our generation law, and the computations would be extremely tedious. By selecting special cases

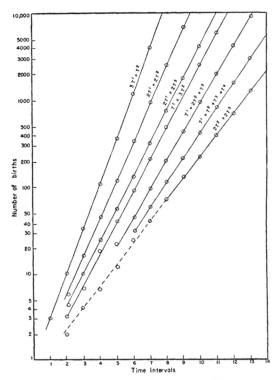

FIG. 1. EXACT VALUES OF POPULATION GROWTH IN
TERMS OF BIRTHS PER UNIT TIME UNDER SEV-
ERAL GENERATION LAWS, WHEN EACH FEMALE
HAS A TOTAL OF FOUR FEMALE OFFSPRING

In each case it is assumed that a single female exists
at time zero and produces her four progeny on or before
her fourth birthday. The plotted points represent exact
values as determined by Thompson's method. To the
extent that the points for any generation law fall on a
straight line in this logarithmic plot, they can be represented by the exponential growth formula (1'), and the
slope of each line is a measure of the intrinsic rate of
natural increase (r).

for study it is sometimes possible greatly to
simplify the procedures. For example, if one is
interested in the case where there is no mortality
during the reproductive span of life and where the
litter size is a constant, say b, the expression for
the generation law (8) can be simplified to:

$$G = bT^\alpha + bT^{\alpha+1} + \cdots bT^\omega = \frac{bT^\alpha - bT^{\omega+1}}{1 - T}.$$

Since one can also write the length of life as

$$1 + T^1 + T^2 + \cdots + T^{\lambda-1} = \frac{1 - T^\lambda}{1 - T},$$

the generating function for the total population
simplifies to

$$\frac{1 - T^\lambda}{1 - T - bT^\alpha + bT^{\omega+1}}.$$

This last formula is much more convenient for
computations than one containing 25 terms or so
in the denominator, but it applies only to a very
special case and is much less convenient than
formula (1'). Consequently, great interest at-
taches to these questions: can (1') be used as a
substitute for (7)? (i.e., does Thompson's method
lead to a geometric progression?) and, if it is so
used, can the constants, particularly r, be interpreted in terms of life-history features?

THE GENERALIZATION OF THOMPSON'S METHOD

Fig. 1 shows the exact values, as determined by
Thompson's method, of the birth series arising
from several generation laws (life-history patterns)
which have in common the feature that in each
case every female produces a total of four female
offspring in her lifetime and completes her reproductive life by the age of four "years." The number
of births is plotted on a logarithmic scale, hence
if it can be represented by formula (1'), $P = Ae^{rx}$
or, logarithmically, $\ln P = \ln A + rx$, the points
should fall on a straight line with slope proportional to r. It is apparent from Fig. 1 that after
the first few time intervals the points in each case
are well represented by a straight line. Therefore,
except in the very early stages, formula (1') does
give a good representation of potential population
growth. The question remains, however, as to
whether we can meet Thompson's objection to (1')
and attach any intelligible significance to the constants of the formula. From Fig. 1 it is obvious
that the lines do not, if projected back to time 0,
indicate exactly the single individual with which
we started. Thus, in these cases the constant A
cannot be precisely P_0 as was the case with non-
overlapping generations.

Before proceeding to interpret the constants of
formula (1') for the case of overlapping generations it will be well to notice one feature of Fig. 1
which is of biological interest. In the literature of
natural history one frequently encounters references to the number of offspring which a female
can produce per lifetime, with the implication
that this is a significant life-history feature. The
same implication is common in the literature
dealing with various aspects of human biology,
where great emphasis is placed on the analysis of
total family size. From Fig. 1 it will be seen that
this datum may be less significant from the standpoint of contributions to future population than
is the age schedule upon which these offspring are

produced. Each life history shown in Fig. 1 represents a total production of four offspring within four years of birth, but the resulting rates of potential population growth are very different for the different schedules. It is clear that the cases of most rapid population growth are associated with a greater concentration of reproduction into the early life of the mother. This is intuitively reasonable because we are here dealing with a compound interest phenomenon and should expect greater yield in cases where "interest" begins to accumulate early. However, the writer feels that this phenomenon is too frequently overlooked in biological studies, possibly because of the difficulty of interpreting the phenomenon quantitatively.

In seeking to reconcile the continuous and discontinuous approaches to potential population growth, let us first note that Thompson's discontinuous method corresponds to an equation of finite differences. We have seen above that the generation law gives us a rule for indefinitely extending the series representing the population size or the number of births in successive time intervals by adding together some of the preceding terms multiplied by appropriate constants. If we let $f_{(x)}$ represent the coefficient of T^x in the expansion of the generating function (7) and, for brevity, write in (8) $V_x = l_x b_x$, then our population series obeys the rule:

$$f_{(x)} = V_\alpha f_{(x-\alpha)} + V_{\alpha+1} f_{(x-\alpha-1)} + \cdots + V_\omega f_{(x-\omega)}, \quad (9)$$

which may be written in the alternative form,

$$f_{(x+\omega)} - V_\alpha f_{(x+\omega-\alpha)} - V_{\alpha+1} f_{(x+\omega-\alpha-1)} - \cdots - V_\omega f_{(x)}, = 0. \quad (10)$$

Thus for our "cliff swallow" example, where we had $G = 2T^1 + 2T^2$ we have

$$f_{(x)} = 2f_{(x-1)} + 2f_{(-2)} \text{ or,}$$
$$f_{(x+2)} - 2f_{(x+1)} - 2f_{(x)} = 0.$$

Formula (10) represents the simplest and best understood type of difference equation, a homogeneous linear difference equation with constant coefficients. It is outside the scope of the present paper to discuss the theory of such equations, which has been given, for example, by Jordan (1950). By the nature of our problem as summarized in formula (9), all of our V_x values are either equal to zero or are positive real numbers and all of the signs of the coefficients in (9) are positive: features which considerably simplify

generalizations. By virtue of these facts it can be shown that there is always a "characteristic" algebraic equation corresponding to (10). This is obtained by writing ρ^x for $f_{(x)}$ and dividing through by the ρ value of smallest index. This gives

$$\rho^\omega - V_\alpha \rho^{\omega-\alpha} - V_{\alpha+1} \rho^{\omega-\alpha-1} \cdots - V_\omega = 0 \quad (11)$$

an algebraic equation which has the roots ρ_1, ρ_2, etc.

The general solution of the corresponding difference equation (10) is

$$f_{(x)} = C_1 \rho_1^x + C_2 \rho_2^x + \cdots + C_n \rho_n^x \quad (12)$$

where the C's are constants to be determined by the initial conditions of the problem. Formula (12) is precisely equivalent to Thompson's method and is a general expression for the number of births or the population size in any future time interval.

As an example we may consider the case where $G = 2T^1 + 2T^2$. The difference equation, as already noted, is $f_{(x+2)} - 2f_{(x+1)} - 2f_{(x)} = 0$ and the characteristic algebraic equation is $\rho^2 - 2\rho - 2 = 0$ which is a quadratic equation with the roots $\rho_1 = 1 + \sqrt{3}$, and $\rho_2 = 1 - \sqrt{3}$. Hence the general solution is $f_{(x)} = C_1(1 + \sqrt{3})^x + C_2(1 - \sqrt{3})^x$. To determine the constants C_1 and C_2 we look at the beginning a of the seriesnd note that we have $f_{(0)} = 1$ and $f_{(1)} = 2$. Substituting these values in the general solution we obtain $C_1 = \dfrac{\sqrt{3} + 1}{2\sqrt{3}}$ and $C_2 = \dfrac{\sqrt{3} - 1}{2\sqrt{3}}$. Therefore, the general expression for the number of births in time interval T^x is

$$f_{(x)} = \frac{\sqrt{3} + 1}{2\sqrt{3}} (1 + \sqrt{3})^x + \frac{\sqrt{3} - 1}{2\sqrt{3}} (1 - \sqrt{3})^x$$

which can be simplified to $f_{(x)} = \dfrac{\rho_1^{x+1} - \rho_2^{x+1}}{\sqrt{3}} = \rho_1^x + \rho_1^{x-1}\rho_2 + \cdots + \rho_2^x$.

In order to have the difference equation (12) correspond to the equation of exponential growth (1′), the ratio between populations in successive time intervals must assume a constant value giving

$$\frac{f_{(x+1)}}{f_{(x)}} = e^r. \quad (13)$$

By the nature of our problem, as already noted, the potential population growth is always positive, so that any limit approached by the ratio $\frac{f_{(x+1)}}{f_{(x)}}$ must be a positive real number.

It is beyond the scope of the present paper to discuss the conditions, for difference equations in general, under which this ratio does approach as a limit the largest real root of the characteristic algebraic equation. (See, for example, Milne-Thompson, 1933, chap. 17). Dunkel (1925) refers to the homogeneous equation with real constant coefficients corresponding to our formulas (10) and (11). The algebraic equation (11) has a single positive root which cannot be exceeded in absolute value by any other root, real or complex. Using (12) to express the ratio between successive terms, we have

$$\frac{f_{(x+1)}}{f_{(x)}} = \frac{C_1\rho_1^{x+1} + C_2\rho_2^{x+1} + \cdots + C_n\rho_n^{x+1}}{C_1\rho_1^{x} + C_2\rho_2^{x} + \cdots + C_n\rho_n^{x}}. \quad (14)$$

If we let ρ_1 represent the root of (11) of greatest absolute value and divide both numerator and denominator of (14) by $C_1\rho_1^{x}$ we obtain

$$\frac{f_{(x+1)}}{f_{(x)}} = \rho_1 \left[\frac{1 + \frac{C_2}{C_1}\left(\frac{\rho_2}{\rho_1}\right)^{x+1} + \frac{C_3}{C_1}\left(\frac{\rho_3}{\rho_1}\right)^{x+1} + \cdots + \frac{C_n}{C_1}\left(\frac{\rho_n}{\rho_1}\right)^{x+1}}{1 + \frac{C_2}{C_1}\left(\frac{\rho_2}{\rho_1}\right)^{x} + \frac{C_3}{C_1}\left(\frac{\rho_3}{\rho_1}\right)^{x} + \cdots + \frac{C_n}{C_1}\left(\frac{\rho_n}{\rho_1}\right)^{x}} \right]. \quad (15)$$

The expressions in parentheses are all less than unity, on the assumption that ρ_1 is the largest root, and the entire expression in brackets approaches unity as x increases. Consequently we have, for x large

$$\frac{f_{(x+1)}}{f_{(x)}} \sim \rho_1 \sim e^r. \quad (16)$$

This then explains the shape of the potential birth and population series as illustrated in Fig. 1. In the very early stages population growth is irregular, because the expressions in (12) and (15) involving the negative and complex roots of (11) are still large enough to exert an appreciable influence. As x increases, the influence of these other roots becomes negligible and the population grows exponentially, conforming to (16). In considering potential population growth we are concerned with the ultimate influence of life-history features, and the equation of geometric progression or compound interest does actually represent the form of potential population growth. We are interested only

in the single positive root of (11) for the purpose of determining the constant r, and this can readily be computed with any desired degree of precision by elementary algebraic methods.

Having established the relationship of formula (13) or (16), it is easy to reconcile Thompson's discontinuous approach to population growth with Lotka's continuous approach, as exemplified by formulas (3), (4), and (5).

Employing formula (9) we may write the ratio between populations in successive time intervals as

$$\frac{(f_{x+1})}{f_{(x)}} = V_\alpha \frac{f_{(x-\alpha+1)}}{f_{(x)}} + V_{\alpha+1}\frac{f_{(x-\alpha)}}{f_{(x)}} + \cdots + V_\omega\frac{f_{(x-\omega+1)}}{f_{(x)}}.$$

Substituting the relationship given by (13), this becomes

$$e^r = V_\alpha e^{-r(\alpha-1)} + V_{\alpha+1}e^{-r\alpha} + \cdots + V_\omega e^{-r(\omega-1)}, \text{ or}$$

$$1 = V_\alpha e^{-r\alpha} + V_{\alpha+1}e^{-r(\alpha+1)} + \cdots + V_\omega e^{-r\omega}.$$

Replacing V_x by its equivalent, $l_x b_x$, this is

$$1 = \sum_{x=\alpha}^{\omega} e^{-rx} l_x b_x. \quad (17)$$

Formula (17) is the precise equivalent in terms of finite time intervals of Lotka's equation (3) for infinitesimal time intervals. In Lotka's equation, as in (17), the limits of integration in practice are α and ω since b_x is zero outside of these limits. Formula (17) was in fact employed by Birch (1948) as an approximation to (3) in his method of determining r for an insect population. The only approximation involved in our derivation of (17) is the excellent one expressed by formula (13); otherwise the formula corresponds to Thompson's exact computational methods. It is hoped that recognition of this fact will make some of the approaches of population mathematics appear more realistic from the biological point of view.

Formulas (4) and (5), originally due to Lotka, are also immediately derivable from the relationship (13). In any time interval, T_x, we may say that the population members aged 0 to 1 are simply the births in that interval, say B_x. The population members aged 1 to 2 are the survivors of the births in the previous interval, that is $l_1 B_{x-1}$, or employing (13), $l_1 B_x e^{-r}$. Quite generally, the population members aged between z and $z + 1$ are the survivors from the birth z intervals previous, or $l_z B_x e^{-rz}$. If λ is the extreme length of life for any population members ($l_{\lambda+1} = 0$) we have for the total population

$$P_x = B_x(1 + l_1 e^{-r} + l_2 e^{-2r} + \cdots$$

$$+ l_\lambda e^{-r\lambda} = B_x \sum_{x=0}^{\lambda} e^{-rx} l_x.$$

The birth rate per individual, β, is B_x/P_x, therefore,

$$1/\beta = \sum_0^\lambda e^{-rx} l_x \tag{18}$$

which is the equivalent in finite time intervals of Lotka's equation (4). Also the proportion, c_z, of the population in the age range z to $z + 1$ is $\dfrac{l_z B_x e^{-rz}}{P_x}$ which is simply,

$$c_z = \beta e^{-rz} l_z. \tag{5}$$

COMPUTATIONAL METHODS

In the following sections we will examine some of the population effects which are the consequences of particular life history patterns. Probably the most significant comparisons are those involving the effects of life-history features on the intrinsic rate of natural increase, r. Of course, any change in r is accompanied by other effects, such as those on the age-structure and on the population birth-rate. However, the intrinsic rate of increase is a parameter of fundamental ecological importance. If a species is exposed to conditions which would favor the ability to outbreed competitors or where exceptional hazards limit the probability that an individual will become established, we might expect to find life-history adjustments tending to increase the value of r. Conversely, if a species has evolved life-history features of a type tending to hold down the intrinsic rate of increase, a fertile field of inquiry may be opened regarding the selective factors to which such a species is subject.

It is probably fairly obvious to anyone that in general a species might increase its biotic potential by increasing the number of offspring produced at a time (litter size), by reducing mortality at least

until the end of active reproductive life, by reproducing oftener, by beginning reproduction at an earlier age, or by minimizing any wastage of environmental resources on sterile members of the population. Any biologist will at once recognize, however, that a great deal of evolution (an extreme case is the evolution of sterility in the social insects) has proceeded in precisely the wrong direction to increase biotic potential by some of these devices. Presumably, this can only mean that the optimum biotic potential is not always, or even commonly, the maximum that could conceivably be achieved by selecting for this ability alone. Comparative life-history studies appear to the writer to be fully as meaningful in evolutionary terms as are studies of comparative morphology or comparative physiology.

Although a great many empirical data on life histories have been accumulated, attempts to interpret these data comparatively have lagged far behind the corresponding efforts in morphology and physiology. The methods exhibited in the preceding parts of the present paper are adaptable for the quantitative interpretation of life history features and, while the number of conceivable life-history patterns is infinite, we propose to examine some of the cases which appear to possess particular ecological interest.

The life-history features with which we are concerned are the age at which reproduction begins (α), the litter size and frequency of reproduction (both summarized by a knowledge of the function b_x, which can also be computed so as to take account of the sex ratio), the maximum age at which reproduction occurs (ω), survivorship (l_x), and maximum longevity (λ). Corresponding to any given set of values for these quantities there is a definite value for the intrinsic rate of natural increase (r) and a definite stable age distribution of the population (c_x). In general, these population features will be altered by any alteration of the life-history features and we wish to examine some of these possible changes quantitatively.

The most efficient way of making the desired computations will vary from problem to problem. Thompson's method (formulas (7) and (8)), could be used to obtain exact population values arising from any life history, but the computations would in many cases be exceedingly laborious and would actually uield no more information about the ultimate course of population growth than would be obtained by solving (11) for the positive root.

In either case it will usually be most efficient to measure time in terms of the shortest interval between the pertinent life-history events with which we are concerned.

Except in very special cases, it is necessary to use iterative methods for obtaining the value of r corresponding to particular life-history patterns. In most cases the solutions are quite rapidly obtained by employing a calculating machine and detailed tables of natural logarithms (e.g., Lowan, 1941) or of the exponential function (e.g., Newman, 1883). In the majority of the cases considered by the writer, the most efficient procedure has been to rewrite formula (17) in the form:

$$e^{r\alpha} = V_\alpha + V_{\alpha+1}e^{-r} + V_{\alpha+2}e^{-2r} + \cdots$$
$$+ V_\omega e^{-r(\omega-\alpha)} \qquad (19)$$

and then to obtain the sum of the series on the right-hand side of (19) for different patterns of variation in the function $l_x b_x = V_x$. This method corresponds exactly to the discontinuous approach, granting only that potential population growth is a geometric progression, and it leads to relatively simple equations in a number of the cases of greatest ecological interest.

A more general approach from the standpoint of formal mathematics can be obtained by rewriting (3) in the form of a Stieltjes integral (Widder, 1940). We may define a maternity function $M(x)$ representing the average number of offspring which an individual will have produced by the time it has attained any age x, and such that its derivative with respect to time is $\left(V_x \dfrac{d}{dx} M(x) = V_x\right)$. We then have

$$\int_0^\infty e^{-rx} dM(x) = 1 \qquad (20)$$

which can represent cases where $V_{(x)}$ is either continuous or discontinuous because the integral vanishes for values where V_x is discontinuous. When V_x can be expressed as a function of time (x), formula (20) is identical with (3) and the use of the Laplace transformation, a procedure of considerable importance in engineering and physical mathematics, makes it possible to avoid the numerical integration and express V_x as a function of r. If V_x is considered as a series of single impulses regularly spaced from α to ω, equation (20) assumes the form (17). Laplace transformations for a number of functions are tabulated by Churchill (1944) and Widder (1947) and, no doubt, there

are cases where this procedure would lead to simpler iterative solutions than those obtained from equation (19). For the cases considered in the present paper, however, the solution of equation (19) generally leads to somewhat simpler results.

In dealing with any particular life-history pattern the computational method of choice may depend upon the types of features to be investigated. The pure numbers α, ω, and λ typically offer no particular computational problems, as they are assigned different values, but this is not always the case with the functions b_x and l_x.

In the cases considered by the writer the intervals between successive periods of reproduction have been considered to be equal. There is no particular difficulty in altering this assumption so as to consider cases where the frequency of reproduction varies with age, but regular spacing seems to be so much more usual in nature as well as representing a limiting case that it seems to merit first consideration. Litter size often does vary with the age of the parent organism, and this fact may introduce complexities into the behavior of the function V_x. In this case also, it appears that the ecologically most interesting cases are those in which the average litter size is a constant. Furthermore, as will become apparent in later sections, the first few litters produced by an organism so dominate its contribution to future population growth that later changes in litter size would have only very minor population consequences. In dealing with empirical data on human populations attempts have been made to express analytically the changes in b_x with age [cf. "Tait's law" that fertility declines in a linear manner (Yule, 1906; Lotka, 1927)] but for the present we shall consider that b_x assumes only the values zero and some constant, b.

The shape of the survivorship (l_x) curve is more difficult to deal with in a realistic manner. Pearl and Miner (1935) originated the classification of survivorship curves which is most employed for ecological purposes (cf. Deevey, 1947; Allee et al., 1949). The "physiological" survivorship curve is the limiting type in which each individual lives to some limit characteristic of the species and the age at death (λ) is regarded as a constant. In this case $l_x = 1$ when $x < \lambda$ and $l_x = 0$ when $x > \lambda$. This is the simplest case for computations, and actual cases are known which approach this type. Furthermore, there are other types of survivorship curves of ecological interest which may be treated in the same manner. In what Deevey (1947) calls Type III there is an extremely heavy early mortality with the few survivors tending to live out a "normal life span." For the computation of r we are only concerned (cf. formula 19) with survivorship during the reproductive span of life, and it appears likely that "Type III" curves can be treated as constant throughout this age range without serious error. Another interesting type of survivorship curve which appears to be consistent with empirical data at least on some wild populations (cf. Jackson, 1939; Deevey, 1947; Ricker, 1948) is that in which a constant proportion of the population dies in each interval of age. This, of course, implies that life expectancy is independent of age, an assumption which cannot in general be considered realistic but which might apply to catastrophic causes of mortality. When this type of l_x curve applies, the V_x values will be in geometric progression and the right side of formula (19) can be summed as easily as in the case where V_x is constant. This case is, therefore, easily dealt with.

The type of survivorship curve usually observed in actual cases is a reverse sigmoid curve, interpreted by Deevey as intermediate between the "physiological" type and the geometric progression. This can be interpreted in various ways as a "wearing-out" curve. Gompertz (1825) attempted to find an analytical form on the assumption that the ability of individuals to "resist destruction" decreases as a geometric progression with age. Elston (1923) has reviewed formulas proposed to represent human mortality; none of these has proved generally applicable, despite great complexity in some cases. Another approach is to assume that some sort of a "vital momentum" (Pearl, 1946) or ability to survive is distributed among the members of the population in the form of a bell-shaped or "normal" frequency distribution. This point of view is a familiar and controversial one in the recent literature on bio-assay problems (Finney, 1947, 1949; Berkson, 1944, 1951) and, at least to the extent that a bell-shaped curve can represent the empirical distribution of ages at death, a probit function or a logit function (Berkson, 1944) can be used to represent l_x.

In the present paper we are concerned primarily with the limiting cases or the *potential* meaning of life-history phenomena. Consequently the writer has chosen to deal with survivorship curves of the physiological type and thus to investigate the ultimate effects of life-history phenomena for a species which is able to reduce mortality during the reproductive part of the life span to a negligible value. Our general conclusions will not be seriously altered even by rather startling drastic alterations of this assumption, and, in any case, our results will indicate the maximum gain which a species might realize by altering its life-history features.

Perhaps the most fundamental type of life-history pattern to be investigated in terms of population consequences is that in which the individuals are assumed to produce their first offspring at the age of α "years" with the mean litter size being a constant, b. A second litter is produced at age $\alpha + 1$ and an additional litter in each subsequent interval of age out to, and including, age ω. The total number of litters produced per individual is then $n = \omega - \alpha + 1$.

We then have, from (19),

$$e^{r\alpha} = b(1 + e^{-r} + e^{-2r} + \cdots + e^{-r(\omega-\alpha)}).$$

The expression in parentheses is a geometric progression the sum of which is $\dfrac{1 - e^{-rn}}{1 - e^{-r}}$. Consequently, the general implicit equation for r under these conditions may be written

$$1 = e^{-r} + be^{-r\alpha} - be^{-r(n+\alpha)} \qquad (21)$$

which may be solved by trial and error by employing a table of the descending exponential function.

Alternative formulas corresponding to (21) may be obtained by the use of the Laplace transformation. In the case where reproduction is considered to occur as a series of regularly spaced impulses, this approach leads to formula (21). Another approach is to consider that $V_x = 0$ when $x < \alpha$, $V_x = b$ when $\alpha \leqq x \leqq \omega$, and $V_x = 0$ when $x > \omega$. The Laplace transformation of a step-function is then employed, leading to the formula

$$\frac{be^{-r\alpha}}{r} - \frac{be^{-r(\omega+1)}}{r} = 1. \qquad (22)$$

Formula (22) and formula (21) would be identical under the condition that $r + e^{-r} = 1$, which is approximately true when r is small. If one desires more nearly to reconcile the continuous and discontinuous approaches in this case, he may note that in formula (21) he is finding the area under a "staircase-shaped" curve with the first vertical step located at $x = \alpha$, whereas in formula (22) he is finding the area under a straight line paralleling the slope of the staircase. It is apparent that the two areas will be more nearly identical if the straight line is started about one-half unit of time earlier. If we substitute in (22) $\alpha - \frac{1}{2}$ for α and $\omega - \frac{1}{2}$ for ω we obtain a formula which gives results for practical purposes identical with those obtained from (21). The formulas are about equally laborious to solve, and the writer has employed (21) for the following computations because of its more obvious relationship to the exact computational methods.

POSSIBLE VALUES OF REPEATED REPRODUCTION (ITEROPARITY)

One of the most significant of the possible classifications of life histories rests on the distinction between species which reproduce only once in a lifetime and those in which the individuals reproduce repeatedly. This being the case, it is very surprising that there seem to be no general terms to describe these two conditions. The writer proposes to employ the term semelparity to describe the condition of multiplying only once in a lifetime, whether such multiplication involves fission, sporulation, or the production of eggs, seeds, or live young. Thus nearly all annual plants and animals, as well as many protozoa, bacteria, insects, and some perennial forms such as century plants and the Pacific salmon, are semelparous species. The contrasting condition will be referred to as iteroparity. Iteroparous species include some, such as small rodents, where only two or three litters of young are produced in a lifetime, and also various trees and tapeworms where a single individual may produce thousands of litters. The distinction between annual and perennial plants is doubtless the most familiar dichotomy separating semelparous and iteroparous species, but general consideration of the possible importance of these two distinct reproductive habits illustrates some points of ecological and evolutionary interest. For purposes of illustration we shall first consider cases where the time interval between reproductive efforts is fixed at one year.

Many plants and animals are annuals. This is true, for example, of many of the higher fungi and seed plants, of insects, and even of a few vertebrates. One feels intuitively that natural selection should favor the perennial reproductive habit because an individual producing seeds or young annually over a period of several years obviously

has the potential ability to produce many more offspring then is the case when reproduction occurs but once. It is, therefore, a matter of some interest to examine the effect of iteroparity on the intrinsic rate of natural increase in order to see if we can find an explanation for the fact that repeated reproduction is not more general.

Let us consider first the case of an annual plant (or animal) maturing in a single summer and dying in the fall at the time of reproduction. We have seen earlier (formula 16, seq.) that if b is the number of offspring produced by such an annual the intrinsic rate of increase would be the natural logarithm of b. We wish to determine by how much this would be increased if the individual were to survive for some additional years, producing b offspring each year. Obviously, an annual species with a litter size of one (or an average of one female per litter in sexual species) would merely be replacing current population and no growth would be possible ($\ln 1 = 0$); therefore, when the litter size is one the species must necessarily be iteroparous.

The most extreme case of iteroparity, and the one exhibiting the absolute maximum gain which could be achieved by this means, would be the biologically unattainable case of a species with each individual producing b offspring each year for all eternity and with no mortality. In this case we have $\alpha = 1$ "year" and, since ω is indefinitely large, the final term $be^{-r(\omega+1)}$ in equation (21) becomes zero. Thus we have

$$r = \ln(b + 1) \qquad (23)$$

which is to be contrasted with $r = \ln(b)$ for the case of an annual. *For an annual species, the absolute gain in intrinsic population growth which could be achieved by changing to the perennial reproductive habit would be exactly equivalent to adding one individual to the average litter size.* Of course, this gain might be appreciable for a species unable to increase its average litter size. The extreme gain from iteroparity for a species with a litter size of two would be ($\ln 3/\ln 2$) or an increase of about 58 per cent, for a species with a litter size of four the increase would be about 16 per cent, but for one producing 30 offspring in a single reproductive period the extreme gain would amount to less than one per cent. It seems probable that a change in life history which would add one to the litter size would be more likely to occur than a change permitting repeated reproduction, which in many cases would necessitate adjustments to survive several seasons of dormancy. It appears that for the usual annual plants and insects with their relatively high fecundity any selective pressure for perennial reproduction as a means of increasing biotic potential must be negligible.

The above conclusion, which appears surprising when first encountered, arouses curiosity as to why iteroparity exists at all. Perhaps some species are physiologically unable to increase their fecundity. This must, however, be unusual and we are led to investigate whether the situation would be different for a species with a prolonged period of development preceding reproduction. One thinks immediately of the giant Sequoias which require a century to mature and begin reproduction but which, once started, produce large numbers of seeds biennially for centuries.

In order to investigate this question we may again compare the intrinsic rate of increase for a single reproduction with that corresponding to an infinite number of reproductions. This procedure will, of course, tend to overestimate the possible

gain from iteroparity although it will set an upper limit, and the first few reproductive periods so dominate the situation that even for very modest litter sizes there is a negligible difference between the results of a very limited number of reproductive periods and an infinite number.

For α not necessarily equal to one, formula (21) gives

$$b = e^{r\alpha} - e^{r(\alpha-1)}, \qquad (24)$$

an implicit equation for r which must be solved by iterative means.

Fig. 2 was constructed from formula (24) to show the relationship between the age at which reproduction begins (α) and the litter size (b) in terms of the possible gain in intrinsic rate of increase which could be achieved by iteroparity. The ordinates represent the proportionate increase in the value of r which could be achieved by changing from a single reproductive effort at age α to an infinite number at ages α, $\alpha + 1$, $\alpha + 2$, etc. The curves all slope upward, indicating that species with long pre-reproductive periods could gain more

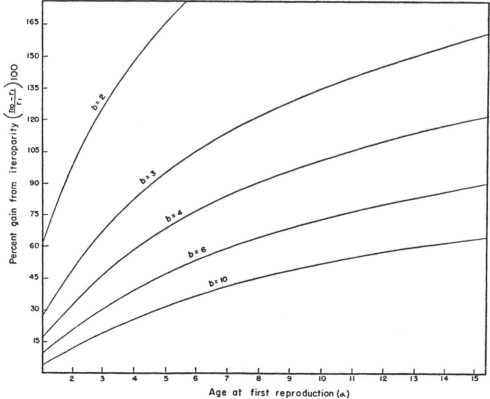

FIG. 2. THE EFFECTS OF LITTER SIZE (b) AND AGE AT MATURITY (α) ON THE GAINS ATTAINABLE BY REPEATED REPRODUCTION

The litter size, b, is the number of female offspring per litter in the case of sexual species.

from iteroparity than forms which mature more rapidly. The tendency of the curves to flatten out with large values of α, however, indicates that the advantages of repeated reproduction increase somewhat less rapidly as the pre-reproductive period is prolonged.

The relationship of iteroparity to litter size is clearly illustrated by Fig. 2. When the litter size is small, as shown by the curve for $b = 2$ (which would correspond to a litter of four individuals when the sex ratio is 1:1), iteroparity can yield important gains in biotic potential, and the possible gains are greater the longer maturity is delayed. The possible advantages diminish quite rapidly as litter size is increased, although it is clear that iteroparity as contrasted with a single reproductive effort would always add something to biotic potential.

Fig. 2 suggests that for semelparous species with large litters there would be very slight selective pressure in favor of adopting the iteroparous habit, and that for iteroparous species with large litters there would be little selection against loss of the iteroparous habit, especially in forms which mature rapidly. On the other hand, in a species which is established as iteroparous there would be slight selection for increasing fecundity or if litter size is relatively large, even against loss of fecundity. This perhaps explains the notoriously low level of viability among the seeds of many trees.

From these considerations it is obvious that when a species could benefit by an increase in the intrinsic rate of natural increase, this advantage might be achieved either by increasing fecundity in a single reproductive period or by adopting the iteroparous habit. A selective advantage would accrue to a mutation altering the life history in either of these directions, and it is an interesting field for speculation as to which type of mutation might be most likely to occur. In this connection it may be interesting to determine the amount of increase in litter size which, for a semelparous species, would be equivalent to retaining the initial litter size but becoming iteroparous.

From (6) we have seen that the intrinsic rate of increase for a semelparous species is defined by $e^{r\alpha} = b$. We wish to find an equivalence factor (E) which will indicate by how much b must be increased to make the value of r for a semelparous species equal to that in formula (21) referring to an iteroparous species. By neglecting the last term in (21) so as to consider the most extreme case of

iteroparity and substituting $Eb = e^{r\alpha}$, we obtain

$$E = \frac{1}{1 - e^{-r}} = \frac{e^{r\alpha}}{b} \qquad (25)$$

where the value of r must be obtained by solving equation (21). When E is plotted against α for various values of b, as shown in Fig. 3, the resulting curves are essentially straight lines.

Fig. 3 illustrates some interesting points bearing on the life histories of organisms, such as tapeworms and many trees, which are iteroparous in addition to producing large litters. From the arrangement of Fig. 2 one might suspect that the iteroparous habit would provide very little advantage to a species that could produce a thousand or so offspring in a single litter, but Fig. 3 indicates that the selective value of iteroparity may be greatly increased when the pre-reproductive part of the life span is prolonged.

A mature tapeworm may produce daily a number of eggs on the order of 100,000 and may continue this for years (Allee et al., 1949, p. 272; Hyman, 1951). With so large a litter size one wonders if iteroparity in this case may not represent something other than an adaptation for increasing biotic potential. Perhaps the probability that a tapeworm egg (or a *Sequoia* seed) will become established may be increased by distributing the eggs more widely in time and space, and this could conceivably be the reason for the iteroparous habit. No definite answer to this problem is possible at present, but Fig. 3 indicates that a knowledge of the length of the life cycle from egg to egg is an essential datum for considering the question. In at least some tapeworms a larva may grow into a mature worm and reproduce at an age of 30 days (Wardle and McLeod, 1952). If this represented the length of the entire life cycle, then Fig. 3 indicates, assuming $b = 100,000$, that a threefold increase in litter size would be the equivalent of indefinite iteroparity. However, with the larval stage in a separate host, the average life cycle must be much longer. If the total cycle requires as much as 100 days, Fig. 3 shows that it would require almost an eight-fold increase in litter size (a single reproductive effort producing 790,167 offspring) to yield the same biotic potential as iteroparity with a litter size of 100,000. Obviously, it is possible, when the life cycle is sufficiently prolonged, to reach a point where any attainable increase in litter size would be less advantageous for potential population growth than a change to the iteroparous

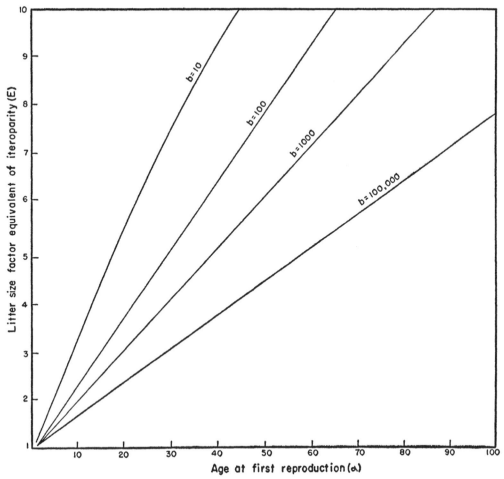

FIG. 3. THE CHANGES IN LITTER SIZE WHICH WOULD BE REQUIRED TO ACHIEVE IN A SINGLE REPRODUCTION THE SAME INTRINSIC RATE OF INCREASE THAT WOULD RESULT FROM INDEFINITE ITEROPARITY

b represents the litter size for an iteroparous species and the ordinate scale (E) represents the factor by which b would have to be multiplied to attain the same intrinsic rate of increase when each female produces only one litter in her lifetime.

habit. Hence a selective pressure can operate in favor of iteroparity even when the litter size is large. It is clear from Fig. 3, noting the greater slope of the lines representing smaller litter sizes, that in these cases the point will be reached more quickly at which the potential gains from iteroparity outweigh those attainable by increasing the litter size.

Man has a life cycle which is rather unusual in that it combines a long pre-reproductive period with a very small litter size; the very conditions under which iteroparity should be most advantageous. Everyone is, of course, aware that multiple births occur in man but with such a low frequency in the population that they are of negligible importance in population phenomena. It is also rather generally accepted that there is a hereditary basis for the production of multiple births. The question

arises as to why increased litter size should not become more common simply as a result of increased contributions to subsequent population resulting from the increase in biotic potential associated with large litters. It should be of interest, therefore, to determine how large a litter would have to be produced in a single reproductive effort to provide an intrinsic rate of increase equal to that resulting from three or more single births.

In the case of man we may rather confidently accept the value $b = \frac{1}{2}$ to signify that the average number of female offspring produced per human birth, and which will ultimately mature, is one-half. Accepting this value means that a mother must on the average produce two "litters" merely to replace herself (to give $r = 0$), so we shall examine the intrinsic rate of increase only for cases where n, the total number of births, is greater than two. To examine the maximum gain attainable by iteroparity we assume that successive births are spaced one year apart and obtain the value of r from formula (21), employing different values of n and α. It is easily seen that the necessary litter size, say b', to give the same value of r by means of a single reproductive effort at age α, would be precisely $e^{r\alpha}$.

The value of r from formula (21) corresponding to three annual births beginning at age 12 is .0312. At the other extreme, if the first of the three births occurs at age 30 we obtain $r = .0131$. The corresponding values of $e^{r\alpha} = b'$ are successively 1.41 and 1.48. Under these conditions *it would require essentially a three-fold increase in litter size to achieve in one reproductive effort the same biotic potential as that obtained from three successive births*. The same conclusion is obtained when we consider larger numbers of births. In the case of man very little could be gained by increasing the litter size by any reasonable amount and it is probable that the biological risk involved in producing multiple

births is more than sufficient to outweigh the very slight gain in biotic potential which could be obtained by this means. This would not be the case if the pre-reproductive period was drastically shortened, so we see that even in the case of man there is an interaction of life-history phenomena such that the importance of any conceivable change can only be evaluated through consideration of the total life-history pattern.

THE EFFECT OF TOTAL PROGENY NUMBER

In the preceding section we compared the two possible means by which an increase in total progeny number might lead to an increase in biotic potential. Our general conclusion was that the relative importance of changes in litter size and changes in the number of litters produced depends upon the rate of maturation. For species which mature early a modest change in litter size might be the equivalent of drastic changes in litter number but the possible value of iteroparity increases as the pre-reproductive part of the life span

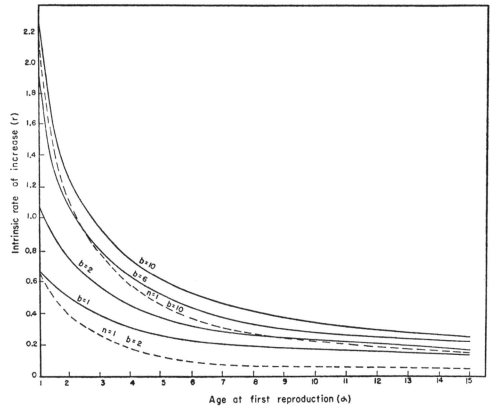

FIG. 4. THE EFFECT OF DELAYED MATURITY ON THE INTRINSIC RATE OF NATURAL INCREASE

The two broken lines represent semelparous species. The solid lines represent indefinitely iteroparous species where each female, after producing her first litter of size b, produces another similar litter in every succeeding time interval.

is lengthened. The importance of discovering the age at which reproduction begins has commonly been overlooked by students of natural history, hence it appears worthwhile to explore the matter further by examining the actual values of the intrinsic rate of natural increase corresponding to specified patterns of reproduction.

Fig. 4 was constructed from formula (21) to show, for several litter sizes, how the intrinsic rate of increase, r, is affected by lengthening the pre-reproductive period, α. Both semelparous ($n = 1$) and indefinitely iteroparous ($n = \infty$) species are illustrated. The striking feature of Fig. 4 is the way in which the lines representing different litter sizes converge as α increases. This occurs whether there is a single reproduction per lifetime or an infinite number, hence it is a general phenomenon. This supplements our earlier conclusions by suggesting that in species where reproductive maturity is delayed there should be relatively slight selection pressure for increased litter size. Here we are referring, of course, to the effective litter size or number of offspring which are capable of maturing. In cases where early mortality is very high, as is known to be the case with many fishes, it might require a tremendous increase in fecundity to produce a very small increase in effective litter size, and such increases might not be very important from the population standpoint. For example, a semelparous species reproducing at age 20 and with an effective litter size of 10 would have, $r = 0.120$. A ten-fold increase in litter size, to $b = 100$, would give $r = 0.231$ or an increase in biotic potential of 92 per cent. Another ten-fold increase to $b = 1000$ would give $r = 0.345$, or a gain of 50 per cent. The diminishing returns attainable by increasing litter size are obvious. For an iteroparous species reproducing first at age 20 and thereafter in each subsequent time interval, the increase in effective litter size from 10 to 100 would give only a 50 per cent increase in biotic potential and a further ten-fold increase in litter size would increase r by only another 35 per cent. In late-maturing species the litter size must be great enough to make it highly probable that *some* of the progeny will mature, but any further increases in fecundity will yield rapidly diminishing returns.

It is also clear from Fig. 4 that for any fixed litter size the biotic potential could be increased by shortening the period of maturation. Any specified amount of decrease in the pre-reproductive period will, however, be most effective for species where this part of the life span is already short.

Fig. 5 illustrates the way in which the two factors of length of the pre-reproductive part of the life span (α) and the number of offspring produced interact to determine the intrinsic rate of natural increase. These values were also computed from formula (21), in this case considering the litter size, b, as a constant with the value one-half. The figure then applies to species which, like man, produce one offspring at a time and where one-half of these offspring are females. Under these conditions it obviously requires two births just to replace the parents, but the population consequences of producing more than two offspring per lifetime vary tremendously with the age at which reproduction begins.

The female of the extinct passenger pigeon presumably produced her first brood consisting of a single egg at the age of one year. From the steep slope of the line representing $\alpha = 1$ in Fig. 5 it is clear that, beyond the minimum of two eggs per average female, several additional eggs produced in successive years would each add very appreciably to the value of r. Accordingly, a relatively slight reduction of the life expectancy for such a species might greatly reduce the biotic potential. The flattening out of the curves in Fig. 5 again illustrates the fact that each litter contributes less to potential population growth than the one preceding it. However, in a case such as that of the passenger pigeon even the seventh and eighth annual "litters" would add appreciable increments to the value of r.

Fig. 5 also shows that as the age at maturity increases the possible gains in biotic potential attainable by producing many offspring rapidly diminish. When $\alpha = 3$, as in the economically important fur-seal, each pup contributes much less to biotic potential than was the case for the eggs of the passenger pigeon. Nevertheless, it is apparent from the figure that if the life expectancy for females should be reduced to seven or eight years (corresponding to the fifth or sixth pup), or less, the species would be in a vulnerable position. The curve is steep in this portion of the graph and relatively slight changes in average longevity could produce disproportionately large population effects.

The lowest curves in Fig. 5 represent ages at maturity falling within the possible range for man. The curves come close together as α increases, so that in this range a change of a year or two in the age at which reproduction begins is less significant than in the case of a species that matures **more**

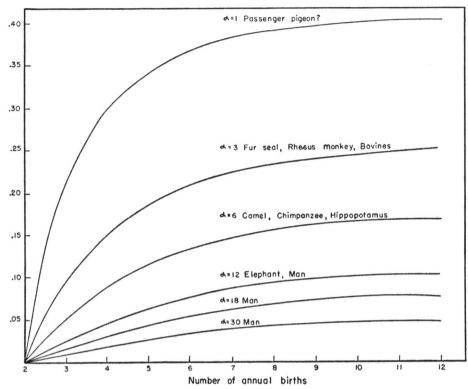

FIG. 5. THE EFFECTS OF PROGENY NUMBER ON THE INTRINSIC RATE OF NATURAL INCREASE WHEN THE
LITTER SIZE IS ONE ($b = \frac{1}{2}$)

The ordinate scale shows the intrinsic rate of increase for species which produce an average of one-half female offspring per litter. For any given total progeny number, the intrinsic rate (r) is seen to be greatly affected by the age (α) at which the first offspring is produced.

rapidly. Furthermore, the curves flatten out rapidly so that families which are very large by ordinary standards actually contribute little more to potential population growth or to future population than do families of quite moderate size. As an explicit illustration consider the intrinsic rate of increase which would result if, on the average, human females produced their first offspring at the age of 20 and had a total of five children spaced at one-year intervals. In this case we would have $r = 0.042$. If, on the other hand, we assume that, instead of producing only five children, the females could live forever producing a child each year we would obtain $r = 0.0887$. Under these conditions we conclude that in terms of biotic potential five children are almost one-half (actually 47 per cent) the equivalent of an infinite number. With larger values of α the effect of very large families would be even further reduced. From these considerations the writer feels that human biologists, as well as other natural historians, often overemphasize the importance of total number of progeny while

underestimating the significance of the age at which reproduction begins. It is impossible to conclude that one segment of the population is contributing more to future population than is some other segment without examining the total life-history pattern. Age at marriage could, in studying human populations, be a more significant datum than total family size.

The foregoing discussion suggests that a species such as man which is characterized by a long period of maturation and a small litter size can exhibit considerable variability in the details of its life history without greatly affecting the intrinsic rate of natural increase. The population consequences to be anticipated if the average age at which reproduction begins were to be altered by a few years or if the average number of progeny per female were slightly altered are much less striking than is the case for many other species. This implies that the intrinsic rate of increase should be relatively constant over the range of possible variations in the life-history features for man.

Such a conclusion would seem to be of both practical and theoretical interest and to merit closer examination.

The intrinsic rate of increase for man would be the rate of compound interest at which a human population, unrestrained by environmental resistance, would grow. We have already noted that Franklin (1751) estimated that a human population could double in 20 years and that this would correspond to $r = .035$. Malthus (1798) estimated that an unrestrained human population such as that of the United States at that time could double in 25 years. Malthus' estimate corresponds to $r = .0277$ which is remarkably close to the value of $r = .0287$ obtained by Lotka (1927) using much more refined methods for estimating the rate of increase prevailing in 1790 in the United States. Pearl and Reed (1920) fitted a logistic curve to the population figures for the United States, and their equation gives the value $r = .03134$. Additional examples of estimates based on empirical data

could be given, but these are sufficient to suggest that the value of the intrinsic rate of increase for man is not far from 0.03.

Fig. 6 was constructed from formula (21) in order to examine the question whether or not the life-history features of man would actually lead us to anticipate the approximate value, $r = 0.03$. Fig. 6 suggests rather definitely that the value $r = 0.02$ is too low, since it falls well below the obvious reproductive capabilities of humans. If females, on the average, had their first child at the age of 12 years (which is possible, cf. Pearl, 1930, p. 223) it would require an average of 2.6 surviving annual births per female to correspond to the rate $r = .02$. This curve is quite flat, so that if the first birth was delayed until the age of 20 years, which seems to be roughly the beginning of the semi-decade of maximum human fertility (Pearl, 1939), three annual births would still be adequate to give $r = .02$. Even if the first birth is delayed until the age of 28 years, this intrinsic rate of increase calls

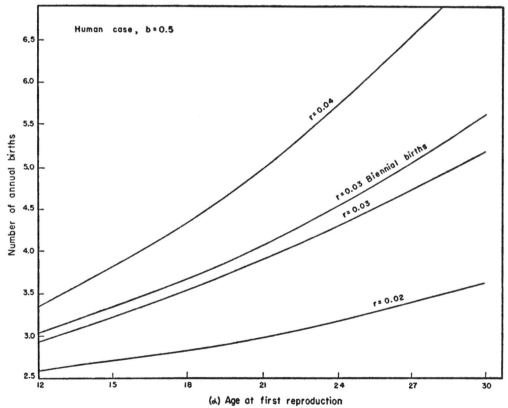

FIG. 6. AVERAGE REPRODUCTIVE PERFORMANCES REQUIRED TO GIVE SPECIFIED VALUES OF r, THE INTRINSIC RATE OF NATURAL INCREASE, IN HUMAN POPULATIONS

Assuming the average number of female offspring per human birth to be one-half ($b = 0.5$), this graph shows the extent to which total progeny number would have to be altered to maintain a specified intrinsic rate of increase while shifting the age at which reproduction begins. The figure also makes it possible to estimate the intrinsic rate of increase for a population when the average reproductive performance per female is known.

for an average progeny number of only 3.5. Clearly, man can easily exceed this average level of performance. On the other hand, the value $r = .04$ seems to require reproductive performance which would be astonishingly high as an average condition. If the first birth occurred when the mother was aged 13 years, an average of 3.5 children per female would suffice to give $r = .04$. However, the curve turns upward and if the first birth was delayed until the age of 20 it would require an average of 4.8 children to obtain this value. A delay of one more year, to the age of 21 for the first child, would increase the necessary mean progeny number to 5.0, while six children would be necessary if the first birth came when the mother was 25 years old. This intrinsic rate of increase then seems to call for exceptional rather than average reproductive performance, and the writer believes that Fig. 6 would lead us to expect that the intrinsic rate of increase for man lies between the limits .02 and .04 and might be estimated at about .03. Of course, the figure makes no allowance for mortality or for spacing the births at intervals of more than one year, so in actual cases the reproductive performance would have to be somewhat greater than indicated. The interval between births, however, is less critical than might be anticipated. The reproductive performances necessary to give $r = .03$ are shown both for one-year spacing and for two-year spacing between births and, to the writer, at least, either of these curves appears to represent a more reasonable picture of average human reproduction than do the cases representing the higher and lower intrinsic rates of increase.

THE POPULATION BIRTH-RATE AS A CONSEQUENCE OF LIFE-HISTORY PHENOMENA

We have noted earlier formulas (4) and (5) which were originally derived by Lotka (1907a, b; Sharpe and Lotka, 1911) and which show that when life-history features remain constant from generation to generation the population will ultimately settle down to a "fixed" or "stable" age distribution and will exhibit a fixed birth rate. We have also noted that this conclusion could be expected intuitively and can be obtained (formula 18) from discontinuous computational methods, once it is established that the potential form of population growth is a geometric progression. These potential consequences appear to provide the best justification for studying the life histories of various species, yet when such studies are conducted

it is common practice not to attempt any interpretation in terms of population phenomena.

It is evident from formulas (3), (4), and (5) that the birth rate and the stable age distribution are tied together with the intrinsic rate of increase and that any extensive discussion of the way in which changes in life-history features would affect these population features might repeat many of the points already covered. Consequently, we shall here note very briefly the relationships between life-history features and the resultant phenomena of birth rates and age structure.

If we consider a "closed" population, which changes in size only through the processes of birth and death, it is apparent that the intrinsic rate of increase, r, in formula (1), $\dfrac{dP}{dx} = rP$, must represent the difference between the instantaneous birth rate and the instantaneous death rate. In practice, however, we are more interested in a finite rate of population change. If we employ formula (1') to express the rate of population growth and consider that the changes result entirely from the birth rate (BR) and the death rate (DR), we obtain:

$$BR - DR = \frac{P_{x+1} - P_x}{P_x} = e^r - 1. \quad (26)$$

A birth rate, β, appropriate to this approach has already been defined by formula (18). However, because we are dealing with finite time intervals the B_x births regarded as occurring at the beginning of some time interval, T_x, should properly be credited to the P_{x-1} individuals living in the previous time interval. The birth rate would, therefore, be:

$$BR = e^r \beta. \quad (27)$$

On the other hand, the death rate should properly be the ratio of the D_x deaths in interval T_x to the total population exposed to the risk of death; that is D_x/P_x. Hence the simple relationship of formulas (5) and (26) can be misleading, especially when there is a rapid population turnover. For example, the birth rate β in Lotka's formula (5) is by definition identical with c_0, the fraction of the population aged between zero and one. Consequently β can never exceed unity no matter how many offspring are on the average produced per individual during a time interval.

In practical population problems the crude birth rate is often observed and employed as a criterion of the state of the population. This practice can be misleading, especially in comparing species which differ widely in their life-history features. It would be redundant to undertake a detailed analysis of the way in which life-history phenomena affect the crude birth rate because, as is evident from formulas (5), (26), and (27), any change which affects the value of *r* also affects the population birth-rate. However, the birth rate is subject to additional influences and these may be briefly examined.

The methods used in computing the value of *r* (formulas 17, 21, and 22) have involved survivorship only out to the age ω at which reproduction ceases. In species having a post-reproductive part of the life span, the crude birth rate corresponding to a given value of *r* will be reduced simply because post-reproductive individuals accumulate and are counted as part of the population on which the computations are based. It is clear, therefore, that

any increase in longevity (λ) which is not accompanied by an increase in ω will tend to lower the observed birth rate; conversely, the birth rate must affect the age structure of the population.

If the maximum longevity for a species is λ, a continued birth rate of 1/λ would just suffice to leave a replacement behind at the time each individual dies. Hence 1/λ represents an absolute minimum for a steady birth rate which is capable of maintaining the population. For example, if it were possible to keep every human female alive for 100 years a birth rate as low as 0.01 or 10 births per thousand population per annum (assuming ½ of the offspring to be females), could theoretically suffice for population maintenance. However, for human females the value of ω is not much beyond 40 years; in general, if a female has not produced a replacement by the age of 40 she is not going to do so. The latter consideration might lead one to

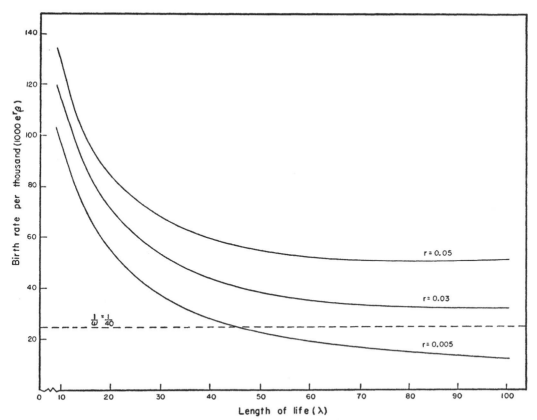

FIG. 7. POPULATION BIRTH RATES CORRESPONDING TO DIFFERENT LIFE HISTORY PATTERNS

The figure shows that the intrinsic rate of natural increase (*r*), the birth rate, and mean longevity (λ) are all interdependent. The broken line represents the minimum birth rate which would maintain a population if females did not survive beyond the age of 40 "years," which is here used as an estimate of the normal age (ω) at which reproduction is concluded. As longevity is increased, slow population growth becomes possible even when birth rates fall below 1/ω. Such low birth rates, however, have sometimes caused unwarranted concern about excessive population increase when abnormal conditions have temporarily reduced the death rate so that there is a large excess of births over deaths.

designate as a minimum maintenance birth rate the value $1/\omega$, or 0.025 in the case of man. But it is clear from the foregoing sections that offspring produced prior to the time that a female reaches the age of 40 will already have begun to "accumulate interest" before the mother can reach the age of 100. The compound interest nature of potential population growth complicates the relationship between birth rate and life-history phenomena and makes it conceivable that populations of species where $\lambda > \omega$ could even continue to grow while exhibiting birth rates lower than $1/\omega$.

No end of interesting combinations of birth rates, death rates, and life-history phenomena might merit consideration, but a single simple example will be selected here to illustrate the general relationship between changing longevity and population birth-rates. If we assume, as before, a physiological type of survivorship such that each individual that is born lives to attain its λth birthday but dies before reaching the age of $\lambda + 1$, we can sum the right-hand side of formula (19) as a geometric progression and combine this with formula (27) to obtain the following expression for the population birth rate (BR):

$$BR = e^r\beta = \frac{e^r - 1}{1 - e^{-r\lambda}}. \qquad (28)$$

Formula (28) shows, as would be expected, that as longevity (λ) is increased, with the other life-history features remaining unchanged, the birth rate will fall and approach as a limit the value $e^r - 1$, which is in accord with (26) when the death rate is set equal to zero.

Fig. 7 was constructed from formula (28) to illustrate the interrelations between BR, r, and λ within a range of values of life history features roughly applicable to man. The condition for a stationary population would, of course, correspond to $r = 0$, while all positive values of r correspond to growing populations. r must become negative when the birth rate falls below $1/\lambda$, and it will be noted that the birth rates below about 20 per thousand which are sometimes observed in human populations (see tabulation in Allee et al., 1949, p. 288) must, unless they represent abnormal temporary phenomena, correspond to populations with very low potential growth rates. The curve for $r = .005$, in fact, does not differ greatly from $1/\lambda$. The curves in Fig. 7 flatten out rapidly for large values of λ, so that drastic and generally unattainable increases in longevity would be re-

quired to make such low birth rates compatible with appreciable population growth.

Looking at these relationships from a different point of view, Fig. 7 shows that a reduction in longevity, such as might result from reducing the life expectancy of game animals, can be expected to result in an increased birth rate even if the intrinsic rate of increase is unchanged. It does not seem worthwhile at present to attempt a quantitative estimate of these relationships because the assumption of a physiological type of survivorship curve is probably not even approximately true for game animals. When more realistic estimates of survivorship are available, however, the type of relationship illustrated in Fig. 7 may assume practical importance.

THE STABLE AGE DISTRIBUTION

The age structure of a population often is a matter of considerable practical concern. In economically valuable species such as timber, game animals, and commercial fishes certain age classes are more valuable than others, and it would be desirable to increase the proportion of the most valuable age classes in a population. Similarly, certain age classes of noxious organisms may be more destructive than others and the relative numbers of these destructive individuals will be governed by life-history phenomena which may conceivably be subject to alteration by control measures. In human populations, also, it is sometimes a matter of concern that the proportion of the population falling within the age limits most suitable for physical labor and military service seems to be below optimum. An article in the New York *Times* for September 24, 1950, headed "population shift in France traced—Study finds too many aged and very young in relation to total of workers" illustrates the potential importance of a knowledge of the age structure of populations.

The mathematical basis for relating the age structure of a population to life-history features was established in Lotka's first paper on population analysis (1907); and in the same year Sundbärg (1907) reached the conclusion that a human population reveals its condition (tendency to grow or decline) through its age structure. These important conclusions have not been sufficiently noted by ecologists. When the mortality factors affecting a population are altered either through natural environmental changes or through human exploitation or attempts at control there will in general result a change in the age structure of the population, and this may be observable even before changes in population size or in birth rates provide evidence of the consequences of the changed mortality factors. The subject of age structure is a large and difficult one because the various combinations of life-history features, birth rates, death rates, and age structure are analogous to a multidimensional figure where a change imposed in any one feature induces changes in all of the others. The subject has been considered most in connection with human populations, and some empirical generalizations have been obtained which may profitably be examined by means of the computational methods we have been employing. For the purpose of illustrating the general character of the relationships involved, one species will serve as well as another.

For illustrative purposes we may proceed as in the preceding sections and consider the stable age distribution for cases where survivorship is of the physiological type. Letting c_x represent the fraction of the population aged between x and $x + 1$, we may employ formulas (5) and (28) directly to obtain, for $x < \lambda$:

$$c_x = \frac{e^{-rx}(1 - e^{r})}{1 - e^{-r\lambda}}. \qquad (29)$$

From formula (29) it is apparent that if the extreme longevity for a population is altered without changing the intrinsic rate of increase, the effect on the age structure will be of a very simple type. An increase in longevity from λ_1 to λ_2 will simply reduce the proportion of the population in each age category below λ_1 by the constant proportion $\frac{1 - e^{-r\lambda_1}}{1 - e^{-r\lambda_2}}$. Consequently, the effect will be most noticeable on the youngest age classes, because these are the largest classes.

Changes in the value of r affect the stable age distribution in a more complex way than do changes in λ, although the general result of increasing the value of r will be to increase the proportion of young in the population, with a corresponding decrease in the proportion of older individuals. Fig. 8 illustrates this effect for three values of r,

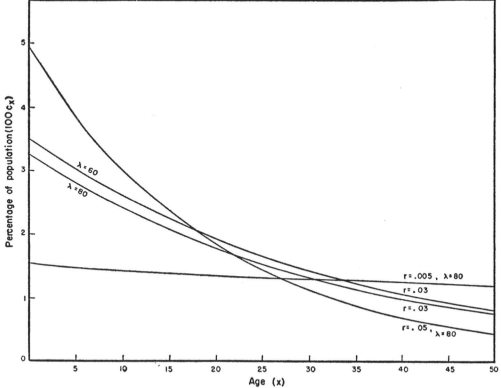

Fig. 8. The Stable Age Distribution, or Proportion of the Population Falling within each Interval of Age, is Shown Here as a Function of the Intrinsic Rate of Increase (r) and the Length of Life (λ)

These relationships have important ecological correlaries. See text for a discussion relating these to the "optimum yield" problem.

assuming that λ in formula (29) remains constant at 80 "years."

The way in which population age-structure is affected by changes in the value of r and λ, as illustrated by formula (29) and Fig. 8, permits some qualitative conclusions which are of interest in connection with the "optimum yield problem" (for discussion, see Allee et al., 1949, p. 377). If man (or some other species) begins exploitation of a previously unexploited population, the age structure will be affected in a definite manner. An obvious result of increased predation will be to decrease the average longevity, corresponding to a decrease in λ. We have seen earlier (Fig. 7) that this will ordinarily have the effect of increasing the population birth rate, and Fig. 8 and formula (29) show that another effect will be to increase the proportion of young in the population, the very youngest age classes being most affected. However, the increased mortality may also affect the value of r. If the population is initially in equilibrium with the capacity of its environment, its total life history pattern will be adjusted to the effective value, $r = 0$. If exploitation is not too intense, it has the effect of making additional environmental resources available to the surviving members of the population and thus of stimulating population increase; the population "compensates" (see Errington, 1946) for the increased mortality by increasing the value of r. Fig. 8 shows that this will have the effect of increasing the proportion of young in the population, thus supplementing the effect of exploitation on λ. On the other hand, the increase in r will have the effect of reducing the proportion of aged individuals in opposition to the effect of reduced λ which is to increase all of the age classes which still persist. It seems clear that the most obvious population consequence of such exploitation will be to increase the proportion of young members of the population. If predation or exploitation becomes still more intense, as in the case of "overfishing" (Russell, 1942), it will reduce the effective value of r, and, of course, still further reduce average longevity. The decrease in r will tend to decrease the proportion of young individuals, but the decrease in λ will tend to increase this proportion. However, both changes will tend to raise the proportion of older individuals in the population, and this combined effect can then be expected to be the most obvious corollary of overfishing. These conclusions, of course, greatly oversimplify a complex phenomenon. In order to make quantitative estimates of these effects it would be necessary to have detailed information about the life-history features, especially survivorship under the conditions of increased predation. Nevertheless, these qualitative conclusions show the type of effect to be expected when populations are subjected to increased predation, and they suggest that observations of the changes in the age structure of populations may provide valuable evidence of over-exploitation, or, from the opposite point of view, of the effectiveness of control measures. Bodenheimer (1938) comments on the fact that ecologists have neglected this important subject.

From Fig. 8 it will be noticed that changes in life-history features produce their greatest effects on the extreme age classes, whereas the curves representing different patterns are close together in the "middle" age range. The same phenomenon is evident, for example, in a graph (Fig. 27) reproduced by Dublin, Lotka, and Spiegelman (1949) from Lotka (1931) to show the age structure of human populations corresponding to stationary, increasing, and decreasing populations. This suggests that the proportion of a population falling in the middle portion of the life span may be relatively independent of factors which produce drastic shifts in the ratio of very old to very young. This was first postulated by Sundbärg (1907), who concluded that it was "normal" for about 50 per cent of a human population to fall in the age range between 15 years and 50 years. Sundbärg distinguishes three primary population types based on the age distribution of the remaining 50 per cent of the population. In the "progressive" type of population there is a strong tendency for increase and the ratio of young (aged under 15) to old (aged over 50) is, by Sundbärg's criteria, about 40 to 10. In the "regressive" type, exhibiting a tendency toward population decrease, the corresponding ratio of young to old is about 20 to 30, while a "stationary" type with the ratio about 33 to 17 shows no particular tendency either to grow or to decrease. When first encountered Sundbärg's conclusion appears surprising, but in actual human populations differing as radically, for example, as those of Sweden and India, the proportion of the population aged between 15 and 50 is remarkably close to 50 per cent (see tabulation in Pearl, 1946, p. 78). It appears to the writer that this conclusion should be of great interest to students of human populations. The age class between 15 and 50 years includes the bulk of the workers and persons of military age, and Sundbärg's conclusion implies that the size of this class relative to the remainder of the population must be determined by life-history features which cannot readily be deliberately controlled.

For other species also, the life span may be meaningfully divided into three primary age classes, pre-reproductive (aged 0 to α), reproductive (aged α to ω), and post-reproductive (aged ω to λ), which differ considerably in their biological significance. If we continue with our assumption of physiological survivorship we can obtain the relative sizes of these three age classes directly from formula (29), and Sundbärg's generalization offers an interesting empirical pattern with which to compare our results. For the relative sizes of the three fundamental age classes formula (29) gives:

$$\left.\begin{array}{ll} \text{Pre-reproductive} & = \dfrac{1 - e^{-r\alpha}}{1 - e^{-r\lambda}} \\[2ex] \text{Reproductive} & = \dfrac{e^{-r\alpha} - e^{-r\omega}}{1 - e^{-r\lambda}} \\[2ex] \text{Post-reproductive} & = \dfrac{e^{-r\omega} - e^{-r\lambda}}{1 - e^{-r\lambda}} \end{array}\right\} \quad (30)$$

By putting $\alpha = 15$ and $\omega = 50$, we may examine the relationship between r and λ for Sundbärg's primary population types.

The generalization that about 50 per cent of a human population normally falls in the age range from 15 to 50 is in accord with formulas (30). When $r = .03$ the value of λ to give a stable age distribution with just 50 per cent in the middle age range would be 59 years, but an increase in longevity to 85 years would only reduce this class to 45 per cent of the population. The same general conclusion applies when r is small. The values $r = .005$ and $\lambda = 63$ years correspond to 55 per cent aged 15 to 50 years, and λ would have to be increased to 81 years to reduce this to 45 per cent. It appears that over the usual range of values of human longevity and potential population growth Sundbärg's generalization is very good. This is shown graphically in Fig. 9. It is noteworthy that when the length of life is about 70 years, Sundbärg's generalization holds over a wide range of values of r. In other words, this ratio of "middle-aged" to total population is quite insensitive to changes in other life history features.

Figure 9 illustrates Sundbärg's population criteria for values of $r > 0$. It is clear that the

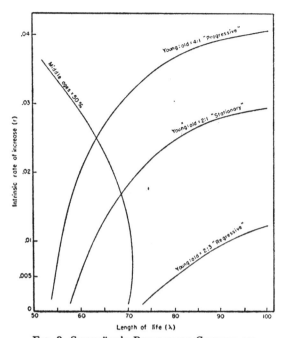

FIG. 9. SUNDBÄRG'S POPULATION CRITERIA IN RELATION TO LIFE-HISTORY FEATURES

The "middle age" range here consists of individuals aged 15 years to 50 years.

"regressive" type of population structure is not consistent with populations possessing a strong tendency to increase and, in fact, for usual longevity figures, will ordinarily correspond to decreasing populations ($r < 0$). On the other hand, the "progressive" type of structure does correspond to large values of r even when λ is not particularly large. The "stationary" type of population structure is less well defined. Any population for which the life-history features correspond to $r > 0$ will tend to grow, and the ratio of two "old" to one "young" can correspond to a large value of r when λ is large. The value of r is, of course, independent of the length of the post-reproductive part of the life span, whereas the age structure is not. For any given value of r, which is determined by the life-history features of individuals aged below ω, the effect of an increase in longevity will be to decrease the proportion of young and, to a lesser extent, the proportion of reproductive members of the population.

At first glance, the type of interactions shown in Fig. 9 might suggest that if the age structure of a population were artificially shifted in the "regressive" direction, for example, through migration or improvements in public health, there would result a reduction in the value of r. This, however, is not the case, since *r* is completely determined by life-history events occurring before age ω. Sharpe and Lotka (1911) showed that a characteristic of the "stable" age distribution is the fact that it will become reestablished after temporary displacements. However, practical problems arise in analyzing actual populations because phenomena such as birth rates and death rates may give a very misleading picture of population trends when the age structure is displaced from the stable type and has not yet had time to reestablish a stable age distribution. From Fig. 7 we have noted that in human populations birth rates on the order of 20 per thousand per annum would in general be little more than sufficient to maintain a population. But if the age distribution is displaced from the stable type in the direction of increased frequency of young individuals, the death rate may be temporarily reduced, so that there is a great excess of births over deaths and the population can grow temporarily even if the age-specific birth rates are too low to maintain the population permanently at a constant size. Under these conditions the population is "aging," and in human populations this phenomenon has a number of interesting and important corollaries (see Dublin, Lotka, and Spiegelman, 1949). Fisher (1930) has considered the problem in a more general sense, noting that the apparent value of *r* given by formula (3) will be incorrect when the age structure is displaced. Fisher suggests the possibility of measuring population size not in terms of the number of individuals present but in terms of "reproductive value," where the "value" of an individual represents his remaining potentialities for contributing to the ancestry of future generations. Some such approach has great potential value for ecological studies of natural populations, but its possibilities in this direction seem not yet to have been explored.

EMPIRICAL APPLICATIONS

In the preceding sections we have been concerned with the influence of specific life-history patterns on the characteristics of populations. In order to examine these effects we have made simplifying assumptions by regarding some of the life-history features as fixed in a certain way while we examined the results of varying other features. While this procedure oversimplifies the biological situation as it exists in actual populations, the writer regards it as a sound way of investigating

the meaning of life-history features. The same general attitude may be traced back to Robert Wallace (1753) in his book which profoundly influenced Malthus. Wallace pointed out that "mankind do not actually propagate according to the rules in our tables, or any other constant rule . . ." but he emphasized that tables of potential population growth are still valuable because they permit us to evaluate the influences restraining population growth.

Wallace, then, was a pioneer in appreciating the potential value of comparisons between empirical and theoretical population phenomena. In modern actuarial practice, population data are subjected to involved mathematical treatments which are sometimes considered to represent biological laws and at other times to be merely empirical equations, but which, in any case, are known to yield results of practical value. In the words of Elston (1923, p. 68):

"... it seems to me that even though there be governing causes of mortality that may result in a true law of mortality, any group of lives studied is so heterogeneous, due to differences in occupation, climate, sanitary conditions, race, physical characteristics, etc., that any formula must in practice be considered to be merely a generalization of what is actually happening."

The number of different combinations of life-history features of the type we have been discussing is essentially infinite, and it is out of the question to make detailed examinations of any great proportion of these from the theoretical standpoint. However, we have seen that certain population features, such as the prevailing age distribution and the intrinsic rate of increase, summarize a great deal of information about the potentialities of the population and its relationship to its particular and immediate environment. As mentioned earlier, the recent ecological literature demonstrates that ecologists are becoming interested in determining such features as the intrinsic rate of increase for non-human populations. These computations may have practical value in dealing with valuable or noxious species, and they possess great theoretical interest for ecologists. For example, the logistic equation has been widely employed to represent population growth in a variety of organisms (Allee et al., 1949; Pearl, 1927); and it has also been attacked (Yule, 1925; and succeeding discussion, Gray, 1929; Hogben, 1931; Smith, 1952), on the grounds that it is too versatile and can be made to fit empirical data that might arise from entirely different "laws" of population growth. This criticism is, of course, directed at the fact that the curve-fitter has three arbitrary constants at his disposal in seeking to obtain a good fit. One of these constants is r, the intrinsic rate of increase. The grounds for accepting or rejecting the logistic equation as a law of population growth (or for seeking some other law) would be greatly strengthened if the value of r was computed directly from observed life-history features and independently of the data on population size.

The computational methods employed in the preceding sections suggest several possible ways of computing the value of r from empirical life-history data. The usual procedure for such computations has been a tedious one based on formula (3) (see Lotka's appendix to Dublin and Lotka, 1925), although Birch (1948) employed formula (17) as an approximation to (3) for his computations. The methods discussed in the preceding sections suggest that a logical procedure for obtaining an empirical value of r would be to observe age-specific birth rates and survivorship under the environmental conditions of interest, and from these to write a "generation law" so that formula (11) can be employed. The single positive root of (11) is e^r and this can be estimated to any desired degree of accuracy without great difficulty even for species where the reproductive life is prolonged. When one is actually solving equation (11) it is strikingly brought to one's attention that the final terms representing reproduction in later life are relatively unimportant in influencing the value of r. This once again reinforces our conclusion that reproduction in early life is of overwhelming importance from the population standpoint, and should be much more carefully observed in field and laboratory studies than has usually been the case.

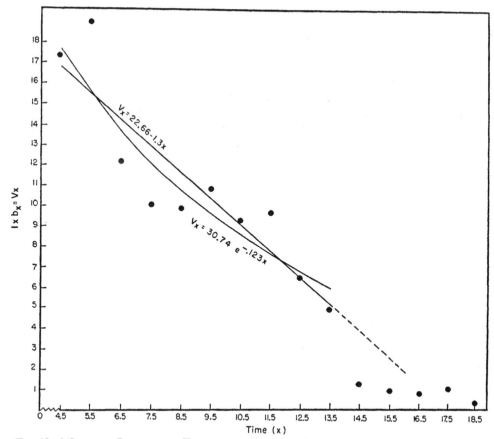

FIG. 10. A STRAIGHT LINE AND AN EXPONENTIAL FUNCTION FITTED TO THE DATA OF BIRCH (1948)

These data are suitable only for illustrating computational methods. Birch observed the age-specific fecundity rates (b_x) for the rice weevil *Calandra oryzae* living under constant conditions of temperature and moisture. He did not observe survivorship (l_x) but carried over the shape of the survivorship curve from the flour beetle *Tribolium confusum*. The black dots represent the product $l_x b_x$.

An alternative approach for obtaining empirical values of r consists of fitting the V_x ($= l_x b_x$) values with mathematical curves of types which make it possible either to obtain the sum of the series on the right-hand side of equation (19) or to employ Laplace transformations to solve equations (3) or (20) by simple iterative methods. Different formulas could be fitted to different sections of the V_x curve, and, when good fits can be obtained with combinations of straight lines, step-functions, exponential functions, and other simple formulas, this procedure will often lead to easy ways of solving for r. As a very simple example of this procedure we may consider the data of Birch (1948), for which he employed formula (17) and, by computations which are given in detail in his paper, obtained the value $r = 0.762$.

The V_x data employed by Birch are shown graphically in Fig. 10 by means of the black dots. The two curves, a straight line and a simple exponential function, were fitted by the method of least squares to the values up to $x = 13.5$. Neither of the functions gives an extremely good fit to the observed data, but, on the other hand, one may question whether the irregularities in the empirical data are not partly artifacts which should be smoothed out when one is attempting to estimate the value of r for a species. In any case, it appears worthwhile to compare Birch's value of $r = .762$ with that obtained by use of the empirically fitted curves.

Using first the exponential curve, if we set $V_x = Ke^{ax}$, formula (19) may be written in the form:

$$Ke^{\alpha(a-r)} = \frac{1 - e^{a-r}}{1 - e^{n(a-r)}}. \qquad (31)$$

When reproduction occurs several times, so that n is fairly large, the denominator in equation (31) becomes for practical purposes unity and may be ignored. In this case we have $\alpha = 4.5$, $K = 30.74$, $a = -.123$, and $n = 9$, so we will ignore the denominator. The equation is easily solved by iterative means, using a table of the exponential function, and the value obtained is $r = .758$, which differs from Birch's value by about one-half of one per cent.

The right-hand side of equation (19) can also be summed for the linear case where $V_x = a + bx$, but in this case the Laplace transformation employed with formula (3) yields a slightly simpler equation:

$$\frac{ar + b}{r^2} = e^{r\alpha}. \qquad (32)$$

Putting $a = 22.66$, $\alpha = 4.5$, and $b = -1.3$ in equation (32), and solving by iterative means we obtain $r = .742$, which differs from Birch's value by nearly three per cent, although this seems to be good agreement in view of the crude approximations employed.

In many practical applications dealing with natural and experimental populations some approximation to the value of r such as those presented above may be all that can be justified by the accuracy of the data. The estimate of r could undoubtedly be improved by fitting different portions of the V_x curve with different functions. Another refinement suggested by observations presented earlier in this paper would be to fit the empirical curves by a method which would give greater weight to the earlier points which have more influence on the value of r than do the later points. Any detailed discussion of these empirical applications would be out of place in the context of the present paper. The writer, however, anticipates that ecologists will in the future devote more attention to the interrelations of life-history features and population phenomena, and it is to be hoped that some of the approaches which have been indicated will accelerate trends in this direction.

SUMMARY

Living species exhibit a great diversity of patterns of such life-history features as total fecundity, maximum longevity, and statistical age schedules of reproduction and death. Corresponding to every possible such pattern of life-history phenomena there is a definitely determined set of population consequences which would ultimately result from adherence to the specified life history. The birth rate, the death rate, and the age composition of the population, as well as its ability to grow, are consequences of the life-history features of the individual organisms. These population phenomena may be related in numerous ways to the ability of the species to survive in a changed physical environment or in competition with other species. Hence it is to be expected that natural selection will be influential in shaping life-history patterns to correspond to efficient populations.

Viewed in this way, comparative studies of life histories appear to be fully as meaningful as studies of comparative morphology, comparative psychology, or comparative physiology. The former type of study has, however, been neglected from the evolutionary point of view, apparently because the adaptive values of life-history differences are almost entirely quantitative. The recent ecological literature does show a trend toward the increasing application of demographic analysis to non-human populations, but the opposite approach of deducing demographic consequences from life-history features has been relatively neglected. The present paper is presented with the hope that this situation can be changed. In other fields of comparative biology it is usual to examine individual characteristics and to regard these as possible adaptations, and the writer believes that life-history characteristics may also be profitably examined in this way.

It is possible by more or less laborious methods to compute the exact size and composition of the population which at any future time would be produced by any given initial population when the life-history pattern of the individual organisms is regarded as fixed. Thus it is possible to make an exact evaluation of the results of changing any life-history feature, and the value of this type of analysis may be apparent to those biologists who distrust the usual demographic procedures.

Starting with exact computational methods, it has been shown that early population growth may exhibit irregularities or cyclic components which are identifiable with negative or complex roots of an algebraic equation, but that these components vanish in time so that potential population growth is ultimately a geometric progression. Having established this fact, it is shown that the exact computational methods and the more convenient approximate methods lead to identical conclusions when considered over the long time scale which is of interest in adaptational and evolutionary considerations.

Some life-history patterns of ecological interest are examined and compared by means of relatively simple formulas derived from a consideration of the form of potential population growth. The results have bearing on the possible adaptive value of genetically induced changes of life-history features. It is suggested that this type of approach may add to the value of life-history studies and that an awareness of the possible meanings of empirical life-history data may aid in planning such studies by insuring that all pertinent information will be recorded. One of the most striking points revealed by this study is the fact that the age at which reproduction begins is one of the most significant characteristics of a species, although it is a datum which is all too frequently not recorded in the literature of natural history.

The number of conceivable life-history patterns is essentially infinite, if we judge by the possible combinations of the individual features that have been observed. Every existing pattern may be presumed to have survival value under certain environmental conditions, and the writer concludes that the study of these adaptive values represents one of the most neglected aspects of biology.

LIST OF LITERATURE

ALLEE, W. C. 1934. Recent studies in mass physiology. *Biol. Rev.*, 9: 1–48.

——, A. E. EMERSON, O. PARK, T. PARK, and K. P. SCHMIDT. 1949. *Principles of Animal Ecology.* Saunders, Philadelphia.

ARCHIBOLD, R. C. 1918. A Fibonacci series. *Amer. math. Monthly*, 25: 235–238.

BANTA, A. M., T. R. WOOD, L. A. BROWN, and L. INGLE. 1939. Studies on the physiology, genetics, and evolution of some Cladocera. *Publ. Carneg. Inst.*, 39.

BERKSON, J. 1944. Application of the logistic function to bio-assay. *J. Amer. statist. Ass.*, 39: 357–365.

——. 1951. Why I prefer logits to probits. *Biometrics*, 7: 327–329.

BIRCH, L. C. 1948. The intrinsic rate of natural increase of an insect population. *J. Anim. Ecol.*, 17: 15–26.

BODENHEIMER, F. S. 1938. *Problems of Animal Ecology.* Oxford Univ. Press, London.

CHAPMAN, R. N. 1928. The quantitative analysis of environmental factors. *Ecology*, 9: 111–122.

——. 1938. *Animal Ecology.* McGraw-Hill, New York.

CHURCHILL, R. V. 1944. *Modern Operational Mathematics in Engineering.* McGraw-Hill, New York.

DEEVEY, E. S., JR. 1947. Life tables for natural populations of animals. *Quart. Rev. Biol.*, 22: 283–314.

DUBLIN, L. I., and A. J. LOTKA. 1925. On the true rate of natural increase as exemplified by the population of the United States, 1920. *J. Amer. statist. Ass.*, 20: 305–339.

——, ——, and M. SPIEGELMAN. 1949. *Length of Life. A Study of the Life Table*, rev. ed. Ronald Press, New York.

DUNKEL, O. 1925. Solutions of a probability difference equation. *Amer. math. Monthly*, 32: 354–370.

ELSTON, J. S. 1923. Survey of mathematical formulas that have been used to express a law of mortality. *Rec. Amer. Inst. Actuar.*, 12: 66–95.

ERRINGTON, P. L. 1946. Predation and vertebrate populations. *Quart. Rev. Biol.*, 21: 144–177, 221–245.

EVANS, F. C., and F. E. SMITH. 1952. The intrinsic rate of natural increase for the human louse, *Pediculus humanus* L. *Amer. Nat.*, 86: 299–310.

FIBONACCI, L. (1857). *Liber Abbaci di Leonardo Pisano publicati da Baldasarre Boncompagni.* Tipografia delle Scienze mathematiche e fisiche, Roma.

FINNEY, D. J. 1947. The principles of biological assay. *J. R. statist. Soc.*, 109: 46–91.

——. 1949. The choice of a response metameter in bio-assay. *Biometrics*, 5: 261–272.

FISHER, R. A. 1930. *The Genetical Theory of Natural Selection.* Oxford Univ. Press, London.

FRANK, P. W. 1952. A laboratory study of intraspecies and interspecies competition in *Daphnia pulicaria* (Forbes) and *Simocephalus vetulus* O. F. Müller. *Physiol. Zool.*, 25: 178–204.

FRANKLIN, B. 1751. Observations concerning the increase of mankind and the peopling of countries. In *The Works of Benjamin Franklin*, ed. by J. Sparks, 1836, Vol. 2: 311–321. Hilliard, Grey, Boston.

GAUSE, G. F. 1934. *The Struggle for Existence.* Williams & Wilkins, Baltimore.

GOMPERTZ, B. 1825. On the nature of the function expressive of the law of human mortality. *Phil. Trans.*, 36: 513–585.

GRAUNT, J. 1662. *Natural and Political Observations Mentioned in a Following Index and Made upon the Bills of Mortality. . . .* Printed by T. Roycroft, London.

GRAY, J. 1929. The kinetics of growth. *Brit. J. exp. Biol.*, 6: 248–274.

HOGBEN, L. 1931. Some biological aspects of the population problem. *Biol. Rev.*, 6: 163–180.

HUMPHREYS, W. J. 1929. *Physics of the Air*, 2nd ed. McGraw-Hill, New York.

HUTCHINSON, G. E. 1948. Circular causal systems in ecology. *Ann. N. Y. Acad. Sci.*, 50: 221–248.

HYMAN, L. H. 1951. *The Invertebrates. Vol. 2. Platyhelminthes and Rhynchocoela. The Acoelomate Bilateria.* McGraw-Hill, New York.

JACKSON, C. H. N. 1939. The analysis of an animal population. *J. Anim. Ecol.*, 8: 238–246.

JORDAN, C. 1950. *Calculus of Finite Differences*, 2nd ed. Chelsea, New York.

KOSTITZIN, V. A. 1939. *Mathematical Biology.* Harrap, London.

KUCZYNSKI, R. R. 1932. *Fertility and Reproduction.* Falcon Press, New York.

——. 1935. *The Measurement of Population Growth.* Sidgwick & Jackson, London.

LESLIE, P. H., and R. M. RANSON. 1940. The mortality, fertility, and rate of natural increase of the vole (*Microtus agrestis*) as observed in the laboratory. *J. Anim. Ecol.*, 9: 27–52.

——, and T. PARK. 1949. The intrinsic rate of natural increase of *Tribolium castaneum* Herbst. *Ecology*, 30: 469–477.

LINNAEUS, C. 1743. Oratio de telluris habitabilis. In *Amoenitates Academicae seu Dissertationes Variae. . . .* Editio tertia curante J. C. D. Schrebero, 1787, Vol. 2: 430–457. J. J. Palm, Erlangae.

LOTKA, A. J. 1907a. Relation between birth rates and death rates. *Science*, 26: 21–22.

——. 1907b. Studies on the mode of growth of material aggregates. *Amer. J. Sci.*, 24: 119–216.

——. 1910. Contributions to the theory of periodic reactions. *J. phys. Chem.*, 14: 271–274.

——. 1925. *Elements of Physical Biology.* Williams & Wilkins, Baltimore.

——. 1927. The size of American families in the eighteenth century and the significance of the empirical constants in the Pearl-Reed law of population growth. *J. Amer. statist. Ass.*, 22: 154–170.

——. 1931. The structure of a growing population. *Human Biol.*, 3: 459–493.

——. 1934. *Théorie Analytique des Associations Biologiques.* Hermann, Paris.

LOWAN, A. N., Technical director. 1941. *Table of Natural Logarithms.* 4 vols. Federal Works Agency, Works Projects Administration for the City of New York. Rept. of Official Project No. 15-2-97-33.

MALTHUS, T. R. 1798. *An Essay on the Principle of Population as it Affects the Future Improvement of Society, with Remarks on the Speculations of Mr. Godwin, M. Condorcet, and Other Writers.* Printed for J. Johnson in St. Paul's Churchyard, London.

MENDES, L. O. 1949. Determinação do potencial biotico da "broca do café"—*Hypothenemus hampei* (Ferr.)—e consideracões sôbre o crescimento de sua população. *Ann. Acad. bras. Sci.*, 21: 275–290.

MILNE-THOMPSON, L. M. 1933. *The Calculus of Finite Differences.* Macmillan, London.

MOLISCH, H. 1938. *The Longevity of Plants.* E. H. Fulling, Lancaster Press, Lancaster, Pa.

NEWMAN, F. W. 1883. Table of the descending exponential function to twelve or fourteen places of decimals. *Trans. Camb. phil. Soc.*, 13: 145–241.

NICHOLSON, A. J., and V. A. BAILEY. 1935. The balance of animal populations. Part I. *Proc. zool. Soc. Lond.*, 1935: 551–598.

PEARL, R. 1925. *The Biology of Population Growth.* Knopf, New York.

——. 1930. *Introduction to Medical Biometry and Statistics*, 2nd ed. Saunders, Philadelphia.

——. 1937. On biological principles affecting populations: human and other. *Amer. Nat.*, 71: 50–68.

——. 1939. *The Natural History of Population.* Oxford Univ. Press, London.

——. 1946. *Man the Animal.* Principia Press, Bloomington, Ind.

——, and L. J. REED. 1920. On the rate of growth of the population of the United States since 1790 and its mathematical representation. *Proc. natl. Acad. Sci. Wash.*, 6: 275–288.

——, and J. R. MINER. 1935. Experimental studies on the duration of life, XIV. The comparative mortality of certain lower organisms. *Quart. Rev. Biol.*, 10: 60–79.

PIERCE, J. C. 1951. The Fibonacci series. *Sci. Monthly*, 73: 224–228.

QUETELET, A. 1835. *Sur l'Homme et le Développment de ses Facultés ou Essai de Physique Sociale.* Bachelier, Imprimeur-Libraire, Paris.

RHODES, E. C. 1940. Population mathematics. *J. R. statist. Soc.*, 103: 61–89, 218–245, 362–387.

RICKER, W. E. 1948. *Methods of Estimating Vital Statistics of Fish Populations.* Indiana Univ. Pubs. Sci. Ser., 15, Bloomington.

ROSS, R. 1911. *The Prevention of Malaria*, 2nd ed. Dutton, New York.

RUSSELL, E. S. 1942. *The Overfishing Problem.* Cambridge Univ. Press, Cambridge.

SADLER, M. T. 1830. *The Law of Population: A Treatise, in Six Books; in Disproof of the Superfecundity of Human Beings, and Developing the Real Principle of their Increase.* Murray, London.

SALT, G. 1936. Experimental studies in insect parasitism. IV. The effect of superparasitism on populations of *Trichogramma evanescens*. *Brit. J. exp. Biol.*, 13: 363–375.

SHARPE, F. R., and A. J. LOTKA. 1911. A problem in age distribution. *Phil. Mag.*, 21: 435–438.

SMITH, F. E. 1952. Experimental methods in population dynamics, a critique. *Ecology*, 33: 441–450.

STANGELAND, C. E. 1904. *Pre-Malthusian Doctrines of Population; a Study in the History of Economic Theory.* Columbia Univ. Stud. Hist. Econ. pub. Law, Vol. 21, no. 3.

SUNDBÄRG, A. G. 1907. *Bevölkerungsstatistik Schwedens 1750–1900.* P. A. Norstedt and Söner, Stockholm.

THOMPSON, D'ARCY W. 1942. *On Growth and Form*, new ed. Cambridge Univ. Press, Cambridge.

THOMPSON, W. R. 1931. On the reproduction of organisms with overlapping generations. *Bull. ent. Res.*, 22: 147–172.

——. 1939. Biological control and the theories of the interactions of populations. *Parasitology*, 31: 299–388.

U. S. Forest Service. 1948. The woody-plant seed manual. *Misc. Publ. U. S. Dep. Agric.*, No. 654.

VERHULST, P. F. 1838. Notice sur la loi que la population suit dans son accroisissement. *Corresp. math. phys., A. Quetelet*, 10: 113–121.

——. 1845. Recherches mathématiques sur la loi d'accroissement de la population. *Mém. Acad. R. Belg.*, 18: 1–38.

VOLTERRA, V. 1927. Variazioni e fluttuazioni del numero d'individui in specie animali conviventi. *Mem. Accad. Lincei*, 324 (ser. sesta). Vol. 2.

——. 1931. Variation and fluctuations of the number of individuals in animal species living together. In *Animal Ecology* (Chapman, R. N.). McGraw-Hill, New York.

WALLACE, R. 1753. *A Dissertation on the Numbers of Mankind in Antient and Modern Times: in which the Superior Populousness of Antiquity is Maintained.* Printed for G. Hamilton and J. Balfour, Edinburgh.

WARDLE, R. A., and J. A. McLEOD. 1952. *The Zoology of Tapeworms.* Univ. of Minnesota Press, Minneapolis.

WIDDER, D. V. 1940. *The Laplace Transform.* Princeton Univ. Press, Princeton.

——. 1947. *Advanced Calculus.* Prentice-Hall, New York.

YULE, G. U. 1906. On the changes in the marriage- and birth-rates in England and Wales during the past half century; with an inquiry as to their probable causes. *J. R. statist. Soc.*, 69: 88–132.

——. 1925. The growth of population and the factors which control it. *J. Roy. statist. Soc.*, 88: 1–58 (related discussions, pp. 58–90).

Edwards, R. Y., and C. D. Fowle. 1955. The concept of carrying capacity. *Transactions of the North American Wildlife and Natural Resources Conference* 20:589–602.

A fundamental term replete in the wildlife literature since Leopold's 1933 text is *carrying capacity*. Regrettably, many terms in the wildlife literature are ambiguous and have a history of misuse. The objective of Edwards and Fowle's early paper on carrying capacity was to review the history of the term's use from Leopold's description (1933:51) through 1955.

In the past six decades the definition has expanded to include ecologically based carrying capacity (i.e., K-carrying capacity, refugium carrying capacity, behavioral carrying capacity, equilibrium carrying capacity) and culturally based carrying capacities (i.e., I-carrying capacity and minimum-impact carrying capacity). Ecologically based carrying capacities describe populations that are maintained at a relatively stable equilibrium by some intrinsic or extrinsic influence. Culturally based carrying capacities describe populations at a steady state density below a resource limitation carrying capacity due to human activity. Leopold initially considered carrying capacity a characteristic of the land, not the population. The discussion continues today, but remarkably, there are still few studies that have been able to measure carrying capacity and apply the concept.

RELATED READING

Caughley, G. 1979. What's this thing called carrying capacity? Pages 2–8 *in* M. S. Boyce, and L. D. Hayden-Wing, editors. North American elk: ecology, behavior, and management. University of Wyoming Press, Laramie, USA.

Hobbs, N. T. 1988. Estimating habitat carrying capacity: an approach for planning reclamation and mitigation for wild ungulates. Pages 3–7 *in* J. Emerick, S. Q. Foster, L. Hayden-Wing, J. Hodgson, J. M. Monarch, and J. Todd, editors. Proceedings III: issues and technology in management of impacted wildlife. Thorne Ecological Institute, Boulder, Colorado, USA.

Hobbs, N. T., and D. M. Swift. 1985. Estimates of habitat carrying capacity incorporating explicit nutritional constraints. Journal of Wildlife Management 49:814–822.

Leopold, A. 1933. Game management. Charles Scribner's Sons, New York, New York, USA.

McCall, T. C., R. D. Brown, and L. C. Bender. 1997. Comparison of techniques for determining the nutritional carrying capacity for white-tailed deer. Journal of Range Management 50:33–38.

McLeod, S. R. 1997. Is the concept of carrying capacity useful in variable environments? Oikos 79:529–542.

Miller, K. V., and J. M. Wentworth. 2000. Carrying capacity. Pages 140–155 *in* S. Demarais, and P. R. Krausman, editors. Ecology and management of large mammals in North America. Prentice Hall, Upper Saddle River, New Jersey, USA.

THE CONCEPT OF CARRYING CAPACITY

R. Y. EDWARDS
British Columbia Forest Service, Victoria;

AND C. DAVID FOWLE
Department of Lands and Forests, Maple, Ontario

The term "carrying capacity" is well-established in the vocabulary of wildlife biologists as well as of many ecologists working in other fields. It is one of those terms often employed without strict consideration of exact meaning which is used to describe a general conception rather than to express an exact idea. As such it is useful, but as our knowledge of factors affecting the survival of animals grows it is wise to examine periodically the terms we use and to consider their adequacy.

In preparing this paper we set out to examine the term "carrying capacity" to determine its meaning and to assess its usefulness in wildlife management. If the term as currently used proved inadequate we hoped to redefine it to give it more useful precision. We have tried not to confine our thinking to herbivores, ungulates or to any particular group of animals, for it should be possible to develop a general concept which can be applied to a wide range of animals. If the term is to have value in wildlife management, it should recognize broad criteria within which must fit particular conditions pertaining to particular species.

Our conclusions can be briefly stated. We find that most definitions of carrying capacity are vague and that some are almost meaningless. There is a difference of opinion in that many have used, and are using, the term as if it applied to food alone, while others use it to denote more than limitation of food and include other factors. We find that carrying capacity is often considered a stable characteristic of environment despite the fact that nearly all limiting factors are known to vary constantly in their influence on populations.

PAST AND PRESENT USE OF THE TERM

In analyzing our thoughts as well as those of associates with whom we have discussed the matter, we have found that the most basic differences of opinion stem from whether carrying capacity has meaning only with reference to food supply or whether its meaning is broader. Most people have assumed the former meaning, especially those working with ungulates. A few have maintained that this is the only meaning possible and have cited Leopold (1948) as the basic reference on this point for wildlife management. His definition states (p. 450) that carrying capacity is:

"The maximum density of wild game which a particular range is capable of carrying."

Here food is not mentioned but elsewhere, in speaking of ungulates, he says (p.54):

". . . . there is so far no visible evidence of any density limit except the carrying capacity of the food."

A somewhat different meaning is implied in another passage (p.135) where he brings the importance of cover and "edge" into the definition. Although Leopold does not provide us with a specific definition it is apparent that he considered food as only one factor determining carrying capacity. This is illustrated by his use of the term "range" which, to him, included a variety of facilities in the environment which are utilized by animals. He says (p.135):

"A range is habitable for a given species when it furnishes places suitable for it to feed, hide, rest, sleep, play, and breed, all within the reach of its cruising radius. Deficiencies in such places are usually seasonal."

The last statement is significant because it shows that Leopold recognized the variability of environment in terms of variations in factors such as weather, season and plant succession, and that dynamic environment is reflected in fluctuations in populations of animals.

A number of authors have emphasized food in considering carrying capacity. Hadwen and Palmer (1922), speaking of reindeer (*Rangifer tarandus*) say that grazing or carrying capacity is (p.29):

". . . . the number of stock which range will support for a definite period of grazing without injury to the range."

Here it is clear that food is the limiting factor under consideration. Trippensee (1948) also emphasizes the importance of food when referring to deer (*Odocoileus*), saying (p.196):

"The carrying capacity of a range, measured in terms of food availability, depends upon two factors; stand age and stand composition."

Elsewhere in his book he avoids the term. Allee *et al.* (1949: 706) also stress food in discussing carrying capacity. Fowle (1950) in reviewing factors controlling deer populations says (p.57):

"The environmental factor which has received the most attention in deer studies is food. Indeed, our concept of carrying capacity for deer centers around the adequacy of the food supply while our criteria of overuse have their basis in the rate of depletion of food supply."

Dasmann (1945:400) regards carrying capacity as:

"... the maximum number of grazing animals of a given class that can be maintained in good flesh year after year on a grazing unit without injury to the range forage growing stock or to the basic soil resource."

The same author modified this definition in a paper in 1948 (p.189) by changing the words "grazing animals" to "foraging animals" and replacing "grazing unit" with "range unit." This definition, confined to foraging animals, represents the modern trend of confining the use of "carrying capacity" to ungulates and other herbivores.

Dasmann's definition adds two concepts to that of Leopold; first, the quality of the animals should be considered, and second, deterioration of range cannot be allowed. Because standards have been established, quality is relatively easy to deal with in domestic stock, but with wildlife species it is more difficult. Perhaps it has yet to receive the attention it deserves in wildlife studies. Fisheries biologists have employed the idea with some success in cases where an environment is found capable of producing a certain aggregate weight of fish (biomass) made up of either a few large fish or many small ones. Reference to the "maximum number" that the waters will hold is meaningless unless the quality (size) of fish required is stipulated. Similarly, a unit of environment may support a large number of deer living at a minimum subsistence level, or a lesser number of healtheir animals.

If a population is sufficiently large to deplete the food supply faster than it is being produced, or injures the environment in some way, the size of future populations will be affected. This is an important concept in Dasmann's definition which is usually assumed in considering carrying capacity. In theory it is simple, recognizing that the animals should be in such numbers that they eat only the annual interest from food plants and none of the principal. Hence the environment should be stocked for perpetual maximum benefit rather than for a short period at its full capacity.

That Dasmann (1948) considers carrying capacity to be a dynamic concept is further illustrated when he says (p.189):

"The number of animals that will take no more than the forage crop in all but poorest growth years is the maximum number a range unit will support on a sustained basis. Since range is dynamic, changing continually with fluctuations in precipitation, temperature, evaporation, and varying use-patterns, no rate of stocking can be considered final."

Leopold (1948:51) is careful to distinguish between "carrying capacity" and "saturation point." The former he regards as "a property of a unit of range" and the latter "a property of a species." However, Graham (1944) refers to both concepts as if they applied to range. He says (p. 60):

"Closely related to cruising radius is the idea of *carrying capacity* of the land, whether it relates to wild species or to domestic animals such as sheep and cattle. Many factors influence the number of animals an area can maintain. The ancients had this in mind when they remarked that 'one hill will not carry two tigers'. A knowledge of the number of animals a habitat can reasonably be expected to support—its *saturation point*—is useful to the land manager."

This statement and the discussion which follows show that Graham was considering all factors, including social intolerances, which can limit the number of animals in an area.

In this connection the definitions of Errington and Hamerstrom (1936:308) are of interest. Speaking of bobwhite quail (*Colinus virginianus*) in winter they say:

"In its simplest form carrying capacity may be said to denote the upper limit of survival possible in a given covey territory as it exists under the most favorable conditions."

Later they state (309):

"The definition of carrying capacity may perhaps be restated as the level beyond which simple predation upon adult birds, their own territorial tolerances, and their tendencies to depart from coverts overcrowded with their own or some other species, do not permit continued maintenance of population."

Reference to "upper limit of survival . . . . under the most favorable conditions" suggests Leopold's phrase "maximum density." Inclusion of territorial tolerance and predation in the second definition gives it the wider meaning inferred in Graham's (1944) statement cited above.

The great importance of social interactions in limiting the number of animals within an area has been illustrated by work on rats (*Rattus norvigecus*) (Calhoun, 1949, 1950, 1952). Hodgdon and Hunt (1953) in defining carrying capacity for beaver also emphasize social tolerances when they say (p.73):

"Carrying capacity, then, is a matter of available food, available water and degree of tolerance one beaver family has for another."

Allen (1954) gives no definition but uses the term frequently.
He states (p.44):

"Within limits a trained observer can make a fair-or-better estimate of what is likely to be a productive area for species he has worked with, but the final proof is *what it is actually supporting*.
The biologist's term for this is 'carrying capacity'."

Again he states (p.132):

".... in the North the carrying capacity of a land unit usually declines during the cold season ...."

Also (p. 259):

"In a given unit of range there probably will be an optimum density level where the population of prey animals will be reasonably safe from predation (i. e., the carrying capacity phenomenon.) But variable factors, such as weather, may alter the condition of this range unit, for better or worse, in a given year."

On page 138 he states:

"Carrying capacity of deer country usually depends upon available food on the winter range."

It is clear that Allen's concept of carrying capacity is broad. He regards predation in relation to carrying capacity much as do Errington and Hamerstrom (1936). While food is acknowledged as the main limiting factor for deer, he recognizes that a number of factors may determine carrying capacity for other species. Of special interest perhaps is the statement from page 44 suggesting that the final proof of carrying capacity is what the area supports. He follows this statement by noting that an area with one quail per acre has a high carrying capacity for quail, but the same area may have a low carrying capacity for turkeys.

These examples will illustrate the general conception of carrying capacity as it has been developed in the literature. There is agreement that for various reasons it is impossible to crowd more than a finite number of animals into any unit of environment. This number will vary with the conditions for life existing in the unit concerned. It is not clear, however, whether all or only some of the many factors tending to limit animal populations should be included as factors determining carrying capacity. Some have stressed the importance of ''quality'' of the animals. It has been recognized that the ability of environment to support population varies from time to time, leading to fluctuations in population. The fluctuating nature of carrying capacity has not, however, been universally recognized, particularly by some earlier writers who inferred that it was a more or less stable attribute of environment.

What is Carrying Capacity?

Our review of the development of the concept of carrying capacity has shown that the original ideas were related to the role of such factors as climate and food-supply in controlling populations. Gradually, however, there has been a shifting of emphasis from a point of view holding that single factors, such as food, are the main controlling factors to a more comprehensive view in which it is recognized that: ''The relationships of a population are with the whole ecosystem (which includes itself) rather than with the environment only.'' (Solomon, 1949:31; Tansley, 1935). The fact that populations do not grow beyond finite limits is the result of the limited capacity of the ecosystem to support organisms.

In view of the complexity of factors influencing populations it is impossible to make a generalization as to the factors which determine carrying capacity since these will vary with time, place and the species concerned. If, however, we include all the factors in the ecosystem which tend to limit populations as factors determining carrying capacity we are led to an apparent paradox; namely that the number of animals present at a certain time in a unit of environment within the main geographic range of the species concerned is in itself a measure of carrying capacity at that time or at a previous time. For example, the number of quail in an area in spring may be a measure of the carrying capacity of the area during the most critical period of the previous winter. Conditions in spring may be favorable to the maintenance of more quail than are present, but in the absence of reproduction or immigration the population remains below carrying capacity.

Hence carrying capacity may be expressed simply by the number of animals in a unit of environment, except when time has been insufficient to enable increase when it is possible and except where the distribution of the animals is such as to leave some parts of inhabitable environment vacant. The *desired* population may be smaller or larger, and it is the business of management to establish the desired levels through manipulation of the appropriate manageable factors in the ecosystem. Our approach leads us to conclude that management must deal with carrying capacities in order to control populations and not, as sometimes thought, with populations in relation to one carrying capacity.

For a given population in a unit of environment there are a number of factors and processes which are potentially capable of placing an upper limit on its size. Several of these may be acting simultaneously as, for example, in a population which is increasing. As the population density increases the food-supply may be gradually reduced, competition for space with associated intraspecific fighting will become keener, and there may be an increase in disease and predation. Some animals will be eliminated in fights, some may die as they are forced out into inhospitable habitat, some may starve, disease may kill some, and predators may take an increasing proportion beyond a certain threshold of density of prey. Severe weather may eliminate others. The net result will be to limit the population, but no one factor may be responsible.

On the other hand, the various factors may act separately in the sense that one may become critical in the manner described by Liebig's Law of the Minimum. Taylor (1934) suggested a somewhat similar idea when he restated the law to apply to environmental factors generally as follows (p.378):

"The growth and functioning of an organism is dependent upon the amount of the essential environmental factor presented to it in minimal quantity during the most critical season of the year, or during the most critical year or years of a climatic cycle."

The role of factors operating at a minimum are also considered by Hubbs and Eschmeyer (1938) who, in speaking of management procedures say (p.21):

"All the essentials of a large fish yield should be provided; none may be omitted, because the lack of any one is sufficient to hold down production."

Later they say (p. 23):

"When the least developed or most limiting factor has been built up and has increased the fish production, some other factor may then become under-developed in terms of the increased population.''

Even here, however, it is virtually impossible to think of a factor acting independently of other factors or processes in the ecosystem. Indeed, as Rübel (1935) points out, factors operating at a minimum will have varying effects depending upon the other factors operating with them.

However, for purposes of management it is useful to approach the concept of carrying capacity from the point of view that populations are, in the last analysis, limited by some factor operating at a minimum. This approach focuses attention upon more or less measurable and manageable factors instead of complex environments regarded as entities. The latter can lead management into useless activity, improving environmental conditions which could be limiting, but which are not limiting in the situation at hand. For example, if 200 deer attempt to live in an area in a year when there is winter food for 100 and shelter at the height of winter for 300, and if we assume further that hunters will harvest 50, it is obvious that improving shelter or restricting hunting will neither maintain nor increase the herd. This can be done only by increasing the food. One detrimental factor alone can limit a population to a low level in what is otherwise superior habitat. Moreover, if the 200 animals are consuming their food supply faster than it is being produced, and the food supply cannot be increased, the aim of management should be to lower the level of critical operation of some other potentially limiting factor so that the herd is reduced to a number in balance with the food supply. Increase of the harvest to 100 would reduce the population to the desired level.

But what of a definition of carrying capacity? It is clear that populations are maintained by periodic recruitment of new organisms through reproduction or immigration. In the intervals between periods of recruitment there is a tendency for populations to decline because of many decimating factors which are operating against them at all times. These vary in the intensity of their effects with both time and density of population. For example, a population of ringnecked pheasants (*Phasianus colchicus*) may be reduced to half its size in winter because of shortage of food and shelter aggravated by severe weather. Again, a dense population of muskrats (*Ondatra zibethica*) may develop in summer only to be reduced later by a reduction in water-level. Here water-level is the critical factor acting in accordance with Liebig's Law. These examples demonstrate that the most critical conditions for survival occur at intervals—in the cases cited, once a year. Thus, there may be definite periods of time in the life of a population in which the most critical controlling factors operate and it is in these critical periods that the population is reduced to minimum.

Hence, for practical purposes we may regard carrying capacity as represented by the maximum number of animals of given species and quaility that can in a given ecosystem survive through the least favorable environmental conditions occurring within a stated time interval. For practical purposes this time interval is usually one year. This number of animals is an expression of the interaction of the properties of the species concerned and the total environment in which it lives. There is, of course, a maximum number of animals that can survive in the unit of environment under conditions existing at any given instant which might be called the current carrying capacity, but this does not seem to be as useful a concept as that for carrying capacity as we have defined it.

Perhaps a definition is not important in itself. The really important thing is to recognize that carrying capacity is not a stable property of a unit of environment but the expression of the interaction of the organisms concerned and their environment. Moreover, the carrying capacity of an ecosystem may fluctuate in response to the ebb and flow of interactions going on within it. The concept developed here requires management to determine, within the framework of local conditions, critical factors and periods in which they operate to limit populations, and to direct its energies into channels which will actually affect the operation of these critical factors.

SUMMARY

A review of the term "carrying capacity" as used in the past and at present reveals serious differences of opinion as to its exact meaning. Some definitions are almost meaningless, some confine the term to food while others include whole environments, and some, particularly earlier workers, consider it as a static attribute while it is known to be dynamic.

We conclude that carrying capacity is determined by the whole environment, and that, with some reservations, the number of animals upon a unit of range is in itself a measure of the carrying capacity of that area.

Factors in the environment should be considered separately. These usually operate as in Liebig's "Law of the Minimum," the most critical factor being the major control on population. The role of management is to recognize and work with critical factors to produce desired populations. "Our approach leads us to conclude that management must deal with carrying capacities in order to control populations and not, as sometimes thought, with populations in relation to one carrying capacity."

ACKNOWLEDGMENTS

For contributing to discussion essential to the evolution of this paper and in some cases for commenting on the paper in its various revisions, we are indebted to: P. J. Bandy, I. McT. Cowan, F. H. Fay, C. J. Guiguet, R. L. Hepburn, H. Lumsden, D. J. Robinson, R. C. Passmore, R. O. Standfield, and M. D. Udvardy. Since our critics did not always agree with us, the views expressed in this paper are, of course, the responsibility of the authors.

LITERATURE CITED

Allee, W. C., A. E. Emerson, O. Park, and K. P. Schmidt
1949.   Principles of animal ecology. W. B. Saunders. Philadelphia.
Allen, D.
1954.   Our wildlife legacy. Funk & Wagnalls. New York.
Calhoun, J. B.
1949.   A method for self-control of population growth among animals living in the wild. Science. 190:333-335.
1950.   The study of wild animals under controlled conditions. Ann. N.Y. Acad. Sci., 51:1113-1122.
1952.   The social aspects of population dynamics. Jour. Mamm., 33:139-159.
Dasmann, W.
1945.   A method for estimating carrying capacity of range lands. Jour. For., 43:401-402.
1948.   A critical review of range survey methods and their application to deer range management. Calif. Fish & Game, 34:189-207.
Errington, P. L. & F. N. Hamerstrom
1936.   The northern bob-whites' winter territory. Iowa State College Agric. & Mech. Arts, Res. Bull., 201.
Fowle, C. D.
1950.   The natural control of deer populations. Proc. 40 Annual Con. Int. Assoc. Game, Fish & Cons. Com. :56-63.
Graham, S. H.
1944.   The natural principles of land use. Oxford. New York.
Hadwen, S., and L. J. Palmer
1922.   Reindeer in Alaska. U.S.D.A. Bull. 1089.
Hodgdon, K. W. and J. N. Hunt
1953.   Beaver management in Maine. Maine Dept. of Inland Fisheries & Game. Game Dev. Bull. No. 3.
Hubbs, C. L. and R. W. Eschmeyer
1938.   The improvement of lakes for fishing. Univ. Mich. Inst. for Fish. Res. Bull. No. 2.
Leopold, A.
1948.   Game management. Chas. Scribner's Sons. New York.
Rübel, E.
1935.   The replaceability of ecological factors and the law of the minimum. Ecology, 16:336-341.
Solomon, M. E.
1949.   The natural control of animal populations. Jour. Animal Ecol., 18:1-35.
Tansley, A. G.
1935.   The use and abuse of vegetational concepts and terms. Ecol., 16:284-307.
Taylor, W. P.
1934.   Significance of extreme and intermittant conditions in distribution of species and management of natural resources, with a restatement of Liebig's Law of Minimum. Ecology, 15:374-379.
Tripennsee, R. E.
1948.   Wildlife management. McGraw-Hill. New York.

DISCUSSION

DISCUSSION LEADER PETERSON: It seems entirely appropriate that we should end this big game session of this Twentieth North American Wildlife Conference with a paper calling for a re-examination of a concept that many of us have used, unfortunately, rather loosely.

The paper is now open for discussion.

MR. PIMLOTT: It has been proposed that this carrying capacity definition be adopted. I feel that would be essential and necessary to adopt new terminology to describe more adequately the relationships of big game populations to their range. In the case of forest big game species, if we cover only the survival of the animal, we have no method of describing whether the animal is damaging the forest reproduction in any area. This is possible under Dasmann's definition of carrying capacity.

I wonder if it would not be more desirable to confine the term more narrowly, rather than to give it such a wide definition that it would have no practical use.

DR. FOWLE: I think I would agree at least in some measure with Mr. Pimlott. Our attempt here, as we indicated, was to try and find some general common ground. We feel that the important thing is that you have to specify what environment you are talking about and what quality of animals you are talking about.

I am sure that everyone here knows that when you say that animals have exceeded their carrying capacity, it's a rather paradoxical statement. If there is a carrying capacity that limits the number of animals, and then you have a whole lot more animals—well, it's a rather absurd conception.

What you really have is another kind of environment. After the supposed carrying capacity has been exceeded, you move to a deteriorated environment which is different from what you had before, and that's why we feel that a dynamic view is necessary.

I would suggest that in the case of big game and forest species generally that it might be desirable to consider desirable rates of stocking and not look at the matter in terms of carrying capacity at all.

DR. IAN MCTAGGART COWAN [British Columbia]: The matter that Dr. Fowle just brought before us, that of the peculiar situation you get into if you define carrying capacity and say that you have an excessive population only arises if you depart from the initial idea that capacity was involved in food. However, I can see considerable merit in the suggestion that Fowle and Edwards have made, but I also suggest that if we use carrying capacity as it is suggested, we need to have something else as well. We need another term that we may call, let's say, ultimate capacity. We need an ideal to shoot at in terms of management, because if you say that this population represents the capacity, and this animal is at capacity, then that immediately removes your ideals, aims, and objectives.

In other words, when you are attempting to manipulate your environment to increase the number of animals, you are shooting at an ideal which you must set as the capacity of this range to support animals when you have done everything humanly possible to improve the range's ability to support those animals.

I would suggest that if we use this, that we must immediately adopt this other idea of an ultimate capacity that can be reached by a good many of us.

DISCUSSION LEADER PETERSON: Perhaps we might employ the word "population optimum." That's equally difficut to define, however.

MR. LOUIS A. KRUMHOLZ [Lerner Marine Laboratory, Bimini, Bahamas]: I hesitated to come here simply because we have batted this thing around in fisheries for quite some time, and this was pretty thoroughly discussed back in the late 1940's when we were talking about carrying capacities in the fish population.

Now I agree, and I think that Mr. Fowle has a good definition of carrying capacity when he says that it should be the population that an environment is capable of maintaining at the most critical time of the year; but I do not believe that carrying capacity fluctuates as he says it does. I would rather say that carrying capacity is pretty much of a stable thing, and the standing crop of animals in the population will fluctuate up and down depending upon the season of the year.

To use a silly example, a quart milk bottle will hold a quart of milk, but if you raise the temperature of that bottle it will hold more than a quart. If you lower it, it will hold less than a quart. The capacity is a quart, but the amount in the bottle will fluctuate as the temperature varies.

Now, food will do the same thing in our area, and I'm firmly convinced from our studies on a good many small lakes that we do know that the number of fish, whether it's total weight in pounds per acre, or however you measure it—is smaller in a lake in the spring of the year, and this applies virtually to all the fresh-water lakes in the world, I imagine—than there are in the fall of the year.

The reason for that is that the fish do not recruit to the bottom of the lake during the winter, but they do in the summer, and this is a change in the size of the standing crop, and not in the carrying capacity of the lake.

DR. FOWLE: I'll have to go and look at a quart of milk, I guess, after that.

That's a very interesting comment, and I thing it brings out something that perhaps we do not emphasize sufficiently. I think Edwards and I would feel that the fish in the pond, if they were fewer in the fall than in the spring—which is certainly the case—the reason for that was that factors had operated against them during the year, and that had there been no mortality factors the numbers of fish would have remained the same.

In other words, we are trying to include everything in the environment, including social interactions of the animals, and perhaps even some of the conceptions that were described here yesterday by Dr. Christisen, and some of the other people who were working with lemmings, and so on. However, I feel that the comments of Dr. Krumholz are well taken. This matter isn't cleared up at all.

DISCUSSION LEADER PETERSON: I think we'll all agree that if he has done nothing else he has stimulated some interest here and some thought, and I see no reason for cutting off the discussion. Do we have a little more?

MR. JAMES BRUCE FALLS [University of Toronto]: I wonder if this new definition is so broad that it includes hunting pressure, and if we are going to adopt such a broad definition whether perhaps this invades the field of already established terms like ''standing crop,'' and ''population.''

I think that perhaps the main idea of the old usage of carrying capacity was standing potential, and perhaps it should be limited to food potential.

DR. FOWLE: We're going to get into a lot more definitions here before we're finished, but I think Dr. Falls has brought out a good point. It simply is a matter of deciding what the definition is going to be. If you don't want to include hunting, you have to think of another term to deal with that.

I still feel that the conception of desirable stocking is one which management should look to. I suggest that there is a slight difference between the definition that we have suggested, in that it perhaps has real biological basis, and it is not perhaps developed as a result of some desire on the part of management; whereas the concept of desirable stocking is something which you make up your mind about and develop procedures which will give you the stocking which you want. It may be much less than the actual number of animals which a segment of the environment will hold, but nevertheless it's what you want, and your program should be geared to that.

DISCUSSION LEADER PETERSON: Who has the next question?

MR. ELSON [St. Andrews]: I'm another one of these fish men. I'd like to make a brief reference to a paper which is going to be given at the forthcoming meeting of the Northeastern Branch of the American Fisheries Society in Atlantic City.

A gentleman from our station in St. Andrews is going to put forth something on the carrying capacity of the small trout streams there. I'm not sure of his definition in terms of words, but I think his practical definition is a very useful one. It will be the mean over a period of eight years of the number of fish which were encountered within the area each year.

It seems to me that these things like carrying capacity are useful particularly if we think of them as a broad band, rather than as a solid line. We have something then that we can work with.

My own particular interest has been with Atlantic salmon. We have had a lot of population work done within a relatively short period of time—I talked a little about it yesterday morning—and there the stream seemed to have a certain carrying capacity for young fish under what might be called natural conditions.

We were able to change that not by raising temperatures or anything like that, but by applying predator control, and then the thing that I think we might call carrying capacity increased by about eight times, so that it is a fluid sort of thing, and I think we should perhaps guard against too hard and fast word definitions to keep these terms useful.

At the high end of the scale I think we do need something like the ultimate carrying capacity that Dr. Cowan mentioned.

DISCUSSION LEADER PETERSON: Any further questions?

DR. W. H. ELDER [Missouri]: I should like to have us remind ourselves in regard to this definition that there is a need to reflect that carrying capacity is the number which a piece of range or a stream will support without deterioration of that range due to the population itself.

MR. KRUMHOLZ: Two years ago I was placed on the Terminology Committee for the Society, and we have had a great deal of difficulty with this term "carrying capacity," as well as "standing crop," "productivity," and so forth. Right now I feel that it's up to us to define this term so that it is pretty much of a standard term and can be used by everybody to mean the same thing. If we have different people believing that it means different things, we might as well redefine the words in the dictionary.

If we adhere to one definition, I think it ought to be that it is the most that any environment will support during the most critical time of the year. If we do that, then we know that we can rely on that environment to support that year after year, unless we have something like a tremendously heavy snow or a tremendous freeze-up in the lake, or there is a heavy ice-cover leading to a diminution of the oxygen or the introduction of undesirable cannibals in the water. Then I think we can say that the carrying capacity went haywire this particular year because of the conditions this particular winter; but we must have a stable definition of the word, and then use another term to definite the population itself, whether its standing crop or something else. I don't care about that, but the term carrying capacity is no more than a unit of measurement which the environment is capable of supporting at the most critical time of the year.

DISCUSSION LEADER PETERSON: Do we have further comments?

MR. CHARLES R. HUNGERFORD [Arizona]: Along with Dr. Cowan's point on the difference between carrying capacity and ultimate carrying capacity, I might ask Dr. Fowle if he includes in the definition the number of animals on an area per unit of area at any given time, how he might define density—if that is the case.

DR. FOWLE: Well, I'm not sure that I get the exact gist of this question. Do you want a definition of "density"?

DR. HUNGERFORD: Yes.

DR. FOWLE: Well, I'll have to look at this definition to see if I'm stuck here.

At first glance, I wouldn't say there is anything the matter with the old mathematical definition of density, being the number of animals per unit of area or volume or whatever it is you are dealing with. Is there a difficulty here that I'm too dense to see?

DR. HUNGERFORD: I believe you included that in your definition. You said any unit at any one time. Your definition starts with, "if at any one time the numbers per unit vary."

DR. FOWLE: No, the definition is that carrying capacity is represented by the maximum number of animals of a given species and quality that can in a given eco-system survive through the least favorable conditions occuring in a stated time interval.

Discussion Leader Peterson: Any further comments?

Mr. Franklin C. Daiber [Delaware]: I don't have any proposed definition for carrying capacity. I would like to make a couple of comments, however.

There have been a couple of ideas expressed here that somewhat confused me as to what is meant by carrying capacity, and one of them is this. We talk about limiting factors, controlling factors, and so forth. We talk about temperature; we talk about nutrition, and so on.

I would just like to call attention to the fact that regardless of how we name it, whether it be temperature or food supply, as long as it has the same effect on the organism to control, to depress, or increase the individual species' population, that it should be considered as a limiting factor.

Now, a controlling factor is something else again. As long as it has some effect on the population, regardless of how it works, I think it should be considered as a limiting factor.

Then in terms of carrying capacity, there was an expression from the man from New Brunswick, and I have heard it also in several fish papers, being a fish man. We talk about the carrying capacity of a fish pond in terms of bluegills, or the carrying capacity of some bay in terms of the number of flounder that are present; but I'm not so sure that that is what we want, because if we are talking about removing predators to increase the size of the flounders population in the bay, we are increasing one population at the expense of the other, and I think we can think of all kinds of examples of control of competitors, or predator control, or some means of direct action.

I think we have to consider the total and not just one population or species.

Discussion Leader Peterson: Do we have any further comments or questions? [There were none.]

I think this paper has been highly successful, and I'm extremely pleased that we got as much comment and thought on it as we have.

Chairman Watkins: I think the fact that nearly 100 of you so long delayed your lunch is rather good testimony that at least this last paper was an interesting one.

I share Dr. Peterson's feeling in that, so I think if there are no further questions, we can consider ourselves adjourned.

Birch, L. C. 1957. The meanings of competition. *The American Naturalist* 91:5–18.

Interspecific competition and intraspecific competition are two ecological processes fundamental to ecosystem function via their effects on animal and plant numbers over time. The debate about the importance of competition in the regulation of wildlife population led to a discussion from the 1930s through the 1950s that resulted in many definitions and interpretations of competition and its relation to animal numbers.

Universal acceptance of ecological terms is rare in the literature, but it is important. When terms are not used properly, their meanings become ambiguous and the scientific meaning is lost. *Competition* is no different. Birch provides a general definition of the word and also explains how it is used in the fields of genetics and evolution.

Birch was concerned because inappropriate use of terms can result in misunderstanding and confusion. This is especially true of *competition*, because writers attribute different things to the word, as Birch points out, "determining both the numbers of animals (the ecological problem) and the perils of animals (the genetic problem) in natural populations." Birch defines the term in four ways after considering an animal's environment in terms of weather: food, a place in which to live, number of other animals of the same species, and number of animals of different species.

The first definition states that "competition occurs when a number of animals (of the same or of different species) utilize common resources the supply of which is short; or if the resources are not in short supply, competition occurs when the animals seeking that resource nevertheless harm one or other in the process." The second definition includes an additional phrase (interference with one species by another), even when there is no demand for a common resource of space or food. Predation is excluded in the second definition. The third definition builds on the others but includes predation. The broadest meaning is the fourth, where competition is synonymous with natural selection.

For each definition, Birch discusses important differences and what the definition means in an ecological context. It is important to understand terms, especially when they have multiple meanings, and it is the responsibility of those using terms to set the parameters with which meaningful discussions can occur. It is ironic that even with Birch's efforts to bring commonality to the word *competition*, it still stirs debate among wildlife biologists and ecologists today.

RELATED READING

Ayala, F. J. 1971. Competition between species: frequency dependence. Science 171:820–824.
Gause, G. F. 1934. The destruction of one species by another. Pages 114–128 *in* G. F. Gause. The struggle for existence. Hafner Publishing, New York, New York, USA.
Grinnell, J. 1917. The niche-relationships of the California thrasher. The Auk 34:427–433.
Hardin, G. 1960. The competitive exclusion principle. Science 131:1292–1297.
MacArthur, R. H. 1958. Population ecology of some warblers of Northeastern coniferous forests. Ecology 39:599–619.
Park, T. 1954. Experimental studies of interspecies competition. II. Temperature, humidity, and competition in two species of *Tribolium*. Physiological Zoology 27:177–238.

Vol. XCI   January–February, 1957   No. 856

THE MEANINGS OF COMPETITION

L. C. BIRCH

Zoology Department, University of Sydney, Australia

INTRODUCTION

"Competition" has come to have at least four meanings in animal ecology, genetics and evolution. Some of these meanings are so ambiguous that the word has largely lost its usefulness as a scientific term. Worse still, it has resulted in much misunderstanding and confusion in some of the writings in these three fields. A clarification seems to be specially important at present because of the importance which some authors attribute to competition in determining both the numbers of animals (the ecological problem) and the perils of animals (the genetical problem) in natural populations.

To distinguish the four meanings of competition in common use, let us consider the environment of an animal as all those things or qualities which influence its chance to survive and reproduce. This is what matters for the purposes of animal ecology and genetics. We are then in a position to break down the environment of an individual animal, in the way suggested by Andrewartha and Birch (1954), into the following components:

Weather .... (I)
Food .... (II)
A place in which to live .... (III)
The number of other animals of the same species .... (IV)
The number of other organisms of different species
    (a) non-predators which utilize the same resources
       (space, food etc.) .... (V)
    (b) non-predators which do not utilize the same re-
       sources .... (VI)
    (c) predators .... (VII)

These components are numbered for reference in classifying the meanings of competition according to breadth of meaning. The first two components are self explanatory. The component "a place in which to live" refers, for example, to a tree-hole of a particular size, containing water of a particular pH with a particular content of organic matter, such as is necessary for the larvae of a tree-hole breeding mosquito to live in. The meaning of the other components will become clearer in the examples given below.

5

The various meanings of competition may be classified according to the number of components of environment which are included within the concept:

Meaning 1 (components: IV, V)
Meaning 2 (components: IV, V, VI)
Meaning 3 (components: IV, V, VI, VII)
Meaning 4 (components: I to VII)

MEANING 1. (COMPONENTS: IV, V)

Competition occurs when a number of animals (of the same or of different species) utilize common resources the supply of which is short; or if the resources are not in short supply, competition occurs when the animals seeking that resource nevertheless harm one or other in the process. This is the strict meaning of competition and the one which corresponds more closely to the etymology of the word, namely "together-seek." Nevertheless there are shades of meaning within this concept which are quite important to distinguish. For example, there is more than one meaning of "short supply." There may be an "absolute shortage" of food, or some other resource, by which is meant that not enough food or suitable places in which to live exist for all the animals present. Or there may be a "relative shortage" of food, even when plenty of food is available, if the animals seeking the resource choose to fight over the same morsel. Two birds may arrive at a carcass at the same time and fight for food even though there are enough other carcasses around for both of them. Further, a distinction can be made within this definition of competition, depending on whether the resource is the sort which is consumed when it is used (for example, food) or not (for example, nesting sites). Nicholson (1954) refers to the first as "scramble" and the second "contest." Examples of "contest" are provided by the aggressive behavior of some birds in seeking watch-posts or singing-places in tree-tops. *Phoenicurus phoenicurus* and *Muscicapa striata* fight for the possession of watch-posts suited to both species. Similarly the species *Sylvia communis* and *Emberiza citrinella* and *Saxicola torquata* in Hungary may fight for the possession of singing posts on the tops of prominent bushes in shrubberies (Udvardy, 1951). An example of "scramble" is provided by Ullyett's (1950) experiments with blowfly larvae. These experiments also illustrate two ways in which non-predators utilising the same resource may influence each other's death-rates. The following summary of some of Ullyett's experiments is paraphrased from Andrewartha and Birch (1954, page 419). Ullyett reared the larvae of the two species of blowflies, *Lucilia sericata* and *Chrysomyia chloropyga*, on a limited quantity of meat. The number of newly hatched larvae placed on 140 grams of meat varied from 100 to 10,000. The death rate was measured for each species at each density, both when the species were reared separately and when they were reared together. When the two species were reared together, the experiments were started with equal numbers of each. So the comparison to be made is between the death-rates, of say, Lucilia when all the other animals in the meat are of the same kind and when every second one belongs to a different species, in this case *Chrysomyia chloropyga*. Ullyett could find no evidence that either species interfered with the other in any way except that they both required exactly the same sort of food. The larvae of *C. chloropyga* in these experiments are non-predators that require to share the same food as *L. sericata* without interfering with it in any other way.

In a parallel series of experiments Ullyett compared *Lucilia sericata* with a different species *Chrysomyia albiceps*, with quite different habits. In these experiments the death-rate of Lucilia was much greater when half the other animals around it were *C. albiceps* than when all were of its own kind. As the densities of the populations increased, the death-rate among Lucilia increased until it reached 100 per cent, that is, none survived to become adult. Although *C. albiceps* undoubtedly needs to share the same food as *L. sericata*, this is a relatively unimportant part of its relationship to Lucilia. Ullyet observed that the larvae of *C. albiceps* attack and eat those of Lucilia. This happens to some extent when both populations are quite sparse but to an increasing degree as the densities increase. Despite these "predatory" habits, the larvae of *C. albiceps* are not to be considered strictly as predators because, in the absence of Lucilia, meat is still their chief source of food. They come into the environment of Lucilia chiefly as non-predators which interfere directly with Lucilia. In the first example one species affects the death rate of the other simply by using its food. The example illustrates what Park (1954) called the "exploitation" component of competition. But in the second example there was an additional direct interference of one species by the other. This is what Park (1954) called the "interference" component of competition. Both types of influence, exploitation and interference, account for high mortality in natural populations of *Lucilia sericata* in South Africa (Ullyett, 1950) and of *Lucilia cuprina* in Australia (Waterhouse, 1947).

An analogous example of direct interference of one species by another in a natural population is provided by Brian's (1952) observations on ants. In a cut-over pine wood *Myrmecia sp.* was unable to maintain itself long in the presence of *Formica sp.* because Formica attacked and destroyed the workers of Myrmecia. The interference of one species by another may take more benign forms, as for example when the humming bird *Calype anna* successfully prevents *Selasphorus sasin* from occupying its chosen territory simply by aggressive displays or by giving chase to the invader. (Pitelka, 1951). Such threats, though not mortally wounding the combatants, are effective in preventing one species from establishing itself in the area of another.

In most of the examples given so far, one species caused an increase in death-rate of the other. But the depressive effect of one species on the rate of increase of another may also be brought about by a decrease in the birth-rate or fecundity. For example, the fecundity of *Trisolium castaneum* was greatly reduced when it was crowded with adults of *T. confusum* as compared with crowding of its own adults. (Birch, Park and Frank, 1951).

All the above examples have been called competition by their authors, though some authors at the same time questioned whether competition was the best word to use (e.g., Ullyett, 1950, p. 117). This meaning is the strict meaning of competition accepted implicitly or explicitly by many ecologists. Other examples of its use will be found in the following references; Clements and Shelford, 1939, p. 159-60; Crombie, 1947, Allee et al, 1949, p. 656, Park, 1954, p. 180; Udvardy, 1951, p. 100 and 104; Andrewartha and Birch, 1954, p. 22.

Darwin has used the word "competition" in this strict sense also as the following quotations from "The Origin of Species" (p. 59, 1876 edition) suggest: "As the species of the same genus usually have, though by no means invariably, much similarity in habits and constitution, and always in structure, the struggle will generally be more severe between them, if they come into competition with each other, than between the species of distinct genera. We see this in the recent extension over parts of the United States of one species of swallow having caused the decrease of another species. The recent increase of the missel-thrush in parts of Scotland has caused the decrease of the song-thrush.... We can simply see why the competition should be most severe between allied forms, which fill nearly the same place in the economy of nature; but probably in no one case could we precisely say why one species has been victorious over another in the great battle of life." In the last part of Chapter III he wrote "Animals come into competition for food or resources." When defining his meaning of the "struggle for existence" (in Chapter 3) Darwin made it clear that he regarded competition as just one of the many components of the "struggle for existence." He certainly did not regard the two terms as synonymous in the way in which we find them used by some modern evolutionists (see Meaning 4).

The concept of competition is also used in palaeontology, though naturally it is much more difficult in this field to adduce evidence of how one species may have aided in the extinction of another. In most examples it is a matter of conjecture. But sometimes evidence is sufficient to provide at least some suggestions as to what might have happened. For example, in writing of the invasion of South America in the Cenozoic by 15 families of North American mammals Simpson (1950 p, 381) says; "To a fauna already large and essentially complete or closed ecologically were added a large number of new forms from the new continent. The enrichment inevitably involved some duplication. No two forms of different origin can have been precisely and fully equal in their needs and capacities, but many were sufficiently similar to be in competition for food and in general, living space. Some native groups held their own and some invading groups became extinct, but some native groups disappeared and (as a rule) were replaced by invaders." As examples of the latter Simpson mentions the disappearance of the old native ungulates and the survival of ungulates of northern origin; many old native rodents became extinct; marsupial carnivores became extinct and placental carnivores survived. Simpson envisages a common demand for similar resources of food and space by the northern and southern forms. And he implies that there was not enough food or space for all. Alternatively they interfered with each other in the process of seeking their common resources. So when a southern carnivore met a northern carnivore they may have fought over a common prey, the victor usually being the northern form. Or it is possible that the northern carnivore was simply able to produce more offspring and carry them through to maturity on a limited amount of food.

Simpson (1953 p, 300) quotes many examples from the fossil record of "two groups of similar adaptive type" which lived together for a time but subsequently one increased while the other decreased to extinction. In these examples he considers it probable that "Competition led to extinction of one group, although of course this may not really have been true in a given case." One of the best documented examples of this type is the replacement of multituberculates by rodents in North America in the early Cenozoic. He suggests that the success of the rodents was associated with their more mechanically efficient incisors, as compared with the multituberculates. If this were the vital factor in their success then we have to visualize a food shortage in which rodents were, in the long run, more successful in getting more than their fair share. The situation might be analogous to the "scramble" of blowflies in the cultures of Ullyett's experiments with *Lucilia sericata* and *Chrysomyia chloropyga* mentioned earlier.

Competition in the sense used in these examples causes (or is presumed to cause) an increased death rate and/or a decreased birth rate, of one of the species. And in the examples in which crowding was confined to members of one species alone, the effect was increased death rate or decreased birth rate of the population, or of certain genotypes in the population. But this may not always be so. Allee et al, (1949, p, 395) gave the following curious example of what they choose to call "cooperative competition." When undiluted with sea water the longevity of spermatozoa of sea urchins is longer than when diluted. Then oxygen consumption is less and their activity is less. "If the available space is much restricted, the competition for it results in an inhibition of movement, a lowering of the rate of oxygen consumption, and a conservation of essential diffusible materials; and all this is accompanied by a decided increase in longevity. Competition here has distinctly beneficial results for all the competitors; it is cooperative, as contrasted with being disoperative." In so far as space and oxygen are in short supply this example might be conceived to conform to the definition of competition at the beginning of this section. But its outcome is so different from all the other examples cited that one wonders whether it is wise to group it with them. The example illustrates the diversity of phenomena which can be given the same name. It would, of course be excluded by any definition of competition which required increased mortality with crowding (see for example Nicholson's usage in Meaning 3).

Within this first meaning of competition we have then various shades of meaning: "contest," "scramble," each of which in turn may be associated with either an "absolute shortage" of resource or a "relative shortage" of resource. And within the meaning of "scramble" the depression of the rate of increase may be due to shortage of resource (that is, starvation if food were the limiting resource to be exploited) or direct interference of one species by the other. "Contest" invariably involves interference of some sort. Each of these situations has been studied in laboratory populations but there is a sparsity of quantitative data from natural populations. However it is now eminently clear that, as Park (1954) points out, prediction about the outcome of competition is a matter of empirical study rather than deduction. Park's (1954) experiments with flour beetles provide a nice demonstration of this point.

MEANING 2 (COMPONENTS: IV, V, VI)

In addition to the phenomena covered by the scope of our first meaning of competition, some authors include an additional item, namely the interference with one species by another (with consequent change in birth rate or death rate) even when there is no demand for a common resource of space or food. But they exclude predation in which one animal eats another for its main source of food.

The following two examples are summarised from the work of Lloyd, Graham and Reynoldson (1940) and Lloyd (1953) on the fauna of sewage beds in England. These beds contain a variety of organisms which have provided scope for some fine studies on the interrelationships of organisms in a more or less confined place. Adults of Psychoda lay most of their eggs near the surface of these beds where two other sewage-bed species are most abundant: Lumbricillus and larvae of Metriocnemus. Lumbricillus tends to pulverize the growth of algae on the surface of the beds. And eggs of the other two species which happen to be in the algae mass tend to be washed away to the bottom of the bed and then out in the effluent. In this way the activities of Lumbricillus reduce the number of the other two species which happen to be in the surface of the bed. In commenting on this the authors state that "the loosening effects of one organism on another are important in the competition." They regard this as an example of competition though the species concerned are not "competing" for any common resource.

Another example which I quote from Lloyd (1943) concerns the two species *Metriocnemus longitarsus* and *M. birticollis*, both of which lay their eggs in the surface of sewage beds. But quite often *M. longitarsus* emerges earlier than the other species and so its eggs are laid earlier, and by the time *M. birticollis* is laying its eggs the larvae of *M. longitarsus* are well established. Both sorts of larvae feed on the algae that encrust the stones on the surface of the bed. There is never any shortage of algae on which they feed. Neither larvae seek out the eggs and larvae of the other species, but any eggs or small larvae which happen to be on the algae which is being eaten will be devoured. When *M. birticollis* lays its eggs after the larvae of *M. longitarsus* are well established, it runs the risk of being killed by the larger larvae of *M. longitarsus*. As a consequence *M. hirticollis* may become quite scarce. Lloyd describes this as competition. In neither of these examples are the two "competing" species seeking a common resource in short supply. The second example may bear some resemblance to what happens in crowded cultures of the blowflies Lucilia and *Chrysomyia albiceps* (see above), where Lucilia destroys Chrysomyia by attacks on its larvae. But there are two important differences. The two species of blowflies seek a common resource in short supply and secondly Lucilia seeks out and destroys larvae of Chrysomyia; its destructiveness is deliberate and not accidental. Now if we are to call accidental destructiveness by the name of competition, then it would be just as logical to say that motor cars compete with *Homo Sapiens* because some men are killed in motor car accidents.

Different again is the following example of interference which Gause (1934) described as competition. When the wild carp, *Cyprinus albus* was introduced into lakes inhabited by *Schizothorax intermidius* in Russia, the latter species was gradually eliminated, largely because the carp eats the spawn of Schizothorax. Gause does not give enough details for us to be certain as to what happened. Whether the carp actively seeks out the spawn of the other species or whether this only occurs when the numbers of both species are high is not stated.

A general definition of competition which covers all the above examples is given by Elton and Miller (1954, p, 475) "Interspecific competition, in the more limited and correct use of the notion, refers to the situations in which one species affects the population of another by a process of *interference,* that is, by reducing the reproductive efficiency or increasing the mortality of its competitor." It will be noted that the definition makes no reference to a common resource. In this respect it is much broader than Meaning 1.

MEANING 3 (COMPONENTS: IV, V, VI, VII)

In addition to including the joint use of a common resource (IV, V), and "interference" of one species by another even when the interfering species do not utilize common resources (VI), some authors include "predation" in their meaning of competition. By predation we mean the use of one animal as the main source of food of another. Predation then includes "parasitism" in the special sense in which entomologists use this term.

This meaning of competition does not include other influences on birth rate and death rate such as the influence of weather or quality of food. The following quotation from Nicholson (1937, p, 103) illustrates this unusual meaning. "The reactions of natural enemies to population changes of their hosts must be regarded as a form of competition, for these reactions decrease the chance of survival of individuals of the host species when the density of the host species is increased, and this is the essential feature of competition in animal communities. Furthermore, competition for food and space, and the interaction of natural enemies and their hosts, can both be rep-

resented by the same fundamental formula and its corresponding exponential curve ... but my remarks about natural enemies will indicate to
you that I am using the word 'competition' in a somewhat wider sense
than that in which it is usually used." Nicholson's criterion of competition
which he gives in this and other papers (For example, Nicholson, 1953, p.
143) is that the chances of survival decrease as density increases. His
definition necessarily excludes Allee's "cooperative competition" (see
above) where crowding increased survival. But it includes predation,
though it is difficult to conceive of any sense in which predator and prey
can be said to compete. There is no sense in which they compete for a
common resource.

MEANING 4 (COMPONENTS: I TO VII)

Here we have the broadest meaning of competition and the meaning which
has least claim to any etymological correctness. Competition is synonymous
with natural selection, that is, the differential survival or fecundity of different species, or genotypes of a species, due to the action of any component of environment. This use of competition is found in the writings of
geneticists and evolutionists more than in those of ecologists. Nevertheless
it was not the sense in which Darwin spoke of competition in The Origin of
Species (see above). Schmalhausen (1949 p. 61) leaves us in no doubt that
this fourth concept is what he means by competition when he writes "elements of competition for the preservation of existence and for propagation
are always present among members of a species; they are expressed in
struggles for food, with enemies and parasites, with severe climatic conditions, with hunger and disease.... In its active form individual competition corresponds to the usual concept of competition for means of subsistence and for propagation.... In its *passive form,* individual competition
includes the struggle of organisms against harmful physical (climate) and
biologic (predators, parasites) factors in order to protect their own lives and
those of their progeny. In this case, competition does not disturb the interests of other individuals of the same population and is not sharpened by an
increase in numbers." The inappropriateness of competition as a term for
these phenomena is evident when we consider the extreme case of a lethal
gene. Such a gene may make some phenotypes lethal at particular temperatures, for example. At these temperatures the organisms with the gene
will tend to be eliminated. They die because their physiological constitution
is not fitted to such temperatures, not because they were competing for anything.

There is another sense in which competition is used commonly in genetics
and evolution. The different genotypes of a species may have different
sexual activities; sometimes females have preferences for mating with
particular males. The males are then said to compete for females (e.g.
Bateman, 1948; Merrell, 1949; Reed and Reed, 1950). Reed and Reed (1950)
found that when a white-eyed male and female of *Drosophila melanogaster*
were kept for 24 hours in a vial with a red-eyed male and female both fe-

males tended to mate with red-eyed males. As a consequence of the selective preference of females there is a rapid increase in the frequency of red-eyed individuals when the two sorts of flies are cultured together in population bottles. In Reed and Reed's experiments the white-eyed individuals had completely disappeared by the 25th generation. Both the authors and Moody (1953, p. 396) in commenting on these experiments state that the white-eyed flies compete with the red-eyed flies. If we regard the female as a prize then there is a sense in which the white-eyed male is unsuccessful in "competition" for the prize though the two sorts of males do not in any sense "contest" for the prize; they do not battle over females as some of the higher animals are supposed to do. But this meaning of competition is different from the strict meaning of competition (Meaning 1) in the following respects. It is not dependent upon crowding or a shortage of resources. It takes place at quite low densities as well as at high ones. There is a peculiar sense in which the white-eyed males suffer from a shortage of suitable females since few females want them. But there is no way of relieving this shortage. They would presumably still be rejected in an infinite population of females. It is quite possible that the frequency of rejection would be less when there were many females and few males if uncopulated females were more receptive than recently copulated ones. If this were the case one might be inclined to include this example within the scope of the strict meaning of competition (Meaning 1) despite the differences already mentioned. My own preference in the matter is to avoid grouping this with the diversity of phenomena which might be strictly called competition and simply call it what it is namely, selective mating.

NATURAL SELECTION WITHOUT "COMPETITION"

The frequency of a genotype in a sexual cross-fertilizing population may be altered by selection which does not involve any shortage of common resources. The genotypes which have a high survival value and which contribute more offspring to the next generation will tend to "swamp" out other less fit genotypes. They will make greater contributions of genes to each successive generation. The greater contribution of a successful genotype may be associated with its higher mating activity, higher fecundity or higher survival. And these advantages may be quite independent of any shortage of resources of food or space. The lethality or semi-lethality conferred on their carriers by some genes are contingent upon components of environment such as temperature, moisture and the quality of food (for examples see Dobzhansky and Spassky, 1944; Dobzhansky, Pavlovsky, Spassky & Spassky, 1955; Dobzhansky, 1956, Da Cunha, 1951). Other genes are deleterious to their carriers in most environments in which the population lives. Genes which confer higher fitness than others will be multiplied and spread in the gene pool until the Mendelian population approaches in constitution its "adaptive norm" for the environment concerned. All this may happen without resources being in short supply. Competition in the strict ecological sense may not be applicable; natural selection may occur without it.

This is not to say that crowding between genotypes and shortage of resources of food and space do not influence their survival value. It simply emphasizes that differences in fitness exist even without such shortages of resources. The influence of shortage of resource, as a consequence of crowding, on the relative fitness of genotypes of a species, is another problem and one which has been little studied either in laboratory populations or in nature. Such studies as have been made are quite illuminating. Some genes confer greater fitness on their carriers (as compared with alternative genes or gene arrangements) at high densities when resources are in short supply. But they are at a disadvantage at low densities. Conversely other genotypes which are at a disadvantage at high densities have greater fitness at low densities. This has been demonstrated for *D. melanogaster* (Lewontin, 1955) and for *D. pseudoobscura* (Birch, 1955).

Natural selection between genotypes thus operates when the population is increasing and when resources may tend to become limiting, and also when the population is declining in numbers. Natural selection does not necessarily become more intense as population density increases, nor is it necessarily reduced as density falls. Sometimes the reverse occurs. An increase in density is not necessarily associated with reduction of amount of resource, nor is a fall in density necessarily associated with an alleviation of a shortage in resource. Many of the changes in numbers of animals from season to season or from one year to another have little to do with the amount of food or space available. The limit of resources may not be reached. Whether or not the limit will be reached sooner or later is a matter of dispute among ecologists. But the dispute need not concern us here, for we want to know what might happen in any fluctuation in numbers we happen to be interested in and not in hypothetical averages. An increase in numbers associated with a decrease in selection and consequent increase in variability seems to be what happened in the population of the butterfly *Melitaea aurina* observed by Ford (1945). The numbers increased because the environment became favorable and at the same time selection was slackened. With rising numbers genotypes of the species which have high fecundity become relatively more frequent in the population. Conversely as numbers of a species decrease, natural selection is not necessarily reduced. Very often the decrease is brought about by unfavorable weather. Those genotypes which are not resistant to heat or cold, or whatever the case may be, will perish, leaving a more homogeneous but resistant population. An example of this is provided by the observations of Thompson, Bell and Pearson (1911) on the decline in numbers during winter of the wasp *Vespa vulgaris* which was associated with a reduction in the wasp's variability.

DISTINCTION BETWEEN THE INFLUENCE OF ONE GENOTYPE ON THE
NUMBERS OF ANOTHER WITHIN A MENDELIAN POPULATION AND
THE INFLUENCE OF ONE SPECIES ON THE NUMBERS
OF ANOTHER SPECIES

Differential selection between genotypes of a species may occur without food or space being limiting. This was illustrated in the previous section. The same holds for selection between species. However it is necessary to make a distinction as between species and as between genotypes when we come to consider the way in which one sort of animal affects the numbers of another sort of animal. If we exclude predation and mutualism, then one species can only affect the numbers of another species when there is either an "absolute" or a "relative" shortage of resource such as food or space. The only exceptions seem to be the special types of "interference" of one species by another which were discussed under the second meaning of competition and the special case of threat discussed in the first meaning of competition. But one genotype can directly affect the numbers of another sort of genotype in the same Mendelian population, even when food and space are not short. It can do this by its greater sexual activity such as a higher mating rate or by selective mating. A species can only reproduce its own species but a genotype may produce others than its own kind. A homozygote which mates with a different homozygote produces all heterozygotes. A homozygote which mates with a heterozygote produces homozygotes and heterozygotes. This difference beyween species and genotypes of a species is quite an obvious one and yet it is sometimes lost sight of in arguments on competition.

There is a second aspect in which competition has a different relevance for ecology as compared with genetics. The ecologist is often unable to distinguish between genotypes within the species he studies. He measures abundance of the species as a whole and not its separate genotypes. Now even though competition in the strict sense of our first meaning may exist, the abundance he measures may not be related to the "intensity of competition." To put the emphasis on competition in such a case is beside the point. Consider a situation in which the number of birds in a wood is dependent upon the number of nesting sites in this wood. Provided the birds do not fight one another in seeking nesting sites, of which we may suppose there are not enough to go around, the number of birds in the wood will be the same as the number of nesting sites. It will be independent of the original number of birds seeking nesting sites and therefore of the "intensity of competition." A study of the "intensity of competition" will only become relevant when the investigator is able to distinguish between genotypes and wants to know the proportion of the different types of birds in the population. Examples of this sort have been discussed in more detail by Andrewartha and Birch (1954). This is another example of the different contexts in which ecologist and geneticist see competition, the one being more concerned with how many animals there are and the other with what kind of animal.

CONCLUSION

Some ecologists and geneticists have expressed doubts about the wisdom of using the one word competition to cover such a variety of phenomena as is included in its many meanings. The point becomes the more cogent when the meaning happens to bear little relation to the use of the word in everyday speech (see, for example, Thompson, 1939; Dobzhansky, 1950;

Ullyett, 1950; Andrewartha and Birch, 1954). Words can be obstacles to clear thinking. Ecology is unfortunately no exception to Shelling's statement, "In science as in life men are governed more by words than by clear concepts." One wonders for example if anything is to be gained by saying that the death of *Chrysomyia chloropyga* was due to competition with Lucilia when they were crowded in the same culture. This statement provides less information than the direct statement of what happened, namely that Chrysomyia larvae died from sheer starvation because there was not enough food for both species. And to say that red-eyed *Drosophila melanogaster* are successful in competition with white-eyed Drosophila tells less than the direct statement that the red-eyed flies increase in frequency in mixed populations because of the preference of females for red-eyed males.

It is not the purpose of this article to propose alternative ways of considering the variety of phenomena which have been called competition. Suggestions along these lines have been made elsewhere by Andrewartha and Birch (1954, p. 399). But whatever personal preferences we may have in this regard it remains important that when the concept of competition is invoked, the particular usage should be clearly understood and stated. This is especially necessary when it is claimed, as it is by some, that competition is centrally important in the dynamics of natural populations and in evolution. We should know what it is that is claimed to be so important. We may then be in a stronger position to investigate the claim. If we choose to retain the use of the word and give it a precise meaning for biology, then the first meaning we have discussed is the logical choice. The additional influences incorporated in the second meaning of competition might better be called simply "interference." As for the third meaning, the additional item is simply predation and might as well be called such. The fourth meaning is synonymous with differential survival and reproduction, and hence with natural selection. But the more logical approach to natural selection is to differentiate those factors which are selective in the first meaning of competition discussed in this article and those which are not. Sometimes the two will operate together. We would then choose to reverse Schmalhausen's scheme, and instead of calling natural selection competition, consider natural selection under two heads; natural selection with competition and natural selection without competition.

SUMMARY

The various meanings of "competition" in ecology, genetics and evolution are grouped into four main meanings which differ in important respects from each other. They grade from a strict meaning to one so broad as to be synonymous with natural selection. In the strict sense competition between animals occurs when a number of animals (of the same or of different species) utilize common resources the supply of which is short; or if the resources are not in short supply competition occurs when the animals seeking that resource nevertheless harm one or another in the process.

Intermediate between the strict and the broad meanings are two others. One includes any interference of one species by another irrespective of whether they use common resources or not, and the other includes predation within the meaning of competition. Each of the four meanings of competition is illustrated with examples.

A distinction is made between the influence of one species on the numbers of another and the influence of one genotype on the frequency of another genotype of the same species. In the former case one species can influence the frequency of another by competition only when there is a shortage of some resource. The one exception is some special cases of threat. The conclusion is drawn that if the term competition is retained in biological writings it should be restricted to the one strict meaning defined above. On the other hand this may be an appropriate time for approaching the problems of how one organism influences the numbers of another in a fresh way without recourse to the concept of competition at all.

ACKNOWLEDGMENTS

This paper has been read and criticized by Professor Th. Dobzhansky, Columbia University, New York, Professor Thomas Park, University of Chicago, Dr. Bruno Battaglia of Padua University, Italy, Dr. R. C. Lewontin, North Carolina State College, and Dr. A. B. da Cunha of the University of São Paulo, Brasil. It is a pleasure to acknowledge their many helpful suggestions most of which, though not necessarily all, have been taken heed of in revising the manuscript. The paper was largely written while I was the recipient of a grant from the Campanha Nacional de Aperfeiçoamento de Pessoal de Nível Superior (Capes), Brasil, for working in the Department of General Biology of the University of São Paulo, Brasil.

LITERATURE CITED

Allee, W. C., Emerson, A. E., Park, T., Park, O., and Schmidt, K. P., 1949, Principles of animal ecology. Philadelphia: W. B. Saunders Co.

Andrewartha, H. G., and Birch, L. C., 1954, The distribution and abundance of animals. University of Chicago Press, Chicago.

Bateman, A. J., 1948, Intra-sexual selection in Drosophila. Heredity, 2: 349-368.

Birch, L. C., 1955, Selection in *Drosophila pseudoobscura* in relation to crowding. Evolution, 9: 389-399.

Birch, L. C., Park, T., and Frank, M. B., 1951, The effect of intraspecies and interspecies competition on the fecundity of two species of flour beetles. Evolution, 5: 116-132.

Brian, M. V., 1952, The structure of a natural dense ant population. J. Anim. Ecol. 21: 12-24.

Clements, F. E., and Shelford, V. E., 1939, Bioecology. New York: John W. Wiley and Sons.

Crombie, A. C., 1947, Interspecific competition. J. Anim. Ecol. 16: 44-73.

Cunha, A. B. da, 1951, Modification of the adaptive values of chromosomal types in *Drosophila pseudoobscura* by nutritional variables. Evolution, 5: 395-404.

Dobzhansky, Th., 1950, Heredity, environment and evolution. Science, III: 161-166.
        1956, A review of some fundamental concepts and problems of population genetics. Cold Spring Harbor Symposium (In Press).

Dobzhansky, Th., and Spassky, B., 1944, Genetics of natural populations XI. Manifestation of genetic variants in *Drosophila pseudoobscura* in different environments. Genetics, 29: 270-290.

Dobzhansky, Th., Pavlovsky, O., Spassky, B., and Spassky, N., 1955, Genetics of natural populations. XXIII. Biological role of deleterious recessives in populations of *Drosophila pseudoobscura*. Genetics, 40: 781-796.

Elton, C. S., and Miller, R. S., 1954, The ecological survey of animal communities with a practical system of classifying habitats by structural characters. J. Ecol. 42: 460–496.

Ford, E. B., 1954, Butterflies. London: Collins.

Gause, G. F., 1934, The struggle for existence. Baltimore: Williams and Wilkins.

Lewontin, R. C., 1955, The effects of population density and composition on viability in *Drosophila melanogaster*. Evolution, 9: 27–41.

Lloyd, L., 1943, Materials for a study in animal competition. II. The fauna of the sewage beds. III. The seasonal rhythm of *Psychoda alternata*, Ann. App. Biol. 30: 47–60.

Lloyd, L., Graham, J. F., and Reynoldson, T. B., 1940, Materials for a study in animal competition. The fauna of sewage beds. Ann. App. Biol. 27: 122–150.

Merrell, D. J., 1949, Selective mating in *Drosophila melanogaster*. *Genetics*, 34: 370–389.

Moody, P. A., 1953., Introduction to evolution. New York: Harper and Brothers.

Nicholson, A. J., 1937, The role of competition in determining animal populations. J. Counc. Sci. Ind. Res. (Australia) 10: 101–6.
    1953, The balance of animal populations. J. Anim. Ecol. 2: 132–178.
    1954, An outline of the dynamics of animal populations. Aust. J. Zool. 2: 9–65.

Park, T., 1954, Experimental studies of interspecies competitions. II. Temperature, humidity and competition in two species of *Tribolium*. Physiol. Zool. 27: 177–238.

Pitelka, F. A., 1951, Ecologic overlap and interspecific strife in breeding populations of Anna and Allen humming birds. Ecology, 32: 641–661.

Reed, S. C., and Reed, E. W., 1950, Natural selection in laboratory populations of Drosophila. II. Competition between a white-eye gene and its wild type allele. Evolution, 4: 34–42.

Schmalhausen, I. I., 1949, Factors of evolution; the theory of stabilizing selection. Philadelphia: The Blakiston Co.

Simpson, G. G., 1950, History of the fauna of Latin America. Amer. Scientist, 38: 361–389.
    1953, The major features of evolution. Columbia University Press. New York.

Thompson, W. R., 1939, Biological control and the theories of interactions of populations. Parasitology, 31: 299–388.

Thompson, R. J., Bell, J., and Pearson, K. 1911, A third cooperative study of *Vespa vulgaris*. Biometrika, 8.

Udvardy, M. D. F., 1951, The significance of interspecific competition in bird life. Oikos, 3: 98–123.

Ullyett, G. C., 1950, Competition for food and allied phenomena in sheep-blowfly populations. Phil. Trans. Roy. Soc. London. B. 234: 77–174.

Waterhouse, D. F., 1947, The relative importance of live sheep and of carrion as breeding grounds for the Australian sheep blowfly *Lucilia cuprina*. Bull. Counc. Sci. Ind. Res. Australia, No. 217.

Holling, C. S. 1959. The components of predation as revealed by a study of small-mammal predation of the European pine sawfly. *The Canadian Entomologist* 91:293–320.

Next to competition, predation is likely the second-most important ecological concept that applies to multiple-species interactions, and it has been a prominent management issue for wildlife biologists. Holling's paper on the process of predation remains relevant today. In this article, Holling presents one of the first experimental studies to examine components of predation (i.e., rates of predations) relative to density of the prey population, density of the predator population, characteristics of the prey and predator, and density and quality of alternate foods. He introduces terms still used in contemporary predator-prey studies, including *functional responses* and *numerical responses.*

Holling's objective is to determine the role predation plays on animal population fluctuation, and he reviews different theories of population density. Many of these concepts form the basis of modern predation theory, including compensatory and non-compensatory predation, the threshold of security, and functional and numerical responses. No study of predation should be undertaken without first reading this foundational paper that describes an outstanding empirical study of the components of predation.

RELATED READING

Errington, P. L. 1945. Some contributions of a fifteen-year local study of the northern bobwhite to a knowledge of population phenomena. Ecological Monographs 15:1–34.

Huffaker, C. B. 1958. Experimental studies on predation: dispersion factors and predator-prey oscillations. Hilgardia 27:343–383.

Leibold, M. A. 1996. A graphical model of keystone predators in food webs: trophic regulation of abundance, incidence, and diversity patterns in communities. The American Naturalist 147:784–812.

Messier, F. 1994. Ungulate population models with predation: a case study with the North American moose. Ecology 75:478–488.

Peterson, R. O. 1999. Wolf-moose interaction on Isle Royale: the end of natural regulation? Ecological Applications 9:10–16.

Rosenzweig, M. L., and R. H. MacArthur. 1963. Graphical representation and stability conditions of predator-prey interactions. The American Naturalist 97:209–223.

Royama, T. 1970. Factors governing the hunting behavior and selection of food by the great tit (*Parus major* L.). Journal of Animal Ecology 39:619–668.

Singer F. J., A. Harting, K. K. Symonds, and M. B. Coughenour. 1997. Density dependence, compensation, and environmental effects on elk calf mortality in Yellowstone National Park. Journal of Wildlife Management 61:12–25.

The Components of Predation as Revealed by a Study of Small-Mammal Predation of the European Pine Sawfly[1]

By C. S. HOLLING

Forest Insect Laboratory, Sault Ste. Marie, Ont.

INTRODUCTION

The fluctuation of an animal's numbers between restricted limits is determined by a balance between that animal's capacity to increase and the environmental checks to this increase. Many authors have indulged in the whimsy of calculating the progressive increase of a population when no checks were operating. Thus Huxley calculated that the progeny of a single *Aphis* in the course of 10 generations, supposing all survived, would "contain more ponderable substance than five hundred millions of stout men; that is, more than the whole population of China", (in Thompson, 1929). Checks, however, do occur and it has been the subject of much controversy to determine how these checks operate. Certain general principles—the density-dependence concept of Smith (1955), the competition theory of Nicholson (1933)—have been proposed both verbally and mathematically, but because they have been based in part upon untested and restrictive assumptions they have been severely criticized (e.g. Andrewartha and Birch 1954). These problems could be considerably clarified if we knew the mode of operation of each process that affects numbers, if we knew its basic and subsidiary components. Predation, one such process, forms the subject of the present paper.

Many of the published studies of predation concentrate on discrete parts rather than the whole process. Thus some entomologists are particularly interested in the effect of selection of different kinds of prey by predators upon the evolution of colour patterns and mimicry; wildlife biologists are similarly interested in selection but emphasize the role predators play in improving the condition of the prey populations by removing weakened animals. While such specific problems should find a place in any scheme of predation, the main aim of the present study is to elucidate the components of predation in such a way that more meaning can be applied to considerations of population dynamics. This requires a broad study of the whole process and in particular its function in affecting the numbers of animals.

Such broad studies have generally been concerned with end results measured by the changes in the numbers of predator and prey. These studies are particularly useful when predators are experimentally excluded from the environment of their prey, in the manner adopted by DeBach and his colleagues in their investigations of the pests of orchard trees in California. This work, summarized recently (DeBach, 1958) in response to criticism by Milne (1957), clearly shows that in certain cases the sudden removal of predators results in a rapid increase of prey numbers from persistently low densities to the limits of the food supply. Inasmuch as these studies have shown that other factors have little regulatory function, the predators appear to be the principal ones responsible for regulation. Until the components of predation are revealed by an analysis of the processes leading to these end results, however, we will never know whether the conclusions from such studies apply to situations other than the specific predator–prey relationship investigated.

Errington's investigations of vertebrate predator–prey situations (1934, 1943, 1945 and 1956) suggest, in part, how some types of predation operate. He has

[1] Contribution from the Dept. of Zoology, University of British Columbia and No. 547, Forest Biology Division, Research Branch, Department of Agriculture, Ottawa, Canada. Delivered in part at the Tenth International Congress of Entomology, Montreal, 1956.

postulated that each habitat can support only a given number of animals and that predation becomes important only when the numbers of prey exceed this "carrying capacity". Hence predators merely remove surplus animals, ones that would succumb even in the absence of natural enemies. Errington exempts certain predator-prey relations from this scheme, however, and quotes the predation of wolves on deer as an example where predation probably is not related to the carrying capacity of the habitat. However logical these postulates are, they are only indirectly supported by the facts, and they do not explain the processes responsible.

In order to clarify these problems a comprehensive theory of predation is required that on the one hand is not so restrictive that it can only apply in certain cases and on the other not so broad that it becomes meaningless. Such a comprehensive answer requires a comprehensive approach, not necessarily in terms of the number of situations examined but certainly in terms of the variables involved, for it is the different reactions of predators to these variables that produce the many diverse predator-prey relations. Such a comprehensive approach is faced with a number of practical difficulties. It is apparent from the published studies of predation of vertebrate prey by vertebrate predators that not only is it difficult to obtain estimates of the density of predator, prey, and destroyed prey, but also that the presence of many interacting variables confuses interpretation.

The present study of predation of the European pine sawfly, *Neodiprion sertifer* (Geoff.) by small mammals was particularly suited for a general comprehensive analysis of predation. The practical difficulties concerning population measurement and interpretation of results were relatively unimportant, principally because of the unique properties of the environment and of the prey. The field work was conducted in the sand-plain area of southwestern Ontario where Scots and jack pine have been planted in blocks of up to 200 acres. The flat topography and the practice of planting trees of the same age and species at standard six-foot spacings has produced a remarkably uniform environment. In addition, since the work was concentrated in plantations 15 to 20 years of age, the closure of the crowns reduced ground vegetation to a trace, leaving only an even layer of pine needles covering the soil. The extreme simplicity and uniformity of this environment greatly facilitated the population sampling and eliminated complications resulting from changes in the quantity and kind of alternate foods of the predators.

The investigations were further simplified by the characteristics of the prey. Like most insects, the European pine sawfly offers a number of distinct life-history stages that might be susceptible to predation. The eggs, laid in pine needles the previous fall, hatch in early spring and the larvae emerge and feed upon the foliage. During the first two weeks of June the larvae drop from the trees and spin cocoons within the duff on the forest floor. These cocooned sawflies remain in the ground until the latter part of September, when most emerge as adults. A certain proportion, however, overwinter in cocoons, to emerge the following autumn. Observations in the field and laboratory showed that only one of these life-history stages, the cocoon, was attacked by the small-mammal predators, and that the remaining stages were inaccessible and/or unpalatable and hence completely escaped attack. These data will form part of a later paper dealing specifically with the impact of small mammal predation upon the European pine sawfly.

Cocooned sawflies, as prey, have some very useful attributes for an investigation of this kind. Their concentration in the two-dimensional environment of the duff-soil interface and their lack of movement and reaction to predators considerably simplify sampling and interpretation. Moreover, the small mammals' habit of making a characteristically marked opening in the cocoon to permit removal of the insect leaves a relatively permanent record in the ground of the number of cocooned sawflies destroyed. Thus, the density of the destroyed prey can be measured at the same time as the density of the prey.

Attention was concentrated upon the three most numerous predators—the masked shrew, *Sorex cinereus cinereus* Kerr, the short-tail shrew, *Blarina brevicauda talpoides* Gapper, and deer mouse, *Peromyscus maniculatus bairdii* Hoy and Kennicott. It soon became apparent that these species were the only significant predators of the sawfly, for the remaining nine species trapped or observed in the plantations were either extremely rare or were completely herbivorous.

Here, then, was a simple predator-prey situation where three species of small mammals were preying on a simple prey—sawfly cocoons. The complicating variables present in most other situations were either constant or absent because of the simple characteristics of the environment and of the prey. The absence or constancy of these complicating variables facilitated analysis but at the expense of a complete and generally applicable scheme of predation. Fortunately, however, the small-mammal predators and the cocoons could easily be manipulated in laboratory experiments so that the effect of those variables absent in the field situation could be assessed. At the same time the laboratory experiments supported the field results. This blend of field and laboratory data provides a comprehensive scheme of predation which will be shown to modify present theories of population dynamics and to considerably clarify the role predators play in population regulation.

I wish to acknowledge the considerable assistance rendered by a number of people, through discussion and criticism of the manuscript: Dr. I. McT. Cowan, Dr. K. Graham and Dr. P. A. Larkin at the University of British Columbia and Dr. R. M. Belyea, Mr. A. W. Ghent and Dr. P. J. Pointing, at the Forest Biology Laboratory, Sault Ste. Marie, Ontario.

FIELD TECHNIQUES

A study of the interaction of predator and prey should be based upon accurate population measurements, and in order to avoid superficial interpretations, populations should be expressed as numbers per unit area. Three populations must be measured—those of the predators, prey, and destroyed prey. Thus the aim of the field methods was to measure accurately each of the three populations in terms of their numbers per acre.

Small-Mammal Populations

Since a complete description and evaluation of the methods used to estimate the density of the small-mammal predators forms the basis of another paper in preparation, a summary of the techniques will suffice for the present study.

Estimates of the number of small mammals per acre were obtained using standard live-trapping techniques adapted from Burt (1940) and Blair (1941). The data obtained by marking, releasing and subsequently recapturing animals were analysed using either the Lincoln index (Lincoln, 1930) or Hayne's method for estimating populations in removal trapping procedures (Hayne, 1949). The resulting estimates of the number of animals exposed to traps were converted to per acre figures by calculating, on the basis of measurements of the home range of the animals (Stickel, 1954), the actual area sampled by traps.

The accuracy of these estimates was evaluated by examining the assumptions underlying the proper use of the Lincoln index and Hayne's technique and by comparing the efficiency of different traps and trap arrangements. This analysis showed that an accurate estimate of the numbers of *Sorex* and *Blarina* could be obtained using Hayne's method of treating the data obtained from trapping with bucket traps. These estimates, however, were accurate only when the populations had not been disturbed by previous trapping. For *Peromyscus*, Lincoln-index estimates obtained from the results of trapping with Sherman traps provided an ideal way of estimating numbers that was both accurate and unaffected by previous trapping.

N. sertifer Populations

Since small-mammal predation of *N. sertifer* was restricted to the cocoon stage, prey populations could be measured adequately by estimating the number of cocoons containing living insects present immediately after larval drop in June. This estimate was obtained using a method outlined and tested by Prebble (1943) for cocoon populations of the European spruce sawfly, *Gilpinia hercyniae* (Htg.), an insect with habits similar to those of *N. sertifer*. Accurate estimates were obtained when cocoons were collected from sub-samples of litter and duff distributed within the restricted universe beneath the crowns of host trees. This method was specially designed to provide an index of population rather than an estimate of numbers per acre. But it is obvious from this work that any cocoon-sampling technique designed to yield a *direct* estimate of the number of cocoons per acre would require an unpractically large number of sample units. It proved feasible in the present study, however, to convert such estimates from a square-foot to an acre basis, by stratifying the forest floor into three strata, one comprising circles with two-foot radii around tree trunks, one comprising intermediate rings with inner radii two feet and outer radii three feet, and one comprising the remaining area (three to five feet from the tree trunks).

At least 75 trees were selected and marked throughout each plantation, and one or usually two numbered wooden stakes were placed directly beneath the crown of each tree, on opposite sides of the trunk. Stakes were never placed under overlapping tree crowns. The four sides of each stake were lettered from A to D and the stake was placed so that the numbered sides bore no relation to the position of the trunk. Samples were taken each year, by collecting cocoons from the area delimited by one-square-foot frames placed at one corner of each stake. In the first year's sample the frames were placed at the AB corner, in the second year's at the BC corner, etc. Different-sized screens were used to separate the cocoons from the litter and duff.

Cocoons were collected in early September before adult sawflies emerged and those from each quadrat were placed in separate containers for later analysis. These cocoons were analysed by first segregating them into "new" and "old" categories. Cocoons of the former category were a bright golden colour and were assumed to have been spun in the year of sampling, while those of the latter were dull brown in colour and supposedly had been spun before the sampling year. These assumptions proved partly incorrect, however, for some of the cocoons retained their new colour for over one year. Hence the "new" category contained enough cocoons that had been spun before the sampling year to prevent its use, without correction, as an estimate of the number of cocoons spun in the year of sampling. A correction was devised, however, which reduced the error to negligible proportions.

This method provided the best available estimate of the number of healthy cocoons per acre present in any one year. The population figures obtained ranged from 39,000 (Plot 1, 1954) to 1,080,000 (Plot 2, 1952) cocoons per acre.

Predation

Small-mammal predation has a direct and indirect effect on *N. sertifer* populations. The direct effect of predation is studied in detail in this paper. The indirect effect, resulting from the mutual interaction of various control factors (parasites, disease, and predators) has been discussed in previous papers (Holling, 1955, 1958b).

The direct effect of predation was measured in a variety of ways. General information was obtained from studies of the consumption of insects by caged animals and from the analysis of stomach contents obtained from animals trapped in sawfly-infested plantations. More particular information was obtained from the analysis of cocoons collected in the regular quadrat samples and from laboratory experiments which studied the effect of cocoon density upon predation.

The actual numbers of *N. sertifer* cocoons destroyed were estimated from cocoons collected in the regular quadrat samples described previously. As shown in an earlier paper (Holling, 1955), cocoons opened by small mammals were easily recognized and moreover could be classified as to species of predator. These estimates of the number of new and old cocoons per square foot opened by each species of predator were corrected, as before, to provide an estimate of the number opened from the time larvae dropped to the time when cocoon samples were taken in early September.

It has proved difficult to obtain a predation and cocoon-population estimate of the desired precision and accuracy. The corrections and calculations that had to be applied to the raw sampling data cast some doubt upon the results and conclusions based upon them. It subsequently developed, however, that a considerable margin of error could be tolerated without changing the results and the conclusions that could be derived from them. In any case, all conclusions based upon cocoon-population estimates were supported and substantiated by results from controlled laboratory experiments.

LABORATORY TECHNIQUES

Several experiments were conducted with caged animals in order to support and expand results obtained in the field. The most important of these measured the number of cocoons consumed by *Peromyscus* at different cocoon densities. These experiments were conducted at room temperature (ca. 20°C) in a screen-topped cage, 10′ x 4′ x 6″. At the beginning of an experiment, cocoons were first buried in sand where the lines of a removable grid intersected, the grid was then removed, the sand was pressed flat, and a metal-edged levelling jig was finally scraped across the sand so that an even 12 mm. covered the cocoons. A single deer mouse was then placed in the cage together with nesting material, water, and an alternate food—dog biscuits. In each experiment the amount of this alternate food was kept approximately the same (i.e. 13 to 17 gms. dry weight). After the animal had been left undisturbed for 24 hours, the removable grid was replaced, and the number of holes dug over cocoons, the number of cocoons opened and the dry weight of dog biscuits eaten were recorded. Consumption by every animal was measured at either four or five different densities ranging from 2.25 to 36.00 cocoons per sq. ft. The specific densities were provided at random until all were used, the consumption at each density being measured for three to six consecutive days. Ideally the size of the cage should remain constant at all densities but since this would have required over 1,400 cocoons at the highest density, practical considerations necessitated a compromise whereby the cage was shortened at the higher densities. In these experiments the total number of cocoons provided ranged from 88 at the lowest density to 504 at the highest. At all densities, however, these numbers represented a surplus and no more than 40 per cent were ever consumed in a single experiment. Hence consumption was not limited by shortage of cocoons, even though the size of the cage changed.

The sources and characteristics of the cocoons and *Peromyscus* used in these experiments require some comment. Supplies of the prey were obtained by collecting cocoons in sawfly-infested plantations or by collecting late-instar larvae and allowing them to spin cocoons in boxes provided with foliage and litter. Sound cocoons from either source were then segregated into those containing healthy, parasitized, and diseased prepupae using a method of X-ray analysis (Holling, 1958a). The small male cocoons were separated from the larger female cocoons by size, since this criterion had previously proved adequate (Holling, 1958b). To simplify the experiments, only male and female cocoons containing healthy, living prepupae were used and in each experiment equal numbers of cocoons of each sex were provided, alternately, in the grid pattern already described.

Three mature non-breeding male deer mice were used in the experiments. Each animal had been born and raised in small rearing cages 12 x 8 x 6 in. and had been isolated from cocoons since birth. They therefore required a period to become familiar with the experimental cage and with cocoons. This experience was acquired during a preliminary three-week period. For the first two weeks the animal was placed in the experimental cage together with nesting material, water, dog biscuits and sand, and each day was disturbed just as it would be if an experiment were in progress. For the final week cocoons were buried in the sand at the first density chosen so that the animal could learn to find and consume the cocoon contents. It has been shown (Holling, 1955, 1958b) that a seven-day period is more than ample to permit complete learning.

THE COMPONENTS OF PREDATION

A large number of variables could conceivably affect the mortality of a given species of prey as a result of predation by a given species of predator. These can conveniently be classified, as was done by Leopold (1933), into five groups:

(1) density of the prey population.
(2) density of the predator population.
(3) characteristics of the prey, e.g., reactions to predators, stimulus detected by predator, and other characteristics.
(4) density and quality of alternate foods available for the predator.
(5) characteristics of the predator, e.g., food preferences, efficiency of attack, and other characteristics.

Each of these variables may exert a considerable influence and the effect of any one may depend upon changes in another. For example, Errington (1946) has shown that the characteristics of many vertebrate prey species change when their density exceeds the number that the available cover can support. This change causes a sudden increase in predation. When such complex interactions are involved, it is difficult to understand clearly the principles involved in predation; to do so we must find a simplified situation where some of the variables are constant or are not operating. The problem studied here presents such a situation. First, the characteristics of cocoons do not change as the other factors vary and there are no reactions by the cocooned sawflies to the predators. We therefore can ignore, temporarily, the effect of the third factor, prey characteristics. Secondly, since the work was conducted in plantations noted for their uniformity as to species, age, and distribution of trees, there was a constant and small variety of possible alternate foods. In such a simple and somewhat sterile environment, the fourth factor, the density and quality of alternate foods, can therefore be initially ignored, as can the fifth factor, characteristics of the predator, which is really only another way of expressing factors three and four. There are thus only two basic variables affecting predation in this instance, i.e., prey density and predator density. Furthermore, these are the only essential ones, for the remainder, while possibly important in affecting the amount of predation, are not essential to describe its fundamental characteristics.

The Basic Components

It is from the two essential variables that the basic components of predation will be derived. The first of these variables, prey density, might affect a number of processes and consumption of prey by individual predators might well be one of them.

The data which demonstrate the effect of changes of prey density upon consumption of cocooned sawflies by *Peromyscus* were obtained from the yearly cocoon quadrat samples in Plots 1 and 2. In 1951, Dr. F. T. Bird, Laboratory of Insect Pathology, Sault Ste. Marie, Ont., had sprayed each of these plots with a low concentration of a virus disease that attacked *N. sertifer* larvae, (Bird 1953). As a result, populations declined from 248,000 and 1,080,000 cocoons per acre, respectively, in 1952, to 39,000 and 256,000 in 1954. Thus predation values at six different cocoon densities were obtained. An additional sample in a neighbouring plantation in 1953 provided another value.

Predation values for *Sorex* and *Blarina* were obtained from one plantation, Plot 3, in one year, 1952. In the spring of that year, virus, sprayed from an aircraft flying along parallel lines 300 feet apart, was applied in three concentrations, with the lowest at one end of the plantation and the highest at the other. An area at one end, not sprayed, served as a control. When cocoon populations were sampled in the autumn, a line of 302 trees was selected at right angles to the lines of spray and the duff under each was sampled with one one-square-foot quadrat. The line, approximately 27 chains long, ran the complete length of the plantation. When the number of new cocoons per square foot was plotted against distance, discrete areas could be selected which had fairly constant populations that ranged from 44,000 to 571,000 cocoons per acre. The areas of low population corresponded to the areas sprayed with the highest concentration of virus. In effect, the plantation could be divided into rectangular strips, each with a particular density of cocoons. The width of these strips varied from 126 to 300 feet with an average of 193 feet. In addition to the 302 quadrats examined, the cocoons from another 100 quadrats were collected from the areas of lowest cocoon densities. Thus, in this one plantation in 1952, there was a sufficient number of different cocoon densities to show the response of consumption by *Sorex* and *Blarina* to changes of prey density.

The methods used to estimate predator densities in each study plot require some further comment. In Plots 1 and 2 this was done with grids of Sherman traps run throughout the summer. In Plot 3 both a grid of Sherman traps and a line of snap traps were used. This grid, measuring 18 chains by 4 chains, was placed so that approximately the same area sampled for cocoons was sampled for small mammals. The populations determined from these trapping procedures were plotted against time, and the number of "mammal-days" per acre, from the start of larval drop (June 14) to the time cocoon samples were made (Aug. 20-30), was determined for each plot each year. This could be done with *Peromyscus* and *Blarina* since the trapping technique was shown to provide an accurate estimate of their populations. But this was not true for *Sorex*. Instead, the number of *Sorex*-days per acre was approximated by dividing the number of cocoons opened at the highest density by the known number consumed by caged *Sorex* per day, i.e. 101. Since the number of cocoons opened at the highest cocoon density was

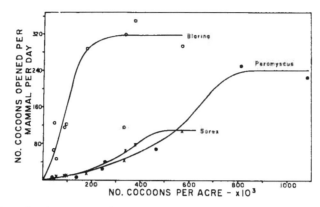

Fig. 1. Functional responses of *Blarina*, *Sorex* and *Peromyscus* in plots 1, 2, and 3.

151,000 per acre, then the number of *Sorex*-days per acre should be 151,000/101 = 1,490. This is approximately 10 times the estimate that was obtained from trapping with Sherman traps. When the various trapping methods were compared, estimates from Sherman trapping were shown to underestimate the numbers of *Sorex* by about the same amount, i.e. one-tenth.

With estimates of the numbers of predators, prey and destroyed prey available, the daily number of prey consumed per predator at different cocoon densities can be calculated. As seen in Fig. 1, the number of cocoons opened by each species increased with increasing cocoon density until a maximum daily consumption was reached that corresponded approximately to the maximum number that could be consumed in any one day by caged animals. For *Sorex* this of course follows from the method of calculation. The rates at which these curves rise differ for the different species, being greatest for *Blarina* and least for *Peromyscus*. Even if the plateaus are equated by multiplying points on each curve by a constant, the rates still decrease in the same order, reflecting a real difference in species behaviour.

The existence of such a response to cocoon density may also be demonstrated by data from the analysis of stomach contents. The per cent occurrence and per cent volume of the various food items in stomachs of *Peromyscus* captured immediately after larval drop and two months later is shown in Table I. When cocoon densities were high, immediately after larval drop, the per cent occurrence and per cent volume of *N. sertifer* material was high. Two months later when various cocoon mortality factors had taken their toll, cocoon densities were lower and

TABLE I

Stomach contents of *Peromyscus* trapped immediately before larval drop and two months later

Time trapped	Approx. no. cocoons per acre	No. of stomachs	Analysis	Plant	*N. sertifer*	Other insects	All insects
June 16–21	600,000	19	% occurrence	37%	95%	53%	100%
Aug. 17–19	300,000	14		79%	50%	64%	86%
June 16–21	600,000	19	% volume	5%	71%	24%	95%
Aug. 17–19	300,000	14		47%	19%	34%	53%

TABLE II

Occurrence of food items in stomachs of *Microtus* trapped before and after larval drop

Time trapped	Plant		*N. sertifer*		All insects	
	No. of stomachs	% occurrence	No. of stomachs	% occurrence	No. of stomachs	% occurrence
before larval drop	25	100%	2	8%	2	8%
after larval drop	29	100%	8	28%	11	38%

N. sertifer was a less important food item. The decrease in consumption of *N. sertifer* was accompanied by a considerable increase in the consumption of plant material and a slight increase in the consumption of other insect material. Plants and other insects acted as buffer or alternate foods. *Microtus*, even though they ate few non-plant foods in nature, also showed an increase in the per cent occurrence of *N. sertifer* material in stomachs as cocoon density increased (Table II). Before larval drop, when cocoon densities were low, the incidence of *N. sertifer* in *Microtus* stomachs was low. After larval drop, when cocoon densities were higher, the incidence increased by 3.5 times. Even at the higher cocoon densities, however, *N. sertifer* comprised less than one per cent of the volume of stomach contents so that this response to changes in prey density by *Microtus* is extremely low.

The graphs presented in Fig. I and the results of the analyses of stomach contents leave little doubt that the consumption of cocooned sawflies by animals in the field increases with increase in cocoon density. Similar responses have been demonstrated in laboratory experiments with three *Peromyscus*. As shown in Fig. 2, the number of cocoons consumed daily by each animal increased with increase in cocoon density, again reaching a plateau as did the previous curves. Whenever the number of prepupae consumed did not meet the caloric requirements, these were met by consumption of the dog biscuits, the alternate food provided. Only one of the animals (A) at the highest density fulfilled its caloric requirements by consuming prepupae; the remaining animals (B and C) consumed

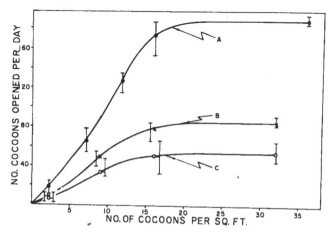

Fig. 2. Functional responses of three caged *Peromyscus* (means and ranges shown).

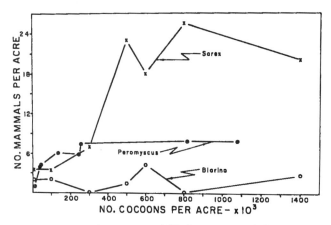

Fig. 3. Numerical responses of *Blarina*, *Sorex* and *Peromyscus*.

less than one-half the number of sawflies they would consume if no alternate foods were present. The cocoons used in experiments involving animals B and C, however, had been spun 12 months earlier than those involving animal A. When the characteristics of the functional response are examined in another paper, it will be shown that the strength of stimulus from older cocoons is less than that from younger cocoons, and that these differences are sufficient to explain the low consumption by animals B and C. The shape of the curves and the density at which they level is very similar for all animals, so similar that multiplying points along any one curve by the proper constant will equate all three. These curves are very similar to the ones based upon field data. All show the same form, the essential feature of which is an S-shaped rise to a plateau.

The effect of changes of prey density need not be restricted exclusively to consumption of prey by individual predators. The density of predators may also be affected and this can be shown by relating the number of predators per acre to the number of cocoons per acre. Conclusions can be derived from these relations but they are tentative. The data were collected over a relatively short period of time (four summers) and thus any relationship between predator numbers and prey density may have been fortuitous. Only those data obtained in plantations over 12 years old are included since small mammal populations were most stable in these areas. The data for the three most important species of predators are shown in the curves of Fig. 3, where each point represents the highest summer population observed either in different plantations or in the same plantation in different years.

The densities of *Blarina* were lowest while those of *Sorex* were highest. In this situation, *Blarina* populations apparently did not respond to prey density, for its numbers did not noticeably increase with increase in cocoon density. Some agent or agents other than food must limit their numbers. Populations of *Peromyscus* and *Sorex*, on the other hand, apparently did initially increase with increase in cocoon density, ultimately ceasing to increase as some agents other than food became limiting. The response of *Sorex* was most marked.

Thus two responses to changes of prey density have been demonstrated. The first is a change in the number of prey consumed per predator and the second is a change in the density of predators. Although few authors appear to recognize the existence and importance of *both* these responses to changes of prey density, they have been postulated and, in the case of the change of predator density, demonstrated.   Thus Solomon (1949) acknowledged the two-fold nature of the response to changes of prey density, and applied the term *functional response* to the change in the number of prey consumed by individual predators, and the term *numerical response* to the change in the density of predators.   These are apt terms and, although they have been largely ignored in the literature, they will be adopted in this paper.   The data available to Solomon for review did not permit him to anticipate the form the functional response of predators might take, so that he could not assess its importance in population regulation.   It will be shown, however, that the functional response is as important as the numerical.

It remains now to consider the effect of predator density, the variable that, together with prey density, is essential for an adequate description of predation. Predator density might well affect the number of prey consumed per predator. Laboratory experiments were designed to measure the number of cocoons opened by one, two, four, and eight animals in a large cage provided with cocoons at a density of 15 per square foot and a surplus of dog biscuits and water.   The average number of cocoons opened per mouse in eight replicates was 159, 137, 141 and 159 respectively.   In this experiment, therefore, predator density apparently did not greatly affect the consumption of prey by individual animals.   This conclusion is again suggested when field and laboratory data are compared, for the functional response of *Peromyscus* obtained in the field, where its density varied, was very similar to the response of single animals obtained in the laboratory.

In such a simple situation, where predator density does not greatly affect the consumption by individuals, the total predation can be expressed by a simple, additive combination of the two responses.   For example, if at a particular prey density the functional response is such that 100 cocoons are opened by a single predator in one day, and the numerical response is such that the predator density is 10, then the total daily consumption will be simply 100 x 10.   In other situations, however, an increase in the density of predators might result in so much competition that the consumption of prey by individual predators might drop significantly. This effect can still be incorporated in the present scheme by adopting a more complex method of combining the functional and numerical responses.

This section was introduced with a list of the possible variables that could affect predation.   Of these, only the two operating in the present study — prey and predator density — are essential variables, so that the basic features of predation can be ascribed to the effects of these two.   It has been shown that there are two responses to prey density.   The increase in the number of prey consumed per predator, as prey density rises, is termed the functional response, while the change in the density of predators is termed the numerical response.   The total amount of predation occurring at any one density results from a combination of the two responses, and the method of combination will be determined by the way predator density affects consumption.   This scheme, therefore, describes the effects of the basic variables, uncomplicated by the effects of subsidiary ones. Hence the two responses, the functional and numerical, can be considered the basic components of predation.

The total amount of predation caused by small mammals is shown in Fig. 4, where the functional and numerical responses are combined by multiplying the number of cocoons opened per predator at each density by the number of effective mammal-days observed.   These figures were then expressed as percentages opened.   This demonstrates the relation between per cent predation and prey density during the 100-day period between cocoon formation and adult emergence.   Since the data obtained for the numerical responses are tentative, some reservations must be applied to the more particular conclusions derived

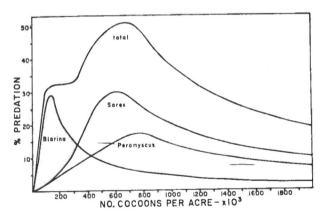

Fig. 4. Functional and numerical responses combined to show the relation between per cent predation and cocoon density.

from this figure. The general conclusion, that per cent predation by each species shows an initial rise and subsequent decline as cocoon density increases holds, however. For this conclusion to be invalid, the numerical responses would have to decrease in order to mask the initial rise in per cent predation caused by the S-shaped form of the functional responses. Thus from zero to some finite cocoon density, predation by small mammals shows a direct density-dependent action and thereafter shows an inverse density-dependent action. The initial rise in the proportion of prey destroyed can be attributed to both the functional and numerical responses. The functional response has a roughly sigmoid shape and hence the proportion of prey destroyed by an individual predator will increase with increase in cocoon density up to and beyond the point of inflection. Unfortunately the data for any one functional response curve are not complete enough to establish a sigmoid relation, but the six curves presented thus far and the several curves to be presented in the following section all suggest a point of inflection. The positive numerical responses shown by *Sorex* and *Peromyscus* also promote a direct density-dependent action up to the point at which predator densities remain constant. Thereafter, with individual consumption also constant, the per cent predation will decline as cocoon density increases. The late Dr. L. Tinbergen apparently postulated the same type of dome-shaped curves for the proportion of insects destroyed by birds. His data were only partly published (1949, 1955) before his death, but Klomp (1956) and Voûte (1958) have commented upon the existence of these "optimal curves". This term, however, is unsatisfactory and anthropocentric. From the viewpoint of the forest entomologist, the highest proportion of noxious insects destroyed may certainly be the optimum, but the term is meaningless for an animal that consumes individuals and not percentages. Progress can best be made by considering predation first as a behaviour before translating this behaviour in terms of the proportion of prey destroyed. The term "peaked curve" is perhaps more accurate.

Returning to Fig. 4, we see that the form of the peaked curve for *Blarina* is determined solely by the functional response since this species exhibited no numerical response. The abrupt peak occurs because the maximum consumption of prepupae was reached at a very low prey density before the predation was "diluted" by large numbers of cocoons. With *Sorex* both the numerical and functional responses are important. Predation by *Sorex* is greatest principally because of the marked numerical response. The two responses again determine the form of the peaked curve for *Peromyscus*, but the numerical response, unlike that of *Sorex*, was not marked, and the maximum consumption of cocoons was reached only at a relatively high density; the result is a low per cent predation with a peak occurring at a high cocoon density.

Predation by all species destroyed a considerable number of cocooned saw-flies over a wide range of cocoon densities. The presence of more than one species of predator not only increased predation but also extended the range of prey densities over which predation was high. This latter effect is particularly important, for if the predation by several species of predators peaked at the same prey density the range of densities over which predation was high would be slight and if the prey had a sufficiently high reproductive capacity its density might jump this vulnerable range and hence escape a large measure of the potential control that could be exerted by predators. Before we can proceed further in the discussion of the effect of predation upon prey numbers, the additional components that make up the behaviour of predation must be considered.

The Subsidiary Components

Additional factors such as prey characteristics, the density and quality of alternate foods, and predator characteristics have a considerable effect upon predation. It is necessary now to demonstrate the effect of these factors and how they operate.

There are four classes of prey characteristics: those that influence the caloric value of the prey; those that change the length of time prey are exposed; those that affect the "attractiveness" of the prey to the predator (e.g. palatability, defence mechanisms); and those that affect the strength of stimulus used by predators in locating prey (e.g. size, habits, and colours). Only those characteristics that affect the strength of stimulus were studied experimentally. Since small mammals detect cocoons by the odour emanating from them (Holling, 1958b), the strength of this odour perceived by a mammal can be easily changed in laboratory experiments by varying the depth of sand covering the cocoons.

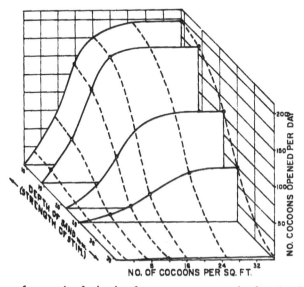

Fig. 5. Effect of strength of stimulus from cocoons upon the functional response of one caged *Peromyscus*. Each point represents the average of three to six replicates.

One *Peromyscus* was used in these experiments and its daily consumption of cocoons was measured at different cocoon densities and different depths of sand. These data are plotted in Fig. 5. Since the relation between depth of sand and strength of stimulus must be an inverse one, the depths of sand are reversed on the axis so that values of the strength of stimulus increase away from the origin. Each point represents the mean of three to six separate measurements. Decreasing the strength of the perceived stimulus by increasing the depth of sand causes a marked decrease in the functional response. A 27 mm. increase in depth (from nine to 36 mm.), for example, causes the peak consumption to drop from 196 to four cocoons per day. The daily number of calories consumed in all these experiments remained relatively constant since dog biscuits were always present as alternate food. The density at which each functional-response curve levels appear to increase somewhat as the strength of stimulus perceived by the animal decreases. We might expect that the increase in consumption is directly related to the increase in the proportion of cocoons in the amount of food available, at least up to the point where the caloric requirements are met solely by sawflies. The ascending portions of the curves, however, are S-shaped and the level portions are below the maximum consumption, approximately 220 cocoons for this animal. Therefore, the functional response cannot be explained by random searching for cocoons. For the moment, however, the important conclusion is that changes in prey characteristics can have a marked effect on predation but this effect is exerted through the functional response.

In the plantations studied, cocoons were not covered by sand but by a loose litter and duff formed from pine needles. Variations in the depth of this material apparently did not affect the strength of the perceived odour, for as many cocoons were opened in quadrats with shallow litter as with deep. This material must be so loose as to scarcely impede the passage of odour from cocoons.

The remaining subsidiary factors, the density and quality of alternate foods and predator characteristics, can also affect predation. The effect of alternate foods could not be studied in the undisturbed plantations because the amount of these "buffers" was constant and very low. The effect of quality of alternate foods on the functional response, however, was demonstrated experimentally using one *Peromyscus*. The experiments were identical to those already described except that at one series of densities an alternate food of low palatability (dog biscuits) was provided, and at the second series one of high palatability (sunflower seeds) was provided. When both foods are available, deer mice select sunflower seeds over dog biscuits. In every experiment a constant amount of alternate food was available: 13 to 17 gms. dry weight of dog biscuits, or 200 sunflower seeds.

Fig. 6 shows the changes in the number of cocoons opened per day and in the amount of alternate foods consumed. The functional response decreased with an increase in the palatability of the alternate food (Fig. 6A). Again the functional response curves showed an initial, roughly sigmoid rise to a constant level.

As cocoon consumption rose, the consumption of alternate foods decreased (Fig. 6B) at a rate related to the palatability of the alternate food. Each line indicating the change in the consumption of alternate food was drawn as a mirror image of the respective functional response and these lines closely follow the mean of the observed points. The variability in the consumption of sunflower seeds at any one cocoon density was considerable, probably as a result of the extreme variability in the size of seeds.

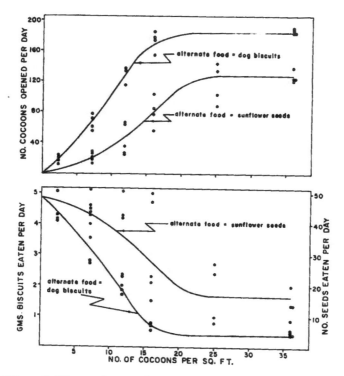

Fig. 6. Effect of different alternate foods upon the functional response of one *Peromys-cus*. A (upper) shows the functional responses when either a low (dog biscuits) or a high (sunflower seeds) palatability alternate food was present in excess. B (lower) shows the amount of these alternate foods consumed.

Again we see that there is not a simple relation between the number of cocoons consumed and the proportion of cocoons in the total amount of food available. This is most obvious when the functional response curves level, for further increase in density is not followed by an increase in the consumption of sawflies. The plateaus persist because the animal continued consuming a certain fixed quantity of alternate foods. L. Tinbergen (1949) observed a similar phenomenon in a study of predation of pine-eating larvae by tits in Holland. He presented data for the consumption of larvae of the pine beauty moth, *Panolis griseovariegata*, and of the web-spinning sawfly *Acantholyda pinivora*, each at two different densities. In each case more larvae were eaten per nestling tit per day at the higher prey density. This, then, was part of a functional response, but it was that part above the point of inflection, since the proportion of prey eaten dropped at the higher density. It is not sufficient to explain these results as well as the ones presented in this paper by claiming, with Tinbergen, that the predators "have the tendency to make their menu as varied as possible and there-fore guard against one particular species being strongly dominant in it". This is less an explanation than an anthropocentric description. The occurrence of this phenomenon depends upon the strength of stimulus from the prey, and the amount and quality of the alternate foods. Its proper explanation must await the collection of further data.

We now know that the palatability of alternate foods affects the functional response. Since the number of different kinds of alternate food could also have

TABLE III

The effect of alternate foods upon the number of cocoons consumed per day by one *Peromyscus*

Alternate food	No. of exp'ts	No. of cocoons opened	
		$\overline{X}$	S.E.$\overline{x}$
none......................................	7	165.9	11.4
dog biscuits...........................	5	143.0	8.3
sunflower seeds........................	8	60.0	6.2
sunflower seeds and dog biscuits.........	8	21.5	4.2

an important effect, the consumption of cocoons by a caged *Peromyscus* was measured when no alternate foods, or one or two alternate foods, were present. Only female cocoons were used and these were provided at a density of 75 per sq. ft. to ensure that the level portion of the functional response would be measured. As in the previous experiments, the animal was familiarized with the experimental conditions and with cocoons for a preliminary two-week period. The average numbers of cocoons consumed each day with different numbers and kinds of alternate foods present are shown in Table III. This table again shows that fewer cocoons were consumed when sunflower seeds (high palatability) were present than when dog biscuits (low palatability) were present. In both cases, however, the consumption was lower than when no alternate foods were available. When two alternate foods were available, i.e., both sunflower seeds and dog biscuits, the consumption dropped even further. Thus, increase in both the palatability and in the number of different kinds of alternate foods decreases the functional response.

DISCUSSION

General

It has been argued that three of the variables affecting predation—characteristics of the prey, density and quality of alternate foods and characteristics of the predators — are subsidiary components of predation. The laboratory experiments showed that the functional response was lowered when the strength of stimulus, one prey characteristic, detected from cocoons was decreased or when the number of kinds and palatability of alternate foods was increased. Hence the effect of these subsidiary components is exerted through the functional response. Now the numerical response is closely related to the functional, since such an increase in predator density depends upon the amount of food consumed. It follows, therefore, that the subsidiary components will also affect the numerical response. Thus when the functional response is lowered by a decrease in the strength of stimulus detected from prey, the numerical response similarly must be decreased and predation will be less as a result of decrease of the two basic responses.

The density and quality of alternate foods could also affect the numerical response. Returning to the numerical responses shown in Fig. 3, if increase in the density or quality of alternate foods involved solely increase in food "per se", then the number of mammals would reach a maximum at a lower cocoon density, but the maximum itself would not change. If increase in alternate foods also involved changes in the agents limiting the numerical responses (e.g. increased cover and depth of humus), then the maximum density the small mammals could attain would increase. Thus increase in the amount of alternate foods could increase the density of predators.

Increase in alternate foods *decreases* predation by dilution of the functional response, but *increases* predation by promoting a favourable numerical response. The relative importance of each of these effects will depend upon the particular problem. Voûte (1946) has remarked that insect populations in cultivated woods show violent fluctuations, whereas in virgin forests or mixed woods, where the number of alternate foods is great, the populations are more stable. This stability might result from alternate foods promoting such a favourable numerical response that the decrease in the functional response is not great enough to lower predation.

The importance of alternate foods will be affected by that part of the third subsidiary component — characteristics of the predators — that concerns food preferences. Thus an increase in plants or animals other than the prey will most likely affect the responses of those predators, like the omnivore *Peromyscus*, that are not extreme food specialists. Predation by the more stenophagous shrews, would only be affected by some alternate, animal food.

Food preferences, however, are only one of the characteristics of predators. Others involve their ability to detect, capture, and kill prey. But again the effect of these predator characteristics will be exerted through the two basic responses, the functional and numerical. The differences observed between the functional responses of the three species shown earlier in Fig. 1 undoubtedly reflect differences in their abilities to detect, capture, and kill. The amount of predation will similarly be affected by the kind of sensory receptor, whether visual, olfactory, auditory, or tactile, that the predator uses in locating prey. An efficient nose, for example, is probably a less precise organ than an efficient eye. The source of an undisturbed olfactory stimulus can only be located by investigating a gradient in space, whereas a visual stimulus can be localized by an efficient eye from a single point in space — the telotaxis of Fraenkel and Gunn (1940). As N. Tinbergen (1951) remarked, localization of direction is developed to the highest degree in the eye. Thus the functional response of a predator which locates prey by sight will probably reach a maximum at a much lower prey density than the response of one that locates its prey by odour. In the data presented by Tothill (1922) and L. Tinbergen (1949), the per cent predation of insects by birds was highest at very low prey densities, suggesting that the functional responses of these "visual predators" did indeed reach a maximum at a low density.

The Effect of Predation on Prey Populations

One of the most important characteristics of mortality factors is their ability to regulate the numbers of an animal — to promote a "steady density" (Nicholson, 1933; Nicholson and Bailey, 1935) such that a continued increase or decrease of numbers from this steady state becomes progressively unlikely the greater the departure from it. Regulation in this sense therefore requires that the mortality factor change with change in the density of the animal attacked, i.e. it requires a direct density-dependent mortality (Smith, 1935, 1939). Density-independent factors can affect the numbers of an animal but alone they can ot *regulate* the numbers. There is abundant evidence that changes in climate, some aspects of which are presumed to have a density-independent action, can lower or raise the numbers of an animal. But this need not be regulation. Regulation will only result from an interaction with a density-dependent factor, an interaction

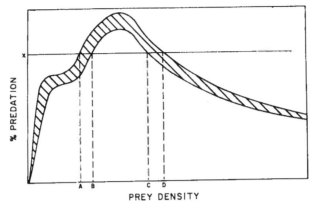

Fig. 7. Theoretical model showing regulation of prey by predators. (see text for explanation).

that might be the simplest, i.e. merely additive. Recently, the density-dependent concept has been severely criticized by Andrewartha and Birch (1954). They call it a dogma, but such a comment is only a criticism of an author's use of the concept. Its misuse as a dogma does not militate against its value as a hypothesis.

We have seen from this study that predation by small mammals does change with changes in prey density. As a result of the functional and numerical responses the proportion of prey destroyed increases from zero to some finite prey density and thereafter decreases. Thus predation over some ranges of prey density shows a direct density-dependent action. This is all that is required for a factor to regulate.

The way in which predation of the type shown in this study can regulate the numbers of a prey species can best be shown by a hypothetical example. To simplify this example we will assume that the prey has a constant rate of reproduction over all ranges of its density, and that only predators are affecting its numbers. Such a situation is, of course, unrealistic. The rate of reproduction of real animals probably is low at low densities when there is slight possibility for contact between individuals (e.g. between male and female). It would rise as contacts became more frequent and would decline again at higher densities when the environment became contaminated, when intraspecific stress symptoms appeared, or when cannibalism became common. Such changes in the rate of reproduction have been shown for experimental populations of *Tribolium confusum* (MacLagan, 1932) and *Drosophila* (Robertson and Sang, 1944). Introducing more complex assumptions, however, confuses interpretations without greatly changing the conclusions.

This hypothetical model is shown in Fig. 7. The curve that describes the changes in predation with changes in prey density is taken from the actual data shown earlier in Fig. 4. It is assumed that the birth-rate of the prey at any density can be balanced by a fixed per cent predation, and that the variation in the environment causes a variation in the predation at any one density. The per cent predation necessary to balance the birth-rate is represented by the horizontal line, x%, in the diagram and variation in predation is represented by the thickness of the mortality curve. The death-rate will equal the birth-rate at two density ranges, between A and B and between C and D. When the densities of the prey are below A, the mortality will be lower than that necessary to balance reproduction and the population will increase. When the densities of the animal are between B and C, death-rate will exceed birth-rate and the populations will decrease. Thus, the density of the prey will tend to fluctuate between densities A and B. If the density happens to exceed D, death-rate will be lower than birth-rate and the prey will increase in numbers, having "escaped" the control exerted by predators. This would occur when the prey had such a high rate of reproduction that its density could jump, in one generation, from a density lower than A to a density higher than D. If densities A and D were far apart, there would be less chance of this occurring. This spread is in part determined by the number of different species of predators that are present. Predation by each species peaks at a different density (see Fig. 4), so that increase in the number of species of predator will increase the spread of the total predation. This will produce a more stable situation where the prey will have less chance to escape control by predators.

Predation of the type shown will regulate the numbers of an animal whenever the predation rises high enough to equal the effective birth-rate. When the prey is an insect and predators are small mammals, as in this case, the reproductive rate of the prey will be too high for predation *alone* to regulate. But if other mortality occurs, density-independent or density-dependent, the total mortality could rise to the point where small mammals were contributing, completely or partially, to the regulation of the insect.

Predation of the type shown will produce stability if there are large numbers of different species of predators operating. Large numbers of such species would most likely occur in a varied environment, such as mixed woods. Perhaps this explains, in part, Voûte's (1946) observation that insect populations in mixed woods are less liable to show violent fluctuations.

I cannot agree with Voûte (1956 and 1958) that factors causing a peaked mortality curve are not sufficient for regulation. He states (1956) that "this is due to the fact that mortality only at low densities increases with the increase of the population. At higher densities, mortality decreases again. The growth of the population is at the utmost slowed down, never stopped". All that is necessary for regulation, however, is a rise in per cent predation over some range of prey densities and an *effective* birth-rate that can be matched at some density by mortality from predators.

Neither can I agree with Thompson (1930) when he ascribes a minor role to vertebrate predators of insects and states that "the number of individuals of any given species (i.e. of vertebrate predators) is . . . relatively small in comparison with those of insects and there is no reason to suppose that it varies primarily in function of the supply of insect food, which fluctuates so rapidly that it is impossible for vertebrates to profit by a temporary abundance of it excepting to a very limited extent". We know that they do respond by an increase in numbers and even if this is not great in comparison with the numerical response of parasitic flies, the number of prey killed per predator is so great and the increase in number killed with increase in prey density is so marked as to result in a heavy proportion of prey destroyed; a proportion that, furthermore, increases initially with increase of prey density. Thompson depreciates the importance of the numerical response of predators and ignores the functional response.

In entomological literature there are two contrasting mathematical theories of regulation. Each theory is based on different assumptions and the predicted results are quite different. Both theories were developed to predict the interaction between parasitic flies and their insect hosts but they can be applied equally well to predator-prey relations. Thompson (1939) assumes that a predator has a limited appetite and that it has no difficulty in finding its prey. Nicholson (1933) assumes that predators have insatiable appetites and that they have a specific capacity to find their prey. This searching capacity is assumed to remain constant at all prey densities and it is also assumed that the searching is random.

The validity of these mathematical models depends upon how closely their assumptions fit natural conditions. We have seen that the appetites of small mammal predators in this study are not insatiable. This fits one of Thompson's assumptions but not Nicholson's. When the functional response was described, it was obvious that predators did have difficulty in finding their prey and that their searching ability did not remain constant at all prey densities. Searching by small mammals was not random. Hence in the present study of predator-prey relations, the remaining assumptions of both Thompson and Nicholson do not hold.

Klomp (1956) considers the damping of oscillations of animal numbers to be as important as regulation. If the oscillations of the numbers of an animal affected by a delayed density-dependent factor (Varley, 1947) like a parasite, do increase in amplitude, as Nicholson's theory predicts (Nicholson and Bailey, 1935), then damping is certainly important. It is not at all certain, however, that this prediction is true. We have already seen that the assumptions underlying Nicholson's theory do not hold in at least some cases. In particular he ignores the important possibility of an S-shaped functional response of the type shown by small mammal predators. If the parasites did show an S-shaped functional response, there would be an *immediate* increase in per cent predation when host density increased, an increase that would modify the effects of the delayed numerical response of parasites emphasized by Nicholson and Varley. Under these conditions the amplitude of the oscillations would not increase as rapidly, and might well not increase at all. An S-shaped functional response therefore acts as an intrinsic damping mechanism in population fluctuations.

Oscillations undoubtedly do occur, however, and whether they increase in amplitude or not, any extrinsic damping is important. The factor that damps oscillations most effectively will be a concurrent density-dependent factor that reacts immediately to changes in the numbers of an animal. Predation by small mammals fulfils these requirements when the density of prey is low. The consumption of prey by individual predators responds immediately to increase in prey density (functional response). Similarly, the numerical response is not greatly delayed, probably because of the high reproductive capacity of small mammals. Thus if the density of a prey is low, chance increases in its numbers will immediately increase the per cent mortality caused by small mammal predation. When the numbers of the prey decrease, the effect of predation will be immediately relaxed. Thus, incipient oscillations can be damped by small-mammal predation.

We have seen that small mammals theoretically can regulate the numbers of prey and can damp their oscillations under certain conditions. Insufficient information was obtained to assess precisely the role of small mammals as predators of *N. sertifer* in the pine plantations of southwestern Ontario, however. Before the general introduction of a virus disease in 1952 (Bird, 1952, 1953), the sawfly was exhausting its food supplies and 70 to 100% defoliation of Scots, jack and red pines was observed in this area. Predators were obviously not regulati-

the numbers of the sawfly. After the virus was introduced, however, sawfly populations declined rapidly. In Plot 1, for example, their numbers declined from 248,000 cocoons per acre in 1952 to 39,000 per acre in 1954. The area was revisited in 1955 and larval and cocoon population had obviously increased in this plot, before the virus disease could cause much mortality. It happened, however, that *Peromyscus* was the only species of small mammal residing in Plot 1 and it is interesting that similar increases were not observed in other plantations where sawfly numbers had either not decreased so greatly, or where shrews, the most efficient predators, were present. These observations suggest that predation by shrews was effectively damping the oscillations resulting from the interaction of the virus disease with its host.

Types of Predation

Many types of predation have been reported in the literature. Ricker (1954) believed that there were three major types of predator-prey relations, Leopold (1933) four, and Errington (1946, 1956) two. Many of these types are merely minor deviations, but the two types of predation Errington discusses are quite different from each other. He distinguishes between "compensatory" and "noncompensatory" predation. In the former type, predators take a heavy toll of individuals of the prey species when the density of prey exceeds a certain threshold. This "threshold of security" is determined largely by the number of secure habitable niches in the environment. When prey densities become too high some individuals are forced into exposed areas where they are readily captured by predators. In this type of predation, predators merely remove surplus animals, ones that would succumb even in the absence of enemies. Errington feels, however, that some predator-prey relations depart from this scheme, so that predation occurs not only *above* a specific threshold density of prey. These departures are ascribed largely to behaviour characteristics of the predators. For example, he does not believe that predation of ungulates by canids is compensatory and feels that this results from intelligent, selective searching by the predators.

If the scheme of predation presented here is to fulfill its purpose it must be able to explain these different types of predation. Non-compensatory predation is easily described by the normal functional and numerical responses, for predation of *N. sertifer* by small mammals is of this type. Compensatory predation can also be described using the basic responses and subsidiary factors previously demonstrated. The main characteristic of this predation is the "threshold of security". Prey are more vulnerable above and less vulnerable below this threshold. That is, the strength of stimulus perceived from prey increases markedly when the prey density exceeds the threshold. We have seen from the present study that an increase in the strength of stimulus from prey increases both the functional and numerical responses. Therefore, below the "threshold of security" the functional responses of predators will be very low and as a result there will probably be no numerical response. Above the threshold, the functional response will become marked and a positive numerical response could easily occur. The net effect will result from a combination of these functional and numerical responses so that per cent predation will remain low so long as there is sufficient cover and food available for the prey. As soon as these supply factors are approaching exhaustion the per cent predation will suddenly increase.

Compensatory predation will occur (1) when the prey has a specific density level near which it normally operates, and (2) when the strength of stimulus perceived by predators is so low below this level and so high above it that there is a marked change in the functional response. Most insect populations tolerate considerable crowding and the only threshold would be set by food limitations. In addition, their strength of stimulus is often high at all densities. For *N. sertifer* at least, the strength of stimulus from cocoons is great and the threshold occurs at such high densities that the functional responses of small mammals are at their maximum. Compensatory predation upon insects is probably uncommon.

Entomologists studying the biological control of insects have largely concentrated their attention on a special type of predator — parasitic insects. Although certain features of a true predator do differ from those of a parasite, both predation and parasitism are similar in that one animal is seeking out another. If insect parasitism can in fact be treated as a type of predation, the two basic responses to prey (or host) density and the subsidiary factors affecting these responses should describe parasitism. The functional response of a true predator is measured by the number of prey it destroys; of a parasite by the number of hosts in which eggs are laid. The differences observed between the functional responses of predators and parasites will depend upon the differences between the behaviour of eating and the behaviour of egg laying. The securing of food by an individual predator serves to maintain that individual's existence. The laying of eggs by a parasite serves to maintain its progenies' existence. It seems likely that the more a behaviour concerns the maintenance of an individual, the more demanding it is. Thus the restraints on egg laying could exert a greater and more prolonged effect than the restraints on eating. This must produce differences between the functional responses of predators and parasites. But the functional responses of both are similar in that there is an upper limit marked by the point at which the predator becomes satiated and the parasite has laid all its eggs. This maximum is reached at some finite prey or host density above zero. The form of the rising phase of the functional response would depend upon the characteristics of the individual parasite and we might expect some of the same forms that will be postulated for predators at the end of this section. To summarize, I do not wish to imply that the characteristics of the functional response of a parasite are identical with those of a predator. I merely wish to indicate that a parasite has a response to prey density — the laying of eggs — that can be identified as a functional response, the precise characteristics of which are unspecified.

The effects of host density upon the number of hosts parasitized have been studied experimentally by a number of workers (e.g., Ullyett, 1949a and b; Burnett, 1951 and 1954; De Bach and Smith, 1941). In each case the number of hosts attacked per parasite increased rapidly with initial increase in host density but tended to level with further increase. Hence these functional response curves showed a continually decreasing slope as host density increased and gave no indication of the S-shaped response shown by small mammals. Further information is necessary, however, before these differences can be ascribed solely to the difference between parasitism and predation. It might well reflect, for example, a difference between an instinctive response of an insect and a learned response of a mammal or between the absence of an alternate host and the presence of an alternate food.

The numerical response of both predators and parasites is measured by the way in which the number of adults increases with increase in prey or host density. At first thought, the numerical response of a parasite would seem to be so intimately connected with its functional response that they could not be separated. But the two responses of a predator are just as intimately connected.

The predator must consume food in order to produce progeny just as the parasite must lay eggs in order to produce progeny.

The agents limiting the numerical response of parasites will be similar to those limiting the response of predators. There is, however, some difference. During at least one stage of the parasites' life, the requirements for both food and niche are met by the same object. Thus increase in the amount of food means increase in the number of niches as well, so that niches are never limited unless food is. This should increase the chances for parasites to show pronounced numerical responses. The characteristics of the numerical responses of both predators and parasites, however, will be similar and will range from those in which there is no increase with increase in the density of hosts, to those in which there is a marked and prolonged increase.

A similar scheme has been mentioned by Ullyett (1949b) to describe parasitism. He believed that "the problem of parasite efficiency would appear to be divided into two main phases, viz.: (a) the efficiency of the parasite as a mortality factor in the host population, (b) its efficiency as related to the maintenance of its own population level within the given area". His first phase resembles the functional response and the second the numerical response. Both phases or responses will be affected, of course, by subsidiary components similar to those proposed for predation—characteristics of the hosts, density and quality of alternate hosts, and characteristics of the parasite. The combination of the two responses will determine the changes in per cent parasitism as the result of changes in host density. Since both the functional and numerical responses presumably level at some point, per cent parasitism curves might easily be peaked, as were the predation curves. If these responses levelled at a host density that would never occur in nature, however, the decline of per cent parasitism might never be observed.

The scheme of predation revealed in this study may well explain all types of predation as well as insect parasitism. The knowledge of the basic components and subsidiary factors underlying the behaviour permits us to imagine innumerable possible variations. In a hypothetical situation, for example, we could introduce and remove alternate food at a specific time in relation to the appearance of a prey, and predict the type of predation. But such variations are only minor deviations of a basic pattern. The major types of predation will result from major differences in the form of the functional and numerical responses.

If the functional responses of some predators are partly determined by their behaviour, we could expect a variety of responses differing in form, rate of rise, and final level reached. All functional responses, however, will ultimately level, for it is difficult to imagine an individual predator whose consumption rises indefinitely. Subsistence requirements will fix the ultimate level for most predators, but even those whose consumption is less rigidly determined by subsistence requirements (e.g., fish, Ricker 1941) must have an upper limit, even if it is only determined by the time required to kill.

The functional responses could conceivably have three basic forms. The mathematically simplest would be shown by a predator whose pattern of searching was random and whose rate of searching remained constant at all prey densities. The number of prey killed per predator would be directly proportional to prey density, so that the rising phase would be a straight line. Ricker (1941) postulated this type of response for certain fish preying on sockeye salmon, and De Bach and Smith (1941) observed that the parasitic fly, *Muscidifurax raptor*, parasitized puparia of *Musca domestica*, provided at different densities, in a similar fashion. So few prey were provided in the latter experiment, however, that the initial linear rise in the number of prey attacked with increase in prey density may have been an artifact of the method.

A more complex form of functional response has been demonstrated in laboratory experiments by De Bach and Smith (1941), Ullyett (1949a) and Burnett (1951, 1956) for a number of insect parasites. In each case the number of prey attacked per predator increased very rapidly with initial increase in prey density, and thereafter increased more slowly approaching a certain fixed level. The rates of searching therefore became progressively less as prey density increased.

The third and final form of functional response has been demonstrated for small mammals in this study. These functional responses are S-shaped so that the rates of searching at first increase with increase of prey density, and then decrease.

Numerical responses will also differ, depending upon the species of predator and the area in which it lives. Two types have been demonstrated in this study. *Peromyscus* and *Sorex* populations, for example, increased with increase of prey density to the point where some agent or agents other than food limited their numbers. These can be termed direct numerical responses. There are some cases, however, where predator numbers are not affected by changes of prey density and in the plantations studied *Blarina* presents such an example of no numerical response. A final response, in addition to ones shown here, might also occur. Morris *et al.* (1958) have pointed out that certain predators might decrease in numbers as prey density increases through competition with other predators. As an example of such inverse numerical responses, he shows that during a recent outbreak of spruce budworm in New Brunswick the magnolia, myrtle, and black-throated green warblers decreased in numbers. Thus we have three possible numerical responses — a direct response, no response, and an inverse response.

The different characteristics of these types of functional and numerical responses produce different types of predation. There are four major types conceivable; these are shown diagramatically in Fig. 8. Each type includes the three possible numerical responses — a direct response (a), no response (b), and an inverse response (c), and the types differ because of basic differences in the functional response. In type 1 the number of prey consumed per predator is assumed to be directly proportional to prey density, so that the rising phase of the functional response is a straight line. In type 2, the functional response is presumed to rise at a continually decreasing rate. In type 3, the form of the functional response is the same as that observed in this study. These three types of predation may be considered as the basic ones, for changes in the subsidiary components are not involved. Subsidiary components can, however, vary in response to changes of prey density and in such cases the basic types of predation are modified. The commonest modification seems to be Errington's compensatory predation which is presented as Type 4 in Fig. 8. In this figure the vertical dotted line represents the "threshold of security" below which the strength of stimulus from prey is low and above which it is high. The functional response curves at these two strengths of stimulus are given the form of the functional responses observed in this study. The forms of the responses shown in Types 1 and 2 could also be used, of course.

The combination of the two responses gives the total response shown in the

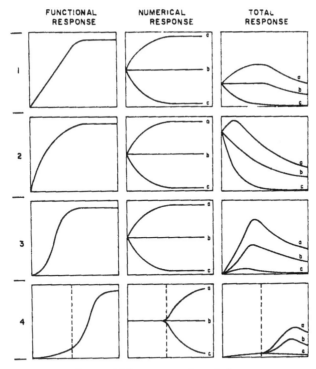

Fig. 8. Major types of predation.

final column of graphs of Fig. 8. Both peaked (curves 1a; 2a; 3a, b, c; 4a, b, c) and declining (1b, c; 2b, c) types of predation can occur, but in the absence of any other density-dependent factor, regulation is possible only in the former type.

This method of presenting the major types of predation is an over-simplification since predator density is portrayed as being directly related to prey density. Animal populations, however, cannot respond *immediately* to changes in prey density, so that there must be a delay of the numerical response. Varley (1953) pointed this out when he contrasted "delayed density dependence" and "density dependence". The degree of delay, however, will vary widely depending upon the rate of reproduction, immigration, and mortality. Small mammals, with their high reproductive rate, responded so quickly to increased food that the delay was not apparent. In such cases the numerical response graphs of Fig. 8 are sufficiently accurate, for the density of predators in any year is directly related to the density of prey in the same year. The numerical response of other natural enemies can be considerably delayed, however. Thus the density of those insect parasites that have one generation a year and a low rate of immigration results from the density of hosts in the preceding generation.

In these extreme cases of delay the total response obtained while prey or hosts are steadily increasing will be different than when they are steadily decreasing. The amount of difference will depend upon the magnitude and amount of delay of the numerical response, for the functional response has no element of delay.

SUMMARY AND CONCLUSIONS

The simplest and most basic type of predation is affected by only two variables — prey and predator density. Predation of cocooned *N. sertifer* by small mammals is such a type, for prey characteristics, the number and variety of alternate foods, and predilections of the predators do not vary in the plantations where *N. sertifer* occurs. In this simple example of predation, the basic components of predation are responses to changes in prey density. The increase in the number of prey consumed per predator, as prey density rises, is termed the functional response. The change in the density of predators, as a result of increase in prey density, is termed the numerical response.

The three important species of small mammal predators (*Blarina*, *Sorex*, and *Peromyscus*) each showed a functional response, and each curve, whether it was derived from field or laboratory data, showed an initial S-shaped rise up to a constant, maximum consumption. The rate of increase of consumption decreased from *Blarina* to *Sorex* to *Peromyscus*, while the upper, constant level of consumption decreased from *Blarina* to *Peromyscus* to *Sorex*. The characteristics of these functional responses could not be explained by a simple relation between consumption and the proportion of prey in the total food available. The form of the functional response curves is such that the proportion of prey consumed per predator increases to a peak and then decreases.

This peaked curve was further emphasized by the direct numerical response of *Sorex* and *Peromyscus*, since their populations rose initially with increase in prey density up to a maximum that was maintained with further increase in cocoon density. *Blarina* did not show a numerical response. The increase in density of predators resulted from increased breeding, and because the reproductive rate of small mammals is so high, there was an almost immediate increase in density with increase in food.

The two basic components of predation — the functional and numerical responses — can be affected by a number of subsidiary components: prey characteristics, the density and quality of alternate foods, and characteristics of the predators. It was shown experimentally that these components affected the amount of predation by lowering or raising the functional and numerical responses. Decrease of the strength of stimulus from prey, one prey characteristic, lowered both the functional and numerical responses. On the other hand, the quality of alternate foods affected the two responses differently. Increase in the palatability or in the number of kinds of alternate foods lowered the functional response but promoted a more pronounced numerical response.

The peaked type of predation shown by small mammals can theoretically regulate the numbers of its prey if predation is high enough to match the effective reproduction by prey at some prey density. Even if this condition does not hold, however, oscillations of prey numbers are damped. Since the functional and numerical responses undoubtedly differ for different species of predator, predation by each is likely to peak at a different prey density. Hence, when a large number of different species of predators are present the declining phase of predation is displaced to a higher prey density, so that the prey have less chance to "escape" the regulation exerted by predators.

The scheme of predation presented here is sufficient to explain all types of predation as well as insect parasitism. It permits us to postulate four major types of predation differing in the characteristics of their basic and subsidiary components.

REFERENCES

Andrewartha, H. G. and L. C. Birch. 1954. The distribution and abundance of animals. *The Univ. of Chicago Press*, Chicago.

Bird, F. T. 1952. On the artificial dissemination of the virus disease of the European saw-fly, *Neodiprion sertifer* (Geoff.). *Can. Dept. Agric., For. Biol. Div., Bi-Mon. Progr. Rept.* 8(3): 1-2

Bird, F. T. 1953. The use of a virus disease in the biological control of the European pine sawfly, *Neodiprion sertifer* (Geoff.). *Can. Ent.* 85: 437-446.

Blair, W. F. 1941. Techniques for the study of mammal populations. *J. Mamm.* 22: 148-157.

Buckner, C. H. 1957. Population studies on small mammals of southeastern Manitoba. *J. Mamm.* 38: 87-97.

Burnett, T. 1951. Effects of temperature and host density on the rate of increase of an insect parasite. *Amer. Nat.* 85: 337-352.

Burnett, T. 1954. Influences of natural temperatures and controlled host densities on oviposition of an insect parasite. *Physiol. Ecol.* 27: 239-248.

Burt, W. H. 1940. Territorial behaviour and populations of some small mammals in southern Michigan. *Misc. Publ. Univ. Mich. Mus. Zool.* no. 45: 1-52.

De Bach, P. 1958. The role of weather and entomophagous species in the natural control of insect populations. *J. Econ. Ent.* 51: 474-484.

De Bach, P., and H. S. Smith. 1941. The effect of host density on the rate of reproduction of entomophagous parasites. *J. Econ. Ent.* 34: 741-745.

De Bach, P., and H. S. Smith. 1947. Effects of parasite population density on rate of change of host and parasite populations. *Ecology* 28: 290-298.

Errington, P. L. 1934. Vulnerability of bob-white populations to predation. *Ecology* 15: 110-127.

Errington, P. L. 1943. An analysis of mink predation upon muskrats in North-Central United States. *Agric. Exp. Sta. Iowa State Coll. Res. Bull.* 320: 797-924.

Errington, P. L. 1945. Some contributions of a fifteen-year local study of the northern bob-white to a knowledge of population phenomena. *Ecol. Monog.* 15: 1-34.

Errington, P. L. 1946. Predation and vertebrate populations. *Quart. Rev. Biol.* 21: 144-177, 221-245.

Fraenkel, G., and D. L. Gunn. 1940. The orientation of animals. Oxford.

Hayne, D. W. 1949. Two methods for estimating population from trapping records. *J. Mamm.* 30: 339-411.

Holling, C. S. 1955. The selection by certain small mammals of dead, parasitized, and healthy prepupae of the European pine sawfly, *Neodiprion sertifer* (Goeff.). *Can. J. Zool.* 33: 404-419.

Holling, C. S. 1958a. A radiographic technique to identify healthy, parasitized, and diseased sawfly prepupae within cocoons. *Can. Ent.* 90: 59-61.

Holling, C. S. 1958b. Sensory stimuli involved in the location and selection of sawfly cocoons by small mammals. *Can. J. Zool.* 36: 633-653.

Klomp, H. 1956. On the theories on host-parasite interaction. *Int. Union of For. Res. Organizations, 12th Congress*, Oxford, 1956.

Leopold, A. 1933. Game management. Charles Scribner's Sons.

Lincoln, F. C. 1930. Calculating waterfowl abundance on the basis of banding returns. *U.S. Dept. Agric.* Circular 118.

MacLagan, D. S. 1932. The effect of population density upon rate of reproduction, with special reference to insects. *Proc. Roy. Soc. Lond.* 111: 437-454.

Milne, A. 1957. The natural control of insect populations. *Can. Ent.* 89: 193-213.

Morris, R. F., W. F. Chesire, C. A. Miller, and D. G. Mott. 1958. Numerical response of avian and mammalian predators during a gradation of the spruce budworm. *Ecology* 39(3): 487-494.

Nicholson, A. J. 1933. The balance of animal populations. *J. Anim. Ecol.* 2: 132-178.

Nicholson, A. J., and V. A. Bailey. 1935. The balance of animal populations. Part 1, *Proc. Zool. Soc. Lond.* 1935, p. 551-598.

Prebble, M. L. 1943. Sampling methods in population studies of the European spruce saw-fly, *Gilpinia hercyniae* (Hartig.) in eastern Canada. *Trans. Roy. Soc. Can.*, Third Series, Sect. V. 37: 93-126.

Ricker, W. E. 1941. The consumption of young sockeye salmon by predaceous fish. *J. Fish. Res. Bd. Can.* 5: 293-313.

Ricker, W. E. 1954. Stock and recruitment. *J. Fish. Res. Bd. Can.* 11: 559-623.

Robertson, F. W., and J. H. Sang. 1944. The ecological determinants of population growth in a *Drosophila* culture. I. Fecundity of adult flies. *Proc. Roy. Soc. Lond.*, B., 132: 258-277.

Solomon, M. E. 1949. The natural control of animal populations. *J. Anim. Ecology* 18: 1-35.

Stickel, L. F. 1954. A comparison of certain methods of measuring ranges of small mammals. *J. Mamm.* 35: 1-15.

Thompson, W. R. 1929. On natural control. *Parasitology* 21: 269-281.

Thompson, W. R. 1930. The principles of biological control. *Ann. Appl. Biol.* 17: 306-338.

Thompson, W. R. 1939. Biological control and the theories of the interactions of populations. *Parasitology* 31: 299-388.

Tinbergen, L. 1949. Bosvogels en insecten. *Nederl. Boschbouue. Tijdschr.* 21: 91-105.

Tinbergen, L. 1955. The effect of predators on the numbers of their hosts. *Vakblad voor Biologen* 28: 217-228.

Tinbergen, N. 1951. The study of instinct. Oxford.

Tothill, J. D. 1922. The natural control of the fall webworm (*Hyphantria cunea* Drury) in Canada. *Can. Dept. Agr. Bull.* 3, new series (Ent. Bull. 19): 1-107.

Ullyett, G. C. 1949a. Distribution of progeny by *Cryptus inornatus* Pratt. (Hym. Ichneumonidae). *Can. Ent.* 81: 285-299, 82: 1-11.

Ullyett, G. C. 1949b. Distribution of progeny by *Chelonus texanus* Cress. (Hym. Braconidae). *Can. Ent.* 81: 25-44.

Varley, G. C. 1947. The natural control of population balance in the knapweed gall-fly (*Urophora jaceana*). *J. Anim. Ecol.* 16: 139-187.

Varley, G. C. 1953. Ecological aspects of population regulation. *Trans. IXth Int. Congr. Ent.* 2: 210-214.

Voûte, A. D. 1946. Regulation of the density of the insect populations in virgin forests and cultivated woods. *Archives Neerlandaises de Zoologie* 7: 435-470.

Voûte, A. D. 1956. Forest entomology and population dynamics. *Int. Union For. Res. Organizations*, Twelfth Congress, Oxford.

Voûte, A. D. 1958. On the regulation of insect populations. *Proc. Tenth Int. Congr. of Ent.* Montreal, 1956.

(Received April 16, 1959)

THE RUNGE PRESS LIMITED
Mailed: Thursday, May 28, 1959

Connell, J. H. 1961. The influence of interspecific competition and other factors on the distribution of the barnacle *Chthamalus stellatus*. *Ecology* 42:710–723.

Interspecific competition and its influence on the distribution and abundance of animals and plants has been the subject of many research projects and resulting papers. Scientists continue to debate the significance of interspecific competition regarding trajectories of plant and animal communities. These perspectives, originating in the 1960s, range from "competition no longer exists" to "competition is the key to ecosystem stability."

It is important to understand how animals are distributed in nature, and it is equally important to understand the role that competition plays. Most biologists agree that interspecific competition is a critical factor affecting distribution of animal and plant species. Before Connell's paper—from his dissertation study of two species of barnacles at Millport, Isle of Cumbrae, Scotland—most evidence for interspecific competition came from laboratory and captive populations. That changed with Connell's classic empirical work, which clearly establishes how strongly interspecific competition affects the distribution and abundance of animals and plants. This and subsequent papers on interspecific competition were instrumental in demonstrating that competition does occur under natural conditions, that it can be empirically examined and evaluated, and that it is the principal factor in determining local distribution. Connell set the standard for determining the ecological importance of interspecific competition.

RELATED READING

Andrewartha, H. G., and L. C. Birch. 1954. The distribution and abundance of animals. University of Chicago Press, Chicago, Illinois, USA.

Bogert, C. M. 1952. Relative abundance, habitats and normal thermal levels of some Virginia salamanders. Ecology 33:16–30.

Connell, J. H. 1961. Effects of competition, predation by *Thais lapillus*, and other factors on natural populations of the barnacle *Balanus balanoides*. Ecological Monographs 31:61–104.

Connell, J. H. 1983. On the prevalence and relative importance of interspecific competition: evidence from field experiments. The American Naturalist 122:661–696.

Mayr, E. 1947. Ecological factors in speciation. Evolution 1:263–288.

Miller, A. H. 1942. Habitat selection among higher vertebrates and its relation to intraspecific variation. The American Naturalist 76:25–35.

Schoener, T. W. 1983. Field experiments on interspecific competition. The American Naturalist 122:240–285.

Wright, S. 1943. Isolation by distance. Genetics 28:114–138.

THE INFLUENCE OF INTERSPECIFIC COMPETITION AND OTHER FACTORS ON THE DISTRIBUTION OF THE BARNACLE *CHTHAMALUS STELLATUS*

Joseph H. Connell

Department of Biology, University of California, Santa Barbara, Goleta, California

Introduction

Most of the evidence for the occurrence of interspecific competition in animals has been gained from laboratory populations. Because of the small amount of direct evidence for its occurrence in nature, competition has sometimes been assigned a minor role in determining the composition of animal communities.

Indirect evidence exists, however, which suggests that competition may sometimes be responsible for the distribution of animals in nature. The range of distribution of a species may be decreased in the presence of another species with similar requirements (Beauchamp and Ullyott 1932, Endean, Kenny and Stephenson 1956). Uniform distribution is space is usually attributed to intraspecies competition (Holme 1950, Clark and Evans 1954). When animals with similar requirements, such as 2 or more closely related species, are found coexisting in the same area, careful analysis usually indicates that they are not actually competing with each other (Lack 1954, MacArthur 1958).

In the course of an investigation of the animals of an intertidal rocky shore I noticed that the adults of 2 species of barnacles occupied 2 separate horizontal zones with a small area of overlap, whereas the young of the species from the upper zone were found in much of the lower zone. The upper species, *Chthamalus stellatus* (Poli) thus settled but did not survive in the lower zone. It seemed probable that this species was eliminated by the lower one, *Balanus balanoides* (L), in a struggle for a common requisite which was in short supply. In the rocky intertidal region, space for attachment and growth is often extremely limited. This paper is an account of some observations and experiments designed to test the hypothesis that the absence in the lower zone of adults of *Chthamalus* was due to interspecific competition with *Balanus* for space. Other factors which may have influenced the distribution were also studied. The study was made at Millport, Isle of Cumbrae, Scotland.

I would like to thank Prof. C. M. Yonge and the staff of the Marine Station, Millport, for their help, discussions and encouragement during the course of this work. Thanks are due to the following for their critical reading of the manuscript: C. S. Elton, P. W. Frank, G. Hardin, N. G. Hairston, E. Orias, T. Park and his students, and my wife.

Distribution of the species of barnacles

The upper species, *Chthamalus stellatus*, has its center of distribution in the Mediterranean; it reaches its northern limit in the Shetland Islands, north of Scotland. At Millport, adults of this species occur between the levels of mean high water of neap and spring tides (M.H.W.N. and M.H.W.S.: see Figure 5 and Table I). In southwest England and Ireland, adult *Chtham-*

alus occur at moderate population densities throughout the intertidal zone, more abundantly when *Balanus balanoides* is sparse or absent (Southward and Crisp 1954, 1956). At Millport the larvae settle from the plankton onto the shore mainly in September and October; some additional settlement may occur until December. The settlement is most abundant between M.H.W.S. and mean tide level (M.T.L.), in patches of rock surface left bare as a result of the mortality of *Balanus,* limpets, and other sedentary organisms. Few of the *Chthamalus* that settle below M.H. W.N. survive, so that adults are found only occasionally at these levels.

Balanus balanoides is a boreal-arctic species, reaching its southern limit in northern Spain. At Millport it occupies almost the entire intertidal region, from mean low water of spring tides (M.L.W.S.) up to the region between M.H.W.N. and M.H.W.S. Above M.H.W.N. it occurs intermingled with *Chthamalus* for a short distance. *Balanus* settles on the shore in April and May, often in very dense concentrations (see Table IV).

The main purpose of this study was to determine the cause of death of those *Chthamalus* that settled below M.H.W.N. A study which was being carried on at this time had revealed that physical conditions, competition for space, and predation by the snail *Thais lapillus* L. were among the most important causes of mortality of *Balanus balanoides.* Therefore, the observations and experiments in the present study were designed to detect the effects of these factors on the survival of *Chthamalus.*

METHODS

Intertidal barnacles are very nearly ideal for the study of survival under natural conditions. Their sessile habit allows direct observation of the survival of individuals in a group whose positions have been mapped. Their small size and dense concentrations on rocks exposed at intervals make experimentation feasible. In addition, they may be handled and transplanted without injury on pieces of rock, since their opercular plates remain closed when exposed to air.

The experimental area was located on the Isle of Cumbrae in the Firth of Clyde, Scotland. Farland Point, where the study was made, comprises the southeast tip of the island; it is exposed to moderate wave action. The shore rock consists mainly of old red sandstone, arranged in a series of ridges, from 2 to 6 ft high, oriented at right angles to the shoreline. A more detailed description is given by Connell (1961). The

other barnacle species present were *Balanus crenatus* Brug and *Verruca stroemia* (O. F. Muller), both found in small numbers only at and below M.L.W.S.

To measure the survival of *Chthamalus,* the positions of all individuals in a patch were mapped. Any barnacles which were empty or missing at the next examination of this patch must have died in the interval, since emigration is impossible. The mapping was done by placing thin glass plates (lantern slide cover glasses, 10.7×8.2 cm, area 87.7 cm²) over a patch of barnacles and marking the position of each *Chthamalus* on it with glass-marking ink. The positions of the corners of the plate were marked by drilling small holes in the rock. Observations made in subsequent censuses were noted on a paper copy of the glass map.

The study areas were chosen by searching for patches of *Chthamalus* below M.H.W.N. in a stretch of shore about 50 ft long. When 8 patches had been found, no more were looked for. The only basis for rejection of an area in this search was that it contained fewer than 50 *Chthamalus* in an area of about 1/10 m². Each numbered area consisted of one or more glass maps located in the 1/10 m². They were mapped in March and April, 1954, before the main settlement of *Balanus* began in late April.

Very few *Chthamalus* were found to have settled below mid-tide level. Therefore pieces of rock bearing *Chthamalus* were removed from levels above M.H.W.N. and transplanted to and below M.T.L. A hole was drilled through each piece; it was then fastened to the rock by a stainless steel screw driven into a plastic screw anchor fitted into a hole drilled into the rock. A hole 1/4″ in diameter and 1″ deep was found to be satisfactory. The screw could be removed and replaced repeatedly and only one stone was lost in the entire period.

For censusing, the stones were removed during a low tide period, brought to the laboratory for examination, and returned before the tide rose again. The locations and arrangements of each area are given in Table I; the transplanted stones are represented by areas 11 to 15.

The effect of competition for space on the survival of *Chthamalus* was studied in the following manner: After the settlement of *Balanus* had stopped in early June, having reached densities of 49/cm² on the experimental areas (Table I) a census of the surviving *Chthamalus* was made on each area (see Figure 1). Each map was then divided so that about half of the number of

TABLE I. Description of experimental areas*

| Area no. | Height in ft from M.T.L. | % of time sub-merged | Population Density: no./cm² in June, 1954 | | All barnacles, undisturbed portion | Remarks |
| | | | Chthamalus, autumn 1953 settlement | | | |
			Undisturbed portion	Portion without Balanus		
MHWS.............	+4.9	4	—	—	—	—
1.................	+4.2	9	2.2	—	19.2	Vertical, partly protected
2.................	+3.5	16	5.2	4.2	—	Vertical, wave beaten
MHWN.............	+3.1	21	—	—	—	—
3a.................	+2.2	30	0.6	0.6	30.9	Horizontal, wave beaten
3b.................	"	"	0.5	0.7	29.2	" " "
4.................	+1.4	38	1.9	0.6	—	30° to vertical, partly protected
5.................	+1.4	"	2.4	1.2	—	" " " " "
6.................	+1.0	42	1.1	1.9	38.2	Horizontal, top of a boulder, partly protected
7a.................	+0.7	44	1.3	2.0	49.3	Vertical, protected
7b.................	"	"	2.3	2.0	51.7	" "
11a.................	0.0	50	1.0	0.6	32.0	Vertical, protected
11b.................	"	"	0.2	0.3	—	" "
12a.................	0.0	100	1.2	1.2	18.8	Horizontal, immersed in tide pool
12b.................	"	100	0.8	0.9	—	" " " " "
13a.................	−1.0	58	4.9	4.1	29.5	Vertical, wave beaten
13b.................	"	"	3.1	2.4	—	" " "
14a.................	−2.5	71	0.7	1.1	—	45° angle, wave beaten
14b.................	"	"	1.0	1.0	—	" " " "
MLWN.............	−3.0	77	—	—	—	—
MLWS.............	−5.1	96	—	—	—	—
15.................	+1.0	42	32.0	—	—	{Chthamalus of autumn, 1954 set-
7b.................	+0.7	44	5.5	3.7	—	{tlement; densities of Oct., 1954.

* The letter "a" following an area number indicates that this area was enclosed by a cage; "b" refers to a closely adjacent area which was not enclosed. All areas faced either east or south except 7a and 7b, which faced north.

Chthamalus were in each portion. One portion was chosen (by flipping a coin), and those Balanus which were touching or immediately surrounding each Chthamalus were carefully removed with a needle; the other portion was left untouched. In this way it was possible to measure the effect on the survival of Chthamalus both of intraspecific competition alone and of competition with Balanus. It was not possible to have the numbers or population densities of Chthamalus exactly equal on the 2 portions of each area. This was due to the fact that, since Chthamalus often occurred in groups, the Balanus had to be removed from around all the members of a group to ensure that no crowding by Balanus occurred. The densities of Chthamalus were very low, however, so that the slight differences in density between the 2 portions of each area can probably be disregarded; intraspecific crowding was very seldom observed. Censuses of the Chthamalus were made at intervals of 4-6 weeks during the next year; notes were made at each census of factors such as crowding, undercutting or smothering which had taken place since the last examination. When necessary, Balanus which had grown until they threatened to touch the Chthamalus were removed in later examinations.

To study the effects of different degrees of immersion, the areas were located throughout the tidal range, either in situ or on transplanted stones, as shown in Table I. Area 1 had been under observation for 1½ years previously. The effects of different degrees of wave shock could not be studied adequately in such a small area

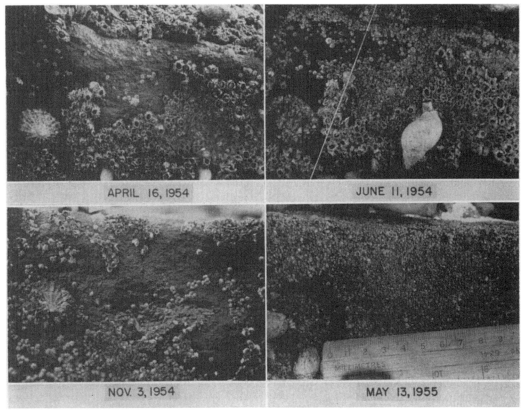

APRIL 16, 1954

JUNE 11, 1954

NOV. 3, 1954

MAY 13, 1955

FIG. 1. Area 7b. In the first photograph the large barnacles are *Balanus,* the small ones scattered in the bare patch, *Chthamalus.* The white line on the second photograph divides the undisturbed portion (right) from the portion from which *Balanus* were removed (left). A limpet, *Patella vulgata,* occurs on the left, and predatory snails, *Thais lapillus,* are visible.

of shore but such differences as existed are listed in Table I.

The effects of the predatory snail, *Thais lapillus,* (synonymous with *Nucella* or *Purpura,* Clench 1947), were studied as follows: Cages of stainless steel wire netting, 8 meshes per inch, were attached over some of the areas. This mesh has an open area of 60% and previous work (Connell 1961) had shown that it did not inhibit growth or survival of the barnacles. The cages were about 4 × 6 inches, the roof was about an inch above the barnacles and the sides were fitted to the irregularities of the rock. They were held in place in the same manner as the transplanted stones. The transplanted stones were attached in pairs, one of each pair being enclosed in a cage (Table I).

These cages were effective in excluding all but the smallest *Thais.* Occasionally small *Thais,* ½ to 1 cm in length, entered the cages through gaps at the line of juncture of netting and rock surface. In the concurrent study of *Balanus* (Con-

nell 1961), small *Thais* were estimated to have occurred inside the cages about 3% of the time.

All the areas and stones were established before the settlement of *Balanus* began in late April, 1954. Thus the *Chthamalus* which had settled naturally on the shore were then of the 1953 year class and all about 7 months old. Some *Chthamalus* which settled in the autumn of 1954 were followed until the study was ended in June, 1955. In addition some adults which, judging from their large size and the great erosion of their shells, must have settled in 1952 or earlier, were present on the transplanted stones. Thus records were made of at least 3 year-classes of *Chthamalus.*

RESULTS

The effects of physical factors

In Figures 2 and 3, the dashed line indicates the survival of *Chthamalus* growing without contact with *Balanus.* The suffix "a" indicates that the area was protected from *Thais* by a cage.

In the absence of *Balanus* and *Thais,* and protected by the cages from damage by water-borne objects, the survival of *Chthamalus* was good at all levels. For those which had settled normally on the shore (Fig. 2), the poorest survival was on the lowest area, 7a. On the transplanted stones (Fig. 3, area 12), constant immersion in a tide pool resulted in the poorest survival. The reasons for the trend toward slightly greater mortality as the degree of immersion increased are unknown. The amount of attached algae on the stones in the tide pool was much greater than on the other areas. This may have reduced the flow of water and food or have interfered directly with feeding movements. Another possible indirect effect of increased immersion is the increase in predation by the snail, *Thais lapillus,* at lower levels.

Chthamalus is tolerant of a much greater degree of immersion than it normally encounters. This is shown by the survival for a year on area 12 in a tide pool, together with the findings of Fischer (1928) and Barnes (1956a), who found that *Chthamalus* withstood submersion for 12 and 22 months, respectively. Its absence below M.T.L. can probably be ascribed either to a lack of initial settlement or to poor survival of newly settled larvae. Lewis and Powell (1960) have suggested that the survival of *Chthamalus* may be favored by increased light or warmth during emersion in its early life on the shore. These conditions would tend to occur higher on the shore in Scotland than in southern England.

The effects of wave action on the survival of *Chthamalus* are difficult to assess. Like the degree of immersion, the effects of wave action may act indirectly. The areas 7 and 12, where relatively poor survival was found, were also the areas of least wave action. Although *Chthamalus* is usually abundant on wave beaten areas and absent from sheltered bays in Scotland, Lewis and Powell (1960) have shown that in certain sheltered bays it may be very abundant. Hatton (1938) found that in northern France, settlement and growth rates were greater in wave-beaten areas at M.T.L., but, at M.H.W.N., greater in sheltered areas.

At the upper shore margins of distribution *Chthamalus* evidently can exist higher than *Balanus* mainly as a result of its greater tolerance to heat and/or desiccation. The evidence for this was gained during the spring of 1955. Records from a tide and wave guage operating at this time about one-half mile north of the study area showed that a period of neap tides had coincided with an unusual period of warm calm weather in April so that for several days no water, not even waves, reached the level of Area 1. In the period

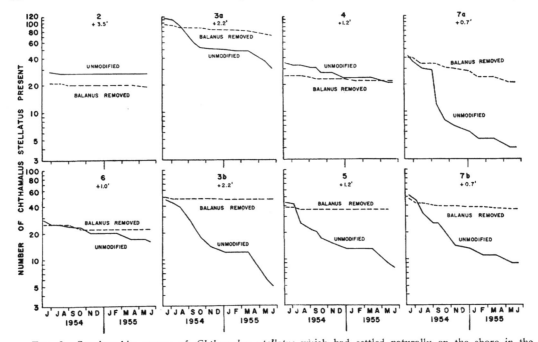

Fig. 2. Survivorship curves of *Chthamalus stellatus* which had settled naturally on the shore in the autumn of 1953. Areas designated "a" were protected from predation by cages. In each area the survival of *Chthamalus* growing without contact with *Balanus* is compared to that in the undisturbed area. For each area the vertical distance in feet from M.T.L. is shown.

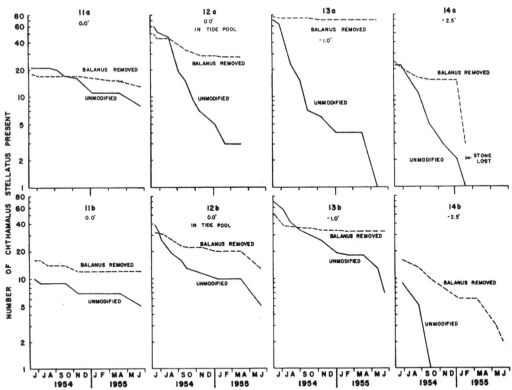

Fig. 3. Survivorship curves of *Chthamalus stellatus* on stones transplanted from high levels. These had settled in the autumn of 1953; the arrangement is the same as that of Figure 2.

between the censuses of February and May, *Balanus* aged one year suffered a mortality of 92%, those 2 years and older, 51%. Over the same period the mortality of *Chthamalus* aged 7 months was 62%, those 1½ years and older, 2%. Records of the survival of *Balanus* at several levels below this showed that only those *Balanus* in the top quarter of the intertidal region suffered high mortality during this time (Connell 1961).

Competition for space

At each census notes were made for individual barnacles of any crowding which had occurred since the last census. Thus when one barnacle started to grow up over another this fact was noted and at the next census 4-6 weeks later the progress of this process was noted. In this way a detailed description was built up of these gradually occurring events.

Intraspecific competition leading to mortality in *Chthamalus* was a rare event. For areas 2 to 7, on the portions from which *Balanus* had been removed, 167 deaths were recorded in a year. Of these, only 6 could be ascribed to crowding between individuals of *Chthamalus*. On the undisturbed portions no such crowding was

observed. This accords with Hatton's (1938) observation that he never saw crowding between individuals of *Chthamalus* as contrasted to its frequent occurrence between individuals of *Balanus*.

Interspecific competition between *Balanus* and *Chthamalus* was, on the other hand, a most important cause of death of *Chthamalus*. This is shown both by the direct observations of the process of crowding at each census and by the differences between the survival curves of *Chthamalus* with and without *Balanus*. From the periodic observations it was noted that after the first month on the undisturbed portions of areas 3 to 7 about 10% of the *Chthamalus* were being covered as *Balanus* grew over them; about 3% were being undercut and lifted by growing *Balanus;* a few had died without crowding. By the end of the 2nd month about 20% of the *Chthamalus* were either wholly or partly covered by *Balanus;* about 4% had been undercut; others were surrounded by tall *Balanus*. These processes continued at a lower rate in the autumn and almost ceased during the later winter. In the spring *Balanus* resumed growth and more crowding was observed.

In Table II, these observations are summarized for the undistributed portions of all the areas. Above M.T.L., the *Balanus* tended to overgrow the *Chthamalus,* whereas at the lower levels, undercutting was more common. This same trend was evident within each group of areas, undercutting being more prevalent on area 7 than on area 3, for example. The faster growth of *Balanus* at lower levels (Hatton 1938, Barnes and Powell 1953) may have resulted in more undercutting. When *Chthamalus* was completely covered by *Balanus* it was recorded as dead; even though death may not have occurred immediately, the buried barnacle was obviously not a functioning member of the population.

TABLE II. The causes of mortality of *Chthamalus stellatus* of the 1953 year group on the undisturbed portions of each area

Area no.	Height in ft from M.T.L.	No. at start	No. of deaths in the next year	PERCENTAGE OF DEATHS RESULTING FROM:			
				Smothering by *Balanus*	Undercutting by *Balanus*	Other crowding by *Balanus*	Unknown causes
2.........	+3.5	28	1	0	0	0	100
3a.........	+2.2	111	81	61	6	10	23
3b.........	"	47	42	57	5	2	36
4.........	+1.4	34	14	21	14	0	65
5.........	+1.4	43	35	11	11	3	75
6.........	+1.0	27	11	9	0	0	91
7a.........	+0.7	42	38	21	16	53	10
7b.........	"	51	42	24	10	10	56
11a.........	0.0	21	13	54	8	0	38
11b.........	"	10	5	40	0	0	60
12a.........	0.0	60	57	19	33	7	41
12b.........	"	39	34	9	18	3	70
13a.........	−1.0	71	70	19	24	3	54
13b.........	"	69	62	18	8	3	71
14a.........	−2.5	22	21	24	42	10	24
14b.........	"	9	9	0	0	0	100
Total, 2- 7..	—	383	264	37	9	16	38
Total, 11-14..	—	301	271	19	21	4	56

In Table II under the term "other crowding" have been placed all instances where *Chthamalus* were crushed laterally between 2 or more *Balanus,* or where *Chthamalus* disappeared in an interval during which a dense population of *Balanus* grew rapidly. For example, in area 7a the *Balanus,* which were at the high population density of 48 per cm², had no room to expand except upward and the barnacles very quickly grew into the form of tall cylinders or cones with the diameter of the opercular opening greater than

that of the base. It was obvious that extreme crowding occurred under these circumstances, but the exact cause of the mortality of the *Chthamalus* caught in this crush was difficult to ascertain.

In comparing the survival curves of Figs. 2 and 3 within each area it is evident that *Chthamalus* kept free of *Balanus* survived better than those in the adjacent undisturbed areas on all but areas 2 and 14a. Area 2 was in the zone where adults of *Balanus* and *Chthamalus* were normally mixed; at this high level *Balanus* evidently has no influence on the survival of *Chthamalus.* On Stone 14a, the survival of *Chthamalus* without *Balanus* was much better until January when a starfish, *Asterias rubens* L., entered the cage and ate the barnacles.

Much variation occurred on the other 14 areas. When the *Chthamalus* growing without contact with *Balanus* are compared with those on the adjacent undisturbed portion of the area, the survival was very much better on 10 areas and moderately better on 4. In all areas, some *Chthamalus* in the undisturbed portions escaped severe crowding. Sometimes no *Balanus* happened to settle close to a *Chthamalus,* or sometimes those which did died soon after settlement. In some instances, *Chthamalus* which were being undercut by *Balanus* attached themselves to the *Balanus* and so survived. Some *Chthamalus* were partly covered by *Balanus* but still survived. It seems probable that in the 4 areas, nos. 4, 6, 11a, and 11b, where *Chthamalus* survived well in the presence of *Balanus,* a higher proportion of the *Chthamalus* escaped death in one of these ways.

The fate of very young *Chthamalus* which settled in the autumn of 1954 was followed in detail in 2 instances, on stone 15 and area 7b. The *Chthamalus* on stone 15 had settled in an irregular space surrounded by large *Balanus.* Most of the mortality occurred around the edges of the space as the *Balanus* undercut and lifted the small *Chthamalus* nearby. The following is a tabulation of all the deaths of young *Chthamalus* between Sept. 30, 1954 and Feb. 14, 1955, on Stone 15, with the associated situations:

Lifted by *Balanus*	: 29
Crushed by *Balanus*	: 4
Smothered by *Balanus* and *Chthamalus*	: 2
Crushed between *Balanus* and *Chthamalus*	: 1
Lifted by *Chthamalus*	: 1
Crushed between two other *Chthamalus*	: 1
Unknown	: 3

This list shows that crowding of newly settled *Chthamalus* by older *Balanus* in the autumn mainly takes the form of undercutting, rather than of smothering as was the case in the spring. The reason for this difference is probably that the *Chthamalus* are more firmly attached in the spring so that the fast growing young *Balanus* grow up over them when they make contact. In the autumn the reverse is the case, the *Balanus* being firmly attached, the *Chthamalus* weakly so.

Although the settlement of *Chthamalus* on Stone 15 in the autumn of 1954 was very dense, 32/cm², so that most of them were touching another, only 2 of the 41 deaths were caused by intraspecific crowding among the *Chthamalus*. This is in accord with the findings from the 1953 settlement of *Chthamalus*.

The mortality rates for the young *Chthamalus* on area 7b showed seasonal variations. Between October 10, 1954 and May 15, 1955 the relative mortality rate per day × 100 was 0.14 on the undisturbed area and 0.13 where *Balanus* had been removed. Over the next month, the rate increased to 1.49 on the undisturbed area and 0.22 where *Balanus* was absent. Thus the increase in mortality of young *Chthamalus* in late spring was also associated with the presence of *Balanus*.

Some of the stones transplanted from high to low levels in the spring of 1954 bore adult *Chthamalus*. On 3 stones, records were kept of the survival of these adults, which had settled in the autumn of 1952 or in previous years and were at least 20 months old at the start of the experiment. Their mortality is shown in Table III; it was always much greater when *Balanus* was not removed. On 2 of the 3 stones this mortality rate was almost as high as that of the younger group. These results suggest that any *Chthamalus* that managed to survive the competition for space with *Balanus* during the first year would probably be eliminated in the 2nd year.

Censuses of *Balanus* were not made on the experimental areas. However, on many other areas in the same stretch of shore the survival of *Balanus* was being studied during the same period (Connell 1961). In Table IV some mortality rates measured in that study are listed; the *Balanus* were members of the 1954 settlement at population densities and shore levels similar to those of the present study. The mortality rates of *Balanus* were about the same as those of *Chthamalus* in similar situations except at the highest level, area 1, where *Balanus* suffered much greater mortality than *Chthamalus*. Much of this mortality was caused by intraspecific crowding at all levels below area 1.

TABLE III. Comparison of the mortality rates of young and older *Chthamalus stellatus* on transplanted stones

| Stone No. | Shore level | Treatment | Number of *Chthamalus* present in June, 1954 | | % mortality over one year (or for 6 months for 14a) of *Chthamalus* | |
			1953 year group	1952 or older year groups	1953 year group	1952 or older year groups
13b	1.0 ft below MTL	*Balanus* removed	51	3	35	0
		Undisturbed	69	16	90	31
12a	MTL, in a tide pool, caged	*Balanus* removed	50	41	44	37
		Undisturbed	60	31	95	71
14a	2.5 ft below MTL, caged	*Balanus* removed	25	45	40	36
		Undisturbed	22	8	86	75

TABLE IV. Comparison of annual mortality rates of *Chthamalus stellatus* and *Balanus balanoides**

Area no.	Height in ft from M.T.L.	Population density: no./cm² June, 1954	% mortality in the next year
Chthamalus stellatus, autumn 1953 settlement			
1	+4.2	21	17
3a	+2.2	31	72
3b	"	29	89
6	+1.0	38	41
7a	+0.7	49	90
7b	"	52	82
11a	0.0	32	62
13a	−1.0	29	99
12a	(tide pool)	19	95
Balanus balanoides, spring 1954 settlement			
1 (top)	+4.2	21	99
1:Middle Cage 1	+2.1	85	92
1:Middle Cage 2	"	25	77
1:Low Cage 1	+1.5	26	88
Stone 1	−0.9	26	86
Stone 2	"	68	94

* Population density includes both species. The mortality rates of *Chthamalus* refer to those on the undisturbed portions of each area. The data and area designations for *Balanus* were taken from Connell (1961); the present area 1 is the same as that designated 1 (top) in that paper.

In the observations made at each census it appeared that *Balanus* was growing faster than *Chthamalus*. Measurements of growth rates of the 2 species were made from photographs of the areas taken in June and November, 1954. Barnacles growing free of contact with each other were measured; the results are given in Table V. The growth rate of *Balanus* was greater than that of *Chthamalus* in the experimental areas; this agrees with the findings of Hatton (1938) on the shore in France and of Barnes (1956a) for continual submergence on a raft at Millport.

TABLE V. Growth rates of *Chthamalus stellatus* and *Balanus balanoides*. Measurements were made of uncrowded individuals on photographs of areas 3a, 3b and 7b. Those of *Chthamalus* were made on the same individuals on both dates; of *Balanus*, representative samples were chosen

	CHTHAMALUS		BALANUS	
	No. measured	Average size, mm.	No. measured	Average size, mm.
June 11, 1954	25	2.49	39	1.87
November 3, 1954..............	25	4.24	27	4.83
Average size in the interval.......		3.36		3.35
Absolute growth rate per day x 100		1.21		2.04

After a year of crowding the average population densities of *Balanus* and *Chthamalus* remained in the same relative proportion as they had been at the start, since the mortality rates were about the same. However, because of its faster growth, *Balanus* occupied a relatively greater area and, presumably, possessed a greater biomass relative to that of *Chthamalus* after a year.

The faster growth of *Balanus* probably accounts for the manner in which *Chthamalus* were crowded by *Balanus*. It also accounts for the sinuosity of the survival curves of *Chthamalus* growing in contact with *Balanus*. The mortality rate of these *Chthamalus*, as indicated by the slope of the curves in Figs. 2 and 3, was greatest in summer, decreased in winter and increased again in spring. The survival curves of *Chthamalus* growing without contact with *Balanus* do not show these seasonal variations which, therefore, cannot be the result of the direct action of physical factors such as temperature, wave action or rain.

Seasonal variations in growth rate of *Balanus* correspond to these changes in mortality rate of *Chthamalus*. In Figure 4 the growth of *Balanus* throughout the year as studied on an intertidal panel at Millport by Barnes and Powell (1953), is compared to the survival of *Chthamalus* at about the same intertidal level in the present study. The increased mortality of *Chthamalus* was found to occur in the same seasons as the in-

creases in the growth rate of *Balanus*. The correlation was tested using the Spearman rank correlation coefficient. The absolute increase in diameter of *Balanus* in each month, read from the curve of growth, was compared to the percentage mortality of *Chthamalus* in the same month. For the 13 months in which data for *Chthamalus* was available, the correlation was highly significant, P = .01.

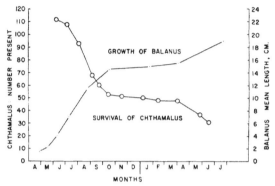

FIG. 4. A comparison of the seasonal changes in the growth of *Balanus balanoides* and in the survival of *Chthamalus stellatus* being crowded by *Balanus*. The growth of *Balanus* was that of panel 3, Barnes and Powell (1953), just above M.T.L. on Keppel Pier, Millport, during 1951-52. The *Chthamalus* were on area 3a of the present study, one-half mile south of Keppell Pier, during 1954-55.

From all these observations it appears that the poor survival of *Chthamalus* below M.H.W.N. is a result mainly of crowding by dense populations of faster growing *Balanus*.

At the end of the experiment in June, 1955, the surviving *Chthamalus* were collected from 5 of the areas. As shown in Table VI, the average size was greater in the *Chthamalus* which had grown free of contact with *Balanus*; in every case the difference was significant (P < .01, Mann-Whitney U test, Siegel 1956). The survivors on the undisturbed areas were often misshapen, in some cases as a result of being lifted on to the side of an undercutting *Balanus*. Thus the smaller size of these barnacles may have been due to disturbances in the normal pattern of growth while they were being crowded.

These *Chthamalus* were examined for the presence of developing larvae in their mantle cavities. As shown in Table VI, in every area the proportion of the uncrowded *Chthamalus* with larvae was equal to or more often slightly greater than on the crowded areas. The reason for this may be related to the smaller size of the crowded *Chthamalus*. It is not due to separation, since *Chthamalus* can self-fertilize (Barnes and Crisp

TABLE VI. The effect of crowding on the size and presence of larvae in *Chthamalus stellatus*, collected in June, 1955

Area	Treatment	Level, feet above MTL	Number of Chthamalus	Diameter in mm Average	Diameter in mm Range	% of individuals which had larvae in mantle cavity
3a......	Undisturbed	2.2	18	3.5	2.7-4.6	61
"	*Balanus* removed	"	50	4.1	3.0-5.5	65
4.......	Undisturbed	1.4	16	2.3	1.8 3.2	81
"	*Balanus* removed	"	37	3.7	2.5-5 1	100
5.......	Undisturbed	1.4	7	3.3	2.8-3.7	70
"	*Balanus* removed	"	13	4.0	3.5-4.5	100
6.......	Undisturbed	1.0	13	2.8	2.1-3.9	100
"	*Balanus* removed	"	14	4.1	3.0-5.2	100
7a & b..	Undisturbed	0.7	10	3.5	2.7-4.5	70
" .	*Balanus* removed	"	23	4.3	3.0-6.3	81

TABLE VII. The effect of predation by *Thais lapillus* on the annual mortality rate of *Chthamalus stellatus* in the experimental areas*

Area	Height in ft from M.T.L.	% mortality of Chthamalus over a year (The initial numbers are given in parentheses) a: Protected from predation by a cage With Balanus	Without Balanus	Difference	b: Unprotected, open to predation With Balanus	Without Balanus	Difference
Area 3..	+2.2	73 (112)	25 (96)	48	89 (47)	6 (50)	83
Area 7..	+0.7	90 (42)	47 (40)	43	82 (51)	23 (47)	59
Area 11..	0	62 (21)	28 (18)	34	50 (10)	25 (16)	25
Area 12 .	0†	100 (60)	53 (50)	47	87 (39)	59 (32)	28
Area 13..	−1.0	98 (72)	9 (77)	89	90 (69)	35 (51)	55

*The records for 12a extend over only 10 months; for purposes of comparison the mortality rate for 12a has been multiplied by 1.2.
†Tide pool.

1956). Moore (1935) and Barnes (1953) have shown that the number of larvae in an individual of *Balanus balanoides* increases with increase in volume of the parent. Comparison of the cube of the diameter, which is proportional to the volume, of *Chthamalus* with and without *Balanus* shows that the volume may be decreased to ¼ normal size when crowding occurs. Assuming that the relation between larval numbers and volume in *Chthamalus* is similar to that of *Balanus,* a decrease in both frequency of occurrence and abundance of larvae in *Chthamalus* results from competition with *Balanus.* Thus the process described in this paper satisfies both aspects of interspecific competition as defined by Elton and Miller (1954): "in which one species affects the population of another by a process of interference, i.e., by reducing the reproductive efficiency or increasing the mortality of its competitor."

The effect of predation by Thais

Cages which excluded *Thais* had been attached on 6 areas (indicated by the letter "a" following the number of the area). Area 14 was not included in the following analysis since many starfish were observed feeding on the barnacles at this level; one entered the cage in January, 1955, and ate most of the barnacles.

Thais were common in this locality, feeding on barnacles and mussels, and reaching average population densities of 200/m² below M.T.L. (Connell 1961). The mortality rates for *Chthamalus* in cages and on adjacent areas outside cages (indicated by the letter "b" after the number) are shown on Table VII.

If the mortality rates of *Chthamalus* growing without contact with *Balanus* are compared in and out of the cages, it can be seen that at the upper levels mortality is greater inside the cages,

at lower levels greater outside. Densities of *Thais* tend to be greater at and below M.T.L. so that this trend in the mortality rates of *Chthamalus* may be ascribed to an increase in predation by *Thais* at lower levels.

Mortality of *Chthamalus* in the absence of *Balanus* was appreciably greater outside than inside the cage only on area 13. In the other 4 areas it seems evident that few *Chthamalus* were being eaten by *Thais.* In a concurrent study of the behavior of *Thais* in feeding on *Balanus balanoides,* it was found that *Thais* selected the larger individuals as prey (Connell 1961). Since *Balanus* after a few month's growth was usually larger than *Chthamalus,* it might be expected that *Thais* would feed on *Balanus* in preference to *Chthamalus.* In a later study (unpublished) made at Santa Barbara, California, *Thais emarginata* Deshayes were enclosed in cages on the shore with mixed populations of *Balanus glandula* Darwin and *Chthamalus fissus* Darwin. These species were each of the same size range as the corresponding species at Millport. It was found that *Thais emarginata* fed on *Balanus glandula* in preference to *Chthamalus fissus.*

As has been indicated, much of the mortality of *Chthamalus* growing naturally intermingled with *Balanus* was a result of direct crowding by *Balanus.* It therefore seemed reasonable to take the difference between the mortality rates of *Chthamalus* with and without *Balanus* as an index of the degree of competition between the species. This difference was calculated for each area and is included in Table VII. If these differences are compared between each pair of adjacent areas in and out of a cage, it appears that the difference, and therefore the degree of competition, was greater outside the cages at the upper shore levels and less outside the cages at the lower levels.

Thus as predation increased at lower levels, the degree of competition decreased. This result would have been expected if *Thais* had fed upon *Balanus* in preference to *Chthamalus*. The general effect of predation by *Thais* seems to have been to lessen the interspecific competition below M.T.L.

Discussion

"Although animal communities appear qualitatively to be constructed as if competition were regulating their structure, even in the best studied cases there are nearly always difficulties and unexplored possibilities" (Hutchinson 1957).

In the present study direct observations at intervals showed that competition was occurring under natural conditions. In addition, the evidence is strong that the observed competition with *Balanus* was the principal factor determining the local distribution of *Chthamalus*. *Chthamalus* thrived at lower levels when it was not growing in contact with *Balanus*.

However, there remain unexplored possibilities. The elimination of *Chthamalus* requires a dense population of *Balanus*, yet the settlement of *Balanus* varied from year to year. At Millport, the settlement density of *Balanus balanoides* was measured for 9 years between 1944 and 1958 (Barnes 1956b, Connell 1961). Settlement was light in 2 years, 1946 and 1958. In the 3 seasons of *Balanus* settlement studied in detail, 1953-55, there was a vast oversupply of larvae ready for settlement. It thus seems probable that most of the *Chthamalus* which survived in a year of poor settlement of *Balanus* would be killed in competition with a normal settlement the following year. A succession of years with poor settlements of *Balanus* is a possible, but improbable occurrence at Millport, judging from the past record. A very light settlement is probably the result of a chance combination of unfavorable weather circumstances during the planktonic period (Barnes 1956b). Also, after a light settlement, survival on the shore is improved, owing principally to the reduction in intraspecific crowding (Connell 1961); this would tend to favor a normal settlement the following year, since barnacles are stimulated to settle by the presence of members of their own species already attached on the surface (Knight-Jones 1953).

The fate of those *Chthamalus* which had survived a year on the undisturbed areas is not known since the experiment ended at that time. It is probable, however, that most of them would have been eliminated within 6 months; the mortality rate had increased in the spring (Figs. 2

and 3), and these survivors were often misshapen and smaller than those which had not been crowded (Table VI). Adults on the transplanted stones had suffered high mortality in the previous year (Table III).

Another difficulty was that *Chthamalus* was rarely found to have settled below mid tide level at Millport. The reasons for this are unknown; it survived well if transplanted below this level, in the absence of *Balanus*. In other areas of the British Isles (in southwest England and Ireland, for example) it occurs below mid tide level.

The possibility that *Chthamalus* might affect *Balanus* deleteriously remains to be considered. It is unlikely that *Chthamalus* could cause much mortality of *Balanus* by direct crowding; its growth is much slower, and crowding between individuals of *Chthamalus* seldom resulted in death. A dense population of *Chthamalus* might deprive larvae of *Balanus* of space for settlement. Also, *Chthamalus* might feed on the planktonic larvae of *Balanus;* however, this would occur in March and April when both the sea water temperature and rate of cirral activity (presumably correlated with feeding activity), would be near their minima (Southward 1955).

The indication from the caging experiments that predation decreased interspecific competition suggests that the action of such additional factors tends to reduce the intensity of such interactions in natural conditions. An additional suggestion in this regard may be made concerning parasitism. Crisp (1960) found that the growth rate of *Balanus balanoides* was decreased if individuals were infected with the isopod parasite *Hemioniscus balani* (Spence Bate). In Britain this parasite has not been reported from *Chthamalus stellatus*. Thus if this parasite were present, both the growth rate of *Balanus*, and its ability to eliminate *Chthamalus* would be decreased, with a corresponding lessening of the degree of competition between the species.

The causes of zonation

The evidence presented in this paper indicates that the lower limit of the intertidal zone of *Chthamalus stellatus* at Millport was determined by interspecific competition for space with *Balanus balanoides*. *Balanus*, by virtue of its greater population density and faster growth, eliminated most of the *Chthamalus* by directing crowding.

At the upper limits of the zones of these species no interaction was observed. *Chthamalus* evidently can exist higher on the shore than *Balanus* mainly as a result of its greater tolerance to heat and/or desiccation.

The upper limits of most intertidal animals are probably determined by physical factors such as these. Since growth rates usually decrease with increasing height on the shore, it would be less likely that a sessile species occupying a higher zone could, by competition for space, prevent a lower one from extending upwards. Likewise, there has been, as far as the author is aware, no study made which shows that predation by land species determines the upper limit of an intertidal animal. In one of the most thorough of such studies, Drinnan (1957) indicated that intense predation by birds accounted for an annual mortality of 22% of cockles (*Cardium edule* L.) in sand flats where their total mortality was 74% per year.

In regard to the lower limits of an animal's zone, it is evident that physical factors may act directly to determine this boundary. For example, some active amphipods from the upper levels of sandy beaches die if kept submerged. However, evidence is accumulating that the lower limits of distribution of intertidal animals are determined mainly by biotic factors.

Connell (1961) found that the shorter length of life of *Balanus balanoides* at low shore levels could be accounted for by selective predation by *Thais lapillus* and increased intraspecific competition for space. The results of the experiments in the present study confirm the suggestions of other authors that lower limits may be due to interspecific competition for space. Knox (1954) suggested that competition determined the distribution of 2 species of barnacles in New Zealand. Endean, Kenny and Stephenson (1956) gave indirect evidence that competition with a colonial polychaete worm, (*Galeolaria*) may have determined the lower limit of a barnacle (*Tetraclita*) in Queensland, Australia. In turn the lower limit of *Galeolaria* appeared to be determined by competition with a tunicate, *Pyura,* or with dense algal mats.

With regard to the 2 species of barnacles in the present paper, some interesting observations have been made concerning changes in their abundance in Britain. Moore (1936) found that in southwest England in 1934, *Chthamalus stellatus* was most dense at M.H.W.N., decreasing in numbers toward M.T.L. while *Balanus balanoides* increased in numbers below M.H.W.N. At the same localities in 1951, Southward and Crisp (1954) found that *Balanus* had almost disappeared and that *Chthamalus* had increased both above and below M.H.W.N. *Chthamalus* had not reached the former densities of *Balanus* except

at one locality, Brixham. After 1951, *Balanus* began to return in numbers, although by 1954 it had not reached the densities of 1934; *Chthamalus* had declined, but again not to its former densities (Southward and Crisp 1956).

Since *Chthamalus* increased in abundance at the lower levels vacated by *Balanus,* it may previously have been excluded by competition with *Balanus.* The growth rate of *Balanus* is greater than *Chthamalus* both north and south (Hatton 1938) of this location, so that *Balanus* would be likely to win in competition with *Chthamalus.* However, changes in other environmental factors such as temperature may have influenced the abundance of these species in a reciprocal manner. In its return to southwest England after 1951, the maximum density of settlement of *Balanus* was 12 per cm^2; competition of the degree observed at Millport would not be expected to occur at this density. At a higher population density, *Balanus* in southern England would probably eliminate *Chthamalus* at low shore levels in the same manner as it did at Millport.

In Loch Sween, on the Argyll Peninsula, Scotland, Lewis and Powell (1960) have described an unusual pattern of zonation of *Chthamalus stellatus.* On the outer coast of the Argyll Peninsula *Chthamalus* has a distribution similar to that at Millport. In the more sheltered waters of Loch Sween, however, *Chthamalus* occurs from above M.H.W.S. to about M.T.L., judging the distribution by its relationship to other organisms. *Balanus balanoides* is scarce above M.T.L. in Loch Sween, so that there appears to be no possibility of competition with *Chthamalus,* such as that occurring at Millport, between the levels of M.T.L. and M.H.W.N.

In Figure 5 an attempt has been made to summarize the distribution of adults and newly settled larvae in relation to the main factors which appear to determine this distribution. For *Balanus* the estimates were based on the findings of a previous study (Connell 1961); intraspecific competition was severe at the lower levels during the first year, after which predation increased in importance. With *Chthamalus,* it appears that avoidance of settlement or early mortality of those larvae which settled at levels below M.T.L., and elimination by competition with *Balanus* of those which settled between M.T.L. and M.H.W.N., were the principal causes for the absence of adults below M.H.W.N. at Millport. This distribution appears to be typical for much of western Scotland.

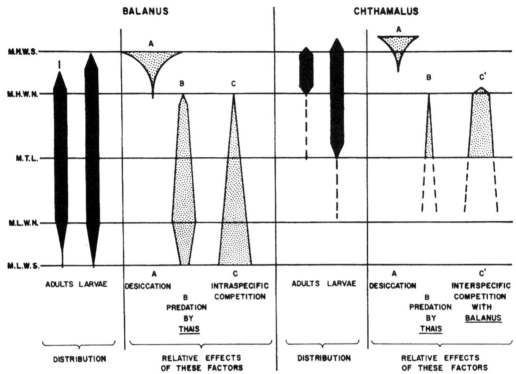

Fig. 5. The intertidal distribution of adults and newly settled larvae of *Balanus balanoides* and *Chthamalus stellatus* at Millport, with a diagrammatic representation of the relative effects of the principal limiting factors.

Summary

Adults of *Chthamalus stellatus* occur in the marine intertidal in a zone above that of another barnacle, *Balanus balanoides*. Young *Chthamalus* settle in the *Balanus* zone but evidently seldom survive, since few adults are found there.

The survival of *Chthamalus* which had settled at various levels in the *Balanus* zone was followed for a year by successive censuses of mapped individuals. Some *Chthamalus* were kept free of contact with *Balanus*. These survived very well at all intertidal levels, indicating that increased time of submergence was not the factor responsible for elimination of *Chthamalus* at low shore levels. Comparison of the survival of unprotected populations with others, protected by enclosure in cages from predation by the snail, *Thais lapillus*, showed that *Thais* was not greatly affecting the survival of *Chthamalus*.

Comparison of the survival of undisturbed populations of *Chthamalus* with those kept free of contact with *Balanus* indicated that *Balanus* could cause great mortality of *Chthamalus*. *Balanus* settled in greater population densities and grew faster than *Chthamalus*. Direct observations at each census showed that *Balanus* smothered,

undercut, or crushed the *Chthamalus;* the greatest mortality of *Chthamalus* occurred during the seasons of most rapid growth of *Balanus*. Even older *Chthamalus* transplanted to low levels were killed by *Balanus* in this way. Predation by *Thais* tended to decrease the severity of this interspecific competition.

Survivors of *Chthamalus* after a year of crowding by *Balanus* were smaller than uncrowded ones. Since smaller barnacles produce fewer offspring, competition tended to reduce reproductive efficiency in addition to increasing mortality.

Mortality as a result of intraspecies competition for space between individuals of *Chthamalus* was only rarely observed.

The evidence of this and other studies indicates that the lower limit of distribution of intertidal organisms is mainly determined by the action of biotic factors such as competition for space or predation. The upper limit is probably more often set by physical factors.

References

Barnes, H. 1953. Size variations in the cyprids of some common barnacles. J. Mar. Biol. Ass. U. K. **32:** 297-304.

———. 1956a. The growth rate of *Chthamalus stellatus* (Poli). J. Mar. Biol. Ass. U. K. **35**: 355-361.

———. 1956b. *Balanus balanoides* (L.) in the Firth of Clyde: The development and annual variation of the larval population, and the causative factors. J. Anim. Ecol. **25**: 72-84.

——— and H. T. Powell. 1953. The growth of *Balanus balanoides* (L.) and *B. crenatus* Brug. under varying conditions of submersion. J. Mar. Biol. Ass. U. K. **32**: 107-128.

——— and D. J. Crisp. 1956. Evidence of self-fertilization in certain species of barnacles. J. Mar. Biol. Ass. U. K. **35**: 631-639.

Beauchamp, R. S. A. and P. Ullyott. 1932. Competitive relationships between certain species of freshwater Triclads. J. Ecol. **20**: 200-208.

Clark, P. J. and F. C. Evans. 1954. Distance to nearest neighbor as a measure of spatial relationships in populations. Ecology **35**: 445-453.

Clench, W. J. 1947. The genera *Purpura* and *Thais* in the western Atlantic. Johnsonia **2**, No. 23: 61-92.

Connell, J. H. 1961. The effects of competition, predation by *Thais lapillus,* and other factors on natural populations of the barnacle, *Balanus balanoides.* Ecol. Mon. **31**: 61-104.

Crisp, D. J. 1960. Factors influencing growth-rate in *Balanus balanoides.* J. Anim. Ecol. **29**: 95-116.

Drinnan, R. E. 1957. The winter feeding of the oystercatcher (*Haematopus ostralegus*) on the edible cockle (*Cardium edule*). J. Anim. Ecol. **26**: 441-469.

Elton, Charles and R. S. Miller. 1954. The ecological survey of animal communities: with a practical scheme of classifying habitats by structural characters. J. Ecol. **42**: 460-496.

Endean, R., R. Kenny and W. Stephenson. 1956. The ecology and distribution of intertidal organisms on the rocky shores of the Queensland mainland. Aust. J. mar. freshw. Res. **7**: 88-146.

Fischer, E. 1928. Sur la distribution geographique de quelques organismes de rocher, le long des cotes de la Manche. Trav. Lab. Mus. Hist. Nat. St.-Servan **2**: 1-16.

Hatton, H. 1938. Essais de bionomie explicative sur quelques especes intercotidales d'algues et d'animaux. Ann. Inst. Oceanogr. Monaco **17**: 241-348.

Holme, N. A. 1950. Population-dispersion in *Tellina tenuis* Da Costa. J. Mar. Biol. Ass. U. K. **29**: 267-280.

Hutchinson, G. E. 1957. Concluding remarks. Cold Spring Harbor Symposium on Quant. Biol. **22**: 415-427.

Knight-Jones, E. W. 1953. Laboratory experiments on gregariousness during setting in *Balanus balanoides* and other barnacles. J. Exp. Biol. **30**: 584-598.

Knox, G. A. 1954. The intertidal flora and fauna of the Chatham Islands. Nature Lond. **174**: 871-873.

Lack, D. 1954. The natural regulation of animal numbers. Oxford, Clarendon Press.

Lewis, J. R. and H. T. Powell. 1960. Aspects of the intertidal ecology of rocky shores in Argyll, Scotland. I. General description of the area. II. The distribution of *Chthamalus stellatus* and *Balanus balanoides* in Kintyre. Trans. Roy. Soc. Edin. **64**: 45-100.

MacArthur, R. H. 1958. Population ecology of some warblers of northeastern coniferous forests. Ecology **39**: 599-619.

Moore, H. B. 1935. The biology of *Balnus balanoides.* III. The soft parts. J. Mar. Biol. Ass. U. K. **20**: 263-277.

———. 1936. The biology of *Balanus balanoides.* V. Distribution in the Plymouth area. J. Mar. Biol. Ass. U. K. **20**: 701-716.

Siegel, S. 1956. Nonparametric statistics. New York, McGraw Hill.

Southward, A. J. 1955. On the behavior of barnacles. I. The relation of cirral and other activities to temperature. J. Mar. Biol. Ass. U. K. **34**: 403-422.

——— and D. J. Crisp. 1954. Recent changes in the distribution of the intertidal barnacles *Chthamalus stellatus* Poli and *Balanus balanoides* L. in the British Isles. J. Anim. Ecol. **23**: 163-177.

———. 1956. Fluctuations in the distribution and abundance of intertidal barnacles. J. Mar. Biol. Ass. U. K. **35**: 211-229.

MacArthur, R. H. and Wilson, E. O. 1963. An equilibrium theory of insular zoogeography. *Evolution* 17:373–387.

Island biogeography as an ecological concept provides the foundation for numerous wildlife management issues, including distribution and abundance of endangered species, species viability within mountain and desert ecosystems, and remnant ecosystems surrounded by an encroaching animal-plant community (e.g., native prairies and grasslands). Prior to the 1960s, numerous theories existed concerning animal species diversity in insular environments, but it was not until MacArthur and Wilson's classic work that ecologically based and tested models were presented to evaluate those theories.

In this seminal paper, MacArthur and Wilson address critical questions that truly started rigorous research concerning island ecology, such as: How are animals distributed? How is distribution related to immigration, time, land mass, and distance to other populations? What are immigration and extinction rates? MacArthur and Wilson present a qualitative and quantitative theoretical framework based on Indio-Australian bird faunas. Because of the prevalence of anthropogenic and natural habitat alteration and fragmentation, their theories have been applied to other flora and fauna around the world. Their views have also been instrumental in designing parks and refuges and in trying to determine the best designs to enhance naturalness.

RELATED READING

Brown, J. H. 1971. Mammals on mountaintops: nonequilibrium insular biogeography. The American Naturalist 105:467–478.

Brown, J. H. 1986. Two decades of interaction between the MacArthur-Wilson model and the complexities of mammalian distributions. Biological Journal of the Linnean Society 28:231–251.

MacArthur, R. H., and E. O. Wilson. 1967. The theory of island biogeography. Princeton University Press, Princeton, New Jersey, USA.

Miller, R. I., and P. S. White. 1986. Considerations for preserve design based on the distribution of rare plants in Great Smoky Mountains National Park, USA. Environmental Management 10:119–124.

Picton, H. D. 1979. The application of insular biogeographic theory to the conservation of large mammals in the northern Rocky Mountains. Biological Conservation 15:73–79.

Simberloff, D. S. 1974. Equilibrium theory of island biogeography and ecology. Annual Review of Ecology and Systematics 5:161–182.

Wilson, E. O., and D. S. Simberloff. 1969. Experimental zoogeography of islands: defaunation and monitoring techniques. Ecology 50:267–278.

© Society for the Study of Evolution.

EVOLUTION

INTERNATIONAL JOURNAL OF ORGANIC EVOLUTION

PUBLISHED BY

THE SOCIETY FOR THE STUDY OF EVOLUTION

| Vol. 17 | DECEMBER, 1963 | No. 4 |

AN EQUILIBRIUM THEORY OF INSULAR ZOOGEOGRAPHY

ROBERT H. MACARTHUR[1] AND EDWARD O. WILSON[2]

Received March 1, 1963

THE FAUNA–AREA CURVE

As the area of sampling A increases in an ecologically uniform area, the number of plant and animal species s increases in an approximately logarithmic manner, or

$$s = bA^k, \qquad (1)$$

where $k < 1$, as shown most recently in in the detailed analysis of Preston (1962). The same relationship holds for islands, where, as one of us has noted (Wilson, 1961), the parameters b and k vary among taxa. Thus, in the ponerine ants of Melanesia and the Moluccas, k (which might be called the *faunal coefficient*) is approximately 0.5 where area is measured in square miles; in the Carabidae and herpetofauna of the Greater Antilles and associated islands, 0.3; in the land and freshwater birds of Indonesia, 0.4; and in the islands of the Sahul Shelf (New Guinea and environs), 0.5.

THE DISTANCE EFFECT IN PACIFIC BIRDS

The relation of number of land and freshwater bird species to area is very orderly in the closely grouped Sunda Is-

lands (fig. 1), but somewhat less so in the islands of Melanesia, Micronesia, and Polynesia taken together (fig. 2). The greater variance of the latter group is attributable primarily to one variable, distance between the islands. In particular, the distance effect can be illustrated by taking the distance from the primary faunal "source area" of Melanesia and relating it to faunal number in the following manner. From fig. 2, take the line connecting New Guinea and the nearby Kei Islands as a "saturation curve" (other lines would be adequate but less suitable to the purpose), calculate the predicted range of "saturation" values among "saturated" islands of varying area from the curve, then take calculated "percentage saturation" as $s_i \times 100/B_i$, where s_i is the real number of species on any island and B_i the saturation number for islands of that area. As shown in fig. 3, the percentage saturation is nicely correlated in an inverse manner with distance from New Guinea. This allows quantification of the rule expressed qualitatively by past authors (see Mayr, 1940) that island faunas become progressively "impoverished" with distance from the nearest land mass.

[1] Division of Biology, University of Pennsylvania, Philadelphia, Pennsylvania.
[2] Biological Laboratories, Harvard University, Cambridge, Massachusetts.

EVOLUTION 17: 373–387. December, 1963   373

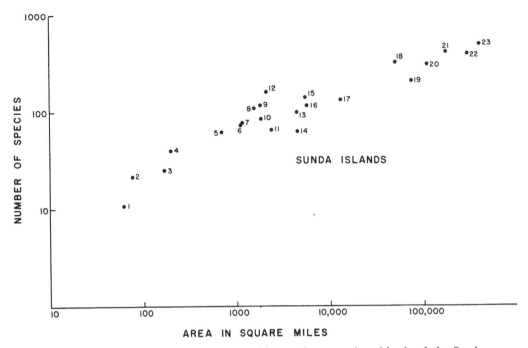

FIG. 1. The numbers of land and freshwater bird species on various islands of the Sunda group, together with the Philippines and New Guinea. The islands are grouped close to one another and to the Asian continent and Greater Sunda group, where most of the species live; and the distance effect is not apparent. (1) Christmas, (2) Bawean, (3) Engano, (4) Savu, (5) Simalur, (6) Alors, (7) Wetar, (8) Nias, (9) Lombok, (10) Billiton, (11) Mentawei, (12) Bali, (13) Sumba, (14) Bangka, (15) Flores, (16) Sumbawa, (17) Timor, (18) Java, (19) Celebes, (20) Philippines, (21) Sumatra, (22) Borneo, (23) New Guinea. Based on data from Delacour and Mayr (1946), Mayr (1940, 1944), Rensch (1936), and Stresemann (1934, 1939).

AN EQUILIBRIUM MODEL

The impoverishment of the species on remote islands is usually explained, if at all, in terms of the length of time species have been able to colonize and their chances of reaching the remote island in that time. According to this explanation, the number of species on islands grows with time and, given enough time, remote islands will have the same number of species as comparable islands nearer to the source of colonization. The following alternative explanation may often be nearer the truth. Fig. 4 shows how the number of new species entering an island may be balanced by the number of species becoming extinct on that island. The descending curve is the rate at which *new* species enter the island by colonization. This rate does indeed fall as the number

of species on the islands increases, because the chance that an immigrant be a new species, not already on the island, falls. Furthermore, the curve falls more steeply at first. This is a consequence of the fact that some species are commoner immigrants than others and that these rapid immigrants are likely, on typical islands, to be the first species present. When there are no species on the island ($N = 0$), the height of the curve represents the number of species arriving per unit of time. Thus the intercept, I, is the rate of immigration of species, new or already present, onto the island. The curve falls to zero at the point $N = P$ where all of the immigrating species are already present so that no new ones are arriving. P is thus the number of species in the "species pool" of immigrants. The shape of the rising curve in the same figure, which represents the

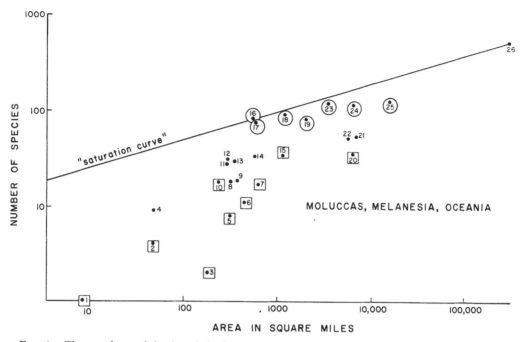

FIG. 2. The numbers of land and freshwater bird species on various islands of the Moluccas, Melanesia, Micronesia, and Polynesia. Here the archiplagoes are widely scattered, and the distance effect is apparent in the greater variance. Hawaii is included even though its fauna is derived mostly from the New World (Mayr, 1943). "Near" islands (less than 500 miles from New Guinea) are enclosed in circles, "far" islands (greater than 2,000 miles) in squares, and islands at intermediate distances are left unenclosed. The saturation curve is drawn through large and small islands at source of colonization. (1) Wake, (2) Henderson, (3) Line, (4) Kusaie, (5) Tuamotu, (6) Marquesas, (7) Society, (8) Ponape, (9) Marianas, (10) Tonga, (11) Carolines, (12) Palau, (13) Santa Cruz, (14) Rennell, (15) Samoa, (16) Kei, (17) Louisiade, (18) D'Entrecasteaux, (19) Tanimbar, (20) Hawaii, (21) Fiji, (22) New Hebrides, (23) Buru, (24) Ceram, (25) Solomons, (26) New Guinea. Based on data from Mayr (1933, 1940, 1943) and Greenway (1958).

rate at which species are becoming extinct on the island, can also be determined roughly. In case all of the species are equally likely to die out and this probability is independent of the number of other species present, the number of species becoming extinct in a unit of time is proportional to the number of species present, so that the curve would rise linearly with N. More realistically, some species die out more readily than others and the more species there are, the rarer each is, and hence an increased number of species increases the likelihood of any given species dying out. Under normal conditions both of these corrections would tend to increase the slope of the extinction curve for large values of N. (In the rare situation in which the species which enter most often as immigrants are the ones which die out most readily—presumably because the island is atypical so that species which are common elsewhere cannot survive well—the curve of extinction may have a steeper slope for small N.) If N is the number of species present at the start, then $E(N)/N$ is the fraction dying out, which can also be interpreted crudely as the probability that any given species will die out. Since this fraction cannot exceed 1, the extinction curve cannot rise higher than the straight line of a 45° angle rising from the origin of the coordinates.

It is clear that the rising and falling curves must intersect and we will denote by $\hat{s}$ the value of N for which the rate of immigration of new species is balanced by

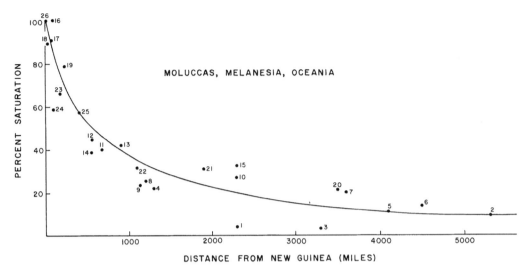

Fɪɢ. 3. Per cent saturation, based on the "saturation curve" of fig. 2, as a function of distance from New Guinea. The numbers refer to the same islands identified in the caption of fig. 2. Note that from equation (4) it is an oversimplification to take distances solely from New Guinea. The abscissa should give a more complex function of distances from all the surrounding islands, with the result that far islands would appear less "distant." But this representation expresses the distance effect adequately for the conclusions drawn.

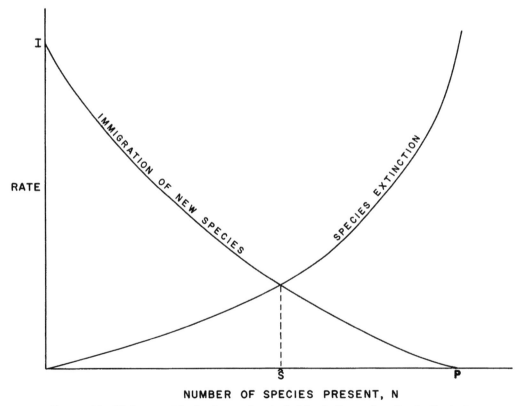

Fɪɢ. 4. Equilibrium model of a fauna of a single island. See explanation in the text.

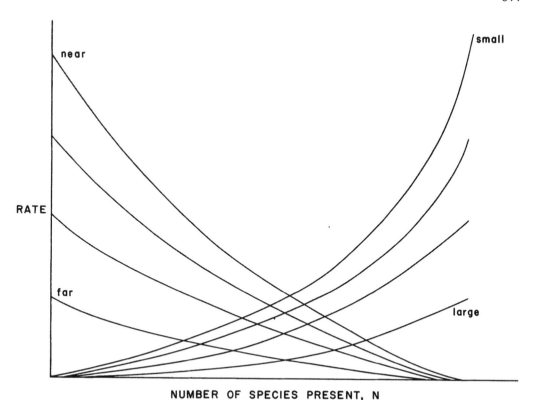

near

RATE

far

small

large

NUMBER OF SPECIES PRESENT, N

FIG. 5. Equilibrium model of faunas of several islands of varying distances from the source area and varying size. Note that the effect shown by the data of fig. 2, of faunas of far islands increasing with size more rapidly than those of near islands, is predicted by this model. Further explanation in text.

the rate of extinction. The number of species on the island will be stabilized at $\hat{s}$, for a glance at the figure shows that when N is greater than $\hat{s}$, extinction exceeds immigration of new species so that N decreases, and when N is less than $\hat{s}$, immigration of new species exceeds extinction so that N will increase. Therefore, in order to predict the number of species on an island we need only construct these two curves and see where they intersect. We shall make a somewhat oversimplified attempt to do this in later paragraphs. First, however, there are several interesting qualitative predictions which we can make without committing ourselves to any specific shape of the immigration and extinction curves.

A. An island which is farther from the source of colonization (or for any other reason has a smaller value of I) will, other things being equal, have fewer species, because the immigration curve will be lower and hence intersect the mortality curve farther to the left (see fig. 5).

B. Reduction of the "species pool" of immigrants, P, will reduce the number of species on the island (for the same reason as in A).

C. If an island has smaller area, more severe climate (or for any other reason has a greater extinction rate), the mortality curve will rise and the number of species will decrease (see fig. 5).

D. If we have two islands with the same immigration curve but different extinction curves, any given species on the one with the higher extinction curve is more likely to die out, because $E(N)/N$ can be seen to be higher [$E(N)/N$ is the slope of the line joining the intersection point to the origin].

E. The number of species found on islands far from the source of colonization will grow more rapidly with island area than will the number on near islands. More precisely, if the area of the island is denoted by A, and $\hat{s}$ is the equilibrium number of species, then d^2s/dA^2 is greater for far islands than for near ones. This can be verified empirically by plotting points or by noticing that the change in the angle of intersection is greater for far islands.

F. The number of species on large islands decreases with distance from source of colonization faster than does the number of species on small islands. (This is merely another way of writing E and is verified similarly.)

Further, as will be shown later, the variance in $\hat{s}$ (due to randomness in immigrations and extinctions) will be lower than that expected if the "classical" explanation holds. In the classical explanation most of those species will be found which have at any time succeeded in immigrating. At least for distant islands this number would have an approximately Poisson distribution so that the variance would be approximately equal to the mean. Our model predicts a reduced variance, so that if the observed variance is significantly smaller than the mean for distant islands, it is evidence for the equilibrium explanation.

The evidence in fig. 2, relating to the insular bird faunas east of Weber's Line, is consistent with all of these predictions. To see this for the non-obvious prediction E, notice that a greater slope on this log-log plot corresponds to a greater second derivative, since A becomes sufficiently large.

THE FORM OF THE IMMIGRATION AND EXTINCTION CURVES

If the equilibrium model we have presented is correct, it should be possible eventually to derive some quantitative generalizations concerning rates of immigration and extinction. In the section to follow we have deduced an equilibrium equation which is adequate as a first approximation, in that it yields the general form of the empirically derived fauna–area curves without contradicting (for the moment) our intuitive ideas of the underlying biological processes. This attempt to produce a formal equation is subject to indefinite future improvements and does not affect the validity of the graphically derived equilibrium theory. We start with the statement that

$$\Delta s = M + G - D, \qquad (2)$$

where s is the number of species on an island, M is the number of species successfully immigrating to the island per year, G is the number of new species being added per year by local speciation (not including immigrant species that merely diverge to species level without multiplying), and D is the number of species dying out per year. At equilibrium,

$$M + G = D.$$

The immigration rate M must be determined by at least two independent values: (1) the rate at which propagules reach the island, which is dependent on the size of the island and its distance from the source of the propagules, as well as the nature of the source area, but not on the condition of the recipient island's fauna; and (2) as noted already, the number of species already resident on the island. Propagules are defined here as the minimum number of individuals of a given species needed to achieve colonization; a more exact explication is given in the Appendix. Consider first the source region. If it is climatically and faunistically similar to other potential source regions, the number of propagules passing beyond its shores per year is likely to be closely related to the size of the population of the taxon living on it, which in turn is approximately a linear function of its area. This notion is supported by the evidence from Indo-Australian ant zoogeography, which indicates that the ratio of faunal exchange is about equal to the ratio of the areas of the source regions (Wilson, 1961). On the other hand, the number of propagules reaching the recipient island probably varies linearly with the angle it subtends with reference to the center of the source region. Only near islands will vary much because of this factor. Finally, the number of propagules reaching the recipient island is most likely to be an exponential function of its distance from the source region. In the simplest case, if the probability that a given propagule ceases its overseas voyage (e.g., it falls into the sea and dies) at any given instant in time remains constant, then the fraction of propagules reaching a given distance fits an exponential holding-time distribution. If these assumptions are correct, the number of propagules reaching an island from a given source region per year can be approximated as

$$\alpha A_i \frac{\text{diam}_i\, e^{-\lambda d_i}}{2\pi d_i}, \qquad (3)$$

where A_i is the area of the source region, d_i is the mean distance between the source region and recipient island, diam_i is the diameter of the recipient island taken at a right angle to the direction of d_i, and α is a coefficient relating area to the number of propagules produced. More generally, where more than one source region is in position, the rate of propagule arrival would be

$$\frac{\alpha}{2\pi} \sum_i \frac{\text{diam}_i}{d_i} A_i e^{-\lambda d_i}, \qquad (4)$$

where the summation is of contributions from each of the ith source regions. Again, note that a propagule is defined as the minimum number of individuals required to achieve colonization.

Only a certain fraction of arriving propagules will add a new species to the fauna, however, because except for "empty" islands at least some ecological positions will be filled. As indicated in fig. 4, the rate of immigration (i.e., rate of propagule arrival times the fraction colonizing) declines to zero as the number of resident species (s) approaches the limit P. The curve relating the immigration rate to degree of unsaturation is probably a concave one, as indicated in fig. 4, for two

reasons: (1) the more abundant immigrants reach the island earlier, and (2) we would expect otherwise randomly arriving elements to settle into available positions according to a simple occupancy model where one and only one object is allowed to occupy each randomly placed position (Feller, 1958). These circumstances would result in the rate of successful occupation decelerating as positions are filled. While these are interesting subjects in themselves, a reasonable approximation is obtained if it is assumed that the rate of occupation is an inverse linear function of the number of occupied positions, or

$$\left(1 - \frac{s}{P}\right). \qquad (5)$$

Then

$$M = \frac{\alpha(1 - s/P)}{2\pi} \sum_i \frac{\text{diam}_i}{d_i} A_i e^{-\lambda d_i}. \qquad (6)$$

We know the immigration line in fig. 4 is not straight; to take this into account we must modify formula 5 by adding a term in s^2. However, this will not be necessary for our immediate purposes.

Now let us consider G, the rate of new productions on the island by local speciation. Note that this rate does not include the mere divergence of an island endemic to a specific level with reference to the stock species in the source area; that species is still counted as contributing to M, the immigration rate, no matter how far it evolves. Only new species generated from it and in addition to it are counted in G. First, consider an archipelago as a unit and the increase of s by divergence of species on the various islands to the level of allopatric species, i.e., the production of a local archipelagic superspecies. If this is the case, and no exchange of endemics is yet achieved among the islands of the archipelago, the number of species in the archipelago is limited to

$$\sum_{i=1}^{\infty} n_i \hat{s}_i, \qquad (7)$$

where n_i is the number of islands in the archipelago of ith area and $\hat{s}_i$ is the number of species occurring at equilibrium on islands of ith area. But the generation of allopatric species in superspecies does not multiply species on single islands or greatly change the fauna of the archipelago as a whole from the value predicted by the fauna–area curve, as can be readily seen in figs. 2 and 3. G, the increase of s by local speciation on single islands and exchange of autochthonous species between islands, probably becomes significant only in the oldest, largest, and most isolated archipelagoes, such as Hawaii and the Galápagos. Where it occurs, the exchange among the islands can be predicted from (6), with individual islands in the archipelago serving as both source regions and recipient islands. However, for most cases it is probably safe to omit G from the model, i.e., consider only source regions outside the archipelago, and hence

$$\Delta s = M - D. \qquad (8)$$

The extinction rate D would seem intuitively to depend in some simple manner on (1) the mean size of the species populations, which in turn is determined by the size of the island and the number of species belonging to the taxon that occur on it; and (2) the yearly mortality rate of the organisms. Let us suppose that the probability of extinction of a species is merely the probability that all the individuals of a given species will die in one year. If the deaths of individuals are unrelated to each other and the population sizes of the species are equal and nonfluctuating,

$$D = sP^{N_r/s}, \qquad (9)$$

where N_r is the total number of individuals in the taxon on the recipient island and P is their annual mortality rate. More realistically, the species of a taxon, such as the birds, vary in abundance in a manner approximating a Barton–Davis distribution (MacArthur, 1957) although the approximation is probably not good for a whole island. In s nonfluctuating species ordered according to their rank (K) in relative rareness,

$$D = \sum_{i=1}^{s} p^{(N_r/s)} \sum_{1=i}^{K} 1/(s-i+1). \qquad (10)$$

This is still an oversimplification, if for no other reason than the fact that populations do fluctuate, and with increased fluctuation D will increase. However, both models, as well as elaborations of them to account for fluctuation, predict an exponential increase of D with restriction of island area. The increase of D which accompanies an increase in number of resident species is more complicated but is shown in fig. 4.

MODEL OF IMMIGRATION AND EXTINCTION PROCESS ON A SINGLE ISLAND

Let $P_s(t)$ be the probability that, at time t, our island has s species, λ_s be the rate of immigration of new species onto the island, when s are present, μ_s be the rate of extinction of species on the island when s are present; and λ_s and μ_s then represent the intersecting curves in fig. 4. This is a "birth and death process" only slightly different from the kind most familiar to mathematicians (cf. Feller, 1958, last chapter). By the rules of probability

$$P_s(t+h) = P_s(t)(1 - \lambda_s h - \mu_s h) \\ + P_{s-1}(t)\lambda_{s-1}h \\ + P_{s+1}(t)\mu_{s+1}h ,$$

since to have s at time $t+h$ requires that at a short time preceding one of the following conditions held: (1) there were s and that no immigration or extinction took place, or (2) that there were $s-1$ and one species immigrated, or (3) that there $s+1$ and one species became extinct. We take h to be small enough that probabilities of two or more extinctions and/or immigrations can be ignored. Bringing $P_s(t)$ to the left-hand side, dividing by h, and passing to the limit as $h \rightarrow 0$

$$\frac{dP_s(t)}{dt} = -(\lambda_s + \mu_s)P_s(t) + \lambda_{s-1}P_{s-1}(t) \\ + \mu_{s+1}P_{s+1}(t). \qquad (11)$$

For this formula to be true in the case where $s = 0$, we must require that $\lambda_{-1} = 0$ and $\mu_0 = 0$. In principle we could solve

(11) for $P_s(t)$; for our purposes it is more useful to find the mean, $M(t)$, and the variance, $\text{var}(t)$, of the number of species at time t. These can be estimated in nature by measuring the mean and variance in numbers of species on a series of islands of about the same distance and area and hence of the same λ_s and μ_s. To find the mean, $M(t)$, from (11) we multiply both sides of (11) by s and then sum from $s = 0$ to $s = \infty$. Since $\sum_{s=0}^{\infty} sP_s(t) = M(t)$, this gives us

$$\frac{dM(t)}{dt} = -\sum_{s=0}^{\infty} (\lambda_s + \mu_s)sP_s(t)$$
$$+ \sum_{s-1=0}^{\infty} \lambda_{s-1}[(s-1)+1]P_{s-1}(t)$$
$$+ \sum_{s+1=0}^{\infty} \mu_{s+1}[(s+1)-1]P_{s+1}(t).$$

(Here terms $\lambda_{-1} \cdot 0 \cdot P_{-1}(t) = 0$ and $\mu_0 \cdot (-1)P_0(t) = 0$ have been subtracted or added without altering values.) This reduces to

$$\frac{dM(t)}{dt} = \sum_{s=0}^{\infty} \lambda_s P_s(t) - \sum_{s=0}^{\infty} \mu_s P_s(t)$$
$$= \overline{\lambda_s(t)} - \overline{\mu_s(t)}. \qquad (12)$$

But, since λ_s and μ_s are, at least locally, approximately straight, the mean value of λ_s at time t is about equal to $\lambda_{M(t)}$ and similarly $\overline{\mu_s(t)} \sim \mu_{M(t)}$. Hence, approximately

$$\frac{dM(t)}{dt} = \lambda_{M(t)} - \mu_{M(t)}, \qquad (13)$$

or the expected number of species in Fig. 4 moves toward $\hat{s}$ at a rate equal to the difference in height of the immigration and extinction curves. In fact, if $d\mu/ds - d\lambda/ds$, evaluated near $s = \hat{s}$ is abbreviated by F, then, approximately $dM(t)/dt = F(\hat{s} - M(t))$ whose solution is $M(t) = \hat{s}(1 - e^{-Ft})$. Finally, we can compute the time required to reach 90% (say) of the saturation value $\hat{s}$ so that $M(t)/\hat{s} = 0.9$ or $e^{-Ft} = 0.1$.
Therefore,

$$t = \frac{2.303}{F}. \qquad (13a)$$

A similar formula for the variance is obtained by multiplying both sides of (11) by $(s - M(t))^2$ and summing from $s = 0$ to $s = \infty$. As before, since $\text{var}(t) = \sum_{s=0}^{\infty} (s - M(t))^2 P_s(t)$, this results in

$$\frac{d\,\text{var}(t)}{dt}$$
$$= -\sum_{s=0}^{\infty} (\lambda_s + \mu_s)(s - M(t))^2 P_s(t)$$
$$+ \sum_{s-1=0}^{\infty} \lambda_{s-1}[(s-1-M(t))+1]^2 P_{s+1}(t)$$
$$+ \sum_{s+1=0}^{\infty} \mu_{s+1}[(s+1-M(t))-1]^2 P_{s+1}(t)$$
$$= 2\sum_{s=0}^{\infty} \lambda_s(s - M(t))P_s(t)$$
$$- 2\sum_{s=0}^{\infty} \mu_s(s - M(t))P_s(t)$$
$$+ \sum_{s=0}^{\infty} \lambda_s P_s(t) + \sum_{s=0}^{\infty} \mu_s P_s(t). \qquad (14)$$

Again we can simplify this by noting that the λ_s and μ_s curves are only slowly curving and hence in any local region are approximately straight. Hence, where derivatives are now evaluated near the point $s = M(t)$,

$$\lambda_s = \lambda_{M(t)} + [s - M(t)]\frac{d\lambda}{ds}$$
$$\mu_s = \mu_{M(t)} + [s - M(t)]\frac{d\mu}{ds}. \qquad (15)$$

Substituting (15) into (14) we get

$$\frac{d\,\text{var}(t)}{dt}$$
$$= 2(\lambda_{M(t)} - \mu_{M(t)})\sum_{s=0}^{\infty} (s - M(t))P_s(t)$$
$$+ 2\left(\frac{d\lambda}{ds} - \frac{d\mu}{ds}\right)\sum_{s=0}^{\infty} (s - M(t))^2 P_s(t)$$
$$+ [\lambda_{m(t)} + \mu_{M(t)}]\sum_{s=0}^{\infty} P_s(t)$$
$$+ \left(\frac{d\lambda}{ds} + \frac{d\mu}{ds}\right)\sum_{s=0}^{\infty} (s - M(t))P_s(t),$$

which, since $\sum_{s=0}^{\infty} P_s(t) = 1$ and

$\sum (s - M(t)) P_s(t) = M(t) - M(t) = 0,$

becomes,

$$\frac{d \operatorname{var}(t)}{dt} = -2\left(\frac{d\mu}{ds} - \frac{d\lambda}{ds}\right) \operatorname{var}(t)$$
$$+ \lambda_{M(t)} + \mu_{M(t)} . \qquad (16)$$

This is readily solved for $\operatorname{var}(t)$:

$$\operatorname{var}(t)$$
$$= e^{-2[(d\mu/ds)-(d\lambda/ds)]t} \qquad (16a)$$
$$\times \int_0^t (\lambda_{M(t)} + \mu_{M(t)}) e^{2[(d\mu/ds)-(d\lambda/ds)]t} \, dt .$$

However, it is more instructive to compare mean and variance for the extreme situations of saturation and complete unsaturation, or equivalently of $t =$ near ∞ and $t =$ near zero.

At equilibrium, $\dfrac{d \operatorname{var}(t)}{dt} = 0$, so by (16) ·

$$\operatorname{var}(t) = \frac{\lambda_{\hat{s}} + \mu_{\hat{s}}}{2\left(\dfrac{d\mu}{ds} - \dfrac{d\lambda}{ds}\right)} . \qquad (17)$$

At equilibrium $\lambda_{\hat{s}} = \mu_{\hat{s}} = x$ say and we have already symbolized the difference of the derivatives at $s = \hat{s}$ by F (cf. eq. [13a]). Hence, at equilibrium

$$\operatorname{var} = \frac{X}{F} . \qquad (17a)$$

Now since μ_s has non-decreasing slope $X/s \leqslant d\mu/ds \big|_{s=\hat{s}}$ or $X \leqslant \hat{s}\, d\mu/ds \big|_{\hat{s}}.$

Therefore, variance $\leqslant \dfrac{\hat{s}\, d\mu/ds}{d\mu/ds - d\lambda/ds}$ or, at equilibrium

$$\frac{\text{variance}}{\text{mean}} \leqslant \frac{d\mu/ds}{d\mu/ds - d\lambda/ds} . \qquad (18)$$

In particular, if the extinction and immigration curves have slopes about equal in absolute value, $(\text{variance}/\text{mean}) \leqslant \frac{1}{2}$. On the other hand, when t is near zero, equation (16) shows that $\operatorname{var}(t) \sim \lambda_0 t$. Similarly, when t is near zero, equations (13) or (14) show that $M(t) \sim \lambda_0 t$. Hence, in a very unsaturated situation, approximately,

$$\frac{\text{variance}}{\text{mean}} = 1 . \qquad (19)$$

Therefore, we would expect the variance/mean to rise from somewhere around $\frac{1}{2}$ to 1, as we proceed from saturated islands to extremely unsaturated islands farthest from the source of colonization.

Finally, if the number of species dying out per year, X (at equilibrium), is known, we can estimate the time required to 90% saturation from equations (13a) and (17a):

$$\frac{2.303}{t} = \frac{X}{\text{variance}}$$

$$t = \frac{2.303 \text{ variance}}{X} \doteq \frac{2.303}{2} \frac{\text{mean}}{X} . \qquad (19a)$$

The above model was developed independently from an equilibrium hypothesis just published by Preston (1962). After providing massive documentation of the subject that will be of valuable assistance to future biogeographers, Preston draws the following particular conclusion about continental versus insular biotas: "[The depauperate insular biotas] are not depauperate in any absolute sense. They have the correct number of species for their area, provided that each area is an isolate, but they have far fewer than do equal areas on a mainland, because a mainland area is merely a 'sample' and hence is greatly enriched in the Species/Individuals ratio." To illustrate, "in a sample, such as the breeding birds of a hundred acres, we get many species represented by a single pair. Such species would be marked for extinction with one or two seasons' failure of their nests were it not for the fact that such local extirpation can be made good from outside the 'quadrat,' which is not the case with the isolate." This point of view agrees with our own. However, the author apparently missed the precise distance effect and his model is consequently not predictive in the direction we are attempting. His model is, however, more accurate in its account of

TABLE 1. *Number of species of land and freshwater birds on Krakatau and Verlaten during three collection periods together with losses in the two intervals (from Dammerman, 1948)*

	1908			1919–1921			1932–1934			Number "lost"	
	Non-migrant	Migrant	Total	Non-migrant	Migrant	Total	Non-migrant	Migrant	Total	1908 to 1919–1921	1919–1921 to 1932–1934
Krakatau	13	0	13	27	4	31	27	3	30	2	5
Verlaten	1	0	1	27	2	29	29	5	34	0	2

relative abundance, corresponding to our equation (10).

THE CASE OF THE KRAKATAU FAUNAS

The data on the growth of the bird faunas of the Krakatau Islands, summarized by Dammerman (1948), provide a rare opportunity to test the foregoing model of the immigration and extinction process on a single island. As is well known, the island of Krakatau proper exploded in August, 1883, after a three-month period of repeated eruptions. Half of Krakatau disappeared entirely and the remainder, together with the neighboring islands of Verlaten and Lang, was buried beneath a layer of glowing hot pumice and ash from 30 to 60 meters thick. Almost certainly the entire flora and fauna were destroyed. The repopulation proceeded rapidly thereafter. Collections and sight records of birds, made mostly in 1908, 1919–1921, and 1932–1934, show that the number of species of land and freshwater birds on both Krakatau and Verlaten climbed rapidly between 1908 and 1919–1921 and did not alter significantly by 1932–1934 (see table 1). Further, the number of non-migrant land and freshwater species on both islands in 1919–1921 and 1932–1934, i.e., 27–29, fall very close to the extrapolated fauna–area curve of our fig. 1. Both lines of evidence suggest that the Krakatau faunas had approached equilibrium within only 25 to 36 years after the explosion.

Depending on the exact form of the immigration and extinction curves (see fig. 4), the ratio of variance to mean of numbers of species on similar islands at or near saturation can be expected to vary between about ¼ and ¾. If the slopes of the two curves are equal at the point of intersection, the ratio would be near ½. Then the variance of faunas of Krakatau-like islands (same area and isolation) can be expected to fall between 7 and 21 species. Applying this estimate to equation (19a) and taking t (the time required to reach 90% of the equilibrium number) as 30 years, X, the annual extinction rate, is estimated to lie between 0.5 and 1.6 species per year.

This estimate of annual extinction rate (and hence of the acquisition rate) in an equilibrium fauna is surprisingly high; it is of the magnitude of 2 to 6% of the standing fauna. Yet it seems to be supported by the collection data. On Krakatau proper, 5 non-migrant land and freshwater species recorded in 1919–1921 were not recorded in 1932–1934, but 5 other species were recorded for the first time in 1932–1934. On Verlaten 2 species were "lost" and 4 were "gained." This balance sheet cannot easily be dismissed as an artifact of collecting technique. Dammerman notes that during this period, "The most remarkable thing is that now for the first time true fly catchers, *Muscicapidae*, appeared on the islands, and that there were no less than four species: *Cyornis rufigastra, Gerygone modigliani, Alseonax latirostris* and *Zanthopygia narcissina*. The two last species are migratory and were therefore only accidental visitors, but the sudden appearance of the *Cyornis* species in great numbers is noteworthy. These birds, first observed in May 1929, had already colonized three islands and may now be called common there. Moreover the *Gerygone*, unmistakable from his gentle note and common along the coast and in the mangrove forest, is certainly a new acquisition." Extinctions are less susceptible of proof but the following evidence is suggestive. "On the other hand two species mentioned by Jacobson (1908) were not found in 1921 and have not been observed since, namely the small kingfisher *Alcedo coerulescens* and the familiar bulbul *Pycnonotus aurigaster*." Between 1919–1921 and 1932–1934 the conspicuous *Demiegretta s. sacra* and *Accipter* sp. were "lost," although these species may not have been truly established as breeding populations. But "the well-known greybacked shrike (*Lanius schach bentet*), a bird conspicuous in the open field, recorded in 1908 and found breeding in 1919, was not seen in 1933. Whether the species had really completely disappeared or only diminished so much in numbers that it was· not noticed, the future must show." Future research on the Krakatau fauna would indeed be of great interest, in view of the very dynamic equilibrium suggested by the model we have presented. If the "losses" in the data represent true extinctions, the rate of extinction would be 0.2 to 0.4 species per year, closely approaching the predicted rate of 0.5 to 1.6. This must be regarded as a minimum figure, since it is likely that species could easily be lost and regained all in one 12-year period.

Such might be the situation in the early history of the equilibrium fauna. It is not possible to predict whether the rate of turnover would change through time. As other taxa reached saturation and more species of birds had a chance at colonization, it is conceivable that more "harmonic" species systems would accumulate within which the turnover rate would decline.

PREDICTION OF A "RADIATION ZONE"

On islands holding equilibrium faunas, the ratio of the number of species arriving from other islands in the same archipelago (G in equation no. 2) to the number arriving from outside the archipelago (M in no. 2) can be expected to increase with distance from the major extra-archipelagic source area. Where the archipelagoes are of approximately similar area and configuration, G/M should increase in an orderly fashion with distance. Note that G provides the best available measure of what is loosely referred to in the literature as adaptive radiation. Specifically, adaptive radiation takes place as species are generated within archipelagoes, disperse between islands, and, most importantly, accumulate on individual islands to form diversified associations of sympatric species. In equilibrium faunas, then, the following prediction is possible: adaptive radiation, measured by G/M, will increase with distance from the major source region and after corrections for area and climate, reach a maximum on archipelagoes and large islands located in a circular zone close to the outermost range of the taxon. This might be referred to as the "radiation zone" of taxa with equilibrium faunas. Many examples possibly conforming to such a rule can be cited: the birds of Hawaii and the Galápagos, the murid rodents of Luzon, the cyprinid fish of Mindanao, the frogs of the Seychelles, the gekkonid lizards of New Caledonia, the Drosophilidae of Hawaii, the ants of Fiji and New Caledonia, and many others (see especially in Darlington, 1957; and Zimmerman, 1948). But there are conspicuous exceptions: the frogs just reach New Zealand but have not radiated there; the same is true of the insectivores of the Greater Antilles, the terrestrial mammals of the Solomons, the snakes of Fiji, and the lizards of Fiji and Samoa. To say that the latter taxa have only recently reached the islands in question, or that they are not in equilibrium, would be a premature if not facile explanation. But it is worth considering as a working hypothesis.

ESTIMATING THE MEAN DISPERSAL DISTANCE

A possible application of the equilibrium model in the indirect estimation of the mean dispersal distance, or λ in equation

(3). Note that if similar parameters of dispersal occur within archipelagoes as well as between them,

$$\frac{G}{M} = \frac{A_1 \operatorname{diam}_1 d_2}{A_2 \operatorname{diam}_2 d_1} e^{\lambda(d_2 - d_1)}, \qquad (20)$$

and

$$\lambda = \ln \frac{A_2 \operatorname{diam}_2 d_1 G}{A_1 \operatorname{diam}_1 d_2 M} \bigg/ (d_2 - d_1), \quad (21)$$

where, in a simple case, A_1, diam_1, and d_1 refer to the relation between the recipient island and some single major source island within the same archipelago; and A_2, diam_2, and d_2 refer to the relation between the recipient island and the major source region outside the archipelago.

Consider the case of the Geospizinae of the Galápagos. On the assumption that a single stock colonized the Galápagos (Lack, 1947), G/M for each island can be taken as equal to G, or the number of geospizine species. In particular, the peripherally located Chatham Island, with seven species, is worth evaluating. South America is the source of M and Indefatigable Island can probably be regarded as the principal source of G for Chatham. Given G/M as seven and assuming that the Geospizinae are in equilibrium, λ for the Geospizinae can be calculated from (21) as 0.018 mile. For birds as a whole, where G/M is approximately unity, λ is about 0.014 mile.

But there are at least three major sources of error in making an estimate in this way:

1. Whereas M is based from the start on propagules from an equilibrium fauna in South America, G increased gradually in the early history of the Galápagos through speciation of the Geospizinae on islands other than Chatham. Hence, G/M on Chatham is actually higher than the ratio of species drawn from the Galápagos to those drawn from outside the archipelago, which is our only way of computing G/M directly. Since λ increases with G/M, the estimates of λ given would be too low, if all other parameters were correct.

2. Most species of birds probably do not disperse according to a simple exponential holding-time distribution. Rather, they probably fly a single direction for considerable periods of time and cease flying at distances that can be approximated by the normal distribution. For this reason also, λ as estimated above would probably be too low.

3. We are using $\hat{S}_G/\hat{S}_M$ for G/M, which is only approximate.

These considerations lead us to believe that 0.01 mile can safely be set as the lower limit of λ for birds leaving the eastern South American coast. Using equation no. 12 in another case, we have attempted to calculate λ for birds moving through the Lesser Sunda chain of Indonesia. The Alor group was chosen as being conveniently located for the analysis, with Flores regarded as the principal source of western species and Timor as the principal source of eastern species. From the data of Mayr (1944) on the relationships of the Alor fauna, and assuming arbitrarily an exponential holding-time dispersal, λ can be calculated as approximately 0.3 mile. In this case the first source of error mentioned above with reference to the Galápagos fauna is removed but the second remains. Hence, the estimate is still probably a lower limit.

Of course these estimates are in themselves neither very surprising nor otherwise illuminating. We cite them primarily to show the possibilities of using zoogeographic data to set boundary conditions on population ecological phenomena that would otherwise be very difficult to assess. Finally, while we believe the evidence favors the hypothesis that Indo-Australian insular bird faunas are at or near equilibrium, we do not intend to extend this conclusion carelessly to other taxa or even other bird faunas. Our purpose has been to deal with general equilibrium criteria, which might be applied to other faunas, together with some of the biological implications of the equilibrium condition.

SUMMARY

A graphical equilibrium model, balancing immigration and extinction rates of species, has been developed which appears fully consistent with the fauna–area curves and the distance effect seen in land and freshwater bird faunas of the Indo-Australian islands. The establishment of the equilibrium condition allows the development of a more precise zoogeographic theory than hitherto possible.

One new and non-obvious prediction can be made from the model which is immediately verifiable from existing data, that the number of species increases with area more rapidly on far islands than on near ones. Similarly, the number of species on large islands decreases with distance faster than does the number of species on small islands.

As groups of islands pass from the unsaturated to saturated conditions, the variance-to-mean ratio should change from unity to about one-half. When the faunal buildup reaches 90% of the equilibrium number, the extinction rate in species/year should equal 2.303 times the variance divided by the time (in years) required to reach the 90% level. The implications of this relation are discussed with reference to the Krakatau faunas, where the buildup rate is known.

A "radiation zone," in which the rate of intra-archipelagic exchange of autochthonous species approaches or exceeds extra-archipelagic immigration toward the outer limits of the taxon's range, is predicted as still another consequence of the equilibrium condition. This condition seems to be fulfilled by conventional information but cannot be rigorously tested with the existing data.

Where faunas are at or near equilibrium, it should be possible to devise indirect estimates of the actual immigration and extinction rates, as well as of the times required to reach equilibrium. It should also be possible to estimate the mean dispersal distance of propagules overseas from the zoogeographic data. Mathematical models have been constructed to these ends and certain applications suggested.

The main purpose of the paper is to express the criteria and implications of the equilibrium condition, without extending them for the present beyond the Indo-Australian bird faunas.

ACKNOWLEDGMENTS

We are grateful to Dr. W. H. Bossert, Prof. P. J. Darlington, Prof. E. Mayr, and Prof. G. G. Simpson for material aid and advice during the course of the study. Special acknowledgment must be made to the published works of K. W. Dammerman, E. Mayr, B. Rensch, and E. Stresemann, whose remarkably thorough faunistic data provided both the initial stimulus and the principal working material of our analysis. The work was supported by NSF Grant G-11575.

LITERATURE CITED

DAMMERMAN, K. W. 1948. The fauna of Krakatau 1883–1933. Verh. Kon. Ned. Akad. Wet. (Nat.), (2) **44**: 1–594.

DARLINGTON, P. J. 1957. Zoogeography. The geographical distribution of animals. Wiley.

DELACOUR, J., AND E. MAYR. 1946. Birds of the Philippines. Macmillan.

FELLER, W. 1958. An introduction to probability theory and its applications. Vol. 1, 2nd ed. Wiley.

GREENWAY, J. G. 1958. Extinct and vanishing birds of the world. Amer. Comm. International Wild Life Protection, Special Publ. No. 13.

LACK, D. 1947. Darwin's finches, an essay on the general biological theory of evolution. Cambridge University Press.

MACARTHUR, R. H. 1957. On the relative abundance of bird species. Proc. Nat. Acad. Sci. [U. S.], **43**: 293–294.

MAYR, E. 1933. Die Vogelwelt Polynesiens. Mitt. Zool. Mus. Berlin, **19**: 306–323.

——. 1940. The origin and history of the bird fauna of Polynesia. Proc. Sixth Pacific Sci. Congr., **4**: 197–216.

——. 1943. The zoogeographic position of the Hawaiian Islands. Condor, **45**: 45–48.

——. 1944. Wallace's Line in the light of recent zoogeographic studies. Quart. Rev. Biol., **19**: 1–14.

PRESTON, F. W. 1962. The canonical distribution of commonness and rarity: Parts I, II. Ecology, **43**: 185–215, 410–432.

RENSCH, B. 1936. Die Geschischte des Sundabogens. Borntraeger, Berlin.

STRESEMANN, E. 1934. "Aves." *In* Handb. Zool.,
W. Kukenthal, ed. Gruyter, Berlin.

————. 1939. Die Vögel von Celebes. J. für
Ornithologie, **87**: 299–425.

WILSON, E. O. 1961. The nature of the taxon
cycle in the Melanesian ant fauna. Amer.
Nat., **95**: 169–193.

ZIMMERMAN, E. C. 1948. Insects of Hawaii.
Vol. 1. Introduction. University of Hawaii
Press.

APPENDIX: MEASUREMENT OF A PROPAGULE

A rudimentary account of how many immigrants are required to constitute a propagule may be constructed as follows. Let η be the average number of individuals next generation per individual this generation. Thus, for instance, if $\eta = 1.03$, the population is increasing at 3% interest rate.

Let us now suppose that the number of descendants per individual has a Poisson distribution. If it has not, due to small birth rate, the figures do not change appreciably. Then, due to chance alone, the population descended from immigrants may vanish. This subject is well known in probability theory as "Extinction probabilities in branching processes" (cf. Feller 1958, p. 274). The usual equation for the probability ζ of eventual extinction (Feller's equation 5.2 with $P(\zeta) = e^{-\eta(1-\zeta)}$, for a Poisson distribution), gives

$$\zeta = e^{-\eta(1-\zeta)}.$$

Solving this by trial and error for the

TABLE 2. *Relation of replacement rate (η) of immigrants to probability of extinction (ζ)*

η	1	1.01	1.1	1.385
ζ	1	0.98	0.825	0.5

probability of eventual extinction ζ, given a variety of values of η, we get the array shown in table 2. From this we can calculate how large a number of simultaneous immigrants would stand probability just one-half of becoming extinct during the initial stages of population growth following the introduction. In fact, if r pairs immigrate simultaneously, the probability that all will eventually be without descendants is ζ^r. Solving $\zeta^R = 0.5$ we find the number, R, of pairs of immigrants necessary to stand half a chance of not becoming extinct as given in table 3. From this it is clear that when η is 1, the propagule has infinite size, but that as η increases, the propagule size decreases rapidly, until, for a species which increases at 38.5% interest rate, one pair is sufficient to stand probability 1.2 of effecting a colonization. With sexual species which hunt for mates, η may be very nearly 1 initially.

TABLE 3. *Relation of replacement rate (η) to the number of pairs (R) of immigrants required to give the population a 50% chance of survival*

η	1	1.01	1.1	1.385
R	∞	34	3.6	1

Paine, R. T. 1966. Food web complexity and species diversity. *The American Naturalist* 100:65–75.

Paine had a substantial influence on the notions of ecosystems, energy, and trophic dynamics as studies of community and ecosystem structure and function began to flourish. From these studies, scientists realized that food web complexity meant system stability. Food webs were not a new concept in Paine's day, but their importance to system stability was. Additionally, as species'—particularly predators'—roles within these ecosystems were studied and identified, biologists discovered that some species were critically important to the stability of the system and contributed to the maintenance of biodiversity. Paine's classic work on Pacific intertidal communities stimulated much thought regarding food webs, keystone species, and stability.

In this overview paper, Paine's primary premise is that food webs are not simple linear functions. Based on his studies of the influence of predators on food webs in marine environments, he reported that local diversity on intertidal rocky bottoms is related to predation intensity in the system, among other factors. Predator efficiency prevents single species from monopolizing important resources. The absence of predators created less diverse systems. In the systems Paine studied (i.e., marine rocky intertidal), the limiting resource was space. Although his work was related to marine systems, it has application to others, and builds the foundation for the idea of a keystone species.

RELATED READING

Hairston, N. G., F. E. Smith, and F. E. Slobodkin. 1960. Community structure, population control, and competition. The American Naturalist 94:421–425.

Henke, S.E., and F. C. Bryant. 1999. Effects of coyote removal on the faunal community in western Texas. Journal of Wildlife Management 63:1066–1081.

Holt, R. D. 1977. Predation, apparent competition, and the structure of prey communities. Theoretical Population Biology 12:197–229.

Hutchinson, G. E. 1959. Homage to Santa Rosalia or why are there so many kinds of animals? The American Naturalist 93:145–159.

MacArthur, R. H., and J. W. MacArthur. 1961. On bird species diversity. Ecology 42:594–598.

Menge, B. A., E. L. Berlow, C. A. Blanchette, S. A. Navarrete, and S. B. Yamada. 1994. The keystone species concept: variation in interaction strength in a rocky intertidal habitat. Ecological Monographs 64:249–286.

Mills, L. S., M. E. Soulé, and D. F. Doak. 1993. The keystone-species concept in ecology and conservation. BioScience 43:219–224.

Paine, R. T. 1995. A conversation on refining the concept of keystone species. Conservation Biology 9:962–964.

Power, M. E., D. Tilman, J. A. Estes, B. A. Menge, W. J. Bond, L. S. Mills, G. Daily, J. C. Castilla, J. Lubchenko, and R. T. Paine. 1996. Challenges in the quest for keystones. BioScience 46:609–620.

Reprinted by permission of University of Chicago Press,
publisher.

Vol. 100, No. 910   The American Naturalist   January–February, 1966

FOOD WEB COMPLEXITY AND SPECIES DIVERSITY

ROBERT T. PAINE

Department of Zoology, University of Washington, Seattle, Washington

Though longitudinal or latitudinal gradients in species diversity tend to be well described in a zoogeographic sense, they also are poorly understood phenomena of major ecological interest. Their importance lies in the derived implication that biological processes may be fundamentally different in the tropics, typically the pinnacle of most gradients, than in temperate or arctic regions. The various hypotheses attempting to explain gradients have recently been reviewed by Fischer (1960), Simpson (1964), and Connell and Orias (1964), the latter authors additionally proposing a model which can account for the production and regulation of diversity in ecological systems. Understanding of the phenomenon suffers from both a specific lack of synecological data applied to particular, local situations and from the difficulty of inferring the underlying mechanism(s) solely from descriptions and comparisons of faunas on a zoogeographic scale. The positions taken in this paper are that an ultimate understanding of the underlying causal processes can only be arrived at by study of local situations, for instance the promising approach of MacArthur and MacArthur (1961), and that biological interactions such as those suggested by Hutchinson (1959) appear to constitute the most logical possibilities.

The hypothesis offered herein applies to local diversity patterns of rocky intertidal marine organisms, though it conceivably has wider applications. It may be stated as: "Local species diversity is directly related to the efficiency with which predators prevent the monopolization of the major environmental requisites by one species." The potential impact of this process is firmly based in ecological theory and practice. Gause (1934), Lack (1949), and Slobodkin (1961) among others have postulated that predation (or parasitism) is capable of preventing extinctions in competitive situations, and Slobodkin (1964) has demonstrated this experimentally. In the field, predation is known to ameliorate the intensity of competition for space by barnacles (Connell, 1961b), and, in the present study, predator removal has led to local extinctions of certain benthic invertebrates and algae. In addition, as a predictable extension of the hypothesis, the proportion of predatory species is known to be relatively greater in certain diverse situations. This is true for tropical vs. temperate fish faunas (Hiatt and Strasburg, 1960; Bakus, 1964), and is seen especially clearly in the comparison of shelf water zooplankton populations (81 species, 16% of which are carnivores) with those of the presumably less productive though more stable Sargasso Sea (268 species, 39% carnivores) (Grice and Hart, 1962).

In the discussion that follows no quantitative measures of local diversity are given, though they may be approximated by the number of species represented in Figs. 1 to 3. No distinctions have been drawn between species within certain food categories. Thus I have assumed that the probability of, say, a bivalve being eaten is proportional to its abundance, and that predators exercise no preference in their choice of any "bivalve" prey. This procedure simplifies the data presentation though it dodges the problem of taxonomic complexity. Wherever possible the data are presented as both number observed being eaten and their caloric equivalent. The latter is based on prey size recorded in the field and was converted by determining the caloric content of Mukkaw Bay material of the same or equivalent species. These caloric data will be given in greater detail elsewhere. The numbers in the food webs, unfortunately, cannot be related to rates of energy flow, although when viewed as calories they undoubtedly accurately suggest which pathways are emphasized.

Dr. Rudolf Stohler kindly identified the gastropod species. A. J. Kohn, J. H. Connell, C. E. King, and E. R. Pianka have provided invaluable criticism. The University of Washington, through the offices of the Organization for Tropical Studies, financed the trip to Costa Rica. The field work in Baja California, Mexico, and at Mukkaw Bay was supported by the National Science Foundation (GB-341).

THE STRUCTURE OF SELECTED FOOD WEBS

I have claimed that one of the more recognizable and workable units within the community nexus are subwebs, groups of organisms capped by a terminal carnivore and trophically interrelated in such a way that at higher levels there is little transfer of energy to co-occurring subwebs (Paine, 1963). In the marine rocky intertidal zone both the subwebs and their top carnivores appear to be particularly distinct, at least where macroscopic species are involved; and observations in the natural setting can be made on the quantity and composition of the component species' diets. Furthermore, the rocky intertidal zone is perhaps unique in that the major limiting factor of the majority of its primary consumers is living space, which can be directly observed, as the elegant studies on interspecific competition of Connell (1961a,b) have shown. The data given below were obtained by examining individual carnivores exposed by low tide, and recording prey, predator, their respective lengths, and any other relevant properties of the interaction.

A north temperate subweb

On rocky shores of the Pacific Coast of North America the community is dominated by a remarkably constant association of mussels, barnacles, and one starfish. Fig. 1 indicates the trophic relationships of this portion of the community as observed at Mukkaw Bay, near Neah Bay, Washington (ca. 49° N latitude). The data, presented as both numbers and total calories consumed by the two carnivorous species in the subweb, *Pisaster ochraceus*,

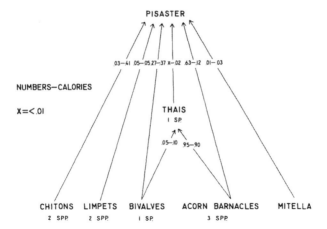

FIG. 1. The feeding relationships by numbers and calories of the *Pisaster* dominated subweb at Mukkaw Bay. *Pisaster*, N = 1049; *Thais*, N = 287. N is the number of food items observed eaten by the predators. The specific composition of each predator's diet is given as a pair of fractions; numbers on the left, calories on the right.

a starfish, and *Thais emarginata*, a small muricid gastropod, include the observational period November, 1963, to November, 1964. The composition of this subweb is limited to organisms which are normally intertidal in distribution and confined to a hard rock substrate. The diet of *Pisaster* is restricted in the sense that not all available local food types are eaten, although of six local starfish it is the most catholic in its tastes. Numerically its diet varies little from that reported by Feder (1959) for *Pisaster* observed along the central California coastline, especially since the gastropod *Tegula*, living on a softer bottom unsuitable to barnacles, has been omitted. *Thais* feeds primarily on the barnacle *Balanus glandula*, as also noted by Connell (1961b).

This food web (Fig. 1) appears to revolve on a barnacle economy with both major predators consuming them in quantity. However, note that on a nutritional (calorie) basis, barnacles are only about one-third as important to *Pisaster* as either *Mytilus californianus*, a bivalve, or the browsing chiton *Katherina tunicata*. Both these prey species dominate their respective food categories. The ratio of carnivore species to total species is 0.18. If *Tegula* and an additional bivalve are included on the basis that they are the most important sources of nourishment in adjacent areas, the ratio becomes 0.15. This number agrees closely with a ratio of 0.14 based on *Pisaster*, plus all prey species eaten more than once, in Feder's (1959) general compilation.

A subtropical subweb

In the Northern Gulf of California (ca. 31° N.) a subweb analogous to the one just described exists. Its top carnivore is a starfish (*Heliaster kubiniji*), the next two trophic levels are dominated by carnivorous gastropods, and the main prey are herbivorous gastropods, bivalves, and barnacles. I have

collected there only in March or April of 1962–1964, but on both sides of
the Gulf at San Felipe, Puertecitos, and Puerta Penasco. The resultant
trophic arrangements (Fig. 2), though representative of springtime conditions and indicative of a much more stratified and complex community, are
basically similar to those at Mukkaw Bay. Numerically the major food item
in the diets of *Heliaster* and *Muricanthus nigritus* (a muricid gastropod),
the two top-ranking carnivores, is barnacles; the major portion of these
predators' nutrition is derived from other members of the community, primarily herbivorous mollusks. The increased trophic complexity presents
certain graphical problems. If increased trophic height is indicated by a
decreasing percentage of primary consumers in a species diet, *Acanthina
tuberculata* is the highest carnivore due to its specialization on *A. angelica*,
although it in turn is consumed by two other species. Because of this, and
ignoring the percentages, both *Heliaster* and *Muricanthus* have been placed
above *A. tuberculata*. Two species, *Hexaplex* and *Muricanthus* eventually
become too large to be eaten by *Heliaster*, and thus through growth join it
as top predators in the system. The taxonomically-difficult gastropod
family Columbellidae, including both herbivorous and carnivorous species
(Marcus and Marcus, 1962) have been placed in an intermediate position.

The Gulf of California situation is interesting on a number of counts. A
new trophic level which has no counterpart at Mukkaw Bay is apparent, in-
terposed between the top carnivore and the primary carnivore level. If
higher level predation contributes materially to the maintenance of diversity, these species will have an effect on the community composition
out of proportion to their abundance. In one of these species, *Muricanthus*,

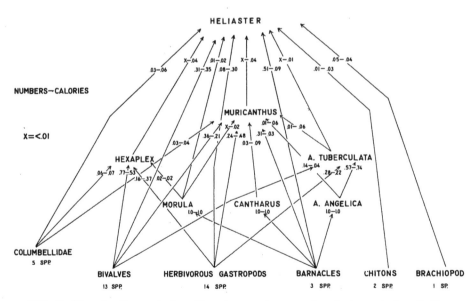

FIG. 2. The feeding relationships by numbers and calories of the *Heliaster*
dominated subweb in the northern Gulf of California. *Heliaster*, N = 2245; *Muricanthus*, N = 113; *Hexaplex*, N = 62; *A. tuberculata*, N = 14; *A. angelica*, N = 432;
Morula, N = 39; *Cantharus*, N = 8.

the larger members belong to a higher level than immature specimens (Paine, unpublished), a process tending to blur the food web but also potentially increasing diversity (Hutchinson, 1959). Finally, if predation operates to reduce competitive stresses, evidence for this reduction can be drawn by comparing the extent of niche diversification as a function of trophic level in a typical Eltonian pyramid. *Heliaster* consumes all other members of this subweb, and as such appears to have no major competitors of comparable status. The three large gastropods forming the subterminal level all may be distinguished by their major sources of nutrition: *Hexaplex*—bivalves (53%), *Muricanthus*—herbivorous gastropods (48%), and *A. tuberculata*—carnivorous gastropods (74%). No such obvious distinction characterizes the next level composed of three barnacle-feeding specialists which additionally share their resource with *Muricanthus* and *Heliaster*. Whether these species are more specialized (Klopfer and MacArthur, 1960) or whether they tolerate greater niche overlap (Klopfer and MacArthur, 1961) cannot be stated. The extent of niche diversification is subtle and trophic overlap is extensive.

The ratio of carnivore species to total species in Fig. 2 is 0.24 when the category Columbellidae is considered to be principally composed of one herbivorous (*Columbella*) and four carnivorous (*Pyrene, Anachis, Mitella*) species, based on the work of Marcus and Marcus (1962).

A tropical subweb

Results of five days of observation near Mate de Limon in the Golfo de Nocoya on the Pacific shore of Costa Rica (approx. 10° N.) are presented in Fig. 3. No secondary carnivore was present; rather the environmental resources were shared by two small muricid gastropods, *Acanthina brevidentata* and *Thais biserialis*. The fauna of this local area was relatively simple and completely dominated by a small mytilid and barnacles. The co-occupiers of the top level show relatively little trophic overlap despite the broad nutritional base of *Thais* which includes carrion and cannibalism. The relatively low number of feeding observations (187) precludes an accurate appraisal of the carnivore species to total web membership ratio.

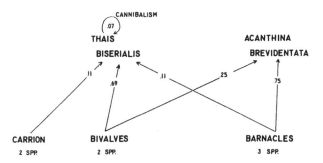

FIG. 3. The feeding relationship by numbers of a comparable food web in Costa Rica. *Thais*, N = 99; *Acanthina*, N = 80.

CHANGES RESULTING FROM THE REMOVAL OF THE TOP CARNIVORE

Since June, 1963, a "typical" piece of shoreline at Mukkaw Bay about eight meters long and two meters in vertical extent has been kept free of *Pisaster*. An adjacent control area has been allowed to pursue its natural course of events. Line transects across both areas have been taken irregularly and the number and density of resident macroinvertebrate and benthic algal species measured. The appearance of the control area has not altered. Adult *Mytilus californianus*, *Balanus cariosus*, and *Mitella polymerus* (a goose-necked barnacle) form a conspicuous band in the middle intertidal. The relatively stable position of the band is maintained by *Pisaster* predation (Paris, 1960; Paine, unpublished). At lower tidal levels the diversity increases abruptly and the macrofauna includes immature individuals of the above, *B. glandula* as scattered clumps, a few anemones of one species, two chiton species (browsers), two abundant limpets (browsers), four macroscopic benthic algae (*Porphyra*-an epiphyte, *Endocladia*, *Rhodomela*, and *Corallina*), and the sponge *Haliclona*, often browsed upon by *Anisodoris*, a nudibranch.

Following the removal of *Pisaster*, *B. glandula* set successfully throughout much of the area and by September had occupied from 60 to 80% of the available space. By the following June the *Balanus* themselves were being crowded out by small, rapidly growing *Mytilus* and *Mitella*. This process of successive replacement by more efficient occupiers of space is continuing, and eventually the experimental area will be dominated by *Mytilus*, its epifauna, and scattered clumps of adult *Mitella*. The benthic algae either have or are in the process of disappearing with the exception of the epiphyte, due to lack of appropriate space; the chitons and larger limpets have also emigrated, due to the absence of space and lack of appropriate food.

Despite the likelihood that many of these organisms are extremely long-lived and that these events have not reached an equilibrium, certain statements can be made. The removal of *Pisaster* has resulted in a pronounced *decrease* in diversity, as measured simply by counting species inhabiting this area, whether consumed by *Pisaster* or not, from a 15 to an eight-species system. The standing crop has been increased by this removal, and should continue to increase until the *Mytilus* achieve their maximum size. In general the area has become trophically simpler. With *Pisaster* artificially removed, the sponge-nudibranch food chain has been displaced, and the anemone population reduced in density. Neither of these carnivores nor the sponge is eaten by *Pisaster*, indicating that the number of food chains initiated on this limited space is strongly influenced by *Pisaster*, but by an indirect process. In contrast to Margalef's (1958) generalization about the tendency, with higher successional status towards "an ecosystem of more complex structure," these removal experiments demonstrate the opposite trend: in the absence of a complicating factor (predation), there is a "winner" in the competition for space, and the local system tends toward simplicity. Predation by this interpretation interrupts the successional process and, on a local basis, tends to increase local diversity.

No data are available on the microfaunal changes accompanying the gradual alteration of the substrate from a patchy algal mat to one comprised of the byssal threads of *Mytilus*.

INTERPRETATION

The differences in relative diversity of the subwebs diagrammed in Figs. 1–3 may be represented as Baja California (45 spp.) >> Mukkaw Bay (11 spp.) > Costa Rica (8 sp.), the number indicating the actual membership of the subwebs and not the number of local species. All three areas are characterized by systems in which one or two species are capable of monopolizing much of the space, a circumstance realized in nature only in Costa Rica. In the other two areas a top predator that derives its nourishment from other sources feeds in such a fashion that no space-consuming monopolies are formed. *Pisaster* and *Heliaster* eat masses of barnacles, and in so doing enhance the ability of other species to inhabit the area by keeping space open. When the top predator is artificially removed or naturally absent (i.e., predator removal area and Costa Rica, respectively), the systems converge toward simplicity. When space is available, other organisms settle or move in, and these, for instance chitons at Mukkaw Bay and herbivorous gastropods and pelecypods in Baja California, form the major portions of the predator's nutrition. Furthermore, *in situ* primary production is enhanced by the provision of space. This event makes the grazing moiety less dependent on the vagaries of phytoplankton production or distribution and lends stability to the association.

At the local level it appears that carnivorous gastropods which can penetrate only one barnacle at a time, although they might consume a few more per tidal interval, do not have the same effect as a starfish removing 20 to 60 barnacles simultaneously. Little compensation seems to be gained from snail density increases because snails do not clear large patches of space, and because the "husks" of barnacles remain after the animal portion has been consumed. In the predator removal area at Mukkaw Bay, the density of *Thais* increased 10- to 20-fold, with no apparent effect on diversity although the rate of *Mytilus* domination of the area was undoubtedly slowed. Clusters (density of 75–125/m²) of *Thais* and *Acanthina* characterize certain rocks in Costa Rica, and diversity is still low. And, as a generality, wherever acorn barnacles or other space-utilizing forms potentially dominate the shore, diversity is reduced unless some predator can prevent the space monopoly. This occurs in Washington State where the shoreline, in the absence of *Pisaster*, is dominated by barnacles, a few mussels, and often two species of *Thais*. The same monopolistic tendencies characterize Connell's (1961a,b) study area in Scotland, the rocky intertidal of northern Japan (Hoshiai, 1960, 1961), and shell bags suitable for sponge settlement in North Carolina (Wells, Wells, and Gray, 1964).

Local diversity on intertidal rocky bottoms, then, appears directly related to predation intensity, though other potential factors are mentioned below. If one accepts the generalizations of Hedgpeth (1957) and Hall

(1964) that ambient temperature is the single most important factor influencing distribution or reproduction of marine invertebrates, then the potential role of climatic stability as measured by seasonal variations in water temperature can be examined. At Neah Bay the maximum range of annual values are 5.9 to 13.3 C (Rigg and Miller, 1949); in the northern Gulf of California, Roden and Groves (1959) recorded an annual range of 14.9 to 31.2 C; and in Costa Rica the maximum annual range is 26.1 to 31.7 C (Anon., 1952). Clearly the greatest benthic diversity, and one claimed by Parker (1963) on a regional basis to be among the most diverse known, is associated with the most variable (least stable) temperature regimen. Another influence on diversity could be exercised by environmental heterogeneity (Hutchinson, 1959). Subjectively, it appeared that both the Mukkaw Bay and Costa Rica stations were topographically more distorted than the northern Gulf localities. In any event, no topographic features were evident that could correlate with the pronounced differences in faunal diversity. Finally, Connell and Orias (1964) have developed a model for the organic enrichment of regions that depends to a great extent on the absolute amount of primary production and/or nutrient import, and hence energy flowing through the community web. Unfortunately, no productivity data are available for the two southern communities, and comparisons cannot yet be made.

PREDATION AND DIVERSITY GRADIENTS

To examine predation as a diversity-causing mechanism correlated with latitude, we must know why one environment contains higher order carnivores and why these are absent from others. These negative situations can be laid to three possibilities: (1) that through historical accident no higher carnivores have evolved in the region; (2) that the sample area cannot be occupied due to a particular combination of *local* hostile physiological effects; (3) that the system cannot support carnivores because the rate of energy transfer to a higher level is insufficient to sustain that higher level. The first possibility is unapproachable, the second will not apply on a geographic scale, and thus only the last would seem to have reality. Connell and Orias (1964) have based their hypothesis of the establishment and maintenance of diversity on varying rates of energy transfer, which are determined by various limiting factors and environmental stability. Without disagreeing with their model, two aspects of primary production deserve further consideration. The animal diversity of a given system will probably be higher if the production is apportioned more uniformly throughout the year rather than occurring as a single major bloom, because tendencies towards competitive displacement can be ameliorated by specialization on varying proportions of the resources (MacArthur and Levins, 1964). Both the predictability of production on a sustained annual basis and the causation of resource heterogeneity by predation will facilitate this mechanism. Thus, per production unit, greater stability of production should be correlated with greater diversity, other things being equal.

The realization of this potential, however, depends on more than simply the annual stability of carbon fixation. Rate of production and subsequent transfer to higher levels must also be important. Thus trophic structure of a community depends in part on the physical extent of the area (Darlington, 1957), or, in computer simulation models, on the amount of protoplasm in the system (Garfinkel and Sack, 1964). On the other hand, enriched aquatic environments often are characterized by decreased diversity. Williams (1964) has found that regions of high productivity are dominated by few diatom species. Less productive areas tended to have more species of equivalent rank, and hence a greater diversity. Obviously, the gross amount of energy fixed by itself is incapable of explaining diversity; and extrinsic factors probably are involved.

Given sufficient evolutionary time for increases in faunal complexity to occur, two independent mechanisms should work in a complementary fashion. When predation is capable of preventing resource monopolies, diversity should increase by positive feedback processes until some limit is reached. The argument of Fryer (1965) that predation facilitates speciation is germane here. The upper limit to local diversity, or, in the present context, the maximum number of species in a given subweb, is probably set by the combined stability and rate of primary production, which thus influences the number and variety of non-primary consumers in the subweb. Two aspects of predation must be evaluated before a generalized hypothesis based on predation effects can contribute to an understanding of differences in diversity between *any* comparable regions or faunistic groups. We must know if resource monopolies are actually less frequent in the diverse area than in comparable systems elsewhere, and, if so, why this is so. And we must learn something about the multiplicity of energy pathways in diverse systems, since predation-induced diversity could arise either from the presence of a variety of subwebs of equivalent rank, or from domination by one major one. The predation hypothesis readily predicts the apparent absence of monopolies in tropical (diverse) areas, a situation classically represented as "many species of reduced individual abundance." It also is in accord with the disproportionate increase in the number of carnivorous species that seems to accompany regional increases in animal diversity. In the present case in the two adequately sampled, structurally analagous, subwebs, general membership increases from 13 at Mukkaw Bay to 45 in the Gulf of California, a factor of 3.5, whereas the carnivore species increased from 2 to 11, a factor of 5.5.

SUMMARY

It is suggested that local animal species diversity is related to the number of predators in the system and their efficiency in preventing single species from monopolizing some important, limiting, requisite. In the marine rocky intertidal this requisite usually is space. Where predators capable of preventing monopolies are missing, or are experimentally removed, the systems become less diverse. On a local scale, no relationship between latitude (10° to 49° N.) and diversity was found. On a geographic scale, an increased stability of annual production may lead to an increased capacity for systems to support higher-level carnivores. Hence tropical, or other, ecosystems are more diverse, and are characterized by disproportionately more carnivores.

LITERATURE CITED

Anon. 1952. Surface water temperatures at tide stations. Pacific coast North and South America. Spec. Pub. No. 280: p. 1-59. U. S. Coast and Geodetic Survey.

Bakus, G. J. 1964. The effects of fish-grazing on invertebrate evolution in shallow tropical waters. Allan Hancock Found. Pub. 27: 1-29.

Connell, J. H. 1961a. Effect of competition, predation by *Thais lapillus*, and other factors on natural populations of the barnacle *Balanus balanoides*. Ecol. Monogr. 31: 61-104.

————. 1961b. The influence of interspecific competition and other factors on the distribution of the barnacle *Chthamalus stellatus*. Ecology 42: 710-723.

Connell, J. H., and E. Orias. 1964. The ecological regulation of species diversity. Amer. Natur. 98: 399-414.

Darlington, P. J. 1957. Zoogeography. Wiley, New York.

Feder, H. M. 1959. The food of the starfish, *Pisaster ochraceus*, along the California coast. Ecology 40: 721-724.

Fischer, A. G. 1960. Latitudinal variations in organic diversity. Evolution 14: 64-81.

Fryer, G. 1965. Predation and its effects on migration and speciation in African fishes: a comment. Proc. Zool. Soc. London 144: 301-310.

Garfinkel, D., and R. Sack. 1964. Digital computer simulation of an ecological system, based on a modified mass action law. Ecology 45: 502-507.

Gause, G. F. 1934. The struggle for existence. Williams and Wilkins Co., Baltimore.

Grice, G. D., and A. D. Hart. 1962. The abundance, seasonal occurrence, and distribution of the epizooplankton between New York and Bermuda. Ecol. Monogr. 32: 287-309.

Hall, C. A., Jr. 1964. Shallow-water marine climates and molluscan provinces. Ecology 45: 226-234.

Hedgpeth, J. W. 1957. Marine biogeography. Geol. Soc. Amer. Mem. 67, 1: 359-382.

Hiatt, R. W., and D. W. Strasburg. 1960. Ecological relationships of the fish fauna on coral reefs of the Marshall Islands. Ecol. Monogr. 30: 65-127.

Hoshiai, T. 1960. Synecological study on intertidal communities III. An analysis of interrelation among sedentary organisms on the artificially denuded rock surface. Bull. Marine Biol. Sta. Asamushi. 10: 49-56.

————. 1961. Synecological study on intertidal communities. IV. An ecological investigation on the zonation in Matsushima Bay concerning the so-called covering phenomenon. Bull. Marine Biol. Sta. Asamushi. 10: 203-211.

Hutchinson, G. E. 1959. Homage to Santa Rosalia or why are there so many kinds of animals? Amer. Natur. 93: 145–159.

Klopfer, P. H., and R. H. MacArthur. 1960. Niche size and faunal diversity. Amer. Natur. 94: 293–300.

———. 1961. On the causes of tropical species diversity: niche overlap. Amer. Natur. 95: 223–226.

Lack, D. 1949. The significance of ecological isolation, p. 299–308. In G. L. Jepsen, G. G. Simpson, and E. Mayr [eds.], Genetics, paleontology and evolution. Princeton Univ. Press, Princeton.

MacArthur, R., and R. Levins. 1964. Competition, habitat selection, and character displacement in a patchy environment. Proc. Nat. Acad. Sci. 51: 1207–1210.

MacArthur, R. H., and J. W. MacArthur. 1961. On bird species diversity. Ecology 42: 594–598.

Marcus, E., and E. Marcus. 1962. Studies on Columbellidae. Bol. Fac. Cienc. Letr. Univ. Sao Paulo 261: 335–402.

Margalef, R. 1958. Mode of evolution of species in relation to their place in ecological succession. XVth Int. Congr. Zool. Sect. 10, paper 17.

Paine, R. T. 1963. Trophic relationships of 8 sympatric predatory gastropods. Ecology 44: 63–73.

Paris, O. H. 1960. Some quantitative aspects of predation by muricid snails on mussels in Washington Sound. Veliger 2: 41–47.

Parker, R. H. 1963. Zoogeography and ecology of some macro-invertebrates, particularly mollusca in the Gulf of California and the continental slope off Mexico. Vidensk. Medd. Dansk. Natur. Foren., Copenh. 126: 1–178.

Rigg, G. B., and R. C. Miller. 1949. Intertidal plant and animal zonation in the vicinity of Neah Bay, Washington. Proc. Calif. Acad. Sci. 26: 323–351.

Roden, G. I., and G. W. Groves. 1959. Recent oceanographic investigations in the Gulf of California. J. Marine Res. 18: 10–35.

Simpson, G. G. 1964. Species density of North American recent mammals. Syst. Zool. 13: 57–73.

Slobodkin, L. B. 1961. Growth and regulation of Animal Populations. Holt, Rinehart, and Winston, New York.

———. 1964. Ecological populations of Hydrida. J. Anim. Ecol. 33 (Suppl.): 131–148.

Wells, H. W., M. J. Wells, and I. E. Gray. 1964. Ecology of sponges in Hatteras Harbor, North Carolina. Ecology 45: 752–767.

Williams, L. G. 1964. Possible relationships between plankton-diatom species numbers and water-quality estimates. Ecology 45: 809–823.

Caughley, G. 1966. Mortality patterns in mammals. *Ecology* 47:906–918.

As with other concepts in this text, understanding population ecology and key population parameters of wildlife is critical when managing wildlife and their habitats. A grasp of mortality patterns is extremely important when examining wildlife interactions (e.g., competition, predation, diseases). Graeme Caughley was a wildlife biologist who made numerous contributions to the field by describing complex interactions clearly. He made theory applicable to wildlife management in his writings.

Life tables, once a common way to examine population dynamics, have been replaced in many cases with more sophisticated models. But life table analysis remains important with some species and in some situations. Herein, Caughley contrasts life tables of seven species and discusses commonalities and the various classifications of mortality patterns. Using life tables and fecundity tables, he shows how mean gestation length, annual mortality, and mean life expectancy can be calculated. By contrasting these parameters with life tables of other species, Caughley demonstrates that most mammalian species have common life tables where "age-specific mortality within species assumes an approximate constant form irrespective of the proximate causes of mortality."

In addition to his presentation of the correct way to use life tables, Caughley also warns of their improper use. In the last sentence of the paper, he asks a question we are still examining today: "Do we yet know enough about mortality patterns in mammals to justify the construction of any system of classification?"

RELATED READING

Cole, L. C. 1954. The population consequences of life history phenomena. Quarterly Review of Biology 29:103–137.

Deevey, E. S., Jr. 1947. Life tables for natural populations of animals. Quarterly Review of Biology 22:283–314.

Hewer, H. R. 1964. The determination of age, sexual maturity, longevity and a life table in the grey seal (*Halichoerus grypus*). Proceedings of the Zoological Society of London 142:593–623.

Taber, R. D., and R. F. Dasmann. 1957. The dynamics of three natural populations of the deer *Odocoileus hemionus columbianus*. Ecology 38:233–246.

Tyndale-Biscoe, C. H., and R. M. Williams. 1955. A study of natural mortality in a wild population of the rabbit *Oryctolagus cuniculus* (L.). New Zealand Journal of Science and Technology B 36:561–580.

Woodgerd, W. 1964. Population dynamics of bighorn sheep on Wildhorse Island. Journal of Wildlife Management 28:381–391.

MORTALITY PATTERNS IN MAMMALS

Graeme Caughley

Forest Research Institute, New Zealand Forest Service, Rotorua, and Zoology Department, Canterbury University, New Zealand

(Accepted for publication December 8, 1965)

Abstract. Methods of obtaining life table data are outlined and the assumptions implicit in such treatment are defined. Most treatments assume a stationary age distribution, but published methods of testing the stationary nature of a single distribution are invalid. Samples from natural populations tend to be biased in the young age classes and therefore, because it is least affected by bias, the mortality rate curve (q_x) is the most efficient life table series for comparing the pattern of mortality with age in different populations.

A life table and fecundity table are presented for females of the ungulate *Hemitragus jemlahicus*, based on a population sample that was first tested for bias. They give estimates of mean generation length as 5.4 yr, annual mortality rate as 0.25, and mean life expectancy at birth as 3.5 yr.

The life table for *Hemitragus* is compared with those of *Ovis aries, O. dalli,* man, *Rattus norvegicus, Microtus agrestis,* and *M. orcadensis* to show that despite taxonomic and ecological differences the life tables have common characteristics. This suggests the hypotheses that most mammalian species have life tables of a common form, and that the pattern of age-specific mortality within species assumes an approximately constant form irrespective of the proximate causes of mortality.

Introduction

Most studies in population ecology include an attempt to determine mortality rates, and in many cases rates are given for each age class. This is no accident. Age-specific mortality rates are usually necessary for calculating reproductive values for each age class, the ages most susceptible to natural selection, the population's rate of increase, mean life expectancy at birth, mean generation length, and the percentage of the population that dies each year. The importance of these statistics in the fields of game management, basic and applied ecology, and population genetics requires no elaboration.

The pattern of changing mortality rates with age is best expressed in the form of a life table. These tables usually present the same information in a variety of ways:

1) Survivorship (l_x) : this series gives the probability at birth of an individual surviving to any age, x (l_x as used here is identical with P_x of Leslie, Venables and Venables 1952). The ages

are most conveniently spaced at regular intervals such that the values refer to survivorship at ages 0, 1, 2 etc. yr, months, or some other convenient interval. The probability at birth of living to birth is obviously unity, but this initial value in the series need not necessarily be set at 1; it is often convenient to multiply it by 1,000 and to increase proportionately the other values in the series. If this is done, survivorship can be redefined as the number of animals in a cohort of 1,000 (or any other number to which the initial value is raised) that survived to each age x. In this way a kl_x series is produced, where k is the constant by which all l_x values in the series are multiplied.

2) Mortality (d_x): the fraction of a cohort that dies during the age interval x, x + 1 is designated d_x. It can be defined in terms of the individual as the probability at birth of dying during the interval x, x + 1. As a means of eliminating decimal points the values are sometimes multiplied by a constant such that the sum of the d_x values equals 1,000. The values can be calculated from the l_x series by

$$d_x = l_x - l_{x+1}$$

3) Mortality rate (q_x): the mortality rate q for the age interval x, x + 1 is termed q_x. It is calculated as the number of animals in the cohort that died during the interval x, x + 1, divided by the number of animals alive at age x. This value is usually expressed as $1,000q_x$, the number of animals out of 1,000 alive at age x which died before x + 1.

These are three ways of presenting age-specific mortality. Several other methods are available— e.g. survival rate (p_x), life expectancy (e_x) and probability of death (Q_x)—but these devices only present in a different way the information already contained in each of the three series previously defined. In this paper only the l_x, d_x and q_x series will be considered.

METHODS OF OBTAINING MORTALITY DATA

Life tables may be constructed from data collected in several ways. Direct methods:

1) Recording the ages at death of a large number of animals born at the same time. The frequencies of ages at death form a kd_x series.

2) Recording the number of animals in the original cohort still alive at various ages. The frequencies from a kl_x series.

Approximate methods:

3) Recording the ages at death of animals marked at birth but whose births were not coeval. The frequencies form a kd_x series.

4) Recording ages at death of a representative sample by ageing carcasses from a population that has assumed a stationary age distribution. Small fluctuations in density will not greatly affect the results if these fluctuations have an average wave length considerably shorter than the period over which the carcasses accumulated. The frequencies form a kd_x series.

5) Recording a sample of ages at death from a population with a stationary age distribution, where the specimens were killed by a catastrophic event (avalanche, flood, etc.) that removed and fixed an unbiased sample of ages in a living population. In some circumstances (outlined later) the age frequencies can be treated as a kl_x series.

6) The census of ages in a living population, or a sample of it, where the population has assumed a stationary age distribution. Whether the specimens are obtained alive by trapping or are killed by unselective shooting, the resultant frequencies are a sample of ages in a living population and form a kl_x series in certain circumstances.

Methods 1 to 3 are generally used in studies of small mammals while methods 4 to 6 are more commonly used for large mammals.

TESTS FOR STATIONARY AGE DISTRIBUTION

Five methods have been suggested for determining whether the age structure of a sample is consistent with its having been drawn from a stationary age distribution:

a) Comparison of the "mean mortality rate," calculated from the age distribution of the sample, with the proportion represented by the first age class (Kurtén 1953, p. 51).

b) Comparison of the annual female fecundity of a female sample with the sample number multiplied by the life expectancy at birth, the latter statistic being estimated from the age structure (Quick 1963, p. 210).

c) Calculation of instantaneous birth rates and death rates, respectively, from a sample of the population's age distribution and a sample of ages at death (Hughes 1965).

d) Comparison of the age distribution with a prejudged notion of what a stationary age distribution should be like (Breakey 1963).

e) Examination of the "l_x" and "d_x" series, calculated from the sampled age distribution, for evidence of a common trend (Quick 1963, p. 204).

Methods a to c are tautological because they assume the sampled age distribution is either a kl_x or kd_x series; method d assumes the form of the life table, and e makes use of both assumptions. These ways of judging the stationary nature of a population are invalid. But I intend something more general than the simple statement that these five methods do not test what they are supposed to test. Given no information other than a single age distribution, it is theoretically impossible to prove that the distribution is from a stationary population unless one begins from the assumption that the population's survival curve is of a particular form. If such an assumption is made, the life table constructed from the age frequencies provides no more information than was contained in the original premise.

Mortality Samples and Age Structure Samples

Methods 4 to 6 for compiling life tables are valid only when the data are drawn from a stationary age distribution. This distribution results when a population does not change in size and where the age structure of the population is constant with time. The concept has developed from demographic research on man and is useful for species which, like man, have no seasonally restricted period of births.

Populations that have a restricted season of births present difficulties of treatment, some of which have been discussed by Leslie and Ranson (1940). Very few mammals breed at the same rate throughout the year, and the stationary age distribution must be redefined if it is to include seasonal breeders. For species with one restricted breeding season each year, a stationary population can be defined as one that does not vary either in numbers or age structure at successive points in time spaced at intervals of 1 yr. The stationary age distribution can then be defined for such populations as the distribution of ages at a given time of the year. Thus there will be an infinite number of different age distributions according to the time of census, other than in the exceptional case of a population having a constant rate of mortality throughout life.

The distribution of ages in a stationary population forms a kl_x series only when all births for the year occur at an instant of time and the sample is taken at that instant. This is obviously impossible, but the situation is approximated when births occur over a small fraction of the year. If a population has a restricted season of births, the age structure can be sampled over this period and at the same time the number of live births produced by a hypothetical cohort can be calculated from the number of females either pregnant or suckling young. In this way a set of data closely approximating a kl_x series can be obtained.

If an age distribution is sampled halfway between breeding seasons, it cannot be presented as a kl_x series with x represented as integral ages in years. With such a sample (making the usual assumptions of stability and lack of bias) neither l_x nor d_x can be established, but q_x values can be calculated for each age interval $x + \frac{1}{2}$, $x + 1\frac{1}{2}$. The age frequencies from a population with a continuous rate of breeding are exactly analogous; they do not form a kl_x series but can be treated as a series of the form

$$k (l_x + l_{x+1}) /2$$

This series does not allow calculation of l_x values from birth unless the mortality rate between birth and the midpoint of the first age interval is known.

Because a sample consists of dead animals, its age frequencies do not necessarily form a mortality series. The kd_x series is obtained only when the sample represents the frequencies of ages at death in a stationary population. Many published samples treated as if they formed a kd_x series are not appropriate to this form of analysis. For instance, if the animals were obtained by shooting which was unselective with respect to age, the sample gives the age structure of the living population at that time; that the animals were killed to get these data is irrelevant. Hence unbiased shooting samples survivorship, not mortality, and an age structure so obtained can be treated as a kl_x series if all other necessary assumptions obtain. Similarly, groups of animals killed by avalanches, fires, or floods—catastrophic events that preserve a sample of the age frequencies of animals during life—do not provide information amenable to kd_x treatment.

A sample may include both l_x and d_x components. For instance, it could consist of a number of dead animals, some of which have been unselectively shot, whereas the deaths of others are attributable to "natural" mortality. Or it could be formed by a herd of animals killed by an avalanche in an area where carcasses of animals that died "naturally" were also present. In both these cases d_x and l_x data are confounded and these heterogeneous samples of ages at death can be treated neither as kd_x nor kl_x series.

Even if a sample of ages at death were not heterogeneous in this sense, it might still give misleading information. If, for instance, carcasses attributable to "natural" mortality were collected only on the winter range of a population, the age frequencies of this sample would provide ages at death which reflected the mortality pattern during only part of the year. But the d_x series gives the proportion of deaths over contiguous periods of the life span and must reflect all mortality during each of these periods.

It has been stressed that the frequencies of ages in life or of ages at death provide life-table information only when they are drawn from a population with a stationary age distribution. This age distribution should not be confused with the stable distribution. When a population increases at a constant rate and where survivorship and fecundity rates are constant, the age distribution eventually assumes a stable form (Lotka 1907 a, b; Sharpe and Lotka 1911). Slobodkin (1962, p. 49) gives a simple explanation as to why this is so. A stable age distribution does not form a kl_x series except when the rate of increase is zero, the season of births is restricted, and the sample is taken at this time. Hence the stationary age distribution is a special case of the stable age distribution.

THE RELATIVE USEFULNESS OF THE l_x, d_x AND q_x SERIES

Most published life tables for wild mammals have been constructed either from age frequencies obtained by shooting to give a kl_x series, or by determining the ages at death of animals found dead, thereby producing a kd_x series. Unfortunately, both these methods are almost invariably subject to bias in that the frequency of the first-year class is not representative. Dead immature animals, especially those dying soon after birth, tend to decay faster than the adults, so that they are underrepresented in the count of carcasses. The ratio of juveniles to adults in a shot sample is usually biased because the two age classes have different susceptibilities to hunting. With such a bias established or suspected, the life table is best presented in a form that minimizes this bias. An error in the frequency of the first age class results in distortions of each l_x and d_x value below it in the series, but q_x values are independent of frequencies in younger age classes. By definition, q is the ratio of those dying during an age interval to those alive at the beginning of the interval. At age y the value of q is given by

$$q_y = d_y / l_y$$

but

$$d_y = l_y - l_{y+1}$$

therefore

$$q_y = (l_y - l_{y+1}) / l_y .$$

Thus the value of q_y is not directly dependent on absolute values of l_x but on the differences between successive values. If the l_x series is calculated from age frequencies in which the initial frequency

is inaccurate, each l_x value will be distorted. However, the difference between any two, divided by the first, will remain constant irrespective of the magnitude of error above them in the series. Thus a q_x value is independent of all but two survivorship age frequencies and can be calculated directly from these frequencies (f_x) by

$$q_x = (f_x - f_{x+1}) / f_x$$

if the previously discussed conditions are met.

The calculation of q from frequencies of ages at death is slightly more complex:

by definition

$$q_y = d_y l_y$$

but

$$l_y = \sum_{x=0}^{\infty} d_x - \sum_{x=0}^{y-1} d_x$$

therefore

$$q_y = d_y / \left(\sum_{x=0}^{\infty} d_x - \sum_{x=0}^{y-1} d_x \right)$$

$$= d_y / \sum_{x=y}^{\infty} d_x$$

but the frequencies of ages at death (f'_x) are themselves a kd_x series and so

$$q_y = f'_y / \sum_{x=y}^{\infty} f'_x .$$

Thus the value of q at any age is independent of frequencies of the younger age classes. Although the calculated value of q for the first age class may be wrong, this error does not affect the q_x values for the older age classes.

The q_x series has other advantages over the l_x and d_x series for presenting the pattern of mortality with age. It shows rates of mortality directly, whereas this rate is illustrated in a graph of the l_x series (the series most often used when comparing species) only by the slope of the curve.

A LIFE TABLE FOR THE THAR,

Hemitragus jemlahicus

The Himalayan thar is a hollow-horned ungulate introduced into New Zealand in 1904 (Donne 1924) and which now occupies 2,000 miles² of mountainous country in the South Island. Thar were liberated at Mount Cook and have since spread mostly north and south along the Southern Alps. They are still spreading at a rate of about 1.1 miles a year (Caughley 1963) and so the populations farthest from the point of liberation have been established only recently and have not yet had time to increase greatly in numbers. Closer to the site of liberation the density is higher (correlated with the greater length of time that animals have been established there), and around the point of liberation itself there is evidence that the population has decreased (Anderson and Henderson 1961).

The growth rings on its horns are laid down in each winter of life other than the first (Caughley 1965), thereby allowing the accurate ageing of specimens. An age structure was calculated from a sample of 623 females older than 1 yr shot in the Godley and Macaulay Valleys between November 1963 and February 1964. Preliminary work on behavior indicates that there is very little dispersal of females into or out of this region, both because the females have distinct home ranges and because there are few ice-free passes linking the valley heads.

As these data illustrate problems presented by most mammals, and because the life table has not been published previously, the methods of treatment will be outlined in some detail.

Is the population stationary?

Although it is impossible to determine the stationary nature of a population by examining the age structure of a single sample, even when rates of fecundity are known, in some circumstances a series of age structures will give the required information. This fact is here utilized to investigate the stability of this population.

The sample was taken about halfway between the point of liberation and the edge of the range. It is this region between increasing and decreasing populations where one would expect to find a stationary population. The animals came into the Godley Valley from the southwest and presumably colonized this side of the valley before crossing the 2 miles of river bed to the northeast side. This pattern of establishment is deduced from that in the Rakaia Valley, at the present edge of the breeding range, where thar bred for at least 5 yr on the south side of the valley before colonizing the north side. Having colonized the northeast side of the Godley Valley, the thar would then cross the Sibald Range to enter the Macaulay Valley, which is a further 6 miles northeast. The sample can therefore be divided into three subsamples corresponding to the different periods of time that the animals have been present in the three areas. A 10×3 contingency test for differences between the three age distributions of females 1 yr of age or older gave no indication that the three subpopulations differed in age structure ($\chi^2 = 22.34$; $P = 0.2$).

This information can be interpreted in two ways: either the three subpopulations are neither increasing nor decreasing and hence are likely to have stationary age distributions, or the subpopulations could be increasing at the same rate, in which case they could have identical stable age distributions. The second alternative carries a

TABLE I. Relative densities of thar in three zones

Zone	Number females autopsied	Mean density index[a]	Standard error
Godley Valley south........	258	2.19	0.56
Godley Valley north........	240	1.67	0.53
Macaulay Valley..........	115	2.66	0.69

$F_{2,56}$ for densities between valleys = 1.74, not significant
[a]Density indices were calculated as the number of females other than kids recorded as autopsied in a zone each day, divided by the number of shooters hunting in the zone on that day.

corollary that the subpopulations would have different densities because they have been increasing for differing periods of time. But an analysis of the three densities gives no indication that they differ (Table I). This result necessitates the rejection of the second alternative.

The above evidence suggesting that the sample was drawn from a stationary age distribution is supported to some extent by observation. When I first passed through the area in 1957, I saw about as many thar per day as in 1963-64. J. A. Anderson, a man who has taken an interest in the thar of this region, writes that the numbers of thar in 1956 were about the same as in 1964 (Anderson, pers. comm.). These are subjective evaluations and for that reason cannot by themselves be given much weight, but they support independent evidence that the population is stationary or nearly so.

Is the sample biased?

A sample of the age structure of a population can be biased in several ways. The most obvious source of bias is behavioral or range differences between males and females. For instance, should males tend to occupy terrain which is more difficult to hunt over than that used by females, they would be underrepresented in a sample obtained by hunting. During the summer thar range in three main kinds of groups: one consists of females, juveniles and kids, a second consists of young males and the third of mature males. The task of sampling these three groupings in the same proportions as they occur throughout the area is complicated by their preferences for terrain that differs in slope, altitude and exposure. Consequently the attempt to take an unbiased sample of both males and females was abandoned and the hunting was directed towards sampling only the nanny-kid herds in an attempt to take a representative sample of females. The following analysis is restricted to females.

Although bias attributable to differences in behavior between sexes can be eliminated by the simple contrivance of ignoring one sex, some age classes of females may be more susceptible than others to shooting. To test for such a difference, females other than kids were divided into two groups: those from herds in which some members were aware of the presence of the shooter before he fired, and those from herds which were undisturbed before shooting commenced. If any age group is particularly wary its members should occur more often in the "disturbed" category than is the case for other age groups. But a χ^2 test ($\chi^2 = 7.28$, df $= 9$, $P = 0.6$) revealed no significant difference between the age structures of the two categories.

The sample was next divided into those females shot at ranges less than 200 yards and those shot out of this range. If animals in a given age class are more easily stalked than the others, they will tend to be shot at closer ranges. Alternatively, animals which present small targets may be underrepresented in the sample of those shot at ranges over 200 yards. This is certainly true of kids, which are difficult to see, let alone to shoot, at ranges in excess of 200 yards. The kids have therefore not been included in the analysis because their underrepresentation in the sample is an acknowledged fact, but for older females there is no difference between the age structures of the two groups divided by range which is not explainable as sampling variation ($\chi^2 = 9.68$, df $= 9$, $P = 0.4$). This is not to imply that no bias exists—the yearling class for instance could well be underrepresented beyond 200 yards—but that

no bias could be detected from a sample of this size.

The taking of a completely representative sample from a natural population of mammals is probably a practical impossibility, and I make no claim that this sample of thar is free of bias, but as bias cannot be detected from the data, I assume it is slight.

Construction of the life table

The shooting yielded 623 females 1 yr old or older, aged by growth rings on the horns. As the sampling period spanned the season of births, a frequency for age 0 cannot be calculated directly from the number of kids shot because early in the period the majority had not been born. In any case, the percentage of kids in the sample is biased.

The numbers of females at each age are shown in Table II, column 2. Although the ages are given only to integral years each class contains animals between ages x yr $-$ ½ month and x yr $+$ 2½ months. Variance owing to the spread of the kidding season is not included in this range, but the season has a standard deviation of only 15 days (Caughley 1965).

Up to an age of 12 yr (beyond this age the values dropped below 5 and were not treated) the frequencies were smoothed according to the formula

$$\log y = 1.9673 + 0.0246x - 0.01036 \, x^2,$$

where y is the frequency and x the age. The linear and quadratic terms significantly reduced

TABLE II. Life table and fecundity table for the thar *Hemitragus jemlahicus* (females only)

1 Age in years x	2 Frequency in sample	3 Adjusted frequency	4 No. female live births per female at age x m_x	5 $1,000 \, l_x$	6 $1,000 \, d_x$	7 $1,000 \, q_x$
0	—	205[a]	0.000	1,000	533	533
1	94	95.83	0.005	467	6	3
2	97	94.43	0.135	461	28	61
3	107	88.69	0.440	433	46	106
4	68	79.41	0.420	387	56	145
5	70	67.81	0.465	331	62	187
6	47	55.20	0.425	269	60	223
7	37	42.85	0.460	209	54	258
8	35	31.71	0.485	155	46	297
9	24	22.37	0.500	109	36	330
10	16	15.04	0.500	73	26	356
11	11	9.64	}0.470	47	18	382
12	6	5.90		29		
13	3		}0.350			
14	4					
15	3					
16	0					
17	1					

[a]Calculated from adjusted frequencies of females other than kids (column 3) and m_x values (column 4).

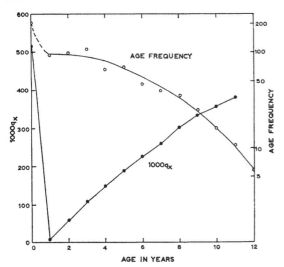

FIG. 1. Age frequencies, plotted on a logarithmic scale, of a sample of female thar, with a curve fitted to the values from ages 1 to 12 yr, and the mortality rate per 1,000 for each age interval of 1 yr $(1,000q_x)$ plotted against the start of the interval.

variance around the regression, but reduction by the addition of a cubic term was not significant at the 0.05 level. There are biological reasons for suspecting that the cubic term would have given a significant reduction of variance had the sample been larger, but for the purposes of this study its inclusion in the equation would add very little. The improved fit brought about by the quadratic term indicates that the rate of mortality increases with age. Whether the rate of this rate also increases, is left open. The computed curve closely fitted the observed data (Fig. 1) and should greatly reduce the noise resulting from sampling variation, the differential effect on mortality of different seasons, and the minor heterogeneities which, although not detectable, are almost certain to be present. The equation is used to give adjusted frequencies in Table II, column 3.

The frequency of births can now be estimated from the observed mean number of female kids produced per female at each age. These are shown in column 4. They were calculated as the number of females at each age either carrying a foetus or lactating, divided by the number of females of that age which were shot. These values were then halved because the sex ratio of late foetuses and kids did not differ significantly from 1:1 (93 ♂ ♂ : 97 ♀ ♀). The method is open to a number of objections: it assumes that all kids were born alive, that all females neither pregnant nor lactating were barren for that season, and that twinning did not occur. The first assumption, if false, would give rise to a positive bias, and the second

and third to a negative bias. However, the ratio of females older than 2 yr that were either pregnant or lactating to those neither pregnant nor lactating did not differ significantly between the periods November to December and January to February $(\chi^2 = 0.79, P = 0.4)$, suggesting that still births and mortality immediately after birth were not common enough to bias the calculation seriously. Errors are unlikely to be introduced by temporarily barren females suckling yearlings, because no female shot in November that was either barren (as judged by the state of the uterus) or pregnant was lactating. Errors resulting from the production of twins will be very small; we found no evidence of twinning in this area.

The products of each pair of values in columns 3 and 4 (Table II) were summed to give an estimate of the potential number of female kids produced by the females in the sample. This value of 205 is entered at the head of column 3. The adjusted age frequencies in column 3 were each multiplied by 4.878 to give the $1,000l_x$ survivorship values in column 5. The mortality series (column 6) and mortality-rate series (column 7) were calculated from these.

Conclusions

Figure 1 shows the mortality rate of females in this thar population up to an age of 12 yr. Had the sample been larger the graph could have been extended to an age of 17 yr or more, but this would have little practical value for the calculation of population statistics because less than 3% of females in the population were older than 12 yr.

The pattern of mortality with age can be divided into two parts—a juvenile phase characterized by a high rate of mortality, followed by a postjuvenile phase in which the rate of mortality is initially low but rises at an approximately constant rate with age.

Table II gives both the l_x and m_x series, and these two sets of values provide most of the information needed to describe the dynamics of the population. Assuming that these two series are accurate, the following statistics can be derived: generation length (i.e. mean lapse of time between a female's date of birth and the mean date of birth of her offspring), T:

$$T = \frac{\Sigma l_x m_x x}{\Sigma l_x m_x} = 5.4 \text{ yr};$$

mean rate of mortality for all age groups, $\bar{q}_x$:

$$\bar{q}_x = 1/\Sigma l_x = 0.25 \text{ per female per annum};$$

life expectancy at birth, e_0:

$$e_0 = \Sigma l_x - \tfrac{1}{2} = 3.5 \text{ yr}.$$

The last two statistics can also be expressed conveniently in terms of the mortality series by

$$\bar{q}_x = 1/\Sigma\,(x+1)\,d_x$$

and

$$e_0 = \frac{\Sigma\,(2x+1)\,d_x}{2}.$$

The relationship of the two is given by

$$\bar{q}_x = 2/(2e_0+1).$$

LIFE TABLES FOR OTHER MAMMALS

The difficulty of comparing the mortality patterns of animals that differ greatly in life span can be readily appreciated. To solve this problem, Deevey (1947) proposed the percentage deviation from mean length of life as an appropriate scale, thereby allowing direct comparison of the life tables of, say, a mammal and an invertebrate. For such comparisons this scale is obviously useful, but for mammals where the greatest difference in mortality rates may be at the juvenile stage the scale often obscures similarities.

By way of illustration, Figure 2 shows $1,000q_x$ curves for two model populations which differ only in the mortality rate of the first age class. When the values are graphed on a scale of percentage deviation from mean length of life the close similarity of the two sets of data is no longer apparent. Thus the use of Deevey's scale for

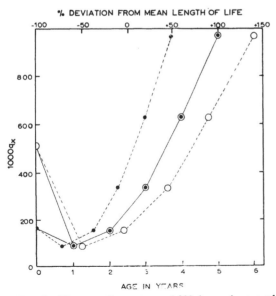

FIG. 2. The mortality rate per 1,000 for each year of life for two model populations that differ only in the degree of first-year mortality. These $1,000q_x$ values are each graphed on two time scales: absolute age in years (continuous lines) and percentage deviation from mean life expectancy (broken lines).

comparing mortality patterns in mammals might result in a loss rather than a gain of information. In this paper, absolute age has been retained as a scale in comparing life tables of different species, although this scale has its own limitations.

Domestic sheep, Ovis aries.—Between 1954 and 1959, Hickey (1960) recorded the ages at death of 83,113 females on selected farms in the North Island of New Zealand. He constructed a q_x table from age 1½ yr by "dividing the number of deaths which have occurred in each year of age by the number 'exposed to risk' [of death] at the same age." An age interval of 1 yr was chosen and the age series 1½, 2½, 3½ etc. was used in preference to integral ages.

The q_x series conformed very closely to the regression: log $q_x = 0.156x + 0.24$, enabling him in a subsequent paper (Hickey 1963) to present the interpolated q_x values at integral ages. He also calculated q for the first year of life from a knowledge of the number of lambs dying before 1 yr of age out of 85,309 (sexes pooled) born alive.

These data probably provide the most accurate life table for any mammal. The $1,000q_x$ curve is graphed in Figure 3.

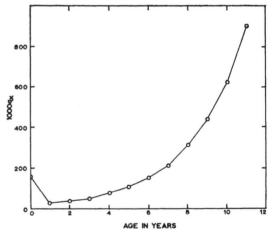

FIG. 3. Domestic sheep: mortality rate per 1,000 for each age interval of 1 yr ($1,000q_v$), plotted against the start of the interval. Data from Hickey (1963).

Dall sheep, Ovis dalli.—During his study on the wolves of Mount McKinley National Park, Murie (1944) aged carcasses of dall sheep he found dead, their ages at death being established from the growth rings on the horns. This sample can be divided into those that died before 1937 and those that died between 1937 and 1941. The former sample was used by Deevey (1947) to construct the life table presented in his classic paper on mortality in natural populations. Kurtén (1953) constructed a life table from the same data, but corrected the underrepresentation of first-year animals resulting from the relatively greater perishability of their skulls by assuming that adult females produce 1 lamb per annum from about their second birthday. Taber and Dasmann (1957) constructed life tables for both males and females from the sample of animals dying between 1937 and 1941, and adjusted both the 0 to 1- and 1 to 2-year age frequencies on the assumption that a female produces her first lamb at about her third birthday and another lamb each year thereafter, that the sex ratio at birth is unity and that the loss of yearlings is not more than 10%.

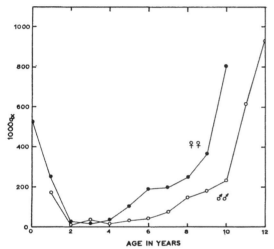

FIG. 4. Dall sheep: mortality rate per 1,000 for each age interval of 1 yr ($1,000q_x$), plotted against the start of the interval. Data from Murie (1944).

Figure 4 shows a version of this table constructed from the pre-1937 sample. The mortality of the first year class has been adjusted by assuming that the sex ratio at birth is unity, that 50% of females produce their first kids at their second birthday and that thereafter 90% produce kids each year. The figure of 50% fecundity at age 2 is borrowed from Woodgerd's (1964) study on the closely related *Ovis canadensis,* and the subsequent 90% fecundity is based on Murie's (1944) statement that twins are extremely rare. To allow for temporarily or permanently barren animals, 10% is subtracted from the potential fecundity.

This life table must be taken as an approximation. As Deevey (1947) has pointed out, the pre-1937 and 1937–41 samples differ significantly in age structure. The obvious conclusion is that the mortality rate by age was changing before and during the period of study. Consequently the age structure of the sample is likely to be only an approximation of the kd_x series. Furthermore,

the q_x values for age 1 yr are likely to have been biased by differential perishability of skulls, but no arbitrary adjustment has been made.

Man.—Most of the life tables available for man show that males have a higher rate of mortality than females. However, Macdonell's (1913) tables for ancient Rome, Hispania and Lusitania suggest that this might not always have been so and that in some circumstance the reverse can be true.

A $1,000q_x$ curve for Caucasian males and females in the United States between 1939 and 1941 is shown in Figure 5. The values are taken from Dublin, Lotka, and Spiegelman (1949).

Rat, Rattus norvegicus.—Wiesner and Sheard (1935) gave the ages at death of 1,456 females of the albino rat (Wistar strain) in a laboratory population. Their table begins at an age of 31 days, but Leslie et al. (1952) calculate from Wiesner and Sheard's data that the probability of dying between birth and 31 days was 0.316. Figure 6 gives a q_x curve constructed from these data.

Short-tailed vole, Microtus agrestis.—The ages at death of 85 males and 34 females were reported by Leslie and Ranson (1940) from a laboratory population of voles kept at the Bureau of Animal Population, Oxford. Frequencies for both sexes were pooled and the data were smoothed by the formula $f_x = f_0 e^{-bx^2}$ where f is the frequency of animals alive age x, and b is a constant. The computed curve closely fitted the data ($P = 0.5$ to 0.7). Figure 6 shows the $1,000q_x$ curve derived from the authors' sixth table.

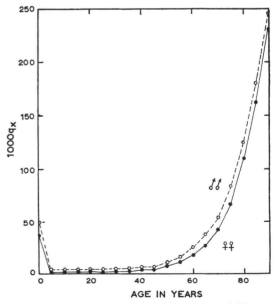

FIG. 5. Man in U.S.: mortality rate per 1,000 per year of age ($1,000q_x$). Data from Dublin et al. (1949).

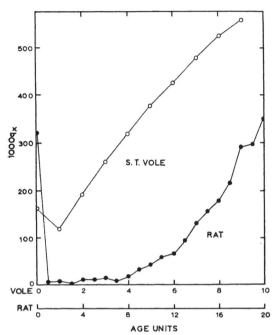

FIG. 6. Short-tailed voles and rats: mortality rate per 1,000 for each age interval $(1,000q_x)$, plotted against the start of the interval. Age interval is 56 days for voles and 50 days for rats. Rat data from Wiesner and Sheard (1935); vole data from Leslie and Ranson (1940).

The pooling of mortality data from both sexes is strictly valid only when the two q_x series are not significantly different. Studies on differential mortality between sexes are few, but those available for man (Dublin et al. 1949, and other authors), dall sheep (Taber and Dasmann 1957, and this paper), the pocket gopher (Howard and Childs 1959) and Orkney vole (Leslie et al. 1955) suggest that although mortality rates certainly differ between sexes, the trends of these age-specific rates tend to be parallel. Consequently, this life table for voles, although based on presumably heterogeneous data, is probably quite adequate for revealing the gross pattern of mortality with age.

Orkney vole, Microtus orcadensis.—Leslie et al. (1955) gave a life table for both males and females in captivity from a base age of 9 weeks. In addition they gave the probability at birth of surviving to ages 3, 6, and 9 weeks, but did not differentiate sexes over this period. The q_x curve given here (Fig. 7) was constructed by calculating survivorship series for both males and females from these data, drawing trend lines through the points, and interpolating values at intervals of 8 weeks.

Proposed life tables not accepted

In the Discussion section of this paper the life tables discussed previously are examined in an

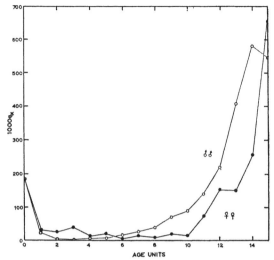

FIG. 7. Orkney vole: mortality rate per 1,000 for each age interval of 56 days $(1,000q_x)$, plotted against the start of the interval. Data from Leslie et al. (1955).

attempt to generalize their form. Only a small proportion of published life tables are dealt with, and any generalization from these could be interpreted as an artefact resulting from selection of evidence.

To provide the reader with the information necessary for reaching an independent conclusion, the published life tables not selected for comparison are listed below with the reason for their rejection. Only those including all juvenile age classes are cited. These tables are rejected only for present purposes because comparison of mortality patterns between species demands a fairly high level of accuracy for individual tables. The inclusion of a table in this section does not necessarily imply that it is completely inaccurate and of no practical value.

Tables based on inadequate data (i.e. less than 50 ages at death or 150 ages of living animals): *Odocoileus hemionus* (Taber and Dasmann 1957), *Ovis canadensis* (Woodgerd 1964);

Probable sampling bias: *Lepus americanus* (Green and Evans 1940), *Rupicapra rupicapra* (Kurtén 1953), fossil accumulations (Kurtén 1953, 1958; Van Valen 1964), *Balaenoptera physalus* (Laws 1962);

Age structure analyzed as a kd_x series: *Sylvilagus floridanus* (Lord 1961), *Odocoileus virginianus* and *Capreolus capreolus* (Quick 1963);

Death and emigration confounded: *Peromyscus maniculatus* (Howard 1949), *Capreolus capreolus* (Taber and Dasmann 1957, Quick 1963);

Sample taken between breeding seasons: *Odocoileus virginianus* (Quick 1963);

Form of life table, or significant portion of it, based largely on assumption: *Callorhinus ursinus* (Kenyon and Scheffer 1954), *Myotis mystacinus* (Sluiter, van Heerdt, and Bezem 1956), *Cervus elaphus* (Taber and Dasmann 1957), *Rhinolophus hipposideros, Myotis emarginatus,* and *Myotis daubentonii* (Bezem, Sluiter, and van Heerdt 1960), *Halichoerus grypus* (Hewer 1963, 1964);

Sample from a nonstationary population: *Sylvilagus floridanus* (Lord 1961);

Inadequate aging: *Gorgon taurinus* (Talbot and Talbot 1963);

Confounding of l_x and d_x data: *Rangifer arcticus* (Banfield 1955).

DISCUSSION

The most striking feature of the q_x curves of species accepted for comparison is their similarity. Each curve can be divided into two components: a juvenile phase where the rate of mortality is initially high but rapidly decreases, followed by a postjuvenile phase characterized by an initially low but steadily increasing rate of mortality. The seven species compared in this paper all produced q_x curves of this "U" or fish-hook shape, suggesting that most mammals share a relationship of this form between mortality rate and age. This conclusion, if false, can be invalidated by a few more life tables from other species. It can be tested most critically by reexamining some of the species for which life tables, although published, were not accepted in this paper. Those most suitable are species that can be adequately sampled, and accurately aged by growth rings on the horns or growth layers in the teeth (chamois, Rocky Mountain sheep, and several species of deer), or those small mammals that can be marked at birth and subsequently recaptured.

High juvenile mortality, characterizing the first phase of the q_x curve, has been reported also for several mammals for which complete life tables have not yet been calculated (e.g. for *Oryctolagus cuniculus* (Tyndale-Biscoe and Williams 1955, Stodart and Myers 1964), *Gorgon taurinus* (Talbot and Talbot 1963), *Cervus elaphus* (Riney 1956) and *Oreamnos americanus* (Brandborg 1955). Kurtén (1953, p. 88) generalized this phenomenon by stating that "the initial dip [in the survivorship curve] is a constitutional character in sexually reproducing forms at least . . .". This phase of mortality is highly variable in degree but not in form. Taber and Dasmann (1957) and Bourlière (1959) have emphasized the danger of

considering a life table of a population in given circumstances as a typical of all populations of that species. Different conditions of life tend to affect life tables, and the greatest differences between populations of a species are likely to be found at the juvenile stage. For example, the rate of juvenile mortality in red deer (Riney 1956) and in man differ greatly between populations of the same species.

The second phase—the increase in the rate of mortality throughout life—is common also to the seven species compared in this paper. However, although the increase itself is common to them, the pattern of this increase is not. Mortality rates have a logarithmic relationship to age in domestic sheep and to a less marked extent in the rat, the Orkney vole, and the dall sheep, whereas the relationship for the thar and the short-tailed vole appears to be approximately arithmetic. However, this difference may prove to be only an artefact resulting from the smoothing carried out on the data from these two species.

Despite these differences, the characteristics common to the various q_x curves dominate any comparison made between them. The similarities are all the more striking when measured against the ecological and taxonomic differences between species. Taxonomically, the seven species represent three separate orders (Primates, Rodentia, and Artiodactyla), and ecologically they comprise laboratory populations (rats and voles), natural populations (thar, dall sheep and man) and an artificial population (domestic sheep). The agents of mortality which acted on these populations must have been quite diverse. Murie (1944) reported that most of the dall sheep in the sample had been killed by wolves; most mortality in the thar population is considered to result from starvation and exposure in the winter; mortality of domestic sheep seems to be largely a result of disease, physiological degeneration, and possibly iodine deficiency in the lambs (Hickey 1963); whereas the deaths in the laboratory populations of voles and rats may be due to inadequate parental care and cannibalism of the juveniles, and perhaps disease and physiological degeneration in the adults. These differences suggest that the q_x curve of a population may assume the same form under the influence of various mortality agents, even though the absolute rate of mortality of a given age class is not the same in all circumstances. This hypothesis is worth testing because it implies that the susceptibility to mortality of an age class, relative to that of other age classes, is not strongly specific to any particular agent of mortality. A critical test would be to compare the life tables of two stationary populations of the same species, where only one population is subjected to predation.

Although no attempt is made here to explain the observed mortality pattern in terms of evolutionary processes, an investigation of this sort could be informative. A promising line of attack, for instance, would be an investigation of what appears to be a high inverse correlation between the mortality rate at a given age and the contribution of an animal of this age to the gene pool of the next generation. Fisher (1930) gives a formula for the latter statistic.

Bodenheimer (1958) divided expectation of life into "physiological longevity" ("that life duration which a healthy individual may expect to live under optimum environment conditions until dying of senescence") and "ecological longevity" (the duration of life under natural conditions). This study suggests that such a division is inexpedient because no clear distinction can be made between the effect on mortality rates of physiological degeneration and of ecological influences.

It is customary to classify life tables according to the three hypothetical patterns of mortality given by Pearl and Miner (1935). These patterns can be characterized as: 1) a constant rate of mortality throughout life, 2) low mortality throughout most of the life span, the rate rising abruptly at old age, and 3) initial high mortality followed by a low rate of mortality. Pearl (1940) emphasizes that the three patterns are conceptual models having no necessary empirical reality, but a few subsequent writers have treated them as laws which all populations must obey. None of these models fit the mortality patterns of the seven species discussed in this paper although Pearl's (1940) later modification of the system provides two additional models (high–low–high mortality rate and low–high–low mortality rate), the first of which is an adequate approximation to these data. For mammals at least, the simple three-fold classification of mortality patterns is both confusing and misleading. The five-fold classification allows greater scope; but do we yet know enough about mortality patterns in mammals to justify the construction of any system of classification?

ACKNOWLEDGMENTS

This paper has greatly benefited from criticism of previous drafts by M. A. Bateman, CSIRO; P. H. Leslie, Bureau of Animal Population; M. Marsh, School of Biological Sciences, University of Sydney; J. Monro, Joint FAO/IAEA Div. of Atomic Energy; G. R. Williams, Lincoln College, and B. Stonehouse, Canterbury University, New Zealand; and B. B. Jones and W. G. Warren of this Institute. The equation for smoothing age frequencies of thar was kindly calculated by W. G. Warren. For assisting in the shooting and autopsy of specimens, I am grateful to Chris Challies, Gary Chisholm, Ian Hamilton, Ian Rogers and Bill Risk.

LITERATURE CITED

Anderson, J. A., and J. B. Henderson. 1961. Himalayan thar in New Zealand. New Zeal. Deerstalkers' Ass. Spec. Publ. 2.

Banfield, A. W. F. 1955. A provisional life table for the barren ground caribou. Can. J. Zool. **33**: 143-147.

Bezem, J. J., J. W. Sluiter, and P. F. van Heerdt. 1960. Population statistics of five species of the bat genus *Myotis* and one of the genus *Rhinolophus*, hibernating in the caves of S. Limburg. Arch. Néerlandaises Zool. **13**: 512-539.

Bodenheimer, F. S. 1958. Animal ecology today. Monogr. Biol. 6.

Bourlière, F. 1959. Lifespans of mammalian and bird populations in nature, p. 90-102. *In* G. E. W. Wolstenholme and M. O'Connor [ed.] The lifespan of animals. C.I.B.A. Colloquia on Ageing **5**.

Brandborg, S. M. 1955. Life history and management of the mountain goat in Idaho. Idaho Dep. Fish and Game, Wildl. Bull. 2.

Breakey, D. R. 1963. The breeding season and age structure of feral house mouse populations near San Francisco Bay, California. J. Mammal. **44**: 153-168.

Caughley, G. 1963. Dispersal rates of several ungulates introduced into New Zealand. Nature **200**: 280-281.

——. 1965. Horn rings and tooth eruption as criteria of age in the Himalayan thar *Hemitragus jemlahicus*. New Zeal. J. Sci. **8**: 333-351.

Deevey, E. S. Jr. 1947. Life tables for natural populations of animals. Quart. Rev. Biol. **22**: 283-314.

Donne, T. E. 1924. The game animals of New Zealand. John Murray, London.

Dublin, L. I., A. J. Lotka, and M. Spiegelman. 1949. Length of life. Ronald Press, New York.

Fisher, R. A. 1930. The genetical theory of natural selection. Clarendon Press, Oxford.

Green, R. G., and C. A. Evans. 1940. Studies on a population cycle of snowshoe hares on the Lake Alexander area. II. Mortality according to age groups and seasons. J. Wildl. Mgmt. **4**: 267-278.

Hewer, H. R. 1963. Provisional grey seal life table, p. 27-28. *In* Grey seals and fisheries. Report of the consultative committee on grey seals and fisheries. H. M. Stationary Office, London.

——. 1964. The determination of age, sexual maturity, longevity and a life table in the grey seal (*Halichoerus grypus*). Proc. Zool. Soc. Lond. **142**: 593-623.

Howard, W. E. 1949. Dispersal, amount of inbreeding, and longevity in a local population of prairie deermice on the George Reserve, Michigan. Contrib. Lab. Vertebrate Biol. Univ. Michigan, 43.

Howard, W. E., and H. E. Childs Jr. 1959. The ecology of pocket gophers with emphasis on *Thomomys bottae mewa*. Hilgardia **29**: 277-358.

Hickey, F. 1960. Death and reproductive rate of sheep in relation to flock culling and selection. New Zeal. J. Agric. Res. **3**: 332-344.

——. 1963. Sheep mortality in New Zealand. New Zeal. Agriculturalist **15**: 1-3.

Hughes, R. D. 1965. On the composition of a small sample of individuals from a population of the banded hare wallaby, *Lagostrophus fasciatus* (Peron & Lesueur). Austral. J. Zool. **13**: 75-95.

Kenyon, K. W. and V. B. Scheffer. 1954. A population study of the Alaska fur-seal herd. United States Dep. of the Interior, Special Scientific Report—Wildlife No. **12**: 1-77.

Kurtén, B. 1953. On the variation and population dynamics of fossil and recent mammal populations. Acta Zool. Fennica **76**: 1-122.

———. 1958. Life and death of the Pleistocene cave bear: a study in paleoecology. Acta Zool. Fennica **95**: 1-59.

Laws, R. M. 1962. Some effects of whaling on the southern stocks of baleen whales, p. 137-158. *In* E. D. Le Cren and M. W. Holdgate [ed.] The exploitation of natural animal populations. Brit. Ecol. Soc. Symp. 2. Blackwell, Oxford.

Leslie, P. H., and R. M. Ranson. 1940. The mortality, fertility and rate of natural increase of the vole (*Microtus agrestis*) as observed in the laboratory. J. Animal Ecol. **9**: 27-52.

Leslie, P. H., U. M. Venables, and L. S. V. Venables. 1952. The fertility and population structure of the brown rat (*Rattus norvegicus*) in corn-ricks and some other habitats. Proc. Zool. Soc. Lond. **122**: 187-238.

Leslie, P. H., T. S. Tener, M. Vizoso and H. Chitty. 1955. The longevity and fertility of the Orkney vole, *Microtus orcadensis*, as observed in the laboratory. Proc. Zool. Soc. Lond. **125**: 115-125.

Lord, R. D. 1961. Mortality rates of cottontail rabbits. J. Wildl. Mgmt. **25**: 33-40.

Lotka, A. J. 1907a. Relationship between birth rates and death rates. Science **26**: 21-22.

———. 1907b. Studies on the mode of growth of material aggregates. Amer. J. Sci., 4th series, **24**: 199-216.

Macdonell, W. R. 1913. On the expectation of life in ancient Rome, and in the provinces of Hispania and Lusitania, and Africa. Biometrika **9**: 366-380.

Murie, A. 1944. The wolves of Mount McKinley. Fauna Nat. Parks U.S., Fauna Ser. 5.

Pearl, R. 1940. Introduction to medical biometry and statistics. 3rd ed. Philadelphia: Saunders.

Pearl, R., and J. R. Miner. 1935. Experimental studies on the duration of life. XIV. The comparative mortality of certain lower organisms. Quart. Rev. Biol. **10**: 60-79.

Quick, H. F. 1963. Animal population analysis, p. 190-228. *In* Wildlife investigational techniques (2nd ed). Wildlife Soc., Ann Arbor.

Riney, T. 1956. Differences in proportion of fawns to hinds in red deer (*Cervus elaphus*) from several New Zealand environments. Nature **177**: 488-489.

Sharpe, F. R., and A. J. Lotka. 1911. A problem in age-distribution. Phil. Mag. **21**: 435-438.

Slobodkin, L. B. 1962. Growth and regulation of animal populations. Holt, Rinehart and Winston, New York.

Sluiter, J. W., P. F. van Heerdt, and J. J. Bezem. 1956. Population statistics of the bat *Myotis mystacinus*, based on the marking-recapture method. Arch. Néerlandaises Zool. **12**: 63-88.

Stodart, E., and K. Myers. 1964. A comparison of behaviour, reproduction, and mortality of wild and domestic rabbits in confined populations. CSIRO Wildl. Res. **9**: 144-59.

Taber, R. D., and R. F. Dasmann. 1957. The dynamics of three natural populations of the deer *Odocoileus hemionus columbianus*. Ecology **38**: 233-246.

Talbot, L. M., and M. H. Talbot. 1963. The wildebeest in western Masailand, East Africa. Wildl. Monogr. 12.

Tyndale-Biscoe, C. H., and R. M. Williams. 1955. A study of natural mortality in a wild population of the rabbit *Oryctolagus cuniculus* (L.). New Zeal. J. Sci. Tech. B **36**: 561-580.

Van Valen, L. 1964. Age in two fossil horse populations. Acta Zool. **45**: 93-106.

Wiesner, B. P., and N. M. Sheard. 1935. The duration of life in an albino rat population. Proc. Roy. Soc. Edinb. **55**: 1-22.

Woodgerd, W. 1964. Population dynamics of bighorn sheep on Wildhorse Island. J. Wildl. Mgmt. **28**: 381-391.

Caughley, G., and L. C. Birch. 1971. Rate of increase. *Journal of Wildlife Management* 35:658–663.

A wildlife population's *rate of increase* is its demographic vigor; it is a reflection of a population's mortality and survival patterns, age structure, and fecundity schedule. Caughley and Birch's objective was to clearly outline the different concepts of rate of increase: observed rate of increase, rate of increase at any given density, and maximum rate of increase. The last of these has also been called the true, the real, the incipient, the inherent, and the intrinsic rate of increase and the innate capacity for increase. The authors demonstrate how the rate should be calculated and how it needs to be considered when examining the dynamics of exploited populations. The paper is important because it summarizes the historical use of the term and explains how the term *rate of increase* should be used. This is another example of how the definition and use of concepts essential for wildlife management are inconsistent in the ecological literature. Understanding rate of increase and the limits of the term's use remain key goals in any wildlife management program.

RELATED READING

Birch, L. C. 1948. The intrinsic rate of natural increase of an insect population. Journal of Animal Ecology 17:15–26.

Caughley, G. 1977. Analysis of vertebrate populations. John Wiley & Sons, New York, New York, USA.

Hairston, N. G., D. W. Tinkle, and H. M. Wilbur. 1970. Natural selection and the parameters of population growth. Journal of Wildlife Management 34:681–690.

Leslie, P. H., and R. M. Ransom. 1940. The mortality, fertility and rate of natural increase of the vole (*Microtus agrestis*) as observed in the laboratory. Journal of Animal Ecology 9:27–52.

Lowe, V. P. W. 1969. Population dynamics of the red deer (*Cervus elaphus* L.) on Rhum. Journal of Animal Ecology 38:425–457.

Rogers, A. 1968. Matrix analysis of interregional population growth and distribution. University of California Press, Berkeley, USA.

RATE OF INCREASE

GRAEME CAUGHLEY, School of Biological Sciences, University of Sydney, Australia

L. C. BIRCH, School of Biological Sciences, University of Sydney, Australia

Abstract: We outline the differences between three notions of *rate of increase*: r, the observed rate of increase; r_s, the rate implied by the prevailing schedules of survival and fecundity; and r_m, the maximum rate at which a population with a stable age distribution can increase in a specified environment. Several mammalogists recently calculated r_s for natural populations under the misapprehension that they were calculating r_m. However, their calculated values are usually defective even as r_s estimates because they have used life tables constructed from age distributions or from distributions of age at death. The estimates are thereby infiltrated by the unrecognized but implicit assumption that $r_s = 0$, and the calculated values of r_s are therefore assumptions retrieved as conclusions. If r_m, the *intrinsic rate of increase*, is to be determined for a natural population of mammals, it is best calculated either by measuring the rate at which a newly established population initially increases or by fitting a curve to the growth of a population after its density has been artificially reduced.

Over about the last 10 years mammalogists have become increasingly interested in population dynamics, and they have begun using concepts and analyses hitherto restricted almost entirely to the demography of man and insects. This is a welcome trend that should be continued, but at the same time a warning is necessary against uncritically using special methods of analysis that may not be appropriate to field studies of mammals. By this we do not imply that the demography of mammals differs essentially from that of insects; we suggest only that some parameters, whose estimation is relevant to questions that entomologists ask, may be extremely difficult to estimate for natural populations of mammals. In addition, the parameters may be irrelevant to the problem in hand. The reverse is also true. Mammalogists ask questions on conservation and harvesting, for example, that entomologists seldom consider, and they must therefore estimate parameters that entomologists ignore.

In this paper we examine some of the ways in which rate of increase can be measured and point out that some mammalogists, using equations that are valid and widely used in insect ecology, are estimating it incorrectly.

KINDS OF RATE OF INCREASE

Because populations tend to grow geometrically, rate of increase is best expressed in exponential form. A population of 100 animals that increases to 200 over a year has been multiplied by 2 or increased by 100 percent, however one prefers to express it. Its exponential rate of increase, r, is given by

$$e^r = 2$$
and $$r = 0.69,$$

e being the base of natural logs, taking the value 2.71828. When the population halves instead of doubles, dropping from 100 to 50 over a year, we say either that the population has multiplied by 0.5 or that it has decreased by 50 percent; hence,

$$e^r = 0.5$$
and $$r = -0.69.$$

The observed rate of increase, r, is the slope of $\log_e$ numbers on time. It can be changed from one unit of time to another by simple multiplication or division. When $r = 0.69$ on a yearly basis, $r = 0.69/365$ when expressed as increase per day. In the material that follows, we give all rates of increase in exponential form. The symbol

658

r should not be confused with r used by Hairston et al. (1970). They employ it to symbolize the statistic we call r_m.

A population's observed rate of increase at a given time is determined by age-specific survival, age-specific fecundity, sex ratio, and age distribution. Sometimes we wish to eliminate the variation in age distribution from the estimate of r because the variation reflects previous influences rather than the population's current ability to increase. If we are interested in the slightly abstract concept of a population's potential to increase at a given density in a given environment, rather than in its actual rate of increase, which is influenced by whatever age distribution has been imposed by events that happened previously, we estimate the rate at which it would increase if it had a *stable age distribution*. This is the distribution of ages that would eventually form if the prevailing life table and fecundity table remained unchanged. Rate of increase at any given density, estimated in such a way as to eliminate the effect of an unstable age distribution, will be symbolized by r_s. Depending on whether survival and fecundity acting together would cause an increase or a decrease, r_s is positive, zero, or negative.

A special case of r_s is provided by the rate of increase at very low density, such that the population has the maximum rates of survival and fecundity attainable in that environment. This rate is r_m, the maximum rate at which a population with a stable age distribution can increase when no resource is limiting. Andrewartha and Birch's (1954) definition of r_m stipulates that the rate is measured in the absence of predators. Although accepting the validity of this constraint, we are inclined to exclude it from a definition of r_m for vertebrates because its inclusion greatly increases the already daunting technical difficulties of estimating r_m in the field. Rate r_m is not a constant for a species but is specific to the particular environment in which it was measured. It is determined by the climate and by the quality of food, shelter, and other environmental influences interacting with the animal's genetically determined capacity to survive and reproduce. Rate r_m can be positive, zero, or negative. Within the range of a species it will tend to be positive, zero at the natural boundary of distribution, and negative beyond it. A negative r_m measured for a particular combination of environmental variables implies that if animals were introduced into such an environment the population would not survive.

Many names have been given r_m. It has been called the true, the real, the incipient, the inherent, and the intrinsic rate of increase or of natural increase, the Malthusian parameter, the natural rate of increase, and the innate capacity for increase. In this paper we will use *intrinsic rate of increase* when we mean r_m.

Estimating Rate of Increase

So far, we have informally defined three variants of rate of increase—r, the observed rate of increase; r_s, the rate implied by the life table combined with the fecundity table; and r_m, the special case of r_s, which obtains when all individuals have access to more food, shelter, water, and other requirements than they need.

Rate r is measured by regressing $\log_e$ of population size (or a linear index of population size) on time. The slope of this line estimates r, which will be positive, zero, or negative depending on whether the population was increasing, stable, or declining over the period it was observed.

Rate r_s must be calculated from a life

Table 1. Two methods of constructing a life table: from the age distribution of living animals (A) and from the distribution of ages at death (B), assuming that the population is stable.

(A)	AGE (x)	NUMBER SAMPLED (f_x)	$\dfrac{f_x}{f_o}$ (l_x)[a]	$l_x - l_{x+1}$ (d_x)[b]	$\dfrac{d_x}{l_x}$ (q_x)[c]
	0	100	1.00	0.50	0.50
	1	50	0.50	0.20	0.40
	2	30	0.30	0.20	0.67
	3	10	0.10	0.10	1.00
	4	0			

(B)	AGE (x)	NUMBER FOUND DEAD (f_x')	$\dfrac{f_x'}{\Sigma f_x'}$ (d_x)	$\sum\limits_{x}^{x} d_x$ (l_x)	$\dfrac{d_x}{l_x}$ (q_x)
	0	50	0.50	1.0	0.50
	1	20	0.20	0.5	0.40
	2	20	0.20	0.3	0.67
	3	10	0.10	0.1	1.00
	4	0			

[a] l_x = probability at birth of surviving to age x.
[b] d_x = probability at birth of dying in age interval x, x + 1.
[c] q_x = probability at age x of dying before age x+1.

table and a fecundity table. In practice, we consider only the female segment of the population and assume that the male segment increases at about the same rate. When l_x is the probability at birth of a female surviving to age x and m_x is the mean number of female offspring she produces in the age interval pivoted at age x,

$$\sum l_x e^{-r_s x} m_x = 1 \qquad (1)$$

and r_s can be estimated by inserting trial values into the equation until it balances. When r_s lies between about plus and minus 0.1, an approximate but reasonably accurate direct solution is possible from

$$r_s = \frac{\log_e R_o}{T_c} \qquad (2)$$

in which $\quad R_o = \sum l_x m_x$

and $\qquad T_c = \dfrac{\sum l_x m_x x}{\sum l_x m_x}.$

These solutions of r_s require a life table (l_x) covering all ages of females. Most life tables of mammals are constructed either from the age distribution of a shot sample or the distribution of ages at death of animals found dead (Deevey 1947, Caughley 1966). These calculations are outlined in Table 1. However, in both cases the construction of life tables from these distributions is based on the assumptions that the population has a stable age distribution and that $r_s = 0$. Consequently, such a life table, although providing a valid approximation to the mortality pattern, cannot be used to calculate r_s. The resultant estimate of r_s will be very close to zero because the calculation simply retrieves an assumption and disguises it as a conclusion. Logically, such a solving returns an estimate of the r_s within parentheses in

$$\sum l_x e^{-r_s x}(e^{-r_s x})m_x = 1 \qquad (3)$$

and not the r_s of equation (1). Equation (1) being true by definition, equation (3) is true only when the parenthesized term equals 1 and therefore only when the value of r_s contained within it is equal to 0. Even when age distribution and fecundity rates are independently calculated, the equation returns an estimate of zero as a logical necessity, and this estimate has no relationship whatsoever to the true value of r_s.

Equations (1) and (2) cannot, therefore, be used on an age distribution or on a distribution of ages at death, but several mammalogists have used them on just such data. The equations have been used incorrectly both to estimate r_s (for example, Lowe 1969, Watson 1969, 1970) and in modified form to test the hypothesis that $r_s = 0$ (for example, Kurtén 1953, 1958, Quick 1963, Hughes 1965, Thomas et al. 1968, Macpherson 1969). In addition, we have recently seen four drafts of papers,

Table 2. Age distributions of female red deer on the island of Rhum—data from Lowe (1969).

YEAR	\multicolumn{14}{c}{AGE IN YEARS}
	0	1	2	3	4	5	6	7	8	9	10	11	12	13
1957	154	129	113	113	81	78	59	65	55	25	9	8	7	2
1958	131	129	129	110	107	69	75	51	59	43	12	5	6	3
1959			123	122										
1960				96	98									
1961					75	66								
1962						52	40							
1963							37	25						
1964								23	18					
1965									19	4				

and know of two others, in which mammalogists have used this tautological analysis. Nothing is wrong with the equations. They can be used to estimate r_s for any species, given appropriate data. However, they do not provide an estimate of r_s if a single age distribution or a single distribution of ages at death is used to give the life table for the calculations.

As an example, part of an array presented by Lowe (1969) of age distributions of female red deer (*Cervus elaphus*) on the island of Rhum, Scotland, is given in Table 2. Lowe compiled these age distributions by aging deer found recently dead, thereby obtaining an estimate of date of birth and period of life for each specimen. Because he found almost all deer that died, he was able to reconstruct the number alive in each age-class in each of several years.

Four life tables could be constructed from the frequencies in Table 2: (1) from the 1957 age distribution if the population had a stable age distribution at the time and if r_s equalled zero; (2) from the 1958 age distribution, given the same assumptions; (3) from the yearly decline between 1957 and 1965 of the cohort born in 1957 (the lower diagonal of frequencies); and (4) from the difference between the number in a given age-class in 1957 and the number in the next age-class a year later.

Although four separate life tables could be constructed, they cannot each be combined with the m_x values of a fecundity table to provide four estimates of r_s. Only life tables 3 and 4 can be used in this way, the others being constructed on the assumption that $r_s = 0$.

Lowe (1969) made two estimates of r_s. The first used the age distribution of 1957, the l_x schedule being constructed as the probability of surviving from 1 year of age. The resultant $r_s = 0.047$ is meaningless both because of the $r_s = 0$ assumed in constructing the life table and because Lowe's calculated l_x is not the probability of surviving from birth, as required by the logic of the analysis. It is a probability of surviving from 1 year of age. His second estimate of r_s uses a life table tracing the fortunes of the cohort that was 1 year of age in 1957, the data being the upper diagonal of frequencies in Table 2. The resultant estimate of $r_s = -0.017$ is again meaningless. Although $r_s = 0$ is no longer an implicit assumption, survival is still incorrectly measured from 1 year of age.

In only very special cases is an estimate of r_s needed in the study of a mammalian population. It is a statistic having little relevance to the problems that a field study is usually expected to solve. In most cases, we need only the observed rate of increase,

r, which we calculate by regressing $\log_e$ numbers on time. Estimating r_s in addition is seldom worth the extra labor. For populations of large mammals, r_s will usually be so close to r that any difference will be swamped by sampling variation. Rate r is useful and relatively easy to calculate; r_s is less likely to be relevant and is more difficult to calculate.

Why, then, have several mammalogists recently tried to estimate r_s? It is apparent from their papers or drafts that in each case they confused r_s with r_m. Now r_m is really worth calculating, particularly if the study is aimed at estimating a sustained-yield harvest. It is a statistic needed to estimate how fast a population will build up after being reduced to any particular density. This induced rate of increase is the cropping rate most likely to hold the population stable at the reduced density. However, r_m cannot be estimated from survival and fecundity schedules unless the population is at minimal density.

We suspect that the confusion has arisen because mammalogists, in extrapolating to mammals the equations used by entomologists, have failed to appreciate that their data are of a different kind from those gathered from insects. Entomologists measure r_m by determining life tables and fecundity tables of insects held at low density in bottles or cages (Birch 1948). They substitute this information into equations (1) or (2) to estimate r_m for a particular combination of temperature and humidity. Leslie and Ranson (1940) estimated a value of r_m for the vole (*Microtus agrestis*) in this manner, but obviously the method cannot be adapted to the study of naturally occurring populations of mammals. In these circumstances, r_m is best approximated by measuring the rate of increase of a newly established population, at which time $r \approx r_s \approx r_m$ in the initial stages.

This opportunity seldom arises, and we are usually forced to calculate r_m indirectly by deducing the rate of increase at minimal density from rate of increase at a much higher density. If the population's pattern of growth is known, a curve of this form can be fitted to successive estimates of the logged size of the growing population, and the curve can be extrapolated backward to minimal population size. The initial slope of this curve is an estimate of r_m.

In fact, the characteristic pattern of growth is seldom known and can only be guessed. Of the multiplicity of curves that might be applicable, we have chosen the logistic curve to demonstrate the general method. Although it is unlikely to provide an exact fit, since the assumptions on which it is based are biologically improbable, it may mimic the growth of some populations closely enough to yield a realistic estimate of r_m. The logistic is a flexible curve; when r_m is below about 0.3 it often provides a good empirical fit to the growth of a population whose dynamics are entirely different from those implied by the derivation of the logistic equation. It can therefore be justified as a first approximation on pragmatic grounds, as long as a tolerable fit is not misinterpreted to indicate that population processes and logistic assumptions are congruent.

Suppose a population that has been fluctuating around a mean of 1,000 animals for some time is suddenly reduced artificially to 800 and that by the following year it has increased to 850 and in the year after that to 890. The intrinsic rate of increase, that rate at which the population would increase if it were reduced to very low density, can be calculated from these totals if we are willing to assume that growth

Table 3. Calculating the intrinsic rate of increase by fitting a logistic curve.

YEAR (t)	NUMBERS (N)	$\frac{K-N}{N}$	$\log_e \frac{K-N}{N}$ (y)
0	800	0.250	−1.39
1	850	0.176	−1.74
2	890	0.123	−2.10

$$-r_m = \frac{\Sigma ty - (\Sigma t)(\Sigma y)/n}{\Sigma t^2 - (\Sigma t)^2/n}$$

$K = 1000$ $n = 3$
$\Sigma t = 3$ $\Sigma y = -5.23$
$\Sigma t^2 = 5$ $\Sigma ty = -5.94$
$\Sigma t^2 - (\Sigma t)^2/n = 2$ $\Sigma ty - (\Sigma t)(\Sigma y)/n = -0.710$
$-r_m = \text{slope} = -0.71/2 = -0.355$; $r_m = 0.355$

is approximately logistic and that the original population size was asymptotic. Symbolizing the prereduction (asymptotic) population size as K and the subsequent totals as N,

$$\log_e \frac{K-N}{N} = a - r_m t$$

and r_m can be estimated by linear regression of the left side of the equation ($\log_e (K-N)/N$) on time t. The slope of this regression estimates $-r_m$. The calculation is summarized in Table 3.

The logistic curve is unlikely to describe the growth of a population, even in an approximate way, if environmental conditions fluctuate markedly between years, or if the population responds slowly to a change in the availability of a resource, or if an artificial reduction in density disrupts the organization of social groups within the population. Estimation of r_m from the logistic curve is at best a stopgap procedure that should be abandoned as soon as a population's pattern of growth is established empirically.

LITERATURE CITED

ANDREWARTHA, H. G., AND L. C. BIRCH. 1954. The distribution and abundance of animals. The University of Chicago Press, Chicago. 782pp.

BIRCH, L. C. 1948. The intrinsic rate of natural increase of an insect population. J. Animal Ecol. 17(1):15–26.

CAUGHLEY, G. 1966. Mortality patterns in mammals. Ecology 47(6):906–918.

DEEVEY, E. S., JR. 1947. Life tables for natural populations of animals. Quart. Rev. Biol. 22(4):283–314.

HAIRSTON, N. G., D. W. TINKLE, AND H. M. WILBUR. 1970. Natural selection and the parameters of population growth. J. Wildl. Mgmt. 34(4):681–690.

HUGHES, R. D. 1965. On the age composition of a small sample of individuals from a population of the banded hare wallaby, *Lagostrophus fasciatus* (Peron & Lesueur). Australian J. Zool. 13(1):75–95.

KURTÉN, B. 1953. On the variation and population dynamics of fossil and recent mammal populations. Acta Zoologica Fennica 76. 122pp.

———. 1958. Life and death of the pleistocene cave bear: a study in paleoecology. Acta Zoologica Fennica 95. 59pp.

LESLIE, P. H., AND R. M. RANSON. 1940. The mortality, fertility and rate of natural increase of the vole (*Microtus agrestis*) as observed in the laboratory. J. Animal Ecol. 9(1):27–52.

LOWE, V. P. W. 1969. Population dynamics of the red deer (*Cervus elaphus* L.) on Rhum. J. Animal Ecol. 38(2):425–457.

MACPHERSON, A. H. 1969. The dynamics of Canadian arctic fox populations. Canadian Wildl. Serv. Rept. Ser. No. 8. 52pp. (French and Russian summaries.)

QUICK, H. F. 1963. Animal population analysis. Pages 190–228. In H. S. Mosby [Editor], Wildlife investigational techniques. 2nd ed. revised. The Wildlife Society, Washington, D.C. 419pp.

THOMAS, D. C., G. R. PARKER, J. P. KELSALL, AND A. G. LOUGHREY. 1968. Population estimates of barren-ground caribou on the Canadian mainland from 1955 to 1967. Canadian Wildl. Serv. Progr. Notes 3. 4pp.

WATSON, R. M. 1969. Reproduction of wildebeest, *Connochaetes taurinus albojubatus* Thomas, in the Serengeti region, and its significance to conservation. J. Reprod. and Fert. Suppl. 6:287–310.

———. 1970. Generation time and intrinsic rates of natural increase in wildebeeste (*Connochaetes taurinus albojubatus* Thomas). J. Reprod. and Fert. 22(3):557–561.

Received for publication February 8, 1971.

Caughley, G. 1974. Bias in aerial survey. *Journal of Wildlife Management* 38:921–933.

Obtaining accurate and precise census estimates has been a major objective of wildlife biologists and agencies. As technology continues to advance, so will our ability to identify and enumerate wildlife populations. Using aircraft to census wildlife is one of those advances and has enhanced our ability to estimate populations of large mammals and waterfowl. Aerial surveys, however, are expensive and can often yield data of questionable use (e.g., with reduced accuracy and precision) because of factors like width of transect, flight speed, altitude, and observers. In summary, not all animals present are seen, resulting in underestimates of population size.

Aerial census of wildlife is actually expanding with the availability of advanced cameras that detect infrared, movements, and other characteristics. Thus, it is essential that aerial censuses are conducted soundly and with minimal bias. Caughley perceived this need, and his objective was to examine the schools of thought related to aerial population estimates. He discusses the importance of recognizing bias in aerial surveys and suggests ways to minimize bias. His work is influential to wildlife management because it has led to other research (e.g., sightability models) that continues to refine aerial surveys. Caughley's methods remain applicable today and will likely remain valuable as even more advanced methodologies of aerially detecting wildlife are developed.

RELATED READING

Anderson, C. R., Jr., D. S. Moody, B. L. Smith, F. G. Lindzey, and R. P. Lanka. 1998. Development and evaluation of sightability models for summer elk surveys. Journal of Wildlife Management 62:1055–1066.

Bartmann, R. M., G. C. White, L. H. Carpenter, and R. A. Garrott. 1987. Aerial mark-recapture estimates of confined mule deer in pinyon-juniper woodland. Journal of Wildlife Management 51:41–46.

Conn, P. B., W. L. Kendall, and M. D. Samuel. 2004. A general model for the analysis of mark-resight, mark-recapture, and band-recovery data under tag loss. Biometrics 60:900–909.

Graham, A., and R. Bell. 1969. Factors influencing the countability of animals. East African Agricultural and Forestry Journal 34:38–43.

Jolly, G. M. 1969. Sampling methods for aerial censuses of wildlife populations. East African Agricultural and Forestry Journal 34:46–49.

Pollock, K. H., and W. L. Kendall. 1987. Visibility bias in aerial surveys: a review of estimation procedures. Journal of Wildlife Management 51:502–510.

Rice, C. G., K. J. Jenkins, and W. Chang. 2009. A sightability model for mountain goats. Journal of Wildlife Management 73:468–478.

BIAS IN AERIAL SURVEY

GRAEME CAUGHLEY, School of Biological Sciences, University of Sydney, New South Wales 2006, Australia

Abstract: Aerial censuses of large mammals are inaccurate because the observer misses a significant number of animals on the transect. The accuracy deteriorates progressively with increasing width of transect, cruising speed, and altitude. Methods of eliminating bias by refining techniques are discussed and rejected; there seems to be no technical solution. An alternative strategy is to measure the bias and correct the estimates accordingly. A method is suggested for estimating bias during an aerial census, the subsequent analysis returning an unbiased estimate of density. No direct measure of true density is needed and little extra effort is involved over that required for a standard aerial survey.

J. WILDL. MANAGE. 38(4):921–933

This paper examines the effect of visibility bias, discusses the means by which the bias arises, and suggests methods by which it might be eliminated from aerial survey estimates of density and population size.

Aerial survey is, at best, a rough method of estimating the size of a population. Most efforts at refinement have been aimed at raising the precision of the estimate by combining impeccable survey design, high sampling intensity, intricate stratification, and powerful methods of analysis. This trend can be traced back to Siniff and Skoog's (1964) superb paper on an aerial census of caribou (*Rangifer tarandus*). Their use of stratified random sampling, with sampling effort allocated proportional to density, contrasted markedly with the crudity of previously reported surveys. Subsequently, Jolly's (1969a) paper on designs and analyses appropriate to aerial survey has encouraged a rigorous and disciplined application of the method. Recent papers following this lead have tended to treat the difficulties of estimating population size from the air largely as constituting a sampling problem, a survey being rated successful or otherwise according to the size of the estimate's standard error. Tacitly, the standard error was treated as a measure of the estimate's accuracy rather than of its repeatability. Underlying the preoccupation with precision there

often lurked an implicit assumption that the estimate is free of bias, that the observers counted all animals on each sampled unit.

A minority of writers have taken another tack. While agreeing with the need for appropriate methods of sampling and analysis, they have argued that the major problem of aerial survey is not so much that the estimates are imprecise as that they are inaccurate. This group has concentrated on detecting how far the mean of survey estimates is displaced from the true population size in an attempt to correct the estimates for bias. Examples of this approach can be found in the papers of Goddard (1967, 1969), Jolly (1969b), Watson et al. (1969a, b), Caughley and Goddard (1972), and Pennycuick and Western (1972).

These approaches are not exhaustive. A third school, recognizing that estimates are biased, treats aerial survey estimates as relative rather than absolute measures of abundance. Methods are rigorously standardized to hold bias constant, the indices thereby obtained allowing accurate estimation of the population's rate of increase. A recent example is Sinclair's (1972) survey design for monitoring long-term trends in ungulate populations of the Serengeti National Park in Tanzania. For such a problem this approach is undoubtedly appropriate, but it is

J. Wildl. Manage. 38(4):1974

Table 1. Accuracy of aerial censusing.

Species	Percent counted	Control	Reference
Wapiti, *Cervus elaphus*	64	poor	Buechner et al. 1951
Moose, *Alces americanus*	77	poor	Edwards 1954
Mule deer, *Odocoileus hemionus*	43	fair	Gilbert & Grieb 1957
Brown bear, *Ursus arctos*	47	good	Erickson & Siniff 1963
African game	88	good	Lamprey 1964
Black rhino, *Diceros bicornis*	29	good	Goddard 1967
Man (fixed wing)	65	good	Watson et al. 1969*d*
Man (helicopter)	75	good	Watson et al. 1969*d*
Simulated game	75	exact	Watson et al. 1969*c*
Indian rhino, *Rhinoceros unicornis*	56	poor	Caughley 1969
Red kangaroo, *Megaleia rufa*	57	fair	Bailey 1971
African large mammals (Mean of 8 spp.)	23	fair	Spinage et al. 1972
Ringed seal, *Phoca hispida* (300 m altitude)	58	good	Smith 1973
Sheep (total count)	74	exact	Caughley (Unpublished)
Sheep (200 m strip)	74	exact	Caughley (Unpublished)
Cattle (total count)	89	exact	Caughley (Unpublished)
Cattle (200 m strip)	39	exact	Caughley (Unpublished)

not appropriate to several other problems routinely faced by the wildlife manager. He is often required to determine offtake rates for game harvesting schemes and for private hunting. The absolute size of the standing crop must therefore be estimated before the permissible rate of harvesting can be translated into a quota of animals. It is to this kind of problem that this paper is addressed. When absolute numbers are estimated the census must be designed to maximize accuracy, even if this is to be achieved at the expense of precision.

I am grateful to S. R. Bleazard and I. S. C. Parker for piloting, to W. G. Swank, R. L. Casebeer, J. Hazam, and R. Sinclair for acting as observers, to A. L. Hodge and particularly E. E. Robinson for mathematical assistance, and to M. Norton-Griffiths, R. H. V. Bell, P. F. Sale, and E. E. Robinson for criticizing a previous draft of this paper. My thanks are due to G. A. Booth, manager of the Zambian F.A.O. Luangwa Valley Project, and to W. G. Swank, manager of the Kenyan F.A.O. Wildlife Management Project, for their help and encouragement. The data presented in this paper were collected when I was a member of the Wildlife Section of F.A.O.

ACCURACY OF ESTIMATES

Graham and Bell (1968) discussed the factors affecting counts of animals from the air and reasoned that most of these would lead to undercounting.

Table 1 demonstrates their conclusion empirically. It lists the number of animals estimated by aerial survey as a percentage of the number estimated by a different, and hopefully more accurate, method. The control estimates are themselves subject to error and a subjective rating of reliability is assigned to each according to the quoted author's description of his method. A rating of "poor" indicates that the control estimate was almost certainly too low, and that the "percent counted" (from the air) is therefore too high.

These data point directly to undercounting as the major problem of aerial survey, and indicate that sources of bias that tend to inflate the estimate are relatively unimportant. Operationally they can be ignored.

SIGHTABILITY MODELS

Sightability may be defined as the probability that an animal within an observer's field of search will be seen by that observer. The probability is determined by the distance between the animal and the observer; by such characteristics of location as thickness of cover, background, and lighting; by such characteristics of the animal as color, size, and movement; and by the observer's eye sight, speed of travel, and level of fatigue. In any specified set of conditions a mean sightability can be assigned to animals positioned at a given distance from the observer, and if values for sufficient distances are available they can be summarized as a sightability curve, the regression of sightability on distance.

Populations are usually surveyed from the air by counting animals on a strip of known width from an aircraft flying a straight line at a constant altitude above the ground. The field of search is defined for the observer by two marks or streamers on the wing strut. Between these he scans a strip of ground which is of constant and known width when the aircraft is at survey altitude. Density is estimated as the number of animals seen on the strip divided by the product of strip width and the length of the flight line.

Within the frame of this design the probability of an animal being seen and counted is determined by a host of influences. The most important are probably strip width, ground speed, and altitude, none of which has a simple relationship to sightability. Although these three influences interact, each can be isolated conceptually by considering its effect when the other two are held constant.

Effect of Strip Width

As strip width is increased: (1) the mean distance between an animal and the observer is increased, thereby reducing mean sightability; (2) the time available to locate, recognize, and count an animal decreases; (3) the amount of eye movement needed to scan the strip increases; and (4) the mean number of obscuring items between the animal and the observer increases.

Effect of Altitude

As altitude is increased: (1) mean distance between the observer and the animal increases; (2) the mean number of obscuring items between the animal and the observer decreases; and (3) the required eye movement decreases.

Effect of Speed

As speed is increased: (1) the time available to locate and count an animal decreases; and (2) the required rate of eye movement increases.

Of these three influences—strip width, altitude, and speed—the first provides four sources of bias, all negative. The second contributes one negative bias and two effects that tend to reduce negative bias. The third contributes two biases that are both negative.

Strip width, even considered in isolation, is probably the most complex influence on the probability of counting an animal. Sightability may or may not have a simple relationship to distance as such, but when

J. Wildl. Manage. 38(4):1974

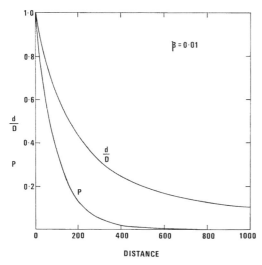

Fig. 1. An exponential decline of sightability P with distance, and the regression it would generate between apparent density relative to true density d/D and transect width.

the concomitant effects of increasing the strip width are taken into account, the chance of a simple theoretical relationship becomes vanishingly small. Nonetheless, the principles underlying the relationship between strip width and estimated density can be investigated in a general way by working forward from a postulated regression of sightability on distance.

Höglund et al. (1967) and Gates (in Eberhardt 1968) suggested that in some circumstances the probability P of seeing an animal positioned at a distance x from the observer approximates the exponential regression $P = e^{-\beta x}$. This simple model is convenient for illustrating the tactics of converting a sightability curve to a regression of estimated density on strip width. When density is calculated as if all animals were counted on the strip the apparent density d, relative to the true density D, can be estimated for any strip width X. This symbol is capitalized to distinguish it from distance x used in relation to sightability. The area under the sightability curve from 0 to

X is solved by integration as $(1 - e^{-\beta X})/\beta$ which is proportional to the number of animals counted on a strip of unit length and of width X. When this expression is divided by X it becomes the apparent density relative to the true density: $d/D = (1 - e^{-\beta X})/\beta X$. Fig. 1 shows an exponential sightability curve parametered by $\beta = 0.01$ and the curve of d/D that it generates.

The regression of apparent density on strip width takes the form $d = D(1 - e^{-\beta X})/\beta X$ which is too complex to lend itself to standard regression techniques for solving D.

The method by which a sightability curve is converted to a regression of apparent density on strip width can be used for any postulated relationship between sightability and distance. Eberhardt (1968) introduced a model more flexible than the exponential, allowing a range of options for the form of the sightability curve. Where, as here, no simple theoretical relationship can be deduced, a flexible empirical curve of this kind has many advantages. Eberhardt's model takes the form $P = 1 - (x/W)^a$ where P is, as before, the probability of seeing an animal at distance x. W is the distance at which the mean probability is zero and a is a constant determining the shape of the curve. When $a > 1$ a curve like the first of Fig. 2 is obtained, when $a = 1$ the regression is linear, and when $a < 1$, the third curve of Fig. 2 is obtained. The three regressions of Fig. 2 have a common constant of $W = 500$ and value of a of 2, 1, and a half respectively.

The corresponding apparent-density equation is again calculated by integrating sightability with respect to distance out to strip width X: $X - X^{(a+1)}/(a+1)W^a$ and dividing by X: $d/D = 1 - X^a/(a+1)W^a$ (when $X < W$). The bound on sightability limits this regression to the range of distance be-

J. Wildl. Manage. 38(4):1974

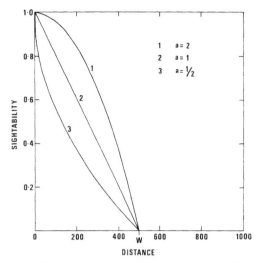

Fig. 2. Three possible forms of the regression of sightability on distance.

	1	a = 2
	2	a = 1
	3	a = ½

Table 2. Censuses of elephants in the South Luangwa National Park, Zambia.

Date	Elephant numbers	Strip width (m) per observer	Biologist
July 1965	9,000	600	P. B. Dean
Sept. 1965	14,000	600	ʺ
June 1966	12,000	600	ʺ
Nov. 1966	10,000	600	ʺ
June 1967	11,000	400	D. R. Patton
Nov. 1969	18,000	400	P. W. Martin
Aug. 1971	28,000	250	L. P. Van Lavieren
Jan. 1973	32,000	100	G. Caughley

tween 0 and W. Beyond $d/D = a/(a + 1)$ (when $X = W$); and $d/D = aW/(a + 1)X$ (when $X > W$). Fig. 3 shows the regressions of d/D on strip width X corresponding to the sightability curves of Fig. 2.

The regression of d/D on strip width that corresponds to the linear case of Eberhardt's sightability model possess interest-

ing properties that may be of practical use. When $a = 1$, the apparent-density equation reduces to $d/D = 1 - X/2W$ (when $X < W$); $d/D = \frac{1}{2}$ (when $X = W$); $d/D = W/2X$ (when $X > W$).

Thus when sightability is linear on distance the regression of apparent density on strip width is also linear out to the bounding distance W: $d = D - DX/2W$, $(X < W)$. It has a y-intercept of D, the true density, and a slope of $-D/2W$. Estimates of both D and W can therefore be calculated by fitting a regression to estimates of apparent density from transects of different widths.

FIELD TRIALS

The relationship between apparent density and strip width, speed, and altitude will be examined empirically in this section against the theoretical background provided previously.

Elephants in Zambia

Table 2 lists aerial survey estimates to the nearest thousand of elephants (*Loxodonta africana*) in the South Luangwa National Park of Zambia, a wooded reserve of approximately 9000 km². These are derived from counts made by the biologists listed in the table, their full reports being filed

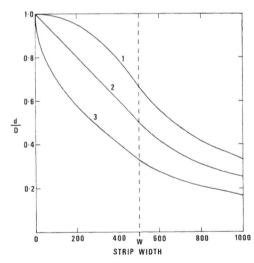

Fig. 3. The regressions of apparent density relative to true density d/D on transect width that would be generated by the three sightability curves of Fig. 2.

J. Wildl. Manage. 38(4):1974

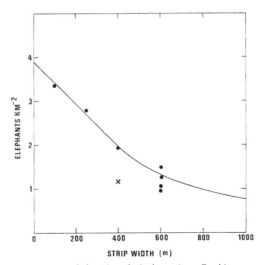

Fig. 4. Estimated densities of elephants in a Zambian national park, graphed against strip width per observer. The point marked by a cross was not used in fitting the regression curve.

with the Zambian Department of Wildlife, Fisheries and National Parks. The censuses up to 1969 utilized a northern boundary to the survey area that was changed thereafter. The 1973 survey was flown according to both designs, thereby producing a correction factor by which earlier survey estimates have been modified.

Superficially, the estimates suggest that elephants have increased at a rate of 15 percent per year between 1965 and 1973. This rate is impossibly high, particularly as there is no evidence of large-scale migration: Hanks and McIntosh (1973) estimate the maximum rate at which an elephant population can increase as about 4 percent a year. A more likely explanation for the range of estimates is suggested by Fig. 4 in which estimated density is graphed against strip width per observer. With the exception of one point, that from Patton's 1967 survey, the density estimates regress smoothly on strip width. If Patton's estimate is ignored, a linear regression through the remaining points provides an equation

of $y = 3.79 - 0.00435X$ that "accounts" for 95 percent of the total variability and shows no significant deviation from linearity ($F < 1$). Visibility bias rather than population increase may well be the dominant influence. If the data do in fact reflect visibility bias, there is a very good reason why Patton's estimate should lie below the regression. On that survey, elephants were counted on a 400-m strip at the same time as buffalo (*Syncerus caffer*) were counted on a strip of 1200 m. Hence an observer's attention was shared simultaneously between two transects, whereas only a single transect per observer was used in the other surveys.

When interpreted in terms of Eberhardt's (1968) linear model for sightability the regression raises a problem. The regression of apparent density d on strip width X takes the form $d = D - DX/2W$ from which true density D is estimated as 3.79 km^{-2} and W, the distance at which sightability is zero, is solved as 436 m. Thus W is lower than the strip width of 600 m used in 1965 and 1966. A second regression was therefore calculated for only those surveys using a strip width equal to or less than 400 m. A regression equation of $y = 3.883 - 0.0048X$ yielded estimates of $D = 3.883$ and $W = 404$ m. The fitted regression line in Fig. 4 is described by this equation out to a strip width of 404 m and thereafter by the theoretical extrapolation of $y = DW/2X$ presented in the previous section. Ninety-five percent of variability in the density estimates is attributable to this regression. It implies a regression of sightability on distance of $P = 1 - (x/404)$.

These data are consistent with a linear regression of sightability on distance. However, the slope of this regression is much steeper than might be expected. Mean sightability drops to zero by 404 m, which

J. Wildl. Manage. 38(4):1974

is intuitively improbable because elephants can be seen considerably further away than that. But the bounding distance of sightability, W, is not quite what it seems. It is the x-intercept of the mean trend of sightability on distance, not a distance beyond which elephants are never seen. W will have a large variance. The sightability of individual elephants will be dispersed around the mean trend and the scatter at the lower end of the regression will provide nonzero sightabilities at distances greatly in excess of W. Hence there is no necessary conflict between solving W as 404 m and sighting an elephant on the horizon.

The two independent variables, time and strip width, are so closely correlated that an attempt to separate their effects on observed density, either by analysis or intuition, is an exercise in futility. The uncontrolled nature of this experiment, and the lack of independent evidence on population trend, forces caution in accepting uncritically the apparent linear regression of sightability on distance. The inferred form of that regression must remain no more than a hypothesis.

Wildebeeste in Kenya

The effects of two influences on apparent density, strip width and altitude, were investigated in the Kadjiado district of Kenya. Wildebeeste (*Connochaetes taurinus*) were counted on three widths of strip from three altitudes on 20 May 1972. Each sampling unit measured 0.915 km², three replicates of this unit being flown consecutively for each combination of strip width and altitude. Viewing conditions were close to optimal. The dark-coated animals showed up well against the short green grass.

Table 3 lists the data as wildebeeste counted per square kilometer. These densities are bimodal, the majority clustering

Table 3. Effect of altitude and strip width on the number of wildebeeste counted per km².

Altitude (m)	Strip width (m)		
	100	200	400
30.5	23.0	6.6	131.1
	65.6	56.8	31.7
	408.7	95.1	74.3
61	359.6	74.3	18.6
	116.9	52.5	360.7
	99.5	92.9	39.3
122	112.6	115.8	1.1
	85.2	44.8	36.1
	80.9	64.5	39.3

around 65 km⁻² in contrast to three above 350 km⁻². The pattern reflects the distribution of wildebeeste during the survey. Most were in small scattered groups but a minority were aggregated into large densely-packed mobs. The three recorded high densities result from a transect passing through one of these mobs. The results provide a faithful record of the distribution pattern and would be analyzed as they stand if the aim were simply to estimate density, but when the purpose is to investigate visibility bias, the three high estimates introduce a perturbation that masks the effects of strip width and altitude. The analysis was therefore restricted to sampling units in which no dense mob was encountered, thereby reducing the number of units from 27 to 24.

The analysis (Table 4) indicates that strip width and altitude jointly accounted for 16 percent of the variability in apparent density but that only a minute fraction of this was associated with altitude. Transformation of densities to logarithms did not change the picture. Again, 16 percent of variability was ascribable to the joint regression, almost all of which was referrable to the effect of strip width. The linear regression was therefore retained for further analysis even though the effect of strip

J. Wildl. Manage. 38(4):1974

Table 4. Regression analysis of variance applied to wildebeeste density y on altitude x_1 and strip width x_2: $y = 93 + 0.005 x_1 - 0.119 x_2$.

Source	SS	df	MS	F
Regression on x_1 and x_2	5,074.41	2		
On x_1 alone	18.34	1		
x_2 isolated from x_1	5,056.07	1	5,056.07	4.11
Regression on x_1 and x_2	5,074.41	2		
On x_2 alone	5,073.46	1		
x_1 isolated from x_2	0.95	1	0.95	0.00
Deviations from joint regression	25,851.17	21	1,231.01	
Total	30,926.57	23		

Table 5. Total counts from the air of elephants in a portion of the Queen Elizabeth National Park, Uganda. Data from Eltringham (1972).

Date (Aug. 1968)	5	7	12	20	
Altitude (m)	300	300	300	150	
Speed (km/h)	180	180	130	180	
Observer	A	A	A	A	B
Morning count	80	88	99	131	125
Afternoon count	93	76	115	104	113
Evening count	105	74		87	87

width just failed to reach significance at the 5 percent level of probability.

Regression of apparent density on strip width alone yielded $y = 93 - 0.119X$ where 93 can be equated with true density D. The regression constant is an estimate of $-D/2W$ from which W is estimated as 391 m. Regression of apparent density on strip widths of 100 m and 200 m, excluding those from strips of 400 m, reduced the estimate of W to 305 m.

Multispecies Counts in Kenya

Pennycuick and Western (1972) investigated the effects of strip width and altitude on their estimates of large mammals in the Masai Amboseli Game Reserve. They demonstrated that significantly fewer animals were counted from an altitude of 152 m than from 91 m, and that apparent densities on a strip of width 371 m were significantly lower than when the width was reduced to 197 m. They investigated the second bias further by dividing the transect into an inner strip of 197 m and an outer

strip of 173 m. Their failure to demonstrate a difference in apparent density between the two strips suggests that the decline in apparent density accruing from increasing the strip width is not a simple function of increasing the mean distance between observer and animal.

Elephants in Uganda

Eltringham (1972) made total counts of elephants in the portion of the Queen Elizabeth National Park lying within North Ankole, Uganda. Standardized flight lines were spaced at intervals of 1500 m. Table 5 gives his counts, excluding one he made in the evening of 12 August when airspeed was increased half-way through the survey. He interpreted these data as indicating that the counts were not influenced by speed or altitude.

These conclusions will be reexamined here. As a preliminary, the data were tested for time effects and day effects by a two-factor analysis of variance on counts of days 5 and 7, these providing the largest subset of data over which speed and altitude were held constant. Neither F ratio approached significance. For purposes of further analysis, counts at different times of the day can therefore be treated as replicates.

The effect of altitude can be examined by comparing observer A's 6 counts on days 5 and 7 with his 3 counts on day 20. The

J. Wildl. Manage. 38(4):1974

counts from 300 m altitude were significantly lower than from 150 m ($t = 1.95$, df $= 7$, one-tailed $P < 0.05$). A 50 percent reduction in altitude was associated with a 25 percent increase in the number of elephants counted.

Likewise, the effect of speed can be examined by t-testing the counts of days 5 and 7 against those of day 12. Again the difference is significant ($t = 2.20$, df $= 6$, one-tailed $P < 0.05$). A 28 percent reduction of speed is associated with a 24 percent increase in elephants.

Unfortunately, the data allow no comparison between high-fast and low-slow surveys. Nor is it possible to isolate a time trend from these counts. The latter restriction forces caution in interpreting the results but, taken at face value, they suggest that sightability was strongly affected by altitude of observation and even more strongly by cruising speed.

DISCUSSION

The evidence presented and reviewed indicates that visibility bias is an important determinant of apparent density. Increase in strip width, altitude, and speed all result in a reduced tally.

The linear decline of apparent density on strip width, a model appearing to fit the elephant counts from Zambia, implies a linear decline of sightability with distance. Such a relationship is theoretically improbable if sightability is determined by the perceived size of an animal. The angle subtended by an object is proportional to the reciprocal of the distance between it and the observer, and hence the area of the observer's retina covered by the image is approximately proportional to the squared reciprocal of distance. If the dominant determinant of sightability were perceived

size we would therefore expect sightability to vary inversely with the square of the distance, or, considering the complexities of visual physiology, to at least approximate this relationship. The data argue strongly against this model. Apparently distance as such is not the most important influence on sightability, a conclusion supported by two additional pieces of evidence. Firstly, Pennycuick and Western (1972) demonstrated that approximately the same number of animals were counted on a strip whether its inner boundary was close to the plane or displaced about 200 m outward. Distance as such could not be shown to affect sightability. Secondly, Fig. 4 indicates that when elephants are counted on a 400-m strip at the same times as buffalo are counted on a 1200-m strip, the elephant tally is depressed. These results suggest that the effect on sightability of widening the strip acts more through time than distance. When the width is doubled the time available for scanning a unit of ground is halved. This mechanism could well generate a linear regression of apparent density on strip width.

That distance as such also plays a part in reducing sightability is attested by the effect of increasing the altitude of observation. Although the effect could not be demonstrated in the wildebeeste experiment it is apparent in Pennycuick and Western's (1972) survey of large mammals and is strongly suggested by Eltringham's (1972) elephant survey.

If, as seems likely, the strongest influence on sightability is the time available for scanning the strip, sightability should be related inversely to the speed of the aircraft. This variable was neglected in all but Eltringham's study. His data suggest that speed has a large effect, but the unbalanced experimental design and the nonrandom order of treatments erodes the confidence that can be placed in this conclusion.

Faced with strong evidence of substantial bias in aerial survey estimates we can seek to improve them by one of two strategies. The first and most obvious is to refine techniques. If sightability falls as strip width, altitude, and speed are increased, a reduction in each of these will improve accuracy. This is likely to provide only a partial solution because there is no present indication that sightability peaks at nonzero values of those variables. Maximum sightability is probably attained when the observer scans a line from a stationary plane on the ground.

The effect of time limitation could be circumvented by photographing the strip, the animals being counted at leisure back at the laboratory. Photography is a considerable aid to aerial survey, being used extensively to census large groups in open country, but it is less efficient than the unaided human eye when cover is available to the animals. An observer can view an animal from several angles as he flies past it. The camera takes one look from one angle and misses any animal obscured at that moment. Infrared photography seems a better bet because it penetrates cover to some extent, but my own view on this technique is pessimistic. In the 5 years since Croon et al. (1968) first investigated its applicability, infrared photography in censusing has improved considerably. However, the most recent demonstration of its effectiveness (Graves et al. 1972) provided results inferior to those returned by a human observer. A suspicious blip could be identified positively as an animal only when its position changed between two photographic runs. From the rate at which this technique has improved since the first crude applications I would hazard a guess that the definition attainable will reach an asymptote no

higher than three times that achieved now. At that level it is still likely to be inferior to the unaided human eye.

There is a strong possibility that no complete technical solution is available through refinement of techniques. The alternative strategy lies in recognizing that our estimates are biased, measuring the magnitude of the bias, and correcting the estimates accordingly. The utility of this approach will be explored next.

CORRECTION FOR BIAS

The meager data available at present indicate that the regression of apparent density on strip width may be approximately linear out to a critical distance. Theoretical considerations suggest that the regression of apparent density on cruising speed will also be linear. The data on the effect of altitude are insufficient to reveal the form of this relationship but for present purposes it will be assumed linear. As a first approximation the regression of apparent density on the three variables can therefore be symbolized as $y = a + b_1X_1 + b_2x_2 + b_3x_3$; where X_1 = strip width; x_2 = ground speed; x_3 = altitude; a = true density; y = apparent density; b = slope of the regression. All three b constants should be negative.

This model will now be given a run on Eltringham's elephant counts (Table 5) to see if it returns plausible values for the constants. Only speed and altitude can be included because strip width was not varied during the experiment. The calculation returns a regression of $y = 205 - 0.420x_2 - 0.146x_3$. It implies that when strip width is maintained at 1500 m while speed and altitude are extrapolated to zero, 205 elephants is the theoretical expected tally. For each kilometer per hour increase of cruising speed above zero the tally will be reduced by 0.420 elephants and for each meter increase in altitude 0.146 elephants will be missed. Extrapolation gives zero sightability at 490 km/h or at 1400 m of altitude. Although the first seems a little low and the second high, both values are within the limits of plausibility.

The only apparent difficulty lies in the estimate of 205 elephants. Eltringham (1972), although allowing that his data could not be used to calculate accuracy because of the lack of a firm standard, discussed differences between his counts in terms of elephants seen to move into and out of the survey area. He considered (p. 304) that "on the whole, therefore, the results suggest that observers were consistent in their counting and any differences can usually be accounted for." Against this optimism stands the extrapolated tally of 205 elephants, about double his average, and the certainty that this total would have been increased by the inclusion of strip width in the calculation.

The discrepancy between the professional judgment of an ecologist in the survey area and the impersonal opinion provided by a partial regression calculated on another continent would have to remain simply an anomaly in the absence of independent evidence. But independent evidence is available. Eltringham's (1972) Fig. 2 is a map of the location of elephant groups seen separately by two observers on the evening of 20 August. Although exact figures cannot be extracted because of mapping inaccuracies and lumping of numbers, an examination of this map suggests that about 35 groups of elephants were seen by one or other observer and 11 groups were seen by both. The following binomial probabilities provide a framework within which these frequencies can be analyzed: p = probability that a given group is seen by a given observer; p^2 = probability that the group is seen by both observers; $2p(1-p)$ = probability that the group is seen by one or other observer; and $(1-p)^2$ = probability that the group is missed by both observers.

We have the number of groups seen by a single observer only, a frequency that will be designated S, and the number, B, seen by both. We have no frequency for M, the number missed by both observers, and no direct information on the value of p.

However p can be estimated from the relationship $2p(1-p)/p^2 = S/B$ from which $p = 2B/(2B + S) = 0.39$. In a similar manner the number of groups missed by both observers can be derived as $M = S^2/4B = 28$. These figures suggest that on average an observer counted only about 40 percent of the groups in the area and that about 28 specific groups were missed by both observers. Note that these figures are of groups, not individuals. No indication of the number of elephants missed from within groups can be extracted from these data. Rough and ready as this analysis is, it favors the figure of 205 elephants against the conclusion that the observers counted a high proportion of the elephants within the survey area.

Experimental Design for Correction of Bias

At this stage it is premature to suggest a survey design based on the regression model because there is no unambiguous information on how the regression surface curves. But the general form of a design can be indicated if linearity is viewed as a special case of a curve. The principles are the same whether the surface is curved or plane. Assuming linearity, the simplest survey design incorporating a correction for bias calls for estimates of apparent density on two widths of strip at two speeds from two altitudes, eight combinations in all. These steps would be followed:

J. Wildl. Manage. 38(4):1974

1. Divide the survey area into strata according to thickness of cover.

2. For a given stratum draw flight lines on a map and assign at random to each one of the eight treatment combinations of strip width, speed, and altitude.

3. Fly the survey, record the number of animals seen on each transect, and calculate an apparent density for each.

4. Calculate separately for each stratum the partial regression of apparent density on strip width, speed, and altitude. Most statistical texts adequately cover this standard analysis.

5. The intercept constant *a* is an estimate of true density which when multiplied by the area of the stratum gives an estimate of numbers.

6. Sum the estimates of numbers over all strata to provide an estimate of total numbers in the survey area.

7. Calculate a standard error of numbers over all strata from the variance of *a* in each stratum.

An alternative method is to first calculate correction factors for each cover type by some modification of the above method and use them to correct the estimates from a routine aerial survey flown at a single combination of speed, altitude, and strip width.

Refinement of the Model

The linear regression method of correcting for bias is a first approximation suggested by inadequate data. It is unlikely to be entirely appropriate. More experimenting is needed to establish the true form of these regressions and whether that form is robust or varies between habitats and animal species. With this information the model can be tuned to provide either a general regression equation, or if generality is not possible, a set of equations from which one can be selected to fit a particular situation.

Having reached this point we can again turn to increasing the precision, as opposed to the accuracy, of the estimate. In any given situation there will be a combination of speeds, strip widths, and altitudes that minimize the variance of the regression constant *a*, but the combination we seek is not necessarily the absolute minimizing set but one that minimizes for a given expenditure of flying time. The necessary information can be obtained experimentally and an optimization calculated by the techniques of linear or nonlinear programming.

LITERATURE CITED

BAILEY, P. T. 1971. The red kangaroo, *Megaleia rufa* (Desmarest), in north-western New South Wales. I. Movements. CSIRO Wildl. Res. 16(1):11–28.

BUECHNER, H. K., I. O. BUSS, AND H. F. BRYAN. 1951. Censusing elk from airplane in the Blue Mountains of Washington. J. Wildl. Manage. 15(1):81–87.

CAUGHLEY, G. 1969. Wildlife and recreation in Nepal. F.A.O. Trisuli Watershed Dev. Rep. 6. 54pp.

———, AND J. GODDARD. 1972. Improving the estimates from inaccurate censuses. J. Wildl. Manage. 36(1):135–140.

CROON, G. W., D. R. McCULLOUGH, C. E. OLSON, JR., AND L. M. QUEAL. 1968. Infrared scanning techniques for big game censusing. J. Wildl. Manage. 32(4):751–759.

EBERHARDT, L. L. 1968. A preliminary appraisal of line transects. J. Wildl. Manage. 32(1):82–88.

EDWARDS, R. Y. 1954. Comparison of aerial and ground census of moose. J. Wildl. Manage. 18(4):403–404.

ELTRINGHAM, S. K. 1972. A test of the counting of elephants from the air. E. Afr. Wildl. J. 10(4):299–306.

ERICKSON, A. W., AND D. B. SINIFF. 1963. A statistical evaluation of factors influencing aerial survey results on brown bears. Trans. N. Am. Wildl. Nat. Resour. Conf. 28:391–409.

GILBERT, P. F., AND J. R. GRIEB. 1957. Comparison of aerial and ground deer counts in Colorado. J. Wildl. Manage. 21(1):33–37.

GODDARD, J. 1967. The validity of censusing black rhinoceros populations from the air. E. Afr. Wildl. J. 5(1):18–23.

J. Wildl. Manage. 38(4):1974

————. 1969. Aerial census of black rhinoceros using stratified random sampling. E. Afr. Wildl. J. 7(1):105–114.

GRAHAM, A., AND R. BELL. 1969 (1968). Factors influencing the countability of animals. E. Afr. Agric. For. J. 34(special issue):38–43.

GRAVES, H. B., E. D. BELLIS, AND W. M. KNUTH. 1972. Censusing white-tailed deer by airborne thermal infrared imagery. J. Wildl. Manage. 36(3):875–884.

HANKS, J., AND J. E. A. McINTOSH. 1973. Population dynamics of the African elephant (*Loxodonta africana*). J. Zool. 169(1):29–38.

HÖGLUND, N., G. NILSSON, AND F. STÅLFELT. 1967. Analysis of a technique for estimating willow grouse (*Lagopus lagopus*) density. Trans. Int. Congr. Game Biol. 8:156–159.

JOLLY, G. M. 1969 (1968)*a*. Sampling methods for aerial censuses of wildlife populations. E. Afr. Agric. For. J. 34(special issue):46–49.

————. 1969 (1968)*b*. The treatment of errors in aerial counts of wildlife populations. E. Afr. Agric. For. J. 34(special issue):50–55.

LAMPREY, H. F. 1964. Estimation of large mammal densities and energy exchange in the Tarangire Game Reserve and the Masai Steppe in Tanganyika. E. Afr. Wildl. J. 2(1):1–46.

PENNYCUICK, C. J., AND D. WESTERN. 1972. An investigation of some sources of bias in aerial transect sampling of large mammal populations. E. Afr. Wildl. J. 10(3):175–191.

SINCLAIR, A. R. E. 1972. Long term monitoring of mammal populations in the Serengeti: census of non-migratory ungulates, 1971. E. Afr. Wildl. J. 10(4):287–297.

SINIFF, D. B., AND D. O. SKOOG. 1964. Aerial censusing of caribou using stratified random sampling. J. Wildl. Manage. 28(2):391–401.

SMITH, T. G. 1973. Population dynamics of the ringed seal in the Canadian eastern Arctic. Fish. Res. Board Can. Bull. 181. 55pp.

SPINAGE, C. A., F. GUINNESS, S. K. ELTRINGHAM, AND M. H. WOODFORD. 1972. Estimation of large mammal numbers in the Akagera National Park and Mutara Hunting Reserve, Rwanda. Terre et Vie 26(4):561–570.

WATSON, R. M., I. S. C. PARKER, AND T. ALLAN. 1969*a*. A census of elephant and other large mammals in the Mkomazi region of northern Tanzania and southern Kenya. E. Afr. Wildl. J. 7(1):11–26.

————, A. D. GRAHAM, AND I. S. C. PARKER. 1969*b*. A census of the large mammals of Loliondo controlled area, northern Tanzania. E. Afr. Wildl. J. 7(1):43–59.

————, G. M. FREEMAN, AND G. M. JOLLY. 1969 (1968)*c*. Some indoor experiments to simulate problems in aerial censusing. E. Afr. Agric. For. J. 34(special issue):56–62.

————, G. M. JOLLY, AND A. D. GRAHAM. 1969 (1968)*d*. Two experimental censuses. E. Afr. Agric. For. J. 34(special issue):60–62.

Accepted 2 July 1974.

J. Wildl. Manage. 38(4):1974

Charnov, E. L. 1976. Optimal foraging, the marginal value theorem. *Theoretical Population Biology* 9:129–136.

Wildlife biologists first began understanding animal foraging and browsing patterns by conducting diet studies and comparing incidence of foodstuffs in the diet with their abundance in the respective habitat. From these classic and essential studies enough empirical evidence was accumulated to improve animal feeding through the next logical step: foraging theory. Foraging theory gained importance when Charnov wrote this seminal paper from his dissertation examining the theory of optimum foraging.

In this paper, Charnov examines why animals use certain areas but avoid others, what they do in areas they use, whether they are influenced more by top-down or bottom-up effects, and numerous other questions related to use and availability of landscapes. Such theory applies to both herbivores and carnivores and to how they search for and secure their food. This short paper laid the groundwork for many studies that examined the questions Charnov addressed and related issues pertaining to foraging and predation. Charnov's work was important to our understanding of how wildlife use their environments and the complexity of factors (i.e., biotic and abiotic) that affect foraging and thus habitat use. His work stimulated others to conduct similar studies.

RELATED READING

Emlen, J. M. 1966. The role of time and energy in food preferences. The American Naturalist 100:611–617.

Krebs, J. R., J. C. Ryan, and E. L. Charnov. 1974. Hunting by expectation or optimal foraging? A study of patch use by chickadees. Journal of Animal Behavior 22:953–964.

Pianka, E. R. 1981. Resource acquisition and allocation among animals. Pages 300–314 *in* C. Townsend and P. Carlow, editors. Physiological ecology: an evolutionary approach to resource use. Blackwell, Oxford, UK.

Pierce, G. J., and J. G. Ollason. 1987. Eight reasons why optimal foraging theory is a complete waste of time. Oikos 49:111–117.

Pyke, G. H. 1984. Optimal foraging theory: a critical review. Annual Review of Ecology and Systematics 15:523–575.

Pyke, G. H., H. R. Pulliam, and E. L. Charnov. 1977. Optimal foraging: a selective review of theory and tests. The Quarterly Review of Biology 52:137–154.

Schoener, T. W. 1969. Models of optimum size for solitary predators. The American Naturalist 103:277–313.

Schoener, T. W. 1969. Optimal size and specialization in constant and fluctuating environments: an energy-time approach. Brookhaven Symposium in Biology 22:103–114.

Reprinted from THEORETICAL POPULATION BIOLOGY   Vol. 9, No. 2, April 1976
All Rights Reserved by Academic Press, New York and London   *Printed in Belgium*

Optimal Foraging, the Marginal Value Theorem

ERIC L. CHARNOV*

*Center for Quan. Science in Forestry, Fisheries, and Wildlife,
University of Washington, Seattle, Washington 98195; and
Institute of Animal Resource Ecology UBC, Vancouver 8, Canada*

Received December 26, 1974

There has been much recent work on foraging that derives hypotheses from the assumption that animals are in some way optimizing in their foraging activities. Useful reviews may be found in Krebs (1973) or Schoener (1971). The problems considered usually relate to breadth of diet (Schoener, 1969, 1971; Emlen 1966; MacArthur, 1972; MacArthur and Painka, 1966; Marten, 1973; Pulliam, 1974; Werner, 1974; Werner and Hall, 1974; Timmins, 1973; Pearson, 1974; Rapport, 1971; Charnov, 1973, 1976; Eggers, 1975), strategies of movement (Cody, 1971; Pyke, 1974; Smith, 1974a, b; Ware, 1975), or use of a patchy environment (Royama, 1970; MacArthur and Pianka, 1966; Pulliam, 1974; Smith and Dawkins, 1971; Tullock, 1970; Emlen, 1973; Krebs, 1973; Krebs, Ryan, and Charnov, 1974; Charnov, Orians, and Hyatt, 1976). The above list of references is provided as a beginning to this fast expanding literature and is far from exhaustive.

This paper will develop a model for the use of a "patchy habitat" by an optimal predator. The general problem may be stated as follows. Food is found in clumps or patches. The predator encounters food items within a patch but spends time in traveling between patches. This is schematically shown in Fig. 1. The predator must make decisions as to which patch types it will visit and when it will leave the patch it is presently in. This paper will focus on the second question. An important assumption of the model is that while the predator is in a patch, its food intake rate *for that patch* decreases with time spent there. The predator *depresses* (Charnov, Orians, and Hyatt, 1976) the availability of food to itself so that the amount of food gained for T time spent in a patch of type i is $h_i(T)$, where the function rises to an asymptote. A hypothetical example is shown in Fig. 2. While it is not necessary that the first derivative of $h_i(T)$ be decreasing for all T (it might be increasing at first if the predator scares up prey upon arrival in a new patch), I will limit discussion to this case since more

* (Present Address) Department of Biology, University of Utah, Salt Lake City, Utah 84112.

129

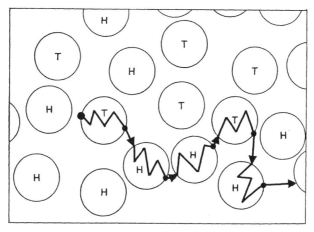

FIG. 1.   A hypothetical environment of two patch types. The predator encounters prey only within a patch, but spends time in traveling between patches. Patches were labeled H or T by the flip of a coin.

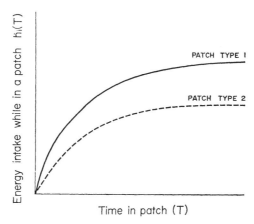

FIG. 2.   The energy intake for T time spent in a patch of type i is given by $h_i(T)$. This function is assumed to rise to an asymptote.

complex functions add essentially nothing new to the major conclusions. The model is also completely deterministic, but this is not a major restriction since qualitatively similar results may be shown to follow from a corresponding stochastic model that considers the foraging as a cumulative renewal process (Charnov 1973 and in preparation). What is important is that if the environment is made up of several "patch types," the types be "mixed up" or rather distributed at random relative to one another (Fig. 1), and that many separate patches are visited in a single foraging bout with little or no revisitation. A patch type has associated with it a particular $h_i(T)$ curve. MacArthur (1972) has termed these assumptions "a repeating environment." One final assumption is necessary. The predator is assumed to make decisions so as to maximize the net rate of energy intake during a foraging bout.

The patch use model

We define as follows:

$P_i =$ proportion of the visited patches that are of type i ($i = 1, 2,..., k$).

$E_T =$ energy cost per unit time in traveling between patches.

$E_{si} =$ energy cost per unit time while searching in a patch of type i.

$h_i(T) =$ assimilated energy from hunting for T time units in a patch of type i minus all energy costs except the cost of searching.

$g_i(T) = h_i(T) - E_{si} \cdot T =$ assimilated energy corrected for the cost of searching.

The time for a predator to use a single patch is the interpatch travel time (t) plus the time in the patch. Let T_u be the average time to use one patch.

$$T_u = t + \sum P_i \cdot T_i.$$

T is now written as T_i to indicate that it may be different for each patch type. The average energy from a patch is E_e.

$$E_e = \sum P_i \cdot g_i(T_i).$$

The net energy intake rate (En) is given by:

$$En = \frac{E_e - t \cdot E_T}{T_u}. \tag{1}$$

En may thus be written as

$$En = \frac{\sum P_i \cdot g_i(T_i) - t \cdot E_T}{t + \sum P_i \cdot T_i}. \tag{2}$$

It is easy to show that (2) is an energy balance equation and that En is the net rate of energy intake. With suitable interpretation of $g_i(T_i)$, (2) is identical to Schoener (1971, Eq. (2)).

The predator is assumed to control which patches it will visit and when it will leave a patch. The t is obviously a function of which patches the predator is visiting and in general should increase as more patch types are skipped over. A simple assumption would have t proportional to the distance between patches divided by the predators speed of movement. It should be noted, however, that there is no good reason to believe that t should be at all related to any of the T_i. The length of time between patches should be independent of length of time the predator hunts within any one (although the reverse statement is not true).

This independence is quite important since when it holds, (2) may be written (from the standpoint of a patch type of interest j) as:

$$En = \frac{P_j \cdot g_j(T_j) + A}{P_j \cdot T_j + B},\tag{3}$$

where A and B are not functions of T_j. A and B are found by equating terms in (2) and (3), naming one patch type as j.

If j is being visited, the predator is assumed to control only T_j. The optimal value of T_j is given by a rather interesting theorem. For some set of patches being visited, write En as En^* when all T_i are at their optimal values. When this is true T_j satisfies the following relation.

$$\frac{\partial g_j(T_j)}{\partial T_j} = En^*, \qquad \text{for all} \quad i = j.\tag{4}$$

The predator should leave the patch it is presently in when the *marginal capture rate in the patch* $(\partial g/\partial T)$ *drops to the average capture rate for the habitat.*

This rule is found by setting $\partial En/\partial T_i = 0$ for all patch types simultaneously. Since we are assuming here that the $\partial h_i(T_i)/\partial T_i$ are always decreasing, so are the $\partial g_i(T_i)/\partial T_i$ and there is a unique set of T_i that fulfills (4). This set represents a maximum as the associated Hessian matrix is negative-definite (Taha, 1971).

A graphical way of showing this result is in Fig. 3. The $g_i(T_i)$ is plotted as a

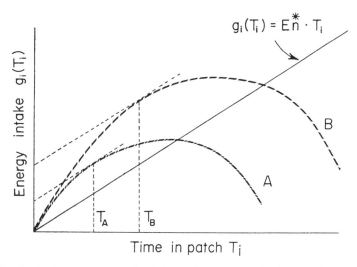

FIG. 3. Optimal use of a patchy habitat. The energy intake functions $g_i(T_i)$, are shown for a habitat with two patch types. If the ray from the origin with slope En^* is plotted, the appropriate time to spend in each patch is found by constructing the *highest* line tangent to a $g_i(T_i)$ curve and parallel to the ray. The lines and the resulting times are shown for the two patch types.

function of T_i for two patch types. If a ray from the origin with slope En^* is then plotted, the optimal T_i are easily found. To find these, simply construct lines with slope En^* and see where they become tangent to the appropriate $g_i(T_i)$ curve. In cases where the $\partial g_i(T_i)/\partial T_i$ are not strictly decreasing with T_i, more than one point of tangency may result. In these cases, the optimal T_i is that associated with the highest line of slope En^*.

DISCUSSION

Two earlier publications (Krebs, Ryan, and Charnov, 1974; Charnov, Orians, and Hyatt, 1976) derived a simplified version of the movement rule given in (4) and discussed some supporting data. Krebs, Ryan, and Charnov 1974 carried out laboratory experiments with chickadees to qualitatively test (4). They defined the time between the last capture and when an individual left a patch (several blocks of wood with mealworms in holes, suspended from an artificial tree) to go to another patch as the "giving up time" (GUT). This was taken to be a measure of the inverse of the capture rate when the bird left the patch (the marginal capture rate). The experimental design consisted of two environments. In the first, the average rate of food intake was high, in the second, it was low. Within each environment there were two or three patch types, each type having a specified number of mealworms. The predictions of the theorem, translated into the GUT measurement, were that (1) GUT should be a constant within an environment across patch types, and (2) GUT should be lower in the rich environment. Both of these predictions were supported by the data. More recent experiments using the Great Tit (*Parus major*) also support the model (R. Cowie, personal communication).

There are little other data that will allow more critical tests of the model although there are many data relating to gross predator behavior relative to clumps of prey. Smith (1974a, b), in some field experiments with thrushes, found that the tendency of a bird to remain in the area where it had already made a capture was greater the lower the overall availability of food in the habitat. The simple tendency for a predator to remain in the area where it was successful has been documented for birds (Tinbergen, Impekoven, and Frank, 1967; Krebs, MacRoberts, and Cullen, 1972); fish (Beukema, 1968); insects (Hafez, 1961; Fleschner, 1950; Laing, 1938; Mitchell, 1963; Dixon, 1959, 1970; Banks, 1957; Richerson and Borden, 1972; Hassell and May, 1973; Murdie and Hassell, 1973). Even unicellular predators exhibit increased frequency of turning following an encounter with food particles, a behavior pattern that results in a more intensive search of the vicinity of the capture (Fraenkel and Gunn, 1940; MacNab and Koshland, 1972).

The rule was used in a slightly different context by Parker and Stuart (1976), whose derivation is independent of the work here. They showed that male dung flies (*Scatophaga stercoraria*) terminate copulation at a time that maximizes the eggs fertilized/unit time for the male.

CONCLUSIONS

A rule is proposed for the movement of an optimal predator through an environment where food is found in patches, and time is expended in movement between patches. The theorem is rather general and should be useful where predators cause the prey in their immediate vicinity to become less available the longer they remain there. It receives some support from both lab and field studies, but has yet to be tested in a quantitative manner.

ACKNOWLEDGMENTS

The ideas presented here have been drawn from a thesis presented to the University of Washington. The thesis work was generously supported by the Ford Foundation (UW) and by funds supplied by the Institute of Animal Resource Ecology (University of British Columbia). I wish to thank D. G. Chapman, W. H. Hatheway, N. Pearson, C. Fowler, C. S. Holling, J. Krebs, G. Pyke, Thomas Schoener, and G. A. Parker. Gordon Orians provided his usual creative input.

REFERENCES

BANKS, C. J. 1957. The behavior of individual coccinellid larvae on plants, *J. Anim. Behav.* **5**, 12–24.

BEUKEMA, J. J. 1968. Predation by the three-spined stickleback *(Gasterosteus aculeatus)*. The influence of hunger and experience, *Behavior* **31**, 1–126.

CHARNOV, E. L. 1973. Optimal foraging: some theoretical explorations, Ph. D. Thesis, Univ. of Wash.

CHARNOV, E. L. 1976. Optimal foraging: attack strategy of a mantid, *Amer. Natur.* to appear.

CHARNOV, E. L., ORIANS, G. H., AND HYATT, K. 1976. Ecological implications of resource depression, Amer. Natur. to appear.

CODY, M. L. 1971. Finch flocks in the Mohave desert, *Theor. Popul. Biol.* **2**, 142–158.

DIXON, A. F. G. 1959. An experimental study of the searching behavior of the predatory coccinellid beetle *Adalia decempunctata, J. Anim. Ecol.* **28**, 259–281.

DIXON, A. F. G. 1970. Factors limiting the effectiveness of the coccinellid beetle, *Adalia bipunctata,* as a predator of the sycamore aphid, *Depanosiphum platonoides, J. Anim. Ecol.* **39**, 739–751.

EGGERS, D. M. 1975. A synthesis of the feeding behavior and growth of juvenile sockeye salmon in the limnetic environment, Ph. D. Thesis, Univ. of Wash.

EMLEN, J. M. 1966. The role of time and energy in food preference, *Amer. Natur.* **100**, 611–617.

EMLEN, J. M. 1973. "Ecology; An Evolutionary Approach," p. 493. Addison–Wesley, Reading, Mass.

FLESCHNER, C. A. 1950. Studies on search capacity of the larvae of three predators of the citrus red mite, *Hilgardia* **20**, 233–265.

FRAENKEL, G. S. AND GUNN, D. L. 1940. "The Orientation of Animals," Oxford Book, New York.

HAFEZ, M. 1961, Seasonal fluctuations of population density of the cabbage aphid *Brevicoryne brassicae* (L.) in the Netherlands and the role of its parasite *Aphidius (Diaretiella) rapae* (Curtis), *Tijdschr. Plantenziekten.* **67**, 445–548.

HASSELL, M. P. AND MAY, R. M. 1973. Stability in insect host-parasite models, *J. Anim. Ecol.* **42**, 693–726.

KREBS, J. R. 1973. Behavioral aspects of predation, in "Perspectives in Ethology," (Bateson, P. P. G., and Klopfer, P. H., Eds.), pp. 73–111. Plenuwm, New York.

KREBS, J. R., MacROBERTS, M. H., AND CULLEN, J. M. 1972. Flocking and feeding in the Great Tit, *Parus major*, an experimental study, *Ibis* **114**, 507–530.

KREBS, J. R., RYAN, J. C., AND CHARNOV, E. L. 1974. Hunting by expectation or optimal foraging? A study of patch use by chickadees, *J. Anim. Behav.* **22**, 953–964.

LAING, J. 1938. Host finding by insect parasites. II. The chance of *Trichogrammae vanescens* finding its hosts, *J. Exp. Biol.* **15**, 281–302.

MacARTHUR, R. H. 1972. "Geographical Ecology," p. 269. Harper and Row, New York.

MacARTHUR, R. H. AND PIANKA, E. R. 1966. On optimal use of a patchy environment, *Amer. Natur.* **100**, 603–609.

MacNAB, R. M. AND KOSHLAND, D. E. 1972. The gradient-sensing mechanism in bacterial chemotaxis, *PNAS* **69**, 2509–2512.

MARTEN, G. 1973. An optimization equation for predation, *Ecology* **54**, 92–101.

MAYNARD SMITH, J. 1974. "Models in Ecology," p. 146. Cambridge Univ. Press, London/ New York.

MITCHELL, B. 1963. Ecology of two carabid beetles, *Bembidion lampros* and *Trechus quadristriatus*, *J. Anim. Ecol.* **32**, 289–299.

MURDIE, G. AND HASSELL, M. P. 1973. Food distribution, searching success, and predator–prey models, *in* "The Mathematical Theory of the Dynamics of Biological Populations," (Barlett, M. S. and Hiorns, R. W. (Eds.), Academic Press, New York/ London.

PARKER, G. A. AND STUART, R. A. 1976. Animal behaviour as a strategy optimizer: evolution of resource assessment strategies and optimal emigration thresholds, *Amer. Natur.* to appear.

PEARSON, N. E. 1974. Optimal foraging theory, Quan. Science Paper No. 39, Center for Quan. Science in Forestry, Fisheries and Wildlife. Univ. of Washington, Seattle, Washington 98195.

PULLIAM, H. R. 1974. On the theory of optimal diets, *Amer. Natur.* **108**, 59–74.

PYKE, G. H. 1974. Studies in the foraging efficiency of animals, Ph. D. Thesis, Univ of Chicago.

RAPPORT, D. J. 1971. An optimization model of food selection, *Amer. Natur.* **105**, 575–587.

RICHERSON, J. V. AND BORDEN, J. H. 1972. Host finding of *Coeloides brunneri* (Hymenoptera: Braconidae), *Canad. Entomol.* **104**, 1235–1250.

ROYAMA, T. 1970. Factors governing the hunting beavior and selection of food by the Great Tit, *Parus major*, *J. Anim. Ecol.* **39**, 619–668.

SCHOENER, T. W. 1969. Models of optimal size for a solitary predator, *Amer. Natur.* **103**, 277–313.

SCHOENER, T. W. 1971. Theory of feeding strategies, *Ann. Rev. Ecol. Syst.* **2**, 369–404.

SMITH, J. N. M. 1974. The food searching behavior of two European thrushes. I. Description and analyses of the search paths, *Behavior* **48**, 276–302; II. The adaptiveness of the search patterns, *Behavior* **49**, 1–61.

SMITH, J. N. M. AND DAWKINS, C. R. 1971. The hunting behaviour of individual Great Tits in relation to spatial variations in their food density, *Anim. Behav.* **19**, 695–706.

TAHA, H. A. 1971. "Operations Research," p. 703. Macmillan, New York.

TIMIN, M. E. 1973. A multi-species consumption model, *Math. Biosci.* **16**, 59–66.

TINBERGEN, N., IMPEKOVEN, M., AND FRANCK, D. 1967. An experiment on spacing-out as a defense against predation, *Behavior* **38**, 307–321.

TULLOCK, G. 1970. The coal tit as a careful shopper, *Amer. Natur.* **104**, 77–80.

WARE, D. M. 1975. Growth, metabolism and optimal swimming speed in a pelagic fish, *J. Fish Res. Bd. Canad.* **32**, 33–41.

WERNER, E. E. 1974. The fish size, prey size, handling time relation in several sunfishes and some implications. *J. Fish Res. Bd. Canad.* **31**, 153–1536.

WERNER, E. E. AND HALL, D. J. 1974. Optimal foraging and size selection of prey by the bluegill sunfish *(Lepomis mochrochirus)*, *Ecology* **55**, 1042–1052.

Printed by the St Catherine Press Ltd., Tempelhof 37, Bruges, Belgium.

McCullough, D. R. 1984. Lessons from the George Reserve, Michigan. Pages 211–242 *in* Halls, L. K., editor. *White-tailed deer: ecology and management*. Stackpole Books, Harrisburg, Pennsylvania, USA.

The landscape for big game harvest management is undergoing serious changes as society alters its views of hunting. Big game managers no longer simply satisfy the needs and desires of hunters. They have to consider all stakeholders who are interested in big game, which makes management increasingly complex. Hunting continues to be an important aspect of wildlife management and requires specific techniques to remain effective. The art and science of harvest management is exemplified in studies from the George Reserve, Michigan.

The George Reserve in southern Michigan has been one of the longest-studied and most well-known sites for learning about the ecology of white-tailed deer in relation to harvest. Such long-term studies are very rare in wildlife ecology, and their results contribute to our understanding of the long-term dynamics of animal populations. Biologists at the University of Michigan began studying the George Reserve's enclosed population in 1928, and McCullough's objective in this chapter was to summarize that work and expand on his classic studies (our definition, not his), which began in 1966.

In addition to information about population increases, recruitment, and productivity rates, McCullough's chapter includes an excellent analysis of exploitation that considers time lags, relationships of productivity to the residual population, and an array of other factors that influence hunting. McCullough also emphasizes the importance of having clear objectives, setting goals, managing deer for non-hunter satisfaction, and defining our roles in society as professional wildlife biologists. The chapter reviews historical hunting models and establishes new models to consider.

RELATED READING

Carpenter, L. H. 2000. Harvest management goals. Pages 192–213 *in* S. Demarais, and P. R. Krausman, editors. Ecology and management of large mammals in North America. Prentice Hall, Upper Saddle River, New Jersey, USA.

Creed, W. A., F. Haberland, B. E. Kohn, and K. R. McCaffery. 1984. Harvest management: the Wisconsin experience. Pages 243–260 *in* Halls, L. K., editor. White-tailed deer: ecology and management. Stackpole Books, Harrisburg, Pennsylvania, USA.

McCullough, D. R. 1979. The George Reserve deer herd: population ecology of a K-selected species. University of Michigan Press, Ann Arbor, USA.

McCullough, D. R. 1996. Spatially structured populations and harvest theory. Journal of Wildlife Management 60:1–9.

McCullough, D. R., and W. J. Carmen. 1982. Management goals for deer hunter satisfaction. Wildlife Society Bulletin 10:49–52.

Peyton, R. B. 2000. Wildlife management: cropping to manage or managing to crop? Wildlife Society Bulletin 28:774–779.

Chapter 10

LESSONS FROM THE GEORGE RESERVE, MICHIGAN

Dale R. McCullough
Professor of Wildlife Biology and Management
Department of Forestry and Resource Management
University of California
Berkeley, California

The George Reserve in southern Michigan has been yielding information on white-tailed deer for more than 50 years. Biologists at the University of Michigan have been studying the deer population there since 1928, when whitetails were introduced by Colonel Edwin S. George, a Detroit industrialist who established the area as a personal estate. I began intensive studies of the population in 1966 and, with the support of the National Science Foundation, these studies have continued to the present.

The George Reserve is a 464-hectare (1,146-acre) area enclosed by a 3.5-meter (11.5-foot) deer-proof fence. It is located in Livingston County, about 7.7 kilometers (3 miles) west of Pinckney, Michigan. It is in a glaciated area where one of the most recent southward advances of the ice shield terminated.

The soils are young and poor in fertility, consisting in the uplands of mainly sands and gravels. It should be kept in mind that productivity of the George Reserve deer population is not the result of superior soils; most soils in the lower Great Lakes and Upper Midwest region are far more fertile that those of the George Reserve. What the Reserve has is diversity, due to recent glaciation. The topography is rough and broken, with rolling uplands, steep-sided eskers and kettle-hole sinks.

The varied topography supports a complex mosaic of vegetation, with five major types and a number of minor ones being recognized (Figure 48). Of the major types, oak/hickory hardwood forest on the uplands predominates (47 percent), while open grasslands, created by the clearing of forests by man in the late 1800s, are second (26 percent). Tamarack swamps are the most common lowland type (about 16 percent), while freshwater marshes cover 8 percent and leatherleaf bogs about 2 percent. The remainder (1 percent) is made up of assorted minor types and open water. For a more complete description of the area and its vegetation, *see* McCullough (1979).

After completing the fence in 1928, Colonel George introduced six whitetails—two bucks and four does (presumed to be pregnant)—from Grand Island in Lake Superior. George then gave the Reserve to the University of Michigan in 1930. The deer population increased rapidly and, by 1933, a minimum of 160 deer were accounted for in a drive count (Hickie 1937, O'Roke and Hamerstrom 1948). Because it is known that drive counts in the early years were low, a revised estimate of more than 220 deer appears more likely (McCullough 1979).

Severe damage to vegetation was being caused by this number of deer. And although the Reserve was managed by the university as a natural area for research, it was apparent that the whitetail population would have to be controlled. Hunting was begun in 1933 and continued thereafter to control the population at desired levels.

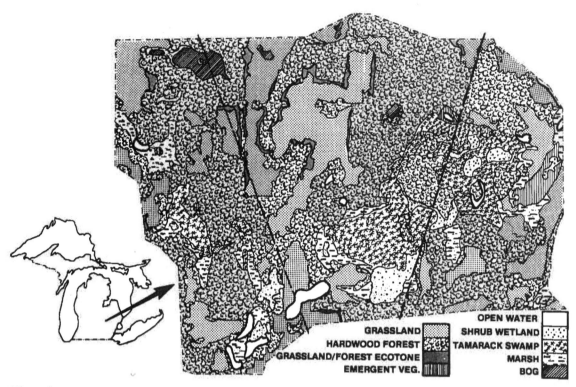

Figure 48. Vegetation of the 464-hectare (1,146-acre) George Reserve in southcentral Michigan (after Roller 1971).

Since 1966, the population size has been altered purposely, in an attempt to determine experimentally the role of density on population dynamics. The population was reduced gradually to about 10 animals in 1975 and then protected from shooting and allowed to increase again, until 212 animals were present in 1980. Since then, the population has been held to 130 animals by shooting.

SEASONAL HABITS

Deer on the Reserve have social systems similar to those of whitetail populations elsewhere in the region (Hirth 1977a). Small groups predominate due to heavy concealment cover. Individuals have home ranges that overlap and usually extend over parts of the major vegetation types (Queal 1962). Deer use all of the vegetational types on the Reserve, shifting about as resources become available and the animals' physiological needs change.

Based on the mean for five years of pellet-group counts, measurements of habitat-use by plant phenological seasons show distinct patterns (McCullough 1982a) that relate primarily to food habits (Figure 49). However, there is substantial variation in use from year to year depending on the acorn crop size, amount of snowfall and other environmental variables. Deer are quite discriminatory, selecting the best food available in any given season.

In early spring, grasses predominate in the whitetail diet, and use of openland habitats is prevalent. As forbs become more available in late spring and early summer, they replace grasses in the diet. As most forbs begin to dry in late summer, the deer switch to green leafy browse and spend much of their time in the wetland types, particularly tamarack swamps (Figure 49). The swamps also are the coolest type when temperatures are highest in July and August. Acorns become available in autumn most years, and use of hardwood forests increases rapidly as acorns become predominant in the diet. If the acorn crop fails, autumn use of hardwood forests is low, as the deer seek out other seeds and fruits, including sumac, hawthorn, lespedesa, apples, grape and chokeberry. These foods are taken in minimal amounts if acorns are available.

In most years acorns disappear by late November. But particularly in years of heavy crops

Oblique aerial view of the central part of the George Reserve. The image area corresponds to the area between the dash lines on Figure 48. Note the rough terrain and interspersion of vegetational types. *Photo by James W. Wheeler; courtesy of The University of Michigan Press, Ann Arbor.*

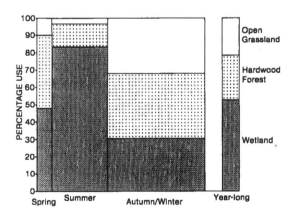

Figure 49. Habitat use of white-tailed deer on the George Reserve by season and yearlong, as indicated by pellet-group counts over five years, 1966–1972 (McCullough 1982a). Yearlong result is weighted by length of seasons. Wetlands include tamarack swamps, leatherleaf bogs and marshes.

they will predominate in the diet into January, and the whitetails will paw through more than 0.3 meter (1 foot) of snow to obtain them. When acorns are gone, deer shift back to grass, which undergoes an autumn flush due to increased rainfall after the drier period in August and September. Grass remains green next to the ground throughout the winter, and deer feed on this material as well as on some green rosettes of forbs at ground level. Browse is used to fill out the winter diet.

These food habits hold as long as snow cover is not present. However, with a snow cover of 7.5 centimeters (3 inches), the deer abandon herbaceous diets and shift almost entirely to evergreen woody browse (Coblentz 1970), primarily redcedar, ground juniper and, to a much lesser extent, leatherleaf. Hardwood forest stands and tamarack swamps are used as thermal cover for bedding, and deer come into the openlands at night to obtain the two species of juniper, of which redcedar is by far the pre-

Redcedar is the most important deer food in winter on the George Reserve when snow depth reaches 7.5 centimeters (3 inches). Whitetail browsing pressure results in a distinct highline on this species. *Photo by Dale R. McCullough.*

ferred. Deer trails after snows form a network connecting the scattered individual redcedar trees, which show a pronounced hedged appearance. Under the coldest conditions, with subzero temperatures and strong winds, deer are restricted to dense tangles of brush in the tamarack swamps.

Does with young occur in social groups separate from male social groups except during the rut (Hirth 1977*a*, Newhouse 1973, McCullough 1982*a*). There also is some evidence that does with young occupy geographic areas separate from those occupied by males (McCullough 1979).

Rutting behavior is shown by bucks from October through December (Hirth 1977*a*), with conception in yearling and adult does occurring in late October and early November (McCullough 1979). Conception of female fawns—which reach sexual maturity at about seven months of age under the good food conditions prevailing at low population densities—occurs in early December (McCullough 1979). Fawns are born in June and July and are hidden in hardwood forest or on the forest-tamarack swamp edge for the first few weeks of life, until they are able to follow their mothers.

MODELS FOR MANAGEMENT

A model is nothing more than a miniature conceptualization of a real-world system (*see* Chapter 9). Models are necessary because the real world is big and complex, and cannot possibly be completely understood. Even after intensive research, many answers elude investigators. The scale over which the typical wildlife agency must operate, and the budget available to do so, are such that only a few statistics can be gathered about the status of a given deer population. Nevertheless, the agency must set regulations for all of the deer population units under its jurisdiction. Models are ways of coping with this problem. But they are always deficient to some extent because, of necessity, they are simplified.

Modern models, because they are invariably mathematical, often are incomprehensible and have created an aura of mystery. Practicing professionals have tended to react to models in one of three ways. First, and most common, is simple bewilderment—not knowing what to think. Second is defensive rejection—considering models as witchcraft or the nasty tricks of mathemagicians. Third is to consider them akin to religion, something to believe in—a miracle that will make the impossible task possible.

Models are tools, nothing more. They differ from religions in that one does not have to die to find out if they are right. They differ from beliefs in that one does not have to accept them in total. Before applying a model to the whole, a separate part can be tested. By themselves, models do nothing. In the hands of a good craftsman, they can produce useful information. Models do not eliminate work; but they can make it easier. On the other hand if a model is misused, as with any decision-making aid it can be dangerous. More than one agency wound has been self-inflicted, and it is inevitable that some nasty gashes will be received from the use of an inadequate model or the misuse of an adequate one.

The first wildlife population models were word models, and they began with Aldo Leopold's *Game Management* (1933). Unfortunately, the wildlife profession accepted this work as a literal description of the real world rather than a word model for how one man thought the real world worked. The genius of Leopold is attested to by the degree to which his wisdom served the wildlife profession so well for so long. It detracts not at all from his genius to observe that, while it served adequately for 30 years, it is inadequate today. However, while Leopold's science has become dated, his philosophy promises to live forever.

Mathematical models have been developed because the symbolic manipulations possible are so powerful and words usually are inadequate to express complex relationships. The computer aided in this development because of its computational power, but a computer simply is a rapid calculator. It is the program that is important, and the computer should receive neither the credit nor the blame for what the programmer does.

There are two different approaches to predictive modeling for white-tailed deer population management. These can be classified broadly as "accounting" models and "black box" models. Accounting models keep track of population size and composition through elaborate bookkeeping of births and deaths. Best known of this kind of model in deer management is ONEPOP (Walters and Gross 1972, Gross et al. 1973) and its offshoots (Bartholow 1981), but there are many other examples (*see* Medin and Anderson 1979, F. M. Anderson et al. 1974).

Accounting models have substantial heuristic value. One can play "what if" games (Pojar 1977, Williams 1981) by changing parameters to test population responses to changes in birth rates or by shifting mortality rates between sex and age classes. But these models have drawbacks as tools for practical management, the most significant of which is their great requirement for data. One needs information on population size, age and sex composition, and birth and death rates by age class. Indeed, if all these data are available, one hardly needs a model.

A second drawback of accounting models is that the rates for sex and age groups are not synchronized within the model. Thus, changes in one parameter are not simultaneously compensated for by changes in other parameters. While some linkages are possible, the programming necessary to integrate the array of categories and variables in the model is overwhelming (Pospahala 1969).

Black box models, on the other hand, assume that a deer population is a far more effective integrator of variables than the population's researchers are. Black box models assume that natural selection over millions of years has shaped responses to perturbations by hunting, habitat change, predation, etc., to the best solution the population is capable of attaining. Thus, a few high-order variables are all that need to be measured, since they represent the population's integration of the total complex of lower-order variables.

Models for whitetail management are useful as practical tools only to the extent that they incorporate the important integrating variables of vegetation/deer/predator systems. One of the failings of the wildlife profession is an erroneous belief that an understanding of the population will somehow emerge from increasingly detailed measurements of plant species composition and pounds of browse, or better estimates of birth and death rates and population size. Having point estimates of these parameters is of some value. But they do not address the fundamental problem—how do changes in each parameter vary in relation to changes in all of the other variables? Deer population response to changes in density integrates all of those variables into numbers of deer with a precision that researchers will never achieve, and that is the bottom line.

Similarly, it is argued whether predators regulate, or deer self-regulate, or hunting is necessary to regulate, without any agreement on what is being argued. What is meant by regulating—holding between limits? Or at a stable point? And if the latter, at what point? I carrying capacity or K carrying capacity (*see* Carrying Capacity)? or somewhere in between? And how does one establish that regulation has occurred? Does any outcome (other than total extinction) mean that regulation has occurred? If not, by what criteria do we determine whether regulation has or has not occurred? If we are to progress it is imperative that our questions be framed the right way. Asking the wrong questions or stating questions in a fuzzy manner diverts attention from real informational needs.

What is important to know is the shape of the productivity curve and the point on it where the population stands. The population density response is independent of the causes of mortality. It makes no difference whatsoever to the survivors in the residual population whether the individuals no longer present fell to bullet, fang, radiator grill or the vagaries of time. If the population's position on the productivity curve is known, the outcome of a higher or lower human hunting harvest can be predicted, and harvests can be changed carefully and incrementally to test those predictions. Model and practice cross-check and reinforce each other. Conversely, current hunting regulations can be fixed and tested to assess how other mortality factors relate to hunter harvest. For example, it can be hypothesized that predators are having an impact on a deer population sufficient to lower the hunting harvest. Thus, if predators are reduced by increasing investment in predator control, the return (deer in the hunter take) can be measured. If the hunting harvest increases as predicted, the cost of the predator-control program can be weighed against the benefit of more deer in the bag. If hunter harvest does not increase, there certainly is no value in predator control for deer management. This is much more useful than knowing exactly how many deer a given predator was killing.

I propose that it is time for management to abandon the quest for the absolute estimate, which is difficult or impossible to obtain and of limited use if known. It is time to concentrate on functional relations that are amenable to discovery through manipulation of controllable variables, such as hunting.

The empirical models I propose in the following discussion are based on this approach. These models are of the black box type known as "stock-recruitment" models. They have been used by fisheries biologists for many years (*see* Ricker 1954, 1975, Beverton and Holt 1957). However, the models presented here are modified somewhat, being derived empirically through data fitting rather than through predetermined mathematical formulations.

RATE OF POPULATION INCREASE

After the introduction of six adult whitetails in the George Reserve in 1928, the population grew to an estimated 222 deer in seven years (McCullough 1979). This is an instantaneous growth rate of 0.516. This population-growth phenomenon was repeated in 1975, when the population was reduced to 10 (primarily fawns) by heavy harvesting and then protected. In 1980, six seasons later, the population was 212 (McCullough 1982*b*). The instantaneous growth rate for this period was 0.509. If the different starting numbers are corrected for, the results of the two periods with deer of different ages

are virtually identical. However, for an opposing view, *see* Van Ballenberge (1983); for a response to that view, *see* McCullough (1983).

It should be noted that although these rates of population growth are among the highest recorded for white-tailed deer, they do not equal the unimpeded rate. With no mortality, and maximum reproductive rate, the population after six years would be 303 for a starting population of 6. For a starting population of 10 it would obviously be still higher. Thus, the rate of growth was actually declining as population increased in a density-dependent fashion (*see* Recruitment Rate, *see also* McCullough 1983).

These growth rates are significant for several reasons. First, they illustrate the capacity of a white-tailed deer population to increase rapidly, even in an environment that inherently is not very fertile, so long as intraspecific competition for resources is minimal. At low densities most does reproduce as fawns, and twins (and even triplets) in fawns have been recorded (Haugen 1975, McCullough 1979). Yearlings commonly have twins, while triplets occur frequently in older females.

Second, the growth rates illustrate that density-dependent effects are subtle but measurable if census methods are reliable. The fact that density effects were present in both growth-rate experiments before high density (indeed before I carrying capacity) was achieved means there is considerable fine-tuning of population-growth response to the quality and quantity of resources available. That the initial introduction was made with adult animals—and that growth would be lowered by the addition of young animals, with their lower reproductive rates—could be put forward as an explanation for the decline in population-growth rate in the years following the experiment's inception. However this explanation was eliminated in the repetition of the experiment, in which an attempt was made to remove all yearlings and adults, leaving primarily fawns. Density-dependence effects, then, are the only reasonable explanation.

That resource availability rather than density *per se* was responsible for the declining population growth rate is demonstrated by a comparison of population densities at different resource states. Impacts on the vegetation were observed—heavy browsing on redcedar being particularly obvious (Hickie 1937, O'Roke and Hamerstrom 1948, Chase and Jenkins 1962)—following the initial increase, and recruitment rates declined in response to vegetation damage. High recruitment rates before vegetation damage and low rates after damage were obtained at the same population size (*see* Figure 51). If density *per se*, rather than resources, had been the cause of the growth-rate decline, the recruitment rate should have been equal at the same density. Recovery of the vegetation occurred during subsequent periods of low deer density. This point will be covered further later in the chapter. Although some changes in major vegetational types have occurred—particularly expansion of the hardwood forest ecotones into the openlands (Roller 1974)—the basic capacity of the area to support deer has not been altered appreciably.

Third, these population growth experiments show that social behavior factors, such as purported social regulation (Wynne-Edwards 1962) or "immature" behavior patterns (Bubenik 1971, 1982), do not play a significant role in white-tailed deer. The first increase began with animals that were two years old or older, and the second began primarily with fawns. The outcome was virtually the same.

Fourth, these experiments reply to the criticism of Smith (1981) about genetic factors being ignored, particularly the effects of inbreeding depression. The George Reserve whitetail population has passed through two genetic bottlenecks and, no doubt, has suffered loss of heterozygosity. That there have been some genetic effects seems likely. For example, George Reserve bucks have an exceptional number of points and other antler characteristics that suggest the influence of the "founder effect" (McCullough 1982c). Because a few individuals establish the population, the genes contributed are limited, and particular characteristics of these initial members are strongly represented in the subsequent population.

However, the George Reserve deer population has had the highest sustained growth recorded for white-tailed deer—not once, but twice. At least 70 percent of fawn does have achieved sexual maturity and recruited young to the population; twins by fawn does have been recorded. Buck fawns commonly exceed 45 kilograms (100 pounds) by autumn; yearling bucks have eight-point antlers; and trophy heads are produced by four years of age (McCullough 1982c). These are the effects of reduced density. And if there was "deleterious" inbreeding, it was overshadowed completely by compensatory processes.

This is not to deny the potential of genetic problems associated with small populations, but rather to question the assumption that they are inevitable, or that they are more important than maintaining the population within the environment's resources—even if that means maintaining a small population.

RECRUITMENT RATE AND PRODUCTIVITY

The dynamics of any population can be described by reproduction, mortality, immigration and emigration over time. Because of the deer-proof fence on the George Reserve, movements as a population variable can be ignored. Gross reproduction was estimated by embryo rates. Although abortion and resorption do occur in deer, they are relatively infrequent and no case of either has been observed on the George Reserve. Furthermore, corpora luteal counts have been virtually equal to embryo counts in the same sample of does (McCullough 1979).

Net reproduction was measured as recruitment. Recruitment is a useful concept in population studies since it is measured at an age when the young are at some threshold of practical importance. In this case, recruitment age was considered to be six months of age—the autumn of the first year, when fawns were large enough to be included in the harvest. The value of recruitment as a concept is that it integrates births and early mortality when death is most probable—each of which is exceedingly difficult to measure in practice—into one measure after survivorship begins to approach that of older animals.

Recruitment rate is the number of recruits per individual in the population producing the recruits. The population producing the recruits will be referred to as the "residual population"—that remaining after hunting and other mortality factors have had their effect. This population was referred to as the "posthunt population" by McCullough (1979). However, because it includes cases where hunting is not present, the term "residual population" is more appropriate.

It is useful to divide mortality occurring after recruitment age into two categories. "Chronic mortality" includes those deaths that are routine and inevitable. They include deaths due to the combined effects of old age, typical parasite loads, nonvirulent diseases, malnutrition,

A deer-proof fence around the George Reserve has made the site a valuable outdoor "laboratory" for study of white-tailed deer populations. *Photo by Doug Fulton.*

debilitating accidents and similar causes. Because they are chronic does not mean that they are density independent. In fact, most are density dependent. The other form is "traumatic mortality," and it includes deaths due to direct intervention of outside forces such as hunting, predators, vehicle collisions and virulent diseases that kill by gross physical damage. These deaths have the characteristic of cutting the lives of individual deer short of what they would have been had only chronic-mortality factors been operating.

Thus, total mortality is comprised of three kinds: prerecruitment, chronic and traumatic. This division is useful because a major source of traumatic mortality, hunting, is under the control of wildlife managers through regulations established for deer management units. Thus, it is necessary to determine how increasing or decreasing hunting-harvest influences traumatic mortality due to other variables, and how total traumatic mortality influences prerecruitment and chronic mortality. The pivotal question is whether hunting mortality is ad-

ditive (that is, in addition to other mortality) or substitutive (that is, compensated for by decreases in other mortality). If it is additive, then the wildlife biologist or resource manager is dealing with an inflexible biological system with little resilience (very prone to overexploitation). Hunting is liable to be, so to speak, the straw that breaks the camel's back, and very conservative approaches are required. If, on the other hand, hunting mortality is compensated for by a reduction in other mortality, then the biologist or manager is dealing with a resilient biological system that is tolerant of exploitation—one in which hunting can be used as a driving variable to milk the proverbial camel for all it is worth.

Net recruitment rate—called "population change" in McCullough (1979)—is defined as the number of recruits minus mortality of animals of recruitment age and older divided by the residual population (Figure 50a). Productivity is the number of fawns born minus the combined mortality of all three kinds at a given residual population size.

The productivity curve of the George Reserve whitetail population is shown in Figure 50b. Maximum sustainable yield is the largest value on the curve (49), and it is obtained at a residual population of 99. It occurs at the inflection point on the net recruitment rate curve, where a linear relationship at lower residual population becomes curvilinear (Figure 50a). It also is at the inflection point of an S-shaped population growth curve, where the increasing curve bends into a decreasing curve (Figure 50d).

CARRYING CAPACITY

Carrying capacity, K, is that residual population at which productivity declines to zero.

K carrying capacity is defined further as the maximum number of animals an environment will support on a sustained basis (that is, without destruction of the vegetation). Indeed, there will be substantial impact, and K is the residual population size causing the maximum defoliation the vegetation is capable of sustaining. Whitetail populations clearly are capable of exceeding K and causing damage to vegetation, as discussed later (*see* Time Lags).

In McCullough (1979), the residual population yielding maximum sustainable yield is referred to as I carrying capacity ("I" for inflection point) to distinguish it from K carrying capacity. There has been considerable confusion in the wildlife literature, and carrying capacity has frequently been defined as that population yielding maximum sustainable yield.

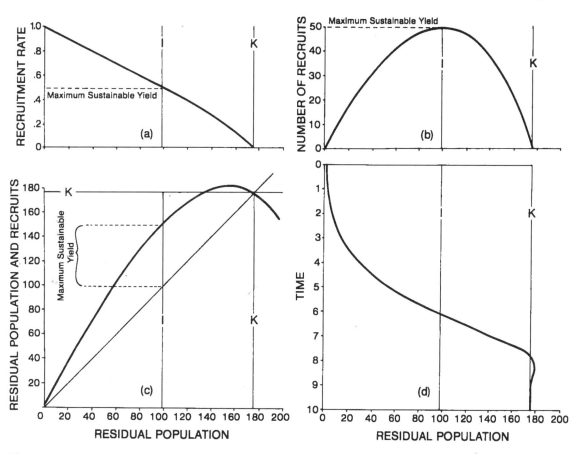

Figure 50. Graphic representation of equilibrium (without time lags) population dynamics for the George Reserve white-tailed deer population in Michigan, showing equivalent values of I and K carrying capacity, based on data gathered from 1952 to 1971: (a) recruitment rate on residual population size; (b) net recruitment or productivity curve; (c) stock-recruitment curve; and (d) population-growth curve over time. K = maximum residual population (density) at which productivity (number of recruits) declines to zero—this is carrying capacity of an area, and represents the maximum number of deer the area will support on a sustained basis without destruction of habitat. I = residual population yielding maximum sustainable yield.

Much of the confusion apparently arose from thinking of the balance of population size and vegetation as representing a given point, when in actuality it is an array of points in a continuum of residual populations from zero to K that represents the functional relationship between the dynamics of deer and the dynamics of vegetation.

In ecology, carrying capacity has been the term applied to K, and this is the correct usage. Referring to residual population yielding maximum sustainable yield as "carrying capacity" clearly is a misnomer. But because of past confusion in the wildlife literature, it seems less confusing at this stage to retain the term and refer to it as I carrying capacity. Curiously, the residual population that yields maximum sustainable yield has not been given a label, for maximum sustainable yield refers to the yield, not the residual population giving the yield. Thus, "I" is an appropriate label for this important residual population on the productivity curve, just as "K" refers to the maximum-equilibrium residual population.

Optimum sustained yield is the yield that a population can sustain that maximizes human benefits in a given case. Because the benefits to be achieved (that is, defining what should be optimized) are derived subjectively, optimum sustained yield depends on the goals of the management program.

IMPACT OF EXPLOITATION

Illustration of the impact on a whitetail population at various levels of exploitation can be made most easily in the simplest case. Thus, no variation from the mean values is assumed, and the productivity curve in Figure 50b is treated as a deterministic model. It can be assumed further that hunting is the only form of traumatic mortality operating on the population, which was true of the George Reserve deer population.

The first point to be made is that with the exception of maximum sustainable yield, which has a single intercept point on the productivity curve at a residual population 99, all smaller values of Y have two intercept points. Thus, a harvest of 30 animals yields a balance point of 38 and 150 residual population, and the same yield can be obtained from two residual populations. This is true of all harvests less than maximum sustainable yield, including zero harvest that is stable at extinction and at K. This phenomenon explains why wildlife populations occasionally are observed to go from a low and stable level to a very much higher and stable level. The reverse of high-to-low also occurs. These are a result of the prevailing mortality rate being temporarily increased or decreased and the population either growing or declining to the opposite balance point where

Recent glaciation created the George Reserve's broken topography and sandy and gravelly soils of low fertility. Oak/hickory forests contribute woody browse and most crops to the deer diet. *Photo by Dale R. McCullough.*

a resumed mortality rate will stabilize it once again (McCullough in press).

This dual balance-point relationship holds promise for management because it means that a given yield can be obtained from a high or low residual population. If the wildlife biologist or resource manager wants to minimize crop damage or deer/vehicle collisions, the lower residual population can be favored. If the desire is to maintain a higher residual population to satisfy protectionists, or so hunters will not worry about the population being overexploited, the high balance point can be favored. This case is treated more fully in McCullough (in press).

However, the next point to be made is that the balance point on the left arm of the productivity parabola (Figure 50b) is highly unstable, while the one on the right arm is highly stable. Therefore, management for the high residual population is forgiving of errors because the population responses are compensatory. At the low population, population responses are destabilizing.

For example, if a fixed harvest of 30 whitetails is being taken at the lower residual population balance point, and 31 are killed accidentally, the residual population will be reduced and an offsetting harvest the following year will be necessary to bring the population to the original residual population. The same is true of a harvest that is accidentally too low, such as 20. Because the higher residual population will recruit additional individuals, a greater correcting harvest will be required to establish the original balance point.

Conversely, on the right arm of the parabola, errors are self-correcting. If too many deer are killed, the reduced residual population has increased recruitment, which tends to force the population back toward the original balance point. If too few are killed, the increased residual population has a lower recruitment, and the population tends to decline back to the original balance point. Therefore, the right arm of the parabola is a highly stable region since the density-dependent population responses tend to correct for errors in management.

Residual populations giving maximum sustainable yield behave like residual populations on the left arm if maximum sustainable yield is exceeded, and like those on the right arm if underharvested. Thus, maximum sustainable yield is unstable in the face of harvests too high, and stable for harvests too low. The rules

for deciding how a population will respond can be stated as follows:

1. If a fixed harvest exceeds maximum sustainable yield, continuing the fixed harvest will lead to population extinction no matter what the residual population.
2. If the residual population is on the left arm of the parabola or at the residual population level that yields maximum sustainable yield, a fixed removal equal to recruitment will stabilize the population at that residual.
3. If the residual population is on the left arm of the parabola, and the fixed harvest exceeds the recruitment, the population will be driven to extinction. If the recruitment exceeds the harvest, the residual population will grow to the balance point on the right arm of the parabola for that fixed harvest.
4. If the residual population is on the right arm of the parabola and if the harvest exceeds the recruitment at that residual population, but is less than maximum sustainable yield (*see* rule 1.), the residual population will decline to the balance point for that fixed harvest on the right arm of the parabola. If the harvest is less than the recruitment, the residual population will grow to the balance point nearer to K.

RELATIONSHIP OF PRODUCTIVITY TO RESIDUAL POPULATION

The relationship of productivity to residual population is best illustrated by a stock recruitment graph (Figure 50c), where the productivity curve represents net recruitment. If human hunting is the sole form of traumatic mortality, the entire productivity needs to be removed to stabilize the residual population at a given value.

From Figure 50c it can be seen that moderate harvests from a white-tailed deer population can result in prehunt populations greater than K and posthunt (residual) populations less than K. At the residual population yielding maximum sustainable yield (that is, I carrying capacity), both the prehunt and the posthunt populations are less than K. This illustrates an important point that has been overlooked by wildlife biologists and resource managers: management for maximum sustainable yield results in fewer deer in the field. This point has resulted in much failure of communication among biologists, hunters and others. Biologists and managers emphasize the high and sustainable harvest that is being supported, while hunters worry about the number of deer and other signs of deer seen in the field while hunting (McCullough 1979). Failure to appreciate that hunting harvests approaching maximum sustainable yield inevitably reduce the deer population has caused loss of credence among hunters, who see that there are fewer deer in the field and therefore assume that antlerless seasons have resulted in overexploitation of the population. Furthermore, because a higher harvest is being taken from a lower population, the average effort of the hunter per deer harvested goes up, which further convinces the hunter that the population is not what it was before antlerless hunting began.

INADEQUACY OF THE HARVESTABLE SURPLUS CONCEPT

The preceding discussion illustrates the deficiency of harvestable surplus as a management strategy. Leopold (1933) defined harvestable surplus as those animals above the replacement population. The replacement population would be equivalent to the residual population, and harvestable surplus equivalent to the recruitment in this context. Because recruitment tends to balance the harvest at any sustainable residual population, the harvestable surplus is a characteristic of the management program practices. Thus, if a white-tailed deer population is lightly harvested, the surplus will be small in the following year because the excess above the replacement population will be small. Because a small surplus suggests that a small harvest should be taken, the concept logically suggests that what has been done in the past should be continued in the future. If a heavy harvest has been taken, recruitment will be high, indicating a high surplus.

Thus, the harvestable surplus concept leads to the conclusion that the right program is being followed, no matter what the program. The productivity-curve approach given here indicates what the potential surplus might be if the population were forced to its maximum productivity (maximum sustainable yield). It also

Deer harvests on the George Reserve between 1953 and 1969 were conducted by drivers pushing deer past shooters on stands. Here a harvest crew takes a break in Camburn Laboratory under a mounted trophy buck taken in the early years. Bucks taken recently, when the population has been low density, exceed this trophy in size and antler quality. *Photo by Doug Fulton.*

indicates removals that exceed maximum sustainable yield and cannot be sustained, as well as the residual population associated with any sustainable harvest.

DETERMINISTIC POPULATION GROWTH

White-tailed deer population growth without destructive vegetative overshoots can be projected using the previous functional relationship between residual population and productivity. Because of the whitetail's high reproductive rate, population growth has a tendency to overshoot K slightly and dampen out to K even without destruction to vegetation (McCullough 1979, in press). Population growth without destruction of vegetation (Figure 50d) shows a substantial difference from the growth rates beginning from very low populations covered in the earlier section, "Rate of Population Increase."

TIME LAGS

There are two types of time lag: those that occur within the constraints of the mean equilibrium values of the productivity curve, and those due to lags in the deer/vegetation system that result in departure from equilibrium conditions (that is, lowering of K by vegetation damage).

The first type, those that follow the mean productivity curve, include lags due to shifts in size of harvest, and temporary nondestructive overshoots of K. Lags due to shifts in harvest occur because the density-dependent response is not instantaneous. For example, if one were stabilizing the residual population on the left arm of the productivity curve with a given harvest and then lowering the harvest, it would take some time for the residual population to grow to the right arm of the parabola and come to equilibrium on that arm with the new harvest.

Nondestructive overshoots (Figure 50d) occur because the mean recruitment curve of the stock-recruitment graph exceeds the horizontal line through K (Figure 50c). Illustrations of population growth curves giving temporary overshoots are shown in McCullough (in press), and seasonal overshoots can be produced by managing for residual populations (138–175) where the recruitment curve is above the horizontal line in Figure 50c.

Destructive overshoots occur where "accumulations" of resources allow population growth that exceeds the equilibrium values. This occurred on the George Reserve following the initial whitetail population increase, which deviated from the mean value at the inflection point of the mean productivity curve (McCullough 1979, in press). Therefore, the critical departure of destructive overshoot occurs not when K is exceeded, but rather when the population growth rate exceeds the mean equilibrium value at I carrying capacity. This departure is shown with reference to recruitment rate in interval B of Figure 51. Note that the same departure was recorded in the experimental increase of the population conducted in recent years, one purpose of which was to study the phenomenon of destructive overshoot.

The harvests imposed in the initial overshoot were too conservative to prevent vegetation destruction, and recruitment rates below the mean values persisted in the following years (interval C of Figure 51). This lowered capacity shows that deer were responding to resources and not to density *per se*, for if the latter were responsible, recruitment should have been the same at the same density. After the residual population on the Reserve was lowered below 80 animals, vegetation recovery occurred, and the recruitment rates returned to the mean value (interval D of Figure 51). Subsequent increases in the residual population (interval E of Figure 51) and decreases in the population as a whole (interval F of Figure 51) followed the mean equilibrium values.

As noted earlier, the recent increase followed the original increase (intervals A and B of Figure 51). The overshoot experiment I currently am conducting is to show that vegetation destruction is not inevitable given the departure of population growth from the mean equilibrium curve at I carrying capacity. Thus, while the initial growth was curbed too slowly, I am trying to show that if sufficient control is exerted to allow time for the population to use up the accumulation of resources remaining from the period of extremely low population, the recruitment rate will decline to the mean equilibrium value without damage to the vegetation and lowering of K. That is, interval B in Figure 51 can be brought down to the equilibrium curve and maintained there without continuing in interval C. That is being done by holding the residual population to 130 animals,

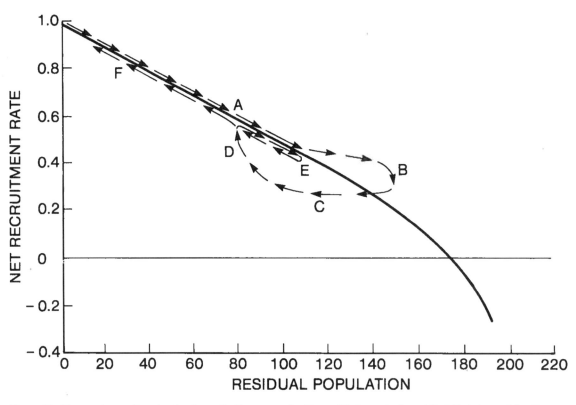

Figure 51. Observed recruitment rates (smoothed) compared with equilibrium rate for white-tailed deer of the George Reserve in Michigan over time, showing time lag. Time intervals are: (A) initial population growth rate following introduction and experimental population growth (1928–1931, 1975–1980); (B) initial population overshoot (1932–1935, 1981–1982); (C) decline in growth rate due to vegetation damage (1936–1946); (D) recovery rate due to population reduction and vegetation recovery; (E) subsequent population increase with observed rate and equilibrium rate comparable (1947–1967); and (F) recent population with observed rate comparable to equilibrium rate (1968–1974).

which is below equilibrium K of 176. However, the recruitment plus the residual population exceeds K on a seasonal basis. This is particularly true early in the overshoot (212 animals in 1980), but the tendency declines as the accumulation of resources is used up. Thus, the first removal required to reduce the population to a residual of 130 was 82 animals. And the hypothesis is that, within the next few years, the required harvest will decline to an average of 44—the equilibrium harvest for a residual population of 130.

STOCHASTIC BEHAVIOR

With the behavior of the whitetail population outlined according to a deterministic model, the model can be complicated to approximate more nearly the conditions of the real world. One of the major problems confronting the wildlife biologist or resource manager is the year-to-year variation in habitat quality due to variable amounts of precipitation, severity of winter, presence or absence of an acorn crop, etc. These variables are beyond managerial control, but whitetail management programs must take them into account. If the environmental stochasticity is small relative to density-dependent responses of the population, management based on empirical models can be followed. This is the case on the George Reserve. If the environmental stochasticity is great relative to the density-dependent response of the environment, an *ad hoc* strategy will be required. That is, the biologist or manager will not be able to develop predictive models based on the density-dependent response of the population and follow a more-or-less consistent management plan. Instead, management decisions will have to be made on a year-by-year basis in response to the environmental factors of the particular year. This does not mean that the deer population does not behave in a density-dependent fashion, or that the conceptual model presented here cannot be used as a general guide to show how the population will respond after an environmental deviation, such as a hard winter, has had its effect. It is just that if extreme conditions occur too frequently, the environmental variation will overshadow the density-dependent response. By analogy to radio communication, what the investigator is dealing with is a signal-to-noise ratio. If the signal-to-noise ratio is high, the signal can be managed, given the level of the noise. If the signal is weak relative to the noise, then response is possible only when the signal is clear, because the noise masks the signal much of the time.

Fortunately, most white-tailed deer populations occur in relatively benign environments and respond well to density-dependent management. However, in the extreme northern and desert fringes of whitetail range, environments can be extreme, and management may well have to be *ad hoc*—or at least require more *ad hoc* interventions than in a density-dependent scheme. Elsewhere I have discussed frequency distributions of good and bad years, and their likely impact on management programs (McCullough 1979, in press).

Stochasticity on the George Reserve can be assigned variance in the regression equation that describes the relationship of recruitment rate to residual population. The confidence bounds on the productivity curve (Figure 50b) are shown in Figure 52, where it can be seen that variation is small at low residual populations but increases rapidly at high residual populations. Good or poor years have a greater impact on high populations of deer.

By analogy, at low deer density the signal (density-dependent effect) is so strong that it greatly exceeds the noise, while at high density the noise has a greater masking effect on the signal. At high densities, most adult whitetail does attempt to reproduce (McCullough 1979), and the potential number of recruits is very high. If a good year occurs, a large number of recruits may be added. Conversely, a poor year may virtually eliminate recruitment in that year, and because the population is on the margin of maintenance costs, many individuals in the residual population will drop below the maintenance level and die of malnutrition. In management terms, this means that populations subject to little hunting that are close to K will show greatest fluctuation, while harvests re-

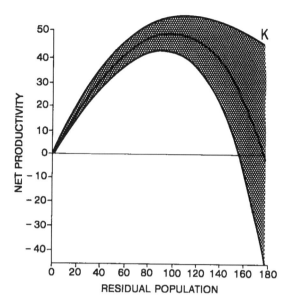

Figure 52. Net productivity of the George Reserve white-tailed deer population in Michigan, based on data from 1952 to 1971, with a 95-percent confidence limit.

ducing the population to lower residual sizes will result in greater stability.

The instabilities previously discussed regarding the left arm of the parabola are intensified by stochasticity (however, *see* Functional Refugia). Because balance at any given residual population on this arm requires exact removal of the number of recruits added, the precision required of management is unrealistic, given that recruitment will vary somewhat even though it is minor compared with that on the right arm (Figure 52).

Because maximum sustainable yield management has instability characteristics similar to those associated with the left arm of the parabola and leads to overexploitation, it is susceptible to excessive removals. However, errors of underharvest are corrected for by density-dependent responses characteristic of the right arm of the parabola. If the mean maximum sustainable yield (49 for the George Reserve deer population) is removed as a fixed harvest over time, extinction of the population is inevitable, because the mean harvest sooner or later will exceed the actual recruitment and move the residual population to the left arm of the productivity parabola. If a disproportionately large recruitment does not occur in the following year, extinction will result.

This characteristic of the population explains why populations overexploited by man show sustaining capability and then apparently collapse suddenly. On the right side of the productivity parabola, the incremental change to removals of maximum sustainable yield, or slightly greater than maximum sustainable yield, are gradual. But once the residual population is reduced to values on the left of maximum sustainable yield, the same fixed harvest that resulted in gradual change in residual populations on the right arm results in rapid extinction.

Simulations of fixed harvests on the George Reserve whitetail population showed that the likelihood of extinction increased as maximum sustainable yield was approached (Figure 53). At maximum sustainable yield, 16 of 20 simulations of 100 years lead to extinction while, with a harvest of 50 (slightly exceeding maximum sustainable yield), all 20 lead to extinction within 100 years. A fixed harvest of 46 was the highest harvest that did not lead to extinction in 20 simulations of 100 years.

If the harvest cannot be controlled and also varies stochastically, probabilities of extinction become even higher. Applying stochastic variation to the harvest in the George Reserve model based on variance in harvest for deer hunting in legal seasons in Michigan with constant regulations (McCullough 1979) resulted in 43 being the highest harvest that did not lead to any extinctions in 20 simulations of 100 years (Figure 53). Harvests of maximum sustainable

yields led to extinction within 100 years for all 20 simulations.

Environmental variation can lead to overshoot of K and vegetation damage in unhunted populations as well as in those whitetail populations hunted very conservatively. Thus, particularly good years may increase the residual population above K. Population adjustments to K accomplished through failure in recruitment have relatively little damaging effect on vegetation even if K is temporarily exceeded, because the young do not live long and are present during the better season of the year. However, adjustments to stochastic overshoots are accomplished by increased mortality of recruited animals. Because these whitetails are larger and have greater reserves, they can persist longer and put correspondingly larger defoliation pressure on the vegetation. Also, several good years in a row can result in relatively great overshoot and destabilize the natural tendencies toward equilibrium between residual population and K. For further treatment of these time lags, *see* McCullough (1979, in press).

NATURAL PREDATORS

Selectivity of natural predators, as previously mentioned, was reviewed for white-tailed deer by McCullough (1979) (*see also* Chapter 8). There is strong evidence that predators take the young, old and unfit. This is not to say that healthy deer are not killed sometimes. In fact they are, and review of wolf predation on deer showed clear evidence that success in taking a greater proportion of prime deer is related to lower deer density (that is, wolves can kill healthier deer by greater effort). However, on the basis of age-class frequency, young, old and unfit are overrepresented in wolf-kill samples. And because large predators have higher reproductive potential than do deer, a vulnerability variable has to be present to account for the long coexistence of prey and predators.

Because human hunting, by reducing whitetail density, increases deer body growth and size, improves health and lowers the age distribution of the population (McCullough 1979), it decreases the overall vulnerability of the deer population to natural predation. Even young deer are less vulnerable because good health of both fawns and does results in rapid growth rates by fawns and alertness on the part of

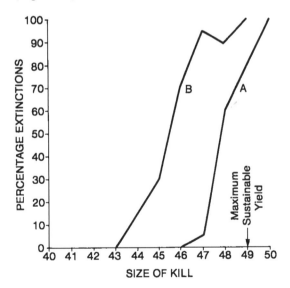

Figure 53. Population extinction rates from simulations of Michigan's George Reserve white-tailed deer model with stochastic recruitment (A) and stochastic recruitment and harvest (B). Percentage of extinctions is based on the percentage of 20 runs that went to extinction within 100 years.

Openlands on the George Reserve contribute high-quality forage to the spring and summer whitetail diet, and also during winter when snow does not cover basal rosettes of forbs and new sprigs of grass that remain green at ground level. *Photo by Doug Fulton.*

mothers. Therefore, heavy hunting by humans results in fewer deer in the population and lowered vulnerability. These changes cannot help but reduce vulnerability to predators and, thus, while hunting by humans and natural predation almost certainly are not entirely substitutive, they are substitutive to some considerable extent. As the proportion of the deer population taken by human hunters is increased, the proportion taken by natural predators should decrease.

The wildlife biologist or manager almost never is in a position to determine the kill of deer by natural predation. It is difficult even in research. However, in managing whitetails the objective should be to derive the empirical relationship between productivity and density. If derived in a management unit where natural predators are present, the empirical productivity curve will reflect the harvest available to hunters with the predator take removed. If human hunting is highly substitutable for natural predation, the slope of the right arm of the productivity curve derived will be steep and, because human hunting is coming at the expense of natural predators, those predators will have to switch to alternative prey, emigrate or decline in population. If human hunting is less substitutable for natural predation, the slope of the productivity curve will be shallow, and hunter harvests will be reduced because of the greater additivity of human and predator take. In either case, managing for maximum substainable yield of hunter harvest on the basis of such an empirical model will manage the deer population for maximum sustainable yield since the take of natural predators already is included in the empirical results. This is the approach I am taking in a current set of experiments on Columbian blacktailed deer at the Hopland Field Station in California, where coyotes and other natural predators take an unknown (and perhaps undeterminable) number of deer.

FUNCTIONAL REFUGIA

An exception to the general rule that the left arm of the productivity curve is unstable is the presence of functional refugia. These may be in the form of an actual designated refuge where hunting is not allowed, an impenetrable vegetation type, a large roadless area or legal restrictions on hunting methods and equipment. The functional refuge sets a constraint on the lower limit to which a population can be reduced (Figure 54). In any event, if the variable under consideration is constrained by a refuge, it may set limitations on what can be accomplished by management. By the same token, a refuge may prevent deer population extinction in the face of large errors in management. Thus, if an unhuntable vegetational type sets a limit on a legal harvest with a one-buck season, increasing the limit to two bucks is not likely to result in any appreciable increase in the kill. Also, because hunters vary in ability (Holsworth 1973), the one-deer bag limit is a refuge effect in itself because good hunters are removed from the hunter population and poor hunters retained as the season progresses.

The Crab Orchard National Wildlife Refuge in Illinois (Hawkins and Klimstra 1970*a*) and Berry College Refuge in Georgia (Kammermeyer and Marchinton 1976*b*) are examples of legal refuges, and the tamarack swamp on the George Reserve is an unhuntable type. In a similar example, Van Etten et al. (1965) reported that it took six hunters 124 hours of hunting even to see a buck in a 2.6-square-kilometer (1-square-mile) enclosure in northern Michigan with seven bucks present. In contrast, Creed and Kubisiak (1973) reported

a harvest of all deer in a Wisconsin enclosure approximately 4,050 hectares (10,000 acres) where escape cover was poor and hunter density high. Swenson (1982) found that mule deer in forested areas were far less vulnerable than those in prairie areas. Northern states, such as Minnesota, Wisconsin, Michigan and Maine, and the eastern Canadian provinces have large roadless areas where obtaining harvests of white-tailed deer is difficult or impossible. Restrictions include prohibition of night spotlighting, baiting, use of dogs, etc. Most refugia are not absolute, but apply to one or a few forms of protection for deer. Thus, a legal refuge may prevent hunter harvest, but natural predation and poaching may continue. And a refugium based on vegetation may prevent a human hunter from ever getting a shot, but legalizing the use of dogs may overcome the effect of the refuge. Peterson (1969) reviewed hunter behavior as it relates to the deer harvest.

GOAL SETTING

Any program of white-tailed deer mangement must start with the definition of what the program is intended to achieve. Who is the audience served? What are ramifications of the program on the deer resource and the satisfactions of the user group or groups? What are the criteria by which the program can be judged as successful or unsuccessful? Will these criteria be acceptable to the user group or groups? What dissatisfactions will be accepted as the cost of allocating the deer resource among user groups? Often, the programs of wildlife biologists and managers have been good in practice; it is the in-depth explanations of deer-population characteristics that have needed firmer foundations.

The frequently heard justification of managers that they managed deer for "the carrying capacity" of the range was a bit of self-delusion, often dressed up by the euphemism "scientific management." While most people can agree that programs resulting in vegetation destruction are self-defeating in the long run, there is a great deal more arbitrariness about impacts on vegetation below K. It is necessary to distinguish defoliation impact from impact so great that it cannot be sustained. Even one deer has some impact, and the impact shows a functional relationship similar to the productivity curve, in that impact increases up to the max-

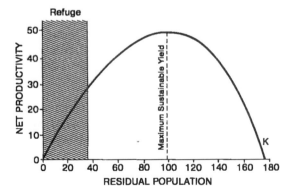

Figure 54. Productivity curve for white-tailed deer on the George Reserve in Michigan, showing the effect of a hypothetical functional refuge.

imum sustainable which, by definition, occurs at K. Some of the least fruitful arguments among big game biologists have revolved about which point estimate of vegetation impact was "correct," when all up to K are arbitrary and can be evaluated only by reference to a predetermined program goal and set of specific objectives. It must be recognized that any point on the residual population axis of Figure 50b from zero to K is a sustainable point (given sufficient skill on the part of the manager), and no scientific method will determine which point is any better than any other. The perception of a given point being best by the manager was a reflection of an unrecognized *a priori* assumption of what was best. It is time to abandon this bit of circular logic.

The goal of a deer management program is arbitrary, not scientific, which may be upsetting news to some wildlife biologists and other resource managers as well as citizens. Science does not make the decision. People do. And the views of various interest groups have to be considered because they cannot be ruled out on scientific grounds. Protectionist groups have recognized intuitively that decisions by biologists and managers may have been biased toward hunter interests, as the overall goal of maintaining sustainable harvests has been pursued (*see* Favre and Olsen 1982). While the protectionists still have a way to go and are diverted by their own shibboleths, they clearly are influencing the goals and objectives used in managing deer populations.

Once the goal has been established for a given management unit, then science—in the form of productivity curves (or some comparable model)—can be used to develop a program to achieve the goal and its associated objectives for the deer population.

There are four major biological goals. First, one wants to avoid extinction of the deer population. Second, one can decide to minimize the residual population to reduce negative impacts of the population. Thus, one might want to reduce the frequency of deer/vehicle collisions, or deer damage to agricultural crops, orchards or forest reproduction. Third, one can decide to maximize the sustainable yield (harvest). And fourth, one can maximize the size of the residual population in the field. Therefore, establishing the optimum sustainable yield is dependent on management program goals, and may range from very low to very high residual populations and sustained yields.

Under circumstances of crop or forest damage, deer/vehicle collisions, etc., it may be desirable to maintain whitetail populations at very low levels, well below I carrying capacity. If a functional refuge is present, more liberal seasons or bag limits, beyond a certain point, are not likely to have an effect. If the degree of control over the deer population is accomplished by the residual population maintained by a refuge, then the management system is easily implemented. If a greater harvest is required, some other hunting methods (depending on the nature of the refuge) may have to be legalized, such as the use of dogs, or hunting at later hours when deer venture out from the refuge. In some cases it may be difficult to motivate hunters to put forth the extra effort required to achieve the desired degree of control.

If one wants to maintain a high residual population for tourists, nature study groups, etc., but also wants some hunting, buck-only hunting is a good strategy. Most sportsmen approve of such programs and find them philosophically satisfying, even though probability of success in taking deer is relatively low, and probability of overshoot of K and vegetation damage is high.

Maximizing the harvest usually would involve manipulating residual populations to just to the right of maximum sustainable yield (Figure 50b) where stability of the population responses is high. Seasons with antlerless hunting will need to be liberal, and the hunter effort per deer taken will need to be high. Deer populations managed in this way are less affected by environmental variations, except, perhaps, in the most extreme years. They respond rapidly to adjustments in management and are most easily modeled for functional relationships. They yield the greatest harvest and, therefore, the highest harvest success rate for hunters. Because residual populations are low, deer damage problems usually are moderate to low or infrequent.

In addition to biological goals, there are social goals. One can manage for deer-hunter satisfaction that may or may not be achieved by management for maximum sustainable yield. One can manage for landowner satisfaction, or for protectionist-group satisfaction. Just as a given program cannot simultaneously achieve all biological goals, it is unlikely that all social goals can be achieved simultaneously. The tough problems confronting wildlife management agencies are the decisions about who gets what—a statement that will come as no surprise. Nevertheless, progress is being made. An important step has been to recognize that landowners and operators accommodate deer on their properties. Thus, more states are realigning deer-hunting regulations to give landowners and/or operators first preference for hunting opportunities in designated management units.

Once decisions have been made, implementation of a new program should be made in

Annual drives, in which a line of people moves across the area, have been conducted on the George Reserve since 1933 to determine whitetail population size. *Photo by Doug Fulton.*

increments. In the first place it should be verified that the direction taken was correct, and conforms to predictions. Second, there is considerable inertia in large public programs, and deviations from past practice cause skepticism among the users. Leadership consists not just of "being out in front," but also in having public backing. Getting too far out in front results in loss of touch with the public. New programs should be geared to a pace the public will follow. Some agencies still are suffering from continuing weak support caused by the attempt to solve a chronic deer overpopulation problem with a massive antlerless kill in the 1950s and 1960s. Changing the reactive attitudes of citizens established two to three decades ago requires a well-founded informational/educational effort based on a fundamentally sound deer management program.

REGULATING THE SEX AND AGE OF THE KILL

Buck-only Versus Any-age-or-sex Hunting

Setting seasons restricted to bucks only is the surest way to minimize the harvest short of no season at all. While there may be good reasons for minimizing the harvest, it is unfortunate that this management approach is applied supposedly in the interest of large harvests. Surely few professions have been cursed by such an intuitively appealing practice—one that is immensely popular with the constituents involved and that works in the diametrically opposite manner from what is intended. It is ironic that those opposing hunting are most incensed by trophy hunting—the hunter's quest for large and inedible antlers—a practice that assures that the fewest deer will die from gunshots.

The failing of buck-only hunting for yielding bucks, much less total deer, can be seen in Figure 55, where the potential harvest curve on residual population size, as shown for the George Reserve deer population, is separated by sex. Sex-ratio variation due to population density, with males predominating at high density (Verme 1965a, McCullough 1979), is included, but the results are approximately the same for a 50:50 ratio.

Because does are spared under buck-only hunting, the residual population grows toward

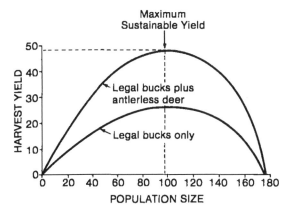

Figure 55. Relationship of white-tailed deer harvests of only bucks to any deer on the George Reserve in Michigan, based on data from 1952–1972.

K. This results in reduction of all recruitment, including recruitment of bucks (Figure 55). Since the harvest is dependent on recruitment, a harvest of only bucks assures a low recruitment of bucks to be harvested.

Maximum sustainable yield of bucks is obtained by the same management as is total maximum sustainable yield, namely by taking harvests of both sexes to reduce the residual population to near I carrying capacity (with due regard to the instability of this point as discussed earlier). Model simulations on the 464-hectare (1,146-acre) George Reserve deer population of buck-only hunting give a buck yield of 2.9 to 6.8 (depending on which model was used), while buck yield at maximum sustainable yield was 26. Furthermore, no antlerless deer take is present in bucks-only hunting, while 23 antlerless deer are taken when the deer population is managed for maximum sustainable yield.

This illustrates a very important point in deer management. Once the residual population has reached or exceeded I carrying capacity, there is no advantage to being selective for sex and age in hunting. To state it another way, the population is quite insensitive to unbalanced sex and age composition. Results from the George Reserve clearly show that most hunters used in earlier years to take the harvest were selective for bucks even when instructed to shoot all deer randomly. This resulted in bucks being taken out of proportion to their presence in the population, but the inevitability of achieving the desired size of harvest required that females be taken. When I assumed the job of doing the harvesting, I selected strictly

for the first clear target. Because of their behavioral vulnerability, yearling bucks were killed in greater poroportion than their prevalence in the population. Most of the other sex and age classes were taken nearly in proportion to their occurrence in the population (McCullough 1979). Despite the different approaches and selectivities of the two harvest systems, the functional outcome was the same.

The constraint of the productivity curve on sex-selective harvest can be illustrated by a further simulation of the George Reserve deer population. Considering a system of harvest that begins with the population being managed for maximum sustainable yield, but then switches to bucks only, with every buck greater than yearling age harvested (Figure 56), the kill of bucks inevitably declines. When it bottoms out, the management system is changed, and all bucks older than yearling age are taken, but does are taken to bring the kill to maximum sustainable yield. Inevitably the size of the buck kill increases until the initial state of maximum sustainable yield is achieved.

Is there evidence that the deer-population response in management units elsewhere shows the same behavior? Yes, virtually universally. As wildlife biologists throughout white-tailed deer range will note, buck-only hunting invariably results in: (1) high residual populations predominantly of females; (2) low overall recruitment rates; and (3) legal bucks comprising 10 percent or less of the population.

These conditions hold for the black-tailed deer at the 2,024-hectare (5,000-acre) Hopland Field Station in California, where only bucks are taken in legal hunts. Given the relation-

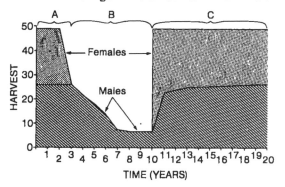

Figure 56. Simulations of multiple sustainable yield harvesting of white-tailed deer on the George Reserve in Michigan (A), switching to buck-only (yearling and older bucks) (B), and then harvesting again at multiple sustainable yield by taking bucks only and completing the harvest with antlerless deer (C).

ships outlined previously, it follows that increased harvests of does should be followed by increased harvests of bucks. By reanalyzing the data of Connolly and Longhurst (1975), who collected various numbers of does for other study purposes that had nothing to do with buck production, I got the relationship shown in Figure 57. There was a highly significant correlation between the number of does taken and the number of bucks taken three years later when recruits of the subsequent residual population first reached legal size with forked-horn antlers. This relationship was strong, despite the fact that the doe harvest varied greatly between years and did not follow a pattern of consistent change. An experiment with incremental increases in doe harvest is one I have begun, and there is every reason to believe that the correlation between doe removal and antlerless harvest will be much higher without the "noise" due to highly variable doe harvest.

The relationship shown in Figure 57 can be used to generate a "productivity curve" of the effect of legal buck harvest on doe removal (Figure 58) that shows maximum sustainable yield of legal bucks to be approximately 47. Because maximum sustainable yield of both sexes occurs at the same point, the combined maximum sustainable yield was 47 bucks and 44 does for a total harvest of 91. Clearly there is much to be gained in the buck harvest by killing does. The harvest of does is in addition to the gain in buck harvest.

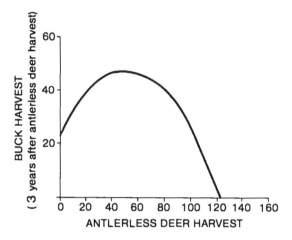

Figure 58. Predicted buck harvest in relation to antlerless harvest of black-tailed deer at Hopland Field Station in California, as derived from Figure 57.

As an aside, it should be noted that on Hopland Field Station, deer density was a far more important variable in controlling productivity than all other variables combined, including coyotes and other natural predators, poaching, drought and variable acorn crops (oaks are abundant on the area). While Hopland Field Station does not represent the extreme in environmental variation that deer face in North America, it certainly is not a benignly stable area. Its Mediterranean climate—mild, wet winters and dry, hot summers—and its vegetation of annual grassland, oak savanna and chaparral certainly present a stark contrast to the George Reserve. That two different species of deer in two very different environments respond similarly to density suggests that these lessons about deer management probably are not endemic to the George Reserve.

Trophy Buck Production

Another irony in the buck-only story is the widely held belief that this scheme of management is the route to production of trophy bucks. While it is true that some trophy bucks are obtained under buck-only hunting, they are relatively few and require a long time to grow to trophy size (McCullough 1979). Management for maximum sustainable yield will produce the greatest number of bucks; and because of rapid growth rate under such management, bucks will reach a large size in a short time.

Trophy bucks are considered those with large, heavy antlers bearing at least four points on

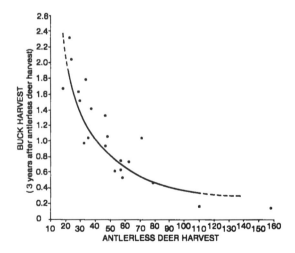

Figure 57. Harvest of Columbian black-tailed deer bucks in relation to antlerless harvest three years earlier at Hopland Field Station in California from 1951 to 1974 (from Connolly and Longhurst 1975). $Y = ax^b = 46.79x^{-1.04}$; $r^2 = 0.88$.

either antler (eight-pointers by eastern count, which includes both antlers). Because antler weight (grams) of George Reserve whitetails is correlated highly with body weight (kilograms) according to the formula $Y = -1695.5 + 28.8 \times (r^2 = 0.57)$ (McCullough 1982c), the management system that produces the largest body size will also produce the largest antlers. Similarly, the number of antler points is related significantly both to antler weight ($Y = 15.7 + 89.15X$, $r^2 = 0.71$) and body weight ($Y = -6.08 + 0.17 \times r^2 = 0.45$). Thus, comparison of maximum sustainable yield management with buck-only management, according to survivorship curves for the George Reserve bucks (McCullough 1979) and assuming selectivity in both systems for the largest bucks and growth rates, shows that management for maximum sustainable yield would yield 26 bucks, of which 12 would be eight-pointers or larger, 10 six-pointers, and 4 three-pointers. Buck-only hunting, by the most liberal calculation, would give 6.8 bucks with eight-point antlers. Note that mean number of points per buck taken is less with management for maximum sustainable yield (weighted $\bar{x} = 6.5$ points), as previously reported for mule deer from Texas (Brownlee 1975). However, absolute numbers by antler-size class as well as averages must be compared under the two systems of management. Declining averages do not show that trophy production and maximum sustainable yield are incompatible, as concluded by Connolly (1981), although the generally lower growth potential of mule deer (McCullough in press) may make the difference less pronounced as compared with that of white-tailed deer.

These conclusions are supported by results from the George Reserve as population size has varied over time (Table 28, Figure 59). Reserve caretaker Lawrence Camburn weighed deer in 1941 and 1942 when the population was about 126. Mean weight of yearling and older bucks was 66.4 kilograms (146.6 pounds). Between 1958 and 1971, the population size averaged 67 whitetails, and yearling and older bucks averaged 71.3 kilograms (157.5 pounds). When the whitetail population was reduced to 10 between 1971 and 1975, the average buck weight rose to 75.9 kilograms (167.6 pounds). Following rapid increase in the whitetail population with protection (*see* Rate of Population Increase), when harvesting was resumed in 1980–1981, average buck weight was 77.6 kil-

ograms (171.4 pounds). Because of the high whitetail densities, weight of bucks harvested the following season (1981–1982) declined to 71.2 kilograms (157.3 pounds). Weight ranges in yearling and adult bucks showed the same patterns as the means (Table 28), as did the

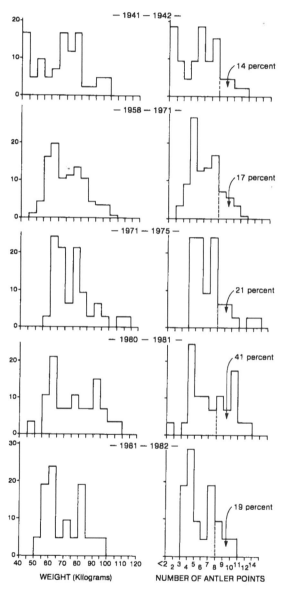

Figure 59. Frequency histograms of fresh kill weights and numbers of antler points of yearling and older white-tailed deer bucks for different periods and population densities on the George Reserve. Population size and buck harvest for periods are as given in Table 28. Because the number of antler points was not recorded for all periods, all results are based on weight according to the equation $Y = -6.08 + 0.17X$ (McCullough 1982c). Less than two antler points—<2—refers to spike antlers averaging less than 15 centimeters (6 inches) in length.

Table 28. Relationship of size and status of the white-tailed deer population to fresh-kill weight of yearling and older bucks on the George Reserve, Michigan.

Years	Mean posthunt population size	Number killed	Mean weight		Standard deviation		Range
			Kilograms	Pounds	Kilograms	Pounds	Kilograms
1941–1942	126[a]	44	66.4	146.4	16.5	36.4	40.3–100.5
1958–1971	67[b]	141	71.3	157.2	11.8	26.0	47.5–101.4
1971–1975	33[c]	35	75.9	167.3	13.5	29.8	59.3–113.6
1980–1981	130[d]	29	77.6	171.1	16.6	36.6	46.2–109.1
1981–1982	± 130[e]	21	71.2	157.0	13.0	28.7	53.4– 98.2

[a]Estimated from corrected drive count (McCullough 1979).
[b]Estimated from reconstructed population (McCullough 1979).
[c]Estimated from drive count. Population reduced from an estimated 60 in 1971–1972 to 10 in 1974–1975.
[d]Estimated from drive count (*see* Rate of Population Increase and Time Lags).
[e]Approximate population size. The most severe winter on record prevented a successful drive count in 1982.

weights of yearling bucks (Figure 59). The first peak in the weight histograms in Figure 59 are predominantly yearlings, and it can be seen from Figure 59 that yearlings in 1941–1942 weighed 40 to 50 kilograms (88–110 pounds). This rose to 55–65 kilograms (121–143 pounds) in the period 1958–1971, to 60–70 kilograms (132–154 pounds) in the period 1971–1975, and fell back to 50–60 kilograms (110–132 pounds) in the periods 1980–1981 and 1981–1982. Note that yearlings during the latter two periods were produced under high densities, while largest bucks harvested in 1980–1981 were born early in the increase experiment, and achieved their growth at low population density. Antlers of yearling animals show the same pattern (Figure 59).

Record bucks were produced not at high densities, which have the oldest individuals, but rather at the lowest densities (1971–1975). The largest buck ever taken (113.6 kilograms [251 pounds]) and the next largest (110.4 kilograms [224 pounds]) were taken in 1971, when total buck harvest was 37. The largest buck of 44 taken in 1941–1942 was 100.5 kilograms (222 pounds). The highest proportion of bucks weighing more than 90.5 kilograms (200 pounds) was obtained at lowest densities, in the period 1971–1975, reflecting low density during the population-growth experiment (1980–1981 harvest) (Table 28). This occurred despite these periods of heavy harvest having the youngest age structure in both the population and the kill. The record buck was five years old and the next largest was three years old. No buck taken in 1971–1975 or 1980–1981 exceeded five years of age and most were yearlings, two- and three-year-olds.

Antler size showed the same relationship, with the greatest proportion of eight-point or better racks obtained in 1971–1975 and 1980–1981 (Figure 59). Furthermore, under heavy exploitation yearling bucks obtain antlers that are much larger than the simple spikes typically associated with this age class. Thus,

The number and trophy quality of George Reserve bucks are determined by the size of the antlerless harvest and its impact on total deer density. *Photo by Doug Fulton.*

Table 28. (continued)

Range	Number of deer more than 90.5 kilograms (200 pounds)	Percentage of deer more than 90.5 kilograms (200 pounds)
Pounds		
88.8–221.6	4	9.1
104.7–223.5	11	7.8
130.7–250.4	5	14.3
101.9–240.5	8	27.6
117.7–216.5	2	9.5

on the George Reserve, when residual populations were well below I carrying capacity (McCullough 1982*c*), eight-point yearlings were common, although the beams of the antlers were substantially lighter than those of older age classes. Because there is a genetic component involved in antler conformation (Harmel 1980), as well as a nutritional component, there may be justification for removing spike-antlered bucks in heavy exploitation programs, since the genetically superior yearlings will have large antlers. And of course the true trophy-rack bucks are those that escape heavy exploitation to become venerable six- to eight-year-olds. Because of great flexibility in the age-sex structure of deer, any workable system to protect bucks until they are older will pay off in trophy antler production under heavy exploitation.

In conclusion, management for maximum sustainable yield has substantial advantage over buck-only harvests in terms of the number of quality bucks produced. There may be considerable public-relations value in billing management for buck-only hunting as trophy management, but a tacit rule of propagandizing is "don't believe it yourself."

RELATIONSHIP OF REGULATIONS TO HARVEST

It is necessary to establish the relationship between regulations and harvest. Does issuance of additional antlerless permits result in the desired increase in harvest? And what is the ratio of permits to harvest? Does the larger hunter population encouraged by antlerless hunting result in an increase in the buck harvest because most hunters will take a buck given the opportunity? Freddy (1982) reported that the number of antlerless animals taken

from a Colorado mule-deer population was correlated with the number of either-sex permits issued, but that antlered-deer harvest was not predictable.

Regression analysis of the whitetail harvest in Michigan (Department of Natural Resources harvest records) that I did for the period 1958–1974—when firearms regulations were fairly consistent—showed variable patterns for the three management regions (Table 29). The size of antlerless harvest was related closely to the number of antlerless hunters in all three regions and, therefore, the number of antlerless hunters to achieve a desired antlerless harvest can be predicted: Region I $\hat{Y} = 1108.6 + 0.223X$; Region II $\hat{Y} = 2074.4 + 0.283X$; and Region III $\hat{Y} = 513.8 + 0.169X$.

However, the relationship of buck harvest to number of buck hunters was correlated significantly only for Region III. In the other two regions buck harvest was correlated more closely to number of antlerless-deer hunters than to number of buck hunters—but not significantly. Probably the major effect was due to the influence of numbers of antlerless-deer permits on the total number of hunters afield, since antlerless permit holders could take a buck instead. In Region I, each antlerless-deer hunter increased the total number of hunters by 0.27 hunter. For example, 5,000 antlerless-deer hunters resulted in 1,350 additional deer hunters in the field. By the same token, each

Table 29. Influence of number of antlerless-deer hunting permits on the white-tailed deer harvest by management region in Michigan: P = probability; r^2 = coefficient of determination.

Region		Buck hunters	Antlerless-deer hunters	Total hunters
Region I[a]				
Buck kill	r^2	0.005	0.178	0.136
	P	0.392	0.247	0.073
Antlerless kill	r^2	0.739	0.804	0.550
	P	<0.001	<0.001	<0.001
Region II[b]				
Buck kill	r^2	0.027	0.184	0.183
	P	0.263	0.176	0.043
Antlerless kill	r^2	0.428	0.799	0.168
	P	0.002	<0.001	0.051
Region III[c]				
Buck kill	r^2	0.827	0.827	0.824
	P	<0.001	<0.001	<0.001
Antlerless kill	r^2	0.484	0.761	0.537
	P	0.009	<0.001	0.005

[a]Upper Peninsula.
[b]Upper part of Lower Peninsula.
[c]Lower part of Lower Peninsula.

antlerless-deer hunter decreased the number of buck hunters by 0.73, which suggests that approximately 73 percent of the hunters who hunted in buck-only seasons obtained antlerless permits when whey were available. Since the time these data were assembled, antlerless hunting in Region I has been curtailed by political intervention.

In Region II, antlerless-deer permits had an even stronger influence on the total hunting population. Each antlerless-deer hunter increased the total number of hunters by 0.55, while 45 percent of the buck-only hunters obtained antlerless-deer permits. This greater effect of numbers of antlerless-deer hunters on total hunters afield resulted in the buck harvest correlated significantly to total hunters afield (Table 29).

In Region III, the total number of hunters increased rapidly during the period covered by the data because of the expanding deer population in southern Michigan. Nevertheless, about 83 percent of the increase was attributable to the number of antlerless-deer hunters, while only 17 percent was due to growth in the number of buck-only hunters. Both buck harvest and antlerless harvest were quite predictable.

While this analysis is cursory, it shows the general approach of analyzing the influence of hunting regulations on hunter numbers and deer harvest by developing predictive tools. Other information on both deer and hunter demography clearly is necessary to interpret results. Hunter dynamics and deer-harvest results are not the same for Region I in the Upper Peninsula of Michigan—a remote and poorly roaded area with a low human population and a stagnant deer population—as for the Lower Peninsula. Similarly, within the Lower Peninsula, the more remote Region II—with its long tradition as a good deer area—is different from the heavily populated, easily accessible Region III with its expanding deer population.

MODELING BUCK HARVEST BY MANIPULATION OF DOE HARVEST

The historic dominance of buck-only hunting means that it may be possible to model populations on the basis of manipulation of doe harvest in order to influence buck harvest, as presently is being done at the Hopland Field Station in California. This assumes that density-dependent responses are present, and are not overshadowed by environmental variation. It makes no assumptions about other mortality factors, except that they behave in some rational fashion. Only the harvests of bucks and does need to be obtained with reasonable precision, and a relative value (or index) will work as well as an absolute value so long as the requirement of precision is met. The total deer population does not need to be known.

In Figure 58, the relationship for Hopland Field Station is presented on the basis of currently available data and suggests that, in the absence of doe harvests, the mean number of bucks taken will be 23. The maximum sustainable yield for bucks and does occurs at the same point, and Figure 58 suggests this should occur at a harvest of 44 does and 47 bucks.

The approach I propose to take is to make the doe harvest under a collecting permit. Examination of reproductive tracts will yield an embryo rate to suggest the reproductive effort. Over time, the kill of does will be incrementally increased. The embryo rate, individual weights, antler size and total buck-kill should increase as doe harvest increases. These data will be the basis of a refined model, probably with lower variance because the harvest of does is incremental and not fluctuating wildly from year to year as in the data presented in Figure 57.

As the buck harvest begins to plateau, maximum sustainable yield is approached. For management purposes, one probably would elect to stop at this point, given that maximum sustainable yield is an unstable point on the left arm of the productivity parabola. On Hopland Field Station, it probably will be desirable to overexploit to verify that maximum sustainable yield has been exceeded (thereby reducing the residual population to the left arm of the productivity parabola temporarily), then decrease the harvest to allow recovery, and once again increase to as near to maximum sustainable yield as seems feasible, given the variance in the new model derived.

Elsewhere (McCullough in press), I have estimated that for mule deer the residual population yielding maximum sustainable yield is about 63 percent of K and will represent 27 percent of the prehunt population, the other 73 percent being the residual population. These values give crude expectations of a residual population (I carrying capacity) of 337 yielding a maximum sustainable yield of 91. Estimated population at K would be 535. These values are slightly below the range of population estimates made by Longhurst, as cited by Connolly (1970), on the basis of mark and recapture, change in ratios and pellet-count methods, although there was great disparity among methods used.

The difference between embryo rates (expected) and harvest rates of individuals born in the same year (observed) will indicate mortality due to factors other than hunting, and serve as a cross-check of total deer-population estimates.

Although the objective is to establish the functional relationship between doe and buck harvests, estimation of absolute values of population size inevitably results from the cross-referencing that becomes possible with data accumulation. These estimates cost little and promise to be at least as reliable as those presently used, which are notoriously difficult and expensive to obtain and lack the cross-checking for consistency that typifies an empirical modeling approach.

STOCK RECRUITMENT MODELS

This approach to empirical modeling has been proposed previously (McCullough 1979). Figure 50c is a stock recruitment model. It is based on the tendency of recruitment to come to an average balance with the harvest. It requires that the harvest be determined and be broken down by new recruits versus older animals. The estimate of recruitment rate is based on dividing the recruits by the older animals taken in the same time-specific hunting-harvest sample. It will work well only if hunting is not selective for sex and age, but will be approximate if the antlerless harvest is relatively liberal. It further requires an estimate of total population, which can be derived from the harvest data (McCullough 1979) or a reasonable reliable index of total population (McCullough 1978).

The approach works best with populations that have been hunted quite conservatively, so that the beginning harvest is small relative to maximum sustainable yield. Therefore, the harvest can be increased incrementally by gradually liberalizing regulations. For further description of the method, *see* McCullough (1979).

MANAGING FOR HUNTER SATISFACTION

Traditionally, whitetails have been managed for deer in the bag, so long as the removal was sustainable, under the assumption that the higher the number and percentage of hunters successful, the greater the hunter satisfaction. However, sociological studies in the 1970s showed that hunter satisfaction was much more complex, and not closely correlated with size of the harvest. This discovery led to the suggestion that the goal of deer management should not be based on yield of deer, but rather on the yield of hunter satisfaction (Hendee 1974).

In principle, this goal is laudable, since the endproduct of recreational hunting is satisfying recreation. In practice it has been difficult to implement. Motivations for hunting are complex and varied, and the sources of satisfaction are equally diverse (Hendee 1972, 1974, Hendee and Schoenfeld 1973, Schole 1973, *see* Chapter 42). Hunters enjoy planning the hunt and build up considerable anticipation of the experience. They enjoy getting away from the routines of home and job and other complexities of modern life. They enjoy being out of doors, close to nature, and living the simple life of hunting cabins or tents, campfires and stoves, sleeping bags, and not shaving. They enjoy camaraderie—to discuss hunts past, present and future. Finally they enjoy the hunt itself, with the potential if not actual ending with the harvesting, dressing, and hauling out of a deer, and the attendant bragging rights.

The wildlife biologist or manager embarking on a program with the goal of hunter satisfaction is confronted with the fact that few of the sources of satisfaction are dependent on the deer-management program. Most are dependent on field experiences of the individual hunter. The wildlife biologist or manager cannot control the satisfaction of the hunter getting away from the home, change in life style, or selection of companions. The program of management can control only a few variables involving the hunt itself. Recent studies at Hopland Field Station in California showed that only 28 percent of total satisfaction was attributable to variables under control of the wildlife manager (McCullough and Carmen 1982). Thus, the manager should not be deluded about the extent to which he can influence hunter satisfaction. However, it is clear that those elements of hunter satisfaction associated with variables that are under control should be incorporated in planning and decision making.

The controllable variables are essentially as follows:

1. Control of hunter density by regulation of number of permits, assignment to hunting units and timing of hunting (that is, weekdays versus weekends, early versus later parts of the hunting season, etc.);
2. Control of size of the deer harvest by length of season, bag limits, and designation of sex and age of legal deer;
3. Control of residual population by regulation of size of the hunt; and
4. Control of harvest success rate through control of hunter density and harvest size.

The response of hunters to hunter densities has been one of the more labile satisfaction variables. It varies tremendously from place to place. Hunters in Maryland were satisfied with extremely high hunter densities (up to 35 per square kilometer: 91 per square mile) (Kennedy 1974), while densities in the West with mule deer hunting are quite low (0.4–8.9 per square kilometer: 1.0–23.1 per square mile) (Miller et al. 1977). It is necessary to keep in mind that hunters are not a uniform population, and several things seem to influence their response to hunter density. People who go hunting for solitude are less likely to want encounters with others than are those who enjoy companionship. Hunters who employ a strategy to hunt are likely to be annoyed by encounters with other hunters who upset that strategy; those who trust to luck probably are not bothered (or are less so) by encounters. It is significant that in the high-density whitetail hunting in Maryland, many hunters thought there were not enough hunters to keep the deer moving (Kennedy 1974). This variable probably is most important in heavy cover where deer are difficult to stalk, and where hunters are relatively unskilled.

A hunter-density experiment on the Sandhill Recreation Area in Wisconsin was particularly instructive on this point. While many hunters objected to frequent encounters with other hunters, some hunters did not, because the high hunter density resulted in moving deer and a higher probability of harvesting a deer (Heberlein et al. 1982). Although frequent encounters lowered hunter satisfaction, the positive aspects of increased harvest caused a disproportionate increase in satisfaction in that

subset of hunters, so that net satisfaction was greater in the most densely used hunting areas.

Reaction of hunters to hunter density is highly conditioned by past history and also by their expectations (Heberlein et al. 1982). Hunters accustomed to high hunter densities find them less objectionable than do those who have not accepted such densities. And it is important to recognize that hunters have practiced selection in their choices of where to hunt. Hunters wanting few encounters select places with lower hunter densities (often with either more rugged hunting conditions or lower deer density or both), while those who do not find encounters with other hunters objectionable hunt in high-density areas.

Although further work is needed, these patterns suggest that hunter-density satisfactions are influenced by human demographic patterns. Higher densities are acceptable in habitats with heavy cover that conceal both deer and hunters, and among relatively unskilled hunters who depend on chance to see deer, which improves as other hunters inadvertently serve as "bird dogs." Hunters who employ skill, whether stalking or selecting carefully considered stands, are less likely to want encounters with other hunters. And, related to the aforementioned hunter characteristics, hunters who place great emphasis on harvesting a deer are more likely to consider high hunter-density acceptable than are those who emphasize quality of the pursuit. It seems likely that the latter are attracted to wilderness hunting, and to bow or muzzleloader seasons if firearm hunting is crowded.

The relationship of size of the deer harvest, percentage of hunter harvest success and size of the residual population is a complex interlinkage of variables. Earlier assumptions that hunters were motivated primarily by taking home a deer were brought up short by resistance to antlerless seasons and adherence to the buck-only philosophy. Hunting traditionally has been a conservative activity, and a majority of hunters have been older, less educated and more rural than the population at large (Hendee and Schoenfeld 1973, Schole 1973). Most studies of hunter satisfaction show that perceptions of the deer population being large and hunting harvest posing no threat to it are the major elements of hunter satisfaction (Langenau et al. 1981, McCullough and Carmen 1982). In a study of hunters at Hopland Field Station, perception of the populaton size being large accounted for nearly all of the satisfaction accounted for (20 of 28 percent), while harvesting a deer accounted for most of the rest. Harvesting a deer remains an important variable for hunter satisfaction in most cases (Schole 1973, Stankey et al. 1973, Gilbert 1977, Langenau et al. 1981), and the possibility of harvesting a deer surely must be present even where expectations of success are not high (Potter et al. 1973, Kennedy 1974).

Management for maximum sustainable yield increases the deer harvest and, assuming hunter density is unchanged, the hunter success ratio. However, as pointed out earlier, this is accompanied by reduced residual populations (Figure 50c). Therefore, if hunters use perception of the size of the deer population as their major reference point for satisfaction, management for increased harvest invariably will result in lower satisfaction. And this probably explains much of the tremendous resistance to antlerless seasons encountered by agencies over the years. Information and educational programs have met with a notable lack of success, and hunter attitudes appear to be formed by peer-group association rather than by information disseminated by management agencies (Shaw 1975). Indeed, there frequently is mistrust and antagonism, and some hunters have charged that wildlife agencies have tried to wipe out the deer population—a belief no less firmly held for being illogical. As a result, high-yield deer-harvest programs in many cases have been constrained by political and social pressure.

A better understanding of the interrelationship of deer productivity, harvest yield and residual deer population may result in better communication among management agencies and hunters. And, as demography of hunters changes, it is likely that hunter attitudes may change as well. The diverse attitudes of hunters toward hunter density demonstrate that hunters are a variable group. And as the hunter population becomes more urban and better educated, shifts are virtually inevitable. After all, Maryland, at one time, was wilderness. Information probably will become more important relative to traditional belief in shaping public attitudes. When probabilities of getting a deer—any deer—get too low, attitudes are likely to change about high-yield management.

I suspect that the backwoods buckhunter is a dying breed. Although they will survive longer in the rural western states, they are doomed by "progress." I further suspect that much of the opposition to antlerless-deer hunting had little to do with the facts under dispute—that biologists in state capitals presumed to know how to manage *their* deer, and general resistance to remote authority may have been the fly in the ointment. The old buckhunters will have modern imitators, but like urban cowboys they will pale by comparison and are not liable to carry the influence the originals did.

The nature and sources of hunter satisfaction are bound to remain labile so long as society and demographic characteristics change. Wildlife management agencies just now are establishing what it is that hunters want (*see* Heberlein and Laybourne 1978), and still have not begun to determine what the vast majority of the population that does not hunt wants from the deer resource. Societal attitudes about wildlife are complex (Kellert 1978, 1979, 1980). These attitudes change over time, and the job is not done when they have been first assessed. A continuing monitoring program is necessary, to update information. Science can assess the deer population, its habitat resource, and also the attitudes and desires of the segments of society to be served. But the real challenge to management is the art of blending the biological properties of the deer population with human uses to optimize or maximize sustained benefits to society.

IS HUNTING NECESSARY?

To many wildlife biologists and resource managers, it is an article of faith that deer populations need to be hunted. That a hunting harvest is sustainable is not the same as being necessary. Certainly, the damage to vegetation that results in lowering K and the productivity curve of a given management unit is a good argument in favor of hunting. Indeed, the George Reserve deer population is a classic case. This research area is managed as a natural area for research, and the deer population is harvested only because of the destruction of vegetation that resulted when it was not controlled. Deer are the only animals (or plants) on the Reserve that are artificially controlled.

However, it does not follow that all deer populations should be or need to be hunted. First, environments that are stable can sustain equilibrium relationships between residual populations and K. Most wildlife biologists and managers can point to situations where deer populations have not been hunted yet do not fluctuate greatly or cause damage to vegetation. Certainly deer reach overpopulation status in some park situations, but the surprising thing is how many parks containing deer populations have no problem.

Second, in extremely fluctuating environments, hunting is not necessary because environmental variation regularly results in the population being below K. The very characteristics that make *ad hoc* management for hunting necessary in such environments make it unnecessary to hunt at all.

Third, hunting in moderately fluctuating environments is not necessary if a good complement of effective natural predators is present. The selectivity of natural predators (more correctly stated as the vulnerability of the prey) is a more exact way to retain equilibrium values of residual populations. Hunting can accomplish the same end, but because of its lack of selectivity, a higher kill is required than for natural predation to achieve the same end. Stated another way, natural predators are better at reducing chronic mortality than are human hunters, because the former remove the vulnerable individuals most likely to succumb to chronic mortality factors. Thus, there is very high substitutability of predator kills for chronic mortality, while for human hunting, chronic mortality is somewhat more additive, although still substitutable to a considerable extent (McCullough 1979).

As professionals, wildlife biologists and managers must distinguish between cases where hunting is necessary and where it is not. It is possible to recognize the legitimate interests and necessary roles of human hunters without becoming apologists or advocates for the recreation. Bias toward hunting in situations where hunting is not necessary can only result in loss of credibility. Professional integrity demands that no side of a controversy be given favor on biological grounds that cannot be justified by the biology of the case under review. If hunters are favored because they pay the costs of management through license fees and special taxes, let that be the justification, and not an indefensible position that hunting is necessary in cases where it is not.

MANAGING FOR NONHUNTER SATISFACTION

Hunters and antihunters appear to be at such opposite philosophical poles that resolution of the conflict between them seems unlikely (Kellert 1978, McCullough 1979). If so, programs that satisfy one group, even to a small degree, are not likely to satisfy the other. It seems to be the unavoidable lot of the agencies that manage deer to suffer the anger of some segments of society in order to satisfy the desires of some other segment.

However, the protectionist philosophy is young and naive, and hopefully will mature with age. It contains a belief in the benevolence of nature that is fostered by being removed, indeed isolated, from nature. A majority of its advocates are urban-born and reared, and thus insulated from the processes of life and death. Nature is what they see in parks or zoos, which is rather reminiscent of Bambi, and their knowledge of wildlife comes from the public media, books and movies. Even their knowledge of domestic animals comes from the little farm or the petting zoo where the fact that these animals grow up to be slaughtered is never mentioned. Their beefsteak comes in a styrofoam tray covered with cellophane. The fact that some calf was bred, born, raised, fattened and then slaughtered to satisfy their wants probably never crossed their minds, or was quickly blotted out if it did. Their thoughts run more to sound nutrition and quality of life—the backyard barbecue with its good food, drink and friends. These people are not in the white cedar swamps in late February or early March when the deer reap the devastating rewards of the protection afforded them. But then, neither are most hunters, many of whom also have some rather quaint beliefs about deer populations.

When deer are starving in the cedar swamps and elsewhere, protectionists want to feed or move them. It often seems that whether such deer live or die is less important than getting the problem out of sight. It is as if doing something, no matter how absurd, will absolve these people of the guilt of their benevolence turning out to be less benevolent than they had expected from their "model" of how deer populations work. This is understandable. Most people mentally blot out things they know to exist that are unpleasant—war, the poor, teenage runaways, organized crime or whatever. Nor do people like to be confronted and forced to face unpleasant realities, particularly if they are in contradiction to philosophical beliefs.

Whether protectionists and antihunters will evolve more realistic views with time remains

The University of Michigan, owner of the George Reserve, holds a Game Breeder's license from the State, and all meat and hides harvested are sold, with the proceeds going to the George Reserve research fund. *Photo by Doug Fulton.*

to be seen. But I am sure they will not if the wildlife profession writes them off as a bunch of bleeding hearts. Certainly the buck-law syndrome of hunters was not overcome by categorizing them as beer-guzzling yahoos with a compulsion to shoot up signs—a characterization that, unfortunately, was appropriate in a discouraging number of cases. Nevertheless, the hunters were right about the decline in deer population that accompanies the liberalization of antlerless seasons, and there was a bit of truth too in the hunter's characterization of wildlife biologists and managers as people brainwashed in universities and who did not know what was going on in the woods.

DEER MANAGEMENT AND PROFESSIONALISM

Many of my colleagues in wildlife management believe that they should have a free hand in managing wildlife—a position I find a little uncomfortable. After all, we are public servants, and if we are not satisfying the public interests with our deer management programs, just what is it we are managing deer populations for? I agree that the public often does not realize the consequences of what they want, but I see it as the job of our profession to give the public not only what it asks for but more— an accurate prediction of the consequences. A medical doctor's job is to make an accurate diagnosis of the illness of a patient and give a reliable prognosis of various treatments. It is the patient's decision what to do about it. It does not reflect on the doctor's professionalism if a patient decides to die of cancer rather than undergo chemotherapy or an operation.

This is the role I advocate for the wildlife profession. Many things have been said about professionalism, but I believe they all are for naught until the profession develops sound diagnostic skills and predictive capabilities about various actions that might be taken. When that is achieved, wildlife management will be rec-

ognized as a scientific discipline by the public and other professionals. In the absence of such capability, no amount of signing of codes of ethics, supporting lobbying efforts or organizing memberships is going to gain wildlife management the respect and stature its participants aspire to.

The art of asking the right questions and framing testable hypotheses must be advanced to improve the mangement of white-tailed deer and other wildlife populations. Solid empirical studies (equivalent to clinical tests) are needed to prove that the wildlife biologists' and managers' prescriptions and remedies work.

Macnab, J. 1985. Carrying capacity and related slippery shibboleths. *Wildlife Society Bulletin* 13:403–410.

Who is John Macnab, and why is he picking on slippery shibboleths? According to the footnote in the article, Macnab is a pseudonym for four wildlife managers from Canada, Australia, Africa, and the U.S. The pseudonym had been used previously in a story about three successful friends who were bored and turned to poaching to provide the adventure they were missing from their lives. The pseudonym is used herein, according to Macnab, for neatness rather than anonymity, and certainly not for poaching. How does the use of Macnab instead of real names contribute to neatness?

The authors (i.e., Graeme Caughley [Australia], A. R. E. Sinclair [Canada], Mike Norton-Griffiths [Africa], and Doug Houston [USA]) are all well-known wildlife professionals. The real purpose of the pseudonym was to allow these biologists to raise issues that needed reevaluation in the practice of wildlife management. The pseudonym was used partly because government employees were discouraged—if not prohibited—from writing contentious scripts, and partly to say it was the issues, not the authors, that count.

The writers' objective was to tackle the slippery terms *carrying capacity, overpopulation, overharvest,* and *overgrazing.* They make the continued plea that a better understanding of these terms will come about only with hypothesis formulation and rigorous scientific testing. If you need proof, look at most publications that use these terms and see if you can determine what they actually mean.

RELATED READING

Anderson, W. E., and R. J. Scherzinger. 1975. Improving quality of winter forage for elk by cattle grazing. Journal of Range Management 28:120–125.

Eberhardt, L. L. 1977. "Optimal" management policies for marine mammals. Wildlife Society Bulletin 5:163–169.

Larkin, P. A. 1977. An epitaph for the concept of maximum sustained yield. Transactions of the American Fisheries Society 106:1–11.

Macnab, J. 1983. Wildlife management as scientific experimentation. Wildlife Society Bulletin 11:397–401.

Macnab, J. 1991. Does game cropping serve conservation? A reexamination of the African data. Canadian Journal of Zoology 69:2283–2290.

Murphy, D. D., and B. R. Noon. 1991. Coping with uncertainty in wildlife biology. Journal of Management 55:773–782.

Sinclair, A. R. E. 1991. Science and the practice of wildlife management. Journal of Wildlife Management 55:767–773.

CARRYING CAPACITY AND RELATED SLIPPERY SHIBBOLETHS

JOHN MACNAB,[1] *P.O. Box 84, Lyneham, A.C.T., Australia 2602*

Carrying capacity—rarely in the field of resources management has a term been so frequently used to the confusion of so many. Close rivals in perplexity include overpopulation, overharvesting, and overgrazing. This paper outlines traditional understanding of these shibboleths, identifies problems that arise from traditional use, and suggests ways of resolving the problems. The past decade has seen changes in the concepts, but these have usually appeared in publications of limited distribution and, consequently, have had minimal impact upon operational wildlife management. The traditional interpretations of the terms, and alternative views, remain largely hypotheses in search of critical field tests. Therefore, we build on the suggestion of

Macnab (1983) that routine wildlife management can be treated profitably as scientific experimentation using appropriate predictions and tests.

CARRYING CAPACITY

For range management, the density of cattle providing maximum sustained production of beef is the carrying capacity of the land. An index of this density is the species composition and growth stages of range plants (Stoddart et al. 1975). These ideas of animal density and related vegetation conditions have also been considered appropriate, for wild ungulates, in natural reserves as criteria to decide on herd management, including culling (e.g., Pengelly 1963, Stelfox 1976, Hanks et al. 1981).

Problems arise when the range management idea of carrying capacity is applied to wildlife management in natural areas, because "carrying capacity" is then used in a context where management objectives are entirely different. In livestock management, commercial

---

[1] John Macnab is the collective pen-name of 4 wildlife managers, 1 based in Canada, 1 in the U.S.A., 1 in Africa, and 1 in Australia. The pseudonym is assumed for neatness rather than for anonymity. The editor is free to reveal their identities to anyone unhappy with this arrangement or who feels slighted by the article.

harvesting is the objective. In wildlife management, harvesting is not always the objective—management in natural areas being an example.

Consider the simplest situation of a population of ungulates increasing in an area where predators, if present, are not regulating the population. The equilibrium reached between herbivores and their food supply represents the maximum sustainable population level. This level has been termed subsistence density (Dasmann 1981), environmental carrying capacity (Clark 1976), *K* carrying capacity (McCullough 1979), and potential carrying capacity (Riney 1982). Rate of increase of both plants and herbivores at *K* carrying capacity (KCC) is, by definition, zero, and the animal side of this equilibrium is often reached through a sequence of increased juvenile mortality, increased age of sexual maturity, decreased birth rate of adults, and increased adult mortality.

Populations of ungulates growing from low densities may overshoot their KCC levels before equilibria are achieved. The sequence of eruption, subsequent crash, and stable equilibrium has been reported by Riney (1964) and Caughley (1970). These authors viewed this dampened oscillation of population growth and the reciprocal downward trajectory of standing plant biomass as the norm, to be expected because the animals passed upward through their equilibrium level before the plants fully adjusted downward to theirs.

Harvest theory suggests that cropping an ungulate population to produce a sustained yield (SY) reduces density and maintains it below the equilibrium unharvested level of KCC (Clark 1976, Holt and Talbot 1978, Beddington 1979, McCullough 1979). The maximum sustained yield (MSY) is harvested from a density in a region between one-half and three-fourths of KCC. This point on the continuum of possible ungulate/vegetation densities has been termed optimum density (Das-

mann 1981), or "I" carrying capacity (ICC) (McCullough 1979). ICC is characterized by a lower standing crop of animals and a higher standing crop of vegetation than at the KCC equilibrium. Note that ICC, in contrast to KCC, is a contrived equilibrium held in place only by human intervention. If that intervention is withdrawn, the system drifts and will stabilize eventually at KCC.

Limited field studies of harvested ungulate populations lend some support to these general conclusions (McCullough 1979, Crête et al. 1981). ICC for the George Reserve whitetailed deer (*Odocoileus virginianus*) is around 0.56 KCC (McCullough 1979). Fowler (1981*a,b*) suggested that density dependent responses in populations of large mammals were expressed only at population levels near KCC, i.e., the maximum yield may be much nearer KCC than the simplified logistic theory suggested. There are remarkably few tests of the theory, however. Additional tests are needed before generalizations can be made.

The difference between KCC and ICC is the main source of confusion when the concept of range management carrying capacity is transferred to management of wild ungulates in natural areas. The equilibrium at ICC, with a comparatively low standing crop of ungulates and a high crop of vegetation, is of primary interest to the game manager whose objective is to maximize the sustained harvest from wild ungulates and to the range manager who attempts to maximize yield from herds of livestock. Indeed, ICC is the only carrying capacity usually recognized by range managers. In contrast, KCC is the equilibrium level usually pertinent to national parks.

A second related problem also exists. A traditional view is that populations are not reduced from KCC when harvested, because the harvest mortality simply substitutes for natural mortality (often called the Harvestable Surplus Model). Confusion arises because some natural mortality will take place in any event, and only some of it will be replaced by harvesting. In such circumstances any harvesting will reduce density below KCC (the ICC Harvest Model).

The full implications of these relationships have not penetrated deeply into either operational wildlife management or range management. There is a recent example of the apparent failure to appreciate these concepts. The sagebrush (*Artemisia* spp.) steppe of western North America supports large populations of mule deer (*O. hemionus*) that experience periodically harsh winters. Prior to the severe winter of 1983–1984, antlerless deer harvests were small to zero in much of eastern Oregon (Ingram 1983, 1984). Consequently, females were at high density, and deer died during the harsh winter (Durbin 1984*a,b*). The managing agency responded by feeding the deer. The implication of this is that the management objective was to maintain high deer densities to obtain bigger harvests later, i.e., offtake was assumed to be proportional to standing crop, the reverse of the theory outlined earlier. This situation should have provided a test of the 2 views, for the ICC Harvest Model predicts that a deer population previously lowered by hunting will provide a larger SY than the high density population, whereas the Harvestable Surplus Model predicts the reverse.

If the ICC model had proved correct, then the people of Oregon had the following management options:

1. Harvest populations to ICC to produce MSY.
2. Manage populations for "optimum sustained recreation"—hunters see large numbers of deer but harvest primarily males (essentially the existing program).
3. Maintain the deer population near KCC to serve as critical winter food for a variety of native carnivores under the policy of preserving all native wildlife species in Or-

egon. Periodic winter mortality is to be expected and desirable.

Finally, we should point out there are variations of the steady state equilibrium model of KCC. First, in extreme environments (e.g., arid and semi-arid regions) vegetation growth and the recruitment and mortality of large herbivores are highly variable, so that the "equilibrium" is more a mathematical abstraction than an operational reality. Secondly, under certain conditions, differences of scale between vegetation responses and ungulate responses may produce stable limit cycles (Peterson et al. 1984). This changes none of the theory, but makes management more difficult.

OVERPOPULATION

The traditional view of overpopulation is that it occurs whenever the population is above ICC. This definition raises problems because the term is dependent upon what kind of carrying capacity is relevant to management. Jewell and Holt (1981) recognized 4 contexts (classes) for overpopulation. There are too many animals if:

I. The animals threaten human life or livelihood.
II. The animals depress the densities of species favored by man.
III. The animals are "too numerous for their own good;" i.e., some animals are periodically in poor condition and undergo natural mortality.
IV. The system of plants and animals is off its equilibrium.

The main problem is that these contexts are often confused with each other. Classes I, II, III are often misidentified as class IV, especially in natural areas. The case histories of "overpopulations" reviewed during a recent workshop (Jewell and Holt 1981) underscored the need to define the sense in which animals are thought to be overabundant, to shape these thoughts into hypotheses, and to test hypotheses by field experimentation.

We note, in reference to class III overpopulation, that natural mortality of ungulates fluctuating around KCC is viewed as "a bad thing" by some people, especially in national parks or nature preserves. Deaths may occur as relatively inconspicuous "chronic" mortality (McCullough 1979) or as more conspicuous episodic mortality associated with severe weather (Coe 1978). Often this mortality is seen as "waste," i.e., the animal did not pass through the human digestive tract or pocketbook. This concept of waste is likely to be an inappropriate perspective from which to view and manage natural ecosystems.

OVERHARVEST

Overharvest, defined as killing too many animals, is also dependent on context. There are 2 types of problems here: (a) an appropriate removal is perceived erroneously as overharvest, (b) an excessive removal is perceived erroneously as an appropriate yield.

In the former case (a), misconceptions arise because of confusion between the "Harvestable Surplus Model" (HSM) and the "ICC Harvest Model." In the HSM, overharvest would occur if a population declines when initially harvested. Under the ICC model, the population would always decline initially when harvested. As stated earlier, these 2 competing models must be adjudicated by suitable field experiments.

The latter case (b), involves MSY harvests. MSY has traditionally been considered a desirable management objective. Recent works (Clark 1976, Larkin 1977, Holt and Talbot 1978) suggest that MSY may be an inappropriate objective because populations harvested for high sustained yields may take longer to recover from environmental disturbance (Beddington and May 1977), particularly when there are strong interactions between species (May et al. 1979); and some animal popula-

tions may have more than 1 equilibrium state (Holling 1973, Peterman et al. 1978), perhaps a higher equilibrium determined by food supply and a lower one determined by predators. In the presence of natural predators, MSY exploitation may drive a herbivore population down to a point where predators are able to drive it lower still, to the new equilibrium point (Smuts 1978, Walker 1981). Heroic measures may then be required to boost it back to higher levels (Peterman 1977). Overall, a policy of MSY may be "brinkmanship"—a small overestimate of population size could result in a continuing harvest set greater than MSY and consequent extinction of the population (Clark 1976, Larkin 1977, Holt and Talbot 1978). Larkin (1977) summarized this danger in an epitaph for MSY:

> Here lies the concept, MSY.
> It advocated yields too high,
> And didn't spell out how to slice the pie.
> We bury it with the best of wishes,
> Especially on behalf of fishes . . . R.I.P.

MSY harvests should be treated with considerable caution, even for the relatively simple ungulate communities typical of the Holarctic. Populations should probably be cropped to produce an "optimum sustained yield" allowing a margin for error in censusing and resultant sustained densities between KCC and ICC (Clark 1976, Eberhardt 1977, Larkin 1977).

Two additional connotations of overharvest are worth mentioning. First, to many accountants, the proper harvest rate for a commercially exploited population is that rate generating the maximum net revenue, discounted to present value, summed across the entire period of harvesting. That rate may result in the population being driven rapidly to commercial extinction (Clark 1976, May 1976, Larkin 1977). The MSY harvest of the biologist may represent a considerable underharvest to the economist. Only when removals exceed the rate at which maximum profit is generated would the population be overharvested economically.

Secondly, preliminary work suggests that certain harvest regimens imposed upon ungulate populations might reduce genetic variability by affecting rates of inbreeding and genetic drift (Ryman et al. 1981). This possibility may have important management implications for small isolated populations, such as those now characteristic of the mountain goat (*Oreamnos americanus*) and mountain sheep (*Ovis canadensis*) in North America. Cropping a high proportion of males under these conditions reduces effective population size (sensu Crow and Kimura 1970) below the actual population size (Ryman et al. 1981) and could affect genetic variability. Conceivably, even low removals could turn out to be genetic overharvest.

OVERGRAZING

The range manager's goal is generally to produce MSY from livestock populations at ICC. Techniques abound for measuring range readiness, production, condition trend, and use. The "ecologically based" or "climax" approach (Dyksterhuis 1949) used to assess range condition on many public lands in western North America is a case in point. Under this system, the presumptive climatic-climax vegetation is the standard ("excellent range condition"), to which other sites are compared and ranked by the proportion of climax species remaining (i.e., good, fair, poor). This work was an important step in providing ecological anchorages in the sea of vegetation conditions that arise with different regimes of livestock grazing. However, with acceptance of these criteria, the best rangeland was judged as that holding only climatic-climax vegetation; in effect the vegetation characteristic of ungrazed or lightly grazed areas became the goal, even though this was apparently not Dyksterhuis' (1949:107) original intent.

Two major problems are evident in this approach:

1. The relationship between range condition and livestock production is not as clear as originally thought. The supposed relationships become murkier still when adopted and used to manage wild ungulates.
2. The concept used as an index for ICC, when applied to natural areas, is inappropriate to measure KCC.

We deal with these in turn.

Some range scientists recognize the shortcomings of the climax approach as it relates to livestock production. These include the observation that various "subclimax" stages may be more stable, more palatable, and more productive than the putative climax (Smith 1978, 1985, Noy-Meir 1981, Risser 1985). Our concern is with management of wild ungulates. Enormous effort has been expended by wildlife biologists in many areas during the past 30 years in measuring the production, use, and morphology of presumptive "key" forage species for wild ungulates on "critical" range areas. Assumed, among other things, is that identification of the plant species and the range sites that actually support the population is straightforward, and that levels of offtake resulting in reduced productivity and mortality of plants are known. The view that climax vegetation represents excellent condition sometimes underpins the interpretation of these measurements. Attempts to relate these plant measurements to either animal density, indices of density and productivity, or to harvest, have been generally unsuccessful or inconclusive (Anderson et al. 1972, Mackie 1976, Kuck 1977) or not attempted. Enough of these efforts have been quietly abandoned to suggest that something went wrong. Most likely the initial assumptions were in error, and both the versatility of the beasts and the ability of many native range plants to withstand repeated seasonal grazing were underestimated

(Gill 1976, McNaughton 1976, 1979, 1983, West et al. 1979, Pellew 1984). Pellew (1984) recently examined the interaction of giraffes (*Giraffa camelopardalis*) with their *Acacia* spp. food supplies in Serengeti National Park. Pellew defined overbrowsing as a reduction of plant productivity caused by unsustainable offtake—and then demonstrated that offtake approaching 85% of the new shoot production of *Acacia* may not have represented overbrowsing! Application of traditional range use criteria would have indicated, incorrectly, that culling of giraffes was long overdue.

The biotic effects of domestic livestock on vegetation at ICC may be quite different from the effects of wild ungulates at KCC in natural areas (problem 2). The climax approach can become a trap when managing wild ungulates, because the vegetation conditions typical of KCC would rank only "fair" or perhaps even "poor" under this system.

A related problem arises through a confusion in the interpretation of what KCC represents in terms of the climatic-climax range criteria. One interpretation is that the KCC equilibrium is a short-term condition where the ungulate population reaches a level appropriate for a "good" range condition, i.e., a high proportion of "decreaser" and "increaser" plants and a low proportion of "invaders." Subsequently, it is assumed that with the high density of animals at KCC, "invaders" take over the range and the ungulate equilibrium density shows a long-term decline, even to extinction. This hypothesis underlies many claims of overgrazing and recommendations for culling. The alternative hypothesis is that the KCC equilibrium already represents the change in vegetation incorporating more "invaders," and has already leveled out. Therefore, it will not show a long-term decline. Many managers assume the first view to be correct without any attempt to test it.

All of these problems can be tackled by appropriate field experimentation. In retrospect,

the plant-based approach to evaluating range condition was uncoupled prematurely from measurements of the density and yield of wild ungulates. The notion that "we don't need to know how many deer there are as long as the range is properly used" has failed. Long-term studies where ungulate densities and offtakes are manipulated experimentally (and measured accurately) must be coupled with work on plant demography and productivity.

DISCUSSION

If there is merit in the suggestion that wildlife management is at its best as scientific experimentation (Macnab 1983), then the shibboleths of wildlife biology will become less slippery the more frequently they are incorporated into hypotheses and subjected to rigorous field tests. Two examples, in definite need of further testing, emerge from the earlier discussions. First, wildlife science is still a long way from having a solid, comprehensive theory for sustained yield harvests of terrestrial mammals. The reasons for this are philosophical (Romesburg 1981, Macnab 1983), technical, and financial. Census techniques sufficiently accurate and precise to measure the changes in animal density imposed by various harvest regimes have become available comparatively recently. Funding for research and for monitoring harvests has increased substantially in many countries. Both of these conditions suggest that the time is right for field testing hypotheses of sustained yield harvest.

Second, the long-term effects of grazing by wild ungulates on the stability, species composition, and productivity of their food supplies are still poorly understood. The concepts of range science are deficient when dealing with the diverse management objectives encountered commonly by wildlife biologists. The need to pose and test research hypotheses by manipulative experimentation, as opposed to application of cookbook methodology, seems especially pertinent in this area.

SUMMARY

Several fundamental concepts used by wildlife managers have evolved considerably during the past decade. These include "carrying capacity" and the derivative notions of overpopulation, overharvesting, and overgrazing. This note outlines the traditional understanding of these terms, identifies problems that arise from traditional views, and suggests ways of resolving some of the problems.

LITERATURE CITED

ANDERSON, A. E., D. E. MEDIN, AND D. C. BOWDEN. 1972. Mule deer numbers and shrub yield—utilization on winter range. J. Wildl. Manage. 36: 571–578.

BEDDINGTON, J. R. 1979. Harvesting and population dynamics. Pages 307–320 *in* R. M. Anderson, B. D. Turner, and L. R. Taylor, eds. Population dynamics: Twentieth Symp. British Ecol. Soc., Blackwell, London.

———— AND R. M. MAY. 1977. Harvesting natural populations in a randomly fluctuating environment. Science 177:463–465.

CAUGHLEY, G. 1970. Eruption of ungulate populations, with emphasis on Himalayan thar in New Zealand. Ecology 51:53–72.

CLARK, C. 1976. Mathematical bioeconomics: the optimal management of renewable resources. J. Wiley & Sons, New York. 352pp.

COE, M. 1978. The decomposition of elephant carcasses in the Tsavo (East) National Park, Kenya. J. Arid Environs. 1:71–86.

CRÊTE, M., R. M. TAYLOR, AND P. A. JORDAN. 1981. Optimization of moose harvest in southwestern Quebec. J. Wildl. Manage. 45:598–611.

CROW, J. F. AND M. KIMURA. 1970. An introduction to population genetics theory. Harper and Row, New York. 591pp.

DASMANN, R. F. 1981. Wildlife biology. Second ed. J. Wiley & Sons, New York. 212pp.

DURBIN, K. 1984*a*. In winter's grip. Oregon Wildl. 39(2):3–8.

————. 1984*b*. Good news and bad news: a winter update. Oregon Wildl. 39(3):3–5.

DYKSTERHUIS, E. J. 1949. Condition and management of range land based on quantitative ecology. J. Range Manage. 2:104–115.

EBERHARDT, L. L. 1977. "Optimal" management policies for marine mammals. Wildl. Soc. Bull. 5: 163–169.

FOWLER, C. W. 1981*a*. Comparative population dynamics in large mammals. Pages 437–455 *in* C. W. Fowler and T. D. Smith, eds. Dynamics of large mammal populations. J. Wiley & Sons, New York.

————. 1981*b*. Density dependence as related to life history strategy. Ecology 62:602–610.

GILL, R. B. 1976. Mule deer management myths and the mule deer population decline. Pages 99–106 *in* G. W. Workman and J. B. Low, eds. Mule deer decline in the west. Symp. Proc. Utah State Univ., Logan.

HANKS, J., ET AL. 1981. Management of locally abundant mammals—The South African experience. Pages 21–55 *in* P. A. Jewell and S. Holt, eds. Problems in management of locally abundant wild mammals. Academic Press, New York.

HOLLING, C. S. 1973. Resilience and stability of ecological systems. Annu. Rev. Ecol. Syst. 4:1–23.

HOLT, S. J. AND L. M. TALBOT. 1978. New principles for the conservation of wild living resources. Wildl. Monogr. 59. 33pp.

INGRAM, R. 1983. 1982 big game seasons. Oregon Wildl. 38(5):3–9.

————. 1984. 1983 big game harvest. Oregon Wildl. 39(5):3–10.

KUCK, L. 1977. The impacts of hunting on Idaho's Pahsimeroi mountain goat herd. Pages 114–125 *in* W. Samuel and W. G. Macgregor, eds. Proc. First Intl. Mountain Goat Symp., British Columbia Fish & Wildl. Branch.

JEWELL, P. A. AND S. HOLT. 1981. Problems in management of locally abundant wild mammals. Academic Press, New York. 361pp.

LARKIN, P. A. 1977. An epitaph for the concept of maximum sustained yield. Trans. Am. Fish Soc. 106:1–11.

MACKIE, R. J. 1976. Evaluation of range survey methods, concepts and criteria (effectiveness of the key browse survey method). Montana Fish and Game Dep. P-R Proj. Rep. W-120-R-G 20pp. (Multilith.)

MAY, R. M. 1976. Harvesting whale and fish populations. Nature 263:91–92.

————, J. R. BEDDINGTON, C. W. CLARK, S. J. HOLT, AND R. M. LAWS. 1979. Management of multispecies fisheries. Science 205:267–277.

MCCULLOUGH, D. R. 1979. The George Reserve deer herd: population ecology of a K-selected species. Univ. Michigan Press, Ann Arbor. 271pp.

MACNAB, J. 1983. Wildlife management as scientific experimentation. Wildl. Soc. Bull. 11:397–401.

MCNAUGHTON, S. J. 1976. Serengeti migratory wildebeest: facilitation of energy flow by grazing. Science 191:92–94.

————. 1979. Grazing as an optimization process: grass-ungulate relationships in the Serengeti. Am. Nat. 113:691–703.

————. 1983. Serengeti grassland ecology: the role of composite environmental factors and contingency in community organization. Ecol. Monogr. 53:291–320.

NOY-MEIR, I. 1981. Responses of vegetation to the abundance of mammalian herbivores. Pages 233–246 *in* P. A. Jewel and S. Holt, eds. Problems in management of locally abundant wild mammals. Academic Press, New York.

PELLEW, R. A. 1984. Food consumption and energy budgets of the giraffe. J. Appl. Ecol. 21:141–159.

PENGELLY, W. L. 1963. Thunder on the Yellowstone. Naturalist 14:18–25.

PETERMAN, R. M. 1977. A simple mechanism that causes collapsing stability regions in exploited salmonid populations. J. Fish Res. Board Can. 34: 1130–1142.

————, W. C. CLARK, AND C. S. HOLLING. 1978. The dynamics of resilience: shifting stability domains in fish and insect systems. Pages 321–341 *in* R. M. Anderson, B. D. Turner, and L. R. Taylor, eds. Population dynamics. Proc. British Ecol. Soc. Symp., Blackwell, London.

PETERSON, R. O., R. E. PAGE, AND K. M. DODGE. 1984. Wolves, moose, and the allometry of population cycles. Science 224:1350–1352.

RINEY, T. 1964. The impact of introductions of large herbivores on the tropical environment. Intl. Union Conserv. Nat. Publ. New Series (4):261–273.

————. 1982. Study and management of large mammals. J. Wiley & Sons, New York. 552pp.

RISSER, P. G. 1985. Range condition analysis: past, present, and future. *In* W. K. Lauenroth and W. A. Laycock, eds. Secondary succession and evaluation of rangeland condition. Springer-Verlag, New York. (In press.)

ROMESBURG, H. C. 1981. Wildlife science: gaining reliable knowledge. J. Wildl. Manage. 45:293–313.

RYMAN, N., R. BACCUS, C. REUTERWALL, AND M. H. SMITH. 1981. Effective poulation size, generation interval, and potential loss of genetic variability in game species under different hunting regimes. Oikos 36:257–266.

SMITH, E. L. 1978. A critical evaluation of the range condition concept. Pages 266–268 *in* D. N. Hyder, ed. Proc. First Intl. Rangeland Congr. Soc. for Range Manage.

————. 1985. Range condition and secondary succession. *In* W. K. Lauenroth and W. A. Laycock, eds. Secondary succession and evaluation of rangeland condition. Springer-Verlag, New York. (In press.)

STODDART, L. A., A. D. SMITH, AND T. W. BOX. 1975. Range management. Third ed. McGraw-Hill Book Co., New York. 532pp.

SMUTS, G. L. 1978. Interrelations between predators, prey, and their environment. Bioscience 28:316–320.

STELFOX, J. G. 1976. Range ecology of Rocky Mountain bighorn sheep. Canadian Wildl. Rep. Ser. No. 39. 50pp.

WALKER, B. H. 1981. Stability properties of semiarid savannas in southern African game reserves. Pages 57–67 *in* P. A. Jewell and S. Holt, eds. Problems in management of locally abundant wild mammals. Academic Press, New York.

WEST, N. E., K. H. REA, AND R. O. HORNISS. 1979. Plant demographic studies in sagebrush-grass communities of southeastern Idaho. Ecology 60: 376–388.

Received 26 February 1985.
Accepted 28 May 1985.

Wilson, E. O. 1985. The biological diversity crisis. *BioScience* 35:700–706.

E. O. Wilson is credited as the first to use the term "biodiversity" in written text. While there are many aspects of nature that society is well aware of, biodiversity is not one of them. Yet there is likely no more formative and important concept in our profession than biodiversity. In this article, Wilson's objective is to understand exactly what the inventory on planet Earth is, as it is the only living world we will ever have.

"The magnitude and cause of biological diversity is not just the central problem of systematics; it is one of the key problems of science as a whole. It can be said that for a problem to be so ranked, its solution must promise to yield unexpected results, some of which are revolutionary in the sense that they resolve conflicts in current theory while opening productive new areas of research. In addition, the answers should influence a variety of related disciplines. They should affect our view of humanity's place in the order of things and open opportunities for the development of new technology of social importance. These several criteria are, of course, very difficult to attain, but I believe the diversity problem meets them all."

Wilson makes a strong case for hiring the necessary taxonomists to complete the job. The effort of classifying and naming all of the world's flora and fauna would not be excessive and could easily be accomplished with current resources. This need becomes critically important when one considers the number of animal and plant species that go extinct annually.

RELATED READING

Andrén, H. 1994. Effects of habitat fragmentation on birds and mammals in landscapes with different proportions of suitable habitat: a review. Oikos 71:355–366.

Ehrlich, P. R., and E. O. Wilson. 1991. Biodiversity studies: science and policy. Science 253:758–762.

Fahrig, L. 2003. Effects of habitat fragmentation on biodiversity. Annual Review of Ecology, Evolution, and Systematics 34:487–515.

Henke, S. E., and F. C. Bryant. 1999. Effects of coyote removal on the faunal community in western Texas. Journal of Wildlife Management 63:1066–1081.

Karr, J. R., and R. R. Roth. 1971. Vegetation structure and avian diversity in several New World areas. The American Naturalist 105:423–435.

Mora, C., D. P. Tittensor, S. Adl, A. G. P. Simpson, and B. Worm. 2011. How many species are there on earth and in the ocean? PLoS Biol 9(8): e1001127. doi:10.1371/journal.pbio.1001127. Accessed 25 June 2012.

Pimm, S. L., G. J. Russell, J. L. Gittleman, and T. M. Brooks. 1995. The future of biodiversity. Science 269:347–350.

Wilson, E. O. 1993. The diversity of life. Penguin Press, London, UK.

Reprinted by permission of The Ecological Society of America
and E. O. Wilson.

The Biological Diversity Crisis

Despite unprecedented extinction rates, the extent of biological diversity remains unmeasured

Edward O. Wilson

Certain measurements are crucial to our ordinary understanding of the universe. What, for example, is the mean diameter of Earth? 12,742 km. How many stars are there in the Milky Way? 10^{11}. How many genes in a small virus particle? 10 (in ϕX174 phage). What is the mass of an electron? 9.1×10^{-28} grams. How many species of organisms are there on Earth? We don't know, not even to the nearest order of magnitude.

Of course, the number of *described* species is so impressive that it might appear complete. The corollary would be that systematics is an old-fashioned science concerned mostly with routine tasks. In fact, about 1.7 million species have been formally named since Linnaeus inaugurated the binomial system in 1753. Some 440,000 are plants, including algae and fungi; 47,000 are vertebrates; and according to one meticulous estimate published in 1985 by R. H. Arnett, 751,012 are insects. The remainder are assorted invertebrates and microorganisms.

But these figures alone grossly underestimate the diversity of life on

Edward O. Wilson is Baird Professor of Science and Curator in Entomology at the Museum of Comparative Zoology, Harvard University, 26 Oxford St., Cambridge, MA 02138. His best-known books include *The Insect Societies, Sociobiology,* and *On Human Nature.* This article has been modified from an article published in *Issues in Science and Technology* [2(1): 20–29, Fall 1985] with permission from the editors.

The pool of diversity is a challenge to basic science and a vast reservoir of genetic information

Earth, and its true magnitude is still a mystery. In 1964 the British ecologist C. B. Williams, employing a combination of intensive local sampling and mathematical extrapolation, projected the number of insect species at three million (Williams 1964). During the next 20 years, systematists described several new complex faunas in relatively unexplored habitats such as the floor of the deep sea. They also began to use electrophoresis and ecological studies routinely, enabling them to detect many more sibling species. A few writers began to put the world's total as high as ten million species.

In 1982 the ante was again raised threefold by Terry L. Erwin (1983) of the National Museum of Natural History. He and other entomologists had developed a technique that for the first time allowed intensive sampling of the canopy of tropical rainforests. This layer of leaves and branches conducts most of the photosynthesis and is clearly rich in species. But it has been largely inaccessible because of its height (a hundred feet or more), the slick surface of the trunks, and swarms of stinging ants and wasps that break forth at all levels. To over-

come these difficulties, a projectile with a line attached is first shot over one of the upper branches. A canister containing an insecticide and swift-acting knockdown agent is then hauled up into the canopy, and the contents are released as a fog by radio command. As the insects and other arthropods fall out of the trees (the chemicals do not harm vertebrates), they are collected in sheets laid on the ground. The numbers of species proved to be far greater than previously suspected because of unusually restricted geographical ranges and high levels of specialization on different parts of the trees. Erwin extrapolated a possible total of 30 million insect species, mostly confined to the rainforest canopy.

If astronomers were to discover a new planet beyond Pluto, the news would make front pages around the world. Not so for the discovery that the living world is richer than earlier suspected, a fact of much greater import to humanity. Organic diversity has remained obscure among scientific problems for reasons having to do with both geography and the natural human affection for big organisms. The great majority of kinds of organisms everywhere in the world are not only tropical, but also inconspicuous invertebrates such as insects, crustaceans, mites, and nematodes. The mammals, birds, and flowering plants of the North Temperate Zone, on which natural history research and popular writing have largely focused, comprise relatively few species. In one aggregate of 25 acres of rainforest in Borneo, for example, about 700 species of trees were identified;[1] there are no more than 700 native tree species in all of North America. Familiarity with organisms close to home gives the false impression that the Linnaean period has indeed ended. But a brief look almost anywhere else (for example the Australian fauna illustrated in Figure 1) shows that the opposite is true.

Why does this lack of balance in knowledge matter? It might still be argued that to know one kind of beetle is to know them all, or at least enough to get by. But a species is not like a molecule in a cloud of molecules. It is a unique population of organisms, the terminus of a lineage that split off thousands or even millions of years ago. It has been hammered and shaped into its present form by mutations and natural selection, during which certain genetic combinations survived and reproduced differentially out of an almost inconceivably large number possible.

In a purely technical sense, each species of higher organism is richer in information than a Caravaggio painting, Bach fugue, or any other great work of art. Consider the typical case of the house mouse, *Mus musculus*. Each of its cells contains four strings of DNA, each of which comprises about a billion nucleotide pairs organized into a hundred thousand structural genes. If stretched out fully, the DNA would be roughly one meter long. But this molecule is invisible to the naked eye because it is only 20 angstroms in diameter. If we magnified it until its width equaled that of a wrapping string to make it plainly visible, the fully extended molecule would be 600 miles long. As we traveled along its length, we would encounter some 20 nucleotide pairs to the inch. The full information contained therein, if translated into ordinary-sized printed letters, would just about fill all 15 editions of the *Encyclopaedia Britannica* published since 1768.

Perhaps because organic diversity is so much greater and richer in history than previously imagined, it has proved difficult to express as a coherent subject of scientific in-

[1]Peter S. Ashton, 1985, personal communication. Arnold Arboretum, Harvard University.

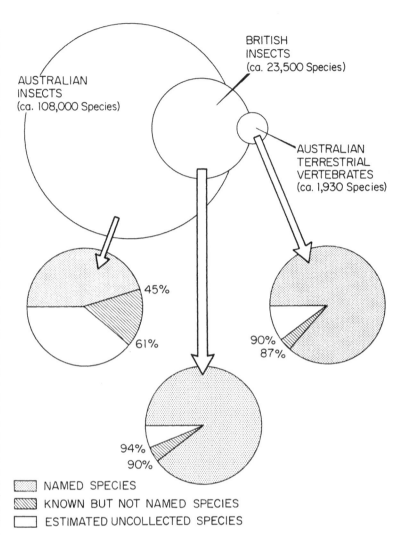

Figure 1. The status of research on diversity is illustrated in this comparison of the estimated sizes of Australian insect, British insect, and Australian terrestrial vertebrate faunas (top) and the levels of taxonomic knowledge about them. Modified from Taylor (1983).

quiry. What *is* the central problem of systematics? Its practitioners, who by necessity limit themselves to small slices of the diversity, understand but seldom articulate a mission of the kind that enspirits particle physics or molecular genetics. For reasons that transcend the mere health of the discipline, the time has come to focus on such an effort. Indeed, if one considers other disciplines that depend directly upon systematics, including ecology, biogeography, and behavioral biology, an entire hierarchy of important problems present themselves.

But one problem stands out, for progress toward its solution is needed to put the other disciplines on a permanently solid basis. For taxonomy, the key question is the number of living species. How many exist in each major group from bacteria to mammals? Where is each found, and how does it differ from related species? I believe that we should aim at nothing less than a full count, a complete catalog of life on Earth. To attempt an absolute measure of diversity is a mission worthy of the best effort from science.

December 1985

701

Green dragonfly (family Libellulidae). Photo: Ronald Bork.

The magnitude and cause of biological diversity is not just the central problem of systematics; it is one of the key problems of science as a whole. It can be said that for a problem to be so ranked, its solution must promise to yield unexpected results, some of which are revolutionary in the sense that they resolve conflicts in current theory while opening productive new areas of research. In addition, the answers should influence a variety of related disciplines. They should affect our view of humanity's place in the order of things and open opportunities for the development of new technology of social importance. These several criteria are, of course, very difficult to attain, but I believe the diversity problem meets them all.

To this end, the problem can be restated as follows: If there are indeed 30 million species, why didn't 40 million evolve, or 2000, or a billion? Many ramifications spring from this ultimate Linnaean question. We would like to know whether something peculiar about the conformation of the planet or the mechanics of evolution itself led to the precise number that does exist. At the next level down, why is there an overwhelming preponderance of insect species on the land, but virtually none of these organisms in the sea? Hot spots of disproportionately high diversity of plants and animals occur within larger rainforests, and we need to know their contents and limits. Would it be possible to increase the diversity of

natural systems artificially to levels above those in nature without destabilizing them? Only taxonomic analysis can initiate and guide research on these and related topics.

The relation of systematics research to other biological disciplines becomes clearer if one considers the way diversity is created. A local community of plants and animals, of the kind occupying a pond or offshore island, is dynamic in its composition. New colonists arrive as old residents die off. If enough time passes, the more persistent populations evolve into local endemic species. On islands as large as Cuba or Oahu, the endemics often split into two or more species able to live side by side. The total play of these forces (immigration, extinction, and evolution leading to species multiplication) determines the global amount of diversity. To understand each of the forces in turn is automatically to address the principal concerns of ecology, biogeography, and population genetics. Our current understanding of the forces is still only rudimentary. The science addressing them can be generously put at about the level of physics as it was in the late nineteenth century.

There is in addition a compelling practical argument for attempting a complete survey of diversity. Only a tiny fraction of species with potential economic importance has been used (Myers 1983, Oldfield 1984). A far larger number,

tens of thousands of plants and millions of animals, have never even been studied well enough to assess their potential. Throughout history, for example, a total of 7000 kinds of plants have been grown or collected as food. Of these, 20 species supply 90% of the world's food and just 3—wheat, maize, and rice—constitute about half. In most parts of the world, this thin reservoir of diversity is sown in monocultures particularly sensitive to insect attacks and disease. Yet waiting in the wings are tens of thousands of edible species, many demonstrably superior to those already in use.

The case of natural sweeteners serves as a parable of untapped resource potential among wild species. A plant has been found in West Africa, the katemfe (*Thaumatococcus daniellii*), that produces proteins 1600 times sweeter than sucrose. A second West African plant, the serendipity berry (*Dioscoreophyllum cumminsii*), produces a substance 3000 times sweeter. The parable is the following: Where in the plant kingdom does the progression end? To cite a more clearly humanitarian example, one in ten plant species contains anticancer substances of variable potency, but relatively few have been bioassayed. Economists use the expression "opportunity costs" for losses incurred through certain choices made over others, including ignorance and inaction. For systematics, or more precisely the neglect of systematics and the biological research dependent upon it, the costs are very high.

Biological diversity, apart from our knowledge of it, is meanwhile in a state of crisis. Quite simply, it is declining. Environmental destruction, a worldwide phenomenon, is reducing the numbers of species and the amount of genetic variation within individual species. The loss is most intense in the tropical rainforests. In prehistoric times, these most species-rich of all terrestrial habitats covered an estimated 5 million square miles. Today they occupy 3.5 million square miles and are being cut down at an annual rate of 0.7%, that is, 25,000 square miles or an area the size of West Virginia. The effect this deforestation has on diversity can be approximated by the following rule of thumb in biogeography. When the area of a habitat is reduced to one-tenth, the number of species that can persist in it indefinitely will eventually decline to one-half. That much habitat reduction has already been passed in many parts of the tropics. The forests of Madagascar now occupy less than ten percent of their original cover, and the once teeming Brazilian Atlantic forests are down to under one percent. Even great wilderness areas are giving way. If present levels of deforestation continue, the stage will be set within a century for the inevitable loss of about 12% of the 700 bird species in the Amazon Basin and 15% of the plant species in South and Central America (Simberloff 1984).

No comfort should be drawn from the spurious belief that because extinction is a natural process, humans are merely another Darwinian agent. The rate of extinction is now about 400 times that recorded through recent geological time and is accelerating rapidly. Under the best of conditions, the reduction of diversity seems destined to approach that of the great natural catastrophes at the end of the Paleozoic and Mesozoic Eras, in other words, the most extreme for 65 million years. And in at least one respect, this human-made hecatomb is worse than any time in the geological past. In the earlier mass extinctions, possibly caused by large meteorite strikes, most of the plant diversity survived; now, for the first time, it is being mostly destroyed (Knoll 1984).

A complete survey of life on Earth may appear to be a daunting task. But compared with what has been dared and achieved in high-energy physics, molecular genetics, and other branches of "big science," it is in the second or third rank. To handle ten million species even with the least efficient old-fashioned methods is an attainable goal. If one specialist proceeded at the cautious pace of an average of ten species per year, including collecting, curatorial work, taxonomic analysis, and publication, about one million person-years of work would be required. Given 40 years of productive life per scientist, the effort would consume 25,000 lifetimes. That is not an excessive investment on a global scale. The number of systematists worldwide would still represent less than ten percent of the current population of scientists working in the United States alone and fall short of the standing armed forces of Mongolia and the population of retirees in Jacksonville. Neither does information storage present an overwhelming problem, even when left wholly to conventional libraries. If each species were given a single double-columned page for the diagnostic taxonomic description, a figure, and brief biological characterization, and if the pages were bound into ordinary 1000-page, six-centimeter-wide hardcover volumes, the 10,000 or so final volumes of this ultimate catalog would fill 600 meters of library shelving. That much is far below the capacity of some existing libraries of evolutionary biology. The library of Harvard's Museum of Comparative Zoology, for example, occupies 4850 meters of shelving.

But I have given the worst scenario imaginable to establish the plausibility of the project. Systematic work can be speeded many times over by new procedures now coming into general use. The Statistical Analysis System (SAS), a set of computer programs currently running in over 4000 institutions worldwide, permits the recording of taxonomic identifications and localities of individual specimens and the automatic integration of data into catalogs and biogeographic maps (La Duke et al. 1984). Other computer-aided techniques rapidly compare species across large numbers of traits, applying unbiased measures of overall similarity, the procedure known as phenetics. Still others assist in sorting out the most likely patterns of phylogeny by which species split apart to create diversity, or cladistics. Scanning electron microscopy has speeded the illustration of insects and other small organisms and rendered descriptions more accurate. The DELTA system, developed and used at Australia's Commonwealth Scientific and Industrial Research Organization, codes data for the automatic identification of specimens (Dallwitz 1980, Taylor 1983). Elsewhere, research is being conducted that might lead to computerized image scanning for automatic description and data recording.

In North America, about 4000 systematists work on 3900 systematics collections (Edwards 1984 and personal communication). But a large fraction of these specialists, perhaps a majority, are engaged only part time in taxonomic research. More to the point, few can identify organisms from the tropics, where both the great majority of species exist and extinction is proceeding most rapidly. Probably no more than 1500 trained professional systematists in the world are competent to deal with tropical organisms. Their number may be declining from decreased professional opportunities, reduced funding for research, and assignment of higher priority to other disciplines (National Research Council 1980). To take one especially striking example, ants and termites make up about one-third of the animal biomass in tropical forests. They cycle a large part of the energy in all terrestrial habitats and include the foremost agricultural pests, which cause billions of dollars of damage yearly. Yet there are exactly eight entomologists worldwide with the general competence to identify tropical ants and termites, and only five of these are able to work at their specialty full time.

It is not surprising to find that the neglect of species diversity retards other forms of biological research. Every ecologist can tell of studies delayed or blocked by the lack

Honeybee *(Apis mellifera)*. Photo: R. Bork.

of taxonomic expertise. In one recent, typical case, William G. Eberhard consulted most of the small number of available (and overworked) authorities to identify South and Central American spiders used in a study of web-building behavior. He was able to obtain determinations of only 87 of the 213 species included, and then only after considerable delay. He notes that "there are some families (e.g., Pholcidae, Linyphiidae, Anyphaenidae) in which identifications even to genus of Neotropical species are often not possible, and apparently will not be until major taxonomic revisions are done. On a personal level, this has meant that I have refrained from working on some spiders (e.g., Pholcidae, one of the dominant groups of web spiders in a variety of forest habitats, at least in terms of numbers of individuals) because I can't get them satisfactorily identified."[2]

If systematics is an indispensable handmaiden of other branches of research, it is also a fountainhead of discoveries and new ideas, providing the remedy for what the biologist and philosopher William Morton Wheeler once called the dry rot of academic biology. Systematics has never been given enough credit for this second, vital role. Every time I walk into a fresh habitat, whether tropical forest, grassland, or desert, I become quickly aware of the potential created by a knowledge of classification. If biologists can identify only a limited number of species, they are likely to gravitate toward them and end up on well-trodden ground; the rest of the species remain a confusing jumble. But if they are well trained in the classification of the organisms encountered, the opportunities multiply. The known facts of natural history become an open book, patterns of adaptation fall into place, and previously unknown phenomena offer themselves conspicuously. By proceeding in this opportunistic fashion, a biologist might strike a new form of animal communication, a previously unsuspected mode of root symbiosis, or a relation between certain species that permits a definitive test of competition theory. The irony is that suc-

[2]Letter from W. G. Eberhard, 1985. Universidad de Costa Rica, Ciudad Universitaria.

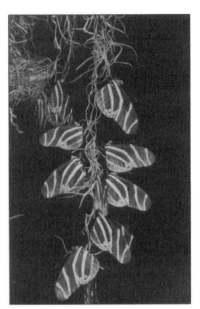

Nocturnal roosting aggregation of zebra butterflies *(Heliconius charitonius)* in Florida. Photo: James L. Castner.

cessful research then gets labeled as ecology, physiology, or almost anything else but its *fons et origo,* the study of diversity.

Systematics is linked in such a manner not only to the remainder of biology but to the fortunes of the international conservation movement, which is now focusing its attention on the threatened environments of the tropics. Plans for systems of ecological reserves have been laid by the International Union for Conservation of Nature and Natural Resources (IUCN), UNESCO, and a growing number of national governments from Australia and Sri Lanka to Brazil and Costa Rica. The aim is to hold on to as many species as possible within the limits imposed by population pressures and the cost of land purchase. The long-term effects of this enterprise can only be crudely predicted until systematics surveys are completed, country by country. In the United States a proposal for a National Biological Survey (NABIS) has been presented to Congress [see p. 686]. The program would establish a survey to describe all the plants and animals, fund basic taxonomic studies to this end, and produce identification manuals, catalogs, and other practical aids (Kosztarab 1984). If multiplied across many countries,

such efforts could bring the full assessment of biological diversity within reach.

Systematics surveys are cost-efficient and provide large proportionate yields with small absolute increments in support. In fiscal 1985, the National Museum of Natural History, the largest organization of its kind in the United States, spent $12.8 million to support 85 scientists engaged partly or wholly in taxonomic studies. In the same year, the Program in Systematic Biology of the National Science Foundation granted $12 million for basic taxonomic research, while other programs in the NSF and Department of Interior provided $13.8 million for support of museum services, studies of endangered species, and other activities related to systematics. The worldwide support for basic tropical biology, including systematics and ecology, is only about $50 million. Just 1000 annual grants of $50,000 devoted to tropical organisms would double the level of support and revitalize the field. To illustrate the difference in scale, the same amount added to the approximately $3.5 billion spent on health-related biology in the United States would constitute an increment of 1.4%, causing a barely detectable change.

In case such an investment, which approximately equals the lifetime cost of one F15 Eagle fighter-bomber, might seem removed from the immediate interests of the United States, let me close with an observation on the importance of biological research to foreign policy. The problems of Third World countries, most of which are in the tropics, are primarily biological. They include excessive population growth, depletion of soil nutrients, deforestation, and the decline of genetic diversity in crop and forest reserves. It is no coincidence that Haiti and El Salvador, which force themselves on our attention at such frequent intervals, are the most densely populated and environmentally degraded countries in the Western Hemisphere, rivaled only by Grenada (another trouble spot) and three other small Caribbean island-nations. Virtually all reports on the subject released by the National Research Council and Office of Technology Assessment during the past ten years agree that the intricate economic and

704

Florida atala *(Eumaeus atala)*. Photo: J. L. Castner.

social problems of tropical countries cannot be solved without a more detailed knowledge of the environment. Increasingly, that must include a detailed account of native faunas and floras.

Congress has addressed this problem in limited degree through the 1980 amendment to the Foreign Assistance Act, which mandates that programs funded through the Agency for International Development include an assessment of environmental impact. In implementing this policy, AID recognizes that "the destruction of humid tropical forests is one of the most important environmental issues for the remainder of this century and, perhaps, well into the next," in part because they are "essential to the survival of vast numbers of species of plants and animals" (Department of State memorandum 1985).

Moving further, AID set up an Interagency Task Force in 1985 to consider biological diversity as a comprehensive issue. In its report to Congress, *US Strategy on the Conservation of Biological Diversity* (AID 1985), the task force evaluated the current activities of the dozen federal agencies that have been concerned with diversity, including the Smithsonian Institution, the Environmental Protection Agency, and AID itself. The most important recommendations made by the group, in my opinion, are those that call for the primary inventory and assessment of native faunas and floras. In fact, not much

else can be accomplished without this detailed information.

AID also supports research programs in which nationals of the recipient countries are principal investigators, and US citizens serve as collaborators. This arrangement is a proven way to build science and technology in the Third World and is particularly well suited to tropical biology. Studies of diversity are best conducted at sites of maximum diversity. They are labor-intensive and require less expensive instrumentation than most kinds of research. Perhaps most important, their relevance to

national identity and welfare are immediately obvious.

To put the matter as concisely as possible, biological diversity is unique in the evenness of its importance to both developed and developing countries and in the cost-effectiveness of its study. The United States would do well to seek a formal international agreement among countries, possibly in the form of an International Decade for the Study of Life on Earth, to improve financial support and access to study sites. To spread technical capability where it is most needed, arrangements can be made to retain specimens within the countries of their origin while training nationals to assume leadership in systematics and the related scientific disciplines.

In *Physics and Philosophy,* Werner Heisenberg suggested that science is the best way to establish links with other cultures because it is concerned not with ideology but with nature and humanity's relation to nature. If that promise can ever be met, it will surely be in an international effort to understand and save biological diversity. This being the only living world we are ever likely to know, let us join to make the most of it.

Acknowledgments

I am grateful to the following persons for valuable advice and assistance given during the preparation of this arti-

Leaf-mimicking mantid *(Choerododid rhombicollis),* Panama. Photo: J. L. Castner.

cle: Peter Ashton, W. L. Brown, David Challinor, John Daly, W. G. Eberhard, S. R. Edwards, J. H. Gibbons, Arturo Gomez-Pompa, A. H. Knoll, P. H. Raven, D. S. Simberloff, G. D. Snell, R. W. Taylor, and P. B. Tomlinson.

References cited

Agency for International Development (AID). 1985. *US Strategy on the Conservation of Biological Diversity: An Interagency Task Force Report to Congress.* Washington, DC.

Arnett, R. H., Jr. 1985. *American Insects.* Van Nostrand Reinhold, New York.

Dallwitz, M. J. 1980. A general system for coding taxonomic descriptions. *Taxon* 29: 41–46.

Edwards, S. R. 1984. The systematics community: priorities for the next decade. *ASC Newsletter* (Association of Systematics Collections) 12(5): 37–40.

Erwin, T. L. 1983. Beetles and other insects of tropical forest canopies at Manaus, Brazil, sampled by insecticidal fogging. Pages 59–75 in S. L. Sutton, T. C. Whitmore, and A. C. Chadwick, eds. *Tropical Rain Forest: Ecology and Management.* Blackwell Scientific Publications, Edinburgh, UK.

Knoll, A. H. 1984. Patterns of extinction in the fossil record of vascular plants. Pages 21–68 in M. H. Nitecki, ed. *Extinctions.* University of Chicago Press, Chicago.

Kosztarab, M. 1984. A biological survey of the United States. *Science* 223: 443.

La Duke, J. C., D. Lank, and T. Sirek. 1984. Utilization of statistical analysis systems

Juvenile Australian katydid (family Tettigoniidae), Seaforth, New South Wales, Australia. Photo: Art Daniel.

(SAS) as a revisionary tool and cataloguing program. *ASC Newsletter* 12(2): 13–14.

Myers, N. 1983. *A Wealth of Wild Species.* Westview Press, Boulder, CO.

National Research Council (NRC). 1980. *Research Priorities in Tropical Biology.* National Academy of Sciences, Washington, DC.

Oldfield, M. L. 1984. *The Value of Conserving Genetic Resources.* US Department of the Interior, Washington, DC.

Simberloff, D. S. 1984. Mass extinction and the destruction of moist tropical forests. *Zh.*

Obshch. Biol. 45: 767–778.

Taylor, R. W. 1983. Descriptive taxonomy: past, present, and future. In E. Highley and R. W. Taylor, eds. *Australian Systematic Entomology: A Bicentenary Perspective.* Commonwealth Scientific and Industrial Research Organization, Melbourne.

US Department of State. 1985. Humid tropical forests: AID policy and program guidance. Memorandum. Washington, DC.

Williams, C. B. 1964. *Patterns in the Balance of Nature.* Academic Press, New York.

Gilpin, M. 1996. Metapopulations and wildlife conservation: approaches to modeling spatial structure. Pages 11–27 *in* D. R. McCullough, editor. *Metapopulations and wildlife conservation.* Island Press, Washington, D.C., USA.

As anthropogenic influences continue to alter natural habitats for wildlife, it becomes increasingly important for us to understand mechanisms that influence population dispersal, movement, extinction, colonization, and connectivity. The term *metapopulation* (i.e., a population of populations) was coined by Levins (1969) when he studied the control of insects in a patchy environment. Subsequent work on mites in subdivided laboratory populations (Huffaker et al. 1963) closely mimicked Levins' model: habitat patches separated by unoccupied space (i.e., nonhabitat) were all equal, and organisms had equal probabilities of dispersing from one patch to any other. As such, the assumption was that some populations would become extinct and others would increase toward carrying capacity. The simplicity of these models and experiments has led to numerous questions related to population persistence at different scales, questions that have since gained importance in wildlife management and conservation.

In this paper, Michael Gilpin provides an introduction and overview of the metapopulation concept and discusses different categories of metapopulation models, applying the models to conservation of wildlife populations. The metapopulation concept is a relatively new way of considering wildlife populations and will alter future management. This chapter is part of a book that originated from a symposium on metapopulations and wildlife conservation held at the first annual meeting of the Wildlife Society in 1994 in Albuquerque, New Mexico.

RELATED READING

Brown, J. H., and A. Kodric-Brown. 1977. Turnover rates in insular biogeography: effect of immigration on extinction. Ecology 58:445–449.

Esler, D. 1999. Applying metapopulation theory to conservation of migratory birds. Conservation Biology 14:366–372.

Hanski, I., and M. Gilpin. 1991. Metapopulation dynamics: brief history and conceptual domain. Biological Journal of the Linnean Society 42:3–16.

Hanski, I., M. Kuussaari, and M. Nieminen. 1994. Metapopulation structure and migration in the butterfly *Melitaea cinxia*. Ecology 75:747–762.

Harrison, S. 1991. Local extinction in a metapopulation context: an empirical evaluation. Biological Journal of the Linnean Society 42:73–88.

Huffaker, C. B., K. P. Shea, and S. G. Herman. 1963. Experimental studies on predation: complex dispersion and levels of food in an acarine predator-prey interaction. Hilgardia 34:305–330.

Levins, R. 1969. Some demographic and genetic consequences of environmental heterogeneity for biological control. Bulletin of the Entomological Society of America 15:237–240.

McCullough, D. R., editor. 1996. Metapopulations and wildlife conservation. Island Press, Washington, D.C., USA.

Wilcove, D. S. 1987. From fragmentation to extinction. Natural Areas Journal 7:23–29.

2

Metapopulations and Wildlife Conservation: Approaches to Modeling Spatial Structure

Michael Gilpin

This chapter is about the spatial structure exhibited by most species populations and, in particular, those populations fragmented by human modification of the landscape. I wish to carry out a general exploration with particular attention to the word "metapopulation," a term with a totally theoretical pedigree. My aim is to provide a conceptual framework that will help wildlife biologists and conservation biologists to understand the general class of spatially structured population-dynamic models—a group of models with sufficient resolution to address some of the decision problems they confront.

As conservationists and species managers, we are concerned with active manipulation and effective protection of biodiversity, which in large part we carry out at the level of single species. Faced with a deteriorating natural biological system, we comprehend as best we can the current state, including threats and possible remedies, and from this we attempt to project the system's future. Employing such projections and alternative scenarios, we engage the biological system, modify its current state, and, we hope, alter its future course of development to an end nearer our desires.

For amelioration, we add or subtract or rearrange individuals of target species; establish new colonies; build captive propagation facilities for sustained supplementation; modify or reconfigure habitat; and affect, positively or negatively, nontarget species that directly or indirectly have an impact on the target species. We do all this using our knowledge, views, and comprehension of the system on whose behalf we are acting and call this broadly interpreted understanding our "model" of the system. Simply put, then, we decide what management actions to take based on our model of the system. Good models are thus the key to good conservation management.

Much of what we do to preserve, conserve, restore, or manage depends on the spatial positions of the species and its members against their natural land-

11

scape and, as well, on the underlying spatial structure of the habitat that supports the system. Stated differently, our models of conservation and management are commonly map-based. Furthermore, our management recommendations are spatially explicit. Reserves are configured with boundaries drawn at *these* locations, not those. Harvesting is carried out with *these* quotas from *these* particular places. Individuals are translocated from here to there. Supplementation and restoration have spatial foci. Position makes a difference.

Having made this point, we must note that many of the population viability analyses (PVAs; Gilpin and Soulé 1986) and reserve design exercises currently being conducted and debated utilize zero-dimensional models and strategies. That is, they are based on demography, on birth–death branching-process models, and on simplified, graphic interpretations from island biogeography theory. I argue this approach to be inadequate, and in the remainder of the chapter I develop the rationale for a contrary view.

Today, the term "map-based" makes one think of geographic information systems (GIS). But a GIS is a passive data structure, rather like a spatial spreadsheet. A GIS holds an abstraction of raw, unprocessed, geo-referenced data. It is a wonderful platform, with a rich tool kit on which to carry out analyses, but this computer representation of data must be distinguished from the actual biological analysis or modeling, which is necessarily based on an ecological understanding of the system being portrayed.

To fix these ideas, consider the following hypothetical case study of a kangaroo rat population threatened by human habitat modification. Ideally we monitor the kangaroo rat's ecology, life history, and population dynamics. We plot densities onto a two-dimensional map of the landscape that also contains information on resources and competitors. We abstract these possibly continuous distributions of densities to discrete local units. Based on possibly inadequate data that parameterized movements and population fluctuations within and between these local units, we search for a possibly disconnected subset of the system that has minimal viability (say, 98 percent survival probability for 100 years), and through this analysis we decide where best to locate a system of public reserves. The progression is:

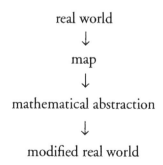

Important decisions take place at each step in the process. In this chapter we are concerned with the set of decisions involved in the map → mathematical-abstraction transformation.

At the outset, I distinguish four categories of metapopulation models with spatial structure: spatially implicit models, spatially explicit models, grid-based (cellular automata) models, and individual-based models. After comparing these models, I will suggest the conditions under which they may effectively be employed.

Spatial Structure: The End Points

The category "spatially structured models" covers models in which positions of populations or their constituent individuals are located, or referenced, against a spatial background. I limit my discussion in this chapter to two-dimensional systems. I base my distinctions on stylized habitat and species density maps contrived for purposes of illustration. In speaking of the term "metapopulation," we must distinguish foreground from background. Following Hanski and Gilpin (1991), the metapopulation is defined at the species level: it is a population of populations. Under this strict usage, the underlying habitat structure is ignored, though this is not something one ever would do in conservation planning. To illustrate spatial structuring, I will consider two extreme types of spatial configuration on opposite sides of the scale: the panmictic population and the Levins metapopulation (Figure 2.1).

Many ecological models are without spatial extension, that is, zero-dimensional. Consider the well-known logistic growth model:

$$\frac{\mathrm{d}n}{\mathrm{d}t} = rn\left(1 - \frac{n}{K}\right)$$

The animal population size is characterized by a single number, a scalar variable, n. There is no indication of where these n individuals are, over what range they extend, or whether there is internal structure to this population. Logically this number should be interpreted as total population size, but it is frequently interpreted as a density, which does imply extension in space. Many multispecies models are structured and interpreted the same way.

Although the issue is normally ducked, it is reasonable to assume, when scalar population state variables are interpreted as densities, that this density is uniform through some range and that, with change in time, the density change is the same throughout the unaltered range of the population (see Figure 2.1). This assumes that the population is extremely well mixed, or panmictic, which implies that the animals in the system can move anywhere in the range of the system within the time step (even if infinitesimal) of the

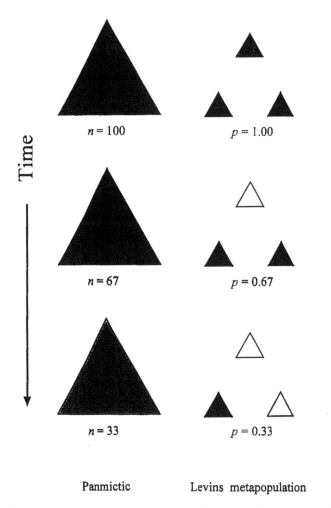

Figure 2.1. The panmictic population is on the left; the three local populations of the Levins metapopulation are on the right. Time is assumed to run down the figure. In each case, a regionwide loss of population is being illustrated. Darker shading corresponds to higher density. Note that the panmictic population maintains spatially uniform density throughout the process. With the Levins metapopulation the internal densities of colonized local populations remain constant, but more go extinct with time.

abstracted process—subject, of course, to the constraint that the density distribution remains uniform. Additionally, this seems to suggest that the habitat quality and other environmental features are the same throughout the domain of this local population.

At the other end of the spectrum of spatial structure is the Levins (1970) metapopulation. I will not discuss the motivations and theoretical background of this system. (The reader is referred to Hanski and Gilpin 1991.) Rather, I want to illustrate the features of the system and compare them graphically to the panmictic system. The Levins metapopulation is composed of local populations (demes), all of which are equal in all aspects. The size (or density) of each local population can be either zero or K, that is, extinct or at carrying capacity, with each local population being internally panmictic. The actual distinction is between absence and occupancy of a patch; consequently, this kind of model is sometimes called an "occupancy model."

Under the Levins metapopulation model, local populations go extinct with a constant probability and realized extinction events are independent from patch to patch. This is tantamount to viewing the environmental stochasticity over the region of the local populations as being independent or uncorrelated. Locally extinct populations may be recolonized by dispersal from extant populations. Recolonization is proportional to the fraction of extant populations in the system, regardless of their position. Dispersal and patch recolonization events are normally assumed to be infrequent. The state of the Levins metapopulation is commonly given by the scalar variable p, which describes the frequency of extant populations.

The most mathematical and idealized version of the Levins metapopulation model is not in fact map-based but an implicit spatial model. That is, the locations and sizes of the patches are ignored. Under these assumptions, the dynamic equation that governs p is

$$\frac{dp}{dt} = cp(1 - p) - ep$$

where c is the parameter for colonization and e is the per-patch parameter for extinction, with an equilibrium value (for $c > e$) of p given by

$$p^{eq} = \frac{(c - e)}{e}$$

Some investigators (for example, Lande 1987; Hanski 1991) have extended this spatially implicit model to give it more realism and more applicability to real-world management.

Later I deal with a spatially explicit, or map-based, version of the Levins metapopulation model in which equal-sized patches are arrayed on a landscape. If dispersal leading to colonization is independent of distance between patches, this spatially explicit form can be approximated by the preceding equations. But it should be recognized that colonization is much more likely to be from nearby occupied patches in such systems.

Figure 2.1 compares the changes of state of the panmictic population with the equivalent change for a Levins metapopulation. For illustration and contrast, both systems show a decline in the total number of animals with time. The panmictic population does this through a uniform decrease of density; the Levins metapopulation does it with a decrease in the number of extant populations. At equilibrium, the panmictic population would have a constant density and the Levins metapopulation would have a stable fraction of occupied patches, all occupied patches having the same internal density. Due to the requirement for ongoing extinction and recolonization in a metapopulation, it is impossible that, at equilibrium, the actual fraction of occupied patches in a finite, spatially explicit Levins metapopulation would be perfectly constant, for the extinction and recolonization events are assumed to be independent. The best that could be anticipated would be a steady-state behavior with a stochastic fluctuation about a constant expectation for the fraction p.

No real-world population or system of local populations fits either of these two extremes. Both models are too stylized and too simplified to be useful for map-based conservation management. We must move to a representation that renders both the landscape and the behavior of animals more faithfully. We must begin by detailing some of the important factors and considerations that govern spatially extended populations. Only then can we characterize which subset of the possible systems are "metapopulational" and which others are better characterized by one of the two other modeling approaches to space—grid or individual—though I must caution at this point that the same real-world system could often, with profit, be modeled with more than one of these alternative approaches.

Here is a partial list of considerations that affect spatially explicit models—that is, realistic features that may need to be accounted for to make accurate future projections:

- Most populations, local or regional, show density variation throughout their range.
- Local populations have densities that can vary over time between zero and some upper limit. This is sometimes called "structure."
- Virtually all fragmentation or patchy population structure exhibits variation in both patch size and patch spacing.
- Most systems have partially correlated environmental variation even within a continuous range.
- Long-term anthropogenic changes can modify underlying habitat structure such as forest succession and spotted owl habitat (Thomas et al. 1990; Smith and Gilpin 1996; Chapter 7 in this volume).
- The edge between the good habitat and the bad habitat, probably corresponding to the spatial limit of the local population, can be of different severities, possibly contributing an "edge effect" (Chapter 10 in this volume).

Distribution, Abundance, and Movement

Figure 2.2 sketches some of the spatial elements we will consider in the analysis to follow. The basic idea is to represent the continuities and discontinuities of spatial variation in density and simultaneously to consider the effect of animal movement on density.

At the map level for population density, we can consider some purely geometrical factors for spacing and patch dimensions as illustrated in Figure 2.3. All four of the configurations lie between panmixia and the Levins metapopulation presented in Figure 2.1. As we will see, these different geometries will have differing interactions with other aspects of spatial structure and individual movement.

Ignoring demographic stochasticity (MacArthur and Wilson 1967; Gilpin 1992)—which, by definition, is independent among local populations—the environmentally driven population dynamics (that is, environmental stochasticity) of the independent units may have different degrees of correlation (see Gilpin 1988, 1990; Harrison and Quinn 1989). Some patterns of correlated environmental stochasticity are illustrated in Figure 2.4. In Figure 2.4*a*, the fluctuations of density (population size) in the three polygonal units are

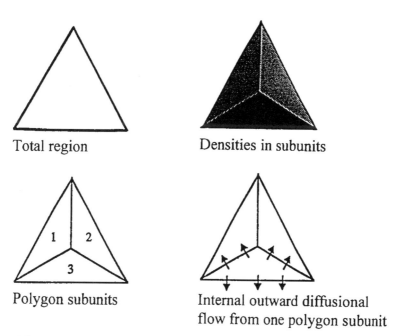

Figure 2.2. The total region is arbitrarily taken to be triangular. The subunits are taken as equally sized smaller triangles. Densities may vary in subunits. The bottom figure shows the localized "diffusional" movement of animals out of one of the cells. The movement of animals out of the system along the lower edge may contribute to an "edge effect."

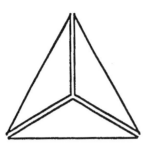

A. Minor fragmentation

B. Unequal separations

C. Unequal sizes

D. Unequal sizes and separations

Figure 2.3. Some geometrical forms between the panmictic and the Levins metapopulation. These forms may or may not be viewed as metapopulation systems, depending on the correlation of the environment and the degree of animal movement between patches.

identical. In Figure 2.4*b,* the correlation is perfect, but the amplitudes are different in the local populations. In Figure 2.4*c,* the three subunits show uncorrelated dynamics. In Figure 2.4*d,* not only are the fluctuations uncorrelated, but patches 1 and 2 at times fluctuate to extinction; presumably their recolonization is mediated from the other patches. The system in Figure 2.4*a,* although spatially disjoint, seems to mimic panmixia. The situation in Figure 2.4*d* behaves more like the Levins metapopulation. The others have an intermediate behavior.

Diffusion of individuals (migration, dispersal, and the like) can modify the environmentally driven dynamics on the separate patches. Typically, a local population with a temporarily or permanently high density may act as a source (or mainland) to a nearby population that is at low density (which

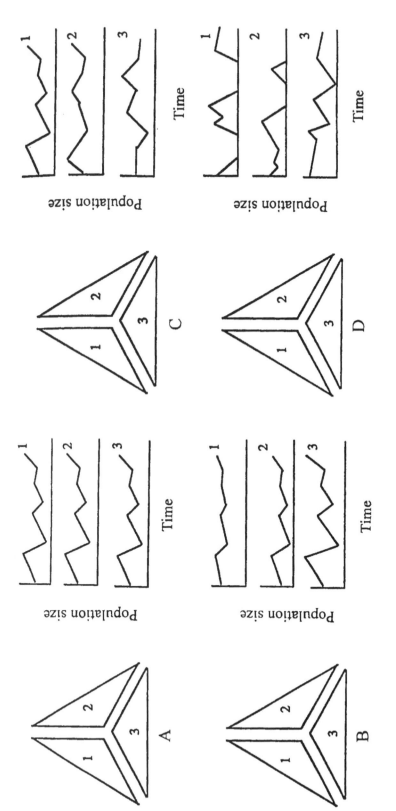

Figure 2.4. Patterns of population change in the three polygonal subunits of a mildly fragmented system. Patterns **A** and **B** show perfect correlation. Pattern **C** shows uncorrelated population changes. Pattern **D** shows uncorrelated change with extinctions. (The recolonization of subunits 1 and 2 is due to immigration of individuals, which is not shown.)

could be called a sink population). Figure 2.5 illustrates some influences of diffusional exchange of individuals. In Figure 2.5*a*, the environment is correlated between the patches and the symmetric diffusion of individuals between these units alters nothing. In Figure 2.5*b*, diffusion partially smooths out the uncorrelated environment. Extreme diffusion could cancel out all irregularities of the environmental stochasticity. Thus, even with fragmentation, the system could most closely resemble a single panmictic population. Figure 2.5*c* illustrates the "rescue effect" first identified by Brown and Kodric-Brown (1977). In this case, the dynamics would drive local populations extinct, but diffusion introduces enough individuals to preclude this extinction. Figure 2.5*d* illustrates source/sink dynamics. Polygon 1 (the sink) has dynamics that would drive it extinct—negative population growth—but the movement of individuals from the stable (source) populations in polygons 2 and 3 rescues this sink population from its fate. Observe that the system depicted in Figure 2.5*b* could also be called source/sink dynamics, but over the time shown, the net movement of animals between pairs of patches shifts back and forth. That is, each patch can act as both source and sink.

Diffusional flow, by strengthening patch-to-patch correlation, can defeat or preclude metapopulation analysis. In the Levins metapopulation, it is uncorrelated dynamics between patches that lead to sudden extinctions, independent from patch to patch. As the distances between local populations become greater, the movement of animals between them lessens and is less likely to produce an effect that mimics panmixia. Rescue and source/sink patterns will also be lessened or eliminated. That is, all else being equal, the greater the separation between patches, the more like a spatially explicit Levins metapopulation the system becomes.

Figure 2.6 shows some situations that have further complications, probably due to more structure in the underlying habitat. In Figure 2.6*a* the population is actually continuous. It is at lower density everywhere other than the three triangular regions, but it is nonetheless connected. In Figure 2.6*b* there are "stepping stones" between the three larger local populations. While either of these cases may be perfectly representative of many natural cases, they are somewhat more difficult to model, especially with a metapopulation approach.

Nonequilibrium Spatially Structured Populations

In a system analysis, one must distinguish between the initial conditions, the transient behavior, and the equilibrium, whether deterministic or stochastic. Just as a panmictic population can exhibit deterministic negative growth, a metapopulation or other spatially structured system can be in collapse mode:

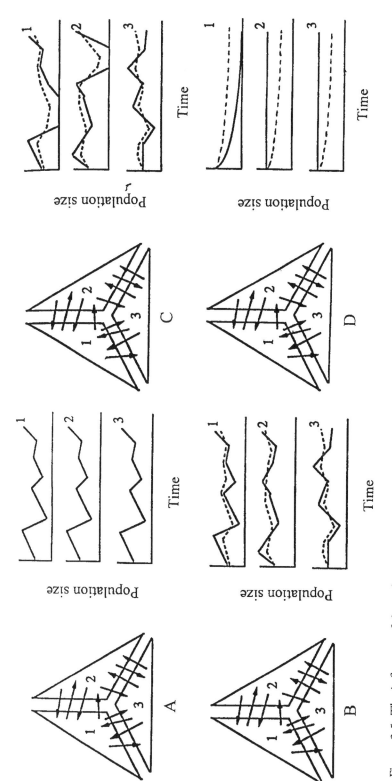

Figure 2.5. The influence of short-distance migration on population dynamics. For each of the three subunits, the density change with time is plotted at the right. The solid lines are the population densities expected without the effects of diffusion; the dashed lines represent the effect of including the immigration due to diffusion. See the text for discussion.

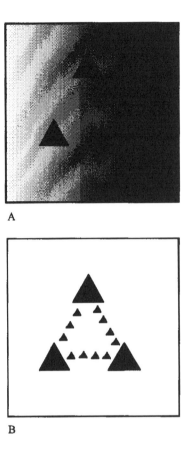

A

B

Figure 2.6. Two spatial variations from the stylized representations considered in the preceding figures. **A** depicts a continuous population with areas of varying population density. **B** shows three large local populations linked by smaller populations.

each local population is subject to some probability of extinction but with a trivial probability of recolonization. That is, the metapopulation system would exhibit little or no turnover of any local patch. From an evolutionary perspective, such systems are uninteresting. They will not exist in the future, and any such systems that existed in the past will not have made it to the present. From a conservation standpoint, however, they represent something that may too often be the case—seriously fragmented species populations cut off from dispersal and recolonization, systems doomed to extinction. Clearly, with such fragmented systems some management of the environment and dispersal must be instituted to reverse the monotonic trends that doom the population to ultimate extinction. A metapopulation model may well suggest fruitful approaches.

Time Scales and Spatial Structure

Time and space are intertwined in the analysis of population dynamics. A decision about the scale of one may limit or preclude certain behavior in the other. If one wanted to talk about human population density changes in the United States following the passage of some piece of trade legislation, for example, one might focus on job creation and ignore demography. That is, the time scale might be short enough to neglect birth and death events. Similarly, some planning horizons may be short enough that metapopulation dynamics—that is, extinction and recolonization—could be ignored to a first approximation. But to the extent that an action will permanently disturb an equilibrium configuration, the spatial implications must be examined. Different species and different landscapes will require different minimum time spans for the emergence of spatially structured population dynamics.

Metapopulation Models

All of the real-world features described above can be incorporated into a general and spatially explicit metapopulation model in which the state variables are the presences and absences of populations on a patch; that is, a vector of 1s and 0s. A more detailed approach would specify the population size on each patch. Such inclusive modeling is well beyond analytical mathematics. Analysis and investigation are possible only through computer-based simulation—repeated simulations that allow characterizations of central tendencies. The idea is that each patch is individually parameterized with a growth rate, a carrying capacity, some features of edge, area and shape, and some function, possibly density dependent, for emigration. A partially or fully correlated environment affects local growth, as does immigration from the set of other occupied patches in the system. A number of such models have been created (Gilpin 1989; Thomas et al. 1990; LaHaye et al. 1994), and many of the chapters in this book employ versions of generalized metapopulation models. The ultimate utility of the metapopulation approach will be determined by the success of these approaches.

Grid-Based and Individual-Based Models

There are two other classes of spatially explicit models that must be compared to the spatially explicit metapopulation model. These are alternative representations of the same underlying biological reality. In a grid-based or cellular automata approach, all spatial units are squares of the same size. Each cell has a state—typically the number of individuals of the target species in that cell. The state values in the cells change with time in a manner that typically depends on inputs and interactions with closely neighboring cells. Depending on the cell size, a patch in a metapopulation model could be composed of multiple grid cells, all of which are contiguous.

In a geo-referenced individual-based model, all individuals are followed spatially and in demography and behavior over the course of their lives. The spatial indexing of animals is based on continuous x and y coordinates. Thus an animal inscribes a trajectory over its landscape during the course of its life. These trajectories are necessarily stochastic. But as animals move and make random encounters, the outcomes of the encounters may depend deterministically on behavioral and demographic states of the two animals. These three spatially explicit approaches are contrasted in Figure 2.7.

I must emphasize that, with the exception of the spatially implicit metapopulation models, all three of these approaches require computers to simulate the dynamics of the system. The grid-based and the individual-based approaches, however, have many more state variables and may require supercomputers or parallel computers for effective implementation. All of these spatially explicit approaches are "data hungry," but each in a different way.

The metapopulation approach is parameterized best with long time series of population sizes on multiple patches. The cell-based grid approach requires high spatial resolution of the population dynamics, but it may permit a more even-handed application of rules to each cell. With the individual-based approach, the population dynamics are a consequence of the parameterization of the individual's demography and various choices of behavior. Thus one could chose between these models solely on the basis of available data and on an a priori understanding of the behavior of the species.

It is beyond the scope of this introduction to go more deeply into the mathematical structure of these three approaches. But it should be noted that each of these different approaches, and also zero-dimensional approaches, could be applied to any system—that is, to any map-based management problem. Different approaches would have different strengths. The metapopulation approach requires the least data and is useful for population viability analysis (Gilpin and Soulé 1986). The grid-based approach might work best for designing reserves (Gilpin 1993; Chapter 10 in this volume). The individual-based approach might be best for situations where environmental mitigation is required throughout the stable range of a species population.

The choice of a modeling strategy initially depends on the character and quality of the data. Typically, in conservation planning for threatened and endangered species, the data are scant and were collected for reasons only tangentially connected to the issue of predicting extinctions or designing reserves. Indeed, the data may be inadequate for any of the spatially explicit

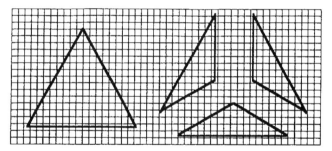

A. Underlying structure

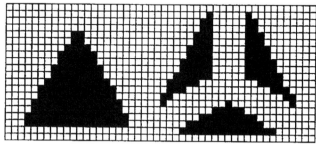

B. Grid model

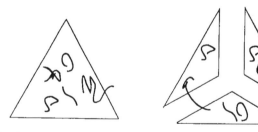

C. Individual-based model

D. Metapopulation model

Figure 2.7. Part **A** shows the map structure of the density of the species. In **B** the structure is decomposed into densities in a number of small, equally sized spatial cells. In **C** the approach is individual-based: the exact movement of each individual is tracked over the landscape. Some individuals remain in their local population throughout their lifetime; others are shown moving between the subunits. **D** is the metapopulation approach: the subunits are abstracted to dimensionless points on a two-dimensional surface.

approaches. Yet real-world conservation and wildlife management are on-going processes in which each element of the effort informs other elements. The initial modeling should guide the second round of data collection. From this point on, there is no excuse not to geo-reference the data. Thus, at some point in the process, the modeling should begin to contain spatial elements that are modeled in one of the fashions outlined above.

REFERENCES

Brown, J. H., and A. Kodric-Brown. 1977. Turnover rates in insular bio-geography: Effect of immigration on extinction. *Ecology* 58:445–449.

Gilpin, M. 1988. Comment on Quinn and Hastings: Extinction in subdi-vided habitats. *Conservation Biology* 2:290–292.

———. 1989. Population viability analysis. *Endangered Species Update* 6:15–18.

———. 1990. Extinction of finite metapopulations in correlated environ-ments. Pages 177–186 in B. Shorrocks and I. R. Swingland, eds., *Living in a Patchy Environment*. Oxford: Oxford University Press.

———. 1992. Demographic stochasticity: A Markovian approach. *Journal of Theoretical Biology* 154:8–18.

———. 1993. A viability model for Stephens' kangaroo rat in western River-side County. Report no. 12, vol. II: Technical reports. Draft habitat con-servation plan for the Stephens' kangaroo rat in western Riverside County, California. Riverside: Riverside County Habitat Conservation Agency.

Gilpin, M. E., and M. E. Soulé. 1986. Minimum viable populations: Processes of species extinction. Pages 19–34 in M. E. Soulé, ed., *Conser-vation Biology: The Science of Scarcity and Diversity*. Sunderland, Mass.: Sinauer Associates.

Hanski, I. 1991. Single species metapopulation dynamics: Concepts, models and observations. *Biological Journal of the Linnean Society* 42:17–38.

Hanski, I., and M. Gilpin. 1991. Metapopulation dynamics: Brief history and conceptual domain. *Biological Journal of the Linnean Society* 42:3–16.

Harrison, S., and J. F. Quinn. 1989. Correlated environments and the persis-tence of metapopulations. *Oikos* 56:293–298.

LaHaye, W. S., R. J. Gutiérrez, and H. R. Akçakaya. 1994. Spotted owl metapopulation dynamics in southern California. *Journal of Animal Ecology* 63:775–785.

Lande, R. 1987. Extinction thresholds in demographic models of territorial populations. *American Naturalist* 130:624–635.

Levins, R. 1970. Extinction. Pages 77–107 in M. Gerstenhaber, ed., *Some Mathematical Questions in Biology.* Providence, R.I.: American Mathematical Society.

MacArthur, R. H., and E. O. Wilson. 1967. *The Theory of Island Biogeography.* Princeton, N.J.: Princeton University Press.

Smith, A. T., and M. Gilpin. 1996. Correlated dynamics in a pika metapopulation. In I. Hanski and M. Gilpin, eds., *Metapopulation Dynamics: Ecology, Genetics and Evolution.* New York: Academic Press.

Thomas, J. W., E. D. Forsman, J. B. Lint, E. C. Meslow, B. R. Noon, and J. Verner. 1990. A conservation strategy for the northern spotted owl. Report of the Interagency Scientific Committee to address the conservation of the northern spotted owl. Portland: USDA Forest Service; USDI Bureau of Land Management/Fish and Wildlife Service/National Park Service.

Habitat

Shelford, V. E. 1912. Ecological succession. IV. Vegetation and the control of land animal communities. *Biological Bulletin* 23:59–99.

℞ Shelford penned this classic study of the relationship between land mammals and succession two decades before Clements (1936) and Gleason (1939), whose papers are also in this volume, initiated the debate over what mechanisms and ecological processes influence succession. Manipulating plant and animal communities by reversing succession is a key tool for managing animal and plant populations and was used well before Leopold (1933) published his text on game management. Shelford discerned the underlying processes that governed succession and wanted to make sure that managers manipulated wildlife habitat with the same understanding.

Habitat is a central theme in wildlife conservation and management that has to be clearly grasped to ensure an effective understanding of wildlife ecology. We must manipulate succession to ensure that the needs of wildlife species are met as animal-plant communities are altered, destroyed, and modified owing to natural and anthropogenic influences. Shelford's paper is one of the first studies to examine terrestrial animal distribution and its association with forest systems, aged from the youngest to the oldest. Changes in diversity were reported to be related to food, cover, soil, temperature, plants and animals, evaporation, and to interactions among these and other factors. Shelford demonstrated that the complexities of animal habitats are linked to habitat fluidity during the process of ecological succession.

RELATED READING

Bray, J. R., and J. T. Curtis. 1957. An ordination of the upland forest communities of southern Wisconsin. Ecological Monographs 27:325–349.

Connell, J. H., and R. O. Slatyer. 1977. Mechanisms of succession in natural communities and their role in community stability and organization. The American Naturalist 111:1119–1144.

Cook, J. G., L. L. Irwin, L. D. Bryant, R. A. Riggs, and J. W. Thomas. 1998. Relations of forest cover and condition of elk: a test of the thermal cover hypothesis in summer and winter. Wildlife Monographs 141:1–61.

Hobbs, N. T., D. L. Baker, G. D. Bear, and D. C. Bowden. 1996. Ungulate grazing in sagebrush grassland: effects of resource competition on secondary production. Ecological Applications 6:218–227.

Horn, H. S. 1974. The ecology of secondary succession. Annual Review of Ecological Systems 5:25–37.

Shelford, V. E. 1913. Animal communities in temperate America. University of Chicago Press, Chicago, Illinois, USA.

Whittaker, R. H. 1953. A consideration of climax theory: the climax as a population and pattern. Ecological Monographs 23:41–78.

Whittaker, R. H. 1962. Classification of natural communities. Botanical Review 28:1–239.

Reprinted by permission of *The Biological Bulletin*.

Vol. XXIII.   *July, 1912.*   *No. 2.*

BIOLOGICAL BULLETIN

ECOLOGICAL SUCCESSION.

IV. Vegetation and the Control of Land Animal Communities.

Victor E. Shelford.

I. Introduction.

In the preceding papers of this series, we have discussed the succession of animals in two types of aquatic habitats with particular reference to fish. While the data presented are only a minor part of those at hand they have served to illustrate some of the principles of succession in aquatic habitats. Discussion of other aquatic situations would enable us to point out many more important facts but we must now pass to land habitats. To illustrate principles here we might discuss the development of either forest or prairie animal communities

59

on sterile mineral soil or in a filling pond. We have selected the succession of forest animal communities on sterile mineral soil and especially those on sand.

The forest conditions on the sand areas at the head of Lake Michigan were once among the best in North America for the study of the problems at hand and in spite of the fact that they are rapidly disappearing and have already been destroyed in some of the localities where the data here presented were collected, there are still various small areas between Indiana Harbor, Ind., and Sawyer, Mich., which taken together present the chief stages of forest development on sand. There are also localities outside the sand area, in which the later stages are to be found on other soils.

The searching of older literature, for possible statements which anticipate ideas here presented would require years and has not been undertaken in any adequate fashion as yet. Such anticipation of ideas and principles is perhaps to be expected but *organization and development are just at their beginnings.* On the plant side some older literature has been brought to attention. Buffon (1742) discovered that poplars precede oaks and beeches in the development of forest (Cowles, '11). Cowles found that in the Lake Michigan sand area cottonwoods precede pines, pines precede black oaks, black oaks precede red oaks, red oaks are usually followed by sugar maple and beech (Cowles '99, '01, '11; Clements, '05; Shantz, '06; Fuller, '11). We are to present certain representative facts concerning the development or succession of animal communities, accompanying and contributing to the causes of plant succession. The data presented are by no means complete as only a small part of the total number of animal species that might be collected from such a series of localities, has been studied. However the data are *adequate* for the purpose of illustrating principles and methods, and of bringing together some of the recent developments in the study of communities of organisms to focus them on the question as to the best method of obtaining and organizing the data of ecology of terrestrial animals.

II. Localities of Study.

The forest development series might be variously divided for purposes of study. The number of stages and variations that might be recognized is considerable. For the purpose just outlined we have selected five stages. Transition areas are present, but only the most important two, namely those between the first and second and the second and third, have been noted. The stages considered are (1) Cottonwood stage; (1–2) the transition between the cottonwoods and the pines; (2) pine stage; (2–3) transition between the pine and oak or mixed pines and oaks together with open places in the oak areas; (3) black oak stage; (4) red oak stage—the red oak associated with the black oak and white oak in the earlier stages and with the shag bark hickory in the later stages; (5) the beech and maple stage.

1. *Cottonwood Stage.*

The cottonwood (*Populus deltoides*) areas are located near the lake shore and the sand is always more or less shifting and rarely with more than traces of humus. The cottonwoods are usually small trees, scattered over the beach ridge, or the lakeward side of the shore dunes as the case may be. Between them are widely scattered bunches of grasses of which *Calamovilfa longifolia* is the most characteristic species and *Ammophila arenaria* usually common. Scattered individuals of *Artemisia canadensis* also occur. The shrubs, which are still more scattered or local, are the beach plum (*Prunus pumila*) and some of the xerophytic willows (*Salix glaucophylla*).

All localities are indicated on the map, p. 62, by letters used as designated below. Two principal cottonwood stations have been studied. One lies to the east, 1*A*, and one to the west, 1*B*, of Pine, Ind. At these points the cottonwood area is about ten rods wide and reaches inland just beyond the crest of the ridge where the plants of transition come in. (For the arrangement of ridges see Fig. 1, p. 137, of "Ecological Succession II.") Two other less fully studied areas have been visited frequently, viz., 1*C*, at Miller, Ind., and 1*D*, at Dune Park. (See map.)

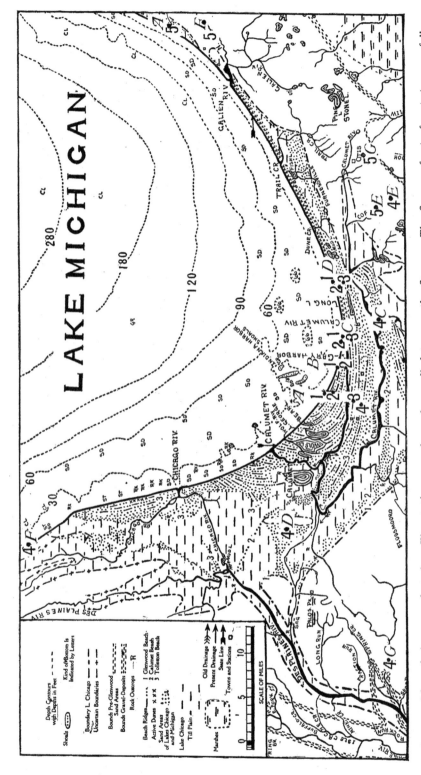

FIG. 1. Map of the area of study. The stations are shown by small dots near the figures. The figures refer to forest stages as follows: 1, cottonwood stage; 2, pine stage; 3, black oak stage; 4, red oak stage; 5, beech stage. The letters standing near are for convenience in reference. A–D refer to stages on sand; E–G on clayey soil. Fuller has studied evaporation at stations 1, 2 and 3D and 5G and in the red oak stage at Palos Park.

1–2. *Mixed Pine and Cottonwood Stage.*

These areas lie on the first ridge from the lake. The sand shows traces of humus which is indicated by a slight darkening. The characteristic grass is the bunch grass (*Andropogon scoparius*) which occurs in dense bunches much more closely set than *Calamovilfa* in the preceding. The juniper (*Juniperus communis*), the bear berry and young pines (*Pinus Banksiana*) are the dominant plants, although old cottonwoods and scattered individuals of all the plants of the preceding stage are present. Two such stages were studied.

2. *Pine Stage.*

The sand here is stable and considerably blackened by humus except in blowouts where the wind keeps it constantly shifting. The trees are scattered stunted pines (*Pinus Banksiana*). The large areas of sand are covered with the recumbent bear berry; scattered specimens of many of the herbaceous plants of the earlier stages are to be found. The New Jersey tea (*Ceanothus americanus*) and the juniper are among the most characteristic shrubs. The cactus (*Opuntia Raffinesquii*) occurs in the older stages only. Three stations have been studied; two, *2A*, and *2B* (adjoining *1A* and *1B*) and *1D*, at Dune Park, Ind.

2–3. *Mixed Pines and Black Oaks.*

The transition area between pine and oak is characterized by the presence of seedlings of the oak, the choke cherry, the fragrant sumac and an abundance of cacti and Liatris. The ground is much darkened. Herbs are much more numerous than earlier but bare spots of relatively clear sand continue, even after the oaks have entirely displaced the pines. Such transition areas proper have been so much disturbed that only the open bare sandy places in the oak areas are here considered as belonging under this head.

3. *Black Oak Stage.*

We now find the sand to be much darkened by humus and locally covered with a dry moss or with dead leaves. Grasses also partly cover the ground. The shrubby undergrowth is made up of blueberry, choke cherry and New Jersey tea. The oaks (*Quercus velutina*) constitute a much thicker stand than the pines but the type which we are considering here has the open places grassy and covered with a growth of vetch (*Tephrosia virginiana*), golden-rod and other Compositæ. The chief points of study were 3*A*, at Miller, Ind., 3*B*, near Clark, Ind., and 3*C*, at Dune Park.

4. *Red Oak Stage.*

This was originally the most abundant type of forest near Chicago but areas which have not been disturbed by man are few in number. On the sand, hickories are rare. The forest is usually made up of black oak, red oak, and white oak. This is true at 4*A*, 4*C*, and 4*D*, of the map. The more mesophytic type is made up of white oak, red oak, and hickory, with the red oak and the hickory dominating. Our type four then, represents a range of conditions. All the stages are included here and all are characterized by the absence of bare sand and other mineral soil, and by the presence of a carpet of leaves and humus which covers the ground. There is a well marked shrubby and herbaceous growth. The characteristic shrubby species, are blueberry, *Viburnum*, *Cornus*, and *Crataegus*. The shrubs are usually quite numerous and make thick stands locally in the forest. The places studied are, 4*A*, at Hessville, Ind., 4*C*, at Liverpool, Ind. and 4*D*, at Beverly Hills in Chicago. These represent white oak, red oak, and black oak on sand. 4*E*, 4*F* and 4*G*, represent areas on till clay which are in something like primeval conditions.

5. *Beech Stage.*

This type is characterized by beech and sugar maple. The mineral soil is covered with several centimeters of humus and a very thick layer of leaves which is often matted together by fungus hyphæ. The number of species of trees is smaller but the number of species of small shrubs is greater than in the preceding stage. The number of individual shrubs however is smaller. Here the trees close the over-head spaces and make a dense shade, while the lower forest is open. The localities of study are, 5*A*, at Sawyer, Mich., on sand, 5*C*, at Wood-ville, Ind., now destroyed, 5*F*, at Sawyer, Mich., and 5*G*, at Otis, Ind., on clay

III. Presentation of Data.

We will begin with the animals of the earliest forest stage and proceed to those of the latest. For purposes of comparison animals must be divided into groups with comparable habitat relations. Warming's ('09, p. 138) division into strata may be employed (see Dahl, '08, p. 11). In any locality we note that there are several levels which may be occupied by animals. Some animals live below the surface of the ground and constitute the *Subterranean stratum*. Others live at the surface of the ground and constitute the *Ground stratum*. Animals inhabiting the herbaceous vegetation, and low shrubs, etc., make up the *Field stratum*. Those living on the shrubs and young trees make up the *Shrub stratum*, and those on the trees, the *Tree strata*. Such a division is essential to any comparison of the animals of different forests, steppes, etc. The ground stratum of one cannot well be compared with the field stratum of another; like strata must be compared.

Many animals invade several strata in connection with their various activities. They should, however, be classed primarily in the stratum in which they breed and secondarily in the stratum or strata in which they feed or forage. The breeding activities are of especial importance to the animals in question, while the feeding and foraging influence other animals. The study of the lower strata of the forest presents no difficulties. The tree stratum however is usually far enough above the ground to make observation difficult. The data at hand are rather incomplete at this point and accordingly the discussion is confined to the lower strata.

Tables I. and II. and the lists following them show the distribution of about 200 species of animals in the forest stages. These tables and lists include chiefly animals that have been encountered in these situations often during several seasons. Where collections are known not to have been representative, they are omitted. For example most of the Diptera collected were pre-

TABLE I.

Showing the distribution of the animals belonging to the subterranean and ground strata of two or more of the animal communities of the forest stages indicated by numbers:—1, cottonwood stage; 1–2, mixed pine and cottonwood; 2, pine stage; 2–3, mixed pine and black oak stage and open places in the black oak forest; 3, the black oak stage, in the later stages white oak occurs; 4, stages containing red oak but not beech and maple (the earlier stages are black oak, white oak and red oak and the later stages white oak, red oak and hickory); 5, beech and maple stage which usually contains some basswood. In the numbered columns, the star indicates that a species is present, F signifies few, C common, A abundant. Strata are indicated by letters at the heads of the columns:—a, subterranean including rotten logs; b, ground; c, vegetation; in these columns F indicates feeding place; B indicates breeding place. Letters in column marked lit. refer to literature in the special bibliography.

Common Name.	Scientific Name.	1	1–2	2	2–3	3	4	5	a	b	c	lit.
Tiger beetle	Cicindela lepida Lee	C	F						B	F		A
Sand spider	Trochosa cinerea Fab.	C							B	B		B
Maritime grasshopper	Trimerotropis maritima Harr.	C	F						B	F		C
Tiger beetle	Cicindela formosa generosa Dj.	F	C	F					B	F		A
Long-horned grasshopper	Psinidia fenestralis Serv.	C	C	C	C				B	F		C
Burrowing spider	Geolycosa pikei Marx.	C	C	C	C				B	F		B
Bembex	Microbembex monodonta Say.	C	C	C	C				B	F	F	C
Bembex	Bembex spinolae Lep.	F	C	F	F				B	F	F	D
Termite	Termes flavipes Koll.	F	F	F	F	F			B	F		E
Wasp	Dielis plumipes Dru.	C	?	*	*				B	F	F	F
Digger wasp	Anophilus divisus Cress.		C	C					B	F	F	D
Sand locust	Agenotetix arenosus Han.		*	*	*				B	F		C
Ant	Lasius niger americanorum Em.		C	F					B	F	F	I
Mottled sand locust	Spharagemon wyomingianum The.		C	C	C				B	F		C
Migratory locust	Melanoplus atlanis Ril.		C	C	C				B	F		C
Tiger beetle	Cicindela scutellaris Lecontei Hald.	F	F	C	C				B	F		A
Locust	Melanoplus angustipennis Dod.		C	C	C				B	F		C
Sixlined lizard	Cnemodophorus sexlineatus Linn.	F	F	C	C				B	F	F	H
Andrenid	Auglochlora confusa Rob.			*	*				B	?		J
Parasitic bee	Epeolus pusillus Cress.			*	*				B	?	F	J
Burrowing bee	Spechodes dicroa Sm.			*	*	*			B	F	F	D
Ammophila	Ammophila procera Klg.			F	C	F			B	F	F	D
Ant lion	Mymeleon sp.?			*	*	*			B	B		D

TABLE I (*continued*).

Common Name	Scientific Name	1	1-2	2	2-3	3	4	5	a	b	c	lit.
Flat larvæ	Pyrochroidæ			F	?	F	C	C	B			L
Snails	Zonites arborens			F	F	F	C	F	B			P
Centipede	Lithobius sp.			F	?	F	C	C	B			
Eyed elater	Alaus oculatus Lin.					F	C	C	B			L
Borer	Orthosoma brunneum Forst.					F	C	C	B			L
Tree frog	Hyla pickeringii Hol.					F	F	F		F		M
" "	Hyla versicolor Lec.					C	F	C		F		M
Snail	Polygyra thyroides Say.					F	C	F	B	F		N
Sow bug	Porcellio rathkei Brandt					F	C	C	B	F	F	O
Snail	Pyramidula alternata Say.						C	C	B	F		P
Snail	Circinaria concava Say.						C	C	B	F		P
Sow bug	Cyclisticus convexus De G.						C*	C*	B	F		O
Centipede	Lyasopetalum laclarium Say.						*	*	B			Q
Centipede	Geophilus rubens Say.						*	*	B			Q
Millipede	Fontaria corrugate Wood.						C	F	B			O
Millipede	Spirobolis marginatus Say.						C	C	B	F		Q
Slug	Philomycus carolinensis Bosc.						C	C	B	F		O
Snail	Pyramidula solitaria Say.						C	C	B			P
Beetle	Galerita janus Fabr.						C	C	B	F		O
Melandrydæ	Penthe promelia Fabr.						C	C*	B			L
Burying beetle	Silpha surinamensis Fabr.						C	C	B	F		L
Harvestman	Liobromum nigripalpi.						C*	C*	B			L
Ant	Camponotus herculeanus pennsylvanicus De G.						C	C	B	F	F	I
Beetle	Cerchus piceus Web.						C	?	B	F		L

served in alcohol and only a few of these could be identified. The flies from one station were pinned and accordingly were all identified so that the inclusion of these in the tables and lists would introduce error. The lists and tables are divided into two groups; the first includes the inhabitants of the ground and subterranean strata, the second of the field, shrub and tree strata.

LIST OF ANIMALS RECORDED IN THE GROUND AND SUBTERRANEAN STRATA OF THE STAGES INDICATED ONLY.

In the first column are common names and in the second scientific names. In the third column *B* indicates breeding; *F*, feeding; *H*, hibernating of animals on the situation indicated in column 4 following. Letters in column *lit.* refer to literature cited in the special bibliography at the end of the paper. Statements made on the authority of others are in italics, those starred are by A. B. Wolcott.

Pine Stage.

1	2	3	4	*lit.*
Bee (Andrenidæ)	*Halictus melumboni* Rob.	*B?*		*J*
Larridæ	*Tachytes texanus* Cres.			*F*
Scoliidæ	*Plesia interrupta* Say			
Ampulicidæ	*Anophilus marginatus* Say		In sand	*F*
Beetle (Elateridæ)	*Cardiophorus cardisce* Say		On sand	*L*
Blue racer	*Coluber constrictor* Lin., Var.	*B*	In sand	*R*
Ground squirrel	*Spermophilus 13-lineatus* Mitch.	*B*	In sand	*S*
Elateridæ	*Alaus myops* Fabr.	*B?*	Under pine bark	

Black Oak Stage.

Elateridæ	*Lacon rectangularis* Say	*B*	Under Opuntia	*
Erotylidæ	*Languria trifasciata* Say	*B*	Under Opuntia	*
Coral winged locust	*Hippiscus tuberculatus* Beau.	*B*	In sand	*C*
Parasitic bee	*Cœlioxys rufitarsus* Smith	*B*	Bee nest	*R*
Psithyridæ	*Psithurus* sp.			
Eumenidæ	*Odynerus anornis* Say			
Hog-nosed snake	*Heterodon platirhinos* Latr.	*B*	In sand	*R*

Red Oak Stage.

Green tiger beetle	*Cicindela sexguttata* Fabr.	*B*	In soil	*A*
White-faced hornet	*Vespa maculata* Lin.	*H*	Rotten wood	*G*
Andrenidæ	*Augochlora pura* Say			
Ant	*Lasius umbratis mixtus aphidicola* Walsh	*B*	Log	*H*
Ant	*Camponotus ligniperda* Latr. *noveboracensis* Fitch	*B*	Log	*H*
Ground beetles	*Pterostichus sayi* Brulle	*B*	Rotten log	*L*
Tenebrionidæ	*Meracantha contracta* Beau.	*B*	Rotten log	*L*
Tenebrionidæ	*Uloma impressa* Mels.	*B*	Rotten log	*L*
Scarabæidæ	*Geotrupes splendidus* Fabr.	*B*	Rotten log	*L*

Staphylinidæ	*Staphylinus violaceus* Grav.	B	Rotten log	L
Elateridæ........	*Melanotus communis* Byl.		Rotten log	L
Slug.............	*Pallifera dorsalis* Bin.	B	Log	P

Beech Stage.

Frog.............	*Rana sylvatica* Le Conte	F	Ground	M
Salamander........	*Plethodon cinereus* Gr.	BF	Under leaves	H
Snail.............	*Polygyra inflecta* Say	BF	Leaves and log	P
Snail.............	*Polygyra oppressa* Say	BF	Leaves and log	
Snail.............	*Polygyra fraudulentia* Pil.	BF	Leaves and log	P
Snail.............	*Polygyra palliata* Say	BF	Leaves and log	
Snail.............	*Pyramidula perspectiva* Say	BF	Leaves and log	P
Snail.............	*Pyramidula solitaria* Say	BF	Leaves and log	P
Ground beetle	*Pterostricus corecinus* Newm.	BF	Leaves and log	L
Beetle.............	*Xylopodus saperaioides* Oliv.	B	Under bark	LV

The distribution of the animals of the field shrub and lower tree strata are shown in Table II. and the lists which follow.

TABLE II.

Showing the distribution of animals recorded from the vegetation in more than one of the animal communities of the forest stages indicated by numbers: 1, the cottonwood stage; 1–2, mixed cottonwood and pine stage; 2, pine stage; 2–3, mixed pine and oak stage and open places in the oak forest; 3, black oak stage, in its later phases white oaks occur; 4, stages containing red oak but not beech and maple; 5, beech and maple stage.

Common Name.	Scientific Name.	1	1–2	2	2–3	3	4	5
(a) Spider (Thomisidæ).	*Philodromus alaskensis* Key	*	*	*				
(b) Butterfly.........	*Anthocharis genuita* Fabr. .		*	*	*			
(c) Spider (Epeiridæ)...	*Epeira domocilorum* Hentz		*	?	?	?	*	*
(dd) Dusky plant bug ...	*Lygus pratensis* Lin.......			*	*			
(d) Phasmidæ........	*Diaphoromera femorata* Say.					*	*	
(e) Assassin bug.......	*Diplodus* sp.............					*	*	*
(f) Spider (Thomisidæ).	*Misumessus asperatus* H		*			*	*	*
(ff) Spider (Dictynidæ).	*Dictyna foliacea* Hentz.					*	*	*
(g) Spider (Epeiridæ)...	*Epeira gigas* Leach.......						C	F
(h) Spider (Theridiidæ) .	*Theridium frondeum* Hentz................						*	*
(i) Bug.............	*Acanthocephala terminalis* Dall................						*	*
(j) Stink bug.........	*Nezara hilaris* Say.......						*	*
(k) Stink bug.........	*Podisus maculiventris* Say..						*	*

The letters below at the left refer to the species opposite which they stand in Table II. and the numbers refer to the forest stages as at the heads of the columns of Table II. The capitals and italics have the same meaning as in the preceding table and list (Table I.).

a—from cottonwoods and juniper (1, 1–2, 3) (*B, K*).

b—from *Arabis lyrata* (*S*).

c—from pine and herbaceous vegetation (*B*) (4) (*K*).

d—from the trunks of various trees.

e—from black oak (3), red oak (4) and from maple (5).

f—Monarda (2–3) and black oak (3) maple (5) (*B, K*).

ff—*F* Monarda (3) (*K*).

g—from undergrowth (4) and beech (5) (*K, B*).

h—from shrubs (4) and young beech (5) (*B, K, T*).

i—shrubs (4) and maple trunk.

j—from red oak trunk (4) and beech trunk (5) (*Linden, Citrus, Gossypium U*).

k—? (4) and beech leaves (5) (*predaceous U*).

dd—herbs (*W*).

The food plant records in the literature are of no great significance ecologically because the character of the leaves of trees growing in open places and in forest and the physical conditions surrounding them are so different that a species feeding on a given tree in the forest might not feed on the same tree in the open and vice versa.

LIST OF ANIMALS RECORDED FROM THE FIELD, SHRUB, AND TREE STRATA OF THE FOREST STAGES NOTED.

All columns and symbols as in the list following Table I.

Cottonwood Stage.

1	2	3	4	Lit.
Chrysomelid beetle	*Disonycha quinquevittata* Say	*B, F*	Willow	*L*
Long-horned borer.....	*Plectodera scalator* Fab.	*B, F*	Cottonwood	*L*
Gall aphid..........	*Pemphigus populicaulis* Fitch	*B, F*	Cottonwood	*Z*
Gall aphid..........	*Pemphigus vagabundus* Walsh	*B, F*	Cottonwood	*Z*

Pine Stage.

Leaf beetle..........	*Nodonota tristis* Oliv.	*F*	Herbs	*L, V*
	Bassareus lativittis Germ.	*F*	Herbs	
Spider (Thomisidæ)....	*Xysticus formosus* Banks		Juniper	*R*
Spider (Attidae).......	*Dendryphantes octavus* Hentz		Juniper	*Y, B, K*
Spider (Theridiidæ)....	*Theridium spirale* Em.		Juniper	*B, K*
Engraver beetle.......	*Ips grandicollis* Eich.	*B, F*	Pine	*V*
Pitch moth..........	*Evetria comstockiana* Fern.?	*B, F*	Locust	*V*

Black Oak Stage.

Thread-waisted wasp ...	*Harpactopus* sp.	*F*	Primrose	*D*
Andrenid..........	*Agapostemon splendens* Lepel	*F*	Primrose	*D*
Spider (Thomisidæ)....	*Philodromus pernix* Black	*F*	Herbs	*K*
Spider (Epeiridæ).....	*Argiope trifasciata* Forsk.	*F*	Herbs	*K*
Sprinkled locust.......	*Chlœaltis conspersa* Har.	*F*	Herbs	*C*
Grasshopper.........	*Schistocera rubignosa* Har.	*F*	Herbs	*C*

Tree cricket	*Œcanthus fasciatus* Fitch.	B, F	Herbs	C
Texas grasshopper	*Scudderia texensis* Scud.	B, F	Herbs	C
Cone-head grasshopper	*Conocephalus ensiger* Scud.	B, F	Herbs	C
Meadow grasshopper	*Xiphidium strictum* Scud.	B, F	Herbs	C
Stink bug	*Euschistus variolarius* Pal.	B, F	Herbs	W
Flower bug	*Tripleps insidiosus* Say.	F	Herbs	T
Fork-tailed larvæ	*Cerura* sp.	B, F	Cherry	V
Fulgorid	*Otiocerus degeeri* Kirby	B, F	Oak	T
Flat bug	*Neuroctenus simplex* Uhl.	B, F	Oak	T
Colydiid beetle	*Ditoma quadriguttata* Say.	F	Oak	V
Prominent larva	*Heterocampa guttivittata* Wlk.	B, F	Oak	V
Prominent larva	*Nadata gibbosa* S. and A.	B, F	Oak	V
Tree hopper	*Telemona querci* Fitch (*monticola*)	B, F	Oak	T
Coreidæ	*Chariesterus autumator* Fabr.	B, F	Oak	T
Jassid	*Typhlocyba querci* var. *bifasciata* Gall.	B, F	Oak	T
Jassid	*Phlepsius irroratus* Say.	B, F	Oak	T

Red Oak Stage.

Rove beetle	*Tachinus pallipes* Grav.	B, F	Mushrooms	D
Spider (Clubionidæ)	*Anyphæna conspersa* Key.	F	Herbs	B, K
Spider (Dictynidæ)	*Dictyna* sp. (juvenile)	F	Herbs	B, K
Spider (Attidæ)	*Mævia niger* Hentz	F	Herbs	Y, B, K
Locustidæ	*Atlanticus pachymerus* Burm.	B	Grass	C
Spider (Epeiridæ)	*Acrosoma gracilis* Wal.	F	Shrubs	B
Spider (Epeiridæ)	*Acrosoma spinea* Hentz	F	Shrubs	B
Spider (Clubionidæ)	*Clubiona* sp.		Shrubs	B
Spider (Epeiridæ)	*Mangora maculata* Key.		Shrubs	K
Jassid	*Scaphodius auronitens* Prov.		Shrubs	
Beetle	*Odontota nervosa* Panz.		Shrubs	
Bug (Nabidæ)	*Coriscus annulatus* Reut.	B	Shrubs	
Spider (Lyngyphiidæ)	*Linyphia phrygiana* Kock.		Shrubs	
Cicada	*Cicada linnei* S. and G.	F	Young maple	
Leaf beetle	*Calligrapha scalaris* Lec.	B F	Young maple	
Stink bug	*Euschistus tristigmus* Say.	B F	Young maple	
Arctiidæ	*Halisdota* sp.	B F	White oak	B
Oak worm	*Anisota senitoria* Sm. and Abb.	B F	White oak	D
White oak gall	*Andricus semiator* Harr.	B F	White oak	Z
Tree cricket	*Œcanthus angustipennis* Fitch.	B F	Red oak	
Katydid	*Cyrptöphyllus perspecivius* L.	B —	Red oak	
Leaf beetle	*Xanthonia 10-notata* Say.	F	Red oak	
Prominent larva	*Symmirista albifrons* S. and A.	B —	Maple	V
Prominent larva	*Datana anguisii* G. and R.	B —	Hickory	
Aphid	*Phylloxera caryæ-caulis* Fitch	B —	Hickory	Z

Beech Stage.

Beetle	*Boletobius cinctus* Grav.		Mushrooms
Fungus beetle	*Boletotherus bifurcus* Fabr.		Shelf fungus
Snout beetle	*Piazurus oculatus* Say.		Shrubs

Cercopidæ (bug)......Clastoptera obtusa Say.	—	F	Hic., map., hazel	T
Tettiginidæ (leaf hop) .Gypona octolineata Fitch.	—	F	Hic., map., beech	J
Leaf hopper.........Jassus obliturus Say.	—	F	Maple	
Ichneumonidæ.......Thalessa atrata Fabr.	B	—	Larvæ	V
Lace wing...........Crysopa rufialbris	B	—	Maple	
Lace bug............Gargaphia tiliæ Walsh.	B	—	Beech	V
Ichneumonidæ.......Trogus vulpinus Cb.	B		Larvæ	D
Pentatomidæ.........Banasa calva Say.		F	Beech	
Lampyrid beetle......Podabrus basilaris Say.	——		Maple	V
Lycosidæ...........Wala mitrata Hentz.			Maple	
Theridiidæ..........Notionella interpres Cam.			Maple	
Lampyridæ..........Telephorus tubercolatus Lec.			(Alder.)	L
Harvestman.........Oligolophus pictus Wood			Maple trunk	—

IV. Discussion of Data.

An examination of Table I. and the lists of ground and subterranean animals, shows that we have on and under the ground a change in species as we pass from the youngest to the oldest stage of forest development. We note also, where data permit estimation of relative abundance, that as we pass from the youngest to the oldest stage, a species is first few in numbers, then common and later decreasing again. Examination of Table II. and the lists of animals inhabiting vegetation, shows the same phenomenon though the delimitation appears somewhat sharper, possibly because these animals are related to plants and the differences in physical conditions are accompanied by quite different plants.

We note that in general, with the change of conditions accompanying the development of forest upon sterile sand or other mineral soil there is also an almost complete change of animal species. This change is comparable to that associated with the development of a stream (Shelford, '11[1]) and the filling of a pond (Shelford, '11[2] and '11[4]). This change in *mores*, if viewed at the oldest point in the environmental series, is *ecological succession*. For example at station 5A where beech forest occurs on sand dunes the cottonwood community has probably been succeeded by the pine community; the pine community by the black oak community; the black oak community by the red oak community which has given way to the present beech community.

This could be discussed as in the cases of the ponds and streams. The discussions already published (l. c.) are sufficient to illustrate the methods and principles. Furthermore the succession of conditions and of the tiger beetles applying to forest development has already been briefly outlined (Shelford, '07). (See Adams, '08, '12.)

V. Causes of Animal Succession and the Control of Animal Communities.

1. The causes of plant succession as summarized by Cowles, '11, may be divided into those related to atmosphere and those related to soil. In the case of animals we recognize also difference in food and materials for abode. *Physical conditions* are believed to be *most* important, as indicated by the great mass of experimental work on animal *behavior*. Representative literature supporting this view is cited in the discussions which follow.

1. *Materials for Abode and Food.*

The former are of great importance. There are the greatest differences between the different forest stages, in this matter. The plants of the later stages are more numerous and the leaves less strongly cutinized, even when the plants belong to the *same* species. The difference in leaf structure may be a factor in limiting the distribution of the phytophaga to a certain part of the range of a species of plant. The leaves, fallen logs, and all conditions in which the animals make their abodes, change as the forest develops. Food has been but little studied and we know little or nothing as to what aspects of the food factor are important. Dahl ('96) has studied the relation of carrion eating animals to their food supply.

2. *Soil.*

Those causes of plant succession which are due to progressive changes of soil, may be briefly summarized from an inspection of the description of stations given above. The chief changes obvious to the eye are an increase of vegetation, of leaf covering and of humus.

(*a*) The last of these changes increases the water holding capacity of the soil, while the other two decrease the evaporation from the soil. The water holding power of different soils is different. It increases with the decrease in size of the soil particles and with the addition of humus which takes up water by imbibition. The amount of water in the soil is usually expressed in terms of per cent. of weight but a soil with 8 per cent. of moisture may not give up water to an organism as readily as another soil with only 2 per cent. It is necessary therefore, to determine the capacity of a soil to retain or give up moisture. This has been determined for a number of soils by Briggs and McLane ('07) and Briggs and Shantz ('12), in terms of what they call the moisture equivalent. The moisture equivalent of a soil is the percentage of water which it can retain in opposition to a centrifugal force 1,000 times that of gravity. This has been determined for a number of soils (l. c., '12, p. 57). The maintenance of turgor in plants is believed to be a purely physical matter. If the roots of a plant are in a mass of soil, the plant gradually reduces the water content until the permanent wilting occurs. The *wilting coefficient* of a soil is the moisture content (in percentage of dry weight) at the time when the leaves of the plant growing in the soil first undergo a permanent reduction in moisture content, as a result of a deficiency of moisture supply. The *moisture equivalent* of a soil is 1.84 times the *wilting coefficient for wheat*, used as a standard plant. Fuller ('12) states that the wilting coefficient of dune sand is about 0.75 per cent. while the usual moisture content of the cottonwood dune sand is two or three times this amount. For the clay soil of the oak-hickory forest, according to McNutt and Fuller ('12) the coefficient is about 8 per cent. These standards of soil moisture indicate the amount of water available to animals through direct contact with the soil or available for evaporation into the air of cavities which they construct for themselves beneath the surface of the soil. The soil inhabiting animals of the cottonwood area live in the presence of a greater amount of available water than do the animals of the oak hickory forest.

(*b*) *Plants and Animals.*—Cowles ('11) mentions the importance of soil bacteria which increase with the increase of the humus, and the development of substances toxic to the plants producing them (Schreiner and Reed, '07). Little is known of the effect of animals upon the soils in which they live but if excretory products ever accumulate in any quantity, they probably have a detrimental effect, especially upon the animals which produce them (Colton '08 and citations). On the other hand, many burrowing animals bury organic material and bring mineral soil to the surface. The digger wasps must add much to the sand by burying many insects for their young. Earth worms appear in the later stages and contribute to soil formation (Darwin). Cowles states further on the authority of Transeau that humus accumulation alters soil aeration.

(c) *Temperature.*—Transeau found that the temperature of bog soil and bog water is below that of other soils and waters. This has however not been observed in the case of dry soils. The differences between soil on the beach at Sawyer, Mich., Aug. 19, 1911, at 3.00 P.M. and in the beech woods near at hand was as follows: Air 20° C., upper one half inch of sand of cottonwood area 38°–39° C., sandy soil of beech woods 19°–20° C., a difference of 19° C. The upper one half inch of bare sand goes as high as 47° C. on the hottest days of summer while the soil in the beech woods is probably always a little cooler than the air at the time of the air maximum. Cottonwood soil temperature on the hottest summer days at about 3.00 P.M. has been found to be as follows:

TABLE II.

SHOWING VARIATION OF SAND TEMPERATURE WITH DEPTH AND MOISTURE CONTENT.
AIR 36° C.

	Dry Sand.	Moist Sand
1.25 cm. below surface	47° C.	32° C.
3–4 cm. below surface	38° C.	31° C.
8–9 cm. below surface	35° C.	29° C.
10–11 cm. below surface	33° C.	———
12–13 cm. below surface	32° C.	27° C.
17–18 cm. below surface	30° C.	———

Simultaneous readings in later forest stages were impracticable. Even where exposed to the sun moist sand is kept at a lower temperature by the evaporation.

3. *Atmosphere.*

Conditions at and above the surface of the soil, *i. e.*, in the ground and field strata.

(a) *Temperature.*—The above data on the temperature of the surface of the soil may be taken to represent essentially the temperature at the surface as well. There are no records of the temperature at various heights above the ground. Noticeable differences within the height of the trees present, are to be expected particularly in the cottonwood and other early stages where much bare sand is exposed.

(b) *Light.*—Animals are either positive or negative to the actinic rays of the spectrum (Congdon, '08, Mast, '11). Considerable work has been done by plant ecologists, on the measurement of light with photographic papers but its bearing on plant problems is questioned by some because the nonactinic portion of the spectrum is most important in the process of photosynthesis. It appears that these measurements are of much greater significance for animals than for plants. Zon and Graves ('11) have brought together the literature and discussed the methods of study (see especially several papers by Wiesner).

The light in the cottonwood stage is more intense than in any other of the habitats that we are to consider. Tests of the light in the beech woods and in the road adjoining, made with a Wynne exposure meter, show the following differences:

Location of Meter.	Time Required to Match Standard Tint.
Beech woods—darkest shadows	1,200 seconds.
Beech woods—medium shadows	180 "
Beech woods—brightest spots	10 "
Road on the north side of woods	3 "

While the above table shows a measurement of the actinic rays only, it indicates that in the beech forest, such rays at least, are diminished in intensity to from 1/3 to 1/400 that of full sunlight. On account of the great amount of reflection from sand, the light in the cottonwood stage is probably double that in the wagon road which is bounded on the south by beech woods and on the north by second growth timber.

(c) *Combinations or Complexes of Factors.*—As we have already pointed out (Shelford, '11), the animal environment is a combination of moisture, temperature, light pressure, materials for abode and food, all of which factors taken together constitute a complex of interdependences. These various factors are so dependent upon one another that any change in one usually affects several others. This property of environmental complexes is what makes ecology one of the most complex of sciences, and experimentation in which the environment is kept normal except for one factor, an ideal rarely realized in practice, even under the best conditions.

The efforts of ecologists, geographers, and climatologists have long been directed toward the finding of a method, of measuring the environment, which shall include a number of the most important environmental factors. De Candolle undertook to base the efficiency of a climate, for supporting plants, upon the mean daily temperatures above 6° C., this temperature being taken as the starting point of plant activity. Merriam has followed this lead and calculated total temperatures for many places in North America and made maps and zones based upon such totals. This system however, has been rejected by botanists and plant ecologists on account of much evidence both experimental and observational, which is quite out of accord with this view. The scheme has not been generally accepted by zoölogists outside of the United States Biological Survey. There is practically no evidence of an experimental sort, for the application of such a scheme to animals. Relative humidity has been suggested as an important index (Walker, '03) but does not properly express the influence of atmospheric humidity upon the animal body (Hann, '03, p. 53). The saturation deficit has also been suggested but does not take temperature into account.

1. *Evaporation.*

"The total effect of air temperature, pressure, relative humidity, and average wind velocity upon a free water surface in the shade or in the sun, is expressed by the amount of water evaporated" (Hann, p. 72). Since temperature in the season without frost is directly due to the sun's rays, light is in part included. In our latitude, clouds in summer slightly decrease the air temperature (Hann, p. 72). In winter however the temperature of cloudy days is higher. The strongest light is usually associated with the greatest evaporation. Yapp ('09) found that the rate of evaporation was directly correlated with temperature and illumination, but most closely correlated with relative humidity. From the standpoint of including many factors, the evaporating power of the air is by far the most inclusive and is therefore by far the best index of physical conditions surrounding animals wholly or partly exposed to the atmosphere. It is not however to be expected that it will hold good for all the factors under all climatic conditions, and for this reason, records of light, temperature, pressure, carbon dioxid, etc., should be made.

(a) *Effect of Evaporation upon Animals.*—In the case of man some observations have been made. According to Pettenkofer and Voit (fide Hann), an adult man eliminates 900 grams of water from his skin and lungs daily. Of this amount 60 per cent. or 540 grams come from the skin alone and changes in relative humidity of only 1 per cent. cause perceptible changes in the amount of evaporation from the skin. If evaporation from the skin and lungs is diminished, the amount of urine is increased, as in many cases are also the secretions of the intestines. Sudden changes in humidity make themselves felt in sudden increased or decreased blood pressure. The less dilute blood of dry climates operates as a stimulant and increases the functions of the nervous system. The consequences are excitement and sleeplessness (Hann, pp. 56–57).

Little has been done on the physiological effect of evaporation or desiccation upon animals. Various writers have found a loss of water associated with hibernation. Greeley obtained the same results with desiccation as with freezing (Greeley, '01; Bachmetjew, '99; Semper, '79, pp. 182–188). The reactions of animals to an atmospheric humidity gradient has probably never been studied. The chief conclusion to be drawn from the literature is that a high rate of evaporation is advantageous to some animals and decidedly detrimental to others. Attempts to keep insects and spiders which live exposed on the prairie vegetation, near Chicago, in the laboratory in screen cages containing vegetation, usually result in the death of the animals within a few hours. On the other hand, the same species will live in glass jars covered or partially covered with glass plates, long after the vegetation which was placed in with them has turned brown and has soured so that it gives off a bad odor. Special investigation would be necessary to determine the cause of this difference in the death rate, yet difference in the rate of evaporation from the animals' bodies is probably an important factor. After long and careful experimental studies dating far back into the history of plant physiology, plant ecologists have come to the conclusion, that the evaporating power of the air is the most satisfactory index of plant environments.

(d) *Evaporation in Forest Animal Habitats.*—Fortunately this has been investigated (Fuller, '11) in the five types of stations, viz., cottonwood, pine, black oak, oak-hickory, and beech. Fuller's first three stations were a little more mesophytic than ours. The data were obtained by using a porous cup atmometer. Evaporation from the atmometer is more nearly like that from an organism than is evaporation from any other device; it was devised by Livingston ('06, '08, '10, '10). It consists of a hollow cup of porous clay 12.5 cm. high, with an internal diameter of 2.5 cm. and a thickness of wall of about 3 mm. It is filled with pure water and connected by means of glass tubing to a reservoir usually consisting of a wide-mouthed glass bottle of one half liter capacity. The water, passing through the porous walls, evaporates from the surface, the loss being constantly replaced from the supply within the reservoir. Readings are made by refilling the reservoir from a graduated burette to a certain mark scratched upon its neck. For convenience in handling a portion of the base of the cup is coated with some impervious substance and before being used in the field, the instrument is standardized by comparing its loss of water with that from a free water surface of 45 sq. cm. exposed under uniform conditions. As a further check against error this standardization is repeated at intervals of six to eight weeks throughout the season (Fuller, '11). In Fuller's work, the bottles were sunk so that the evaporating surface of the instrument was 20–25 cm. above the surface of the soil.

Figure 2 shows the results of a season's study by Fuller. "The graph for the pine dunes is decidedly lower and more regular in its contour than that of the association which it succeeds. Its four nearly equal maxima would indicate that

80   VICTOR E. SHELFORD.

within its limits there was throughout the summer season a
continuous stress rather than a series of violent extremes.   On
the whole it shows a water demand of little more than half of

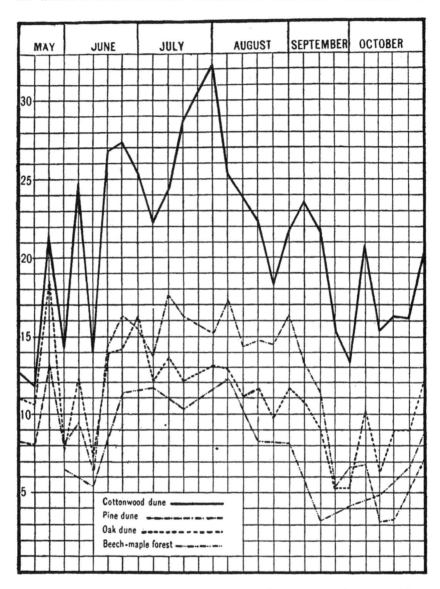

FIG. 2.   Mean daily evaporation rates (cc. per day) in the ground stratum of four
of the animal communities.   (Courtesy of G. D. Fuller and the *Botanical Gazette*.)

that occurring in the cottonwood dunes.   Its greatest divergence
is plainly due to the evergreen character of its vegetation and is
seen on its low range in May and the first part of June, and again

in October when it falls below that of the oak dunes and is even less than that of the beech maple forest. This would give good reasons for expecting to find within this association truly mesophytic plants whose activities are limited to the early spring.

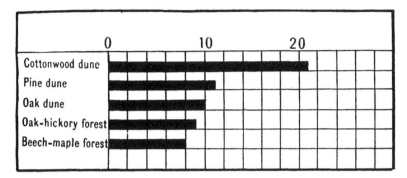

FIG. 3. Showing the comparative evaporation rates (cc. per day) in the ground stratum of the different animal communities from May to October. (Courtesy of Mr. G. D. Fuller.)

Evaporation in the various associations varies directly with the order of their occurrence in the succession (Figs. 3, 4). The differences in the rate of evaporation in the various plant asso-

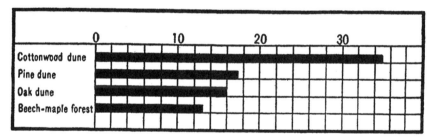

FIG. 4. Showing the comparative evaporation rates (cc. per day) in four of the animal communities on the basis of the maximum amount per day for any week from May to Oct. (Courtesy of Mr. G. D. Fuller and the *Botanical Gazette*.)

ciations studied are sufficient to indicate that the atmospheric conditions are most efficient factors in causing succession." (Fuller, '11.)

A comparison of Fuller's data with the tables and lists of animals shows that the distribution and succession of animals is *clearly correlated* with the *evaporating power of the air*. Further comparison with the description of stations (p. 61) shows that the evaporating power of the air may be taken, in this case, as an index of the materials for abode, etc.

4. *Influence of Physiography and Vegetation upon Animal Habitats.*

In some cases the evaporating power of the air is apparently largely controlled by the vegetation and in others largely by physiographic conditions while as a rule both physiographic conditions and vegetation play important rôles. The importance of the combined effect of physiographic conditions and vegetation is well shown on the steep clay bluffs of Lake Michigan. For example, at Glencoe, Ill., erosion has rendered the bluff steep and brought the ground water near the surface in some places (Shelford, 11[4]). Forest animals occur among the shrubs and under the dead sweet clover (Fig. 5).

TABLE IV.

Showing forest animals in the early stages of forest development of a clay bluff of Lake Michigan. Subterranean and ground strata, 1; bare clay, 2; sweet clover, 3; shrubs, goldenrod, etc., 4; sapling stage; animals same as in 5, the oak-hickory forest.

Common Names.	Scientific Names.	1	2	3	4	5
Tube weaver.	*Aglena nœvia* Wal.	×	×			
Lycosid	*Pardosa lapidicina* Em.	×	×			
Carolina locust.	*Dissostiera carolina* Linn.	×	×			
Mud dauber.	*Pelopœus cementarius* Dru.	×	×			
Tiger-beetle larvæ.	*Cicindela purpurea limbalis* Klg. . .	×	×			
Sow bugs.	*Porcellio rathkei* Brandt.		F	C	A	A
Centipede.	*Geophilus* sp.		×	×	×	×
Snail.	*Polygyra thyroides* Say.		×	×	×	×
Snail	*Pyramidula alternata* Say.		×	×	×	×
Snail.	*Polygyra monodon* Rach.			×	×	×
Tiger-beetle larvæ.	*Cicindela sexguttata* Fabr.			×	×	×
Snail.	*Polygyra albolabris* Say.			×	×	×
Slug.	*Phylomycus carolinensis* Bosc.			×	×	×
Yellow-margined millipede.	*Fontaria corrugate* Wood			×	×	×
Centipede.	*Lyasopetalum lactarium* Say.			×	×	×

All of the species beginning with *Geophilus* are commonly found in the oak-hickory forest. On the covered bluff however, where the moisture content of the soil is great and the dense sweet clover and the shrubs make a good covering we find these animals associated with the earliest stages of vegetation development. Shade and moisture here appear to be the determining factors. We note here then that the forest floor conditions are in advance of the forest while on the dry well-drained sand they lag behind in succession.

Some investigators have questioned the importance of vegetation to animals and we note here that the distributions of plant and animal species are not always correlated. If one

FIG. 5. The bluff of Lake Michigan at Glencoe, Ill., showing several stages of forest development. To the right of an imaginary line *a–b* are small areas of the habitats shown in Table IV., in columns 1 and 2. Within the triangle *a–b–c* are areas of the same habitat invaded by shrubs under which are found forest animals. To the left of *a–c* is an area of shrubs and saplings which has a full quota of forest floor animals. (Reprinted from the *Journal of Morphology*.)

refers to *species* of *plants* and *species* of *animals* then the vegetation very often is not correlated with the distribution of the animals. If on the other hand one means that the plants are controllers of physical conditions, then vegetation can be said to be of very great importance.

5. *Stratification of Conditions.*

An inspection of the tables and the discussion following them shows that different animals which do not burrow into the ground inhabit different levels of the forest. For example *Acrosoma spinea* Hentz builds its web 1–3 ft. above the ground while *Acrosoma gracile* Wal. builds 4–6 ft. above the ground (see Dahl, '08).

TABLE V.

EVAPORATION FROM POROUS CUP EVAPORIMETERS IN DIFFERENT STRATA OF A SUMMER DRY MARSH, CAMBRIDGESHIRE, ENGLAND, DURING THREE PERIODS BETWEEN JULY 9 AND SEPTEMBER 8, 1907.

(Yapp, '09, p. 299 and 294.)

Year.	Height above Ground.	Ratio of Evapor.	Temperature.			
			Mean Max.	Mean Min.	Mean.	
1907	5 ft. 6 in. to 4 ft. 6 in.	100.00	22.1	6.6	16.5	Well above vegetation.
1907	2 ft. 2 in.	32.8	23.0	—	—	A little above the mid height.
1907	.5 in.	6.6	18.0	7.1	14.1	
1907	soil	——	12.7	11.2	11.8	

The above table shows marked differences in the rate of evaporation, considerable differences in temperature at the different levels and both due largely to vegetation. Differences in light are also to be expected. Sherff ('12, p. 420) has found conditions similar to the above by a two months' study of evaporation on Skokie marsh near Chicago. The evaporation there was three times as great at a height of 1.95 m. as at the surface of the soil in among the plants of Phragmites. Mr. Harvey has also secured similar (unpublished) results on the prairie at Chicago Lawn, Chicago.

It has been long recognized that there are distinct growth-form strata in nearly all plant formations, pelagic algæ formations being a possible exception. The data of Sherff and Yapp indicate differences in conditions in the strata of grass formations and associations. Greater differences are to be expected in the different strata of forests and shrub covered areas. Mr. Fuller informs us that there are marked differences in the structure of leaves at different levels of the same forest tree.

6. *Apparent Anomalous Distribution.*

Are physical conditions sometimes similar when vegetation and landscape aspect are very different? That they are is clearly suggested when we compare the forest and the shrub covered bluff where forest animals occur. Plants grow from seeds only under a very limited range of conditions. However if trees are given a few years' growth under favorable conditions they will be successful under a great range of conditions. The great age to which trees often live and the slowness with which they grow makes it possible for conditions to change while the trees still live on with changes only in leaf structure. It is to be expected that the distribution of animals is correlated with the occurrence of seedlings or of quick growing plants or at least with leaf structure types rather than strictly with species of trees. These facts suggest that there are two types of cases in which physical conditions and forest conditions are not in accord. In the first case atmospheric conditions become favorable for forest animals before any woody plants have been able to grow, in the second, woody plants remain after conditions have become unfavorable for forest animals; both are due to lagging behind of vegetation; both are very local and of minor significance.

A comparison of the data of Yapp (Table V.) and Transeau (Fig. 6) shows a difference between the evaporation of the lower

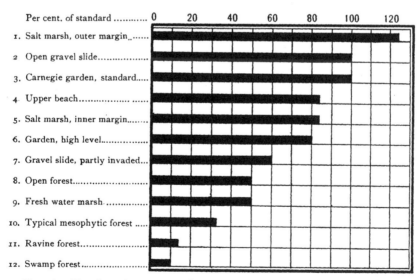

FIG. 6. Showing the comparative evaporation rates (c.c. per day) in the ground stratum of several animal habitats on Long Island during July and August. (After Transeau, courtesy of the *Botanical Gazette*.)

stratum of a marsh and the free atmosphere above, comparable
to that found by Transeau between the lowest stratum of the
mesophytic forest and the open gravel slide. An inspection of
Transeau's data (Fig. 6) on evaporation indicates several obvious
cases of similarity; *e. g.*, we note that the rate of evaporation
is about the same in the fresh water marsh and the open forest.
The data at present available do not justify definite conclusions,
yet it may be suggested that there are various stations in strata
of the different plant communities where the conditions of the
physical factors are essentially identical, but where the necessary
materials of abode, especially those used in breeding, are sometimes wanting. Their absence alone is sufficient to prevent
animals of specialized habits and structures from taking possession of situations otherwise entirely suitable.

The reasons for the wide distribution of some animals in the
forest stages which we are considering are no doubt various.
For example *Zonites arborens* (Table I, p. 67) is rare in the
early stages and is confined to the lower and moister localities.
If *Epeira domicilorum* is a species of stable physiological makeup
we can offer no explanation for its peculiar distribution (Table
II, p. 69). A species may have its critical period in the early
spring when the leaves are off the trees and the condition of the
atmosphere similar in all (see Fig. 2) stages or may live at higher
levels in the denser and older stages, and thus be surrounded by
similar atmospheric conditions, but we are not warranted in
assuming either of these causes here.

Another striking feature of the distribution of many beetles,
bugs and spiders, and Orthoptera is the fact that they are found
in open woods, edges of woods, on the vegetation of marshes
and over the water of small ponds in which vegetation is growing.
In this way many species are found to occur in what at first
appear to be very unlike situations. *Lygus pratensis, Tripleps
insiduosus*, and *Euschistus variolarius* which occur on the vegetation of the margins of swamps, of the black oak forest dunes and
on prairies and agricultural lands may serve as examples. Shull
('11) has pointed out similar facts as one of the difficulties in
the way of ecological classification of Orthoptera and Thysanoptera. Such species as the bugs mentioned above are said to

occur "everywhere," although they are rarely found in moist woods or in any situation in which they are not fully exposed to the sun and may always live in similar conditions.

There are great differences between open prairies and closed forests. Shimek ('11) found that the evaporation in the undisturbed groves in eastern Iowa during July and August was very much less than that in the prairies adjoining. From the free surfaces of pans set in the ground so that the water which they contained was level with the surface of the soil, the evaporation of the groves was about 27 per cent. of that of the prairie; with cup evaporimeters about 37 per cent. and with Piche evaporimeters about 47 per cent. This is about the same as the difference on Long Island between the inner side of Transeau's salt marsh dominated by grass-like plants and his mesophytic forest. Sherff ('12) found the evaporation in a marsh forest to be a little less than that in the beech maple and from 1.8 to 2.6 times as great as in the lowest stratum of a marsh. While differences and similarities of physical conditions are sufficient to account for many peculiarities of ecological distribution, it must be recognized that the *same species may occur under different conditions and show difference in mores* (Bohn, '09, Allee, '12).

5. *Agreement of Plant and Animal Communities.*

Before discussing the problem of agreement between plant and animal communities, it is necessary to state what is meant by agreement. According to present developments of the science of ecology *plant and animal communities may be said to be in full agreement when the growth form of each stratum of the plant community is correlated with the conditions selected by the animals of that stratum.* Questions of agreement are primarily questions for experimental solution. Two types of disagreement are to be expected. We may illustrate the first by a bog or marsh community. Considering plants rooted in the soil we note that water is secured from the soil by the roots and is lost through the leaves and twigs. Accordingly since bog soil is unfavorable, due to the presence of toxins or to other causes, plants growing in it do not secure water easily even when the quantity of soil water is great. *Such plants have xerophytic structures (which tend to check the loss of water) developed far beyond the requirements of the atmospheric conditions surrounding their vegetative parts.* It is improbable that the animals inhabiting a bog-vegetation field-stratum would *select* atmospheric conditions such as produce equally xerophytic structures under *favorable soil* conditions. We may therefore expect disagreement. The smaller plants such as fungi, algæ, etc., are related to the strata of soil and atmosphere exactly as the smaller animals and as *much disagreement* is to be expected between such plants and the rooted vegetation as between the rooted vegetation and animals. It must also be noted that the xerophytic structures of the plants of *unfavorable* soils may have important influence upon ectophytic plants and animals and in part counteract the effect of favorable atmospheric conditions.

The second type of disagreement is represented by cases in which the vegetation is said to lag behind. We have noted that on the clay bluff, conditions become favorable for inconspicuous plants and forest animals as soon as the growth of the pioneer vegetation gives shade to the soil. In other cases woody vegetation remains in situations where the conditions have become unfavorable for it and the less conspicuous plants and some of the animals have disappeared. We may expect lack of accord within and between plant and animal communities under such conditions. In these cases, however, conditions are only *temporarily out of adjustment*, due to rapid physiographic changes and we note from the data presented that plant and animal communities are usually in agreement. The exceptions are often apparent only and due to the emphasis of *species* instead of *mores* and *growth form*.

VI. GENERAL DISCUSSION.

At this point we may note certain aspects of the basis for the organization of ecology into a science. It is possible to characterize the communities of the forest in physiological terms though we cannot be as definite as is to be desired, until *mores* have been studied in detail. Taking the communities one by one and stratum by stratum we may note obvious characters.

A. Pioneer Communities.

The communities of the cottonwood, pine and black oak stages may be designated as pioneer because of the presence of bare mineral soil.

1. *Subterranean and Ground Strata.*—(*a*) The cottonwood community is characterized by animals which breed and spend the dark and cloudy days chiefly below the surface of the sand. They are very largely diurnal and predatory, are exceedingly swift and wary. The burrowing spider (*Geolycosa pikei*) is one of the few nocturnal animals.

(*b*) The pine community is characterized by similar *mores*, but is to be distinguished from the preceding by the presence of many animals which prefer sand that is less shifting and which is slightly *darkened by humus* (Shelford, '07). Animals requiring "cover," such as the lizard, the blue racer, a few ground squirrels, etc., give character because of their absence from earlier and later communities.

(*c*) The black oak community represents the climax of diversity of the subterranean and ground strata. The bare sand *mores* continue in the open spaces, which we have designated as transition areas. Leaf cutters are now present while among the burrowers, the root borers (Prionids and Lucanids) work on the roots of the decaying trees. The behavior differences between this and the preceding communities are differences of detail which, for the making of deductions, would require much careful study.

2. *Field and Shrub Strata.*—The field and shrub strata of the cottonwood, pine and oak communities are less easily characterized. The cottonwoods of the beach are far less commonly infested with aphid galls than are trees of the same species growing in less exposed situations. Furthermore we have never found any of the lepidopterous larvæ such as *Basilarchia archippus* near the beach. Animals living exposed upon the trees are few in number. The same general conditions obtain on and among the pines but spiders are more numerous. On the black oak the number of phytophaga is increased and the number of galls appears to be greater than in the later stages; the inhabitants of the herbaceous vegetation are chiefly those found in

open situations such as prairies and roadsides, where the physical conditions are similar. Some animals of the same species which make up the black oak community were taken from a roadside and after being mixed with the inhabitants of the shrubs of the beech forest, were placed in a light gradient. Soon the insects and spiders of the two communities separated sharply from each other, the beech-inhabiting species going to the darkest end while the roadside species all crowded to the light.

B. Later Communities.

With the coming in of red oak true forest with the mineral soil largely covered with humus and leaves is present and very different *mores* obtain. The diurnal diggers are practically absent. Snails, beetles, grasshoppers, spiders and myriopods living under bark, decaying wood, and leaves, avoiding strong light and requiring moisture, are the chief types. The *mores* are typically forest in character. The differences between these and the later stages are those of detail and degree which need careful study. In general with a lessening in the severity of the conditions, there is a proportional increase in the use of the vegetation as a place of abode.

In the field and shrub strata, we note that the animals of the cottonwood, pine and oak stages are characteristic of open dry situations, requiring or tolerating strong light, while those animals of the red oak, hickory and beech stage are negatively phototactic to light of the same intensity, as shown by mixing the animals in a gradient.

The animals of the tree stratum are few and scattered in the cottonwood, pine and black oak stage while animals enclosed in galls or cases are common if not dominant. In the red oak, hickory and beech stage phytophaga are often gregarious and numerous. The vegetation is used more and more for a breeding place as the forest increases in denseness. Groups such as orthoptera, beetles, bees and wasps, are represented more and more by species which make use of the vegetation as forest development goes on. The tree strata of all the forest stages are characterized by species given to frequenting a limited number of kinds of trees.

C. General Considerations.

We note that the distribution of animal species which occur chiefly on a particular plant species or on closely related species of a group, do *not often occur everywhere* that the plant or plants occur, and if they do there is a marked difference in the number of individuals. Such phenomena appear to be matters of common observation among naturalists. While they are still subjects for investigation, there is much evidence that the local distribution of the phytophaga is not that of the food plant or plants but is limited to a certain portion of the local range of the plant, by differences in the physical conditions, or the growth form of the plant or both. The food plants of phytophaga having a number of food plants are usually those growing in associations. The fauna of trees growing in different communities or under different conditions are probably *commonly different*. The differences in the *mores* of the communities outlined above are clearly correlated with factors known to be of importance in the behavior and physiology of animals. These are materials for abode, soil moisture, light and the condition of the atmosphere.

The more important features of the environment of an animal are selected through its reactions, which are probably innate or instinctive (Wheeler, '10, p. 159; Shelford, '11[4], pp. 556–582; Hancock, '11, p. 327; Herrick, '05, p. 201). Different species usually select different habitats or different strata in the same habitat. It is well known among naturalists and experimenters that *different species* usually have *different mores* (Brehm, '96, p. 73).

Animals of the same species show behavior differences in different habitats (Jennings, '06, Ch. XXI.; Shelford, '11[4], p. 584; Allee, '12). Bohn found, that the sea anemones living near the surface of the sea where the wave and tide action are strongest, showed more marked rhythms of behavior in relation to tide than those living lower down where the action of the tide and waves is less marked (Bohn, '10, p. 156; Holmes, '11, p. 155). These rhythms disappeared slowly when the animals were removed from the tide to the aquarium. Many such cases are probably to be found in the natural history literature. For example the chipmunk differs in behavior under different conditions (Wood,

'11, p. 523).   Abbot ('70, p. 104) makes a similar statement about fish.   It is apparent then that one species may have *several mores* (Bohn, '10 et al.).   Different species may sometimes have *identical mores;* these cases are usually separated geographically (Shelford, '11, p. 32; '11², p. 147; '11⁴, p. 604).   In addition to these relations, the relation of ecology to species is largely a matter of language, names being necessary as a means of referring to animals.

Animal ecology has very much in common with plant ecology. Diatoms, flatworms and many other marine animals and plants meet the same conditions in the same or similar ways (Loeb, '06, p. 121; Bohn, '10, p. 156; Holmes, '11, p. 155).   Sessile animals, such as reef-forming corals, show growth form differences (Woods-Jones, '10) under different conditions, just as sessile plants do. Comparable plants and animals show comparable responses. The physiological life history aspect of plant ecology (Ganong, '07) is parallel with the same phenomenon in animals, but the activities of motile animals correspond roughly to the growth-form phenomena in sessile plants (Shelford, '11⁴, p. 593).   Results of study of the environment are equally applicable to plants and to animals.   Since *mores* and *growth-form* are correlated with the *environment* much progress can be made by the *study* of the environment; in fact, study of the environment is necessary for progress.

On the other hand the study of the environment must be accompanied by experiments designed to determine the relative importance of the different factor *to animals,* or the results, like so many of our meteorological records, will prove to be of questionable value for the purpose for which they are intended.   In the case of the forest animal communities which we have studied, experiments must be undertaken to determine the physiological relations of animals to materials for abode, soil moisture, light and the condition of the atmosphere before the subject can progress beyond the suggestive stage which this paper necessarily represents.

Ecology or ethology of single isolated species is a very old branch of biological study.   The developments of the last twenty years have been in the direction of organization of these isolated facts into a science on the basis of *mores*, including *habitat preferences*. The similarities between the response phenomena of plants and animals have led in the direction of the organization of a branch of biological science embracing both plants and animals. It is this organization, or the possibility of organization, which we are attempting to introduce here. The experimental work cited above is adequate to indicate the lines along which further investigation should be directed and that *the mores problem, which includes the habitat preference problem, is the central problem of ecology.*

VII. SUMMARY.

1. The development of forest on sand or other mineral soil is accompanied by an almost complete change of animal *species* and probably by a complete change of animal *mores* (pp. 67–72).

2. Forest development is accompanied by marked changes in soil and physical factors; animal distribution is more closely correlated with differences in *physical factors* than with species of plants (pp. 73–82).

3. For animals living in the soil, the moisture equivalent, or the wilting coefficient for a standard plant, is the best index of the moisture available to the animals (p. 74).

4. The rate of evaporation or the evaporating power of the air is probably the best index of the conditions of the atmosphere (p. 77).

5. The rate of evaporation, temperature, etc., have been found to be very different in the different communities and also different in the different strata of the same communities. The amount of evaporation in animal communities is directly related to their order of occurrence in succession (p. 81).

6. Plant and animal communities are divisible into strata which represent vertical differences in physical conditions. The bodies of many plants occupy several strata but their vegetative parts are usually in some particular stratum. Land animals are comparable to smaller non-rooted plants such as algæ, lichens and fungi. Many animals carry on different activities in different strata, but are to be classed primarily with the stratum in which they breed (p. 84).

7. Succession of all the animals of the forest communities under consideration is comparable in principle to that in ponds. Succession is due to an increment of changes in conditions produced by the plants and animals living at a given point. Animals through their effect upon the soil play an important though minor part in the process (pp. 73, 75).

8. The various animal species are arranged in these communities in an orderly fashion and the *dominating animal mores* are correlated with the *dominating conditions* (pp. 81, 89–90).

9. Taxonomic (structural) species usually have distinct *mores*, though the same species often has different *mores* under different conditions, and different species may have the same *mores*. *Species* and *mores* are therefore not synonymous (pp. 91–92).

10. Ecology considers together *mores* that are alike or similar in their larger characters (p. 92).

HULL ZOÖLOGICAL LABORATORY,
UNIVERSITY OF CHICAGO,
May 1, 1912.

VII. ACKNOWLEDGMENTS AND BIBLIOGRAPHY.

1. *Acknowledgments.*

The writer is indebted to Mr. Geo. D. Fuller for data and assistance in correlating the work on environmental analysis with that on animal distribution and to Dr. H. C. Cowles for reading the manuscript. He is also indebted to the following persons for identifications and for advice concerning the groups in which they are specialists: Mr. Frank C. Baker, Mollusca; Mr. Nathan Banks, spiders; Mr. O. F. Cook, Myriopods; Mr. Wm. J. Gerhard, Hemiptera; Dr. Joseph L. Hancock, Orthoptera; Mr. Chas. A. Hart, general entomology; Mr. S. F. Hildebrand and Dr. S. E. Meek, vertebrates; Professor H. F. Wickam and Mr. A. B. Wolcott, beetles.

2. *Special Bibliography.*

Literature on the species included in the lists and tables is arranged in the order of first citation. It is not intended to be complete but only to give a lead for finding more literature.

(A) **Shelford, Victor E.**
'08 Life Histories and Larval Habits of the Tiger Beetles. Linn. Soc. (London) Jour. Zoöl., Vol. 30, pp. 157–184.

(B) **Emerton, J. H.**
'02 Common spiders. Boston.

(C) **Hancock, J. L.** As below.

(D) **Smith, J. B.**
'09 Insects of New Jersey. (27th. Rep. of State Board of Agric.) Second edition,—Rept. N. J. State Museum, 1909.

(E) **Marlatt, C. L.**
'02 The White Ant. U. S. Dept. of Agric. Div. of Entomology. Circular 50, 2d Series.

(F) **Peckham, G. W., and Peckham, E. G.**
'98 Instincts and Habits of Solitary Wasps. Wis. Geol. and Nat. Hist. Surv. Bull. 2, Scientific Series 1.

(G) **Howard, L. O.**
'02 Insect Book. New York.

(H) **Ruthven, Alex.**
'11 Amphibians and Reptiles. A Biological Surv. of the Sand Dune Region on the South Shore of Saginaw Bay. Mich. Geol. and Biol. Surv. Publication 4, Biol. Ser. 2, p. 257.

(I) **Wheeler, W. M.** As below.

(J) **Robertson, C.**
99 Flowers and Insects, XIX; Bot. Gaz., XXVIII, 27–45.

(K) **Banks, Nathan.**
'10 Catalogue of Nearctic Spiders. U. S. Nat. Mus. Bull. 72.

(L) **Blatchley, W. S.**
'10 Coleoptera or Beetles of Indiana. Bull. 1, Ind. Dept. of Geol. and Nat. Resources.

(M) **Dickerson, M. C.**
'07 Frog Book. New York.

(N) **Baker, H. B.**
'10 Mollusca: Biological Survey of the Sand Dune Region on the South Shore of Saginaw Bay, Mich. L. c., Ruthven, p. 121.

(O) **Richardson, H.**
'05 Monograph on the Isopods of North America. Bull. 54 U. S. Nat. Mus.

(P) **Baker, F. C.**
'02 The Mollusca of the Chicago Area: Gasteropoda. Chicago Acad. of Sciences. Nat. Hist. Surv. Bull., III., Part II.

(Q) **Bollman, C. H.**
'93 The Myriapoda of North America. U. S. Nat. Mus. Bull. 46.

(R) **Ditmars, R. L.**
Reptile Book. New York.

(S) **Shull, Chas.**
'07 Life History of Habits of *Anthocharis olympia* Edw., Ento. News, Vol. XIX., pp. 73–82.

(T) **Wirtner, P. M.**
'04 Preliminary List of the Hemiptera of Western Pennsylvania. Ann. Carnegie Mus., Vol. III., 133–228.

(U) Kirkaldy, G. W.

'09 Catalogue of the Hemiptera, Heteroptera. Vol. I., Cimicidæ. Berlin.

(V) Felt, E. P.

'06 Insects Affecting Park and Woodland Trees. N. Y. State Education Dept. Mus. Memoir, 8. 2 vols. with good index of species and bibliography in text.

(W) Forbes, S. A.

'05 Insects Injurious to Indiana Corn. 23d Rep. Ill. State Entomologist.

(X) Scudder, S. H.

'89 Butterflies of Eastern U. S. and Canada. Cambridge, Mass.

(Y) Peckham, G. W., and Peckham, E. G.

'95 Sense of Sight in Spiders with Some Observations on Color Sense. Wis. Ac. of Sci., Vol. X., pp. 231–261.

(Z) Beutenmüller, Wm.

'04 Insect Galls within 50 Miles of New York. Guide Leaflet, No. 16, Am. Mus. Nat. Hist.

3. *General Bibliography.*

Abbot, C. C.

'70 Notes on Fresh-water Fishes of New Jersey. Am. Nat., Vol. IV., pp. 99–117.

Adams, C. C.

'08 Isle Royale as a Biotic Environment. An Ecological Survey of Isle Royale, Lake Superior. State Biol. Surv. (Mich.). Published by the Geol. Surv. Lansing.

'08 Ecological Succession of Birds. L. c., pp. 121–154.

'08 The Coleoptera of Isle Royale and their Relation to North American Centers of Dispersal. L. c., pp. 157–191.

'08 Notes on Isle Royale Mammals and their Ecological Relations. L. c., pp. 389–96.

'12 A Hand Book for Students of Animal Ecology (outline). Ill. St. Ac., Trans., Vol. III, 1911, p. 33.

'12 Ecological Studies of Prairie and Forest (summary). Ill. St. Ac., Trans., Vol. III, 1911, p. 33.

Allee, W. C.

'12 Physiological States and Rheotaxis in Isopods. Jour. Exp. Zoöl. July.

Bachmetjew, P.

'99 Ueber die Temperatur der Insekten nach Beobachtungen in Bulgarien. Zeit. f. wiss. Zool., Bd. 66, pp. 521–604.

Baker, F. C.

'02 Mollusca of the Chicago Area. Chi. Acad. of Sci., N. H. Surv., Bull. III., Part II.

'10 Ecology of the Skokie Marsh Area. Bull. Ill. St. Lab. of Nat. Hist., Vol. VIII., pp. 441–96.

Bohn, G.

'10 Naissance de l'Intelligence. Paris (Biblioteque de Philosophie scientifique).

Brehm, A. E.

'96 From North Pole to Equator. (London.) Translation by the Thomsons' introduction with bibliography.

Briggs, L. J., and McLane, J. W.

'07 The Moisture Equivalents of Soils. Bull. 45, Bureau of Soils, U. S. Dept. of Agric.

Briggs, L. J., and Shantz, H. L.
'12 The Wilting Coefficients for Different Plants and their Indirect Determination. Bull. 230, Bureau of Plant Industry, U. S. Dept. Agric.

Brown, Wm.
'10 Evaporation and Plant Habitats in Jamaica. Plant World, XIII., pp. 268-72.

Buffon, G. L. L.
1742 Mémoire sur la culture des forêts. Hist. Acad. Royal. Sci. Paris, pp. 233-46. (Fide Cowles, '11.)

Clements, F. E.
'05 Research Methods of Ecology. Lincoln, Nebr.

Colton, H. S.
'08 Some Effects of Environment on Growth of *Lymnæa Columella* Say. Proc. Ac. Nat. Sci. Phila., pp. 410-48.

Congdon, E. D.
'08 Recent Studies upon the Locomotor Responses of Animals to White Light. Jour. Comp. Neur. and Psych., Vol. 18, pp. 309-28.

Cowles, H. C.
'99 The Ecological Relations of the Vegetation on the Sand Dunes of Lake Mich. Bot. Gaz., XXVII., pp. 95-117, 169-202, 281-308, 361-91.
'01 The Plant Societies of the Vicinity of Chicago. Bull. 2, Geog. Soc. Chicago. Also Bot. Gaz., Vol. XXXI., pp. 73-108, 145-182.
'01 The Influence of Underlying Rocks on the Character of the Vegetation. Bull. Am. Bur. Geog., Vol. II., 1901, pp. 163-176, 376-388.
'11 The Causes of Vegetational Cycles. Bot. Gaz., Vol. XLI., pp. 161-83. Also Ann. Ass. Am. Geog., Vol. I.
'11 A Text-book of Botany. Vol. II., Ecology. New York.

Dachnowski, A.
'08 The Toxic Properties of Bog Water and Bog Soil. Bot. Gaz., Vol. 46, pp. 130-143.

Dahl, F.
'96 Lebenweise Wirbeloser Aasfresser. Sit. K. Preuss. Ak. Wiss. Berlin, pp. 17-30. (Abstract, Jour. Roy. Micro. Soc., 1896, p. 617.)
'03 Winke fur ein Wissenschaftliches Sammeln von Thiere. S. B. d. Ges. Naturforschender Freunde zu Berlin, pp. 444-476.
'03 Kurze Anleitung zum wissenschaftlichen Sammeln und Konservieren von Thiere. Leipzig.

Darwin, Chas.
'92 Earthworms and Vegetable Mould. London.

Dureau de la Malle, A. J. C. A.
'25 Mémoire sur l'alternance ou sur ce problème:—la succession alternative dans la reproduction des especes végétales vivant en sociétés est-elle une loi générale de la nature? Ann. Soc. Nat., Vol. I., 5, pp. 353-81. (Fide Cowles.)

Fuller, G. D.
'11 Evaporation and Plant Succession. Bot. Gaz., Vol. LII., pp. 195-208.
'12 Soil Moisture in the Cottonwood Dune Association of Lake Michigan. Bot. Gaz., Vol. LIII., pp. 512-514.

Ganong. W. F.

'07 Organization of the Ecological Investigation of the Physiological Life-histories of Plants. Bot. Gaz., XLIII., 341–344.

Gleason, H. A.

'08 Ecological Relations of the Invertebrate Fauna of Isle Royale, Mich. State Biol. Surv., pp. 57–78. (Pub. by Geol. Surv. Lansing.)

'10 The Vegetation of the Inland Deposits of Illinois. Bull. Ill. St. Lab. Nat. Hist., Vol. IX., pp. 23, 174.

Greeley, A. W.

'01 On the Analogy between the Effect of Loss of Water and Lowering of Temperature. Am. Jour. Phys., Vol. VI., No. 2.

Hancock, J. H.

'11 Nature Sketches in Temperate America. Chicago.

Hann, J.

'03 Hand Book of Climatology. Part I. Translation R. de C. Ward. N. Y.

Hart, C. A., and Gleason, H. A.

'07 On the Biology of the Sand Areas of Illinois. Bull. Ill. St. Lab. of Nat. Hist., Vol. VII., pp. 137–242.

Herrick, F. H.

'05 The Home-life of Wild Birds. A New Method of the Study and Photography of Birds. New York.

Holmes, S. J.

'11 Evolution of Animal Intelligence. New York.

Jennings, H. S.

'06 Behavior of the Lower Organisms. N. Y. Bibliography.

Livingston, B. E.

'06 The Relation of Desert Plants to Soil Moisture and to Evaporation. Carnegie Inst. of Wash., Publication 50.

'08 Evaporation and Plant Habitats. Plant World, Vol. IX., pp. 1–10.

'10 A Rain Correcting Atmometer for Ecological Instrumentation. Plant World, Vol. XIII., pp. 79–82.

'10 Operation of the Porous Cup Atmometer. Plant World, Vol. XIII., pp. 111–19.

Loeb, J.

'06 Dynamics of Living Matter. New York.

McNutt, W., and Fuller, G. D.

'12 The Range of Evaporation and Soil Moisture in the Oak-Hickory Forest Association of Ill. Trans. Ill. Ac. of Sci., 1912.

Mast, S. O.

'11 Light and the Behavior of Organisms. New York.

Morse, A. P.

'04 Researches on North American Acrididæ. Carnegie Inst. of Wash. Pub. No. 18.

Ruthven, A. G.

'06 An Ecological Survey in the Porcupine Mts. and Isle Royale, Mich. Pub. by the State Board of Geol. Surv. Lansing.

Semper, K.

'81 Animal Life. New York.

Shantz, H. L.

'06 A Study of the Vegetation of the Mesa Region East of Pikes Peak. Part I., The Bouteloua Formation. Bot. Gaz., Vol. 42, p. 179.

Shelford, V. E.

'07 Preliminary Note on the Distribution of the Tiger Beetles (*Cicindela*) and its Relation to Plant Succession. Biol. Bull., Vol. XIV., pp. 9–14.

'10 Ecological Succession of Fish and its Bearing on Fish Culture. Ill. St. Ac. Trans., Vol. II., pp. 108–10.

'11[1] Ecological Succession. I. Stream Fishes and the Method of Physiographic Analysis. Biol. Bull., Vol. XXI., pp. 9–35.

'11[2] Ecological Succession. II. Pond Fishes. Biol. Bull., Vol. XXI., pp. 127–51.

'11[3] Ecological Succession. III. A Reconnaissance of its Causes in Ponds with Particular Reference to Fish. Biol. Bull., Vol. XXII., pp. 1–38.

'11[4] Physiological Animal Geography. Jour. of Morph. (Whitman Volume), Vol. XXII., pp. 551–617.

Sherff, E. E.

'12 The Vegetation of Skokie Marsh. Bot. Gaz., Vol. LIII., pp. 415–435.

Shimek, B.

'11 The Prairies. Bull. Lab. Nat. Hist. State Univ. Iowa, Apr., 1911, pp. 169–240.

Shull, A. F.

'10 Thysanoptera and Orthoptera; A Biological Survey of the Sand Dune Region on the South Shore of Saginaw Bay, Mich. Mich. Biol. and Geol. Surv. Pub. 4, Biol. Ser. 2, pp. 137–232.

Transeau, E. N.

'08 The Relation of Plant Societies to Evaporation. Bot. Gaz., XLV., pp. 217–31.

'10 A Simple Vaporimeter. Bot. Gaz., XLIX., pp. 459–60.

Walker, A. C.

'03 Atmospheric Moisture as a Factor in Distribution. S. E. Nat., VIII., pp. 43–47.

Warming, E.

'09 Ecology of Plants, an Introduction to the Study of Plant Communities. Oxford. (Translation by Percy Groom.)

Wheeler, W. M.

'10 Ants, Their Structure, Development and Behavior. New York.

Yapp, R. H.

'09 Stratification of the Vegetation of a Marsh and its Relation to Evaporation and Temperature. Ann. Bot., Vol. XXIII., pp. 275–319.

Zon, R., and Graves, H. S.

'11 Light in Relation to Tree Growth. U. S. Dept. of Agric. Forest Service Bull. 92. General summary and bibliography.

Wood, E. F.

'10 A Study of the Mammals of Champaign County, Ill. Bull. Ill. State Lab. of Nat. Hist., VIII., Art. V., pp. 501–613.

Woods-Jones, F.

'10 Coral and Atolls. London.

Clements, F. E. 1936. Nature and structure of the climax. *Journal of Ecology* 24:252–284.

Understanding successional processes, regardless of the ecosystem you are working in, is essential for conducting sound, sustained wildlife management. Manipulating successional stages to match a particular species' needs is a fundamental strategy of wildlife habitat management. It is, therefore, important that the process of succession be fully understood. In most cases, wildlife managers fight succession to set it back to earlier seres (e.g., grasslands, shrubland).

Two key papers introduced wildlife biologists to succession: Clements and Gleason (1939). In this work, Clements' objective is to present a solid overview of the terms used in vegetation studies—*climax*, *biome*, *habitat*, *ecotones*, *dominants*, *associations*, and others—and provide biological explanations for each. Furthermore, human influences on vegetation (e.g., fire, overgrazing) have led to even more terms that are useful in ecology, including *proclimax*, *subclimax*, *disclimax*, *preclimax*, *postclimax*, and others that describe climaxes. Clements presents a method for classifying form and function of climax, as necessary as Linnaeus' system of biologic classification of plants and animals.

RELATED READING

Connell, J. H., and R. O. Slatyer. 1977. Mechanisms of succession in natural communities and their role in community stability and organization. The American Naturalist 111:1119–1144.

Holdridge, L. R. 1967. Life zone ecology. Tropical Science Center, San José, Costa Rica.

Horn, H. S. 1974. The ecology of secondary succession. Annual Review of Ecological Systems 5:25–37.

Odum, E. P. 1969. The strategy of ecosystem development. Science 164:262–270.

Shelford, V. E. 1913. Animal communities in temperate America. University of Chicago Press, Chicago, Illinois, USA.

Watt, A. S. 1947. Pattern and process in the plant community. The Journal of Ecology 35:1–22.

Whittaker, R. H. 1953. A consideration of climax theory: the climax as a population and pattern. Ecological Monographs 23:41–78.

Whittaker, R. H. 1962. Classification of natural communities. Botanical Review 28:1–239.

NATURE AND STRUCTURE OF THE CLIMAX

By FREDERIC E. CLEMENTS

(Carnegie Institution of Washington, Santa Barbara, California)

(With Plates VI—XI)

CONTENTS

INTRODUCTION

MORE than a century ago when Lewis and Clark set out upon their memorable journey across the continent of North America (1803–6), they were the first to traverse the great climaxes from deciduous woods in the east through the vast expanse of prairie and plain to the majestic coniferous forest of the north-west. At this time the oak-hickory woodland beyond the Appalachians was almost untouched by the ax except in the neighborhood of a few straggling pioneer settlements, and west of the Mississippi hardly an acre of prairie had known the plow. A few years later (1809), Bradbury states that the boundless prairies are covered with the finest verdure imaginable and will become one of the most beautiful countries in the world, while the plains are of such extent and fertility as to maintain an immense number of animals. It appears probable that at this time no other grassland in the world exhibited such myriads of large mammals belonging to but a few species.

The natural inference has been that the prairies were much modified by the grazing of animals and the fires of primitive man, and this has been reinforced by estimates of the population of each. Seton (1929) concludes that the original number of bison was about 60 million with a probable reduction to 40 million by 1800, and that both the antelope and white-tailed deer were equally abundant, while elk and mule-deer each amounted to not more than 10 million at the maximum. However, these were distributed over a billion or two acres, and the average density was probably never more than a score to the square mile. Estimates of the Indian tribes show the greatest divergence, but it seems improbable that the total population within the grassland ever exceeded a half million. The general habit of migration among the animals further insured that serious effects from overgrazing and trampling were but local or transitory, while the influence of fires set by the Indians was even less significant in modifying the plant cover. As to the forests, those of the north-west were still primeval and in the east they were yet to be changed over wide areas by lumbering and burning on a large scale.

THE CLIMAX CONCEPT

The idea of a climax in the development of vegetation was first suggested by Hult in 1885 and then was advanced more or less independently by several investigators during the next decade or so (cf. Clements, 1916; Phillips, 1935). It was applied to a more or less permanent and final stage of a particular succession and hence one characteristic of a restricted area. The concept of the climax as a complex organism inseparably connected with its climate and often continental in extent was introduced by Clements (1916). According to this view, the climax constitutes the major unit of vegetation and as such forms the basis for the natural classification of plant communities. The relation between climate and climax is considered to be the paramount one, while the intimate bond between the two is emphasized by the derivation of the terms from the same Greek root. In consequence, under this concept climax is invariably employed with reference to the climatic community alone, namely, the formation or its major divisions.

At the outset it was recognized that animals must also be considered members of the climax, and the word *biome* was proposed for the purpose of laying stress upon the mutual roles of plants and animals (Clements, 1916*b*; Clements and Shelford, 1936). With this went the realization that the primary relations to the habitat or ece were necessarily different by virtue of the fact that plants are producents and animals consuments. On land, moreover, plants constitute the fixed matrix of the biome in direct connection with the climate, while the animals bear a dual relation, to plants as well as to climate. The outstanding effect of the one is displayed in reaction upon the ece, of the other in coaction upon plants, which constitutes the primary bond of the biotic community.

Because of its emphasis upon the climatic relation, the term climax has come more and more to replace the word formation, which is regarded as an exact synonym, and this process may have been favored by a tendency to avoid confusion with the geological use. The designation "climatic formation" has now and then been employed, but this is merely to accentuate its nature and to distinguish it from less definite usages. Furthermore, climax and biome are complete synonyms when the biotic community is to be indicated, though climax will necessarily continue to be employed for the matrix when plants alone are considered.

Nature of the climax

This theme has been developed in considerable detail in earlier works (Clements, 1916, 1920, 1928; Weaver and Clements, 1929), as well as in a recent comprehensive treatment by Phillips (1935), and hence a summary account of the major features will suffice in the present place. These may be conveniently grouped under the following four captions, i.e. unity, stabilization and change, origin and relationship, and objective tests.

Unity of the climax

The inherent unity of the climax rests upon the fact that it is not merely the response to a particular climate, but is at the same time the expression and the indicator of it. Because of extent, variation in space and time, and the usually gradual transition into adjacent climates, to say nothing of the human equation, neither physical nor human measures of a climate are adequately satisfactory. By contrast, the visibility, continuity, and sessile nature of the plant community are peculiarly helpful in indicating the fluctuating limits of a climate, while its direct response in terms of food-making, growth and life-form provides the fullest possible integration of physical factors. Naturally, both physical and human values have a part in analyzing and in interpreting the climate as outlined by the climax, but these can only supplement and not replace the biotic indicators.

It may seem logical to infer that the unity of both climax and climate should be matched by a similar uniformity, but reflection will make clear that such is not the case. This is due in the first place to the gradual but marked shift in rainfall or temperature from one boundary to the other, probably best illustrated by the climate of the prairie. In terms of precipitation, the latter may range along the parallel of 40° from nearly 40 in. at the eastern edge of the true prairie to approximately 10 in. at the western border of the mixed grassland, or even to 6 in. in the desert plains and the Great Valley of California. Such a change is roughly 1 in. for 50 miles and is regionally all but imperceptible. The temperature change along the 100th meridian from the mixed prairie in Texas to that of Manitoba and Saskatchewan is even more striking, since only one association is concerned. At the south the average period without killing frost is about 9 months, but at the north it is less than 3, while the mean annual temperatures are 70 and 33° F. respectively. The variation of the two major factors at the extremes of the climatic cycle is likewise great, the maximum rainfall not infrequently amounting to three to four times that of the minimum.

The visible unity of the climax is due primarily to the life-form of the dominants, which is the concrete expression of the climate. In prairie and steppe, this is the grass form, with which must be reckoned the sedges, especially in the tundra. The shrub characterizes the three scrub climaxes of North America, namely, desert, sagebrush, and chaparral, while the tree appears in three subforms, coniferous, deciduous, and broad-leaved evergreen, to typify the corresponding boreal, temperate, and tropical climaxes. The life-form is naturally reflected in the genus, though not without exceptions, since two or more forms or subforms, herb or shrub, deciduous or evergreen, annual or perennial, may occur in the same genus. Hence, the essential unity of a climax is to be sought in its dominant species, since these embody not only the life-form and the genus, but also denote in themselves a definite relation to the climate. Their reactions and coactions are the most controlling both in kind and amount, and thus they determine the conditions under which all the remaining species are associated with them. This is true to a less degree of the animal influents, though their coactions may often be more significant than those of plants.

Stabilization and change

Under the growing tendency to abandon static concepts, it is comprehensible that the pendulum should swing too far and change be overstressed. This consequence is fostered by the fact that most ecological studies are carried out in settled regions where disturbance is the ruling process. As a result, the climax is badly fragmented or even absent over wide areas and subseres are legion. In all such instances it is exceedingly difficult or entirely impossible to strike a balance between stability and change, and it becomes imperative to turn to regions much less disturbed by man, where climatic control is still paramount. It is likewise essential to employ a conceivable measure of time, such as can be expressed in human terms of millennia rather than in eons. No student of past vegetation entertains a doubt that climaxes have evolved, migrated and disappeared under the compulsion of great climatic changes from the Paleozoic onward, but he is also insistent that they persist through millions of years in the absence of such changes and of destructive disturbances by man. There is good and even conclusive evidence within the limitations of fossil materials that the prairie climax has been in existence for several millions of years at least and with most of the dominant species of to-day. This is even more certainly true of forests on the Pacific Coast, owing to the wealth of fossil evidence (Chaney, 1925, 1935), while the generic dominants of the deciduous forests of the Dakota Cretaceous and of to-day are strikingly similar.

It can still be confidently affirmed that stabilization is the universal tendency of all vegetation under the ruling climate, and that climaxes are characterized by a high degree of stability when reckoned in thousands or even millions of years. No one realizes more clearly than the devotee of succession that change is constantly and universally at work, but in the absence of civilized man this is within the fabric of the climax and not destructive of it. Even in a country as intensively developed as the Middle West, the prairie relicts exhibit almost complete stability of dominants and subdominants in spite of being surrounded by cultivation (cf. Weaver and Flory, 1934). It is obvious that climaxes display superficial changes with the season, year or cycle, as in aspection and annuation, but these modify the matrix itself little or not at all. The annuals of the desert may be present in millions one year and absent the next, or one dominant grass may seem prevailing one season and a different one the following year, but these changes are merely recurrent or indeed only apparent. While the modifications represented by bare areas and by seres in every stage are more striking, these are all in the irresistible process of being stabilized as rapidly as the controlling climate and the interference of man permit.

In brief, the changes due to aspection, annuation or natural coaction are superficial, fleeting or periodic and leave no permanent impress, while those of succession are an intrinsic part of the stabilizing process. Man alone can destroy the stability of the climax during the long period of control by its climate, and he accomplishes this by fragments in consequence of a destruction that is selective, partial or complete, and continually renewed.

Origin and relationship

Like other but simpler organisms, each climax not only has its own growth and development in terms of primary and secondary succession, but it has also evolved out of a preceding climax. In other words, it possesses an ontogeny and phylogeny that can be quantitatively and experimentally studied, much as with the individuals and species of plants and animals (*Plant Succession*, 1916, pp. 181, 342). Out of the one has come widespread activity in the investigation of succession, while interest in the other lingers on the threshold, chiefly because it demands a knowledge of the climaxes of more than one continent. With increasing research in these, especially in Europe and Asia, it will be possible to test critically the panclimaxes already suggested (Clements, 1916, 1924, 1929), as well as to determine the origin and relationships of the constituent formations.

This task will also require the services of paleo-ecology for the reconstruction of each eoclimax, which has been differentiated by worldwide climatic changes into the existing units of the panclimax or panformation. As it is, there can be no serious question of the existence of a great hemispheric clisere constituted by the arctic, boreal, deciduous, grassland, subtropical and tropical panclimaxes. Desert formations for the most part constitute an exception and may well be regarded as endemic climaxes evolved in response to regional changes of climate (Clements, 1935).

It is a significant fact that the boreal formations of North America and Eurasia are more closely related than the coniferous ones of the former, but this seeming anomaly is explained by the greater climatic differences that have produced the forests of the Petran and Sierran systems. The five climaxes concerned are relatively well known, and it is possible to indicate their relationships with some assurance, and all the more because of their parallel development on the two great mountain chains. In the case of deciduous forest and grassland, only a single formation of each is present in North America, and the problem of differentiation resolves itself into tracing the origin and relationship of the several associations. It has been suggested that the mixed prairie by virtue of its position, extent and common dominants represents the original formation in Tertiary times, an assumption reinforced by its close resemblance to the steppe climax (Clements, 1935). It is not improbable that the mixed hardwoods of the southern Appalachians bear a similar relation to the associations of the modern deciduous forest (Braun, 1935).

Tests of a climax

As has been previously indicated, the major climaxes of North America, such as tundra, boreal and deciduous forest, and prairie, stand forth clearly as distinct units, in spite of the fact that the prairie was first regarded as comprising two formations, as a consequence of the changes produced by overgrazing. The other coniferous and the scrub climaxes emerge less distinctly because of the greater similarity of life-form within each group, and hence it is necessary to appeal to criteria derived from the major formations just mentioned. This insures uniformity of basis and a high degree of objectivity, both of which are qualities of paramount importance for the natural classification of biomes. In fact, entire consistency in the application of criteria is the best warrant of objective results, though this is obviously a procedure that demands a first-hand acquaintance with most if not all the units concerned and over a large portion of their respective areas.

The primary criterion is that afforded by the vegetation-form, as is illustrated by the four major climaxes. The others of each group, such as coniferous forest or scrub, are characterized also by the same form in the dominants, but this is not decisive as between related climaxes and hence recourse must be taken to the other tests. The value of the life-form is most evident where two climaxes of different physiognomy are in contact, as in the case of the lake forest of pine-hemlock and the deciduous forest of hardwoods. The static view would make the hemlock in particular a dominant of the deciduous formation, but the evidence derived from the vegetation-form is supported by that of phylogeny and by early records of composition and timber-cut to show that two different climaxes are concerned. Secondary forms or subforms rarely if ever mark distinctions between climaxes, but do aid in the recognition of associations. This is well exemplified by the tall, mid and short grasses of the prairie and somewhat less definitely by the generally deciduous character of the Petran chaparral and the typically evergreen nature of the Sierran.

As would be expected, the most significant test of the unity of a formation is afforded by the presence of certain dominant species through all or nearly all of the associations, but often not in the role of dominants by reason of reduced abundance. Here again perhaps the best examples are furnished by prairie and tundra, though the rule applies almost equally well to deciduous forest and only less so to coniferous ones because of a usually smaller number of dominants. For the prairie, the number of such species, or *perdominants*, found in all or all but one or two of the five associations is eight, namely, *Stipa comata*, *Agropyrum smithi*, *Bouteloua gracilis*, *Sporobolus cryptandrus*, *Koeleria cristata*, *Elymus sitanion*, *Poa scabrella* and *Festuca ovina*. Even when a species is lacking over most of an association, as in the case of *Stipa comata* and the true prairie, it may be represented by a close relative, such as *S. spartea*, which is probably no more than a mesic variety of it. As to the three associations of the deciduous forest, a still larger number of dominant species occur to some degree in all; the oaks comprise *Quercus borealis, velutina, alba, macrocarpa, coccinea, muhlenbergi, stellata* and *marilandica,* and the hickories, *Carya ovata, glabra, alba* and *cordiformis.*

It was the application of this test by specific dominants that led to the recognition of the two climaxes in the coniferous mantle of the Petran and

Sierran cordilleras. The natural assumption was that such a narrow belt could not contain more than one climax, especially in view of its physiognomic uniformity, but this failed to reckon with the great climatic differences of the two portions and the corresponding response of the dominants. The effect of altitude proved to be much more decisive than that of region, dominants common to the montane and subalpine zones being practically absent, though the rule for the same zone in each of the two separate mountain systems. Long after the presence of two climaxes had been established, it was found that Sargent had anticipated this conclusion, though in other terms (1884, p. 8).

As would be inferred, the dominants of related associations belong to a few common genera for the most part. Thus, there are a dozen species of *Stipa* variously distributed as dominants through the grassland, nearly as many of *Sporobolus*, *Bouteloua* and *Aristida*, and several each of *Poa*, *Agropyrum*, *Elymus*, *Andropogon*, *Festuca* and *Muhlenbergia*. In the deciduous forest, *Quercus*, *Carya* and *Acer* are the great genera, and for the various coniferous ones, *Pinus*, *Abies* and *Picea*, with species of *Tsuga*, *Thuja*, *Larix* and *Juniperus* hardly less numerous.

The perennial forbs that play the part of subdominants also possess considerable value in linking associations together, and to a higher degree in the deciduous forest than in the prairie, owing chiefly to the factors of shade and protection. Over a hundred subdominants belonging to two score or more of genera, such as *Erythronium*, *Dicentra*, *Trillium*, *Aquilegia*, *Arisaema*, *Phlox*, *Uvularia*, *Viola*, *Impatiens*, *Desmodium*, *Helianthus*, *Aster* and *Solidago*, range from Nova Scotia or New England beyond the borders of the actual climax to Nebraska and Kansas. Across the wide expanse of the prairie climax, species in common are only exceptional, these few belonging mostly to the composites, notably *Grindelia squarrosa*, *Gutierrezia sarothrae*, *Artemisia dracunculus* and *vulgaris*. On the other hand, the number of genera of subdominants found throughout the grassland is very large.

The greater mobility of the larger mammals in particular renders animal influents less significant than plants as a criterion, but several of these possess definite value and the less mobile rodents even more. The antelope and bison are typical of the grassland climax, the first being practically restricted to it, while jack-rabbits, ground-squirrels and kangaroo-rats are characteristic dwellers in the prairie, as is their chief foe, the coyote.

The remaining criteria are derived from development directly or indirectly, though this is less evident in the case of the ecotone between two associations. Here the mixing of dominants and subdominants indicates their general similarity in terms of the formation, within which range their preferences assign them to different associations. The evidence from primary succession is of value only in the later stages as a rule, since initial associes like the reed-swamp may occur in several climaxes. With subseres, however, all or nearly all the stages are related to the particular climax and such seres denote a

17–2

corresponding unity in development. This is especially true of all subclimaxes and most evidently in the case of those due to fire. More significant still are postclimaxes in both grassland and forest. For example, the associes of species of *Andropogon*, which is subclimax to the oak-hickory forest, constitutes a postclimax to five out of the six associations of the prairie. On the other hand, the community of *Ulmus*, *Juglans*, *Fraxinus*, etc., found on flood-plains through the region of deciduous forest, forms a common subclimax to the three associations.

In addition to such ontogenetic criteria, phylogeny supplies tests of even greater value. This is notably the case with the two associations of the montane and subalpine coniferous forests of the west, though perhaps the most striking application of this criterion is in connection with the lake forest of pine-hemlock. Though the concrete evidence for such a climax recurs constantly through the region of the Great Lakes to the Atlantic, it is fragmentary and there is no evident related association to the westward. However, the four genera are represented by related species in the two regions, namely, *Pinus strobus* by *P. monticola*, *P. banksiana* by *P. contorta*, *Tsuga canadensis* by *T. heterophylla*, *Larix laricina* by *L. occidentalis*, and *Thuja occidentalis* by *T. plicata*, though the last two genera have changed from a subclimax role in the east to a climax one in the west. As suggested earlier, phylogenetic evidence of still more direct nature is supplied by the mixed prairie with the other enclosing associations and by the remnants of a virgin deciduous forest that exhibits a similar genetic and spatial relation to the associations of this climax (cf. Braun, 1935).

Finally, it is clear that any test will gain in definiteness and accuracy of application whenever dependable records are available with respect to earlier composition and structure. These may belong entirely to the historical period, as in the case of scientific reports or land surveys, they may bridge the gap between the present and the past as with pollen statistics, or they may reach further back into the geological record, as with leaf-impressions or other fossils (Chaney, 1925, 1933; Clements, 1936). Two instances of the scientific record that are of the first importance may be given as examples. The first is the essential recognition by Sargent of the pine-hemlock climax under the name of the northern pine belt (1884), at a time when relatively little of this had been logged, by contrast with 90 per cent. or more at present (cf. also Bromley, 1935). The second is an account, discovered and communicated by Dr Vestal, of the prairies of Illinois as seen by Short in *ca.* 1840. This is of heightened interest since its discovery followed little more than a year after repeated field trips had led to the conclusion that all of Iowa, northern Missouri and most of Illinois were to be assigned to the true prairie,[1] a decision

---

[1] The true prairie is characterized by the three eudominants, *Stipa spartea*, *Sporobolus asper* and *S. heterolepis*. The presence of tall-grasses in it to-day, particularly *Andropogon furcatus* and *nutans*, is the mark of the disclimax due to the varied disturbances associated with settlement.

Phot. 1. Sierran subalpine climax: Consociation of *Tsuga mertensiana*; Crater Lake, Oregon.

Phot. 2. Proclimax of Sagebrush (*Artemisia tridentata*), a disclimax due to overgrazing of mixed prairie; climatically a preclimax.

CLEMENTS—Nature and Structure of the Climax

Face p. 261

confirmed for Illinois by Short's description, and supported by the more general accounts of Bradbury (*ca.* 1815) and Greeley (1860).

CLIMAX AND PROCLIMAX
Essential relations

In accordance with the view that development regularly terminates in the community capable of maintaining itself under a particular climate, except when disturbance enters, there is but one kind of climax, namely, that controlled by climate. This essential relation is regarded as not only inherent in all natural vegetation, but also as implicit in the cognate nature of the two terms. While it is fully recognized that succession may be halted in practically any stage, such communities are invariably subordinate to the true climax as determined by climate alone. From the very meaning of the word, there can not be climaxes scattered along the developmental route with a genuine climax at the end. There is no intention to question the reality of such pauses, but only to emphasize the fact that they are of a different order from the climax.

While it is natural to express new ideas by qualifying an old term, this does not conduce to the clearest thinking or the most accurate usage. Even more undesirable is the fact that the meaning of the original word is gradually shifted until it becomes either quite vague or hopelessly inclusive. At the hands of some, climax has already suffered this fate, and fire, disease, insects, and human disturbances of all sorts are assumed to produce corresponding climaxes (cf. Chapman, 1932). On such an assumption corn would constitute one climax, wheat another, and cotton a third, and it would then become imperative to begin anew the task of properly analyzing and classifying vegetation.

In the light of two decades of continued analysis of the vegetation of North America, as well as the application of the twin concepts of climax and complex organism by workers in other portions of the globe and the strong support brought to them by the rise of emergent evolution and holism (Phillips, 1935), the characterization of the climax as given in *Plant Succession*, in 1916, still appears to be both complete and accurate. "The unit of vegetation, the climax formation, is an organic entity. As an organism, the formation arises, grows, matures and dies. Its response to the habitat is shown in processes or functions and in structures that are the record as well as the result of these functions. Furthermore, each climax formation is able to reproduce itself, repeating with essential fidelity the stages of its development. The life-history of a formation is a complex but definite process, comparable in its chief features with the life-history of an individual plant. The climax formation is the adult organism, of which all initial and medial stages are but stages of development....A formation, in short, is the final stage of vegetational development in a climatic unit. It is the climax community of a succession that terminates in the highest life-form possible in the climate concerned."

To-day this statement would need modification only to the extent of substituting "biome" for climax or formation and "biotic" for vegetational. This characterization has recently been annotated and confirmed by Phillips' masterly discussion of climax and complex organism, as cited above, a treatise that should be read and digested by everyone interested in the field of dynamic ecology and its wide applications.

Proclimaxes

As a general term, proclimax includes all the communities that simulate the climax to some extent in terms of stability or permanence but lack the proper sanction of the existing climate. Certain communities of this type were called potential climaxes in *Plant Succession* (p. 108; 1928, p. 109), and two kinds were distinguished, namely, preclimax and postclimax. To avoid proposing a new term in advance of its need, subclimax was made to do double duty, denoting both the subfinal stage of succession, as well as apparent climaxes of other kinds. This dual usage was criticized by Godwin (1929, p. 144) and partially justified by Tansley in an appended note on the ground just given. However, this discussion made it evident that a new term was desirable and proclimax was accordingly suggested (Clements, 1934). While this takes care of the use of subclimax in the second sense noted above, it is better adapted by reason of its significance to apply to all kinds of subpermanent communities other than the climax proper. However, there is still an important residuum after subclimax, preclimax and postclimax have been recognized, and it is proposed to call these *disclimaxes*, as indicated later.

The proclimax may be defined as any more or less permanent community resembling the climax in one or more respects, but gradually replaceable by the latter when the control of climate is not inhibited by disturbance. Besides its general function, it may be used as a synonym for any one of its divisions, as well as in cases of doubt pending further investigation, such as in water climaxes. The four types to be considered are subclimax, disclimax, preclimax and postclimax.

Subclimax

As the stage preceding the climax in all complete seres, primary and secondary, the subclimax is as universal as it is generally well understood. The great majority of such communities belong to the subsere, especially that following fire, owing to the fact that disturbance is to-day a practically constant feature of most climaxes. Fire and fallow are recurrent processes in cultivated regions generally and they serve to maintain the corresponding subsere until protection or conversion terminates the disturbance. Though the subclimax is just as regular a feature of priseres, these have long ago ended in the climax over most of the climatic area and the related subclimax communities are consequently much restricted in size and widely scattered. Smallness is

Phot. 3. Fire subclimax of dwarf *Pinus* and *Quercus*; the "Plains",
New Jersey Pine-barrens.

Phot. 4. Subclimax of consocies of *Aristida purpurea* in field
abandoned 15 years; Great Plains.

CLEMENTS—NATURE AND STRUCTURE OF THE CLIMAX

Face p. 263

naturally a characteristic of nearly all subclimaxes, the chief exceptions being due to fire or to fire and logging combined, but by contrast they are often exceedingly numerous.

Because of its position in the succession, the subclimax resembles the preclimax in some respects and in a few instances either term may be properly applied. The distinction between subclimax and disclimax presents some difficulty now and then, as the amount of change necessary to produce the latter may be a matter of judgment. This arises in part also from the structural diversity of formation and association, as a consequence of which the dominants of a particular type of subsere vary in different areas. When there is but a single dominant, as in many burn subclimaxes, no question ensues, but if two or more are present, the decision between subclimax and disclimax may be less simple, as is not infrequent in scrub and grassland.

Examples of the subclimax are legion, the outstanding cases being mostly due to fire, alone or after lumbering or clearing. Most typical are those composed of "jack-pines" or species with closed cones that open most readily after fire. Each great region has at least one of these, e.g. *Pinus rigida, virginiana* and *echinata* in the east, *P. banksiana* in the north, *P. murrayana* in the Rocky Mountains, and *P. tuberculata, muricata* and *radiata* on the Pacific Slope. *Pinus palustris* and *taeda* play a similar role in the "piney" woods of the Atlantic Gulf region, as does *Pseudotsuga taxifolia* in the north-west. The characteristic subclimaxes of the boreal forest are composed of aspen (*Populus tremuloides*), balsam-poplar (*P. balsamifera*), and paper-birch (*Betula papyrifera*), either singly or in various combinations. Aspen also forms a notable subclimax in the Rocky Mountains, for the most part in the subalpine zone. Prisere subclimaxes are regular features of bogs and muskeags throughout much or all of the boreal and lake forests, the three dominants being *Larix laricina, Picea mariana* and *Thuja occidentalis*, often associated as zonal consocies. Where pines are absent in the region of the deciduous forest, two xeric oaks, *Quercus stellata* and *marilandica*, may constitute a subclimax, and this role is sometimes assumed by small trees, *Sassafras, Diospyrus* and *Hamamelis* being especially important.

Subclimaxes in the grassland are composed largely of tall-grasses, usually in the form of a consocies. In the true prairie, this part is taken by *Spartina cynosuroides* and in the desert plains by *Sporobolus wrighti*, while *Elymus condensatus* plays a similar role in the mixed prairie and in portions of the bunch-grass prairies. The function of the tall Andropogons is more varied; they are typically postclimax rather than subclimax, though they maintain the latter relation along the fringe of the oak-hickory forest and in oak openings. They occupy a similar position at the margin of the pine subclimax in Texas especially, and hence they are what might be termed "sub-subclimax" in such situations. Beyond the forest and in association with *Elionurus, Trachypogon*, etc., they appear to constitute a faciation of the coastal prairie. Chaparral proper is to be regarded as a climax, but with a change of species it extends into the montane and even the subalpine zone and there constitutes a fire subclimax. In the foothills of southern California, the coastal sagebrush behaves in like manner where it lies in contact with the chaparral.

The disposition of seral stages below the subclimax that exhibit a distinct retardation or halt for a longer or shorter period is a debatable matter. It is entirely possible to include them among subclimaxes, but this would again fail in accuracy and definiteness and hence lead to confusion. The decision may well be left to usage by providing a term for such seral or sub-subclimax communities as persist for a long or indefinite period because of continued or recurrent edaphic control or human disturbance. By virtue of its significance, brevity and accord with related terms, the designation "serclimax" is suggested, with the meaning of a seral community usually one or two stages before the subclimax, which persists for such a period as to resemble the climax in this one respect. For reasons of brevity and agreement, the connecting vowel is omitted, but the *e* remains long as in sere.

For the most part, serclimaxes are found in standing water or in saturated soils as a consequence of imperfect drainage. The universal example is the reed-swamp with one or more of several consocies, such as *Scirpus, Typha, Zizania, Phragmites* and *Glyceria*: this is typical of the lower reaches of rivers, of deltas and of certain kinds of lakes, the great tule swamps of California affording outstanding instances. Another type occurs in coastal marshes in which *Spartina* is often the sole or major dominant, while sedge-swamps have a wider climatic range but are especially characteristic of northern latitudes and high altitudes. The Everglades of Florida dominated by *Cladium* constitute perhaps the most extensive example of the general group, though *Carex* swamps often cover great areas and the grass *Arundinaria* forms jungle-like cane-brakes through the south. Among woody species, *Salix longifolia* is an omnipresent consocies of sand-bars and river-sides, but the most unique exemplar is the cypress-swamp of the south, typified by *Taxodium*. In boreal and subalpine districts the distinctive serclimax is the peat-bog, moor or muskeag, more or less regularly associated with other seral communities of *Carex* and usually of *Larix* or *Picea* also in the proper region.

Frequent burning may retard or prevent the development of the normal fire subclimax and cause it to be replaced by a preceding stage. This may be a scrub community or one kept in the shrub form by repeated fires, but along the Atlantic and Gulf Coasts it is usually one of *Andropogon virginicus*, owing to its sufferance of burning. The so-called "balds" of the southern Appalachians are seral communities of heaths or grasses initiated and maintained primarily by fire. Finally, there are serclimaxes of weeds, especially annuals, in cultivated districts, and a somewhat similar community of native annuals is characteristic of wide stretches in the desert region.

Phot. 5. Serclimax of *Taxodium*, *Nyssa* and *Quercus* forming a
cypress swamp; Paris, Arkansas.

Phot. 6. Disclimax of *Bouteloua*, *Muhlenbergia* and *Opuntia*, due to
overgrazing of mixed prairie, Great Plains.

CLEMENTS—NATURE AND STRUCTURE OF THE CLIMAX

Face p. 265

Disclimax

As with the related concepts, the significance of this term is indicated by a prefix, *dis-*, denoting separation, unlikeness or derogation, much as in the Greek *dys*, poor, bad. The most frequent examples of this community result from the modification or replacement of the true climax, either as a whole or in part, or from a change in the direction of succession. These ensue chiefly in consequence of a disturbance by man or domesticated animals, but they are also occasionally produced by mass migration. In some cases, disturbance and the introduction of alien species act together through destruction and competition to constitute a quasi-permanent community with the general character of the climax. This type is best illustrated by the *Avena-Bromus* disclimax of California, which has all but completely replaced the bunch-grass prairie.[1] A similar replacement by *Bromus tectorum* has more recently taken place over large areas of the Great Basin, while *Poa pratensis* has during the last half-century steadily invaded the native hay-fields and pastures of the true prairie, an advance first noted by Bradbury in 1809. An even more striking phenomenon is the steadily increasing dominance of *Salsola* over range and crop land in the west, and this is imitated by *Sisymbrium* and *Lepidium* in the north-west. It is obvious that all cultivated crops belong in the same general category, but this point hardly requires consideration.

Probably the example most cited in North America is that of the short-grass plains, which actually represent a reduction of the mixed prairie due to overgrazing, supplemented by periodic drouth. Over most of this association, the mid-grasses, *Stipa*, *Agropyrum*, etc., are still in evidence, though often reduced in abundance and stature, but in some areas they have been practically eliminated. Similar though less extensive partial climaxes of short-grasses characterise pastures in the true and both pasture and range in the coastal prairie, the dominants regularly belonging to *Bouteloua*, *Buchloe* or *Hilaria*. Of essentially the same nature is the substitution of annual species of *Bouteloua*

---

[1] The grassland climax of North America comprises six well-marked associations (*Plant Indicators*, 1920; *Plant Ecology*, 1929). The mixed prairie, so-called because it is composed of both mid-grasses and short-grasses, is more or less central to the other five and is regarded as ancestral to them. To the east along the Missouri and Mississippi Rivers, it has become differentiated into the true prairie formed by other species of mid-grasses pertaining mostly to the same genera, and this unit is flanked along the western margin of the deciduous forest by a proclimax of tall grasses, chiefly *Andropogon*. Southward the true prairie is replaced by coastal prairie, which in the main occupies the Gulf region of Texas and Mexico and is constituted by similar dominants but of different species. The desert plains are characterized primarily by species of *Bouteloua* and *Aristida*, which range from western Texas to the edge of the deserts of Mexico and Arizona. In the north-west the short-grasses disappear and the Palouse prairie of eastern Washington and adjacent regions is formed by mid-grasses of the bunch-grass life-form, among which *Agropyrum spicatum* is the eudominant. The same life-form signalizes the California prairie, found from the northern part of the state southward into Lower California, but its especial character is derived from endemic species of *Stipa*. As indicated in the discussion, the short-grass plains, composed of *Bouteloua*, *Buchloe*, and *Carex*, are not climatic in nature, and this statement applies likewise to the tall-grass meadows of *Andropogon* mentioned above.

and *Aristida* in the desert plains for perennial ones of the same genera, which is a case of short-grasses being followed by still shorter ones.

In other instances, the effect of disturbance is to produce a community with the appearance of a postclimax, when the life-form concerned is that of an undershrub or tall grass. This is notably the case in the mixed prairie when overgrazing is carried to the point of breaking up the short-grass sod and permitting the dominance of *Artemisia frigida* or *Gutierrezia sarothrae*. In essence, the wide extension of sagebrush (*Artemisia tridentata*) and of creosote-bush (*Larrea tridentata*) is the same phenomenon, though each of these is a climax dominant in its own region. In the case of *Opuntia*, the peculiar life-form suggests an important difference, but the numerous species behave in all significant respects like other shrubs, though with the two advantages of spines and ready propagation.

The communities of tall-grasses formed by species of *Andropogon* originally presented some difficulty, since these naturally have all the appearance of a postclimax to the prairie. Probably the greater number are to be assigned to this type, but the evidence from reconnaissance and record indicates that in the true prairie and especially the eastern portion, *Andropogon furcatus* in particular now constitutes a disclimax due to pasturing, mowing and in some measure to fire also (Clements, 1933). A characteristic disclimax in miniature is to be found in the "gopher gardens" of the alpine tundra, where coaction and reaction have removed the climax dominants of sedges and grasses to make place for flower gardens of perennial forbs. "Towns" of prairie-dogs and kangaroo-rats often produce similar but much more extensive communities.

Selective cutting not infrequently initiates disclimaxes, as may likewise the similar action of other agents such as fire or epidemic disease. The most dramatic example is the elimination of the chestnut (*Castanea dentata*) from the oak-chestnut canopy, but of even greater importance has been the extreme reduction and fragmentation of the lake forest through the overcutting of white pine. Finally, what is essentially a disclimax may result from climatic mass migration, such as in the Black Hills of South Dakota has brought together *Pinus ponderosa* from the montane climax of the Rocky Mountains and *Picea canadensis* from the boreal forest.

Preclimax and postclimax

These related concepts were first advanced in *Plant Succession* (1916, 1928) and have since been discussed in *Plant Ecology* (1929) and in the organisation of the relict method (1934). They are both direct corollaries of the principle of the clisere, the spatial series of climaxes that are set in motion by a major climatic shift, such as that of the glacial epoch with its opposite phases. The clisere is most readily comprehended in the case of high ranges or summits, such as Pikes Peak where the entire series of climaxes is readily visible, and is what Tournefort described on Mount Ararat in his famous journey of 1700.

However, this is but an expression of the continental clisere in latitude, which achieves perhaps its greatest regularity in North America. A similar relation is characteristic of the longitudinal disposition of climaxes in the temperate zone between the two oceans, the portion from deciduous forest through prairie to desert being the most uniform.

With the exception of the two extremes, arctalpine and tropical, each climax has a dual role, being preclimax to the contiguous community of socalled higher life-form and postclimax to that of lower life-form. This may be illustrated by the woodland climax, which is postclimax to grassland and preclimax to montane forest. The arctic and alpine tundras exhibit only the preclimax relation, to boreal and subalpine forest respectively, since a potential lichen climax attains but incomplete expression northward or upward. While the general primary relation is one of water in terms of rainfall and evaporation, temperature constantly enters the situation and at the extremes may be largely controlling, as in the tundra especially. However, in our present imperfect knowledge of causal factors it is simpler and more definite to determine rank by position in the cliseral sequence, each community higher in altitude or latitude being successively preclimax to the preceding one. This relation is likewise entirely consistent in the clisere from deciduous forest to desert, as it is among the associations of the same climax, though in both these cases the zonal grouping may be more or less obscured.

Wherever concrete preclimaxes or postclimaxes occur, either between climaxes or within a single one, they are due to the compensation afforded by edaphic situations. The major examples of the latter are provided by valleys, especially gorges and canyons, long and steep slope-exposures, and by extreme soil-types such as sand and alkali. The seration is a series of communities produced by a graduated compensation across a valley and operating within a formation or through adjacent ones, while the ecocline embraces the differentiation brought about by shifting slope-exposures around a mountain or on the two sides of a high ridge. In the case of such soils as sand or gravel at one extreme and stiff clay at the other, the edaphic adjustment may sometimes appear contradictory. Thus, sand affords a haven for postclimax relicts in the dry prairie and for preclimax ones in the humid forest region, while the effect of heavy soils is just the reverse. However, this is readily intelligible when one recalls the peculiar properties of such soils in terms of absorption, chresard and evaporation (Clements, 1933).

Preclimax

Since they occupy the same general antecedent position with respect to the climax, it is necessary to distinguish with some care between subclimax and preclimax, especially in view of the fact that they often exhibit the same lifeform. However, this is not difficult when the priseres and subseres have been investigated in detail, as the actual composition and behavior of the two communities are usually quite different. Moreover, in the first, reaction leads to the entry of the climax dominants with ultimate conversion, while in the second the compensation by local factors is rarely if ever to be overcome within the existing climate, short of man-made disturbance.

Preclimaxes are most clearly marked where two adjacent formations are concerned, either prairie and forest or desert and prairie. Examples of the first kind are found in the grassy "openings" and oak savannahs of the deciduous forest and in the so-called "natural parks" along the margin of the montane and boreal forests. They are also well developed on warm dry slope-exposures or xeroclines in the Rocky Mountains. In the one, compensation is usually afforded by a sandy or rocky soil, in the other by a local climate due to insolation. Desert climaxes regularly bear the proper relation to circumjacent grassland, but this is somewhat obscured by the shrub life-form, which would be expected to characterize the less xeric formation. This may be explained, however, by the wide capacity for adaptation shown by such major dominants as *Larrea tridentata* and *Artemisia tridentata*, a quality that is lacking in most of their associates. Left stranded as relict communities in desert plains and mixed prairie by the recession of the last dry phase, they have profited by the overgrazing of grasses to extend across a territory much larger than that in which they are climax. Here they have all the appearance of a postclimax, especially in the case of *Larrea*, which commonly attains a stature several times that found in the desert. However, since this is the direct outcome of disturbance in terms of grazing, it is better regarded as a disclimax, particularly since the climax grasses still persist in it to some degree.

Within the same formation, the more xeric associations or consociations are preclimax to the less xeric ones. This is the general relation between the oak-hickory and beech-maple associations of the deciduous forest, the former occupying in the latter the warmer drier sites produced by insolation or type of soil. A similar relation may obtain in the case of faciations, the *Quercus stellata-marilandica* community often being a border of marginal preclimax to the more mesic oak-hickory faciations. Such preclimaxes naturally persist beyond the limits of the association proper as relicts in valleys or sandy soils and then assume the role of postclimaxes to the surrounding grassland, a situation strikingly exemplified in the "Cross Timbers" of Texas. In the montane forest of the Rockies, the consociation of *Pinus ponderosa* is preclimax to that of *Pseudotsuga taxifolia*, and a similar condition recurs in all forests where there is more or less segregation of consociations.

In the mixed prairie, fragments of the desert plains occur all along the margin as preclimaxes, the most extensive one confronting the Colorado Valley, where it is at the same time postclimax to the desert. The mixed prairie constitutes relicts of this type where it meets the true prairie. The most frequent examples are provided by *Bouteloua gracilis* and *Sporobolus cryptandrus*, though as with all the short-grasses in this role, grazing has played some part.

Phot. 7. Postclimax of *Quercus, Juglans, Ulmus, Fraxinus*, etc.,
Canadian River, near Oklahoma City.

Phot. 8. Postclimax of tall-grasses, *Andropogon, Calamovilfa* and
Panicium in sandhills; Thedford, Nebraska.

CLEMENTS—NATURE AND STRUCTURE OF THE CLIMAX

Postclimax

As a general rule, postclimax relicts are much more abundant than those that represent preclimaxes, owing in the first place to the secular trend toward desiccation in climate and in the second to the large number of valleys, sandhills and sandy plains, and escarpments in the grassland especially. Postclimaxes of oak-hickory and of their flood-plain associates, elm, ash, walnut, etc., are characteristic features of the true and mixed prairies, holding their own far westward in major valleys but limited as outliers on ridges and sandy stretches to the eastern edge. However, the compensation afforded by the last two is incomplete as a rule and the postclimax is typically reduced to the savannah type. The latter is an almost universal feature where forest, woodland or chaparral touches grassland, owing to the fact that shrinkage under slow desiccation operates gradually upon the density and size of individuals. Savannah is derived from the reduction of deciduous forest along the eastern edge of the prairie, of the aspen subclimax of the boreal forest along the northern, and of the montane pine consociation, woodland or chaparral on the western and southern borders, recurring again on the flanks of the Sierras and Coast Ranges in California. On the south, the unique ability of the mesquite (*Prosopis juliflora*) to produce root-sprouts after fire, its thorniness, palatable pods and resistant seeds have permitted it to produce an extensive savannah that often closely simulates a true woodland climax.

As would be expected, a point is reached in the reduction of rainfall westward in the prairie where sand no longer affords compensation adequate for trees. In general this is along the isohyet of 30 in. in the center and south, and of about 20 in. in the north. Southward from the parallel of 37° the further shrinkage of the oak savannah may be traced in the "shinry", which dwindles from four or five feet to dwarfs only "shin" high. With these are associated tall-grasses, principally *Andropogon* and *Calamovilfa* in the form *gigantea*. To the north of this line, the shin oaks are absent and the tall-grasses make a typical postclimax that extends into Canada, though the compensatory influence of sand is still sufficient to permit an abundance of such low bushes as *Amorpha, Ceanothus, Artemisia filifolia* and *Yucca*, as well as depauperate hackberry and aspen. In the vast sandhill area of central Nebraska, the tall-grass postclimax attains its best development, which is assumed to reflect the climate when the prairies were occupied by the Andropogons and their associates some millions of years ago. The gradual decrease to the rainfall of the present has led to the tall-grasses finding refuge in all areas of edaphic compensation, not only in sand but likewise on foothills and in valleys, and in addition along the front of the deciduous forest.

STRUCTURE OF THE CLIMAX
Community functions

The nature of community functions and their relation to the structure of climax and sere have been discussed in considerable detail elsewhere (*Plant Succession*, 1916, 1928; *Plant Ecology*, 1929; *Bio-ecology*, 1936), and for the present purpose it may well suffice to emphasize the difference in significance between major or primary and minor or secondary functions. The former comprise aggregation, migration, ecesis, reaction, competition, cooperation, disoperation, and coaction, together with the resulting complexes, invasion and succession. Any one of these may have a profound effect upon community structure, but the driving force in the selection and grouping of life-forms and species resides chiefly in reaction, competition, and coaction. Migration deals for the most part with the movement and evolution of units under climatic compulsion, and succession with the development and regeneration of the climax in bare or denuded areas.

In contrast to these stands the group of minor functions that are concerned with numbers and appearance or visibility as it may be termed. The first is annuation, in accordance with which the abundance of any species may fluctuate from dry to wet phases of the various climatic cycles or the growth differ in terms of prominence, the two effects not infrequently being combined. For the grassland, a season of rainfall more or less extreme in either direction often emphasizes one dominant at the expense of others, though the balance is usually redressed by the following year, while in the desert in particular the swing in number of annuals may be from almost complete absence to seasonal dominance, again with one or few species taking the major role. Aspection is mainly the orderly procession of societies through each growing season, more or less modified by changes in number ensuing from annuation. Hibernation and estivation merely affect seasonal appearance and are forms of aspection, with the temporary suspension of coaction effects. While usually applied to the animal members of the biome, it is obvious that plants exhibit certain responses of similar nature. Diurnation is likewise best known in the case of animals, especially nocturnal ones, but it is exhibited also by the vertical movement of plankton and in different form by the opening and closing of flowers and the "sleep" movements of leaves.

Roles of constituent species: dominants

The abundant and controlling species of characteristic life-form were long ago termed dominants (Clements, 1907, 1916), this property being chiefly determined by the degree of reaction and effective competition. In harmony with the concept of the biome, it has become desirable to consider the role of animals likewise; since their influence is seen chiefly in coaction by contrast to the reaction of plants, the term *influent* has been applied to the important species of land biomes (cf. Clements and Shelford, 1936). It is an axiom that the life-form of the dominant trees stamps its character upon forest and woodland, that of the shrub upon chaparral and desert, and the grass form on prairie, steppe and tundra. There are seral dominants as well as climax ones, and these give the respective impresses to the stages of prisere and subsere. Finally, there are considerable differences in rank or territory even among the dominants of each formation. The most important are those of wide range that bind together the associations of a climax; to these the term *perdominant* (*per*, throughout) may well be applied. In contrast to these stand the dominants more or less peculiar to each association, such as beech or chestnut in their respective communities and *Sporobolus asper* in the true and *Stipa comata* in the mixed prairie, for which *eudominant* may be employed.

Subdominants regularly belong to a life-form different from that of the dominants and are subject to the control of the latter in a high degree, as the name indicates. They are best exemplified by the perennial forbs, though biennials and annuals may serve as seral subdominants; all three may be actual dominants in the initial stages of succession and especially in the subsere. The term *codominant* has so far had no very definite status; it is hardly needed to call attention to the presence of two or more dominants, since this is the rule in all cases with the exception of consociation and consocies. In contrast to the types mentioned stands a large number of secondary or accessory species that exhibit no dominance, which may be conveniently referred to as *edominants*, pending more detailed analysis.

Influents

As indicated previously, the designation of *influent* is applied to the animal members of the biome by virtue of the influence or coaction they exert in the community. The significance of this effect depends much upon the life-form and to a large degree upon the size and abundance of the species as well, and is seen chiefly in the coactions involved in food, material, and shelter. Influents may be grouped in accordance with these properties, or they may be arranged with respect to distribution and role in climax or sere, or to time of appearance (Clements and Shelford, 1936). For general purposes it is perhaps most convenient to recognize subdivisions similar to those for dominants and with corresponding terms and significance. Thus, a *perfluent* would occur more or less throughout the formation, while the *eufluent* would be more or less typical or peculiar to an association. *Subfluents* would mark the next lower degree of importance, roughly comparable to that of subdominant, while minute or microscopic influents of still less significance might well be known as *vefluents*.

Climax and seral units

No adequate analysis of vegetation or of the biome is possible without taking full account of development. As the first step, this involves a distinction between climax communities proper and those that constitute the successional movement toward the final stages. The two groups differ in composition, stability, and type of control, but they agree in the possession of dominants, subdominants, and influents. These primary differences made it desirable to recognize two series of communities, viz. climax and seral and to propose corresponding terms, distinguished by the respective suffixes, *-ation* and *-ies* (Clements, 1916). These have gradually come into use as the feeling for dynamic ecology has grown and bid fair to constitute a permanent basis for all such studies. It is not supposed that they embrace all the units finally necessary for a complete system, but their constant application to the great climaxes of North America for nearly two decades indicates that they meet present needs in the matter of analysis.

Not all communities can be certainly placed in the proper category at the outset, but the number of doubtful cases is relatively small and few of these present serious difficulty under combined extensive and intensive research. This statement, however, presupposes an experience sufficiently wide and long to permit distinguishing between climaxes and the various types of proclimax, as well as recognizing the characteristic features of subclimaxes in particular. Comparative studies over a wide region are indispensable and the difficulties will disappear to the degree that this is achieved. While ecotones and mictia necessarily give rise to some questions in this connection, these in turn are resolved by investigations as extensive as they are detailed.

Climax units

In the organization of these, four types of descending rank and importance were distinguished within the formation, namely, association, consociation, society, and clan. Like the formation itself, the first two were based upon the dominant and its life-form, while the last were established upon the subdominant and its different life-form. It was recognized at the time that the association contained within itself other units formed by the dominants (cf. *Plant Indicators*, 1920, pp. 107, 276), and two further divisions, *faciation* and *lociation*, with corresponding seral ones, *facies* and *locies*, were suggested and submitted to Prof. Tansley for his opinion as to their desirability. These have been tested in the course of further field studies and have now and then been used in print, though the complete series was not published until 1932 (cf. Shelford, 1932). The climax group now comprises the following units, viz. association, consociation, faciation, lociation, society, and clan. At the beginning, it was intended to replace society by *sociation* for the sake of greater uniformity in terms, but the former had attained such usage that the idea was relinquished. However, the use of society in quite a different sense by students of social relations, especially among insects, again raises the question of the desirability of such a substitution, in view of the growing emphasis upon bioecology (cf. also Du Rietz, 1930; Rübel, 1930).

Phot. 9. Association of mixed prairie, *Stipa, Agropyrum, Bouteloua*, etc.: Monument, Colorado.

Phot. 10. Foothill faciation of the desert-plains association, *Bouteloua eriopoda, B. gracilis, B. hirsuta, B. filiformis*, etc.: Safford, Arizona.

CLEMENTS—NATURE AND STRUCTURE OF THE CLIMAX

Face p. 273

Association.

Under the climax concept this represents the primary division of the biome or formation, and hence differs entirely from the generalized unit of the plant sociologists, for which the term *community* is to be preferred. Each biome consists regularly of two or more associations, though the lake forest and the desert scrub embody two apparent exceptions, each seeming to consist of one association only. However, these are readily explained by the fact that the western member of the former has been obscured by the expansion of montane and coast forests in the north-west, while one or more additional associations of the desert climax occur to the southward in Mexico, and apparently in South America also.

The number of associations in a particular formation is naturally determined by the number of primary differences and these in turn depend upon the presence of eudominants. Just as the unity of the formation rests upon the wide distribution of several major dominants or perdominants, so the association is also marked by one or more dominants peculiar to it, and often as well by differences in the rank and grouping of dominants held in common. Thus, in the true prairie association, the eudominants are *Stipa spartea*, *Sporobolus asper* and *heterolepis*; for the desert plains, *Bouteloua eriopoda*, *rothrocki* and *radicosa* and *Aristida californica*, while *Stipa comata* and *Buchloe* take a similar part in the mixed prairie. In the deciduous climax, the characteristic dominants of one association are supplied by the beech and hard maple, of a second by chestnut and chestnut-oak, though the oak-hickory association, of wider range and greater complexity, is comparatively poor in eudominants by contrast with the number of species.

The structural and phyletic relations of the associations of a climax are best illustrated by the grassland, which is the most highly differentiated of all North American formations, largely as an outcome of its great extent. The most extensive and varied unit is the mixed prairie, which occupies a generally median position with respect to the other five associations of this climax. Originally, it derived its dominants from three separate regions, *Stipa*, *Agropyrum* and *Koeleria* coming from Holarctica, *Sporobolus* from the south, and the short-grasses from the Mexican plateaux, and it still exhibits the closest kinship with the Eurasian steppe. It contains nearly all the genera that serve as dominants in the related associations, while many of the eudominants of these have all the appearance of direct derivatives from its species, as is shown by *Stipa*, *Sporobolus*, *Poa*, and *Agropyrum*. The evolution of both species and communities is evidently in response to the various subclimates, that of the true prairie being moister, of the coastal warmer as well; the desert plains are hotter and drier, the California prairie marked by winter rainfall and the Palouse by snowfall.

Consociation.

In its typical form the consociation is constituted by a single dominant, but as a matter of convenience the term is also applied to cases in which other dominants are but sparingly present and hence have no real share in the control of the community. It has likewise been convenient to refer in the abstract to each major dominant of the association as a consociation, though with the realization that it occurs more frequently in mixture than by itself. In this sense it may be considered a unit of the association, though the actual area of the latter is to be regarded as divided into definite faciations. Consociation dominants fall into a more or less regular series with respect to factor requirements, especially water content, and often exhibit zonation in consequence. This is a general feature of mixed prairie where *Agropyrum smithi* and *Stipa comata* are the chief mid-grasses, the former occupying swales and lower slopes, the latter upper slopes and ridges.

The consociation achieves definite expression over a considerable area only when the factors concerned fluctuate within the limits set by the requirements of the dominant or when the other dominants are not found in the region. The first case may be illustrated by *Pinus ponderosa* in the lower part of the montane forest and by *Adenostoma fasciculatum* in the Sierran chaparral, while the second is exemplified by *Picea engelmanni* in the Front Range of Colorado, its usual associate, *Abies lasiocarpa*, being absent from the district. In rolling terrain like that of the prairie, each consociation will recur constantly in the proper situation but is necessarily fragmentary in nature. Such behavior is characteristic of dominants with a postclimax tendency, as with *Stipa minor* and *Elymus condensatus* in swales and lower levels of the mixed prairie.

Faciation.

This is the concrete subdivision of the association, the entire area of the latter being made up of the various faciations, except for seral stages or fragments of the several consociations. Each faciation corresponds to a particular regional climate of real but smaller differences in rainfall/evaporation and temperature. It may be characterized by one or two eudominants, such as *Hilaria jamesi* and *Stipa pennata* in the southern mixed prairie, but more often it derives its individuality from a sorting out or a recombination of the dominants of the association. As is evident, the term is formed from the stem *fac-*, show, appear, as seen in *face* and *facies*, and the suffix *-ation*, which denotes a climax unit.

During the past decade, much attention has been given to the recognition and limitation of faciations on the basis of the presence or absence of a eudominant, such as *Hilaria*, *Buchloe*, or *Carex*, or a change in the rank or grouping of common dominants, like *Stipa*, *Agropyrum*, *Sporobolus* or *Bouteloua*. In the prairie this task has been complicated by overgrazing, cultivation and related disturbances, while selective lumbering and fire have added to the difficulties, in the deciduous forest especially. In general, temperature appears to play the leading part in the differentiation of faciations, since they usually fall into a sequence determined by latitude or altitude, though rainfall/ evaporation is naturally concerned also. The mixed prairie exhibits the largest number, but it is approached in this respect by the deciduous forest as a consequence of wide extent and numerous dominants. Over the Great Plains from north to south, the successive faciations are *Stipa-Bouteloua, Bouteloua-Carex, Stipa-Agropyrum-Buchloe, Bouteloua-Buchloe, Hilaria-Stipa-Bouteloua,* and *Agropyrum-Bouteloua.* However, the short-grass communities are to be regarded as disclimaxes wherever the mid-grasses have been eliminated or nearly so, a condition that fluctuates in relation to dry and wet phases of the climatic cycle.

Lociation.

In its turn, the lociation is the subdivision of the faciation, the term being derived from *locus,* place, as indicating a general locality rather than a large region. Nevertheless, a lociation may occupy a relatively extensive territory up to a hundred miles or more in extent, by comparison with several hundred for the faciation. It is characterized by more or less local differences in the abundance and grouping of two or more dominants of the faciation. These correspond to considerable variations in soil, contour, slope-exposure or altitude, but all within the limits of the faciation concerned. As a consequence, lociations are very often fragmented, recurring here and there as alternes with each other, and frequently with proclimaxes of various types. Like most climax units, they have been modified by disturbance in some degree, and this fact must be constantly kept in mind in the task of distinguishing them from subclimax or disclimax.

A detailed knowledge of the faciation is prerequisite to the recognition of the various lociations in it. The number for a particular faciation naturally depends upon the extent of the latter and the number of dominants concerned. Consequently, lociations are more numerous in the faciations of the mixed prairie and desert plains, of the chaparral and the oak-hickory forest. As would be expected, they are often most distinct in ecotones and in districts where there is local intrusion of another dominant. In correspondence with their local character, it is important to eliminate or diminish superimposed differences through restoration of the original cover by means of protection enclosures and thus render it possible to disclose the true composition.

Society.

This term has had a wide range of application, but by dynamic ecologists it has generally been employed for various groupings of subdominants, of which those constituted by aspects or by layers are the most important. In addition, there is a host of minor communities formed by cryptogams in the ground layer or on host-plants and other matrices. The soil itself represents a

18–2

major layer, divisible into more or less definite sublayers. Animals regularly assume roles of varying importance in all of these, especially the insects, arachnids and crustacea, and hence most if not all societies comprise both subdominant plants and subinfluent animals. It is doubtful whether animals form true societies independently of their food-plants or those used for materials or shelter, but this is a question that can be answered only after the simplest units, namely, family and colony, have been recognized and coordinated in terms of their coactions.

In view of what has been said previously, it seems desirable to employ society as the general term for all communities of subdominants and subinfluents above the rank of family and colony, much as community is the inclusive term for all groupings of whatsoever rank. This then permits carrying out the suggestion made two decades ago that the major types of societies be set apart by distinctive names. In accordance, it is here proposed to call the aspect society a *sociation* and the layer society a *lamiation*, while the corresponding seral terms would be *socies* and *lamies*. Many of the societies of cryptogams and minute animals would find their place in these, particularly so for those of the surface and soil layers, but many others take part in a miniature sere or *serule*, such as that of a moldering log, and may best receive designations that suggest this relation.

Sociation. Wherever societies are well developed, they regularly manifest a fairly definite seasonal sequence, producing what have long been known as aspects (Pound and Clements, 1898). As phenomena of the growing season, these were first distinguished as early spring or prevernal, vernal proper, estival, and serotinal or autumnal, but there may also be a hiemal aspect, especially for animals, in correspondence with an actual and not merely a calendar winter as in California.

Sociations are determined primarily by the relation between the life cycle of the subdominants and the seasonal march of direct factors, temperature in particular. So far as the matrix of plants is concerned, the constituent species may be in evidence throughout the season, but they give character to it only during the period of flowering, or fruiting in the case of cryptogams. They are present largely or wholly by sufferance of the dominants, and they are to be related to the reactions of these and competition among themselves rather more than to the habitat factors as such. In grassland and desert, they are often more striking than the dominants themselves, sometimes owing to stature but chiefly as an effect of color and abundance, and they may also attain much prominence in woods with the canopy not too dense.

Sociations are usually most conspicuous and best developed in grassland, four or even five distinct aspects occurring in the true prairie from early spring to autumn. In the mixed prairie these are usually reduced to three, and in the desert plains and desert proper, to two major ones, summer and winter, in which however there may be subaspects marked by *sations*, as indicated

Phot. 11. Sociation of *Erigeron* and *Psoralea*, estival aspect; Belmont
Prairie, Lincoln, Nebraska.

Phot. 12. Lamiation of mid-herbs, *Laportea*, *Physostegia*, *Impatiens*, etc.,
in Oak-Hickory forest.

CLEMENTS—NATURE AND STRUCTURE OF THE CLIMAX

Face p. 277

later. Short seasons due to increasing latitude or altitude afford less opportunity, and the tundra, both alpine and arctic, usually exhibits but two aspects, the sociations however taking a conspicuous role as in the prairie. In woodland the number and character of the sociations depend largely upon the nature of the canopy, and for deciduous forest the flowery sociations regularly belong to the spring and autumn aspects, when the foliage is either developing or disappearing.

It is convenient to distinguish sociations as simple or mixed with respect to the plant matrix in accordance with the presence of a single subdominant or of two or more. However, when animals are included in the grouping, such a distinction appears misleading and may well be dropped. The word "mixed" would be more properly applied to plant-animal societies were it not for the fact that this appears to be the universal condition. Since seasonal insects are legion, many of the societies in which they take part are best denoted as sations.

Lamiation. The term for a layer society is derived from the stem *lam-*, seen in lamina and lamella. As is well known, layers are best developed in forests with a canopy of medium density, so that under the most favorable conditions as many as five or six may be recognized above the soil. In such instances, there are usually two shrub stories, an upper and lower, often much interrupted, followed by tall, medium and low forb layers, and a ground community of mosses, lichens and other fungi, and usually some delicate annuals (cf. Hult, 1881; Grevillius, 1894). The soil population is perhaps best treated as a single unit, though it may exhibit more or less definite sublayers. When the various layers beneath the dominants are distinct, each is regarded as a lamiation, but in many cases only one or two are sufficiently organized to warrant designation, e.g. shrub or tall-herb lamiation.

Layers are often reduced to a single lamiation of low herbs in the climax forest, especially of conifers, and even this may be entirely lacking in dense chaparral. Two or three layers of forbs may be present in true prairie in particular, the upper lamiation being much the most definite and often concealing the grass dominants in the estival aspect, but the structure of grassland generally reflects the greater importance of sociations. Climaxes of sagebrush and desert scrub exhibit no proper lamiations, owing to the interval between the individual dominants, but the herbaceous societies of the interspaces show something of this nature.

As a rule, well-developed lamiations also manifest a seasonal rhythm, corresponding to aspects or to subaspects. These constitute recognizable groupings in the lamiation and for the sake of determination and analysis may be termed *sations*. This word is a doublet of season, both being derived from the root *sa-*, sow, hence grow or appear. Because of the frequent interplay of aspects and layers, the sation may for the present be employed for the subdivision of both sociation and lamiation, and especially where seasonal species of invertebrates play a conspicuous role.

Clan. This is a small community of subordinate importance but commonly of distinctive character. It is marked by a density that excludes all or nearly all competing species, in consequence of types of propagation that agree in the possession of short offshoots. Extension is usually by bulb, corm, tuber, stolon or short rootstock, each of which produces a more or less definite family grouping; in fact, most clans are families developed in the climax matrix and sometimes with a blurred outline in consequence. Clans of a particular species such as *Delphinium azureum* or *Solidago mollis* are dotted throughout the respective sociation, often in large numbers, and contribute a distinctive impress much beyond their abundance. Like all units, but small ones especially, they are subject to much fluctuation with the climatic cycle, as a result of which they may pass into societies or be formed by the shrinkage of the latter.

Seral units.

The concepts of dominance and subdominance apply to the sere as they do to the climax, as does that of influence also, and the corresponding sets of units bear the same general relation to each other. Each of the four major units is the developmental equivalent of a similar community in the climax series and this is likewise true of the various kinds of societies. They constitute the successive stages of each sere, both primary and secondary, including the subclimax, where they often achieve their best expression. It has also been customary to employ seral terms for preclimax and postclimax, and this appears to be the better usage for disclimax and proclimaxes in general. From the fragmentary nature of bare areas and suitable water bodies in particular, seral communities are often but partially developed and one or more units will be lacking in consequence. Thus, the reed-swamp associes is frequently represented by a single one of its several consocies and the minor units are even more commonly absent.

The associes is the major unit of every sere, the number being relatively large in the prisere and small in the subsere. The universal and best understood examples are those of the hydrosere, in which *Lemna, Potamogeton, Nuphar, Nymphaea, Nelumbo* and others form the consocies of the floating stage, and *Scirpus, Typha, Phragmites*, etc., are the dominants of the reed-swamp or amphibious associes. As already indicated, every consocies may occur singly and often does when the habitat offers just the proper conditions for it or the others have failed to reach the particular spot. When the ecial range is wider, various combinations of two or three dominants will appear, to constitute corresponding facies. Locies are less definitely marked as a rule, except in swamps of vast extent, but are to be recognized by the abundance of reed-like dominants of lower stature, belonging to other species of *Scirpus*, to *Heleocharis, Juncus*, etc. Both facies and locies seem to be better developed in sedge-swamp with its larger number of dominants, though the Everglades with the single consocies of *Cladium* form a striking exception.

The tree-swamps of the south-eastern United States contain a considerable number of consocies, such as *Taxodium distichum, Nyssa aquatica, biflora* and *ogeche, Carya aquatica, Planera aquatica, Persea palustris* and *borbonia, Magnolia virginiana, Fraxinus pauciflora, profunda* and *caroliniana,* and *Quercus nigra.* These are variously combined in several different facies, though a more detailed and exact study of the swamp sere may show the presence of two woody associes, distinguished by the depth or duration of the water. As with other scrub communities, the heath associes of peat-bog and muskeag comprises a large number of dominants and presents a corresponding wealth of facies and locies.

In the hydrosere of the deciduous forest, the typical subclimax is that of the flood-plain associes, composed of species of *Quercus, Ulmus, Fraxinus, Acer, Betula, Juglans, Celtis, Platanus, Liquidambar, Populus* and *Salix* for the most part. There are at least three well-marked facies, namely, northern, central and southern, each with a number of more or less distinct locies. The swamp associes or subclimax of the lake and boreal forests consists of *Larix laricina, Picea mariana* and *Thuja occidentalis,* occurring often as consocies but generally in the form of zoned facies. A large number of fire subclimaxes appear in the form of consocies, as with many of the pines, but associes are frequent along the Atlantic Coast, as they are in the boreal climax, where aspens and birches are chiefly concerned. The number of shrubs and small trees that play the part of seral dominants in the deciduous climax is much larger, producing not merely a wide range of associes but of facies and locies as well. More than a dozen genera and a score or so of species are involved, chief among them being *Sassafras, Diospyrus, Asimina, Hamamelis, Prunus, Ilex, Crataegus* and *Robinia.* The subclimax of the xerosere is constituted for the most part by species of *Quercus,* forming an eastern, a south-eastern and a western associes, the last with two well-marked facies, one of *stellata* and *marilandica,* the other of *macrocarpa* and *Carya ovata.*

Among postclimax associes, those of grassland and scrub possess a large number of dominants and exhibit a corresponding variety of facies and locies, together with fairly definite consocies. In the sandhills of Nebraska, the tall-grasses concerned are *Andropogon halli, furcatus* and *nutans, Calamovilfa longifolia, Eragrostis trichodes, Elymus canadensis* and *Panicum virgatum*; some of these drop out to the northward and others to the south, thus producing at least three regional facies. The mesquite-acacia associes of the south-west possesses a larger number of dominants and manifests a greater variety of facies through its wide area, and this is likewise true of the coastal sagebrush of California.

With reference to seral societies, it must suffice to point out that these are of necessity poorly developed in the initial stages of both hydrosere and xerosere, as the dominants are relatively few. Even in the reed-swamp, true layers are the exception, being largely restricted to such subdominants as

Alisma, Pontederia, Hydrocotyle and *Sagittaria*, which are found mostly in borders and intervals. However, in extensive subclimaxes and postclimaxes the situation is quite different. The tall-grass associes of sandhills is often quite as rich in saties and lamies as the true prairie, while the various subclimaxes of the several great forest types may equal the latter in the wealth of subdominants for each season and layer, the actual communities being very much the same.

Serule.

This term, a diminutive of *sere*, has been employed for a great variety of miniature successions that run their short but somewhat complex course within the control of a major community, especially the climax and subclimax. They resemble ordinary seres in arising in bare spots or on matrices of different sorts, such as earth, duff, litter, rocks, logs, cadavers, etc. Parasites and saprophytes play a prominent and often exclusive role in them, and plants and animals may alternate in the dominant parts. The organisms range from microscopic bacteria and worms to mites, larvae and imagoes on the one hand and large fleshy and shelf fungi on the other. The most important of these in terms of coaction and abundance are known as *dominules* (Clements and Shelford, 1936), with *subdominule* and *edominule* as terms for the two degrees of lesser importance. On the same model are formed *associule, consociule*, and *sociule* in general correspondence with the units of the sere itself. In addition there are families and colonies of these minute organisms, which are essentially similar to those of the initial stages of the major succession. Up to the present, little attention has been devoted to the development and structure of serules, but they are coming to receive adequate consideration in connection with bio-ecological problems. Many of the coactions, however, have long been the subject of detailed research in the conversion of organic materials.

RANK AND CORRESPONDENCE OF UNITS

The following table exhibits the actual units of climax and sere, as well as their correspondence with each other. However, for the complete and accurate analysis of a great climax and especially the continental mass of vegetation, it is necessary to invoke other concepts, chiefly that of the proclimax and of communities mixed in space or in time. The several proclimaxes have been characterized (pp. 262–8), and the ecocline and seration briefly defined (p. 267). To these are to be added the *ecotone* and *mictium*, both terms of long standing, the former applied to the mixing of dominants between two units, the latter to the mixed community that intervenes between two seral stages or associes. Finally, there will be the several types of seres in all possible stages of development, the prisere in the form of hydrosere, xerosere, halosere or psammosere in regions less disturbed and a myriad of subseres in those long settled.

TABLE OF CLIMAX AND SERAL UNITS

	Eoclimax	
	Panclimax	
Climax		*Sere*
	Climax (formation)	
Association		Associes
Consociation		Consocies
Faciation		Facies
Lociation		Locies
Sociation		Socies .
Lamiation		Lamies
Sation		Saties
Clan		Colony
		Family
	Serule	
	Associule	
	Consociule	
	Sociule	

As indicated previously, the word *community* is employed as a general term to designate any or all of the preceding units, while *society* may well be used to include those of the second division, i.e. sociation, etc. These are characterized by subdominants in contrast to the dominants that mark the first group. It has also been pointed out that the entire area of the association is divided into faciations and that the consociation is the relatively local expression of complete or nearly complete dominance on the part of a single species. The clan corresponds to the family as a rule, but in some cases resembles the colony in being formed by two species.

Families and colonies may also appear in climax communities, but this is regularly in connection with the serule.

Panclimax and eoclimax

The comprehensive treatment of these concepts is reserved for the succeeding paper in the present series, but it is desirable to characterize them meanwhile. The *panclimax* (παν, all, whole) comprises the two or more related climaxes or formations of the same general climatic features, the same life-form and common genera of dominants. The relationship is regarded as due to their origin from an ancestral climax or *eoclimax* (ἠώs, dawn), of Tertiary or even earlier time, as a consequence of continental emergence and climatic differentiation. In the past, eoclimaxes formed a series of great biotic zones in the northern hemisphere with the pole as a focus, and this zonal disposition or clisere is still largely evident in the arrangement of panclimaxes at the present. It is striking in the case of the arctic tundra and taiga or boreal forest, fairly evident for deciduous forest and prairie-steppe, and somewhat obscure for woodland and chaparral-macchia, while the position of deserts is largely determined by intervening mountain ranges. This is true likewise of grassland in some degree, and taken with the former broad land connection between North and South America explains why both prairie and desert panclimax contain at least one austral formation.

In the light of what has been said earlier, it is readily understood that panclimax and panformation are exact synonyms, as are eoclimax and eoformation. Panbiome and eobiome are the corresponding terms when the biotic community is taken as the basis for research.

Prerequisites to research in climaxes

It would be entirely superfluous to state that the major difficulty in the analysis of vegetation is its complexity, were it not for the fact that it is too often taken as the warrant for the static viewpoint. This was embodied in the original idea of the formation as a unit in which communities were assembled on a physiognomic basis, quite irrespective of generic composition and phyletic relationship. It is not strange that this view and its corollaries should have persisted long past its period of usefulness, since this is exactly what happened with the artificial system of Linnaeus, but the time has come to recognize fully that a natural system of communities must be built just as certainly upon development and consequent relationship as must that of plant families. Complexity is an argument for this rather than against it, and especially in view of the fact that the complexity discloses a definite pattern when the touchstone of development is applied to it.

Though the mosaic of vegetation may appear to be a veritable kaleidoscope in countries long occupied by man, the changes wrought upon it are readily intelligible in terms of the processes concerned. As emphasized previously, the primary control is that of climate, in a descending scale of units that correspond to formation, association, and faciation. Upon this general pattern are wrought the more circumscribed effects of physiography and soil, and both climatic and edaphic figures are overlaid and often more or less completely obscured with a veneer applied by disturbance of all possible kinds. Even above this may be discerned the effect, transient but nonetheless apparent, of such recurrent changes as annuation and aspection. Moreover, the orderly pattern of climate is complicated by great mountain ranges so that such climaxes as tundra and taiga occur far beyond their proper zone, and the effect is further varied by the relative position of the axis.

The migrations of climaxes in the past are a prolific source of fragmentary relicts, the interpretation of which is impossible except in terms of dynamics. This is likewise true of savannah, which represents the shrinkage of forest and scrub under a drying climate and is then usually further modified by fire or grazing. Fragmentation from this and other causes is characteristic of every diversified terrain and reaches its maximum when human utilization enters the scene upon a large scale. Somewhat similar in effect though not in process is the reduction of number of dominants by distance, with the consequence that an association of several may be converted into a consociation of one. Such a shrinkage naturally bears some relation also to climate and physiography, especially as seen in the glacial period, and finds its best illustration in the general poverty of dominants in the coniferous and deciduous climaxes of Europe, by contrast with those of eastern Asia and North America. A similar contrast obtains between the grassland of Asia and of North America, the latter being much richer in dominants, while South America approximates it closely in this respect.

On the part of the investigator, the difficulties in the way of an extensive and thoroughgoing study of climaxes are usually more serious. They arise partly from the handicap too often set by state or national boundaries and partly from the limitations of funds and time. They are also not unrelated to the fact that it is easiest to know a small district well and to assume that it reflects larger ones with much fidelity. As a consequence, it is impossible to lay too much stress upon the need for combining intensive and extensive methods in the research upon climaxes, insofar as their nature, limits and structure are concerned. The detailed development in terms of primary and secondary succession lends itself much more readily to local or regional investigation, but even here a wider perspective is essential to accurate generalization.

REFERENCES

Bradbury, J. "Travels in the Interior of North America, *ca.* 1815." In **Thwaites'** *Early Western Travels,* 5, 1904.

Bromley, S. W. "The original forest types of southern New England." *Ecol. Mon.* **5**, 61–89, 1935,

Chaney, R. W. "A comparative study of the Bridge Creek flora and the modern redwood forest." *Publ. Carneg. Instn,* No. 349, 1925.

Chaney, R. W. and E. I. Sanborn. "The Goshen Flora of West-central Oregon." *Publ. Carneg. Instn,* No. 439, 1933.

Chapman, H. H. "Is the longleaf type a climax?" *Ecology,* **13**, 328–34, 1932.

Clements, F. E. *Plant Physiology and Ecology,* New York, 1907.

Clements, F. E. *Plant Succession,* Washington, 1916.

Clements, F. E. "Development and structure of the biome." *Ecol. Soc. Abs.* 1916.

Clements, F. E. *Plant Indicators,* Washington, 1920.

Clements, F. E. "Phylogeny and classification of climaxes." *Yearb. Carneg. Instn,* **24**, 334–5, 1925.

Clements, F. E. *Plant Succession and Indicators,* New York, 1928.

Clements, F. E. "The relict method in dynamic ecology." This JOURN. **22**, 39–68, 1934.

Clements, F. E. "Origin of the desert climate and climax in North America." This JOURN. **24**, 1936.

Clements, F. E. and E. S. Clements. "Climate and climax." *Yearb. Carneg. Instn,* **32**, 203, 1933.

Clements, F. E. and V. E. Shelford. *Bio-ecology,* 1936.

Du Rietz, G. E. "Classification and nomenclature of vegetation." *Svensk. Bot. Tid.* **24**, 489, 1930.

Godwin, H. "The subclimax and deflected succession." This JOURN. **17**, 144, 1929.

Greeley, H. *An Overland Journey to California in* 1859, New York, 1860.

Grevillius, A. Y. "Biologisch-physiologische Untersuchungen einiger Schwedischen Hainthälchen." *Bot. Z.* **52**, 147–68, 1894.

Hult, R. "Forsök til analytisk behandling af växformationerna." *Medd. Soc. Faun. Flor. Fenn.* **8**, 1881.

Hult, R. "Blekinges vegetation. Ett bidrag till växformationernas utvecklingshistorie." *Medd. Soc. Faun. Flor. Fenn.* **12**, 161, 1885. *Bot. Zbl.* **27**, 192, 1888.

Lewis, M. and W. Clark. Journal, 1803–1806. In **Thwaites'** *The Original Journals of the Lewis and Clark Expedition,* 1904–05.

Phillips, J. "The biotic community." This JOURN. **19**, 1–24, 1931.

Phillips, J. "Succession, development, the climax and the complex organism: an analysis of concepts. Part I." This JOURN. **22,** 554–71, 1934.

Phillips, J. "Succession, development, the climax and the complex organism: an analysis of concepts. Part II." This JOURN. **23,** 210–46, 1935.

Phillips, J. "Succession, development, the climax and the complex organism: an analysis of concepts. Part III." This JOURN. **23,** 488–508, 1935.

Rübel, E. *Pflanzengesellschaften der Erde,* Bern-Berlin, 1930.

Sargent, C. S. *Report on the Forests of North America (exclusive of Mexico),* Washington, 1884.

Seton, E. T. *Lives of Game Animals,* **3,** New York, 1929.

Shelford, V. E. "Basic principles of the classification of communities and habitats and the use of terms." *Ecology,* **13,** 105–20, 1932.

Short, C. W. "Observations on the botany of Illinois." *West. J. Med. Surg.* **3,** 185, 1845.

Tansley, A. G. "Editorial note." This JOURN. **17,** 146–7, 1929.

Tansley, A. G. "The use and abuse of vegetational concepts and terms." *Ecology,* **16,** 284–307, 1935.

Tournefort, J. P. *Relation d'un Voyage du Levant,* Paris, 1717.

Weaver, J. E. and **F. E. Clements.** *Plant Ecology,* New York, 1929.

Weaver, J. E. and **E. L. Flory.** "Stability of climax prairie and some environmental changes resulting from breaking." *Ecology,* **15,** 333–47, 1934.

Stewart, G., and S. S. Hutchings. 1936. The point-observation-plot (square-foot density) method of vegetation survey. *Journal of the American Society of Agronomy* 28:714–722.

The wildlife literature is replete with techniques to better measure habitats and populations. If biologists are to understand wildlife, they will need to better understand the animal's habitat and be able to measure it in a meaningful way. One of the key components of habitat is vegetation, which is time-consuming to measure properly. Two of the major decisions that need to be made when measuring vegetation are what to measure and how to measure it. In other words, what are the important attributes of vegetation to the animal?

The point-observation-plot method has numerous advantages over the other techniques that were used when measuring vegetation. It is easy to learn, saves time, allows for randomization and replication, is suitable for statistical analysis, and can be modified to record other data about the landscape. This paper is an example of the transition from descriptive studies of vegetation to those that are more quantitative, which improved hypothesis testing and our knowledge of the natural world.

RELATED READING

Block, W. M., K. A. With, and M. L. Morrison. 1987. On measuring bird habitat: influence of observer variability and sample size. Condor 89:241–251.

Cottam, G., and J. T. Curtis. 1956. The use of distance measures in phytosociological sampling. Ecology 37:451–460.

Daubenmire, R. 1958. A canopy-coverage method of vegetational analysis. Northwest Science 33:43–64.

Etchberger, R. C., and P. R. Krausman. 1997. Evaluation of five methods for measuring desert vegetation. Wildlife Society Bulletin 25:604–609.

Hurlbert, S. H. 1978. The measurement of niche overlap and some relatives. Ecology 59:67–77.

James, F. C., and H. H. Shugart, Jr. 1970. A quantitative method of habitat description. Audubon Field Notes 24:727–736.

Mueller-Dombois, D., and H. Ellenberg. 1974. Aims and methods of vegetation ecology. John Wiley and Sons, New York, New York, USA.

Rice, E. L., and W. T. Penfound. 1955. An evaluation of the variable-radius and paired-tree methods in the blackjack–post oak forest. Ecology 36:315–320.

Reprinted by permission.

THE POINT-OBSERVATION-PLOT (SQUARE-FOOT DENSITY) METHOD OF VEGETATION SURVEY[1]

George Stewart and S. S. Hutchings[2]

FOR years ecological and range studies have been impeded by the need for a more accurate and time-saving method of making an inventory of plant cover on a given area. At a range-management conference in 1927, 40 to 50 range administrators and researchers attempted to estimate vegetation by the large-plot method. Independent estimates of vegetation in the same plot by experienced range men showed considerable variation, and, as a whole, seemed high. In 1930, the senior author began to devise a method of inventorying the plant cover that was aimed to correct the weaknesses of the methods then in use.

This paper explains the working of the new point-observation-plot method of vegetation survey which has evolved over a period of 4 years. The following formal description is deemed necessary because this method is already highly useful and has been tried sufficiently by many agencies concerned with the measurement of range plant growth to show its practicability. Its application extends throughout the fields of range management, pasture management, agronomy, and soil erosion. It provides definitely quantitative data instead of merely qualitative.

The new method is founded on the technic long used by plant breeders to obtain reliable preliminary field tests of large numbers of strains. It includes randomization and replication of plots and, therefore, lends itself readily to statistical analysis. In addition, it has been modified to include some phases of the timber survey and certain refinements of the range reconnaissance survey long used by the Forest Service and other organizations. The junior author, who has had considerable experience in range reconnaissance, was instrumental in adapting certain principles of this work to the new method.

NEED FOR A NEW METHOD

The point-observation-plot method, as the name indicates, utilizes a series of replicated plots from which is recorded the kind and amount of vegetation on a small area at a particular point. The charted quadrat, in a way, furnishes these data, but this method is too time-consuming to be used widely for surveys. Also, the meter quadrat, sometimes used in the area list method, is often too small to

[1]Contribution from the Intermountain Forest and Range Experiment Station, Forest Service, U. S. Dept. of Agriculture, Ogden, Utah. Received for publication July 27, 1936.

[2]Senior and Assistant Forest Ecologists, respectively. Acknowledgments are gratefully made to G. D. Pickford and J. F. Pechanec of the Intermountain Forest and Range Experiment Station for help in developing and in testing the point-observation-plot method; to Dr. W. P. Cottam of the University of Utah for testing it; and to Director R. E. McArdle and Dr. D. F. Costello of the Rocky Mountain Forest and Range Experiment Station for introducing this method widely in their region, both to officials of the Forest Service and to those of other agencies.

contain the 20 to 50 or more individual plants of major species which must be included to overcome the differences of individual plants. The large-estimate-plot method, in part, compensates for drawbacks encountered in the use of the quadrat, but both lack elasticity and speed.

In both the quadrat and the large-estimate-plot methods, the areas to be studied are ordinarily "selected" because of the slow process of taking data. This practice prevents randomization and replication. On the other hand, the point-observation-plot method provides for randomization and replication, insures a closer scrutiny of the plant cover than is the case on the large-estimate plot, a greater ease and reduction of error in compilation, and an immense saving in time.

THE PLOT

In size, the plot commonly used in the new method is usually 100 square feet in area, though on sparse desert vegetation often it has seemed best to use 200 square feet. The size may vary within reasonable limits, as it needs only to be large enough to be representative of the major forage species, small enough to be viewed readily from above, and to require only a few minutes to observe and record the data. In all plant associations so far studied, except in the desert shrub, a plot area of 100 square feet has proved most satisfactory.

For convenience and speed, the circular plot, so far used in tests with the new method, has many advantages. Following is the method used in measuring off a plot to be studied: Since the radius of a 100 square-foot circle is 5.64 feet, a light chain of that length or a strong string, without stretch, is used to fasten two pegs together which are used to describe the desired circle. One peg is driven in the ground to serve as a pivot about which the other turns in tracing the circumference. Sometimes a straight piece of wood or light metal with a peg at the end is used for this purpose; or, in taller or brushy vegetation, surveyor pins, pegs, or other easily visible reliable markers are sometimes used. The error incurred in designating the plot is small when reasonable care is exercised in the operation.

THE ESTIMATE

Previous to beginning the inventory, a decision is reached as to the classes of vegetation for which data are to be recorded. For some purposes, all grasses may be considered as one group, all range weeds another, and all shrubs another. Usually, owing to differences in palatability, the more abundant species of each plant class will be listed separately. This may be done to whatever degree of refinement is desired, i.e., there may be as few or as many groups according to the information desired and as the time available may warrant. This decision requires careful thought lest the surveyor overburden himself with too many details.

With these groups or species in mind, the estimator visualizes how much current growth of a given vegetation group equals 1 square foot of plant cover. This done, he counts up the number of square feet of plant cover of that species or plant group and records it on an especially prepared form. He then makes the estimate for other species or plant groups and records the data. Usually, it is sufficiently accurate to record the ground cover of each species or group to the nearest 1/4 square foot. The scarcer species may have only a small volume on one plot, too little to equal 1/4 square foot. This condition is indicated by a "trace" (T) unless it is not desired to record such information. The estimate of fractions of a square foot of ground cover has limits. For example, ¼ square foot is the middle point between ⅛ and ⅜, ½ square foot between ⅜ and ⅝, ¾ square foot between ⅝ and ⅞, and 1 square foot between ⅞ and 1⅛. The ½ square foot may be the smallest fractional area of plant cover needed on an extensive survey, in which case the limits are ¼ square foot below and ¼ above each figure to be recorded. Sometimes for very careful work it may be desirable to take data to ⅛ square foot, making the limits $\frac{1}{16}$ and $\frac{3}{16}$ square foot, etc.

PLOT SYSTEM AND REPLICATION

When the data are taken on one plot, the estimator moves in a predetermined direction by a distance interval also prearranged. Whether the interval is 100 feet, 0.1 mile, or 0.5 mile, depends on the intensity of the survey, inasmuch as the method is adjustable to any intensity desired. When the interval has been measured or paced according to the working plan, the point at the end of the interval mechanically becomes the center of the next plot, and the circumference of the plot is drawn in as previously described. In order to provide for randomization and replication, data are gathered from a large number of plots spaced at mechanical intervals.

For example, to compare the amount of plant growth on a protected range with that on a heavily grazed one, the worker estimates the particular vegetation in which he is interested on 10, 20, 30, 50, or 100 plots on each area and compares the number of square feet of this kind of vegetation found on each plot of the two series studied. Often the area of the protected or moderately used range is smaller than that of the heavily used one. This condition is of little consequence if care be taken to compare only that part of the heavily used range which is truly comparable with the "protected" area, i.e., comparisons should be made between areas on which the vegetation would have been essentially equal under equal treatment.

COMPARISON MUST BE RELIABLE

True comparability is highly important in making plant depletion inventories, because wide variations in productivity have always existed on different areas on the original undisturbed range. Comparisons must, therefore, be real and not specious. Minor differences will occur on any areas, but consequential ones can and must be eliminated or the comparisons are of little value; therefore, great care has been taken in this respect in the inventories made by the Intermountain Forest and Range Experiment Station.

Frequently a fence marks the boundary between a moderately used and a heavily used range. An examination will usually reveal whether the range areas on the two sides of the fence have the same kind of vegetation, have comparable soil and moisture conditions, and are essentially equal in other respects. A series of plots in a line on one side of the fence is paralleled by a similar series on the other side. The two series are both far enough from the fence to omit the zone of animal concentration that almost invariably occurs along a fence. This is ordinarily not more than 40 to 100 feet, usually less. The mechanical interval between plots is the same on both "transects", as such lines of plots are designated.

At the Intermountain Forest and Range Experiment Station, two systems have been used in laying out the position of plots, (1) one for obtaining a comparison of forage on various range units; and (2), more important, one for setting up a network of plots over a given range to obtain a forage inventory, or to determine the volume of plant cover on a pasture or on a watershed. This second system is also used for establishing permanent plots from which data are to be taken in later years. However, since the first-named phase of this plot system, that of making comparative surveys, is the one so far used outside the Intermountain region, it is described first.

COMPARATIVE SURVEYS AND TYPE MAPS

Forage inventories made in 1932 and 1933 with the new method extended into approximately 80 typical range areas in Idaho, Nevada, Utah, and southwestern Wyoming. They afforded the only approximately accurate method of estimating the condition of range vegetation which has, since settlement, developed from uncontrolled grazing. In 1935 the Rocky Mountain Forest and Range Experiment Station adopted this plot system and further extended the survey into Colorado and Wyoming and into western Kansas and Nebraska. In 1936 the Forest Service, in cooperation with other agencies, plans to take many thousands of plot estimates. A survey of Utah ranges will be completed, as well as surveys in Idaho in some of the more important areas which were omitted in the 1932 study. The surveyor, by drawing in the type lines when he passes from one plant association to another, can, with little extra labor, also have, when the estimating is finished, a type map of the vegetation. The point-observation-plot method, as a new tool and because of the rapidity of its operation, has made such surveys possible.

COMPILING THE DATA

Since the original data are taken in square feet on plots of 100 square feet in area, they read in percentages directly without calculation or adjustment. Data-sheet forms (Fig. 1) are drawn up to accommodate 10 plots on one sheet, inasmuch as at least 10 plots are taken in one protected area or in one grazed area. When the number of plots taken exceeds 10, a multiple of 10 is used.

Clerks are able to compile the data quickly on adding machines and to mark off one decimal point to the left; whereas, the quadrat and large-estimate-plot methods require technicians to work through

SAMPLE PLOT - Transect

Location ENCLOSURE NO. 1 FREMONT COUNTY BURNING PROJECT
Transect No. #1 Date JULY 9 1936 Examiner RALPH JENSEN
Type SAGEBRUSH GRASS

Plot Number	1	2	3	4	5	6	7	8	9	10	TOTAL DENSITY
Density	8 1/2	3 3/4	5 3/4	10 3/4	8 3/4	8 1/2	6 3/4	14 1/4	6 1/2	9	82 1/2
Species											
GRASSES											
Agropyron dasystachyum	1/4		1/4	1/4	1/2	1 3/4	3/4	1	1/2	1/2	6 1/4
Danthonia unispicata		1/4									1/4
Festuca idahoensis			1/4	1 1/4	1/4		1/4	1/4	1/2		2 3/4
Koeleria cristata										1/4	1/4
Poa secunda		1/2	3/4			1/2	1/4				2
Poa nevedensis				1/4		1/4		1/4	1/4	1/4	1 1/4
Stipa columbiana			1/4	1 1/2	1/4	1/2	1/4	1 1/4	2 1/4	2	8 1/4
Stipa comata				3/4		1/2		1 3/4			3
GRASSLIKE PLANTS											
Carex douglasii				1/4			3/4		1/2	1/4	1 3/4
Carex filifolia				1/2					1/4		3/4
WEEDS											
Achillea lanulosa			1/4			1/4					1/2
Antennaria microthecum										1/4	1/4
Eriogonum heracleoides						1/2					1/2
Gutierrezia sarothrae		1/2									1/2
Polygonun douglasii				1/4				1/4			1/2
Sphaeralcea munroana		1/4									1/4
SHRUBS											
Artemesia tridentata	8 3/4	2	3 1/2	5 1/4	7 1/2	4	3 1/4	9 1/4	2 1/4	5	50 1/4
Leptodactylon pungens		1/4									1/4
Prushia tridentata					1/4	1/4	1 1/4	1/2		1/2	2 3/4
Symphoricarpus oreophilus			1/4								1/4

FIG. 1.—Data sheet used by the Intermountain Forest and Range Experiment Station in taking field data and in compiling data regarding range surveys made by the point-observation-plot method. The major species for which data are to be gathered are listed at the left. The number of square feet of each species is recorded in the correct narrow column. Each sheet holds data from 10 plots, which are averaged in the column at the extreme right. This column is then added to give the total mean density for 10 plots in this plant type. When data are taken from 20 or more plots, two or more sheets are used and these summarized on an additional sheet. All figures thus recorded are in percentages. Machine addition and pointing off of a decimal are all the calculations required.

an intricate system of cross-multiplications and divisions. The compiled figures for quadrats and large-estimate plots usually bear three to five decimal points, whereas frequently none and at most two decimal points are required in making computations with the point-observation-plot system. Thus, the chances of making errors are greatly reduced.

One simple mathematical addition and the pointing off of a decimal place keep the mean figure used much nearer to the observed data than is possible with the intricately "calculated" figures derived by the other methods. This use of nearly "real" figures is a scientific advantage not to be overlooked.

FORAGE INVENTORIES AND PERMANENT PLOTS

A second use of the point-observation-plot method is in making a forage inventory of an experimental range or pasture before it has been grazed or at intervals during grazing. On the Desert Experimental Range such a study was made by means of a network of 100 plots to the section, equidistant in cardinal directions. A survey was made of each of the 16 320-acre pastures by means of 128 plots in 16 lines of 8 plots each. A similar survey will be repeated each season.

Many hundreds of permanent plots have been established by this method at the Intermountain Station. Some are grouped in a network on comparative areas inside and outside of enclosures, others are in transect lines across a grazed zone, and still others radiate outward from watering places around which progressive forage depletion is thought to be taking place. When fortified with chart quadrats and brush plots,[3] a better picture of the kind, the growth, and the changes in vegetation can be obtained in a limited time by the point-observation-plot method than by any other yet described.

EROSION SURVEYS

Repeated observations by this new method at a large number of systematically located plots will also give a representative picture of surface soil erosion conditions, insofar as these can be determined on a small plot. Such a picture has been obtained very successfully on the Davis County flood area just north of Salt Lake City. Variance calculations established that the amount of vegetation on small plots is significantly related to erosion conditions of the surface soil. The point-observation-plot method, by providing replicated and randomized samples, is a valuable addition to the technic of conducting a survey of surface erosion.

METHOD EASY TO LEARN AND OPERATE

During more than 4 years in testing this method it has been demonstrated that several estimators can make approximately the same estimations for given plots, provided they practice together for 2 or 3 half-days and check with each other occasionally. It can also

---

[3] PICKFORD, G. D., and STEWART, GEO. Coordinate method of mapping low shrubs. Ecol., 16:257–261. 1935.

be determined, within narrow limits, whether the estimate is essentially correct. For this purpose, vegetation is cut below the surface of the ground with a knife, a sharpened shovel, or some other tool and moved gently together so as to cover the ground. The plants are not jammed together, but placed where small uncovered angles, when viewed from above, are compensated for by similar small angles where the leaves overlap. Thus the space occupied by each class or species can actually be measured. If this procedure is followed carefully once or twice a day for a few days, most men educated in plant science will soon be able to make estimates that are consistent with each other and that are close to the true value. Consistency and correctness result from the use of a small plot on which the estimator can look directly down at the vegetation. The possibility of testing the estimate against its true value also counts for consistency and correctness. The inherent error in this estimate can be ignored because it can be kept small if care is exercised.

Two examples will also bring out concretely the ease with which men trained in plant science, but entirely unacquainted with the point-observation-plot method, learned and applied it on an extensive scale.

1. A professor of botany, working as a temporary employee of the Intermountain Forest and Range Experiment Station, was shown one afternoon in 1933 how to operate the method. He made estimates alone during the next forenoon, and in the afternoon checked with the authors. He then took six above average CCC boys, coached them for 2 or 3 days, and then added six more to his crew. Repeated careful checks showed that only general supervision by the authors was required. Data were taken from more than 11,000 plots in about 3 months. During the following winter all of the data from these surveys were compiled, checked, and rechecked by 25 untrained CWA workers supervised by one technically trained man.

2. In July of 1935, when the method was in a late stage of development and somewhat simplified, an ecologist with a doctor's degree, also a temporary employee of the Intermountain Station, did his first field work on range studies. For about half of the month he helped in a range survey by the point-observation-plot method. In September he went to another region and conducted a range survey without further help, except for the Intermountain field-data forms and instructions for compilation. He supervised a number of trained Forest Service personnel, none of whom had ever used the method before, in taking, during September and October, data on more than 4,000 range plots. During October and early November, office help, with a minimum of technical supervision, compiled the data and summarized them to show the relative depletion of plant cover on national forests, on state and private lands, and on public domain.

The progress made by these two men is in part explained by their technical training and high ability. However, examples are on record of range men of junior grade who, in 2 half-days, learned the method and thereafter made successful surveys with only one or two checkings by experienced users.

SAVES TIME

After 2 weeks' experience, one technical man working alone can take data at the rate of 20 to 50 plots in a day, instead of at the rate of 2 to 6 chart quadrats a day for two men, or of 2 to 4 large-estimate plots a day for one man. Compilation, consisting of machine addition and one simple division, often the mere pointing off of a decimal, is also about 5 to 10 times as rapid as that for the other methods. This rapidity makes possible the taking of data from a large number of plots with the minimum of help.

MAKES POSSIBLE RANDOMIZATION AND REPLICATION

It is not the details of the plot or the system, important and useful as they are, that the authors regard as most valuable, however. It is the system of point observation, the randomized location of plots at mechanical intervals, and the full replication of plot data that makes it both biologically and statistically sound. Lack of randomization in the other methods prevents plots from being representative, and lack of full replication prevents them from being statistically reliable when used as units for a vegetation survey.

LENDS ITSELF TO STATISTICAL ANALYSIS

Numerical observations of several species of plants on a series of randomized and replicated plots, either with or without observations concerning erosion conditions, make possible statistical analyses of the data. Statistical studies can be made by the probable-error method, by analysis of variance, or by correlation coefficients, depending on the nature of the problem and the number and sort of numerical data. For example, Stewart and Keller[4] show that the point-observation-plot survey provides the ecologist with a means of making his observations definitely quantitative, instead of merely qualitative.

POSSIBLE VARIATIONS AND IMPROVEMENTS

Though a profound step in advance of systems used heretofore, the point-observation-plot method is not perfect. As yet, neither it nor the other methods have been so refined as to permit the proper adjustments for differences in height of plant growth which obviously influence the volume of forage produced. Two studies are in progress at the Intermountain Forest and Range Experiment Station in an effort to devise a suitable modification in the new method to allow for volume of forage. It is to be expected that other workers will develop improvements in the method, because the system by its very nature is highly adjustable.

CONCLUSIONS

After 4 years' trial over a wide territory, the following conclusions regarding the point-observation-plot method seem warranted:

[4]STEWART, GEO., and KELLER, WESLEY. A correlation method for ecology as exemplified by studies of native desert vegetation. Ecol., 17:—. 1936.

1. This method has proved superior to the chart quadrat and to the large-estimate-plot method of making comparative surveys of range vegetation. It should be given a trial in range reconnaissance.

2. It provides for randomization and replication of plots which make it a sound method, both biologically and statistically. In supplying soundness, and therefore, reliability, the new method provides a basis for a vegetation survey which does away with the necessity of "selecting" plots.

3. It is easy to learn and to operate. Its time-saving qualities so greatly reduce the labor cost of providing replication and randomization that these two essential requirements of a good survey and experimental grazing can be fully met. Forage inventories on both range and farm pastures can be made by the new method with a reasonable labor cost.

4. Surface soil erosion conditions can also be readily studied by using the point-observation-plot method. Replicated representative samples, made possible by this new method, will add much to the reliability of erosion surveys.

5. Data obtained by the method are subject to statistical analyses.

SUMMARY

1. A new plot method of vegetation survey, known as the point-observation-plot method, has been devised by ecologists of the Inter-mountain Forest and Range Experiment Station. This method has many distinct advantages over the ones formerly used. It has been nick-named the "square-foot-density" method because data are recorded in that unit.

2. The point-observation plot is usually 100 square feet in area, and for convenience is circular, though there is nothing essential in the shape of plot. The plot is marked off by drawing a circle around a central point which is located mechanically and in no way selected. Locating the plot mechanically without selection gives it randomness — an absolutely essential point in sampling an area.

3. One man can take data on from 20 to 50 plots in a day, thereby making possible full replication of observations—a second absolutely essential quality to statistically reliable data. Plots are replicated 10 or 20 (or other multiples of 10) times to afford representativeness and replication. Such replication may be in a line transect of plots or in a network of plots more or less equi-distant in one or more directions.

4. Separate workers can easily check with each other, i.e., one worker can duplicate another's data, if they practice together for a short time before beginning the survey.

5. Comparisons can be readily made of "protected" and heavily grazed areas, or of other range conditions.

6. The method is suitable for depletion surveys, for forage inventories, and for permanent plots; also for studies in range and pasture management, in erosion surveys, in agronomy, and in ecological observations.

7. Furthermore, it makes possible a statistical analysis of the data by probable errors, by analysis of variance, or by the correlation method.

Dalke, P. D. 1937. The cover map in wildlife management. *Journal of Wildlife Management* 1:100–105.

Mapping out management areas, study areas, or other areas of interest and identifying key features on those maps is a requisite to most wildlife management plans. Dalke's paper proposes the logical next step: applying vegetation and other habitat features to a practical map for wildlife management based on the ecological succession of plant associations. Dalke's objective was to accurately depict habitat for wildlife on any scale, and this short publication formed the basis for an explosion of map development—including computer applications using geospatial data, GIS, and GPS—in natural resources management. Maps are readily available today in a variety of formats, from hard copies to those available online. Furthermore, they are becoming increasingly sophisticated as remote sensing continues to evolve. Although the technology available to us is greatly advanced compared to what was available to Dalke in 1937, the basic concept remains the same.

RELATED READING

Beasom, S. L., E. P. Wiggers, and J. R. Giardino. 1983. A technique for assessing land surface ruggedness. Journal of Wildlife Management 47:1163–1166.

Copeland, H. E., K. E. Doherty, D. E. Naugle, A. Pocewicz, and J. M. Kiesecker. 2009. Mapping oil and gas development potential in the US Intermountain West and estimating impacts to species. PLoS One 4(10): e7400. doi: 10.1371/journal.pone.0007400. Accessed 25 June 2012.

Hepinstall, J. A., and S. A. Sader. 1997. Using Bayesian statistics, thematic mapper, and breeding bird survey data to model bird species probability of occurrence in Maine. Photogrammetric Engineering and Remote Sensing 63:1231–1237.

Kiesecker, J. M., H. Copeland, A. Pocewicz, and B. McKenney. 2010. Development by design: blending landscape-level planning with the mitigation hierarchy. Frontiers in Ecology and the Environment 8:261–266.

Knick, S. T., and J. T. Rotenberry. 1998. Limitations to mapping habitat use areas in changing landscapes using the Mahalanobis distance statistic. Journal of Agricultural, Biological, and Ecological Statistics 3:311–322.

Mead, R. A., T. L. Sharik, S. P. Prisely, and J. T. Heinen. 1981. A computerized spatial analysis system for assessing wildlife habitat from vegetation maps. Canadian Journal of Remote Sensing 7:34–40.

O'Neil, T. A., P. Bettinger, B. G. Marcot, B. W. Luscombe, G. T. Koeln, H. J. Bruner, C. Barrett, J. A. Pollock, and S. Bernatas. 2005. Application of spatial technologies in wildlife biology. Pages 418–447 *in* C. E. Braun, editor. Techniques for wildlife investigations and management. Sixth edition. The Wildlife Society, Bethesda, Maryland, USA.

THE COVER MAP IN WILDLIFE MANAGEMENT[1]

Paul D. Dalke

The methods employed in making cover maps vary from those showing only the broad types of cover to intricate systems for the most part impractical. Wight ('34) developed a system of cover mapping for the purpose of wildlife management based upon the ecological succession of plant associations. The correlation of the major plant associations into an orderly arrangement of ecological succession is the working basis for this system of cover mapping. This system has been used successfully in Michigan for the past five years and its simplicity and ease of interpretation are rapidly bringing about its use in several other sections of the country.

In regions where the ecological succession of plant communities has been determined, much of the data necessary for a cover mapping legend are already available. Nichols ('13a, '13b, '14, '15, '16) and Lutz ('28) in discussing the vegetation of Connecticut and upland succession in southern New England, present data that have been very helpful in the defining of types for a mapping system. Cover types based upon ecological succession are now used in Connecticut and are well adapted to other eastern localities.

The U. S. Geological Survey topographical sheets have served as the basis for all maps of the state since Connecticut was first surveyed some 40–50 years ago. Since then, many changes have taken place in the landscape. The State Planning Board initiated an aerial mapping program for the entire state about three years ago. The aerial maps (Plate 10) have proved very valuable and for many purposes, superior to the U. S. Geological Survey topographical sheets. The aerial maps have the advantage of being available in several sizes, roads and streams are accurately shown, and wooded areas of both conifers and hardwoods are easily recognized. They also show in farm territory, the large number of fences and stone walls that could otherwise be mapped only with much difficulty. The aerial mosaic prints not only solve the problem of accurately locating fences and stone walls, but facilitate the identification of the boundaries of seepage or periodic swamp areas regardless of the kind or amount of cover occurring on them. Since seepage areas are used mostly for pastures, the cover is usually either shrubby, herbaceous, or with a light mixture of tree species (Plate 10).

The mosaic maps are available for the entire state and are now used by several state departments. The most practical size seems to be the 19"×25" sheets with a scale of 1" = 1,200' or 4.4" = 1 mile. This size mosaic print represents five minutes of latitude and longitude. The scale gives good results in cover mapping and because of the small size of fields in the farm-game areas, it

[1] The author wishes to acknowledge the helpful criticisms and suggestions of Prof. H. M. Wight, Dr. R. E. Trippensee, and Dr. P. F. English, in the preparation of this paper.

100

Scale 1″ = 1200′

Aerial map—an excellent base for cover maps
(Courtesy Fairchild Aerial Surveys, Inc.)

is our experience that a field map on a smaller scale is impractical.

The location of each mosaic print is easily accomplished by the aid of a small guide to all the U. S. G. S. quadrangles occuring within the state. In addition to the latitude and longitude marks on the corners of each mosaic, the quadrangle number and the number within the quadrangle appears in the lower right hand corner. Each U. S. G. S. quadrangle requires nine of the mosaic prints, which are numbered from left to right as follows:

1	2	3
4	5	6
7	8	9

In the preparation of base maps for field use (Plate 11) tracings are first made of the aerial mosaic maps. The principal features of the landscape are outlined in ink and the boundaries of the fields and woodland are traced in pencil. Brown line prints are then made for use as base filed maps. Where boundaries of fields, woodland, or other areas need correcting, the pencil lines can be easily erased and the correction made on the office tracing at the time the cover map symbols are put on. The resulting cover map is then correct in all details. Symbols for roads, trails, fences,[2] streams, etc., conform to accepted map standards.

Connecticut is not included in the rectangular system of land surveys and in order to facilitate the recording of

field data for management purposes, a grid of lines running north-south and east-west has been arbitrarily drawn on each base field map (Plate 11). The resulting squares are designated by numbers and letters and represent approximately 20 acres each. Frequently it is important to indicate an exact spot within a 20-acre area. These squares, which have been ruled off as solid lines are further divided into-5-acre areas by dotted lines as shown in Plate 11. It is possible then to confine the recording of field observations within a definite 5-acre unit. For the convenience of the field worker, 5-acre units within each 20-acre square are numbered as follows:

1	2
3	4

The complete notation for the exact location of any observation can be written as a fraction, the numerator expressing the location within any given area and the denominator indicating the number of the quadrangle and the sheet number within the quadrangle. $\frac{(6\ B\ 4)}{(21-4)}$ represents the location of a small pond in the southwest portion of the map in Plates 11 and 12.

COVER SYMBOLS

To facilitate the preparation of cover type legends the major plant associations are divided into those successions originating from land or upland types (xerosere) and those originating from lakes, ponds, streams, or swamp types (hydrosere). In addition to this broad classification, there is the seepage area, a periodic or permanent swamp originating from springs.

[2] In the area illustrated in this paper, practically all the fences are stone walls. Because of the small size of fields, the conventional symbols for stone walls would occupy too much space. A solid black line was used on this particular map as a stone wall symbol.

Plant Succession Symbols
Origin from land (xerosere)

A4 Herbs
A5 Shrubs, deciduous
A5e Shrubs, evergreen
A6 Gray birch
A7 Red cedar
A8 Pitch pine
A9 Aspen
A10 White pine

A11 Pine or spruce plantation
A12 Hardwood plantation
A13 Oak-hickory-red maple-ash
A14 Hemlock
A15 Mixed hardwoods and conifers
(oak-hickory-birch-maple-pine-hemlock)
A16 Beech-birch-maple-hemlock.

Origin from open water (lakes, ponds, and streams; hydrosere)

Marsh Phase

B1 Submerged vegetation (pond weeds)
B2 Floating vegetation (water lilies)
B3 Emergent vegetation (bulrushes)
B4 Sedges-grasses
B5 Shrub-herb-sedge

Bog Phase

E1 Sedge-sphagnum-loosestrife
E2 Shrubs (heaths)-sphagnum
E3 Tamarack-spruce

Wooded Swamp Phase

B6 White cedar or tamarack
B7 White cedar plantation
B8 Red maple-yellow birch-black ash

Stream Flood Plain

P1 Emergent vegetation
P2 Mixed herbaceous
P3 Shrub (willow-dogwood-alder)
P4 Willow-cottonwood-elm-silver maple

Seepage-Swamps (periodic or permanent spring swamps)

C1 Sedge-grass-herbs
C2 Alder-dogwood-willow

C3 Spirea-shrubby cinquefoil
C4 Red maple-elm-birch-blue-beech hickory

Other Symbols

Fence row plant densities		*Timber Density*		*Understory or shrub density*	
Clear or light	|	Scattered	|	Scattered	—
Medium	||	Medium	||	Medium	=
Heavy	|||	Heavy	|||	Heavy	≡

Timber Age

0– 10 years—1
11– 20 years—2
21– 40 years—3
41– 60 years—4
61– 80 years—5
81–100 years—6
101–120 years—7
A—all ages

Cropped Land

al—alfalfa
cn—corn
gr—garden
pot—potatoes
t—tobacco
cr—cereals
ve—vegetables

*Vegetation in
Fence Rows*

s—shrubs
t—trees
h—herbs
m—mixed

Plantation species and miscellaneous

wc—white cedar
rp—red pine
wp—white pine
scp—Scotch pine
ns—Norway spruce
ws—white spruce
wo—white oak
wa—white ash

c—cropped land
or—orchard
p—pastured
x—scattered conifers
⊗—scattered apple trees
tp—tulip poplar
ro—red oak
sm—sugar maple

UPLAND COVER TYPES

The upland succession is designated by a capital A followed by a number which represents a type or association. The cover types A1, A2, and A3, have been omitted because they concern

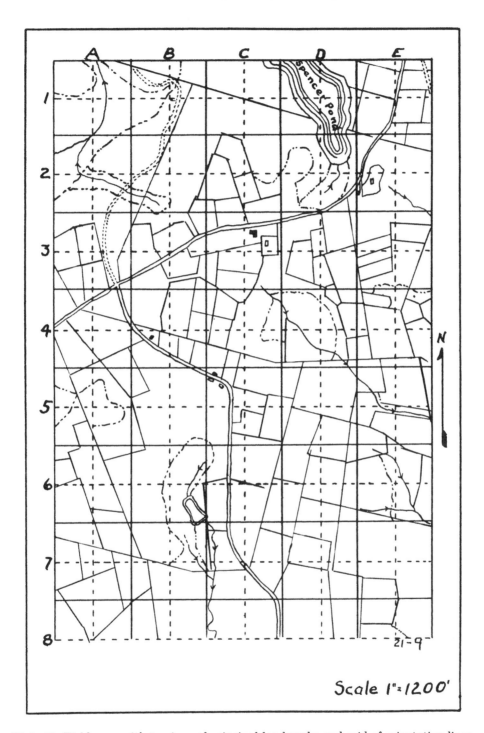

Plate 11. Field map with tracings of principal landmarks and grid of orientation lines.

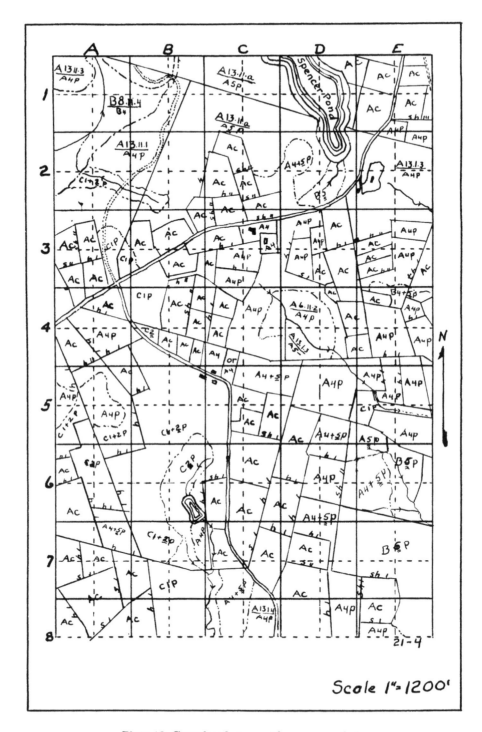

Plate 12. Completed map with cover symbols.

only the crustose and foliose lichens and mosses. For more detailed ecological work, however, they may be included.

The A4 type includes all herbaceous growth. The pasture and abandoned field are representative of this type. Areas in the A4 type are usually pastured, in which case the small letter p is added to the symbol.

The symbols for shrubs (A5 and A5e), indicate deciduous and evergreen species, respectively. In Connecticut the latter may be ground juniper, mountain laurel, or rhododendron. In areas where two species are about equally represented as for example gray birch and red cedar, the symbols are combined as A6&7. In areas where one species forms at least 80% of the stand, only one is indicated, unless occurring as an understory. In wooded areas both the timber species in the overstory, and shrub species in the understory are shown by symbol. The former occupying the numerator of the fraction and the latter the denominator, as for example $\frac{A\,13}{A\,10}$ or $\frac{A\,13}{A5e}$.

There are numerous plantations of red, white, and Scotch pine, Norway and white spruce both on state and private lands. These plantations are designated by the symbol A11 with a subscript to indicate the species. Smaller areas of hardwood species are also to be found on state lands and these are designated by the symbol A12 with a sub-script for the species symbol.

The end series in the upland types is the climax forest usually composed of American beech, yellow birch, hard maple, and hemlock. On certain sites, the end series appears to be the oak-hickory association. The composition of the climax forest varies considerably in different parts of the state.

SWAMP COVER TYPES

The plant succession initiated from open water, marshes, or streams is all classified under swamp cover types. The term swamp is used in a generic sense and includes both the herb, shrub, and tree stages of the hydrarch succession. For convenience in cover mapping, the swamp is sub-divided into the marsh phase (which contains only the herbaceous and shrubby species), the wooded swamp phase, the bog phase, and the stream flood plain.

The B1, B2, and B3 cover types are of importance in waterfowl management. The B1 type includes only the submerged vegetation such as pond weeds. The B2 type is composed of the species whose leaves float on the surface such as waterlilies. The B3 type is characterized by emergent vegetation represented by bulrushes, pickerel weed and cattails. The B4 cover type of sedges and grasses is common and constitutes some of the best winter pheasant roosting sites. The B5 type is also useful as cover for pheasants if not too much overgrown with shrubs.

The vegetation in a wooded swamp in the southeastern portion of the state may be either the coast white cedar or the typical red maple association. The B6 type contains tamarack if located in the northern part of the state, and coast white cedar if in the southeastern part of the state. The B8 type of red maple, yellow birch, and black ash is common throughout the state.

The bog phase of the swamp succession is classified into three types. The E1 type is made up mostly of sedge, sphagnum, and loosestrife. Following this initial stage is the E2 type or the shrubs and sphagnum. The shrubs in this type are mostly the heaths or ericaceous plants. The end succession, E3, of the characteristic bog vegetation is found in the tamarack-spruce association.

The succession of vegetation along the flood plains of the larger rivers such as the Connecticut and Housatonic is designated by the symbols, P1, P2, P3, and P4. While these cover types are very restricted in size and their occurrence is rather rare, they deserve a place in an ecological classification of plant types.

The seepage-swamp is the most common type of swamp in Connecticut and owes its existence to the relation between the topography and the ground water level. Seepage-swamps are usually situated in shallow depressions and on side hills where the water runs off as fast as it seeps to the surface. Where the ground water level is sufficiently high to allow a flow of water the year round, the swamp is known as a permanent spring- or seepage-swamp. Where the ground is saturated with water at certain seasons and dry at others, a periodic seepage-swamp exists.

The initial stage in either the periodic, or permanent, spring swamp is the sedge, grass and herbaceous association (C1). The shrub stage (C2) which follows is similar to the lake-swamp shrub stage and is characterized by swamp alder, dogwood-poison sumac-willow and spirea. In the northwestern portion of the state, a separate shrub association is very common and is designated as the spirea-shrubby cinquefoil type (C3). The tree stage (C4) of the seepage-swamp may contain any of the species characteristic of a permanent swamp such as red maple, American elm, yellow birch and also blue beech, hickory, tulip-tree, and sycamore.

The character of the vegetation growing along fence rows or stone walls is designated by a system of vertical lines (|, ||, and |||) indicating a sparse, medium, or heavy vegetation. Symbols of s, t, and h are used to show whether the vegetation is shrubby, arboreal, or herbaceous. In case all three types of plants occur, the letter m indicates a fence row of mixed types. In farm-game areas, the fence row often forms one of the important elements of permanent cover and should therefore be indicated on the cover map.

Timber density also is indicated by vertical lines (|, ||, |||) accompanied by the type symbol and age class. Timber age is expressed by age classes and indicated on the cover map by arabic numbers. For example, the 0–10 year age class is denoted by 1, 11–20 by 2, 21–40 year age class by 3, and so on.

From a wildlife management aspect, the younger age classes are the most important. Food is more abundant and the vegetation provides better protective cover. It is for these reasons that the first 20 years of a stand are divided into the 0–10 and 11–20 year age groups. The remainder of the age classes are separated by 20 year intervals. Reproduction, and underbrush or shrub density is expressed by 1, 2, or 3 horizontal lines such as B6≡.

The tree association symbols indicate first the type, then density and age class such as B8.11.4. This type frequently contains an understory of shrubs, so the complete symbol appears as $\dfrac{B8.11.4.}{B6=}$. The uses of the symbols are demonstrated in Plate 12. This map is a tract of typical farm-game territory in the highlands of eastern Connecticut. The entire area, 6,400 acres in size and managed as a State-regulated shooting ground, was cover-mapped during November and December in an actual field time of 40 hours, or an average of two sections per day. Cover mapping in Connecticut can be most ef-

ficiently done in the autumn after the leaves have fallen. Visibility is greatly increased and fence rows and shrub areas reveal their true cover value.

The method just outlined has a specific use only for southern New England. The principles of plant succession, however, may be employed as a basis for cover mapping in all sections of the country. The method is sufficiently flexible to permit any degree of accuracy or detail in making up the cover map. As methods of estimating available game food are perfected, such information should be included on the cover map.

LITERATURE CITED

LUTZ, HAROLD L.
   1928. Trends and Silvicultural Significance of Upland Forest Succession in Southern New England. Yale Univ. School of Forestry Bull. No. 22.
NICHOLS, G. E.
   1913a. The vegetation of Connecticut. I. Phytogeographical Aspects. Torreya, Vol. 13, No. 5. pp. 89–112.
   1913b. The Vegetation of Connecticut. II. The Virgin Forests. Torreya, Vol. 13, No. 9, pp. 199–215.
   1914. The Vegetation of Connecticut. III. The Plant Societies on Uplands.

Torreya, Vol. 14, No. 10, pp. 167–194.
   1915. The Vegetation of Connecticut. IV. Plant Societies in Lowlands. Bull. Torrey Bot. Club. V. 42, No. 4, pp. 169–217.
   1916. The Vegetation of Connecticut. V. Plant Societies Along Rivers and Streams. Bull. Torrey Bot. Club. V. 43, No. 5, pp. 235–264.
WIGHT, H. M.
   1934. The Cover Map and Game Census in Pheasant Management. Trans. Twentieth American Game Conference, pp. 334–339.

Paul D. Dalke
Connecticut Cooperative
Wildlife Research Station
State College
Storrs, Conn.

Gleason, H. A. 1939. The individualistic concept of the plant association. *American Midland Naturalist* 21:92–110.

The second name that every ecologist thinks of when you say "succession" is Gleason (the first should be Clements). Gleason's theories concerning succession are based on the interaction of the individual plants and animals, whereas Clements viewed succession as changes in one community over another. Gleason's landmark paper asks two key questions about succession: What is the best way to classify vegetation on the land? What factors have to be considered in making classifications?

Plant associations are commonly used in the literature to describe vegetation, often very loosely or without thought as to what a plant association is. Gleason reviews the three basic theories of what constitutes a plant association. One theory is that a plant association is a quasi-organism composed of individual plants and animals held together by a band of interdependence. The second theory is that the association is not an organism but a series of separate similar units repeated in numerous situations. The first views the association as an organism, the second as a species. The third theory sees the plant association as a temporary and fluctuating phenomenon dependent on surrounding vegetation and the environment. Gleason puts forth his reasoning for this individualist concept of the plant association and outlines why "a precisely logical classification of communities is not possible."

RELATED READING

Bray, J. R., and J. T. Curtis. 1957. An ordination of the upland forest communities of southern Wisconsin. Ecological Monographs 27:325–349.

Crocker, R. L. 1952. Soil genesis and the pedogenic factors. Quarterly Review of Biology 27:139–168.

Horn, H. S. 1974. The ecology of secondary succession. Annual Review of Ecological Systems 5:25–37.

Odum, E. P. 1969. The strategy of ecosystem development. Science 164:262–270.

Shelford, V. E. 1913. Animal communities in temperate America. University of Chicago Press, Chicago, Illinois, USA.

Watt, A. S. 1947. Pattern and process in the plant community. The Journal of Ecology 35:1–22.

Whittaker, R. H. 1953. A consideration of climax theory: the climax as a population and pattern. Ecological Monographs 23:41–78.

Whittaker, R. H. 1962. Classification of natural communities. Botanical Review 28:1–239.

Whittaker, R. H. 1970. Communities and ecosystems. Macmillan, New York, New York, USA.

Reprinted by permission of *American Midland Naturalist.*

The Individualistic Concept of the Plant Association

H. A. Gleason

The units of vegetation, which form the subject of our discussions this week, were just as easily visible to primitive man as they are to us today. They were equally visible to the sages of Greece and Rome two thousand years ago, to our European ancestors a thousand years ago. Through all these long years they were neglected by scientists, whose thoughts were turned in different directions, but they were recognized by the laity. As a result the languages of Europe all contain a number of terms which refer, usually rather loosely, to vegetation.

Certainly every botanist who studied plants in their natural habitat during the eighteenth century was familiar with many types of vegetation, but to them he gave little thought, possibly considering that they were not proper subjects for scientific investigation. To see vegetation units is one thing; to take cognizance of their existence, to investigate their nature, to become aware that they have structure and behavior which may be analyzed, to formulate a philosophy in explanation of them, are entirely different matters.

Four our purposes, and with the admission that many desultory references to vegetational units may be found in the works of earlier authors, we may say that the first definite, scientific discussion of the subject may be attributed to Grisebach, and we may repeat his often quoted words: Ich möchte eine Gruppe von Pflanzen, die einen abgeschlossenen Charakter trägt, wie eine Wiese, ein Wald, u.s.w., eine pflanzengeographische Formation nennen. This was in 1838, exactly a century ago, and we may congratulate ourselves that in this conference we are celebrating the centenary of the association-concept.

During the next sixty years, the study of vegetational units was sporadic, mostly superficial, and usually purely descriptive in nature. True, we can sometimes see in the literature of this period statements which may be interpreted as applying to the underlying philosophy of the association, but in most cases such statements are casual or accidental and do not indicate that the authors had given deep thought to the fundamental nature of the vegetation which they described.

Not until the advent of the twentieth century did botanists turn their minds seriously to the consideration of underlying questions. Since that time we have made great progress. We have developed methods for the exact observational study of the association. We have recognized conditions and processes in their development, their existence, and their disappearance, and these conditions and processes are quite unlike anything in the life-history of an individual plant or animal, so that our recognition of them has required the development of new habits of thought. We have described in modern terms vegetation from nearly all parts of the world. We have developed

92

systems of classification, by which the units of vegetation may be orderly arranged. We have invented and brought into accepted use a new terminology by which these conditions, processes, structures, and concepts may be described and discussed.

There is, however, one important question which has not yet been settled to the satisfaction of all concerned. This is the fundamental question, basic to all our work: What *is* a plant association?

Out of the thousands of pages of literature which have been used in expounding various views on the matter, three well marked theories may be chosen, and all others may be regarded as merely variants from them. These three are:

1. The association is an organism, or a quasi-organism, not composed of cells like an individual plant or animal, but rather made up of individual plants and animals held together by a close bond of interdependence; an organism, or a quasi-organism, with properties different from, but analogous to, the vital properties of an individual, including phenomena similar to birth, life, and death, as well as constant structural features comparable to the structures of the individual.

2. The association is not an organism, but is a series of separate similar units, variable in size but repeated in numerous examples. As such, it is comparable to a species, which is also composed of variable individuals. Under this view, the association is considered by some to be a concrete entity, merely divided into separate pieces, while by others the association as a whole is regarded as a mental concept, based on the common characters of all its separate pieces, and capable of typification by one or more of those pieces which most nearly approach the average or ideal condition.

3. The vegetation-unit is a temporary and fluctuating phenomenon, dependent, in its origin, its structure, and its disappearance, on the selective action of the environment and on the nature of the surrounding vegetation. Under this view, the association has no similarity to an organism and is scarcely comparable to a species.

In the original paper, in which this idea was presented to the botanical public in 1926, it was called the individualistic concept, a term which may well be continued.[1]

Whether fortunately or unfortunately, my own work during these twelve years has been wholly taxonomic. Nevertheless, observation of numerous plant associations during my field work in this period has merely intensified my own belief in the fundamental truth of the individualistic concept. The exposition of the concept which follows is merely a re-statement of the subject in different terms; it is in no way different in principles or conclusions from the first presentation in 1926.

The argument for the individualistic concept rests on a series of theses,

---

[1] Gleason, H. A. The individualistic concept of the plant association. Bull. Torrey Bot. Club **53**(1):7-26, 1926.

each of which is so obvious, so well known, so universally understood and accepted by all ecologists, that none of them requires prolonged discussion.

1. Every species of plant has reproductive powers in excess of its need.

The land surface of the world is already fully occupied by plants. Room for additional plants is made available only by the death of plants now existing. If seed germination were always perfect, if there was no mortality among plants before reaching their reproductive stage, it would be necessary for each existing plant to produce only one seed in order to perpetuate its species and to maintain the existing number of individuals. On the contrary, every species of plants produces a considerable number of seeds or other propagating bodies, often yielding an annual crop over a long period of years. Since the world is full of plants, it is a fact that, on the average, only one of them comes to a state of full maturity, but huge numbers of seeds or other propagules are capable of growth and ready to grow if favorable conditions are offered. A well kept lawn, for example, may produce very few weeds during the course of the summer, but if the lawn is plowed, the same expanse will promptly develop an astonishing crop of weeds of many species. The bottoms of drained ponds, the first season after drainage, produce many plants of terrestrial species. These and other examples of the same sort are so well known and so conspicuous that we may safely state that the surface of the world is heavily planted with an excess of seeds, most of which never develop, but many of which will develop if a favorable opportunity is offered.

2. Every species of plant has some method of migration.

The means of migration are well understood and require no discussion. The effectiveness of migration is often not appreciated. The distance to which some seeds may be carried by currents of water, by wind, or by birds is known to some degree, but the migration of other less favorably adapted species is also remarkable. Such seeds may be carried by rodents or ants or washed away by heavy rains. It is a demonstrable fact for all plants, without regard to their methods of migration, that more seeds are finally deposited near the parent plant than are carried to a great distance, the number decreasing, roughly speaking, inversely as the square of the distance. But effectiveness of migration is increased by various kinds of accidents and also by the known longevity of seeds in many species. We may therefore conclude, and our conclusion is supported by direct evidence familiar to all of us by personal experience, not only that the world is heavily planted with seeds, as stated under our first thesis, but also that these seeds come not only from the existing plants of the immediate vicinity, but also from plants at some distance. In any unit of vegetation the potentiality of plant production includes not only the natural species of this unit, but numerous species not now found in it and derived from other vegetational units of different character.

3. The environment in any particular station is variable.

Probably the simplest instance of variability which may be mentioned is light. Each day starts at midnight with light at or near zero. A little before sunrise the light curve begins to rise, reaches a theoretical maximum at noon and drops to zero shortly after sunset. The length of the curve varies with the season, being longest in summer and shortest in winter; the amplitude of the curve varies in precisely the same way, reaching its maximum on June 22 and its minimum six months later. Any and all of the 365 curves which constitute a year's cycle may be subject to irregular reductions in amplitude by cloudy weather. Nor is the quality of the light constant, but varies somewhat with the altitude of the sun and the condition of the atmosphere. These are all relatively simple matters, but there are also variations due to sun-spots, and there may have been and in fact may still be variations due to changes in the inclination of the earth's axis or the eccentricity of its orbit, or to other causes even more remote and far slower in their action. Locally, for many plants the light is changed again by the shade of taller plants, and this shade varies in its effectiveness from hour to hour, from season to season, and as the shading trees grow taller, from year to year.

The variation in temperature of the surrounding air shows similar irregular variations superposed on cyclic progressions in a way quite similar to the variation in light. Soil moisture is more irregular, with abrupt rises followed by longer periods of decrease, and with great variability in the amplitude of the curves. Available soil moisture is another question, based primarily on the total soil moisture, but complicated by matters of temperature, acidity, and other abstruse conditions. Similar conditions hold for every other factor of the environment taken individually and with all of them taken collectively: they are complex and variable to the last degree.

A second class of environmental variations may be called fluctuations. They are illustrated by our irregular alternation of cold and warm, of dry and wet years, of late and early seasons.

Still a third class of variations is important in its effect on plant life, and includes cumulative environmental changes which progress over a period of years or centuries or ages. Such, for example, are the silting up of a pond, the deepening and widening of a ravine by erosion, the exhaustion of soil fertility by percolation of rain water, the accumulation of humus, the increase in temperature following the retreat of a glacier, or the decrease in rainfall in the rain-shadow of a mountain range during its elevation. Although excessively slow, the cumulative effect of these environmental changes is ultimately profound.

All three of these classes of environmental variation are in operation simultaneously in every situation. The first class is regular and predictable; the second class is irregular and unpredictable; the third class is slow and often immeasurable. The amplitude of the fluctuations in the second class is normally much greater than the steady progression of the third class. A single year of deficient rainfall may cause a greater change in the depth of water in a pond than fifty years of silting, or may have more effect on the crops of Montana than five thousand feet of additional elevation of the Rocky Mountains. If we assume that we are now approaching another advance of continental glaciers, the annual drop in temperature associated with it is far less than the fluctuation from a warm to a cold year. Nevertheless, the only fundamental difference between the three classes is the factor of time.

4. The development of a vegetational unit depends on one or the other of two conditions, the appearance of new ground or the disappearance of an existing association.

The appearance of new ground is a matter of very little importance. More land may be added around the coasts of our continents by further elevation of the coastal plain or the building up of the shores by coastal deposits. More land may be added in supra-alpine regions by reduction of altitude through erosion. More land can not be added by increasing rainfall in deserts: our deserts are already fully occupied by plant life; some deserts merely support less plant life than others.

In the vast majority of cases, an association appears on the ground previously occupied by a different association. It makes no difference whether the earlier association is removed by the slow processes of ordinary succession or suddenly by some cataclysm. If the existing vegetation is destroyed by the axe, by fire, by a landslide or by hot lava, some plants go first and some go last. If the vegetation is changed by a slow process of succession, some of the original inhabitants still go first and some linger. In both cases, some plants of the next association, the pioneers, appear first and others are delayed. The only essential difference between them is the factor of time, greatly shortened in the first case, often greatly prolonged in the second.

The various factors which collectively constitute the environment of a plant may be separately measured and diagrammed, although often with much difficulty, but no one has ever succeeded in reducing to a single statement or a single equation the total environment of any plant. For practical purposes the only measure of the environment is its result, as expressed in plant life. An oak tree may easily live to be three hundred years old. Let us see what this implies concerning its relation to the environment. It means that through this whole period, in spite of its extraordinary variability, the environment has never once exceeded the limits tolerated by the living protoplasm of the tree. The weather has never been too hot or too cold, too wet or too dry; the soil has never been too acid or too alkaline; no environmental factor has ever been too much or too little, or if the limits have sometimes been surpassed, it was for a period too short to be fatal.

The clearest example of temporary excess is the annual cold winter, during which the activities of the oak sink so low that we call the tree dormant. The oak is not only able to adapt itself to the variations in environment, but is able to vary its own life processes enormously to meet such a critical condition as winter cold. Not every plant meets the emergency in the same way. Some herbs actually die, but are fortunate in producing a seed which is not killed by cold and which produces another generation the following summer. There are still other plants whose seeds are also killed by the cold. These plants do not live with the oak, or, of they do, they must be re-established every summer by fresh seeds from a more favorable climate. Such plants are familiar as cultivated in our gardens. The canna and dahlia may flourish not far from the oak, but they disappear forever if not removed by us to a more favorable climate for the winter and replaced the following spring.

Now leaving these trite illustrations, which have been used merely to recall to our minds what we all know from common experience, let us consider the question from a broader standpoint. We at once arrive at the general theorem, that each plant seizes and uses the particular time-period during which the environment is in a condition suitable to it. When the environment passes these limits, the plant dies. For the old oak, this limit has not been reached for the three hundred years of its life and, so far as we can imagine, not for still longer periods in the past when preceding generations of oaks were growing on the same spot, nor will it be surpassed for many years in the future, during which the descendants of the present tree will be living there. For the dahlia, the limit is reached before the plant has completed a single generation: it is cut off in October while it still bears young leaves and unopened buds. During the same summer, the Galinsoga which infests our gardens in this vicinity may produce several generations. If our lives were measured by days instead of years, we can imagine an ecologist saying, "I have seen three generations of Galinsoga on this one spot of ground. Evidently we are dealing with a stable environment and Galinsoga will live here forever." He would be wrong. With our knowledge of vegetational conditions actually extending over about three centuries, we now say, "Oaks have occupied this spot of ground for three hundred years. Evidently we have here a stable environment and the oak will live here forever." If our lives were seventy centuries instead of seventy years, would we not see that we were again wrong? Like the short-lived Galinsoga, which utilizes the time period which we call summer, the long-lived oak is utilizing a longer time-period, and the time has been, and in the future will be again, when the race of oaks can no longer live on this particular spot.

During the time-period when a species may occupy a station, diverse effects are exerted upon the plants by the variable environment. Variations of the first type, which are regular, periodic, and uniform, such as night and day, winter and summer, act uniformly and unavoidably upon all plants. The physiological processes of every plant are adjusted to meet them. While their investigation forms an important part of plant physiology, their effect on plant distribution and their interest to the plant ecologist are negligible.

Variations or fluctuations of the second type, including such phenomena as cold and warm years, of dry and wet periods, of late and early seasons, have a pronounced ecological effect. This is evidenced in agriculture and horticulture by years of heavy or light yield, by the abundance or lack of pests and parasites, by the lateness or earliness of bloom or fruit, by the losses due to late frosts in spring or early frosts in autumn. If such fluctuation is maintained for even a few years, its results become so grave as to be of national importance, as shown recently by the effect of only four years of deficient rainfall in our Plains States.

The effect of these fluctuations on natural vegetation is precisely the same qualitatively, but less pronounced quantitatively, since crop plants are deliber-ately introduced into a region, while natural vegetation has been adjusted to the fluctuations through previous experience. In natural vegetation, the effect of fluctuation in the environment is seen in phenological phenomena, in the amount of annual growth, in the vigor of individuals, in the number of seeds produced, in the number of seedlings produced, and therefore in the relative number of individuals. These fluctuations rarely last very long, seldom affect-ing more than a single year; they therefore rarely cause the disappearance of a species or the appearance of a new one.

That fluctuations of environment affect the relative number of individuals of a species is a fact which has rarely been demonstrated in research in pure ecology, since accurate quantitative studies of vegetation have seldom been repeated in the same area over a series of years. I have personal records from the same area taken at varying intervals over twenty years and can assure you that they show conspicuous variations in the number of individuals of some species, and scarcely none for other species. We may assume that the physiological processes and environmental demands of the latter are so broadly adjusted that no fluctuations have ensued of sufficient magnitude to affect the plants. In the former group we may also assume that their physiological processes are more strictly defined or that the normal environment is already off the optimum for the species, so that fluctuations do interfere demonstrably with the number of individuals. Abundant records of this phenomena are available through the careful observations on our grazing lands. Here the environmental fluctuation is largely in the number of animals feeding on the plants, but the effect is precisely the same. Differences in grazing lead imme-diately to the reduction of certain species and the multiplication of others.

The rare disappearance of a species because of environmental fluctuation is probably due to the fact that the fluctuations of any one year have been re-peated at various times in the past, and species which would be exterminated by them have already been removed. Nevertheless, the long period of drought recently ended in the western states will probably have this effect in many instances, although I am not able to cite any single example at the present time.

The relatively rare appearance of a new species for the same cause, is due primarily to the time-factor involved in plant migration. Environmental fluc-tuations are of relatively short duration; migration of plants is slow and before additional species can reach the spot, migration has been stopped by the return of the environment to normal. If next winter were to be omitted and summer temperatures should continue for another twelve months without interruption, we still could not expect to find mangroves along the shores of the north Atlantic. Sufficient time must always be allowed for the reaction of plants to any environmental change.

The variations of the third type, the slow-moving, long-continuing environ-mental changes caused by physiographic processes, geological developments, or climatic changes, cause no directly observable or measurable effect on vegeta-tion, at least in most cases. In these days the effect of erosion during a single year can sometimes be measured or estimated but it has taken a century for it to make an impression on our minds. The silting of a pond with the result- ant changes in plant life is also fairly rapid, so that it has sometimes been observed and recorded by a single person during his own lifetime. Even in these relatively rapid examples, and of course in the slower variations of cli- mate, the effect of the change is always masked by the wider amplitude and quicker action of the fluctuations just discussed. Nevertheless, these fluctua- tions are based on a norm, and if the norm itself varies the amplitude of the fluctuations tend always to abate in one direction and to extend in another.

Suffice it then to repeat that on every spot of ground the environment is continually in a state of flux, and that the time-period in which a certain environmental complex is operative is seized on by the particular kinds of plants which can use it. The vegetation of every spot of ground is therefore also continually in a state of flux, showing constant variations in the kinds of species present, in the number of individuals of each, and in the vigor and reproductive capacity of the plants.

With this idea of the time-period in our minds, let us look back to the origin of some hypothetical association, which is now populated by three spe- cies, designated X, Y, and Z. This association originated in one of two ways, on new and denuded ground bare of plants, or in replacement of a preceding association.

In the former case, of denuded or new ground, species X, Y, and Z must have been derived from some neighboring area or areas, and these must have been within migrating distance. In this neighboring area species X, Y, and Z may or may not have been living together. In these original stations the environment was variable, their seed production was variable, and the migration of these seeds was controlled largely by those inexplicable factors which we call chance. Some times, possibly seldom or possibly frequently, their seeds were carried to this particular spot of ground which we are considering. On this spot of ground the environment was also variable, and in this variation a certain time-period began which permitted the growth of species X, Y, and Z. Their seeds happened to be available, as association of X, Y, and Z appeared, and the time-period during which it may continue has not yet ended.

If our hypothetical association appeared in succession to some preceding association, the circumstances are still similar. The plants of the earlier asso- ciation, which we may designate A, B, and C, eventually reached the end of their time-period and disappeared. Their disappearance may or may not have been simultaneous. The particular environmental changes which facilitated the entrance of X, Y, and Z may not have been the changes which caused the loss of A, B, and C. The former may have entered first, so that at one time the association consisted of all six species, or entry and disappearance may have been more or less serial, so that the association passed through suc- cessive stages of A, B, C; X, B, C; X, Y, C; and finally X, Y, Z.

In either case, whether the association developed *de novo* or as a stage in succession, the variations of the environment may have been such that the permanent entry of X, Y, and Z was preceded by shorter time-periods during which one or some of them was temporarily present. In short, any number of different sequences may be imagined, all based on variation in the environment and on the species then available through the chances of migration. As a further generalization we may also postulate that species J, K, and L, although adapted to the new environment, did not enter because their nearest stations were beyond the reach of migration; that M, N, and O did not enter because of the chance failure of migration, but may be expected to appear at any time; and that P, Q, and R actually migrated into the area but were unable to utilize the environment.

I have seen this variable result of the interaction of migration and environment beautifully illustrated along the shore of a lake, where winter storms and ice action keep the shore denuded of permanent vegetation. Each spring some storm washes on the beach a long strip of flotsam, twigs, fragments of bark, and dead leaves, and in it seedlings develop. Some years the strip of drift is marked by hundreds of seedlings of willow, at other times by seedlings of elm, and again by seedling red maples. In each case the environment is the same. The difference is caused by migration. If the flotsam is deposited at the particular time when maple seeds are falling in quantity, maple seedlings appear. A difference of a week or two in time results in willows, or in elms, or in no plant life at all.

Under ordinary circumstances, the annual fluctuations of environment are not sufficient to cause any profound or permanent change in the vegetation. Their effects are shown in the size and vigor of individuals or by changes in the relative number of individuals of different species. Only when the fluctuations reach a certain maximum or minimum do old species disappear or new species appear. But in the meantime the slow long-term variations in climate are still in operation and are cumulatively important in environmental selection.

An excellent example of these conditions became apparent during my early work in northern Michigan. Thousands of acres were then covered with a virgin forest of the so-called climax type, with beech, maple, and hemlock as the dominant trees. Of beech and maple there were millions of young trees, ranging from seedlings to tall specimens of the second story of the forest. But there were no seedling hemlocks, except in rare cases; all the hemlocks were mature trees, mostly 300-350 years old. About the beginning of the seventeenth century there had been some sort of an environmental change. Since that time the old hemlock trees had not died, but new hemlock trees had not been established. Another century or less would probably have caused the disappearance of the old trees, and the hemlock would have almost disappeared from the forest. Was that change of three centuries ago a mere fluctuation which has not yet returned to the normal, or was it the accretion of a climatic change to such a stage that it permanently passed the limit of tolerance of the hemlock?

In spite of this exception, I am quick to recognize that the effects of environmental change is less apparent in eastern America and western Europe, where most of our ideas on the association-concept have developed, than in certain other parts of the world. Our vegetation is composed very largely of perennial species, and even the herbs are generally of long duration. That means that these species are already adjusted to all ordinary fluctuations of the environment, so that the vegetation persists with relatively little change through cycles of high or low temperature, or of excessive or deficient rainfall.

But in other parts of our country, where the vegetation is largely annual, fluctuations in environment are followed immediately by a reaction in the plant life. In the Mohave desert, rains at the proper season induce a great development of the desert annuals, which sometimes carpet large expanses of the land with a sheet of color. Species appear which have not been detected before and rare species become common. The next year, if it fails to receive rain at the critical time, may see the display replaced by mere bare areas of desert.

For ordinary plants, rooted to one place in the ground, environmental selection depends wholly on excess seed production and migration. But many animals, gifted with powers of voluntary locomotion, are continually selecting their environment or being selected by it. No such animal is restricted to any biotic association unless it alone affords him his necessary environment. The scarlet tanager, an inhabitant of the forest canopy, appears in my garden in town when cherries are ripe, and disappears when they are eaten. Migratory birds choose a new environment twice a year. The turkey buzzard of our southern states, except in its nesting, belongs to no biotic association, but enters any one in which it finds a temporary food supply.

The development of a new biota on any spot is therefore the result of environmental selection from the available population of adjacent areas. Similarly, the continuation on any spot of ground of what we have been pleased to call a biotic association is also due to the interaction of the same causes, available population and environmental selection.

While we may all admit the validity of this statement, one argument will immediately come to mind which will seem at first to minimize its importance, namely, that the changes in environment which I have mentioned are so slight that they produce relatively little effect. It is a fact that the majority of individuals of perennial plants continue through environmental fluctuation, because of the well known fact that a mature plant can successfully withstand conditions which would be fatal to seedlings or would prevent the germination of seeds. It is also a fact that plants are themselves a part of the environment, and their persistence always tends to reduce the effect of a variation in the physical environment. It is not until a reduction in their vigor or in their number has broken their environmental control that the full effect of an environmental change can be seen. The individualistic concept does not postulate annual changes of great magnitude, it merely emphasizes that a fixed and definite vegetational structure does not exist.

Another objection which may be brought forward is that the argument of a continuously fluctuating or steadily progressing environment, or both, can scarcely be reconciled with the sudden successional transitions from one association to another. This has already been answered in part, when it was stated that mature plants can withstand conditions impossible for their seedlings. When finally, as the result of cumulative changes, the environmental control of the mature plants is broken, the consequent change in the total vegetation is rapid and profound. In this connection it must also be remembered that the theory of sudden steps in succession is largely based on the abrupt transitions in space between different associations. As a matter of fact, these narrow transition lines are more apt to indicate a high degree of stability of vegetation than a rapid rate of succession. . In each of the adjoining associations variation in vegetational structure is always active, along the lines stated by me already, but the absence of cumulative progressive changes in environment, together with a high degree of environmental control by the plants, leads to narrow and well-defined lines between them.

All the statements which have been made so far concerning environmental variability have purposely been restricted to variation in one spot during the course of time.

Variability of environment may also be correlated with space. In time-variability, the environment changes from one time to another on the same spot. In space-variability, environment differs from spot to spot at the same moment. The idea is so axiomatic, so familiar to all, that it scarcely needs any discussion, yet some generalizations may be desirable before we attempt to relate the condition to the individualistic concept.

Space variations are in many ways analogous to time-variations, both in their nature and in their effect. Analogous to the fluctuations of environment in time, which fluctuations are comparatively uniform over a reasonable amount of space, are the abrupt fluctuations in space. These, as is well known, are mostly correlated with the soil, in its chemical and physical properties, its elevation, slope and exposure. It must be repeated that these are fluctuations in space, and emphasized that they are relatively constant on any one spot through a reasonable period of time. Such fluctuations are often of extreme importance to plant life. In any region, they tend to show a wider amplitude of variability than do time-fluctuations.

There are also gradual variations in environment, so gradual, in fact, that their effect can not be observed by comparison of adjacent regions. Such are the variation in mean temperature, in average rainfall, in length of growing season, in length of day and night. Between two adjacent regions, the difference in these conditions is so slight that they are concealed by the far greater and locally more important fluctuations mentioned in the preceding paragraph. But they are cumulative, and it is probably impossible to draw a line 500 miles long anywhere in the United States, certainly in the north-south direction, along which these climatic variations do not cumulate to such an extent as to cause a marked difference in the vegetation at the two ends of the lines.

The selective effect of space-variation on plants is very similar to that of time-variation. To any spot of ground come seeds of various species of plants. Those seeds which find in the spot the environmental conditions which they demand germinate and as adult plants become an integral part of the association. The population varies from place to place in a manner and to an extent determined by the variation in the environment. The population is chosen by environmental selection from the available immigrants.

There is nevertheless one important difference. In a time-series, the surrounding population remains relatively constant. From year to year it affords the same group of species from which immigrants may be drawn. In a space-series, the surrounding plant population may vary greatly from one end to the other. Not only is the environment different, but the immigrating species are also different and the resulting environmental selection is different.

Furthermore, time is always necessary for an environmental variation to exert its full effect on plant life. Fluctuations in time are often of short duration and have passed before their full effects can be realized. Fluctuations in space, on the contrary, are of relatively long duration, and on each spot the full effect of environmental selection can be reached.

In summary it may be stated that environment varies constantly in time and continuously in space; environment selects from all available immigrants those species which constitute the present vegetation, and as a result vegetation varies constantly in time and continuously in space.

Those who disagree with the individualistic concept will very properly raise at this time certain questions, based on facts which at first thought seem to invalidate the whole concept. Before these questions are stated here, so that their obvious implications may be refuted, two general statements may be introduced.

First, an association, or better one of those detached pieces of vegetation which we may call a community, is a visible phenomenon. As such it has dimensions and area, and consequently boundary. While its area may be large, the community is nevertheless a very tangible thing, which may be mapped, surveyed, photographed, and analyzed. Over this area it maintains a remarkable degree of structural uniformity in its plant life. Homogeneity of structure, over a considerable extent, terminated by definite limits, are the three fundamental features on which the community is based. Without these three features, Grisebach would never have published his statement of a century ago; without them, all our studies of synecology would never have been developed. Also, besides its extent in space, every community has a duration in time. Uniformity, area, boundary, and duration are the essentials of a plant community.

Second, every community occupies a position in two series of environmental variation. In the space-series, as the community exists *here*, in this spot, it is part of a space-variation, and its environment differs from the adjacent communities. In the time-series, as the community exists *now*, at this time, it is part of a time-variation and in its environment differs from the communities which preceded it or will follow it.

The individualistic concept postulates a continuous variation in space and time. How can we reconcile this with the admitted uniformity in space and time?

In any community of reasonable extent, the variation of rainfall, temperature, length of day, and similar factors from one end to the other is extremely small. Not only is their effect proportionately small, but this effect is overshadowed by the much greater seasonal fluctuation. Soil, also is often uniform over the whole community, and when it it not uniform but varies significantly within small distances, we are prone to overlook its effects and to classify the variable vegetation in a single community. More important still, the dominant plants, which are distributed over the whole area of the community, exert such a uniform effect on the other species that discrepancies in the physical environment are more or less smoothed out or obliterated.

Nevertheless, it is difficult or impossible to find in any community two quadrats which are precisely similar. The community is a complex or mosaic of slight irregularities, all of which blend into an entirety of apparent homogeneity. We have all known of this lack of perfect uniformity, and have endeavored to evade it by developing the concept of a minimal area, but we have failed to realize its significance as indicating the general variability of vegetation.

Cumulative progressive changes in environment are generally so slow in their development that they make no pronounced effect from one year to the next, while the wider swings of fluctuating environment are so short in their duration that their full effect is not experienced. Also, comparatively few observers have kept careful records of vegetational change over a series of years. Furthermore, as in space variation, environment control by plants tends to overshadow environmental variation. Nevertheless, succession, which is merely vegetational change, is accepted by all as a fact; exact statistical records, when available, do show continuous variations in structure, and in many locations complete vegetational changes have occurred within the experience of a single observer.

The postulated uniformity of the community is therefore far from absolute. A community is uniform, either in space or in time, only to a reasonable degree. This uniformity is sufficient to enable us to recognize the community and to accept it as a unit of vegetation, while its variability, although slight, is sufficient to indicate the impossibility of considering any such area of vegetation as a definitely organized unit.

If vegetation varies continuously in space, how can we explain the abrupt transitions from one community to another, which are so conspicuous a feature of natural vegetation in many regions?

Abrupt transitions are in every case correlated with abrupt variations in the environment or with abrupt differences in the immigrating plant population. Some abrupt changes in environment are due to physical conditions, notably the soil, which may change notably within a short distance. The other changes are due to environmental control of the physical factors by the plant life itself. These account for most of the abrupt transitions in the eastern states, or in any other region where a dense vegetation is possible. The sharp demarcation of zones around a pond or bog, for example, is caused almost entirely by vegetational control.

Abrupt differences in immigration exist only in areas which have recently been disturbed, such as an abandoned field, a lake shore recently worked over by waves, a ballast heap or a tract of newly filled ground. In such places the accidents of immigration often lead to the temporary establishment of distinct patches of vegetation, each characterized by one or a few species. Continued migration tends to smooth out these irregularities in a short time; environmental fluctuation favors certain species over others; denser growth leads to environmental control by certain species, and in the course of a few years such patchwork vegetation has blended to a relatively homogeneous community.

If vegetation varies continuously in space, how can we account for the repetition of the same vegetation in many separate communities?

The answer to this question is simple. There is no exact repetition of the same vegetation from one community to the next. There is an approximate repetition only.

It is a fact that in any region several to many examples of vegetation may be found in which the differences are so slight that they are not observed, or if observed are considered as unimportant and negligible. In any community absolute homogeneity is impossible, and the observed heterogeneity may well be due to chance. But in a single community, if it is large enough, differences between two ends may be discovered. Every ecologist who has undertaken quantitative analysis of vegetation will probably agree with this statement. I once examined two adjacent sections, each an exact mile square, of virgin hardwood forest. The soil was uniform and level and there was no surface drainage system, nor any indication of wetter and drier parts. Careful quantitative studies in each section showed conspicuous differences, not important differences, to be sure, which could not be explained by any visible feature. At the Biological Station of the University of Michigan, the aspen association, with a single continuous community some six miles long, exhibits demonstrable variation from one end to the other, with no visible reason.

Between two different communities, not too far removed from each other, the observable differences in structure are of essentially the same degree of magnitude as these fluctuations within the same community, and one may easily tend to credit them also to the effect of chance. But part of them, possibly only a small part, is due to the space-variation in the environment, and another part, again possibly small, is due to a difference in the available plant population upon which environmental selection operates. If a series of communities are observed at successively greater distances, these differences cumulate, so that those at the ends of the series may be strikingly different, although connected by imperceptible or apparently negligible intermediates.

One short-coming of our ecology has been that our field work has generally been confined to a small area. We have investigated and described all the associations in a small area, instead of trying to trace a single association over its whole extent. In any small area, environmental variation, essentially repeated in many spots, produces several well-marked types of environment, each characterized by a similar vegetation. We justifiably draw the conclusion, from this limited evidence, that association-types are definite. But as soon as we extend our observations, we begin to realize that each separate community is merely one minute part of a vast and ever-changing kaleidoscope of vegetation, a part which is restricted in its size, limited in its duration, never duplicated except in its present immediate vicinity and there only as a coincidence, and rarely if ever repeated.

In other words, the similarities between adjacent communities, which have led to the views that the association is analogous to a species, or analogous to an organism, are not perfect similarities. They are caused by *nearly* similar environmental selection, intensified by *nearly* similar environmental control, from a nearly similar population. In addition to the imperfections of similarity caused by chance, and largely masked by them, are other variations of a cumulative nature. These, increasing in importance and in conspicuousness as more distant communities are considered, finally lead to vegetation of such unlike nature that they would never be classed in the same association-type.

In my original paper on this subject I mentioned as an example the alluvial forests along the Mississippi River and its tributaries over a stretch of about a thousand miles from its mouth to the northward. From mile to mile these forests show no considerable change; over a space of a hundred miles the changes may or may not be of ecological importance and the more important differences are apparently due to lack of time for sufficient migration to smooth out the variation. Yet these differences cumulate. One by one species disappear; one by one other species appear, and by the time one has reached, say, Indianapolis, there has been an almost complete change in the appearance and composition of the forest. Or, if the observer swings more to the west and travels up the Missouri and the Platte, he can see the disappearance of species one by one without corresponding replacement, until the forest is reduced to a fringe of willows and finally disappears completely in western Nebraska.

Within the state of Michigan, the beech-maple climax forest, always considered to be a definite, well distinguished association-type, exhibits profound changes from one end of the state to the other.

I also venture to say, without personal experience to verify the opinion, that even more remarkable transitions might be discovered elsewhere. For example, the forests of the foothills of the Rocky Mountains in Colorado, composed there largely of *Pinus ponderosa*, might be traced northward with similar gradual variation, thence eastward along the northern boundary of the grassland in Canada, and again southward to the forests of Illinois, and lead us to the extraordinary conclusion that the *Pinus ponderosa* forests of Colorado represent the same association as the *Quercus velutina* forests of Illinois and the aspen groves of Manitoba.

Over such distances as the three I have mentioned the flora from which any area may be populated changes greatly. Such environmental factors as temperature, rainfall, and length of growing season also vary greatly and to an extent that completely surpasses the local fluctuatione between any adjacent communities. With the vegetation determined by environmental selection from the available plant population, and with both of these underlying features altered, obviously the resultant vegetation must also be entirely changed.

It must be remembered that we admit the essential uniformity of vegetation within a single community, and the frequent striking uniformity between adjacent communities. But the fact that these small cumulative differences do exist is basically important in the consideration of the general concept of the plant-association. They indicate that each community, and for that matter each fraction of one, is the product of its own independent causative factors, that each community in what we now choose to call an association-type is independent of every other one, except as a possible source of immigrating species. With no genetic connection, with no dynamic connection, with only superficial or accidental similarity, how can we logically class such a series of communities into a definite association-type? Truly the plant community is an individualistic phenomenon.

Every species of plant and animal migrates, whether as a mature individual, as many species of animals, or as a reproductive body, as the vast majority of plants. Among animals, migration is simetimes selective in its direction and goal, as illustrated by birds which follow definite routes to definite established breeding grounds. With other animals and with all plants, migration is purely fortuitous. It progresses by various means, it brings the organisms into variout places and to varying distances, but only those organisms which have reached a favorable environment are able to continue their life. Into this favorable environment other species also immigrate, and from all of the arrivals the environment selects those species which may live and dooms the others.

In this migration each migrating body acts for itself and moves by itself, almost always completely independent of other species. The idea of an association migrating *en masse* and later reproducing itself faithfully is entirely without foundation. Those cases in which there is a semblance of such a condition are caused by the proximity of the original association and the advantage which its species therefore have in migration. Even then, certain species always precede and certain others lag behind.

Summary

1. The ordinary processes of migration bring the reproductive bodies of a single plant or a species of plant into many places.

2. The ordinary processes of migration bring the reproductive bodies of various plants into the same place.

3. Of the various species which reach one spot of ground, the local environ-

ment determines which may live, depending on the individual physiological demands of each species separately.

4. On every spot of ground, the environment varies in time, and consequently the vegetation varies in time.

5. At any given time, the environment varies in space, and consequently vegetation varies in space.

6. A piece of vegetation which maintains a reasonable degree of homogeneity over an appreciable area and a reasonable permanence over a considerable time may be designated as a unit community. Within such an area and during such a period similarity in environmental selection tends toward similarity in vegetation.

7. Since every community varies in structure, and since no two communities are precisely alike, or have genetic or dynamic connection, a precisely logical classification of communities is not possible.

Although a summary is generally considered to close a discussion, it nevertheless seems desirable to indicate very briefly the relation of the individualistic concept to other current philosophies of the plant community.

The individualistic concept is totally at variance with the idea that the association is an organism, represented by many individuals, and also does not admit an analogy or homology between the species and the association. While affirming the existence of definite communities, characterized by reasonable uniformity over a considerable area terminated by a definite boundary, the concept denies that all vegetation is thus segregated into communities.

The concept is by no means opposed to the recognition of the synusia, or union, defining it as a group of plants whose physiological demands are so similar that they are regularly selected by the same environment and consequently regularly live together.

If classification of communities on the basis of floristic resemblance is rejected, and if broader areas of vegetation are searched for characters indicating genetic or dynamic similarity, these are first found in the broad floristic group now known as a biome. The plants of a biome, living together now as their ancestors have lived together and evolved together in the past, represent an environmental selection of a broad type, while lociation and faciation within the biome indicate the variability of the environment and the irregularity of migration.

DISCUSSION

Lippmaa: Though the theses of the speaker are acceptable the same can not be said of his conclusion that it is "impossible to choose any one phase of the vegetation as typical." The environment does not vary *continuously* in time and space. Abrupt changes caused by the differences in rock and soil composition, exposure, water supply, etc. are not uncommon. To such changes as well as those factors listed by the speaker effecting the development of vegetation must be added other eminently important ones such as the competition between the species composing the community and the time necessary for the stabilization of the community.

As regards "the exact repetition of similar vegetation" it is clear, that the units used in phytosociology are of a different kind than those of taxonomy. For the fundamental unit of taxonomy—the species—could hardly be considered "exact" by the chemist.

Gleason: Most of these remarks do not differ essentially from anything which I have said. The first topic regarding variation may require brief explanation. The curve, in which periods of slow and rapid variation alternate, still represents continuous variation.

In my early work I tried to read all the works on descriptive ecology dealing with American vegetation. I could scarcely find any two regions which contained the same associations. Conard's paper illustrated the same variability since he reported a Brometum in western Europe and England which has twenty-nine variants. These are almost certainly an expression of such cumulative differences in structure. In every small region, association-individuals may be found which are sufficiently alike to encourage us to classify them together, but extension of observation over greater areas shows that the small observed differences cumulate into important differences.

Emerson: It is useful to break down the larger categories into their units. It is also useful to build the units into larger wholes. I think by the same logic could be proven that the concept of the individual and of the species presents the same difficulties as the concept of the association. Is it not true that the biological integration of units is as real as the units?

Gleason: It was brought out in several places in my paper that the biological control by existing plants is an important feature of the environment.

Emerson: The speaker showed that each community is an individual, but I think it is also necessary to put them together again.

Gleason: The species are present, and these groupings of plants into communities exist. What I hoped to bring out here was that the particular grouping of plants is not found duplicated elsewhere. There are similarities among stands, or slight differences from stand to stand, and if you trace an association those little differences are cumulative.

Emerson: That is also true of the individual. A research taxonomist talks about species and races. Does the concept of a human race have meaning?

Gleason: I think the concept of race does have a meaning which is merely complicated by the intermediate forms, but we have no genetic relation or connection between these plant communities.

Emerson: One of the first principles of genetics is the reassortment of genes and chromosomes. I do not think the only reality is the individual and that the associations are not real integrated units.

Steiger: I agree with everything the speaker said about these variations in time and species, but all that is true of every living being. Plants are living beings and as such they do not possess the same degree of exactness, as NaCl for instance. Apparently we need a terminology.

Gleason: A classification of plant communities is certainly needed to enable us to talk about them conveniently.

Steiger: However, I would not compare an association with an individual. Even the comparison with the species is limited.

Spieth: As a taxonomist I do not like to hear associations compared with species. Any herring caught is going to have 60-65 vertebrae but it can not be predicted how many it is going to have. I think that is a very fundamental difference between the two.

Conard: I certainly enjoyed the speaker's analysis very much and I do not see any way of opposing his logic. Comparison of the association with the species has been stoutly opposed, but I feel that the same logic rules out the use of such terms as "species" or "human being". We are too different to be classified. It seems to me both from my own personal feeling and from reading the results of the work of others where the association as understood in the international sense has been used that there has been nothing in ecological inquiry which has been so fertile and so productive of results as the idea of the association. It is therefore so useful that whether logical or not, I am for it.

Gleason, Jr.: One can not deny the objective reality in a stand of vegetation any more than the objective reality in the daisy. But if all the daisies in the world were brought to some one location, that collection would be something very different from the concept of *Chrysanthemum Leucanthemum.* The latter is conceptual and purely subjective and very different from the other which is a specific collection of all the individuals which we have classified. There is a difference between human beings and *Homo sapiens.* The former have an objective reality, but the latter does not exist except in the mind of man. The same is true of any classification of vegetation. The vegetation exists and each individual stand exists, but the classificatory category exists only in the mind of man.

Emerson: I can not allow these remarks to go unchallenged. I think that the species has objective reality as well as the individual. If you say that "Homo sapiens" is a figment of the mind, I think you will be forced by logic to say that we are all individually figments of the mind.

Gleason, Jr.: I am certainly an aggregation of cells and a reality but the category obtained by classifying those cells is different from that aggregation, and is purely subjective. Modern logic has recognized the great importance of differentiating between a class and its components. The two have an entirely different logical value, and must be treated differently. Other treatment brings about the confusion of terms on which many well known paradoxes rest.

Emerson: The individual is a population of cells.

Gleason, Jr.: There is a distinction between a population of cells and a class of cells. A class of cells is an abstraction characterized by a quality, which is held in common by all those cells. That I think is fundamental.

Gleason: One can combine things of different nature and still have a concrete reality. When one classifies things he arranges them according to similarities. Philosophers agree that the development of a classificatory group leads always to an abstract concept.

MacGinitie: It seems to me that the speaker will have some trouble in limiting his defense from the other side. Why did he stop where he did? I do not like it for that reason.

Wecker, S. C. 1964. Habitat selection. *Scientific American* 211:109–116.

One of the most crucial yet difficult aspects of evaluating an animal's home range is not necessarily determining habitat use within that range but how each habitat is selected by the animal. Evaluating habitat selection is an exciting area in wildlife management. Differentiating the importance of specific sites within the home range, however, is difficult. Wildlife management has evolved through a series of stages: formulating laws, predator control, reservation of landscapes, exotic management, and environmental control. Today a sixth step, habitat management, could be added. We have the same basic objectives in 2012 regarding preserving habitat as we had in 1900 with our attempts to conserve and preserve wildlife. As our urban areas continue to grow and expand, it is critical to know the value of all habitats to wildlife species; otherwise, extinction is likely. Hopefully we will be as successful in maintaining and preserving habitats as we were in rescuing big game and other wildlife in the 1900s.

Wecker's study was classic in that it linked behavioral concepts to habitat selection. Wecker wanted to determine how learning, experience, and heredity influenced habitat selection. His experiments led to a better understanding of wildlife habitat and paved the way for numerous studies of it. His work showed that habitat selection was behavioral and had learned and innate components.

RELATED READING

Bell, R. H. V. 1969. The use of the herbivorous layer by grazing ungulates in the Serengeti. British Ecological Society Symposium 10:111–124.

Carson, R. G., and J. M. Peek. 1987. Mule deer habitat selection patterns in north-central Washington. Journal of Wildlife Management 51:46–51.

Hall, L. S., P. R. Krausman, and M. L. Morrison. 1997. The habitat concept and a plea for standard terminology. Wildlife Society Bulletin 25:173–182.

Jones, J. 2001. Habitat selection studies in avian ecology: a critical review. The Auk 118:557–562.

Morrison, M. L., B. G. Marcot, and R. W. Mannan. 2006. Wildlife-habitat relationships: concepts and relationships. Third edition. Island Press, Washington, D.C., USA.

Wecker, S. C. 1963. The role of early experience in habitat selection by the prairie deer mouse *Peromyscus maniculatus bairdi*. Ecological Monographs 33:307–325.

Wiens, J. A., and J. T. Rotenberry. 1981. Habitat associations and community structure of birds in shrubsteppe environments. Ecological Monographs 51:21–41.

HABITAT SELECTION

How does an animal choose its environment? Experiments with mice that live either in fields or in forests indicate that both heredity and learning have played a role in the evolution of this behavior

by Stanley C. Wecker

*Mid pleasures and palaces
'though we may roam,
Be it ever so humble,
there's no place like home.*

If animals were capable of understanding verse, this sentiment would doubtless have as much meaning for the denizens of a rotting log as it does for the inhabitants of the most fashionable suburb. One need only visit the countryside to perceive that the plants and animals in a natural community, like their human counterparts, are not scattered haphazardly over the landscape. Each organism tends to be restricted in distribution by its behavioral and physiological responses to the environment. It follows that living things must be able to locate favorable places in which to live. Their methods of doing so are so numerous and varied, however, that it is difficult to generalize about the selection of habitat.

On the one hand, many small organisms of otherwise low mobility have evolved means for utilizing air and water currents in the dispersion of members of their species. Spores, seeds, ballooning spiders and a surprisingly large number of insects drift in the upper reaches of the atmosphere, and a wide variety of planktonic forms ride the waves of the waters below. Occasionally terrestrial organisms accidentally cross long stretches of sea on pieces of driftwood, and live fish have been transported from pond to pond by hurricanes. The end result of this passive and essentially random dissemination of individuals is that a small number of them eventually reach areas conducive to continued survival and reproduction.

For the majority of animals, on the other hand, choosing a habitat is a more active process. This does not imply that most species can make a critical evaluation of the entire constellation of factors confronting them. More probably they react automatically to certain key aspects of their surroundings. For example, a wide variety of animals, ranging from single-celled protozoans to beetles and salamanders, often select their habitat at least in part by orientation along physicochemical gradients in the environment. These include such factors as temperature, moisture, light and salinity.

Another form of behavior that results in habitat selection is the choice of egg-laying sites by insects. Among certain beetles, butterflies and wasps the gravid female instinctively selects a plant or an animal host that will satisfy the requirements of the developing larva, whether or not the needs of the larva coincide with her own. Among the birds that live in shrubbery or forest the choice of habitat has been found to be associated with the height, spacing and form of the vegetation. Even when the overall character of the vegetation is appropriate for a species, a deficiency of specific environmental cues, such as song perches and nest sites, may exclude the species from an area within its range. The British ornithologist David Lack has called this phenomenon a "psychological factor" in habitat selection. Among the higher forms of life such factors may be fully as important as stimuli more directly related to physiological tolerances.

Although many ecologists have investigated the physical and biological factors that cause mammals to occupy certain habitats and avoid others, little is known about the role of psychological factors. One genus that is particularly well suited for the study of these factors is *Peromyscus*, the deer mouse. To this hardy little mammal almost every conceivable ecological situation, ranging from desert to tropical rain forest, from barren tundra to windblown mountaintops, is home. One species, *P. maniculatus*, is among the most variable of all North American rodents. It has 66 subspecies, which are found in so many habitats that a leading ecologist has remarked that probably no environmental change short of the inundation of the entire continent would eliminate all of them! In spite of this variability, however, in the sense of ecological adaptation the species has just two principal types: the long-tailed, long-eared forest forms, and the smaller short-tailed, short-eared grassland forms.

The prairie deer mouse of the Middle Western and Plains states (*Peromyscus maniculatus bairdi*) is a strictly field-dwelling subspecies that avoids all forested areas, even those with a grassy floor. Studies comparing the food preferences and the requirements for temperature and moisture of this subspecies and a closely related woodland form, *P. m. gracilis*, have not revealed any physiological differences of sufficient magnitude to account for the difference in their choice of habitat. It has therefore been concluded that the absence of the prairie deer mouse from forested areas within its geographic range is primarily a behavioral response to its environment.

The first experimental attempt to identify the environmental cues that cause these mice to choose a place to live was undertaken in 1950 by Van T. Harris, then working with Lee R. Dice at the University of Michigan Laboratory of Vertebrate Biology. Harris presented individual prairie and woodland deer mice with a choice between a laboratory "field" and a laboratory "woods." Each type of mouse exhibited a clear preference for the artificial habitat more closely resembling its natural en-

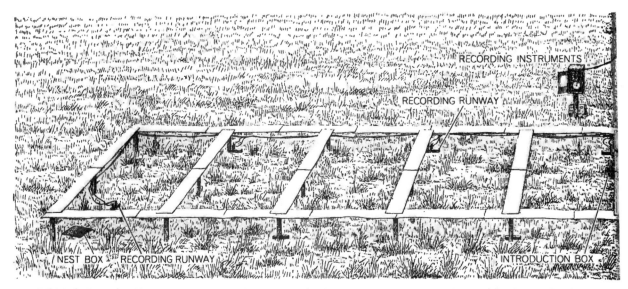

EXPERIMENTAL ENCLOSURE for testing habitat preference of prairie deer mice is 100 feet long and 16 feet wide. Five of its 10 compartments are in a field (*left*) and five are in an oak-hickory woodlot. For testing, each mouse is placed in the introduction box near the middle. It can go from there into either the field half or the woods half of the enclosure. Each partition has a run-

vironment. Since the physical conditions throughout the experimental room were uniform, Harris concluded that the mice were reacting to the character of the artificial vegetation. Moreover, laboratory-reared animals with no outdoor experience chose the "correct" artificial habitat as readily as the wild mice did. Harris therefore decided that this behavior was innate.

These experiments were not, however, designed to test the possibility that learning might also be involved. It has recently been established that early experience is of greater importance in the development of adult behavior than had once been thought [see "Early Experience and Emotional Development," by Victor H. Denenberg; SCIENTIFIC

AMERICAN, June, 1963]. Since young prairie deer mice are normally born and reared in open fields, one would expect their early experience to reinforce any innate preference for this habitat.

These considerations raise two questions: (1) Does learning actually play a role in habitat selection by *Peromyscus*? (2) Can an innate preference for field conditions be overridden by early experience in a different environment? In order to investigate these problems I constructed a 100-foot-long outdoor pen on the University of Michigan's Edwin S. George Reserve, 26 miles northwest of Ann Arbor. The project was initiated with the support of the Department of Zoology and the Muse-

um of Zoology and was carried out under the auspices of the Laboratory of Vertebrate Biology and its director, Francis C. Evans.

The long axis of the experimental pen crosses a relatively sharp boundary between an open field and an oak-hickory woodlot. The enclosure is divided into 10 compartments, five of which are in the field and five in the woods. There are two underground nest boxes, one at the end that extends farthest into the woods, the other at the end that extends farthest into the field. A third underground box in the middle of the enclosure serves as a chamber for introducing mice. Small metal runways leading from one compartment to another allow the animals to go anywhere with-

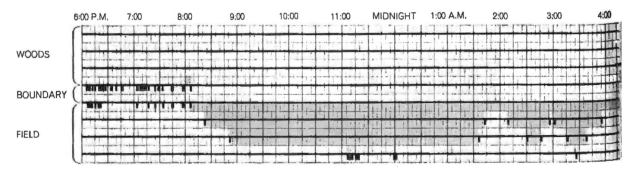

RECORD MADE BY MOUSE in two nights shows preference for field. The mouse, from laboratory stock, was in the field for 10 days when quite young, then lived in laboratory for 56 days before test. Daylight hours are omitted here because the mouse was quiet in the field nest box. The eight horizontal lines (*black*) were traced by

pens connected with the various treadles in the enclosure, ranging in order from the woods nest box at top to field nest box at bottom. Short vertical lines along the tracings are "blips" made when mouse crossed treadle. Just after 6:00 P.M. (*far left*) mouse leaves introduction box, runs back and forth across treadles at

110

way at one end that enables the mouse to go from one compartment to the next. Two of the seven runways with recording treadles are labeled. Nest boxes, both of which have treadles, are in the last compartments at left and right. The instruments that make permanent records of movements of each mouse (*see bottom of these two pages*) are in box at top, just to left of center.

in the entire fenced area. A centrally located electric device records the time at which a mouse passes through the runways and enters the nest boxes. I place each mouse in the experimental enclosure alone and leave it there until it has nested in the same habitat for two consecutive days.

Prairie deer mice are nocturnal and are inactive during the day. I decided, therefore, that it would be most meaningful to consider the length of an animal's active and inactive periods in each environment (woods and field) as separate measures of habitat selection. Three other categories of measurement provide further data for comparing an animal's response to the woods with its response to the field. The five categories used in

this study, then, are (A) *time active*, or time spent outside the nest boxes in woods and field respectively; (B) *time inactive*, or time spent nesting in woods or field; (C) *rate of travel*, or the speed at which a mouse moves about in each of the two habitats; (D) *activity*, or the frequency with which a mouse changes compartments or enters nest boxes in woods or field; and (E) *average penetration* (in feet) into either of the two habitats each time a mouse crosses the boundary between them. In all categories except rate of travel the higher score for woods or field is taken to indicate habitat preference. In the case of rate of travel it was assumed that a mouse travels more slowly in the preferred habitat; presumably the animal

is less subject to stress in its normal environment.

In the course of the study I tested six groups of prairie deer mice, one mouse at a time, in the enclosure. Observing the 132 mice occupied the spring, summer and fall of two successive years. The two control and four experimental groups were each characterized by a different combination of two variables: hereditary background and pretest experience. The hereditary distinction was between field-caught mice (and their immediate offspring) and individuals selected from a laboratory stock. The experience was provided in the field, in the woods and in the laboratory.

The first group to be tested consisted

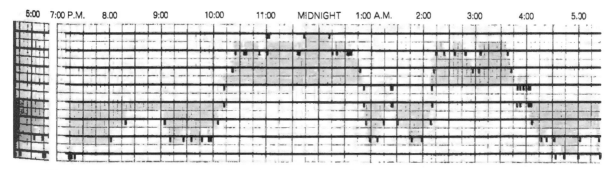

habitat boundary. After brief foray into woods (7:58 P.M.) it returns to field and gradually moves all the way to the nest box (11:07 P.M.). It goes in and out, then moves back several times toward habitat boundary but never crosses it. At 5:16 A.M. it enters field nest box (*end of first night's record*) and remains throughout the day. Record for second night shows two long periods in the woods, including two entries into the woods nest box. It was usually assumed that the mouse went at least halfway to the next treadle after crossing a treadle, as shown by colored shading. Actually the mouse could have been anywhere between the two treadles.

INTRODUCTION BOX opens into the runway (*left*) that crosses the habitat boundary. The two recording treadles for the boundary can be seen in this runway. Tiny door at end of exit tube opens outward only. The two nest boxes resemble the introduction chamber.

of individuals recently caught in old fields of the Edwin S. George Reserve, where earlier studies had clearly demonstrated the strong affinity of prairie deer mice for the field environment. My assumption was that the reactions of these adult animals would provide a basis for evaluating any unnatural effects of the enclosure itself. Accordingly I designated the eight males and four females in the group as Control Series I. At the end of the test it was obvious from all five measurements that the mice much preferred the field half of the enclosure [*see upper illustration on opposite page*]. From this I concluded that the testing situation permitted the animals to exercise their normal habitat preference.

If this preference is innate, field mice reared in the laboratory should also choose the field environment. In order to evaluate this hypothesis I tested seven males and six females from the prairie deer mouse colony of the University of Michigan Mammalian Genetics Center. These were Control Series II. The entire laboratory stock, designated *Peromyscus maniculatus bairdi* Washtenaw (for Washtenaw County), was descended from 10 pairs of animals trapped in the vicinity of Ann Arbor by Harris in 1946. According to the records the 13 individuals of Control Series II were 12 to 20 generations removed from any field experience. Their performance in the enclosure contrasted sharply with that of Control Series I [*see lower illus-

tration on opposite page*]. In three of the five categories more Control Series II individuals preferred the woods to the field! The most that can be said of the group as a whole, however, is that it did not demonstrate a well-defined preference for either habitat.

In its laboratory environment the *bairdi* Washtenaw stock has been subjected to different selective pressures from those encountered in fields. Combinations of genes that are advantageous to prairie deer mice in nature, such as those affecting response to the environment, would in the laboratory probably not be selected for and might even be selected against. One can therefore assume that the field and laboratory populations used in my experiments had genetically diverged. Since other investigators have shown that such divergence in laboratory stocks can lead to morphological changes, it seems reasonable to assume that behavioral modifications will arise also. I suggest that these contributed to the highly variable habitat response of the mice in Control Series II. It is of considerable interest that the marked preference for fields displayed by their ancestors has been lost in only 12 to 20 generations.

Thus the data from Control Series II neither support nor refute Harris' contention that the habitat preference of prairie deer mice is normally determined by heredity. The next experiment provided a more rigorous evalua-

tion. For this test I caught more wild field mice and bred them in the laboratory. The offspring, which were separated from their parents shortly after weaning, lived in laboratory cages for an average of about two months. Then I tested eight males and four females in the enclosure as Experimental Series I. None had had any previous outdoor experience.

Among these mice there was no reason to anticipate hereditary modifications of the type postulated for the laboratory stock. Thus if habitat preference is genetically determined, the behavior of the mice in Experimental Series I should approximate that of Control Series I. As the records indicate [*see top illustration on page 114*] these animals did display a pronounced affinity for the field half of the enclosure. Obviously prior experience in this environment is not a necessary prerequisite for habitat selection.

Since the animals were reared by field-caught parents, however, it is possible that some form of noninherited social interaction brought about the results. Unfortunately I have had no opportunity to evaluate this possibility, but other investigators have failed to find evidence in prairie deer mice for transfer of behavioral traits from generation to generation through learning. Litters reared by foster parents do not reveal any consistent indication of maternal, paternal or joint parental influ-

112

ence. It seems likely that, as Harris concluded, the habitat preference of wild populations of prairie deer mice is an expression of an innate pattern of behavior. The pattern may be elicited by certain key environmental stimuli, but it apparently does not depend on a period of habituation to the environment for its expression. This does not mean, however, that early experience has no effect on the selection of habitat by an adult animal. It seems reasonable to assume that a young deer mouse's normal association with open fields will reinforce its innate preference for this environment.

In order to ascertain the role such experience plays, I allowed pairs of laboratory animals to rear litters in a 10-by-10-foot pen constructed in the field. Located a short distance from the main enclosure, this area was divided into two compartments, each of which contained a number of nest boxes. Mice that had mated in the laboratory were moved into the nest boxes soon after they had borne litters and before the eyes of the young had opened. After an average of 31 days in the field pen 13 of the offspring were tested in the main enclosure. I labeled this group Experimental Series II.

These 13 mice—eight males and five females—displayed a well-defined preference for the field habitat. Since the laboratory stock of Control Series II had not particularly preferred the field, the highly contrasting behavior of the offspring of such stock can only be explained by their field experience. Although the laboratory animals have apparently lost the innate preference of the subspecies for fields, they have retained a capacity for learning that enables them to exercise habitat selection if they are exposed to the field environment at an early age. Whether or not early experience in a different environment would reverse normal habitat affinities, however, remained to be determined.

Accordingly field-caught prairie deer mice were allowed to raise litters in a 10-by-10-foot pen in the woods. Subsequently I tested seven of the woods-reared offspring in the large enclosure. These mice—six males and one female—constituted Experimental Series III. The two weeks of woods experience did not noticeably influence their selection of habitat. In all five categories of measurement a majority of the mice exhibited the normal field preference. It thus appears that early experience in the "wrong" environment is not enough to override the innate habitat response. Since learning assumes a more impor-

tant role in the development of a well-defined habitat preference by laboratory animals, it seemed possible that early experience in the woods might lead mice from laboratory stock to prefer the woods habitat.

In order to determine if this was the case I transferred to the woods pen litters born to *bairdi* Washtenaw females. Nine of these offspring, six males and three females, were subsequently tested in the main enclosure as Experimental Series IV. As a whole, in spite of their 24 days of woods experience, these animals did not demonstrate a pronounced tendency to select the woods half of the enclosure. On the other hand, neither did they display any special preference for the field half. One must therefore conclude that prairie deer mice can only learn to respond to

environmental cues associated with the field habitat.

To summarize the six experiments, four groups (Control Series I and Experimental Series I, II and III) consistently selected the field half of the enclosure, whereas the other two (Control Series II and Experimental Series IV) did not exhibit a well-defined habitat preference. All the individuals in the four groups that preferred the field environment had either field-caught parents or field experience or both. The other two groups were offspring of laboratory animals and had had no contact with the natural field environment prior to testing in the enclosure.

The data warrant the following conclusions: (1) The choice of the field environment by *P. m. bairdi* is normally

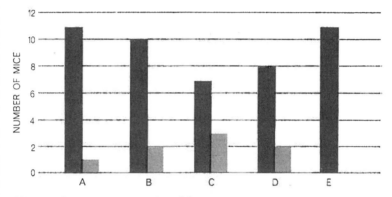

CONTROL SERIES I, consisting of 12 adult mice trapped in an open field, showed a clear preference for the field. Gray bars indicate choice of field for each criterion of measurement (*A* through *E*); colored bars denote a preference for the woods. In this and the five bar graphs that follow, a pair of bars for some criteria does not add up to the total number of mice in the group. This results from a failure in the recording apparatus, or from the fact that an animal's score was the same for both woods and field, or because the mouse spent all its time in one habitat, making the comparisons in categories C, D and E impossible.

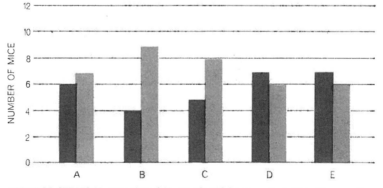

CONTROL SERIES II, consisting of 13 animals of laboratory stock, 12 to 20 generations away from the field, preferred the woods according to three categories of measurement. As a whole, however, this group cannot be said to have selected either half of the enclosure. Categories of measurement are (*A*) percent of time active in woods or field, (*B*) percent of time inactive, (*C*) rate of travel in woods or field, (*D*) activity in field or woods and (*E*) average penetration in feet by a mouse into woods or field from the habitat boundary.

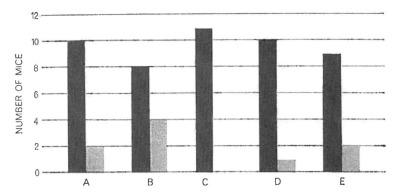

EXPERIMENTAL SERIES I, 12 mice, were first-generation offspring of field stock, reared in the laboratory. In all five measurements of habitat selection, they chose the field.

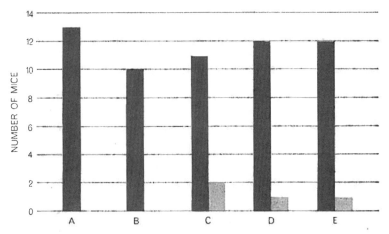

EXPERIMENTAL SERIES II, 13 animals, were laboratory stock reared in a pen in the field. By all five criteria of measurement they displayed a strong preference for the field.

determined by heredity. (2) Early field experience can reinforce this innate preference, but it is not a prerequisite for subsequent habitat selection. (3) Early experience in other environments (woods or laboratory) cannot override the normal affinity of field stock for the field habitat. (4) Confinement of the *bairdi* Washtenaw stock in the laboratory for 12 to 20 generations has apparently reduced hereditary control over the habitat response. This genetic change has markedly increased the behavioral variability of these animals when tested in the enclosure. (5) The laboratory stock did retain an innate capacity for learning from early field experience to respond positively to stimuli associated with this environment. Experience in the woods, however, did not cause them, on the whole, to select the woods habitat.

These results indicate that both heredity and experience can play a role in determining the preference of the prairie deer mouse for the field habitat,

which raises an interesting question. Since the same affinity for fields can be learned by each generation, why has natural selection produced an apparently parallel, genetically determined response?

According to the British zoologist C. H. Waddington, evolutionary changes that increase hereditary control are advantageous because they tend to limit the number of possible ways an organism can respond to a particular environmental stimulus. This is beneficial because natural selection favors only those responses conducive to survival. Therefore, as long as the environment remains relatively stable, the population as a whole will eventually become genetically adjusted to the ecological situation it is most likely to encounter and best able to exploit. The innate preference of prairie deer mice for the field environment represents such an adjustment. Why, then, does the mouse retain what appears to be an independent mechanism for habitat selection based

on learning? Furthermore, if we are dealing with two independent mechanisms, why should relaxation of natural selection under laboratory conditions remove one and not the other?

I would suggest that the innate pattern of habitat selection is not independent of the learned pattern but rather is really an extension of the learned pattern. This idea derives support from the observations on *bairdi* Washtenaw stock: the laboratory animals have not lost any innate habitat preference but learn to select the "correct," or field, half of the enclosure after being reared in a pen in the field. Presumably a certain number (X) of "field-adapting" genes would give the prairie deer mouse the ability to learn to respond positively to the field environment; a larger number of such genes (X plus Y) could make this behavior innate. After 12 to 20 generations in the laboratory the mouse reverts from the X-plus-Y genotype back to the X genotype.

The behavioral evolution from learned to innate response can be explained as an example of the "Baldwin effect," originally called organic selection when postulated in 1896 by J. M. Baldwin of Princeton University. Recently George Gaylord Simpson of Harvard University has redefined the process to explain how individually acquired, nongenetic adaptations may, under the influence of natural selection, be replaced in a population by similar hereditary characteristics.

As an alternative to accepting the old Lamarckian doctrine that acquired characteristics can be directly inherited, one might apply Simpson's interpretation of the Baldwin effect to the prairie deer mouse situation as follows: As the mice became physiologically and morphologically adapted to existence in the grasslands, patterns of behavior based on some form of learning (homing, for example) tended to confine individuals to the field environment. These patterns, although not exclusively hereditary as such, were still advantageous in that they restricted the animals to the habitat best suited for their survival and reproduction. Then chance mutation created genetic factors that facilitated the development of behavior patterns whose effects resembled those acquired through learning. Finally, since natural selection favored these factors, they spread through the population.

Waddington believes, however, that the Baldwin effect, with its emphasis on chance mutation, involves an oversimplification that ignores the role of the environment in determining the manner in

114

which particular combinations of genes will be expressed. For example, climate or some other aspect of the environment may determine what color certain animals will be, the animals themselves having a genetic potential for more than one color. Waddington maintains that natural selection operates not in favor of genes whose effects happen *by chance* to parallel acquired (nongenetic) adaptations but in favor of factors that control the capacity of an individual to respond to its surroundings. The interaction of organism and environment has the effect of reducing the number of different pathways for genetic expression, thus facilitating the production of better-adapted individuals. The more thorough this "canalization" of developmental possibilities is, the more likely it will be that favorable combinations of genes already present in the population in low frequency will find expression. Once expressed, these combinations of genes can be acted on by natural selection. Since they are favorable, the number of individuals bearing them will ultimately increase. Waddington terms this process the "genetic assimilation" of a character that is initially acquired, or nongenetic.

The results of experiments I am now conducting suggest that the *bairdi* Washtenaw stock learns to respond to the field environment very quickly and may indeed exhibit what the British zoologist W. H. Thorpe has called habitat imprinting. If imprinting is actually operating, one would expect the adult habitat response to be determined during a critical period early in the life of the animal, probably shortly after the young mouse first leaves the nest. It is significant, therefore, that young laboratory animals receiving only 10 days of early field experience still have a marked preference for that environment when they are tested in the enclosure, even after two months of confinement in laboratory cages! On the other hand, exposure of adult laboratory animals to the field environment for as long as 59 days does not cause them to develop a well-defined habitat preference.

In view of the above, it appears that one result of selection for an increased number of "field-adapting" genes has been to shift the development of the behavior patterns involved in habitat selection to earlier and earlier periods in the life of the individual. Obviously survival is enhanced by recognition of a favorable environment over successive generations through learning. It would be even more advantageous to restrict

learning capacity to include only those cues associated with the favorable environment and to reduce to an absolute minimum the time required for such learning. Finally, the necessity for learning could be eliminated altogether by selection for sets of genes that endow an individual with the capacity for making an adaptive response to the critical stimuli as soon as the stimuli are encountered. In this context a hypothetical imprinting stage may have been an important preliminary to the ultimate genetic assimilation of the habitat response of the prairie deer mouse. Indeed, the behavioral differences among the various groups of mice tested during my investigation could be taken to reflect different steps in an evolutionary sequence leading from behavior largely dependent on learning to the development of an innate pattern of control.

This sequence might have occurred as follows: (1) habitat restriction through social factors and homing, (2) recognition of the field environment through learning, (3) learning capacity reduced to cues associated with the field habitat, (4) imprinting to the field environment through exposure very early in life, (5) innate determination of the habitat response.

So far no one has identified the specific cues by which a prairie deer mouse recognizes the field environment. Fortunately the results of my investigations suggest a unique approach to this problem. Young laboratory animals do not develop a well-defined habitat preference in the absence of early field experience. It should therefore be meaningful to expose them in the laboratory to single stimuli designed to simulate different aspects of the natural field en-

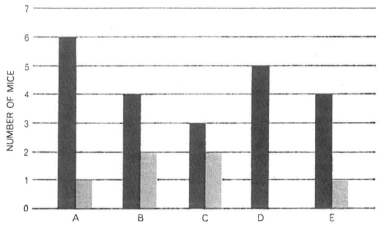

EXPERIMENTAL SERIES III, seven mice, offspring of adults trapped in the field, were conditioned by rearing in a pen in the woods. They tended to choose the field habitat.

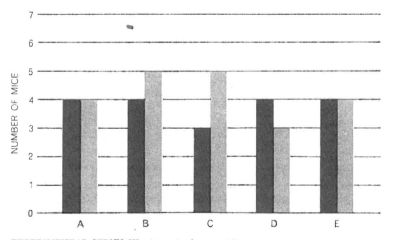

EXPERIMENTAL SERIES IV, nine animals, were laboratory stock reared in the pen in the woods. Overall they appeared to have no particular preference for either of the habitats.

vironment. These stimuli include the sight, odor and touch of field vegetation, with artificial grass as the touch stimulus. Groups of animals, each group exposed to only one such factor, could then be tested in the experimental enclosure. In fact, I am now conducting such tests, but the data are not yet extensive enough to warrant any conclusions.

Having considered the role of evolution in habitat selection, I should like to discuss briefly the part that habitat selection plays in the evolutionary process. A diversity of habitat prefer-ences within a species favors survival by making the species more adaptable to environmental change. Such a diversity, however, might be expected to lead to genetic divergence by selective processes similar to those already described. Nevertheless, most biologists do not believe that new species can arise in this way unless some form of geographical isolation occurs. In Michigan the ranges of the prairie deer mouse and the woodland deer mouse overlap, but there is no evidence of intergradation, or interbreeding, between the two types. It is known, however, that these subspecies did not develop side by side, because they were formerly isolated geographically. Indeed, the two forms came into close contact only during the past century, when the clearing of forests by man enabled the prairie field mouse to extend its range northward.

Both Harris' experiments and mine provide evidence that the observed difference in habitat preference of these subspecies forms the basis for their continued segregation. As Ernst Mayr of Harvard University points out, ecological differences between two such overlapping forms are to be expected, since competition would otherwise prevent both from coexisting in the same area.

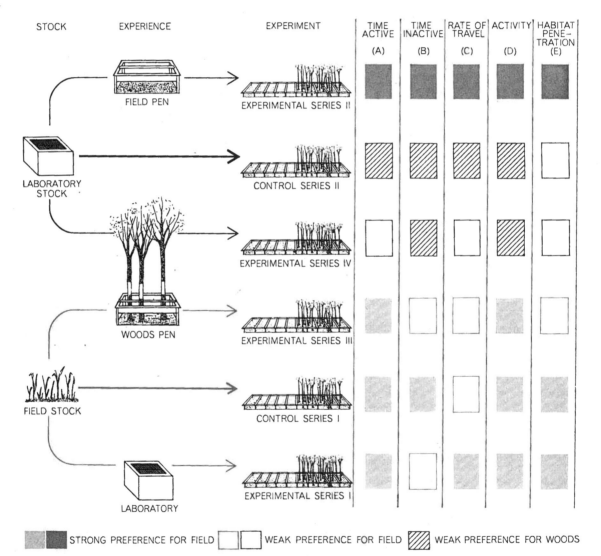

SUMMARY OF OBSERVATIONS reveals preferences of prairie deer mice from various backgrounds. Mice captured in the field were placed in the experimental pen as "Control Series I" (*thick colored arrow*). Offspring of field-caught mice were given conditioning in laboratory cages or in a pen in the woods before testing (*thin colored arrows*). Laboratory stock were tested in the experimental enclosure as "Control Series II" (*thick black arrow*), and their offspring were conditioned in woods or field pens before testing (*thin black arrows*). Results of the tests on the six groups are given in right half of the diagram. The five categories of measurement are explained in the text. Degree of preference is indicated. Results are based on mean response of each group.

116

Johnson, D. H. 1980. The comparison of usage and availability measurements for evaluating resource preference. *Ecology* 61:65–71.

As wildlife science progresses, some of the important advances gradually gain acceptance and are eventually used in the profession. Other ideas and concepts quickly gain nearly universal acceptance once the ideas are presented to the scientific community. Johnson's paper falls into the latter category.

Understanding the components of the different habitats for species of wildlife and how those components are used by wildlife is a central theme in the wildlife literature. Information about habitat use and availability is difficult to obtain, especially when we examine use by gender and age group; in different vegetation associations, topographies, and climate regimes; and with other biotic and abiotic conditions. Johnson's objective in this paper is to present a new approach for examining wildlife habitat use by considering the hierarchical nature of selection: the species' physical or geographical range (i.e., first-order selection), the home range of the species (i.e., second-order selection), use of habitat components within the home range (i.e., third-order selection), and the actual procurement from those components available at the site (i.e., fourth-order selection). Johnson's proposed method has been widely used in studies of habitat, allowing researchers to gain further insights into wildlife-habitat relationships.

RELATED READING

Block, W. M., K. A. With, and M. L. Morrison. 1987. On measuring bird habitat: influence of observer variability and sample size. Condor 72:182–189.

Fretwell, S. D., and H. L. Lucas. 1969. On territorial behavior and other factors influencing habitat distribution in birds. Acta Biotheoretica 19:16–36.

Guthery, F. S. 1997. A philosophy of habitat management for northern bobwhites. Journal of Wildlife Management 61:291–301.

Morrison, M. L., B. G. Marcot, and R. W. Mannan. 2006. Wildlife-habitat relationships: concepts and applications. Third edition. Island Press, Washington, D.C., USA.

Robbins, C. S, D. K. Dawson, and B. D. Dowell. 1989. Habitat area requirements of breeding forest birds of the Middle Atlantic states. Wildlife Monographs 103:1–34.

Schooley, R. L. 1994. Annual variation in habitat selection: patterns concealed by pooled data. Journal of Wildlife Management 58:367–374.

Van Horne, B. 1983. Density as a misleading indicator of habitat quality. Journal of Wildlife Management 47:893–901.

Ecology, 61(1), 1980, pp. 65–71
© 1980 by the Ecological Society of America

THE COMPARISON OF USAGE AND AVAILABILITY MEASUREMENTS FOR EVALUATING RESOURCE PREFERENCE[1]

DOUGLAS H. JOHNSON

United States Fish and Wildlife Service, Northern Prairie Wildlife Research Center,
Jamestown, North Dakota 58401 USA

Abstract. Modern ecological research often involves the comparison of the usage of habitat types or food items to the availability of those resources to the animal. Widely used methods of determining preference from measurements of usage and availability depend critically on the array of components that the researcher, often with a degree of arbitrariness, deems available to the animal. This paper proposes a new method, based on ranks of components by usage and by availability. A virtue of the rank procedure is that it provides comparable results whether a questionable component is included or excluded from consideration. Statistical tests of significance are given for the method.

The paper also offers a hierarchical ordering of selection processes. This hierarchy resolves certain inconsistencies among studies of selection and is compatible with the analytic technique offered in the paper.

Key words: availability; food habits; habitat selection; preference; resource utilization; selection; usage.

INTRODUCTION

Central to the study of animal ecology is the usage an animal makes of its environment: specifically, the kinds of foods it consumes and the varieties of habitats it occupies. Many analytic procedures have been devised to treat data on the usage of such resources, particularly in relation to information on their availability to the animal, for the purpose of determining "preference." The objectives of this report are to describe the problem of determining preference by comparing usage and availability data, to illustrate a serious shortcoming in the routine application of most procedures for comparing these data, and to suggest a new method that resolves this difficulty. The proposed technique results in a ranking of the components on the basis of preference, and permits significance tests of the ranking.

Many investigators who use analytic procedures to handle usage and availability data fail to recognize the conditional nature of inferences drawn by comparing usage to availability. Conclusions about whether an individual component is used above, in proportion to, or below its availability are critically dependent upon the array of components the investigator deems available to the animal. This decision is often made somewhat arbitrarily by the investigator. The following contrived example will illustrate the point.

Suppose an investigator collects a fish, and finds that its stomach contains food items A, B and C in the percentages shown in Table 1(A) under "Usage." A sample of the animal's feeding site at the time the fish was collected reveals that the items were present in the proportions shown under "Availability." Many investigators would conclude that Item A is avoided, because usage was less than availability, while Items B and C are preferred, because usage exceeded availability. But suppose another investigator, equally familiar with the biology of the fish, does not believe that Item A is a valid food item (perhaps he thinks it is ingested only accidentally while the animal is consuming other foods). He would then consider the data in Table 1(B), obtained by deleting Item A from the analysis. Now, although Item C is still deemed preferred, the assessment of Item B has changed from preferred to avoided.

Conclusions are not apt to be drawn from one fish, but whatever conclusions are reached about the preference or avoidance of any particular component of the environment depend markedly upon the array of components deemed by the investigator to be available to the animal. To the extent that the decision is arbitrary, so will be the conclusions drawn from the analysis. This inconsistency can result from the use of any of the standard methods, e.g., the forage ratio (Williams and Marshall 1938, Hess and Rainwater 1939), its modifications (Jacobs 1974, Chesson 1978), the index of electivity (Ivlev 1961), the difference (Swanson et al. 1974, Gilmer et al. 1975), or contingency tables (Hanson and Labisky 1964, Buchler 1976). Some authors have recognized the difficulty. Bartonek and Hickey (1969) noted that their decision to measure only items they considered as potential foods was subjective. Sugden (1973:28–29) mentioned that "the presence of other items will influence the rating for a given item. When the available food includes mostly unimportant items measured in the habitat, other items will be given a higher rating." Certain other authors

[1] Manuscript received 23 May 1978; revised 1 May 1979; accepted 8 May 1979.

TABLE 1. Example illustrating results of comparing usage and availability data when a common but seldom-used item is included 1(A) and when excluded 1(B) from consideration.

Item	Usage (%)	Availability (%)	Conclusion	Rank		
				Usage	Availability	Difference
			(A)			
A	2	60	Avoided	3	1	+2
B	43	30	Preferred	2	2	0
C	55	10	Preferred	1	3	−2
			(B)			
B	44	75	Avoided	2	1	+1
C	56	25	Preferred	1	2	−1

(e.g., Ivlev 1961, Chamrad and Box 1968) have been circumspect about interpreting usage-availability data, but many others (e.g. Hess and Swartz 1940, Bellrose and Anderson 1943, Jones 1952, Van Dyne and Heady 1965) have termed a component "preferred" if its usage exceeded its availability, and "avoided" if the reverse was true.

For the sequel, we define the following terms: The *abundance* of a component is the quantity of that component in the environment, as defined independently of the consumer. The *availability* of that component is its accessibility to the consumer. The *usage* of a component by the consumer is the quantity of that component utilized by the consumer in a fixed period of time. The *selection* of a component is a process in which an animal actually chooses that component. Usage is said to be *selective* if components are used disproportionately to their availability. The *preference* of a consumer for a particular component is a reflection of the likelihood of that component being chosen if offered on an equal basis with others. In theory, components can be ranked from "most preferred" to "least preferred." Preference is ordinarily claimed to be independent of availability, but is generally defined by reference to the choice made at equal availabilities (e.g., Pirnie 1935, Ellis et al. 1976).

A PROPOSED METHOD

The method that I suggest for analyzing usage-availability data yields rankings of items by preference with the following properties: (1) significance tests can be made for differences in preference among items; and more important, (2) the method gives largely comparable results whether the analysis includes or excludes doubtful items.

As a measure of preference, I propose using the difference between the rank of usage and the rank of availability. Call this difference t_{ij}, where i indexes the component and j indexes the individual animal. The differences can be averaged across animals, to obtain a mean for the ith component. Averages for different components can then be compared to deter- mine which are more preferred. If components are ordered by these average differences, the ranking will be from least preferred to most preferred.

Returning to the one-animal example previously considered, with Item A included, Table 1(A), the differences in the ranks of usage and availability are +2, 0, and −2 for Items A, B, and C, respectively. Should Item A be excluded from the analysis, Table 1(B), B and C have values +1 and −1, respectively. Although the values themselves change, the difference between B and C remains 2, suggesting that C is preferred to B, regardless of whether A is included or excluded. We thus avoid absolute statements about preference.

Standard methods (e.g., forage ratio, Ivlev's index of electivity) can also be used to develop rankings in order of preference. Indeed, Ivlev (1961) recognized that preference values indicate only the relative value of a component in comparison to others, and Chesson (1978) did likewise. But many authors go much further and make absolute statements about preference and avoidance. The proposed method discourages this by using ranks, which by their nature represent relative values.

Furthermore, the loss of information resulting from the use of ranks of usage and availability, instead of the measured values, is of less consequence than might be supposed (Lehmann 1975). First of all, statistical methods based on ranks are nearly as efficient as methods based on the original data even when all the assumptions necessary to treat the original data hold (e.g., measurements are exact, their distribution is normal). Moreover, if the assumptions are not met, the rank methods have considerable advantages of efficiency and validity. And we have good reason to doubt the strict propriety of availability measurements. Sampling procedures used to determine availability values for the various components may not faithfully reflect the true availabilities to the animal under study (Savage 1931, Landenberger 1968, Bartonek and Hickey 1969, Sugden 1973, and Mitchell 1975). Thus, availability values are measured inexactly and methods based on ranks are to be preferred.

General formulation

Let X_{ij} be some measure of usage of component i by individual j, and Y_{ij} be a measure of the availability of component i to individual j, where $i = 1, 2, \ldots, I$ (I = number of components) and $j = 1, 2, \ldots, J$ (J = number of individuals). The values need not be scaled to be percentages. Take r_{ij} to be the rank of X_{ij} within j (animal) and s_{ij} the rank of Y_{ij} within j. The difference in these ranks, $t_{ij} = r_{ij} - s_{ij}$, is a measure of preference for component i by individual j.

It is a simple step to average the t_{ij} across individuals, obtaining $\bar{t}_i = J^{-1} \sum_{j=1}^{J} t_{ij}$. A ranking of components in order of increasing $\bar{t}_i$ will then indicate the relative preference of the components by the entire sample of animals.

To draw statistical conclusions about the differences among components, we invoke the following model:

$$t_{ij} = \mu + \alpha_i + \beta_j + \epsilon_{ij}, \tag{1}$$

where

μ is the overall mean,
α_i is the effect due to component i ($i = 1, \ldots, I$),
β_j is the effect due to animal j ($j = 1, \ldots, J$),
ϵ_{ij} is the random error term,

and

$$\sum_i \alpha_i = \sum_j \beta_j = 0.$$

Because the t_{ij}'s are differences in ranks within individuals, they sum to zero across i:

$$\sum_{i=1}^{I} t_{ij} = 0 \text{ for all } j, \tag{2}$$

which implies $\mu = 0$, $\beta_j = 0$, and $\sum_{i=1}^{I} \epsilon_{ij} = 0$, all j. (3)

Thus the model (1) reduces to

$$t_{ij} = \alpha_i + \epsilon_{ij}.$$

Interest lies in the null hypothesis that

$$\alpha_1 = \ldots = \alpha_I \, (=0), \tag{4}$$

that is, all components are equally preferred. Should that hypothesis be rejected in favor of the alternative that some components are more preferred than others, we would then wish to know which of the components are preferred to which others (the problem of multiple comparisons).

The distributional properties of our statistic are needed to test the null hypothesis. The average $\bar{t}_i$ equals the difference in the averages of the ranks:

$$\bar{t}_i = \bar{r}_i - \bar{s}_i.$$

It can be shown (e.g., by the method of Haigh 1971)

that under general conditions $\bar{r}_i$ and $\bar{s}_i$ are normally distributed in large samples. Thus, their difference is also asymptotically normal, which allows us to employ the heavy statistical artillery developed for normal variables.

We assume the error terms $[\epsilon_{ij}]$ are distributed with zero mean, and independently between animals. Within an animal, however, error terms are (slightly) correlated (they sum to zero by Eq. 3), so standard analysis of variance techniques are inappropriate. A procedure that allows for correlations of error terms within animals is Hotelling's T^2 (e.g., Anderson 1958), which is used to test the hypothesis that a multivariate normal vector of means is equal to a specified vector (in the present case, a vector of zeroes).

Let

$$v_{ik} = (J - 1)^{-1} \sum_{j=1}^{J} (t_{ij} - \bar{t}_i)(t_{kj} - \bar{t}_k)$$

be the covariance between components i and k. (A computational note: Because of Eq. 2, the variance-covariance matrix for all components is singular. The following calculations are made by deleting one component from the analysis. The same answer ensues regardless of which component is deleted.) Let V be the $(I - 1) \times (I - 1)$ covariance matrix, $V = [v_{ik}]$. Then the statistic

$$F = \frac{J(J - I + 1)}{(J - 1)(I - 1)} \sum_{i=1}^{I-1} \sum_{k=1}^{I-1} \bar{t}_i \bar{t}_k u_{ik},$$

where u_{ik} is the designated element of the inverse matrix of V and $U = [u_{ik}] = V^{-1}$, is distributed under the null hypothesis (Eq. 4) as Snedecor's F with $I - 1$ and $J - I + 1$ degrees of freedom.

Should the calculated statistic be larger than the tabled F value at some assigned significance level, the investigator will likely be interested in finding the source of the heterogeneity among the α's. This is the multiple comparisons problem, which has been attacked by a number of procedures. (See review by O'Neill and Wetherill 1971.) In the example that follows, I chose to use the Bayesian decision procedure developed by Waller and Duncan (1969). It is rather simple to apply, solves the dilemma of whether to use experimentwise or comparisonwise error rates, and has performed nicely in comparative studies (Carmer and Swanson 1973).

Waller and Duncan suggested declaring significant a difference between two means if the difference exceeds WS_d, where S_d is the standard error of the difference and W is a function of the number of means under comparison (in our case $I - 1$), the degrees of freedom ($J - I + 1$), and the F statistic obtained earlier. The dependence of W on F is the characteristic feature of the Waller-Duncan method; its use reduces the chance of a Type I error by demanding a large

TABLE 2. An example of wetland usage* and availability† data for 2 birds and 12 wetland classes.

| Wetland class | Measured values | | | | Rank | | | |
| | Bird 5198 | | Bird 5205 | | Bird 5198 | | Bird 5205 | |
	Usage	Avail-ability	Usage	Avail-ability	Usage	Avail-ability	Usage	Avail-ability
1/2	0.0	0.1	0.0	0.4	10.5	11	9.5	12
3/8	10.7	1.2	0.0	1.4	4	7	9.5	6
9	4.7	2.9	21.0	3.5	6	5	2	4
10	20.1	0.8	0.0	0.4	3	9	9.5	11
11/14	22.1	20.1	5.3	1.2	2	2	6	7.5
15	0.0	1.4	10.5	4.9	10.5	6	4.5	3
17/20	2.7	12.6	0.0	1.0	7.5	3	9.5	9
31/34	29.5	4.7	15.8	5.1	1	4	3	2
35	0.0	0.0	10.5	0.7	10.5	12	4.5	10
36/38	2.7	0.2	36.8	1.8	7.5	10	1	5
39	7.4	1.1	0.0	1.2	5	8	9.5	7.5
Open	0.0	54.9	0.0	78.3	10.5	1	9.5	1
Total	99.9	100.0	99.9	99.9				

* Usage = percentage of recorded locations in each wetland class.
† Availability = percentage of wetland area in a bird's home range in each wetland class.

difference if F is small, and reduces the chance of a Type II error by requiring a less marked difference if F is large.

Application to real data

The procedure described above is illustrated by some habitat usage and availability data collected by Gilmer et al. (1975). Data for 2 of their 24 radio-marked adult Mallards (*Anas platyrhynchos*) are displayed in Table 2. For each bird, a measure of usage is the percentage of locations recorded in each of 12 wetland classes, including "open water." (For this example, certain of the wetland classes used by Gilmer et al. have been combined.) Availability is taken to be the percentage of a wetland area in an individual bird's home range constituted by each wetland class. Interest lies in determining which classes of wetlands are favored, in the sense of receiving more intensive use by the Mallards.

It is apparent (Table 2) that the availability of open water far surpasses its usage. For this reason, usage of the other classes tends to exceed availability, which would suggest, if caveats about absolute statements were disregarded, that most of the other wetland classes were "preferred," whereas open water was avoided. In fact, in their original analysis, Gilmer et al. (1975) excluded most of the available open water from consideration. It is readily seen that the question of inclusion/exclusion is germane in this application.

To apply the new procedure, we first take the ranks of usage and availability values within each bird. Ranks for the two birds are shown in Table 2, where open water is included. (Results for "open water excluded" are not shown.) Next, for each bird, we take the difference between the rank of usage and the rank of availability. Averaging across all 24 birds in the complete sample yields the average differences shown in Table 3.

The hypothesis test outlined earlier yields the F-statistics $F = 20.28$ (df = 11 and 13) when open water is included and $F = 8.68$ (df = 10 and 14) when excluded. Both values are highly significant ($P < .001$), leading us to reject the null hypothesis that all wetland classes are used with equal intensity. We now seek to determine the significant differences in preference among the wetland classes.

To declare a difference significant, it must exceed in absolute value WS_d, where W is obtained from tables in Waller and Duncan (1969) and S_d is the standard error of a difference between two means. For example, if $d = \bar{t}_i - \bar{t}_k$, then $s_d{}^2 = \mathrm{var}\,(\bar{t}_i) + \mathrm{var}\,(\bar{t}_k) - 2\,\mathrm{cov}\,(\bar{t}_i, \bar{t}_k)$. To determine W the investigator must select a value for K, the Type I to Type II error seriousness ratio. We use $K = 100$, which Waller and

TABLE 3. Average differences between ranks of wetland class usage and the availability of that class.

| Wetland class | Average difference in ranks | |
	Open water included	Open water excluded
1/2	−2.44	−1.94
3/8	−3.29	−2.42
9	1.50	2.02
10	−1.33	−0.98
11/14	.52	1.31
15	3.60	3.96
17/20	2.94	3.35
31/34	−1.19	−0.19
35	−2.88	−2.38
36/38	−2.58	−1.81
39	−1.54	−0.94
Open	6.69	—

Duncan concluded to be closely analogous to the usual Type I significance level of $P = .05$. Looking in Table A2 of Waller and Duncan, with 11 means under consideration, 13 error degrees of freedom, and $K = 100$, we can interpolate for $F = 20.28$ between the values for $F = 10$ and $F = 25$. The appropriate value is $W = 1.93$. Thus, any difference $d = \bar{t}_i - \bar{t}_k$ between wetland classes i and k in the "open water included" portion of Table 3 is declared significant if

$$\frac{|d|}{S_d} > 1.93.$$

Following the same procedure for "open water excluded," we find the critical value to be 2.00.

The wetland classes may then be ordered to sort out the significant differences. With open water included, we get

3/8 35 36/38 1/2 39 10 31/34 11/14 9 17/20 15 open

Wetland classes underscored by the same line are deemed not significantly different, while lack of a common underscore indicates that the habitat classes differ significantly. The results when open water is excluded are as follows:

35 3/8 1/2 36/38 10 39 31/34 11/14 9 17/20 15

Notice particularly that the conclusions reached about the relative preference of each habitat are substantially similar in the two cases, a desirable feature of the method.

Discussion

It is clear that conclusions reached from usage-availability studies depend on the investigator's notion of what components are available to the animal. This dependency is more pervasive than it may first appear. In habitat studies, as an example, the usage of particular habitat types is compared with the availability of each type within the animal's home range, or perhaps within the study area defined by the investigator. But the very fact that the animal has its home range where it does, or that it occurs within the study area, is itself indicative that the animal has already made a selection. The analogous situation appears in feeding studies, where the presence of an animal at its feeding site suggests that it selected that site in part because of the food items available there. Comparing usage values to the availabilities within the home range, or at the feeding site, may well be misleading.

To recognize this hierarchical nature of selection, the concept of selection order can be introduced. A selection process will be of higher order than another if it is conditional upon the latter. As an example, selection of habitat types within a home range of an animal is of higher order than selection of the home range, because the availability of each habitat type is determined by the selection of the home range. Similarly, selection of food items is of higher order than selection of feeding site, for the site delimits the array of food items available to be selected.

A natural ordering of selection processes can be identified. *First-order selection* can be defined as the selection of physical or geographical range of a species. Within that range, *second-order selection* determines the home range of an individual or social group. *Third-order selection* pertains to the usage made of various habitat components within the home range. Finally, if third-order selection determines a feeding site, the actual procurement of food items from those available at that site can be termed *fourth-order selection*. Although it is no doubt possible to divide these selection orders more finely, those defined above should suffice for most applications.

The concept of selection order has been implicitly recognized in the ecological literature. Owen (1972) noted that "selection can be exercised at difference scales"; he contrasted selection of vegetative zones for feeding sites (third order) and selection within zones of plant species or parts of plants (fourth order). Wiens (1973) recognized different levels or scales of distributional patterns among breeding birds, and identified geographic range (first order), local site and plot patterns in territories (second order), and patterns of utilization (third order).

This hierarchy of selection has a unifying nature. Habitat usage studies and investigations of feeding are no longer qualitatively distinct; they are simply of different orders. The question of inclusion/exclusion of components also resolves itself in this context. The components available depend upon the order of selection being considered. Related to this, it is easy to avoid the fallacy of absolute claims, such as saying that a food item is avoided by an animal because only 50% of the animal's consumption consisted of that item, whereas it made up 90% of the items available at the feeding site. The animal may indeed have chosen that site because the item was abundant there. Absolute statements about preference or avoidance should be guarded against. Relative statements are possible because their nature invokes the concept of selection order.

The ranking approach has been used earlier, for example by Landenberger (1968), who found a hierarchy of preferences to be well defined and consistent among replicates. Consistency of preference rankings has also been found by investigators employing paired-choice experimental designs (e.g., Thompson 1965, Mulkern 1967). A ranking of components is all that can be expected from an analysis of usage relative to availability. Indeed, a ranking may be all that is desired: many of the models developed in optimal foraging theory rely on rank orders of food types from most preferred to least preferred (e.g., Pyke et al. 1977).

Another consideration is that preference is reflected in selection, which can occur only when the component is relatively scarce A component vital to the consumer may be so abundant that the consumer need only use small amounts of it to satisfy its requirements. Thus, usage is less than availability, but a conclusion that the component is of little value may not be warranted (Maitland 1965).

The method of comparing usage and availability data presented here possesses several desirable features. First, it places the components in order according to preference, an ordering consistent with the hierarchical selection model proposed herein. The method is relatively insensitive to the inclusion/exclusion of doubtful components. Results are less subjective, in the sense of being affected by possibly arbitrary decisions made by the investigator. Second, because the method employs the ranks of usage and availability measurements, these measures need not be estimated exactly or without bias. Finally, the method yields tests of significance, which permit statistical comparisons among the components.

A FORTRAN program to perform the calculations described in this report is available.[2]

ACKNOWLEDGMENTS

I am grateful to David S. Gilmer for providing the source data used in the example, to Ell-Piret Multer for bibliographic assistance, and to Alan B. Sargeant and Erik Fritzell for suggestions on the manuscript.

LITERATURE CITED

Anderson, T. W. 1958. An introduction to multivariate statistical analysis. John Wiley and Sons, New York, New York, USA.

Bartonek, J. C., and J. J. Hickey. 1969. Selective feeding by juvenile diving ducks in summer. Auk 86:443–457.

Bellrose, F. C., Jr., and H. G. Anderson. 1943. Preferential rating of duck food plants. Illinois Natural History Survey Bulletin 22:417–433.

Buchler, E. R. 1976. Prey selection by Myotis lucifugus (Chiroptera: Vespertilionidae). American Naturalist 110:619–628.

Carmer, S. G., and M. R. Swanson. 1973. An evaluation of ten pairwise multiple comparison procedures by Monte Carlo methods. Journal of the American Statistical Association 68:66–74.

Chamrad, A. D., and T. W. Box. 1968. Food habits of white-tailed deer in south Texas. Journal of Range Management 21:158–164.

Chesson, J. 1978. Measuring preference in selective predation. Ecology 59:211–215.

Ellis, J. E., J. A. Wiens, C. F. Rodell, and J. C. Anway. 1976. A conceptual model of diet selection as an ecosystem process. Journal of Theoretical Biology 60:93–108.

Gilmer, D. S., I. J. Ball, L. M. Cowardin, J. H. Riechmann, and J. R. Tester. 1975. Habitat use and home range of mallards breeding in Minnesota. Journal of Wildlife Management 39:781–789.

Haigh, J. 1971. A neat way to prove asymptotic normality. Biometrika 58:677–678.

Hanson, W. R., and R. F. Labisky. 1964. Association of pheasants with vegetative types in east-central Illinois. Transactions of the North American Wildlife and Natural Resources Conference 29:295–306.

Hess, A. D., and J. H. Rainwater. 1939. A method for measuring the food preference of trout. Copeia 3:154–157.

Hess, A. D., and A. Swartz. 1940. The forage ratio and its use in determining the food grade of streams. Transactions of the North American Wildlife Conference 5:162–164.

Ivlev, V. S. 1961. Experimental ecology of the feeding of fishes. Yale University Press, New Haven, Connecticut, USA.

Jacobs, J. 1974. Quantitative measurement of food selection. Oecologia (Berlin) 14:413–417.

Jones, N. S. 1952. The bottom fauna and the food of flatfish off the Cumberland coast. Journal of Animal Ecology 21:182–205.

Landenberger, D. E. 1968. Studies on selective feeding in the Pacific starfish Pisaster in southern California. Ecology 49:1062–1075.

Lehmann, E. L. 1975. Nonparametrics: statistical methods based on ranks. Holden-Day, San Francisco, California, USA.

Maitland, P. S. 1965. The feeding relationships of salmon, trout, minnows, stone loach and three-spined sticklebacks in the River Endrick, Scotland. Journal of Animal Ecology 34:109–133.

Mitchell, J. E. 1975. Variation in food preferences of three grasshopper species (Acrididae: Orthoptera) as a function of food availability. American Midland Naturalist 94:267–283.

Mulkern, G. B. 1967. Food selection by grasshoppers. Annual Review of Entomology 12:59–78.

O'Neill, R., and G. B. Wetherill. 1971. The present state of multiple comparison methods (with discussion). Journal of the Royal Statistical Society B33:218–250.

Owen, M. 1972. Some factors affecting food intake and selection in white-fronted geese. Journal of Animal Ecology 41:79–92.

Pirnie, M. D. 1935. Michigan waterfowl management. Michigan Department of Conservation, Lansing, Michigan, USA.

Pyke, G. H., H. R. Pulliam, and E. L. Charnov. 1977. Optimal foraging: a selective review of theory and tests. Quarterly Review of Biology 52:137–154.

Savage, R. E. 1931. The relation between the feeding of the herring off the east coast of England and the plankton of the surrounding waters. [Great Britain] Fishery Investigations Series, Number 2. Volume 12. Ministry of Agriculture and Fisheries, London, England.

Sugden, L. G. 1973. Feeding ecology of pintail, gadwall, American widgeon and lesser scaup ducklings in southern Alberta. Canadian Wildlife Service Report Series 24. Canadian Wildlife Service, Ottawa, Canada.

Swanson, G. A., M. I. Meyer, and J. R. Serie. 1974. Feeding ecology of breeding blue-winged teals. Journal of Wildlife Management 38:396–407.

Thompson, D. Q. 1965. Food preferences of the meadow vole (Microtus pennsylvanicus) in relation to habitat affinities. American Midland Naturalist 74:76–86.

Van Dyne, G. M., and H. F. Heady. 1965. Botanical composition of sheep and cattle diets on a mature annual range. Hilgardia 36:465–492.

[2] See National Auxiliary Publications Service document #03648 for 21 pages of supplementary material. For a copy of this document, contact the author or order from ASIS/ NAPS, Microfiche Publications, P.O. Box 3513, Grand Central Station, New York, New York 10017 USA.

Waller, R. A., and D. B. Duncan. 1969. A Bayes rule for the symmetric multiple comparisons problem. Journal of the American Statistical Association **64**:1484–1503.

Wiens, J. A. 1973. Pattern and process in grassland bird communities. Ecological Monographs **43**:237–270.

Williams, C. S., and W. H. Marshall. 1938. Duck nesting studies, Bear River Migratory Bird Refuge, Utah, 1937. Journal of Wildlife Management **2**:29–48.

Human Dimensions

Kellert, S. R. 1976. Perceptions of animals in American society. *Transactions of the North American Wildlife and Natural Resources Conference* 41:533–546.

There are three main components of wildlife management; the first two are animal biology and wildlife habitat. In the 1970s and 1980s, the third component, human dimensions, was gaining ground and growing in popularity. Given current trends in societal involvement in wildlife management issues, the human dimensions component of wildlife management is becoming just as important as the other two, if not more so.

One of the most influential individuals to bring human dimensions to the forefront of wildlife management was Stephen Kellert. In this paper, Kellert provided managers and biologists with one of the first views into understanding "people's motivations for involvement in animal-related activities . . . including hunting, observing, and scientific studies, and how human demography was important in views and behaviors." This was the first paper to clearly classify the attitudes of Americans toward wild animals into nine categories: naturalistic, ecologistic, humanistic, moralistic, scientistic, aesthetic, utilitarian, dominionistic, and negativistic. The characteristics of these 9 attitudes were related to 11 social-demographic and 12 animal-activity groups.

The information Kellert shared led to a paradigm shift in wildlife management by allowing managers to better understand members of the public and their relation to wildlife. Since the publication of this work there have been numerous studies that have expanded on, and modified, Kellert's work on the many aspects of human dimensions.

RELATED READING

Conover, M. R. 1998. Perceptions of American agricultural producers about wildlife on their farms and ranches. Wildlife Society Bulletin 3:597–604.

Crank, R. D., S. E. Hygnstrom, S. R. Groepper, and K. M. Hams. 2010. Landowner attitudes toward elk management in the Pine Ridge region of northwestern Nebraska. Human–Wildlife Interactions 4:67–76.

Gentile, J. R. 1987. The evolution of antitrapping sentiment in the United States: a review and commentary. Wildlife Society Bulletin 15:490–503.

Lauber, T. B., and B. A. Knuth. 1997. Fairness and moose management decision-making: the citizens' perspective. Wildlife Society Bulletin 25:776–787.

Manfredo, M. J., D. C. Fulton, and C. L. Pierce. 1997. Understanding voter behavior on wildlife ballot initiatives: Colorado's trapping amendment. Human Dimensions of Wildlife 2:22–39.

Martin, S. R. 1997. Specialization and differences in setting preferences among wildlife viewers. Human Dimensions of Wildlife 2:1–18.

Reprinted by permission of the Wildlife Management Institute.

Perceptions of Animals in American Society

Stephen R. Kellert

Behavioral Sciences Study Center,
Yale University School of Medicine,
New Haven, Connecticut

This paper will summarize the results of a study on American attitudes toward animals which has been conducted over the past three years.[1] One of the primary goals of this study was to devise a typology of attitudes toward animals which hopefully would provide a better understanding of people's motivations for involvement in animal-related activities such as hunting, pet ownership, bird watching, animal welfare, scientific study, and so on. The other major goal of the study was to determine the distribution of these attitudes within the general American population. Thus two separate stages were implied in carrying out the investigation: first, formulating a typology of basic attitudes toward animals and, second, determining their apportionment among major population groups (for example, age, sex and education) and among major animal activity groups (for example, hunters, humanitarians and wildlifers).[2]

Methodology

The initial investigation—aimed at developing a typology of attitudes toward animals—focused on the views of people specifically interested in animals in some significant way. Studying a select, atypical group (composed only of persons involved with animals) in order to generate understandings relevant to a broader population was considered the most appropriate method at this stage for revealing fundamental aspects of contemporary human-animal relationships.

Although restricted, the population to be investigated still offered a wide variety of people for study, including animal artists, bird watchers, breeders, conservationists, ecologists, farmers, horsemen, humanitarians, hunters, pet owners, preservationists, scientists, vegetarians, veterinarians, writers and zoo personnel. Names of possible subjects were obtained through consultation with over 30 animal-related organizations. More than 700 persons were recommended, and 65 were eventually interviewed. There were 16 women and 49 men, averaging 45 years of age, and representing every section of the country. Each of the 65 was interviewed for approximately one hour. The first half of the

[1]This study was contracted by the Fish and Wildlife Service of the U.S. Department of the Interior. Special thanks are extended to Dr. Robert I. Smith, Paul Breer, Steven Schwager and Carolyn Kellert.

[2]Space restrictions prevent detailed presentation of the methodology and results of these two studies. A complete description of the first study can be found in a Fish and Wildlife Service report entitled "From Kinship to Mastery," by the author. The national study methods and results will be available at the end of June, also in a report to the Fish and Wildlife Service.

interview was open-ended and covered development of interest and present involvement with animals. The second half included only close-ended questions on a variety of animal, environmental and social issues.

Based on the typology of attitudes toward animals developed through the initial investigation, the follow-up study focused on the views and behaviors of the general American public. A close-ended, structured questionnaire was administered to 553 randomly selected Americans in 45-minute personal interviews which covered five areas—attitudes toward animals and the natural world, attitudes toward man's social world, a knowledge-of-animals quiz, respondents' animal-related activities, and basic social characteristics. The questionnaire was designed and pretested by the author. The final list of 79 attitude questions was gleaned from over 1000 considered, including more than 240 pretested. The sample selection and personal interviews were carried out by the Gallup Organization, using a "quota" rather than a "probability" random sampling method. The quota method randomly selects geographical areas (e.g., city blocks, rural neighborhoods) instead of individuals. Although less efficient and reliable than the probability method, the quota technique was considered most appropriate in view of the study's exploratory nature and the comparative costs involved.

The results of this two-stage investigation are presented in the following two sections of this paper. The first section defines and describes nine basic attitudes toward animals as well as their intercorrelations and their approximate frequency of occurrence in the American population. The second section presents findings on the distributions of these attitudes among various social-demographic and animal-activity groups.

A Typology of Attitudes[3] Toward Animals

Nine basic attitudes toward animals were identified and labelled as the naturalistic, ecologistic, humanistic, moralistic, scientistic, aesthetic, utilitarian, dominionistic and negativistic attitudes.

The Naturalistic Attitude

Although the naturalistic attitude is associated with an interest in all animals, its most outstanding characteristic is a profound attraction to wildlife and to the outdoors in general. The naturalistically oriented have affectionate feelings for pets but tend to regard them as inferior to wild animals. A primary satisfaction is in direct, personal contact with wilderness; and, in this regard, wildlife is valued particularly for the opportunities it provides for activity in the natural environment. An occasional manifestation of this attitude is an atavistic reward derived from experiencing wilderness as an escape from the perceived pressures and deficiences of modern industrial life.

The Ecologistic Attitude

The ecologistic attitude is also primarily oriented toward wildlife and natural settings, but typically is more intellectual and detached. This attitude views the

[3]Space does not permit a thorough discussion of the concept of attitude. It is broadly defined here as a distinguishable patterning of related ideas (cognitive notions), feelings (emotional-affective notions), and beliefs (cultural value notions).

natural environment predominantly as a system of interdependent parts. Rather than focusing on individual animals, wild or domesticated, the major emphasis and affection is for species of animals in their natural habitats. The perspective is often marked by considerable knowledge of animals, although this interest tends to concentrate more on behavioral relations of animal species than on their physical or biological properties.

While adhering to the notion that man is just another animal species, as ultimately dependent on the natural environment as any other, the ecologistic attitude tends to be concerned with protecting the environment primarily for the sake of humankind. Associated with this view is an interest in modifying modern society's impact on the natural world, although typically by compromising between the values of practical human advantage and protection of natural habitats.

The Humanistic Attitude

The humanistic attitude is distinguished by strong personal affection for individual animals, typically pets rather than wildlife. The pet animal is viewed as a friend, a companion, a member of the family. The love of the humanistically oriented person for animals can often be compared to that felt for human beings. Although not specifically interested in wildlife, the humanistically oriented often extend their empathy for pet animals to a general concern for the well-being of all animals, wildlife included. This concern for animal welfare originates less in any general ethical philosophy or in any particular concern for animal species than in an identification with the experience of individual animals extended from pets to wildlife.

The Moralistic Attitude

The most striking feature of the moralistic attitude is its great concern for the welfare of animals, both wild and domesticated. Rather than deriving from strong affection for individual animals (the humanistic point of view) or from consideration for animal species (the ecologistic attitude), the moralistic attitude is typically more philosophical. It is based on ethical principles opposing the exploitation and the infliction of any harm, suffering or death on animals. The moralistically oriented object to activities involving the killing of animals (e.g., hunting, trapping), and they also oppose many practices involving exploitation of animals (e.g., rodeos, cage zoos, horse racing). There is a tendency to perceive a kinship, a sense of equality, between humans and animals.

The Scientistic Attitude

The scientistic point of view is characterized by an objective, intellectualized, somewhat circumscribed perspective of animals. Animals are regarded more as physical objects for study than as subjects of affection or moral concern. There is typically little personal attraction to pets, wildlife or the natural environment among the scientistically oriented. Animals are usually perceived as the means for acquiring specific knowledge (mainly physiological, biological and taxonomic), and as offering opportunities for problem-solving. The affective relationship is one of emotional detachment, with curiosity often constituting the primary motivation for interest in animals.

Perceptions of Animals in American Society   535

The Aesthetic Attitude

The aesthetic attitude also tends to be associated with emotional detachment, but with a central interest in the beauty or symbolic properties of animals. Although many people possess a feeling for the physical attractions of animals, the aesthetically oriented base their interest almost exclusively on this artistic appeal. As an example, the aesthetically oriented tend to be attracted to animal sporting activities involving considerable artistic display such as animal showmanship, fox hunting and bullfighting. For the most part, they remain aloof from the living animal, enjoying it more as an object of beauty (in paintings, sculpture, movies) or of symbolic significance (in poetry, children's stories, cartoons).

The Utilitarian Attitude

The primary characteristic of the utilitarian attitude is the perception of animals in terms of their practical or profitable qualities—largely for their material benefit to humans. The utilitarian attitude is not necessarily marked by a lack of affection or interest in animals, although such feelings are usually subordinated to the more predominant interest in the usefulness of animals. While many utilitarian-oriented persons own pets, for example, most believe they should be trained for specific tasks and not kept just as companions or friends. Persons with a utilitarian attitude tend to be indifferent to issues of animal welfare which do not affect the animal's performance or practical value

The Dominionistic Attitude ʹ

A sense of superiority and a desire to master animals are defining features of the dominionistic attitude. Animals are mainly regarded from the perspective of providing opportunities for dominance and control, and expressions of prowess and skill in competition with animals are typically emphasized. Considerable attachment to animals may accompany the dominionistic attitude, but usually in the context of dominating them as, for example, in rodeos, trophy hunting and obedience training.

The Negativistic Attitude

A number of quite distinctive attitudes are included within the negativistic category, with the common feature being a desire to avoid animals. Typical of the negativistic attitude are such feelings as indifference, dislike, fear and superstition. This viewpoint is quite often marked by a fundamental sense of separation and alienation from the natural world. For many negativistically oriented persons, an utter gulf in emotion and spirit distinguishes animals from humans. The negativistic attitude is obviously very much people-centered, involving little, if any, sense of empathy or kinship with animals and the nonhuman world.[4]

The nine attitudes have been presented as ideal types and should be regarded as conceptual constructs of general human tendencies rather than specific descriptions of actual behavior. Most people typically possess more than one attitude toward animals, feeling and behaving a certain way in one situation while manifesting a different attitude under other circumstances. Additionally, when

[4]The author has not attempted to imply that any attitude is intrinsically superior or inferior to any other, but has sought merely to describe the qualities characteristic of each.

individuals express a particular attitude, rarely do they exhibit every characteristic of this attitude. In other words, not only do people have multiple attitudinal orientations, but they also vary considerably in the intensity of their commitments. In general, however, it is possible to identify in most individuals predominant characteristics of a primary attitude toward animals, with elements of secondary and tertiary attitudes present as well.[5]

Related to the expression of multiple but hierarchical attitudes in individuals is the question of which animal attitudes tend to cluster with one another. Based on impressions and on a correlation matrix computed in the national study, the most typical affiliations and antagonisms of the nine attitudes are presented in the chart below. (Key identifying terms for each attitude are also presented as a crude summary index.)

Attitude	Key Identifying Terms	Highly Correlated with	Most Antagonistic toward
Naturalistic	Wildlife exposure, contact with nature	Ecologistic, humanistic	Negativistic
Ecologistic	Ecosystem, species interdependence	Naturalistic, scientistic	Negativistic
Humanistic	Pets, love for animals	Moralistic	Negativistic
Moralistic	Ethical concern for Animal welfare	Humanistic	Utilitarian, dominionistic, scientistic, aesthetic, negativistic
Scientistic	Curiosity, study, knowledge	Ecologistic	None
Aesthetic	Artistic character and display	Naturalistic	Negativistic
Utilitarian	Practicality, usefulness	Dominionistic	Moralistic
Dominionistic	Mastery, superiority	Utilitarian, negativistic	Moralistic
Negativistic	Avoidance, dislike, indifference, fear	Dominionistic, utilitarian	Moralistic, humanistic, naturalistic

[5]The validity of Classifying people on seven of the nine attitudes was partially tested during the course of the first study through statistical examination of the subjects' responses to the close-ended questions. (The dominionistic and negativistic attitudes were not considered in the first study.) Because assignment of subjects to attitude categories was based on responses to the open-ended section of the interview, it was possible to examine responses to the close-ended questions without this analysis being tautological. The results of this statistical analysis—referred to as the multiple discriminant function analysis—roughly affirmed (1) the correct classification of nearly all respondents and (2) the general distinctiveness of all the attitudes from one another (with the partial exception of the ecologistic attitude).

Perceptions of Animals in American Society   537

The relative popularity of the attitudes can be examined in terms of both their prevalence (total number of cases in the population) and their incidence (rate of new cases within a given time period). Prevalence figures provide a good idea of absolute frequency, while incidence suggests historical changes and trends. Tentative prevalence statistics from the national study suggest that the humanistic, the utilitarian, and the "indifference" component of the negativistic attitude are the most common attitudes among Americans, while the most uncommon seem to be the scientistic, the aesthetic and the ecologistic. Considering incidence during the past 10 years, the impression is that the utilitarian attitude is decreasing in popularity along with the negativistic, while the naturalistic, humanistic and ecologistic viewpoints appear to be substantially increasing. National study data comparing different age, educational and urban-rural groups generally corroborates this change, although considerably more analysis is needed before a trend can be definitely substantiated.

Distribution of the Attitudes in the General American Population

The remainder of the paper deals with the distribution of the nine attitudes within major social-demographic and animal-activity groups in the general American population. Unfortunately, these findings must be largely confined to six attitude types, as indices of the aesthetic, scientistic and, to a lesser degree, ecologistic attitudes were not particularly reliable or valid. Problems with these three attitudes probably relate to their relative infrequency in the population exacerbated by a sample size of 553. Interpretation of the findings was largely based upon three overlapping though different criteria: individual attitude cross-tabulations, attitude scale cross-tabulations, and regression analysis (analysis of variance and multiple regression). Attitude scales were created largely by examining the results of a statistical technique referred to as cluster analysis.

The national study findings must be regarded as preliminary and tentative. This was largely an exploratory study based on a limited sample size and involving rather large numbers of often complex variables. These are not exact relationships but tentative approximations of the underlying situation. Additionally, it should be emphasized that these findings are gross generalizations of large population groups. Individual person differences cannot be inferred from such results, only the statistical likelihood that a particular attitude may occur in a particular group. Further analysis may reveal the presence of unconsidered "X" factors that largely account for found relationships between attitudes and groups of people.

Findings on the distribution of the attitude types within 11 social-demographic groups are briefly reviewed below. For convenience, the 11 variables have been divided into four major groups: basic ascriptive variables (age, sex, race); socio-economic variables (education, occupation, income); geographic variables (childhood residence, present residence, section of the country); and familial variables (marital status, number of children).

Basic Ascriptive Variables

Age. By far the most significant age differences were between persons 18 to 29 (people under 18 were not sampled) and those 65 and older. The 65 and older population was significantly less naturalistic and somewhat less moralistic than those 18 to 29. Furthermore, the elderly population was significantly more utilitarian and negativistically oriented.

Regarding naturalistic differences, the 65+ group was far less interested in wildlife, wilderness and spending time in the outdoors. Age differences were, in fact, the largest found on the naturalistic dimension. Moralistic differences between age groups were not as great but did reveal 18 to 29 people to be considerably more interested in problems of animal welfare and significantly more opposed to hunting than older people. The 18 to 29 group also expressed a more ecologistic attitude, though differences here were not great. Exemplary of the prominent utilitarian attitude of elderly persons, in contrast to those 18 to 29, was support for such activities as predator control, commercial activity at the expense of wildlife, and a preference for work animals over pets. Differences on the negativistic dimension were revealed in considerably less affection and desire for personal contact with animals among the 65 and older population.

Sex. Significant differences were found between males and females on at least seven of the nine attitudes (moralistic, humanistic, naturalistic, utilitarian, dominionistic, scientistic and ecologistic). Sex and education, in fact, were consistently the two most important social differentiators of people's attitudes toward animals and the natural world.

Sex differences were especially impressive on the moralistic and humanistic dimensions. Females were far more concerned with protecting animals from suffering, and much more inclined to express strong loving feelings toward pet animals. Naturalistic differences between the sexes were manifest in the much greater desire among males for direct contact with wildlife and the outdoors. Males were also significantly more utilitarian and expressed more acceptance of such activities as killing animals for meat, predator control, and harvesting wildlife for fur. On the dominionistic dimension, females were less interested either in mastering animals for sporting purposes or in training them to perform specific tasks. Finally, although the scientistic and ecologistic attitudes were somewhat poorly measured, males generally scored higher on these dimensions, especially regarding knowledge of animals.

Race. Racial differences between blacks and whites were quite striking on the negativistic, naturalistic and moralistic attitudes. Additionally, noteworthy differences on the dominionistic attitude were revealed by black males. These differences remained significant after all other demographic variables had been taken into account, but small sample sizes for blacks render the findings somewhat tentative at this time.

The more negativistic attitude of blacks was manifested by less animal interest and affection as well as more fear of animals, wildlife in particular. Naturalistic differences were expressed in greater wildlife interest and desire for personal contact with wilderness settings among whites. Not surprisingly, given the negativistic differences, blacks were generally less interested in moralistic issues of animal welfare and exploitation. Black males (although not black females) were significantly more dominionistic, generally admiring expressions of superior physical strength over animals.

Socioeconomic Status

Education. As indicated, education and sex were the most consistent social differentiators of people's views toward animals. Educational differences were most striking when comparing persons with a less than eighth grade education to those with some or completed college. The most significant educational findings were on the negativistic, utilitarian and dominionistic attitudes, while important differences were also recorded on the ecologistic and naturalistic dimensions.

Negativistic differences between educational groups were particularly remarkable—in fact, the single biggest distinction on any attitude dimension by any demographic variable. People with low education manifested far more fear, lack of affection, and disinterest in animals. Although these feelings were most apparent with regard to wildlife, they were also evident in views toward domesticated animals (e.g., expressing fear of stray dogs, perceiving cats as vicious and supporting extermination of all pests). Higher educational groups—especially the college-educated—tended to be less utilitarian toward animals. In contrast, the less than eighth grade group expressed much greater support for such utilitarian activities as trapping, hunting for meat, and the value of conquering and taming the wilderness. Dominionistic differences were reflected in a more authoritarian relation to animals among lower educational groups. Concern with wildlife and wilderness among the college educated accounted for their higher naturalistic scores. This interest in wildlife, along with a more extensive knowledge of animals, was the basis for the somewhat increased ecologistic orientation among college-educated people. Relatively insignificant differences were found among educational groups on the moralistic attitude. Apparently concern for animal welfare and the environment is more likely to be manifested in an ecologistic than a moralistic attitude among higher educational groups.

Occupation. Although a number of interesting occupational differences were revealed in the item and scale analyses, many of these differences were minimized in the multiple regression analyses when the influence of other factors was considered. Nevertheless, some potentially significant findings were revealed among occupational groups on the utilitarian, moralistic, dominionistic, naturalistic and negativistic attitudes.

Farmers, as expected, were by far the most utilitarian group. This pattern was revealed in strong support for such activities as the use of steel leg-hold traps (one of the few occupational groups to support this practice), predator control (the only occupational category to unanimously support this activity), pest extermination (also 100% support), and agreement with the notion that animals exist largely for the benefit of man. The one occupational category consistently contrasting with the views of farmers were students who generally expressed a highly moralistic attitude. Students not only opposed many of the above mentioned activities but also objected to rodeos, trophy and sport (though not meat) hunting, and the killing of wild animals for their fur. Students contrasted markedly with farmers and also unskilled workers who were much more dominionistically oriented. Students, as well as professionals, business executives and skilled workers, scored high on the naturalistic attitude. A discernible though not strong negativistic attitude was found among unskilled and clerical workers, especially regarding feelings of indifference and fear of animals.

Income. Perhaps the most significant finding on the income variable was its relative unimportance. On few attitudes did income differences prove noteworthy. Insofar as predicting views toward animals and the natural world, income appears to have only marginal relevance, in sharp contrast to the other socioeconomic indicators (education and occupation).

There was a moderate tendency to find lower income groups with a more negativistic attitude toward animals. Higher income people tended to be slightly more naturalistic and less utilitarian in their outlook than lower income persons.

Geographical Variables

Childhood residence. The most outstanding finding on this urban-rural dimension was the much greater dominionistic orientation of those reared in rural areas of less than 2000 population. In general, rural childhood background was related to a greater sense of superiority and a lack of affinity with animals. Additionally, rural dwellers were significantly more utilitarian-oriented than city residents and supported such practices as hunting for meat, the use of steel traps and the killing of wild animals for their fur. Those raised in cities of 1 million plus were more moralistic, objecting to these practices as well as to rodeos and predator control. Not surprisingly, noncity-dwellers were more naturalistically oriented; what was unexpected was that the strongest naturalistic attitude was found in persons from towns of 10,000 to 50,000 population. One could surmise that those raised in small towns not only had the opportunity for outdoor and wildlife exposure but also the noncommercial involvement to allow this type of appreciative interest.

Present residence. Findings on population of present residence followed the same pattern as childhood residence and therefore do not require much further discussion. Perhaps the most interesting finding was that, although childhood and present residence results were similar, the former was consistently stronger. This finding supports the developmental hypothesis that childhood environment is most important in the formation of attitudes toward animals.

Section of the country. Differences in this variable were not great although they were consistent across a number of indicators. The Rocky Mountain, South Central and West Central States seemed to be the most dominionistic and utilitarian in their attitudes. A general perspective of human superiority and of the practical use of animals was common in these areas. Additionally, the Rocky Mountain States projected the greatest naturalistic interest. In terms of affection for pet animals, the Mid-Atlantic States were generally more humanistically oriented, contrasting with the South Central and South Eastern States which were more negativistically oriented toward both wildlife and domesticated animals.

Familial Variables

Marital status. Differences among marital status groups did not usually remain significant when other variables were taken into account. This was particularly the case for widows on the negativistic dimension, and for single persons on the naturalistic, where variations were mostly a function of age. One rather significant difference on this variable was between single and married persons on the

Perceptions of Animals in American Society 541

utilitarian dimension. Married persons generally revealed pragmatic perceptions of animals, whereas single persons more frequently objected to utilitarian exploitation (for example, predator control and the raising of fur-bearing animals). Single persons scored consistently higher on the humanistic dimension, partially supporting the "human-substitute" hypothesis that unmarried people often own pets as a way of compensating for the absence of other people.

Number of children. Differences on this variable were not impressive. Two tendencies were for families of five and more children to be more utilitarian, and for persons without children to be more humanistic. The latter finding further supports the human-substitute theory as one motivation for having pets.

Summary of the Demographic Findings

As previously indicated, three types of data were considered in assessing the significance of relationships between social-demographic and attitudinal variables—individual item cross-tabulations, scale score cross-tabulations, and multivariate analysis (multiple regression and analysis of variance). Figure 1 roughly summarizes the assessed significance of these findings for all the demographic variables described (without indicating the direction of these relationships.).

Relation of Animal Activities to Attitudes Toward Animals

A number of behavioral relations to animals were partially built into the attitude scales, and thus some of the following findings will not be surprising and may be somewhat tautological. Nevertheless, there are some illuminating results in relation to both activities and attitudes.

Hunters were categorized according to their primary reason for hunting—for close contact with nature, for meat or for sport. *Nature hunters* were predictably the most naturalistically oriented, expressing great interest in wildlife, getting out in the woods, and seeing wilderness left unspoiled and unexploited. Nature hunters also scored quite high on the humanistic and ecologistic variables, indicating their basic affection and concern for animals and the natural environment. *Meat hunters*, in contrast, were not particularly naturalistically oriented. As anticipated, they scored highest on the utilitarian dimension. *Sport hunters*, somewhat unexpectedly, manifested a rather strong negativistic attitude toward animals. Apparently interest and affection for animals is not typical of those who report sporting enjoyment as their primary reason for hunting. Sport hunters were very dominionistically oriented, suggesting that displays of skill, expressions of prowess, competition, and mastery over the animal were important motivations in this activity. Finally, it is relevant to note that, whereas sport hunters were among the lowest scoring groups on the knowledge of animals quiz, nature hunters scored higher than any other social demographic or animal activity group.

Differences in predominant attitudes of the three hunting groups underscore the value of distinguishing between attitudes toward and activities with animals although sometimes particular attitudes may be closely associated with particular activities. In a similar example, regarding opposition to hunting, at least three relevant attitudes can be cited—the moralistic, the humanistic and the ecologis-

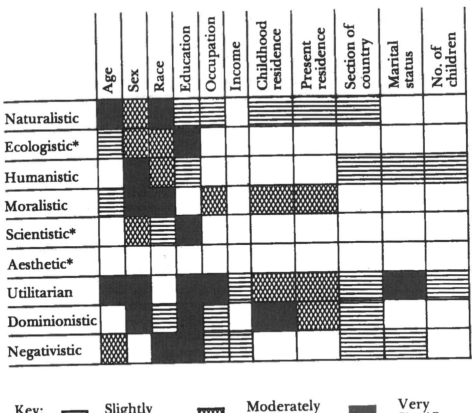

Key: ▤ = Slightly Significant Differences   ▧ = Moderately Significant Differences   ■ = Very Significant Differences

* Scale and item indices fair to poor in national study.

Figure 1. Summary of Social Demographic Findings by Attitudes

tic. The first would object to hunting on the grounds that killing in the pursuit of pleasure is inherently wrong. The humanistically oriented identify with the experience of the individual animal and typically object out of sympathy with its suffering. Finally, the ecologically oriented, although not always opposed to hunting, voice strenuous objections when a species is endangered.

Two other activity groups fundamentally interested in wildlife and the outdoors were backpackers and bird watchers. *Backpackers*, as expected, manifested strong naturalistic attitudes although less than nature hunters. Additionally, backpackers were both very humanistically and moralistically oriented, expressing (quite unlike the nature hunters) a fair degree of opposition to hunting and to the harvesting of wildlife for fur. *Bird watchers* were both very naturalistically and ecologically oriented, manifesting strong interest and concern for the condition of wildlife and the natural environment.

Two groups interested in domesticated animals were pet owners and rodeo enthusiasts. *Pet owners* were divided into two groups—those who own pets primarily for companionship and those who own pets mainly for work, protec-

Perceptions of Animals in American Society   543

tion or sport (combined together because of small sample sizes). The *companion pet owner* was, of course, very humanistically oriented. This activity group was also one of the most moralistically oriented, opposing such activities as raising animals for fur, predator control, and, to a lesser degree, sport hunting. *Work, protection, sport pet owners*, in dramatic contrast, were among the most negativistically oriented groups examined in the study. They manifested strong dominionistic and utilitarian attitudes also. Apparently, pet ownership in this group was almost exclusively motivated by practical considerations and did not involve much personal affection for pet animals.

Rodeo enthusiasts manifested a relatively unusual combination of dominionistic and humanistic attitudes. Although supporting the value of dominating and mastering animals, they still revealed strong affection for pet animals. Additionally, rodeo enthusiasts were rather utilitarian-oriented, believing in the virtue of work animals, predator control, wilderness exploitation, and the killing of wildlife for fur.

Vegetarians and zoo enthusiasts were among the most moralistically oriented groups. *Zoo enthusiasts*, for example, strongly objected to rodeos, killing animals for fur, and utilitarian exploitation of animals in general. This group also revealed strong humanistic attitudes and, somewhat surprisingly, were not particularly naturalistic or ecologistic in their attitude toward animals. *Vegetarians* were even more moralistically oriented, expressing opposition to a wide range of activities including hunting, trapping, predator control, neutering of dogs, and much medical investigation involving the killing of animals. Additionally, they were one of the few moralistically oriented groups which were not also strongly humanistic, embodying a general philosophical concern for animal welfare divorced from any personal affection for animals.

Two of the more utilitarian-oriented groups were animal farmers and trappers. *Animal farmers*, in addition to their utilitarian outlook, were also quite dominionistically oriented. *Trappers* were among the most negativistic of all groups studied, revealing a general disinterest and lack of affection for animals. Additionally, trappers were highly dominionistic, supporting such activities as cock fighting, training animals through strong physical force, and trophy hunting.

Figure 2 summarizes, in a manner similar to the previous figure, the various relationships described between animal activity groups and attitudes toward animals.

Summary

A wide range of results from a study of American attitudes toward animals has been presented. A typology of nine attitudes toward animals was described and related to 11 social-demographic and 12 animal-activity groups. A number of policy implications derive from this data; but, as this matter has not yet been sufficiently considered, speculation will be deferred.

Discussion

MR. CHARLES NEWLING [U.S. Army Corps of Engineers]: Dr. Kellert, I found the differentiation between the type of hunters rather incredible. Did your data suggest any way to reconcile differences between hunters and nonhunters, consumptive and nonconsumptive use of wildlife?

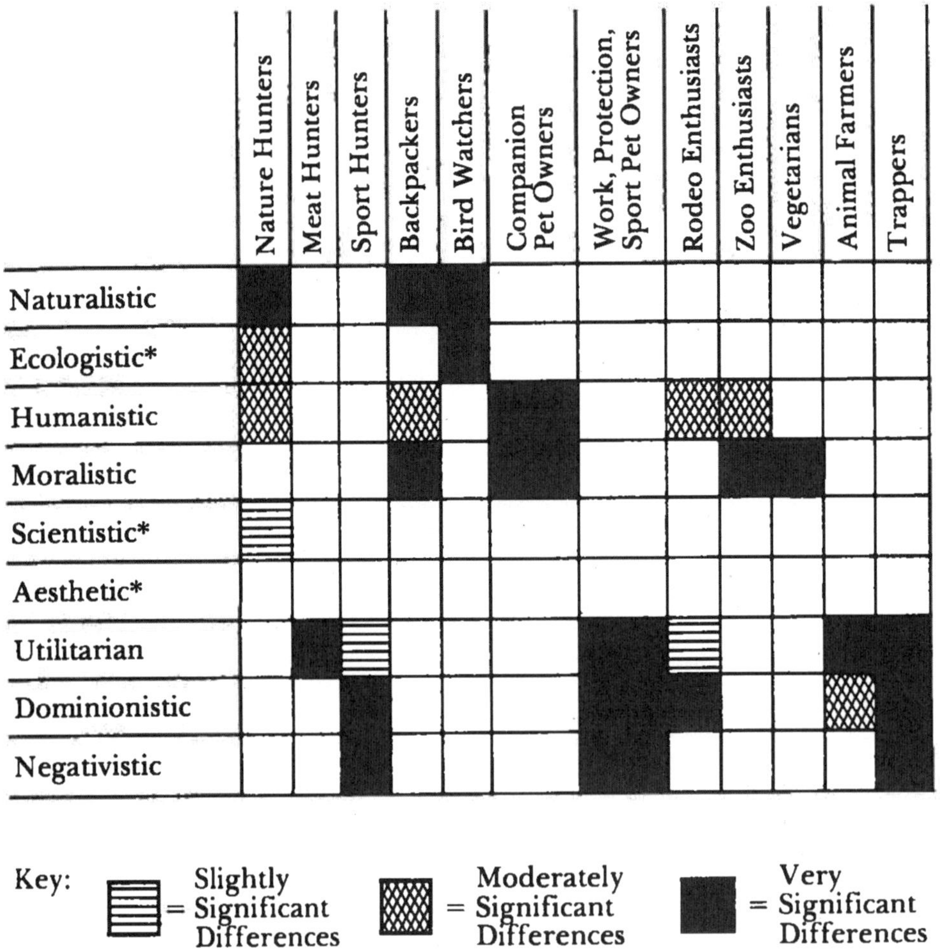

	Nature Hunters	Meat Hunters	Sport Hunters	Backpackers	Bird Watchers	Companion Pet Owners	Work, Protection, Sport Pet Owners	Rodeo Enthusiasts	Zoo Enthusiasts	Vegetarians	Animal Farmers	Trappers
Naturalistic												
Ecologistic*												
Humanistic												
Moralistic												
Scientistic*												
Aesthetic*												
Utilitarian												
Dominionistic												
Negativistic												

Key:
▤ = Slightly Significant Differences
▨ = Moderately Significant Differences
■ = Very Significant Differences

* Scale and item indices fair to poor in national study.

Figure 2. Summary of Animal Activity Group Findings by Attitudes

DR. KELLERT: I don't know if any data necessarily suggests any way for reconciliation. One of the problems is that many of these attitudes start from fundamentally different premises, and often these groups talk about one another rather than with one another.

If there is an opportunity for reconciliation, or compromise, it is probably in the ecologistic attitude. The two that oppose one another are the moralistic and the utilitarian groups, and I think there are elements which can be related to the ecologistic type and in that attitude type there may be a possibility for dialogue.

MR. LARRY HARRIS [University of Florida]: Based on similar research in Florida, we established a higher proportion of the anti-hunting segment is constituted of alienated former hunters. I wonder if you can address this issue that we commonly see, that the anti-hunters are Bambi lovers. Perhaps that is not true.

DR. KELLERT: It has not been my finding, although I think that is quite interesting. I have not found that reformed hunters were the most anti-hunting group in the study. The two primary sources for anti-hunting represent more or less a composite of the feeling that killing of animals is inherently wrong and involves a great deal of suffering; and I think the other is guilt by association phenomenon, where sentiment does enter, where opposition to guns and violence is such that the hunter is often found in that position.

Perceptions of Animals in American Society 545

CHAIRMAN CLUSEN: Let me ask one question. You said you had not raised strong questions of policy statements, and you may not want to answer this. Nevertheless, the obvious question as far as the public is concerned would be what does this all mean and what kinds of land or territory would or could be set aside for wildlife preservation or hunting? Is there anything in your analysis or your statistics which indicated attitudes toward this.

DR. KELLERT: That is a difficult question to answer.

There are a number of specific findings which may have relevance toward education and conservation of multiple resources, but they have not been developed in that category.

Dickson, J. G., R. N. Conner, and K. T. Adair. 1978. Guidelines for authorship of scientific articles. *Wildlife Society Bulletin* 6:260–261.

Scientists beginning their career are well versed in the scientific process but often have not addressed the issue of authorship for papers with more than a single author. Who should be the senior author? Who qualifies for coauthorship, and how should they be listed following the senior author? These are important questions, because authorship is a critical aspect of scientific work and it recognizes the research team. Regrettably, without a standard from which to base authorship, the process can become contentious.

Several papers in other fields of study have addressed this issue, but it was Dickson et al. who provided the first useful guidelines to assist wildlife biologists in deciding authorship. They divided research investigations into five areas (i.e., conception of the idea, study design, data collection, data analysis, manuscript preparation), and recommended that all authors of a paper be involved in some aspect of manuscript preparation and at least one other area of investigation.

Dickson et al.'s guidelines are straightforward and are similar to those in other disciplines. Authorship is serious business and should be discussed at the beginning of any research project. This paper is a good place from which to start those discussions.

RELATED READING

Bennett, D. M., and D. M. Taylor. 2003. Unethical practices in authorship of scientific papers. Emergency Medicine 15:263–270.

Iammarino, N. K., T. W. O'Rourke, R. M. Pigg, and A. D. Weinberg. 1989. Ethical issues in research and publication. Journal of School Health 59:101–104.

Laurance, W. F. 2006. Second thoughts on who goes where in author lists. Nature 442:26.

Schmidt, R. H. 1987. A worksheet for authorship of scientific articles. Bulletin of the Ecological Society of America 68:8–10.

Shewan, L. G., and A. J. S. Coats. 2010. Ethics in the authorship and publishing of scientific articles. International Journal of Cardiology 144:1–2.

Smith, R. 1997. Authorship: time for a paradigm shift? British Medical Journal 314:992.

Tscharntke, T., M. E. Hochberg, T. A. Rand, V. H. Resh, and J. Krauss. 2007. Author sequence and credit for contributions in multiauthored publications. PLoS Biol 5(1): e18. doi:10.1371/journal.pbio.0050018. Accessed 25 June 2012.

Weltzin, J. F., R. T. Belote, L. T. Williams, J. K. Keller, and E. C. Engel. 2006. Authorship in ecology: attribution, accountability, and responsibility. Frontiers in Ecology and the Environment 4:435–441.

GUIDELINES FOR AUTHORSHIP OF
SCIENTIFIC ARTICLES

The move toward interdisciplinary and team research imposes special requirements for guidelines that delineate authorship rights of the scientists involved. Each contributing individual of a research team should be accurately recognized.

We propose some simple guidelines and recommend their adoption *before* research projects begin. Adoption of standards should prohibit unauthorized or unethical use of data and should enhance maintenance of professional ethics.

Research investigations can be divided into 5 basic areas: conception, design (procedures), data collection, data analysis, and manuscript preparation. The relative importance of each area varies considerably among studies. Some investigations are innovative and emphasize an original idea, rather than extensive data collection. In other investigations, the idea of the research is not unique and most of the effort is in data collection and analysis. In some studies intricate research design or establishment of sophisticated measurement procedures requires the most effort. Every researcher is aware of the work involved in the writing phase of a project.

Determination of authors and sequence of authorship of manuscripts should be based on contributions in each facet of the research. All authors should have made significant contributions in writing (at least a review of the manuscript) and in at least 1 additional area of investigation.

Much wildlife research is being conducted by universities and research unit employees of state, federal, and other cooperative agencies. In federal agencies, many studies involve a researcher dominant in all areas except data collection which is often done by technicians. The researcher should be the sole author in this situation unless the technician is allowed and accepts responsibility in more areas.

At universities, much research involves a professor/graduate student relationship with varying contributions from each. If the student is given and accepts responsibility in all areas he should be first author of manuscripts, or sole author if the professor has not made substantial contributions to the study in at least 2 of the 5 areas. If the professor conceives and designs a project and is instrumental in other areas, he should be first author. If the student is supposedly conducting original research, the situation should be discouraged where the student achieves his main goal (the degree), but the major professor becomes the sole author. If the student has not contributed sufficiently to merit authorship, he or she probably has not fulfilled the degree requirements of original research.

Directors of research units or laboratories should not automatically be authors of research publications from the research organizations (CBE Style Manual Committee 1978:8); nor should authorship be automatically tied to source of funding for research. All authors should be active participants in the actual workings of a project.

In situations at universities and research agencies where the person contributing the most to a research study has no intention of publishing the results (usually because of departure from the organization), another person involved in the study should be able to assume senior authorship if he or she writes the manuscript. In this situation the person contributing the most should be offered the chance to review the manuscript and be a junior author.

Adoption of these or similar guidelines will help ensure proper recognition of the contributions of each researcher. It would also ensure that study results would be published by some member of a research team with adequate recognition for the contributions of others. At the very least, proper prior planning would avoid situations where contributions of some remain unrecognized.

LITERATURE CITED

CBE STYLE MANUAL COMMITTEE. 1978. Council of Biology Editors style manual: a guide for authors, editors, and publishers in the biological sciences. 4th ed. Council of Biology Editors. 265pp.

JAMES G. DICKSON AND RICHARD N. CONNER, USDA Forest Service, Southern Forest Experiment Station, Nacogdoches, TX 75962, AND KENT T. ADAIR, School of Forestry, Stephen F. Austin State University, Nacogdoches, TX 75962.

Received 24 July 1978.
Accepted 11 October 1978.

Ehrlich, P. R., and E. O. Wilson. 1991. Biodiversity studies: science and policy. *Science* 253:758–762.

Biologists estimate that there are 10 to 30 million species of flora and fauna in the world, but less than 2 million have been given scientific names. Although scientists have struggled for centuries to estimate and catalogue the world's plants and animals, we are not even close to completing the job. In this paper, Ehrlich and Wilson outline what is known about the world's biodiversity, explain why species naturally become extinct (i.e., extinction proneness), discuss causes of human-caused extinction (i.e., habitat destruction), and examine why society should be concerned. They briefly discuss what should be done in the policy arena to preserve biodiversity, arguing that action needs to start immediately.

Many have considered the consequences of mass extinctions; this paper serves as a wake-up call. "The indispensable strategy for saving our fellow living creatures and ourselves in the long run, is, as the evidence compellingly shows, to reduce the scale of human activities. The task of accomplishing this goal will involve a cooperative worldwide effort unprecedented in history. Unless humanity can move determinedly in that direction, all of the efforts now going into in situ conservation will eventually lead nowhere, and our descendents' future will be at risk." This is a challenge that cannot be overlooked.

RELATED READING

Czech, B., P. R. Krausman, and P. K. Devers. 2000. Economic associations among causes of species and endangerment in the United States. BioScience 50:593–601.

Miller, J. R. 2005. Biodiversity conservation and the extinction of experience. Trends in Ecology and Evolution 20:430–434.

Pimm, S. L., and P. Raven. 2000. Biodiversity: extinction by numbers. Nature 403:843–845.

Soule, M. E. 1991. Conservation: tactics for a constant crisis. Science 253:744–749.

Swanson, T. 1994. The economics of extinction revisited and revised: a generalised framework for the analysis of the problems of endangered species and biodiversity losses. Oxford Economic Papers 46:800–821.

Thomas, C. D., A. Cameron, R. E. Green, M. Bakkenes, L. J. Beaumont, Y. C. Collingham, B. F. N. Erasmus, M. Ferreira de Siqueira, A. Grainger, L. Hannah, L. Hughes, B. Huntley, A. S. van Jaarsveld, G. F. Midgley, L. Miles, M. A. Ortega-Huerta, A. T. Peterson, O. L. Phillips, and S. E. Williams. 2004. Extinction risk from climate change. Nature 427:145–148.

Article

Biodiversity Studies: Science and Policy

PAUL R. EHRLICH AND EDWARD O. WILSON

Biodiversity studies comprise the systematic examination of the full array of different kinds of organisms together with the technology by which the diversity can be maintained and used for the benefit of humanity. Current basic research at the species level focuses on the process of species formation, the standing levels of species numbers in various higher taxonomic categories, and the phenomena of hyperdiversity and extinction proneness. The major practical concern is the massive extinction rate now caused by human activity, which threatens losses in the esthetic quality of the world, in economic opportunity, and in vital ecosystem services.

FROM LINNAEUS TO DARWIN TO THE PRESENT ERA OF cladograms and molecular evolution, a central theme of biology has always been the diversity of life. A new urgency now impels the study of this subject for its own sake: just as the importance of all life forms for human welfare becomes most clear, the extinction of wild species and ecosystems is seen to be accelerating through human action (1). The dilemma has resulted in the rise of biodiversity studies: the systematic examination of the full array of organisms and the origin of this diversity, together with the methods by which diversity can be maintained and used for the benefit of humanity. Biodiversity studies thus combine elements of evolutionary biology and ecology with those of applied biology and public policy. They are based in organismic and evolutionary biology in the same manner that biomedical studies are based in molecular and cellular biology. They include the newly emergent discipline of conservation biology but are even more eclectic, subsuming pure systematic research and the practical applications of such research that accrue to medicine, forestry, and agriculture, as well as research on policies that maximize the preservation and use of biodiversity. In biodiversity studies, the systematist meets the economist and political scientist. In this article we will present some of the key issues that newly link these two principal domains.

Species Formation

A rich medley of models has been constructed to account for the origin of species by reproductive isolation. Two broad categories have been substantiated by empirical evidence. The first is poly-

ploidy, the multiplication of entire chromosome numbers within individual species or within hybrids of species, a process that isolates the new breed from its ancestor in one step. This instantaneous mode has generated 40% of contemporaneous plant species and a much smaller number of animal species (2). Of comparable importance is geographic (or allopatric) speciation, the origin of intrinsic isolating mechanisms in two or more daughter populations while they are isolated by a geographic barrier, such as a sea strait, desert basin, or mountain range. Evidence of this two-step process, which occurs widely in plants and animals, has been documented minutely, often to the level of the gene, in birds, mammals, and a few groups of insects such as drosophilid flies and butterflies (3).

The diversification processes of polyploidy and geographic isolation are generally appreciated because they follow an easily traced pathway of measurable steps. Other modes of speciation are more difficult to conceive and test, but this does not mean they do not occur widely. Perhaps the most common is nonpolyploid sympatric speciation, in which new species emerge from the midst of parental species even when individuals of both populations are close enough to intermingle during part of their life cycles. The dominant process of this category, at least the one most persuasively modeled and documented, is by intermediate host races. Members of the parental species feed upon and mate in the vicinity of one kind of plant; they give rise to an alternate host race that shifts to a second species of host plant growing nearby; the two races, thus isolated by their microhabitat differences, diverge further in other traits that reinforce reproductive isolation. Sympatric speciation may play a key role in the origin of the vast numbers of insects and other invertebrates specialized on hosts or other types of microhabitats. The early stages are difficult to detect, however, and few studies have been initiated in the invertebrate groups most likely to display them (4).

Certain forms of speciation can thus occur very rapidly, within one to several generations. And when species meet, they can displace one another genetically within ten or fewer generations, reducing competition and the likelihood of hybridization (5). A question of central importance is the impact of high speciation rates on standing diversity. Although the probability of extinction of species within a particular group at a particular place (say, the anole lizards of Cuba) eventually rises with the number of species, the number of species should increase with greater speciation rates at all levels up to equilibrium. But does it really? And if so, in which groups and to what degree?

Current Levels of Biodiversity

Also in an early stage, and surprisingly so, is the elementary taxonomic description of the world biota. At the present time approximately 1.4 million species of plants, animals, and microorganisms have been given scientific names (1, 6). Terrestrial and freshwater species diversity is greater than marine diversity. The

P. R. Ehrlich is Bing Professor of Population Studies and Professor of Biological Sciences at Stanford University, Stanford, CA 94305. E. O. Wilson is Baird Professor of Science and Curator in Entomology at Harvard University, Museum of Comparative Zoology, Cambridge, MA 02138–2902. This article is based on their Crafoord Prize lectures given at the Royal Swedish Academy of Sciences, Stockholm, 26 September 1990.

overwhelming elements are the flowering plants (220,000 species) and their coevolutionary partners, the insects (750,000 species). The reverse is the case at the highest taxonomic levels, with all of the 33 living animal phyla present in the sea and only 17, or half, present on land and in fresh water (7).

Known species diversity is only a small fraction of actual species diversity, especially in the invertebrates and microorganisms. In this century the class Insecta has always been considered the most speciose group at the class level. As early as 1952, Sabrosky estimated that the number of living species is as high as 10 million (8). In 1982, Erwin found that beetle diversity in Neotropical trees, revealed in samples knocked down by insecticidal fogs, suggest far higher levels of insect and other arthropod diversity in tropical rain forests than had previously been estimated for the entire world fauna and flora (9). His figure, 30 million, was reached by extrapolating from counts of beetle species (1200) in a Panamanian tree species through estimates of total arthropod diversity per tree species to the percentages of species limited to each tree species to the total of tree species in tropical rain forests. Stork (10) reassessed this bold extrapolation, and in essence agreed with it, adding data of his own from Indonesian forests to produce a possible range of 10 to 80 million tropical forest arthropods. The most sensitive parameter remains the degree to which species of beetles and other arthropods are found uniquely on individual tree species.

In fact, because the life of the planet remains mostly unexplored at the species and infraspecies levels, systematists do not know the species diversity of the total world fauna and flora to the nearest order of magnitude. It is easily possible that the true number of species is closer to 10^8 than 10^7. Relatively little effort has been expended on nematodes, mites, or fungi, each highly diverse and containing undescribed species that could easily range into the hundreds of thousands or millions. Bacteria, with only about 4000 described species, remain a terra vitae incognita because of the astonishingly small amount of research devoted to their diversity, as opposed to the genetics and molecular biology of select species.

Hyperdiversity

Certain taxa are hyperdiverse, that is, they contain more species, genera, or higher ranked groups within them than expected by a null model of random assortment (11). Examples include arthropods among animal phyla, insects among arthropods, rodents among mammalian orders, orchids among monocotyledonous plant families, *Sciurus* among the genera of Sciuridae (squirrels), and so forth. It can be expected in a Darwinian world, where chance and opportunism prevail, that production of great diversity depends to substantial degree on special adaptations allowing penetration of multiple niches, such that each hyperdiverse group has its own magic key. For example, the ants appear to have expanded by virtue of fungistatic secretions, series-parallel work operations, and a highly altruistic worker caste (12). But recent research has also begun to identify properties possessed by many groups: small size, permitting fine niche subdivision (7, 13, 14); phytophagy and parasitism with specialization on hosts (15); specialized life stages that allow species to occupy multiple niches; entry into new geographic areas with subsequent adaptive radiation and preemption; and greater dispersal ability, promoting the colonization of empty areas. Southwood neatly summarized the likely causes of the extreme hyperdiversity of insects as "size, metamorphism, and wings" (13).

Hyperdiversity also occurs in certain habitats and geographical areas. The strongest trend worldwide is the latitudinal diversity gradient, with group after group reaching its maximum richness in the tropics and most particularly in the tropical rain forests and coral reefs. (Exceptions include conifers, salamanders, and aphids.) The hyperdiversity of continental rain forests is legendary. Gentry found about 300 tree species in single-hectare plots in Peru (16), to be compared with 700 native tree species in all of North America. A single tree in the same area yielded 43 species of ants in 26 genera, about equal to the ant fauna of the entire British isles (17). Explaining the latitudinal diversity gradient has proven an intractable problem. But clues exist which when pieced together suggest the possibility of a general explanation, involving climatic stability and extreme biological specialization and niche division (18).

Natural Extinction

One of the qualities reducing diversity in particular groups is extinction proneness, which renders populations vulnerable to environmental change and reduces taxonomic groups to one or a very few threatened species. A threatened or endangered species (the two grades commonly employed by conservationists) is one with a high probability of extinction during the next few years or decades. The principal demographic properties contributing to the status are a low maximum breeding population size and a high coefficient of variation in that size (19). When the breeding size drops to a hundred or less, the likelihood of extinction is enhanced still further by inbreeding depression (20).

The overall natural extinction rate (at times other than mass extinction episodes) estimated from fossil data to the nearest order of magnitude is 10^{-7} species per species year (21). This estimate refers to true extinction, from the origin of a species to the extinction of that species and any species descended from it (altogether, called the clade) and excludes "pseudoextinction," the evolution of one species into another. Wide variation exists among major taxonomic groups in the longevity of clades. Mesozoic ammonoid and Silurian graptolite clades lasted only 1 million to 2 million years, whereas most other Paleozoic and Mesozoic invertebrate clades lasted closer to 10 million years (21). In general, planktonic and sessile marine animals, including corals and brachiopods, have had higher extinction rates than mobile benthic animals such as gastropods and bivalves (22). Using anatomical evidence from fossils and comparisons with related living species, paleobiologists have begun to infer the determinants of clade longevity by relating the adaptations of the organisms to maximum population size, population fluctuation, and dispersal ability (23).

Human-Caused Extinction

Biodiversity reduction is accelerating today largely through the destruction of natural habitats (1). Because of the latitudinal diversity gradient, the greatest loss occurs in tropical moist forests (rain forests) and coral reefs. The rate of loss of rain forests, down to approximately 55% of their original cover, was in 1989 almost double that in 1979. Roughly 1.8% of the remaining forests are disappearing per year (24). By the most conservative estimate from island biogeographic data, 0.2 to 0.3% of all species in the forests are extinguished or doomed each year (25). If two million species are confined to the forests, surely also a very conservative estimate, then extinction due to tropical deforestation alone must be responsible for the loss of at least 4000 species annually.

But there may well be 20 million or more species in the forests, raising the loss tenfold. Also, many species are very local and subject to immediate extinction from the clearing of a single habitat isolate, such as a mountain ridge or woodland patch (26). The absolute extinction rate thus may well be two to three orders of magnitude greater than the area-based estimates given above. If current rates of clearing are continued, one-quarter or more of the species of organisms on Earth could be eliminated within 50 years—and even that pessimistic estimate might be conservative (25). Moreover, for the first time in geological history, plants are being extinguished in large numbers (27).

Another data set illuminating the urgency of dealing with the extinction problem measures the human impact on global net primary productivity (NPP) (28); global NPP is roughly the total food supply of all animals and decomposers. Almost 40% of all NPP generated on land is now directly used, coopted, or forgone because of the activities of just one animal species—*Homo sapiens*.

Since the overwhelming majority (possibly more than 90%) of species now exists on land, the 40% human appropriation there alone shows why there is an extinction crisis. Furthermore, the human population is projected to double in the next half-century or so—to more than 10 billion people. Most ominous of all, the widely admired Brundtland Report speaks of a five- to tenfold increase in global economic activity needed during that period to meet the demands and aspirations of that exploding population (29). If anything remotely resembling that population-economic growth scenario is played out, with an acceleration of habitat destruction, most of the world's biodiversity seems destined to disappear.

Why Should We Care?

The loss of biodiversity should be of concern to everyone for three basic reasons (1, 30). The first is ethical and esthetic. Because *Homo sapiens* is the dominant species on Earth, we and many others think that people have an absolute moral responsibility to protect what are our only known living companions in the universe. Human responsibility in this respect is deep, beyond measure, beyond conventional science for the moment, but urgent nonetheless. The popularity of ecotourism, bird-watching, wildlife films, pet-keeping, and gardening attest that human beings gain great esthetic rewards from those companions (and generate substantial economic activity in the process).

The second reason is that humanity has already obtained enormous direct economic benefits from biodiversity in the form of foods, medicines, and industrial products, and has the potential for gaining many more. Wheat, rice, and corn (maize) were unimpressive wild grasses before they were "borrowed" from the library and developed by selective breeding into the productive crops that have become the feeding base of humanity. All other crops, as well as domestic animals, have their origins in the genetic library, as do many medicines and various industrial products, including a wide variety of timbers (1, 30). Throughout the world almost a quarter of all medical prescriptions are either for chemical compounds from plants or microorganisms, or for synthetic versions or derivatives of them (31). One plant compound, quinine, is still a mainstay of humanity's defense against its most important disease, malaria.

Biodiversity is a precious "genetic library" maintained by natural ecosystems. But the potential of the library to supply such benefits has barely been tapped. Only a tiny portion of plant species has been screened for possible value as providers of medicines (31), and although human beings have used about 7000 plant species for food, at least several times that number are reported to have edible parts (1).

The third reason, perhaps the most poorly evaluated to date, is the array of essential services provided by natural ecosystems, of which diverse species are the key working parts. Ecosystem services include maintenance of the gaseous composition of the atmosphere, pre-

venting changes in the mix of gases from being too rapid for the biota to adjust. In Earth's early history, photosynthesizing organisms in the seas gradually made Earth's atmosphere rich in oxygen. Until there was enough oxygen for an ozone shield to form, the land surface was bathed in ultraviolet-B radiation. Up to some 450 million years ago life was confined to the seas. Only with the protection of the ozone shield were plants, arthropods, and amphibians able to colonize the land.

Significant alteration of the atmosphere has signaled the arrival over the past few decades of *Homo sapiens* as a global force, one capable of destroying most of biodiversity. As a result of human activities (32), the ozone shield has thinned by as much as 5% over Europe and North America (33), and there is some evidence that the surface intensity of ultraviolet-B radiation has increased there (34). Each spring the shield is now reduced over the Antarctic by approximately 50%. The global impact of the human economy is even more evident in the prospect of climatic change in response to increasing concentrations of greenhouse gases (35).

The organisms in natural ecosystems influence the climate in ways other than the role they play in regulating atmospheric gases. The vast rain forests of Amazonia to a large degree create the moist conditions that are required for their own survival by recycling rainfall. But as the forest shrinks under human assault, many biologists speculate that there will be a critical threshold beyond which the remaining forest will no longer be able to maintain the climate necessary for its own persistence (36). Deforestation and the subsequent drying of the climate could have serious regional effects in Brazil outside of Amazonia, conceivably reducing rainfall in important agricultural areas to the south. There also appear to be regional effects on climate when semi-arid regions are desertified (37), but their extent remains unknown.

The generation and maintenance of soils is another crucial service supplied most efficiently by natural ecosystems. Soils are much more than fragmented rock; they are themselves complex ecosystems with a rich biota (38). The elements of biodiversity in soil ecosystems are crucial to their fertility—to their ability to support crops and forests.

Many green plants enter into intimate relationships with mycorrhizal fungi in the soil. The plants nourish the fungi, which in turn transfer essential nutrients into the roots of the plant. In some forests where trees appear to be the dominant organisms, the existence of the trees is dependent upon the functioning of these fungi. On farms, other microorganisms play similar critical roles in transferring nutrients to crops such as spring wheat.

Organisms are very much involved in the production of soils, which starts with the weathering of the underlying rock. Plant roots can fracture rocks and thus help generate particles that are a major physical component of soils; plants and animals also contribute CO_2 and organic acids that contribute to the weathering of parent rock. More importantly, many species of small organisms, especially bacteria, decompose organic matter (shed leaves, animal droppings, dead organisms, and so on), releasing carbon dioxide and water into the soil and leaving a residue of humus, or tiny organic particles. These are resistant to further decomposition, help maintain soil texture and retain water, and play a critical role in soil chemistry, permitting the retention of nutrients essential for plant growth.

Soil ecosystems themselves are the main providers on land of two more essential ecosystem services: disposal of wastes and cycling of nutrients. Decomposers break wastes down into nutrients that are essential to the growth of green plants. In some cases, the nutrients are taken up more or less directly by plants near where the decomposers did their work. In others, the products of decomposition circulate through vast biogeochemical cycles before being reincorporated into living plants.

Another critical service provided by natural ecosystems is the control of the vast majority of species that can attack crops or domestic animals. Most of those potential pests are herbivorous insects, and the control is provided primarily by numerous species of predacious and parasitic insects that naturally feed upon them.

While natural ecosystems are providing crop plants with stable climates, water, soils, and nutrients, and protecting them from pests, they also often pollinate them. Although honeybees, essentially domesticated organisms, pollinate many crops, numerous other crops depend on the services of pollinators from natural ecosystems. One such crop is alfalfa, which is most efficiently pollinated in cooler areas by wild bees.

The biodiversity in natural ecosystems also supplies people with food directly—most notably with a critical portion of their dietary protein from fishes and other marine animals. This service is provided by oceanic ecosystems in conjunction with coastal wetland habitats that serve as crucial nurseries for marine life.

The ecosystem services in which biodiversity plays the critical role are provided on such a grand scale and in a manner so intricate that there is usually no real possibility of substituting for them, even in cases where scientists have the requisite knowledge. In fact, one could conclude that virtually all human attempts at large-scale inorganic substitution for ecosystem services are ultimately unsuccessful, whether it be introductions of synthetic pesticides for natural pest control, inorganic fertilizer for natural soil maintenance, chlorination for natural water purification, dams for flood and drought control, or air-conditioning of overheated environments. Generally, the substitutes require a large energy subsidy, thereby adding to humanity's general impact on the environment, and are not completely satisfactory in even the short run (39).

It is important to note that in supplying ecosystem services the species and genic diversity of natural systems is critical. One might assume that one grass or tree species can function as well as any other in helping control the hydrologic cycle in a watershed, or that one predator will be as good as another in controlling a potential pest. But, of course, organisms are generally highly adapted to specific physical and biotic environments—and organic substitutions, like inorganic ones, are likely to prove unsatisfactory.

In sum, much of biodiversity and the quality of ecosystem services generated by it will be lost if the epidemic of extinctions now under way is allowed to continue unabated.

Public Policy

Many steps can be taken to preserve biodiversity, if the political will is generated. Perhaps the first step, which would be seen as especially extreme by Americans, would be to cease "developing" any more relatively undisturbed land. Every new shopping center built in the California chaparral, every hectare of tropical forest cut and burned, every swamp converted into a rice paddy or shrimp farm means less biodiversity.

In rich countries, stopping the more destructive forms of "development" is relatively simple in principle. Age structures are such that population shrinkage in most rich nations could be achieved with little effort (a few are already in that desirable mode). When new facilities are needed, they should replace deteriorating old ones. Forestry should be placed on a sustainable basis with careful attention to the conservation of precious reserves of old growth. And much more scientific effort and public support should go into biodiversity studies, including the cataloging of the genetic library and national biological inventories (1, 31).

In poor nations, the task is both more urgent and vastly more difficult. It cannot be accomplished immediately, and will not be accomplished at all without massive assistance from the rich. For instance, stopping the expansion of cropland and pasture into virgin areas cannot be accomplished unless birth rates can be dramatically lowered and the development of sustainable high-yield agricultural systems is backed by land reform and a sound agricultural infrastructure and economy. In many cases, new social and economic systems must be developed in which preservation of biodiversity and its sustainable exploitation go hand in hand. The social, political, economic, and scientific barriers to achieving the goal are so formidable that nothing less than the kind of commitments so recently invested in the Cold War could possibly suffice to accomplish it. And we are 45 years late in starting.

But ending direct human incursions into remaining relatively undisturbed habitats would be only a start. Simultaneously, global cooperative efforts to reduce anthropogenic impacts on ecosystems, such as those directed at a reduction of emissions of greenhouse gases and ozone-destroying compounds, must be greatly enhanced. They are much more likely to be successful if population growth can be halted and the cessation of forest destruction can be achieved.

Finally, because humanity already occupies so much of Earth's surface, substantial effort should be directed at making areas already used by people more hospitable to other organisms. Those efforts can range from the substitution of game ranching for cattle and sheep ranching in many areas to the substitution of native vegetation for European-style lawns in desert cities.

If there is to be any chance of abating the loss of biodiversity, action must be taken immediately. The essential tactics of conservation are being developed within conservation biology, as a subdiscipline of biodiversity studies. The indispensable strategy for saving our fellow living creatures and ourselves in the long run, is, as the evidence compellingly shows, to reduce the scale of human activities. The task of accomplishing this goal will involve a cooperative worldwide effort unprecedented in history. Unless humanity can move determinedly in that direction, all of the efforts now going into in situ conservation will eventually lead nowhere, and our descendents' future will be at risk.

REFERENCES AND NOTES

1. See, for example, P. R. Ehrlich and A. H. Ehrlich, *Extinction: The Causes and Consequences of the Disappearance of Species* (Random House, New York, 1981); E. O. Wilson and F. M. Peter, Eds., *Biodiversity* (National Academy Press, Washington, DC, 1988); C. C. Black *et al.*, *Loss of Biological Diversity: A Global Crisis Requiring International Solutions* (National Science Foundation, Washington, DC, 1989); W. V. Reid and K. R. Miller, *Keeping Options Alive: The Scientific Basis for Conserving Biodiversity* (World Resources Institute, Washington, DC, 1989).
2. That 40% of living plant species originated by polyploidy is the estimate widely accepted by plant systematists; the figure is likely to be higher if the earlier origins of stocks giving rise to living species are also considered.
3. D. J. Futuyma, *Evolutionary Biology* (Sinauer, Sunderland, MA, ed. 2, 1986); D. Otte and J. A. Endler, Eds., *Speciation and Its Consequences* (Sinauer, Sunderland, MA, 1989).
4. G. L. Bush, *Annu. Rev. Ecol. Syst.* 6, 399 (1975); C. A. Tauber and M. J. Tauber, in *Speciation and Its Consequences* D. Otte and J. A. Endler, Eds. (Sinauer, Sunderland, MA, 1989), pp. 307–344.
5. W. H. Bossert and E. O. Wilson, in *The Genetics of Colonizing Species*, H. G. Baker and G. L. Stebbins, Eds. (Academic Press, New York, 1964), pp. 7–24; J. Diamond, S. L. Pimm, M. E. Gilpin, M. LeCroy, *Am. Nat.* 134, 675 (1989).
6. E. O. Wilson, *Issues Sci. Technol.* 2 (Fall), 20 (1985).
7. R. M. May, *Science* 241, 1441 (1988).
8. C. W. Sabrosky, *Insects, The Year of Agriculture, 1952* (U.S. Department of Agriculture, Washington, DC, 1952), pp. 1–7.
9. T. L. Erwin, *Coleopt. Bull.* 36, 74 (1982); *Bull. Ent. Soc. Am.* 30, 14 (1983).
10. N. E. Stork, *Biol. J. Linn. Soc.* 35, 321 (1988).
11. K. P. Dial and J. M. Marzluff, *Syst. Zool.* 38, 26 (1989).
12. E. O. Wilson, *Success and Dominance in Ecosystems: The Case of the Social Insects* (Ecology Institute, Oldendorf-Luhe, Germany, 1990).
13. T. R. E. Southwood, in *Diversity of Insect Faunas*, L. A. Mound and N. Waloff, Eds. (Blackwell, London, 1978), pp. 17–40.
14. K. P. Dial and J. M. Marzluff, *Ecology* 69, 1620 (1988).
15. P. R. Ehrlich and P. H. Raven, *Evolution* 18, 586 (1965); C. Mitter, B. Farrell, B. Wiegmann, *Am. Nat.* 132, 107 (1988).
16. A. H. Gentry, *Proc. Natl. Acad. Sci. U.S.A.* 85, 156 (1988).
17. E. O. Wilson, *Biotropica* 19, 245 (1987).
18. H. L. Sanders, *Am. Nat.* 102, 243 (1968); F. Grassle, *Trends Ecol. Evol.* 4, 12

(1989); G. C. Stevens, *Am. Nat.* **133**, 240 (1989); E. R. Pianka, *Trends Ecol. Evol.* **4**, 223 (1989); H. R. Pulliam, *Am. Nat.* **132**, 652 (1988).
19. S. L. Pimm, H. L. Jones, J. M. Diamond, *Am. Nat.* **132**, 757 (1988).
20. K. Ralls, J. D. Ballou, A. Templeton, *Conserv. Biol.* **2**, 185 (1988).
21. D. M. Raup, in *Biodiversity*, E. O. Wilson, Ed. (National Academy Press, Washington, DC, 1988), pp. 51–57.
22. M. L. McKinney, *Nature* **325**, 143 (1987).
23. S. M. Stanley, *Paleontology* **20**, 869 (1977); *Bull. Am. Acad. Arts. Sci.* **40**, 29 (1987); P. D. Taylor, *Hist. Biol.* **1**, 45 (1988).
24. N. Myers, *Deforestation Rates in Tropical Forests and Their Climatic Implications* (Friends of the Earth, London, 1989).
25. E. O. Wilson, *Sci. Am.* **261**, 108 (September 1989). Other, higher estimates are reviewed by W. V. Reid and K. R. Miller [*Keeping Options Alive: The Scientific Basis for Conserving Diversity* (World Resources Institute, Washington, DC, 1989)].
26. J. M. Diamond, *Discover* **11**, 55 (1990).
27. A. H. Knoll, in *Extinctions*, M. H. Nitecki, Ed. (Univ. of Chicago Press, Chicago, 1984), pp. 21–68.
28. P. M. Vitousek, P. R. Ehrlich, A. H. Ehrlich, P. A. Matson, *BioScience* **36**, 368 (1986).
29. World Commission on Environment and Development, *Our Common Future* (Oxford Univ. Press, New York, 1987).
30. N. Myers, *A Wealth of Wild Species* (Westview Press, Boulder, CO, 1983).
31. T. Eisner, *Issues Sci. Technol.* **6** (Winter), 31 (1989).
32. See summary in S. Roan, *Ozone Crisis: The 15 Year Evolution of a Sudden Global Emergency* (Wiley, New York, 1989).
33. R. A. Kerr, *Science* **247**, 1297 (1989).
34. M. Blumthaler and W. Ambach, *ibid.* **248**, 206 (1990).
35. S. H. Schneider, *Global Warming* (Sierra Club Books, San Francisco, 1989); D. Abrahamson, Ed., *The Challenge of Global Warming* (Island Press, Washington, DC, 1989); P. R. Ehrlich and A. H. Ehrlich, *Healing the Planet* (Addison-Wesley, Reading, MA, in press).
36. J. Shukla, C. Nobre, P. Sellers, *Science* **247**, 1322 (1990).
37. J. Charney, P. H. Stone, W. J. Quirk, *ibid.* **187**, 434 (1975).
38. For more details on soils and technical citations on what follows, see P. R. Ehrlich, A. H. Ehrlich, J. P. Holdren, *Ecoscience: Population, Resources, Environment* (Freeman, San Francisco, 1977).
39. P. R. Ehrlich and H. A. Mooney, *BioScience* **33**, 248 (1983).
40. This article is dedicated to some of the key individuals who have influenced our thinking on biodiversity: W. L. Brown, E. Mayr, C. D. Michener, and R. R. Sokal; and the late J. H. Camin, R. W. Holm, and R. H. MacArthur. For helpful comments on the manuscript we are indebted to G. C. Daily, J. M. Diamond, A. H. Ehrlich, T. Erwin, J. Harte, M. E. Harte, C. E. Holdren, J. P. Holdren, T. Lovejoy, J. P. Meyers, S. Conway Morris, D. M. Raup, P. H. Raven, and M. E. Soulé.

Decker, D. J., and L. C. Chase. 1997. Human dimensions of living with wildlife—a management challenge for the 21st century. *Wildlife Society Bulletin* 25:788–795.

In real estate it is all about location, location, location. Most wildlife professionals have been told early in their education that wildlife management is all about people, people, and people. The importance of including people in wildlife management and conservation has been expressed since the early years of the profession but has received serious attention only in the past few decades. The human dimensions of wildlife management and conservation (i.e., identifying what people think and do regarding wildlife, understanding why, and incorporating that insight into policy and decision-making processes and programs) will continue to gain importance as the human population increases and wildlife habitats are altered and reduced.

In many cases, human dimensions may be more important than biology when problems have to be addressed in a social context. The objective of this paper is to examine the changes in managing people–wildlife conflicts and to provide guidance so the wildlife professional can address future issues while maintaining responsibilities to wildlife and society. The profession has evolved from a top-down authoritative model of management where biologists make most decisions about wildlife management toward a co-managerial/delegatorial model where management is shared with stakeholders. However, Decker and Chase warn, "in trying to be good public servants, wildlife managers should be cautious not to become servantile to public opinion by relying too heavily on opinion polls to determine what they ought to do . . . it would be heading down a road toward devaluation of professional judgment and perhaps even to abrogation of professional responsibility." To be effective, the profession has to find the correct balance between the two extremes. The authors do not advocate that wildlife managers should simply defer to the public to make decisions; instead, managers can use human dimensions knowledge to aid their decision making.

RELATED READING

Chase, L. C., M. Schusler, and D. J. Decker. 2000. Innovations in stakeholder involvement: what's the next step? Wildlife Society Bulletin 28:208–217.

Manfredo, M. J. 1989. Human dimensions of wildlife management. Wildlife Society Bulletin 17:447–449.

Manfredo, M. J., and A. D. Bright. 1991. A model for assessing the effects of communication on recreationists. Journal of Leisure Research 23:1–20.

Riley, S. J., D. J. Decker, L. H. Carpenter, J. F. Organ, W. F. Siemen, G. F. Mattfeld, and G. Parsons. 2002. The essence of wildlife management. Wildlife Society Bulletin 30:585–593.

Human dimensions of living with wildlife—a management challenge for the 21st century

Daniel J. Decker and Lisa C. Chase

Problems addressed by wildlife management have changed dramatically during the 20th century. Some species have emerged from a period of scarcity to a state of overabundance. Wildlife managers now face many situations marked by an urgent, growing demand to reduce conflicts between people and species of wildlife that were scarce just a few decades ago. Managers are finding that they must attempt to work within a complex interface of biological and sociological forces. They are dealing with the difficulties of managing wildlife and people to optimize benefits to a society that is living with wildlife, and experiencing the diverse benefits and problems associated with such intimacy. Managers working at this interface are applying their ingenuity, experimenting with approaches to achieve success in a milieu of diverse and often conflicting stakeholder expectations and wildlife acceptance capacities (Decker and Purdy 1988). We believe developing the strategies to integrate-informed stakeholder input and involvement in decision-making is 1 of the greatest challenges facing wildlife management as we move into the 21st century.

We first define human dimensions of wildlife management and clarify our notion of "solutions" to people–wildlife problems. We then categorize and describe approaches that managers have taken to incorporate human dimensions into decision making in their search for solutions to people–wildlife problems. We use the evolution of white-tailed deer (*Odocoileus virginianus*) management in New York as a case study demonstrating our points, and also refer to other North American examples. We suggest next steps for incorporating human dimensions into solutions for people–wildlife problems, reflecting on

what that entails for the future of professional wildlife management.

Human dimensions of wildlife management

Repeatedly, wildlife managers dealing with people–wildlife conflicts report that the human dimensions of such situations are the most difficult to understand and manage. Therefore, it is not surprising that interest in the human dimensions of wildlife management has grown steadily from the late 1960s to the present (Manfredo et al. 1996). The wildlife management profession, dominated since its inception by scientific expertise in the biological and ecological dimensions of management, has gradually broadened its view of wildlife management to include a scientifically-based understanding of people as an essential part of the management equation (Decker et al. 1992).

Human dimensions of wildlife management as an academic and applied focus of interest and activity has been defined variously by different authors. Manfredo et al. (1996:54) defined human dimensions as "an area of investigation which attempts to describe, predict, understand, and affect human thought and action...." Decker and Lipscomb (1991) described human dimensions of wildlife management as identifying what people think and do regarding wildlife, understanding why, and incorporating that insight into policy and management decision-making processes and programs. Human dimensions specialists maintain that while traditional biological considerations are essential, managing people and manag-

Authors' address: Human Dimensions Research Unit, Department of Natural Resources, Cornell University, Ithaca, NY 14853, USA.

Key words: citizen participation, human dimensions, people–wildlife conflicts, problem wildlife, public input, public involvement, stakeholder approach, wildlife management

Wildlife Society Bulletin 1997, 25(4):788–795   **Peer edited**

ing the decision-making process itself are equally important for dealing with people-wildlife problems today (Decker and Richmond 1995).

What is a people-wildlife problem? It is potentially any situation where: (1) the behavior of people negatively impacts wildlife (this includes human impacts on habitat); (2) the behavior of wildlife creates a negative impact for some stakeholders, or is perceived by some stakeholders to impact themselves or others adversely; or (3) the wildlife-focused behavior of some people creates a negative interaction with other people, often in the form of a values clash. Thus, a people-wildlife problem can involve a people-wildlife interaction or a people-people interaction (i.e., a controversy), or both.

The real dilemma for managers is that different stakeholders have different expectations of their interactions with wildlife, and, based on these, they develop different acceptance capacities that cannot be met simultaneously at the same place and time. The primary challenge for the manager is reconciling these competing interests, or stakes, in the wildlife resource. That is not a profound observation to the experienced wildlife manager who has faced this daunting challenge many times; what is exciting now is that some wildlife managers are gaining a reasonable measure of success in achieving this seemingly impossible goal.

Solving people–wildlife problems

What does it mean to solve people-wildlife problems? From a human dimensions perspective, a people-wildlife interaction or people-people interaction problem can be considered solved only when the stakeholders involved believe it to be so. That is, the condition of being solved is not the result of a declaration following an objective assessment made by wildlife managers alone.

The definition of "solution" we propose relative to management of people-wildlife interactions emphasizes the way a problem is addressed:

"The method or process of solving a problem." (The American Heritage Dictionary 1985:1164).

Experience has shown that there is seldom 1 "answer" to a people-wildlife problem. Rather, a range of more or less acceptable management objectives and actions exists. For this reason, we prefer to think about solutions as being tied primarily to the process used to make decisions, especially the extent of stakeholder input and involvement. To manage people-wildlife interactions successfully, stakeholders must be considered throughout all phases. This makes managers' work difficult because the nature of stakeholders—the kinds of stakes and backgrounds

of people holding stakes—is diverse. Furthermore, stakeholders' beliefs are changeable as they gain information and experience with the situation. Thus, one must be careful of assumptions about stakeholder beliefs and attitudes—they are not static (Enck and Decker, in press). Fortunately, this same characteristic represents the potential for educational communication to modify beliefs and attitudes.

The art and challenge of managing people-wildlife problems is to reach an enduring, biologically sound agreement among people with competing stakes in the resource; e.g., balancing some stakeholders' desires for benefits with others' expectations of relief from the negative impacts of wildlife. In their efforts to find solutions, wildlife managers can choose from an array of approaches, ranging from an authoritative, "expert" approach to those that provide formally for a sharing of responsibility between the wildlife agency and communities (and the diverse stakeholders they represent). The wisdom of managers' selection and quality of execution have great bearing on achieving an acceptable solution.

Accumulating experience is demonstrating that stakeholder input in decisions and some level of stakeholder involvement in the decision-making process are necessary to achieve success in managing people-wildlife problems and resolving associated people-people conflicts. Research has shown that stakeholders are more likely to consider a public issue resolved or problem solved acceptably when they have had a voice in the decision-making process (Lind et al. 1990).

Seeking solutions—wildlife managers' approaches to stakeholder input and involvement

Citizen participation has been familiar to some wildlife agencies for many years, but the intent and extent of stakeholder input and involvement in management over the last 10 years has represented something of a paradigm shift for wildlife management. In this section we outline a typology of approaches that characterize most of the ways wildlife managers tend to address public input and involvement. The 5 types of approaches we have identified are: authoritative, passive-receptive, inquisitive, transactional, and co-managerial or delegatorial (Fig. 1). Each approach is described and some are illustrated with examples of their implementation.

Authoritative "expert" approach. This "top-down" approach is a vestige of the time when managers served a narrow constituency, with which they normally personally identified and shared values (Decker et al. 1996). Major value differences seldom

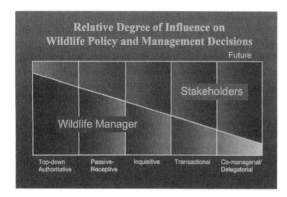

Fig. 1. Management approaches to stakeholder input and involvement.

were at issue; whoever had the greatest knowledge about a people-wildlife interaction, with knowledge confined largely to biological expertise, usually carried the day. In this simple human dimensions system, an authoritative approach by wildlife managers (biological experts) could work because there were few recognized groups of stakeholders in decisions. Today several kinds of stakeholders are interested in most people-wildlife issues, they hold diverse values, and they are willing to advocate actively their preferred outcome in a management decision, through political and legal means if necessary.

Under the authoritative approach, "game" management created some notable successes continent wide (U.S. Fish and Wildl. Serv. 1987). In New York for example, white-tailed deer management provided excellent hunting opportunities in much of the state. But in some instances, the "expert" approach was ineffective. For example, about 30 years ago in the Adirondack Region of northern New York, hunters did not agree with state biologists regarding population objectives. Deer managers, convinced they were right, held their ground. However, hunters, working through their local representatives to state government, successfully relieved the agency of real management authority in the Adirondack Region, with the result that 1/3 of New York's deer range has not been effectively managed for more than a quarter of a century (Decker et al. 1985).

Passive–receptive approach. Over time, as populations of some species such as deer and elk rebounded due to protection and favorable habitat changes, farmers, ranchers, and forest landowners gained importance as stakeholders. Managers listened to their concerns and gave weight to these stakeholders in decisions. In this passive–receptive approach managers are receptive to stakeholder input, but do not seek it systematically. Importantly,

when making decisions, wildlife managers—not stakeholders—determine the weight given to various concerns voiced by stakeholders.

Of the 5 management approaches presented here, this approach requires perhaps the least initiative on the part of the wildlife manager. The initiative for input comes largely from the stakeholder. The passive–receptive manager listens intently, makes paper or mental notes, and takes information offered voluntarily by stakeholders into consideration when making decisions.

For example, deer management in the agricultural region of central and western New York operated in the passive–receptive mode up until the mid- to late 1970s. Managers there gave considerable weight to crop damage concerns expressed as complaints to the agency, keeping the deer population well below any notion of range carrying capacity (Brown et al. 1978).

Inquisitive approach. As wildlife biologists gained expertise and precision manipulating wildlife populations (especially deer), they began to seek more knowledge of stakeholders. Managers' need for such information was the impetus for systematic inquiry and the advent of human dimensions research, ushering in the inquisitive approach to gaining stakeholder input.

Managers began taking an inquisitive approach because they recognized that: (1) relying solely on unsolicited input can be selective and lead to a biased perspective, and (2) they had little information about the new, nontraditional stakeholders that were emerging and placing new demands on wildlife management. In an inquisitive approach, managers seek input from a broad array of stakeholders and from multiple members of any single stakeholder group. They employ systematic surveys to become more scientific in their efforts to understand stakeholders.

The inquisitive approach has yielded many benefits. Going back to New York's experience, we learned from several studies that tolerance of some species of wildlife was greater than had been assumed. With this knowledge in hand, actions were taken to allow populations to build, yielding greater net benefits from the wildlife resource. For example, deer management in central and western New York changed as we learned farmers were more tolerant of deer than had been generally thought. Deer populations were allowed to increase significantly, and farmers' acceptance of deer was systematically monitored (Decker et al. 1985). After deer populations increased, the need for effective hunter participation in the Deer Management Permit system was critical, so understanding hunter motivations was needed to ensure these volunteer "tools" of management took enough deer to control populations. Hunter input was sought to determine how the Department of Environmental Conservation (DEC) might better encourage participation, such as offering 2 permits per hunter in some areas so at least 1 is used (Decker and Connelly 1989, Enck and Decker 1991).

Transactional approach. During the mid-1980s to the present, the wildlife management establishment has increasingly focused on new people-wildlife problems. Farmers and rural landowners have been joined by other vocal groups who expect relief from costs associated with living with wildlife. The greatest increase in people-wildlife problems seems to be in urban and suburban environments, where residents are finding that living with burgeoning deer and Canada goose (*Branta canadensis*) populations, trash-seeking black bears (*Ursus americanus*), and pet-hunting mountain lions (*Puma concolor*) has its drawbacks. As a result, managers are dealing with many new stakeholders who have diverse beliefs and attitudes about wildlife management. Situations are encountered where the usual management prescriptions may not be logistically feasible or socially acceptable.

In such situations, top-down approaches are out. Passive-receptive, and even inquisitive approaches by themselves fall short because, in addition to understanding stakeholder beliefs and attitudes, managers also need to weigh the breadth of likely consequences from management alternatives such that they and stakeholders find resulting objectives and actions acceptable. This is a daunting task given the diversity of views encountered about the importance of various consequences.

The 1990s has been a period of experimentation in stakeholder approaches to solve difficult people-wildlife problems. Managers have discovered they need to get out of the middle in conveying the values of every stakeholder group to every other group. In their attempts to balance competing interests, they have been pulled in every direction. This has become an untenable role. Now, managers seem to be seeking processes where stakeholders engage one another directly (Fig. 2). They are applying their ingenuity to this need and creating a new paradigm for managing people-wildlife problems in many situations.

A recent development in the evolution of wildlife managers' approaches to stakeholder input in decision-making is the transactional approach. In this approach, managers initiate and implement processes where stakeholders not only articulate their stakes to one another, rather than through the manager as an intermediary (which is the situation in either the passive-receptive or inquisitive approaches), but also essentially negotiate acceptable weights for their own and others' stakes. Although these processes may be time-consuming and costly in the short-term, some managers believe the up-front costs are worthwhile over the long-term. Stakeholder participants in such approaches frequently can reach consensus about an appropriate objective and course of action (Nelson 1992). This is not simply a process where representatives of various stakeholder interests get together and vote; rather, consensus is accomplished through education, discussion, and debate. Depending on the nature of the decision and the confidence of the wildlife manager (management agency), the transactional approach might even permit the stakeholders involved to make a decision within some sideboards or bounds determined by the wildlife manager. Typically the negotiation of weights for the various stakes

Fig. 2. Managers have found they need to get out of the middle in conveying the values of every stakeholder group to every other group. Instead, managers are creating processes where stakeholders engage one another directly while the manager maintains control over the process and provides expert advice. Drawing by A. Campbell.

is a qualitative process, not one where weights are explicitly expressed quantitatively. Wildlife managers in this context are managing the process of involvement, purposefully intervening with educational communication about biological and socioeconomic knowledge pertinent to the issue to ensure informed discussion and decisions.

For example, DEC staff instituted Citizen Task Forces around 1990 as an integral part of deer management in New York. Conceptually, New York's deer managers designed an approach that got them out of the middle, but left them in charge of managing a process and playing the role of expert and responsive public servant. Essentially, DEC managers provide a forum and some ground rules that allow representatives of various stakeholder groups to deal with one another in the negotiation of the weights their respective stakes will have in the decision. Consensus about deer population objectives is the goal. Citizen Task Forces in New York exemplify the transactional approach, where—ideally—all important stakeholders are included, an educational element is involved to ensure informed discussion, and an external, skilled facilitator handles the meetings. Discussion among stakeholders is encouraged to achieve mutual understanding of stakes involved. Participants determine management objectives, within sideboards set by the wildlife manager (Nelson 1992, Stout et al. 1996).

The transactional approach has gained favor for a variety of applications in New York and elsewhere. For example, during the late 1980s in west-central Montana, a transactional approach grew out of grassroots concerns about elk (*Cervus elaphus*) management. Landowners, hunters, conservationists and the state wildlife agency created a "team" to address several related issues. Central among the subjects addressed was competition for forage between cattle and elk. Hunting access to private lands and elk herd quality also were concerns. The task force named themselves the "Devil's Kitchen Management Team" and committed themselves to the difficult goal of building consensus around a "win-win or no deal" philosophy. Among the positive outcomes of the Devil's Kitchen Management Team's efforts was the Montana Fish and Wildlife Commission's approval of the team's proposal to liberalize hunting regulations, thereby increasing benefits to hunters and landowners in the region (Dagget 1995, Baumeister et al. 1996). The group continues to meet, and it has addressed additional wildlife concerns, including deer and mountain lion management (M. Aderhold, Mont. Dep. Fish, Wildl. and Parks, pers. commun.).

In the Minnesota River Valley we found another ex-

Living with elk in Estes Park, Colo.

periment in stakeholder involvement. White-tailed deer were becoming a major nuisance in 4 municipalities bordering the Minnesota Valley National Wildlife Refuge and the Fort Snelling State Park. Deer management was at an impasse until the creation of the Minnesota Valley Deer Management Task Force in 1989. The Task Force represented a variety of citizen stakes and governmental interests. This was an ambitious group, meeting nearly 20 times over 1 1/2 years. Despite the range of views represented around the table, task force members were able to resolve many of their differences about management objectives and even about the actual measures to be employed to achieve herd reduction. They issued a report, complete with recommendations, which the Minnesota Department of Natural Resources is currently implementing (McAninch and Parker 1991).

To summarize, the wildlife management profession has experienced some significant developments in dealing with the people part of people–wildlife problems. New techniques have been developed and added to the manager's tool chest of citizen-input approaches to meet changing needs and varied conditions. What might the future hold?

Co-managerial or delegatorial approach. Although it is difficult and professionally risky to predict what lies on the horizon for wildlife management, given the dynamic human dimensions environment in which management occurs, it is possible that the challenges of living with wildlife may lead to adoption of yet another approach to public involvement in wildlife management. If several trends continue—(1) more people-wildlife problems (often location specific), (2) greater public expectations for tailored solutions, (3) broader management responsibilities for agencies, and (4) continuing limitations on agency funds and personnel—it is conceivable that the combination of these pressures could lead to a situation where agencies will need to share or delegate authority for management to stakeholders, especially those in affected communities. This would signal another significant paradigm shift for wildlife management.

Development and widespread adoption of a co-managerial or delegatorial approach would require rethinking the role of state wildlife agencies and their professionals, as well as the acceptance by local communities for greater responsibility in solving local wildlife problems. This approach would recognize the need, value, and difficulty of tailoring decisions and actions to meet local conditions. It also would recognize the complex breadth and variability of stakeholder values that must be addressed in most management decisions. The co-managerial or delegatorial approach would require the establishment of operational guidelines for partners, as well as oversight, accountability, and evaluation processes. It also would necessitate educational communication programs for stakeholders on a level seldom seen in contemporary wildlife management. Decision-making bodies representing local stakeholders would have to adopt receptive, inquisitive, and transactional elements in this approach. The role of the wildlife management agency might include providing expertise, managing processes, training community participants, approving community management plans, and monitoring programs. Agencies might be working more extensively with stakeholders in local communities, collaborating with them to develop guidelines, standards, criteria, and requirements for local community management efforts.

The prospect of such an approach raises several philosophical questions for wildlife management, but 1 practical question stands out. Is it reasonable to expect all the knowledge and skills needed to address complex people-wildlife problems to be embodied in a single individual—"super wildlife manager?" Likely not. To operationalize the co-managerial or delegatorial approach will require a professional management team made up of experts in several areas, such as wildlife biology, human dimensions research, citizen participation, educational communication, and others. Perhaps community wildlife management teams will develop, made up of professionals with diverse expertise operating on a regional basis within a state. Furthermore, for this approach to work, local community "partners" must accept greater responsibility. Concomitant with the desire for tailored management, communities should expect to share costs, participate in implementation, and share accountability for outcome.

This future is already here for some agencies in some places. For example, a co-managerial approach was taken to resolve a difficult situation involving cross-cultural communications and regulation in Alaska's Yukon Delta Region. By the 1980s, populations of 4 species of migratory waterfowl—cackling geese (*Branta canadensis* minima), white-fronted geese (*Anser* spp.), emperor geese (*Anser canagica*), and black brant (*Branta bernicla*)—were in serious decline in the region. Their nesting areas in western Alaska were the homeland of native people for whom these birds were an important part of traditional subsistence harvest, especially during spring. Local communities as well as state and federal wildlife authorities in the Pacific Flyway recognized the need to collaborate to reverse the population declines of these waterfowl. A primary action in this case was educational communication about the impacts of spring harvest on waterfowl populations. Importantly, the educational thrust was implemented primarily by local communities, and monitoring, research, and enforcement have become cooperative efforts between the native people and government wildlife authorities. Since the Goose Management Plan was agreed to in 1984, populations of cackling geese and white-fronted geese have prospered, and black brant and emperor geese populations have remained stable (Brelsford 1995). Co-management seems to be working in the Yukon Delta.

A caveat

Although the approaches to solving people-wildlife problems we have outlined emphasize public input and involvement, they do not advocate uninformed public decision-making. We are concerned that a pitfall awaits the wildlife management profession if we do not look ahead and examine where we are going in terms of "responsiveness" to the public. That is, managers need to be mindful that human di-

News media coverage depicting the controversy of living with wildlife.

mensions knowledge aids decision-making, but seldom, if ever, in itself reveals what should be done in a particular situation. In most situations, wildlife managers must avoid any temptation to use only stakeholder preferences as the basis for decisions. Our concern about this is heightened as we observe the spread of a "customer service and satisfaction" attitude among wildlife agencies. To a large extent, this orientation is laudable, but we fear that without sufficient discourse about the ramifications of this attitude, customer service could shift to an extreme, "give 'em what they want" approach.

In trying to be good public servants, wildlife managers should be cautious not to become servantile to public opinion by relying too heavily on opinion polls to determine what they ought to do. Surveys and opinion polls should not become in effect surrogate referenda where what the majority want, the manager will automatically try to provide. Though some feel that is a justifiable course of action, it would be heading down a road toward devaluation of professional judgment and perhaps even to abrogation of professional responsibility.

These concerns notwithstanding, the future is promising for wildlife managers with the right attitudes and skills to develop acceptable approaches to solving people–wildlife problems and associated people–people conflicts. Community level "solutions" seem possible, based on experiences to date. McAninch and Parker (1991:435) concluded from their experiences with deer management in Minnesota that:

"Open communication and discussion has also resulted from the deliberations of the DMTF [Deer Management Task Force] and will continue to produce dialogue where none existed previously. We are optimistic that these exchanges will lead to better and more timely resolution of wildlife management problems in other situations. Perhaps when problems can be limited to a particular area, and stakeholders are identified and given an equal opportunity to participate in the input process, wildlife management programs might best be formulated by community residents and local staff from the resource agencies."

Conclusions

No single, sure-fire secret to success exists in the business of managing people and wildlife for peaceful coexistence across the broad and varied landscape of people–wildlife interactions. There is no pat "solution" in the form of a single technique that can be prescribed to solve the variety of people–wildlife problems faced by wildlife managers. Rather the solution is in adopting an approach that is sufficiently robust to encompass the breadth of stakeholder interests that exist and accommodate their ever-changing nature.

Certainly wildlife managers' experiences are indicating some key ingredients for success. We have concluded that 1 is essential: Initiate and manage stakeholder involvement in wildlife management from situation analysis through implementation and evaluation, and continuing through subsequent adaptive iterations of the management process. Regardless of which general approach is adopted for development and application in a specific situation, it should include:

- determining management objectives and selecting management actions that involve stakeholders in an inquiry or process that discovers and applies stakeholder weights in decisions;
- involving stakeholders in evaluating management action implementation and resulting outcomes;
- using ongoing stakeholder input for adaptive management.

People–wildlife problems get "fixed" to the extent diverse stakeholder interests, complementary and oppositional, are accommodated by the process that yields decisions about management objectives and actions to be taken. We believe overall stakeholder satisfaction with these objectives and actions, and the outcomes achieved therefrom, is largely based on their satisfaction with the process developed to incorporate their concerns in decision-making and the degree to which their ever-changing beliefs and attitudes are accurately reassessed and integrated into an adaptive management approach. Thus, satisfaction with wildlife management in many people–wildlife conflicts may be as much related to wildlife managers' approaches to stakeholder input as it is to outcomes in terms of objective impacts on the people–wildlife interactions of concern. Recognition of this connection will be important to meet successfully many of the human dimensions challenges of wildlife management in the future.

Acknowledgments. This article is based on a presentation given during the plenary session for the Third Annual Conference of The Wildlife Society, October 1–5, 1996, Cincinnati, Ohio. The authors acknowledge with appreciation the review comments and suggestions on draft manuscripts of this paper offered by colleagues in the Human Dimensions Research Unit. Preparation of this paper was supported in part by the Division of Fish, Wildlife, and Marine Resources, New York State Department of Environmental Conservation through Federal Aid Grant WE-173-G and the Cornell University Agricultural Experiment Station through Hatch Project NYC147403.

Literature cited

BAUMEISTER, T. R., H. SALWASSER, AND A. L. PRESTON. 1996. The role of elk conservation in sustainable development: past, present, and future on Montana's Rocky Mountain Front. Trans. North Am. Wildl. and Nat. Resour. Conf. 61:301-313.

BRELSFORD. T. 1995. Co-management under the Yukon-Kuskokwim goose management plan in Western Alaska. Presentation at Conference of Circumpolar Aboriginal People and Co-management Practice: Current Issues in Co-management and Environmental Assessment. Arctic Inst. of North Am. and Joint Secretariat Inuvialuit Renewable Resour. Comm., Inuvik, Northwest Territories, Can.

BROWN, T. L., D. J. DECKER, AND C. P. DAWSON. 1978. Willingness of New York farmers to incur white-tailed deer damage. Wildl. Soc. Bull. 6:235-239.

DAGGET, D. 1995. Beyond the rangeland conflict: toward a west that works. Gibbs Smith, Publ., Layton, Ut. 104pp.

DECKER, D. J., T. L. BROWN, AND G. F. MATTFELD. 1985. Deer population management in New York: using public input to meet public needs. Pages 185-196 *in* S. L. Beasom and S. F. Roberson, eds. Game harvest management. Caesar Kleberg Wildl. Res. Inst., Kingsville, Tex.

DECKER, D. J., T. L. BROWN, N. A. CONNELLY, J. W. ENCK, G. A. POMERANTZ, K. G. PURDY, AND W. F. SIEMER. 1992. Toward a comprehensive paradigm of wildlife management: integrating the human and biological dimensions. Pages 33-54 *in* W. R. Mangun, ed. American fish and wildlife policy: the human dimension. Southern Illinois Univ. Press, Carbondale.

DECKER, D. J., AND N. A. CONNELLY. 1989. Motivations for deer hunting: implications for antlerless deer harvest as a management tool. Wildl. Soc. Bull. 17:455-463.

DECKER, D. J., C. C. KRUEGER, R. A. BAER, JR., B. A. KNUTH, AND M. E. RICHMOND. 1996. From clients to stakeholders: a philosophical shift for fish and wildlife management. Hum. Dimensions Wildl. 1:70-82.

DECKER, D. J., AND J. LIPSCOMB. 1991. Toward an organizational philosophy about integrating biological and human dimensions in management. Hum. Dimensions Perspect. Ser. No. 1, Occas. Pap., Colorado Div. Wildl., Denver. 3pp.

DECKER, D. J., AND K. G. PURDY. 1988. Toward a concept of wildlife acceptance capacity in wildlife management. Wildl. Soc. Bull. 16:53-57.

DECKER, D. J., AND M. E. RICHMOND. 1995. Managing people in an urban deer environment: the human dimensions challenges for managers. Pages 3-10 *in* J. B. McAninch, ed. Urban deer: a manageable resource? Proc. Symposium 55th Midwest Fish and Wildlife Conference, 12-14 December 1993, St. Louis, Mo. North Cent. Sect., The Wildl. Soc., West LaFayette, Ind.

ENCK, J. W., AND D. J. DECKER. 1991. Hunter's perspectives on satisfying and dissatisfying aspects of the deer-hunting experience in New York. Hum. Dimensions Res. Unit Publ. 91-4. Dep. Nat. Resour., N.Y. State Coll. Agric. and Life Sci., Cornell Univ., Ithaca. 83pp.

ENCK, J. W., AND D. J. DECKER. In Press. Examining assumptions in wildlife management: a contribution of human dimensions inquiry. Hum. Dimensions Wildl.

LIND, E. A., R. KANFER, AND P. C. EARLEY. 1990. Voice, control, and procedural justice: instrumental and noninstrumental concerns in fairness judgements. J. Personality and Social Psychol. 59:952-959.

MANFREDO, M. J., J. J. VASKE, AND L. SIKOROWSKI. 1996. Human dimensions of wildlife management. Pages 53-72 *in* A. W. Ewert, ed. Natural resource management: the human dimension. Westview Press, Inc., Boulder, Colo.

McANINCH, J. B., AND J. M. PARKER. 1991. Urban deer management programs: a facilitated approach. Trans. North Am. Wildl. and Nat. Resour. Conf. 56:428-436.

NELSON, D. 1992. Citizen task forces on deer management: a case study. Northeast Wildl. 49:92-96.

STOUT, R. J., D. J. DECKER, B. A. KNUTH, J. C. PROUD, AND D. H. NELSON. 1996. Comparison of three public-involvement approaches for stakeholder input into deer management decisions: a case study. Wildl. Soc. Bull. 24:312-317.

THE AMERICAN HERITAGE DICTIONARY. 1985. Second ed. Houghton Mifflin Co., Boston, Mass. 1568pp.

U.S. FISH AND WILDLIFE SERVICE. 1987. Restoring America's wildlife 1937—1987. U.S. Gov. Printing Off., Washington D.C. 394pp.

Daniel (Dan) J. Decker is an associate professor in the Department of Natural Resources at Cornell University, where he is co-leader of the Human Dimensions Research Unit. He is currently serving as associate director for research for the Cornell Agricultural Experiment Station and the College of Agriculture and Life Sciences. He received his B.S., M.S., and Ph.D. from Cornell, where he has been involved in studies of the human dimensions of wildlife management for more than 20 years. ***Lisa C. Chase*** is a Ph.D. candidate studying resource policy and management in the Department of Natural Resources at Cornell University. Her research interests include public involvement in wildlife management. She received her B.S. from the University of Michigan and her M.S. from Cornell University.

Decker, D. J., and T. A. Gavin. 1987. Public attitudes toward a suburban deer herd. *Wildlife Society Bulletin* 15:173–180.

This is one of the first peer-reviewed papers that used data on public attitudes to manage suburban deer. Decker and Gavin's work serves as an example of how human dimensions can effectively be incorporated into management. Since the late 1980s, management agencies across the U.S. have increasingly incorporated public attitudes into management plans and decisions. As humans and wildlife associate more because of human encroachment into wildlife habitat, conflicts around vegetation damage, vehicle collisions, disease, and other issues will continue to arise. These conflicts have created a new era where a broader, vocal public increasingly uses the political process to serve its interests. Thus, managers are faced with the complex issue of deciding the extent to which public attitudes and desires should be included in management decisions.

Effective wildlife management cannot overlook the critical need to incorporate human dimensions into wildlife management and planning at the initial stages of program development. This idea of incorporating the public into wildlife management is not new; it had been discussed by Leopold (1933:423) when he wrote, "twenty centuries of 'progress' have brought the average citizen a vote, a national anthem, a Ford, a bank account, and a high opinion of himself, but not the capacity to live in high density without befouling and denuding his environment, nor a conviction that such capacity, rather than such density, is the true test of whether he is civilized. The practice of game management may be one of the means of developing a culture which will meet this test." Including the public as part of the plan is how the test will be passed.

RELATED READING

Arnstein, S. A. 1969. A ladder of citizen participation. Journal of American Institutional Planners 35:216–224.

Guynn, D. E. 1997. Miracle in Montana—managing conflicts over private lands and public wildlife issues. Transactions of the North American Wildlife and Natural Resources Conference 62:146–154.

Leopold, A. 1933. Game management. Charles Scribner's Sons, New York, New York, USA.

Loker, C. A., D. J. Decker, and S. J. Schwager. 1999. Social acceptability of wildlife management actions in suburban areas: three cases from New York. Wildlife Society Bulletin 27:152–159.

Osherenko, G. 1988. Can comanagement save Arctic wildlife? Environment 30:6–34.

Reiter, D. K., M. W. Brunson, and R. H. Schmidt. 1999. Public attitudes toward wildlife damage management and policy. Wildlife Society Bulletin 27:746–758.

PUBLIC ATTITUDES TOWARD A SUBURBAN
DEER HERD

DANIEL J. DECKER, *Department of Natural Resources, Cornell University, Ithaca, NY 14853*

THOMAS A. GAVIN, *Department of Natural Resources, Cornell University, Ithaca, NY 14853*

Management problems associated with expanding populations of white-tailed deer (*Odocoileus virginianus*) in suburban areas of the eastern United States are increasing (Flyger et al. 1983). According to the U.S. Bureau of the Census (1985), we can expect human populations to continue to grow by 50 million by the year 2000, a trend that portends even more suburban deer problems. In these areas of high human density the impact of a small herd of deer can be significant. The coexistence of deer and humans in residential areas results in potential conflicts, including deer damage to plantings, deer-vehicle collisions, and disease transmission from deer to humans or their domestic pets. Natural areas and "greenbelts"

(e.g., bird sanctuaries, county parks, or wooded stream corridors) often exacerbate the problem by permitting deer to penetrate suburban areas more easily. Such sites provide refuge where deer may spend most of their time, but from which they can access nearby residential properties.

A new challenge for wildlife managers in some suburban areas of the eastern United States is understanding the role of deer in transmitting Lyme disease to humans. Lyme disease, which has only been documented in North America since 1970 (Spielman et al. 1985), is caused by a spirochete that is transmitted by the deer tick (*Ixodes dammini*) (Burgdorfer et al. 1982); the adult stage of the tick uses white-

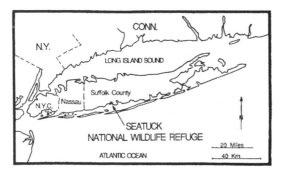

Fig. 1.   Location of study area in Islip (Long Island), New York.

tailed deer as a primary host. This arthritis-like ailment occurs on Long Island, and the deer tick is now common in coastal areas of New England and the mid-Atlantic states (Spielman et al. 1985). For example, in 1984 there were 2,053 cases of Lyme disease diagnosed for humans in the State of New York, with most occurring in Westchester and Suffolk counties (E. Bosler, N.Y. State Dep. Health, pers. commun.). These various deer-human interactions present a complicated and challenging situation for wildlife managers because conventional approaches to managing deer in rural settings are typically unworkable in suburban settings. For example, even where recreational hunting could be used safely, this approach often is not acceptable to suburban residents (Flyger et al. 1983). On the other hand, education of the public, always an important aspect of deer management, may be even more salient to management in suburban than in rural areas. But, experience has shown that education programs developed for rural applications may not be effective in suburban areas (San Julian 1983). This is not surprising in view of the findings of surveys of the American public (Kellert 1976) that indicate substantial differences between rural and urban or suburban residents' attitudes toward wildlife and people's use of wildlife. In response to these realities of the management environment, wildlife managers may have to intervene to change public attitudes, or at least en-

sure that they are based on facts about particular situations. Careful situation analyses must be conducted to ensure that the attitudes and values of the affected publics are understood (O'Donnell and Van Druff 1983). Our purpose is to provide insights for management of deer in suburban areas by describing the attitudes and concerns of residents in 1 suburban community about deer in their neighborhood.

STUDY AREA

The study was conducted in a residential area of Islip, Suffolk County, New York, adjacent to the Seatuck National Wildlife Refuge (SNWR) on Long Island. Residences in this area are primarily single-family dwellings, and include a few large homes on 0.5–2.0 ha, although most houses are on 0.1-ha lots. SNWR is an 80-ha site bounded by Champlin Creek, Great South Bay, and the Scully Audubon Sanctuary on the east, south, and west, respectively (Fig. 1). The refuge contains about 30 deer, which is a density of $375/km^2$.

Movements of the deer have been studied using radiotelemetry since March 1984. Some deer moved through the north end of the SNWR onto private property at night during fall and winter of 1984–1985, expanding the area used by about 10%. Three adult males moved east by swimming across Champlin Creek, where they remained off the refuge for several weeks. Information on deer movements also was used to delineate zones of deer-human contact of varying intensity adjacent to the SNWR and aided interpretation of responses from different areas within the neighborhood. An additional feature of this site was the insular nature of the deer herd; deer-human interactions in the neighborhood and locally derived perceptions about deer could be attributed primarily to this herd.

As in other coastal regions of the Northeast, Lyme disease is a growing concern in the study area. The etiology, symptoms, and occurrence of Lyme disease have been publicized widely. Feature articles about the disease have appeared in local (e.g., *Newsday*) and national (e.g., *The New York Times*) media. Public information meetings sponsored by state and local health departments have been held in Islip, as well.

METHODS

Names and mailing addresses of all residential property owners in the study area were obtained using property-tax records. Each of the 605 people represented an Islip household near SNWR. A self-administered, mail-back questionnaire was designed to characterize residents and their properties. Residents were queried about their experiences with deer damage and other deer-related concerns as well as their attitudes about having deer in their neighborhood. The mail survey was implemented in early spring 1985, using up to 3 follow-up mailings to nonrespondents following a modification of Dillman's (1978) Total Design Method for mail questionnaire surveys. A telephone interview was conducted with all nonrespondents who lived within the area used by refuge deer, as determined by radiotelemetry, to assess nonresponse bias. The telephone interview solicited information on the key variables of the mail questionnaire: nonrespondents' duration of residence in Islip, deer sightings, concerns about deer, damage to plantings and gardens, attitudes about deer, preferences for future deer population trends, age, sex, education, and wildlife-related recreational activities.

The study area also included residences outside the zone of deer movements, so that a future resurvey following a period of anticipated herd expansion would include residences that sustained damage for the first time since the original survey. In this paper we concentrate on the segment of the total survey population who indicated they had personal experience with deer (i.e., saw deer or deer sign on their property during the previous year, or ever saw a deer in the vicinity of their property). Data were analyzed using the SPSSX (SPSS, Inc. 1983) and SAS (SAS Inst. Inc. 1982) computer packages. Logistic regression analysis was used to model experiential and other influences on residents' attitudes about the presence of deer in their neighborhood.

A measure for classifying people based on their attitudes about wildlife (Purdy et al. 1984) was included in the questionnaire. A set of 18 attitude statements was evaluated independently by a respondent using a 5-point Likert Scale, where 1 and 2 are "strongly agree" and "agree," respectively, and 4 and 5 are "disagree" and "strongly disagree," respectively. Factor analysis was conducted on the attitude scale. Then, scores for each statement in the scale were standardized to Z-scores, with a mean of 0 and a variance of 1. The Z-score of each item was weighted by the item's factor loading coefficient for each dimension of the attitude scale derived from the factor analysis. These weighted Z-scores were then averaged for each dimension to yield a dimension score for every individual respondent. The range of these weighted Z-scores must fall between −1 and 1. Therefore, a score for an attitude dimension that is preceded by a negative sign denotes a negative or disagree position, whereas a score without the negative sign denotes a positive or agree position. Scores close to 0 should be interpreted as neutral.

RESULTS AND DISCUSSION

Survey Response

The survey of 605 households had 13 undeliverable questionnaires and 406 returned questionnaires, for an adjusted response rate of 68.5% of deliverable questionnaires. Of the respondents, 300 had personal experience with deer on or near their residence; the following analyses include only the responses of these Islip residents and the useable responses for a variable may be less than 300 due to item nonresponse. Results of the nonrespondent telephone interviews indicated that nonrespondents were similar to respondents for all key attitudinal and profiling variables; thus no nonresponse bias was indicated.

Characteristics of Residents

Islip residents were typically middle-aged (median = 47 years old) and 75% had some college education; 44% were 4-year college graduates. The ratio of males to females was 55:45. These people reported some household involvement in wildlife-related recreational activities during the year preceding the survey: 81% fed birds, 81% observed wildlife, and 38% photographed wildlife. Twenty percent of the respondents fed deer and 16% hunted in 1984–1985. Almost all of these residents (95%) maintained shrubs and other woody ornamentals on their properties; many also had flower gardens (71%), vegetable gardens (40%), and fruit trees (37%). Forty-nine percent of Islip residents either saw deer or evidence of deer on their property. One-fourth of these residents sustained deer damage to shrubs and other plantings (see Decker and Gavin [1985] for details). Most respondents (70%) had lived in their neighborhood for ≥5 years, but few (8%) had lived there >30 years (mean = 14.7 years, median = 11.0 years).

Residents' Attitudes About Deer

Islip residents had a variety of concerns about deer, particularly deer-car collisions and Lyme disease (Table 1). Nevertheless, only 9% of them considered deer a nuisance and preferred having no deer in their neighborhood, compared to 57% who enjoyed having deer in their neighborhood and considered them an esthetic re-

Table 1.   Deer-related concerns of Islip, New York, residents (*n* = 299) based on data collected in 1985.

Concern	Percent having concern[a]	Percent rating as primary concern
Deer-car collision	57	41
Lyme disease transmission	53	50
Damage to vegetable garden	12	1
Damage to yard plantings	30	4
Personal injury from deer	10	4

[a] Multiple responses were permissible; therefore this column does not sum to 100%.

source. A minority (29%) enjoyed having a few deer in their neighborhood but worried about disease or damage that might be caused by deer. Those who had experienced deer damage recently were less positive; 40% indicated they could enjoy a few deer, but worried about damage and disease, whereas 20% considered deer a nuisance and preferred having no deer in their neighborhood.

An indication of people's overall attitudes about deer in their neighborhood is their preference for trends in the deer population. Seventy-two percent of the Islip residents surveyed had sufficiently positive attitudes about the deer in their neighborhood to propose maintaining numbers at or increasing them above current levels. However, a higher proportion of those people who had recently experienced deer damage to plantings wanted the population reduced compared to those without such experience (52% vs. 18%).

Residents' preferences for future trends in the deer population differed for those with dif-

ferent primary concerns about deer. For respondents who had deer-car collisions or other personal injury as a primary concern, the majority preference was for maintaining the deer population at current levels (Table 2). However, those who had either Lyme disease or plant damage as their primary concern were more likely to favor a decrease in deer numbers.

Residents' Attitudes Toward Wildlife

Factor analysis of the 18 Likert-scale attitude statements identified 4 dimensions of wildlife attitudes (Table 3). Labels used to describe these dimensions were similar to those used in previous studies (Purdy et al. 1984, Decker 1985) that included this set of attitude statements: wildlife problem tolerance, economic uses of wildlife, noneconomic uses of wildlife (Types A and B). (See Table 3 for examples of noneconomic uses, Types A and B.)

Islip residents' concerns about deer, opinions about the presence of deer, and preferences for deer population trends were consistent with their basic attitudes about wildlife. People who reported personal hazard (deer-car collision or personal injury) as a primary concern typically reflected more positive attitudes about tolerating wildlife problems and about noneconomic uses of wildlife (Type A) than those who had Lyme disease or damage to plantings as a primary concern (Table 4). People who were concerned about Lyme disease and reported that deer use their property indicated partic-

Table 2.   Effects of the primary deer-related concern of Islip, New York, residents on the percentage of residents preferring specified trends in a suburban deer population based on data collected in 1985.

Residents' primary concern	*n*	Percent preferring deer population trend				
		Moderate increase	Slight increase	Stable	Slight decrease	Moderate decrease
Car collision and personal hazard	96	28.1	8.3	54.2	3.1	6.3
Lyme disease	107	6.5	6.5	39.3	8.4	39.3
Garden and ornamental damage	16	6.3	0.0	25.0	12.5	56.2
Overall	219	16.0	6.8	44.8	6.4	26.0

Table 3. Responses of Islip, New York, residents to statements reflecting attitudes toward wildlife and people's use of wildlife, organized by 4 dimensions identified by factor analysis. Data were collected in 1985.

Attitude dimension Statement topic	Strongly agree	Agree	Neither agree nor disagree	Disagree	Strongly disagree
Problem tolerance					
Nuisance	21.4	47.8	10.9	13.4	6.5
Damage	16.2	39.4	16.5	19.4	8.5
Disease	9.6	33.8	14.2	19.9	22.4
Personal hazard	19.7	48.2	13.0	10.9	8.1
Economic uses					
Trapping	0.7	1.1	9.9	12.0	76.3
Recreational hunting	6.1	5.7	9.0	9.3	69.9
Hunt for meat	4.6	10.2	11.7	11.3	62.2
Economic benefit	8.6	29.4	37.8	10.4	13.8
Renewable resource	28.6	37.1	18.9	6.1	9.3
Type A noneconomic uses					
Environmental quality monitor	54.7	38.2	3.9	2.1	1.1
Existence	56.8	39.9	2.9	0.4	0.0
Ecological role	52.6	42.8	3.5	0.0	1.1
Learning subject	54.9	39.6	3.6	0.7	1.0
Behavior study	36.7	46.3	14.1	1.8	1.1
Type B noneconomic uses					
Vicarious experience	33.1	41.9	21.8	1.4	1.8
Nonextractive use	35.3	41.7	17.6	3.2	2.2
Art subject	47.6	39.6	11.3	1.1	0.4
Social action	15.5	37.7	39.2	5.4	2.2

ularly low tolerance of problems caused by wildlife.

People who believed deer had "esthetic value" were more positive in their attitudes about problem-tolerance and noneconomic uses of wildlife (Type A) than other respondents (Table 4). Those who considered deer to be a "nuisance" had extremely negative attitudes about problem tolerance and noneconomic uses of wildlife (Type A). As respondents preferences for more deer decreased, so did the standardized scores reflecting their attitudes for wildlife problem tolerance and noneconomic uses of wildlife (Type A). Conversely, those people who wanted an increase in the deer population expressed high tolerance of wildlife problems and positive attitudes about noneconomic uses of wildlife (Type A). Thus, attitudes about problem tolerance and noneconomic uses of wildlife (Type A) were the dimensions that

best distinguished among respondents' attitudes about deer and their preferences for population trends.

What Might Effective Programming Accomplish?

A prerequisite to developing a management or educational program is understanding how residents' attitudes about wildlife and concerns about deer influence their preferences for deer population trends. Logistic regression was used to estimate the changes in trend preferences that could be expected to occur if attitudes were altered and concerns alleviated by education or by management actions.

The best logistic regression model of the probability of preferring the deer population to decrease (dependent variable) included 2 independent variables (both significant at $P \leq 0.02$). This model was:

Table 4.   Standardized wildlife-attitude-dimension scores[a] for Islip, New York, residents with various concerns, attitudes, and opinions based on data collected in 1985.

Category Topic	Problem tolerance	Economic uses	Type A[b]	Type B[c]
			Noneconomic uses	
Primary concern				
Car collision and personal hazard	0.16	0.03	0.11	−0.02
Lyme disease	−0.24	−0.04	−0.15	−0.07
Garden and ornamental damage	−0.26	0.01	<−0.01	0.03
Attitude about deer				
Esthetic	0.28	<0.01	0.20	0.09
Enjoy-worry	−0.33	0.02	−0.09	−0.07
Nuisance	−0.67	0.01	−0.77	−0.14
Don't care	−0.28	0.06	−0.43	−0.29
Deer population trend preference				
Moderate increase	0.50	0.02	0.32	0.19
Slight increase	0.09	−0.13	0.31	<0.01
No change	0.07	0.04	0.03	−0.03
Slight decrease	−0.16	0.14	−0.01	0.09
Moderate decrease	−0.54	−0.02	−0.42	−0.10

[a] A score that is preceded by a negative sign denotes a negative or disagree position, whereas a score without the negative sign denotes a positive or agree position. The range of scores is between −1 and 1. Scores close to 0 should be interpreted as neutral.
[b] See Table 3 for examples of noneconomic uses Type A.
[c] See Table 3 for examples of noneconomic uses Type B.

$$\log \frac{(P_i)}{(1 - P_i)} = -5.26 + 2.59\text{OPIN} + 0.92\text{PROB} + u,$$

where P_i = probability that a resident will prefer the deer population to decrease,

OPIN = residents' opinions of the esthetic value of deer,

PROB = residents' score on the problem-tolerance dimension of the attitude scale,

u = error term.

The R statistic for the model was 0.65. The model correctly predicted the response of 84.8% of respondents, including 69.5% of those who preferred a decrease. The best predictive statistic (Harrell 1983), the fraction of concordant pairs of predicted probabilities and responses, was 0.89. That is, of all possible pairs of respondents that included one who wanted a decrease and one who did not, the model predicted correctly for 89% of the pairs a higher probability of wanting a decrease for the respondent who actually gave this response.

Imagine that an education program to alleviate residents' concerns about Lyme disease, deer damage, and deer-car collisions could be implemented, and that it achieved moderate success. As a result, let us suppose the proportion of Islip residents who did not enjoy deer unconditionally dropped from the 38% who were actually in this category to 20% (i.e., those who could enjoy a few deer but worried about damage and disease would be 15% and those who considered deer a nuisance would be 5%). From the model we could expect this change in opinions (OPIN) about the presence of deer to result in a drop from 28 to 12 in the percentage of people who wanted a decrease in the deer herd. Within the limitations of the model, the scenario described above helps put realistic expectations on the extent of change that might result from an information program. The cost of conducting such an educational campaign could then be weighed against any reduction in the cost of management due to a change in public attitudes.

IMPLICATIONS AND CONCLUSIONS

We believe that wildlife management has entered an era where our programs are addressing the interests of a broader, more vocal public, which is increasingly likely to use the political process effectively to serve its interests. For example, in 1984 a small group of Islip residents enlisted the support of their U.S. Congressman to "request" the U.S. Fish and Wildlife Service to build a deer-proof fence along the north boundary of SNWR to prevent deer from moving into the adjacent neighborhood. The fence was completed in spring 1985. It may be indicative of the future that so few deer (30–35) can generate so much concern and stimulate as much effort in research and management.

Traditional approaches in wildlife management, and the assumptions upon which they are based, may not be appropriate for deer management in suburban areas. This study has shown that residents of 1 suburban area generally enjoyed having deer in their neighborhood, not unlike the perspective found repeatedly in studies of rural landowners in New York (Brown et al. 1978, Brown and Decker 1979, Decker et al. 1983). Like their rural counterparts, Islip suburbanites recognized that they paid a price for tolerating deer: damage to landscape plantings and vegetable gardens, hazard of deer-vehicle collisions, and potential transmission of disease. Additionally, most residents of Islip near SNWR did not want more deer, but neither did they want fewer. Thus, this public was not extreme in its view of deer, even though Lyme disease had received considerable public attention. However, Islip residents were unlikely to support a traditional approach to population control, given their negative view of hunting and lack of personal participation in hunting. This does not mean they would oppose a program of population control, but that managers would be ill-advised to apply a traditional recreational hunt in the Islip situation without an effective educational effort preceding such an activity.

A major responsibility and challenge for managers is ensuring that management action is consistent with public attitudes and site-specific situations. Here again, an effective educational program is required. But managers are faced with another, possibly more difficult problem—deciding to what extent the attitudes and desires of the public should influence the management approach. Should affected publics simply "vote" through their responses to questionnaires and have management agencies formulate policies accordingly? Should public preferences be ignored in deference to professional judgment? Or, should managers attempt to change public attitudes that seem to reflect little knowledge of the facts of the management situation?

After answering these questions and obtaining information about public attitudes, the wildlife manager has the monumental task of integrating these data to arrive at a decision. Currently, we are in need of an easily understood process, which is defensible and comprehensive, to incorporate disparate public opinions into wildlife management programming decisions. Such a process should integrate sociological, political, economic, and biological elements (Krueger et al. In press). The need for such a process will grow as we face new situations where experience alone will not suffice as a basis for the development of wildlife management programs.

SUMMARY

An important aspect of managing deer in suburban areas is understanding residents' attitudes about deer. In suburban Islip, New York, residents who had contact with the deer herd associated with the Seatuck National Wildlife Refuge generally expressed positive attitudes about deer in their neighborhood; the majority

180   *Wildl. Soc. Bull.* 15(2) 1987

enjoyed deer and considered deer an esthetic resource. Residents' enthusiasm for deer was tempered by negative experiences with deer, such as ornamental and garden damage, and by concerns about Lyme disease. Overall, nearly three-fourths of Islip residents surveyed wanted deer numbers in their neighborhood to be increased or maintained at current levels. However, those residents who were concerned with Lyme disease and plant damage were more likely to want the number of deer to decrease than increase, suggesting that if these concerns became more widespread the prevalent future deer population trend preference may be a decrease. This could present managers with a dilemma because Islip residents generally held a negative view of hunting. Thus, either a nontraditional approach to deer population management or an effective educational effort preceding a hunt would be required. This study has provided evidence that the major task for managers in suburban areas is formulating management actions that are consistent with public attitudes regarding specific situations.

Acknowledgments.—The authors acknowledge the assistance of T. S. Litwin and M. C. Capkanis of the Seatuck Res. Prog., and R. Spaulding of the U.S. Fish and Wildl. Serv. for logistical and financial support. T. L. Brown, E. E. Langenau, D. L. Leedy, and 2 anonymous reviewers provided helpful comments on previous drafts of this paper. N. Connelly processed data and E. Bowmaster and C. Westbrook typed drafts. This study was funded by the Cornell Agric. Exp. Stn. (Hatch Proj. 147442) and the Seatuck Res. Prog., Cornell Lab. Ornithol.

LITERATURE CITED

BROWN, T. L., AND D. J. DECKER. 1979. Incorporating farmers' attitudes into management of white-tailed deer in New York. J. Wildl. Manage. 43:236–239.

———, ———, AND C. P. DAWSON. 1978. Willingness of New York farmers to incur white-tailed deer damage. Wildl. Soc. Bull. 6:235–239.

BURGDORFER, W., A. G. BARBOUR, S. F. HAYES, J. L. BENACH, E. GRUNWALDT, AND J. P. DAVIS. 1982. Lyme disease—a tick-borne spirochetosis? Science 216:1317–1319.

DECKER, D. J. 1985. Agency image: a key to successful natural resource management. Trans. Northeast Sect. Wildl. Soc. 41:43–56.

———, AND T.A. GAVIN. 1985. Public tolerance of a suburban deer herd: implications for control. Proc. East. Wildl. Damage Control Conf. 2:192–204.

———, N. SANYAL, T. L. BROWN, R. A. SMOLKA, JR., AND N. A. CONNELLY. 1983. Reanalysis of farmer willingness to tolerate deer damage in western New York. Proc. East. Wildl. Damage Control Conf. 1:37–45.

DILLMAN, D. A. 1978. Mail and telephone surveys. John Wiley & Sons, Inc., New York, N.Y. 325pp.

FLYGER, V., D. L. LEEDY, AND T. M. FRANKLIN. 1983. Wildlife damage control in eastern cities and suburbs. Proc. East. Wildl. Damage Control Conf. 1:27–32.

HARRELL, F. E., JR. 1983. The logist procedure. Pages 181–202 *in* SUGI supplemental library user's guide. SAS Inst. Inc., Cary, N.C.

KELLERT, S. R. 1976. Perceptions of animals in American society. Trans. North Am. Wildl. and Nat. Resour. Conf. 41:533–545.

KRUEGER, C. C., D. J. DECKER, AND T. A. GAVIN. In press. A concept of natural resource management: an application to unicorns. Trans. Northeast. Fish and Wildl. Conf. 42.

O'DONNELL, M. A., AND L. W. VAN DRUFF. 1983. Wildlife conflicts in an urban area: occurrence of problems and human attitudes toward wildlife. Proc. East. Wildl. Damage Control Conf. 1:315–323.

PURDY, K. G., D. J. DECKER, AND T. L. BROWN. 1984. Standardizing basic wildlife attitudes and values: data acquisition methods. Outdoor Recre. Res. Unit Publ. 84-3, Cornell Univ., Ithaca. 30pp.

SAN JULIAN, G. J. 1983. The need for animal damage control. Proc. East. Wildl. Damage Control Conf. 1:313–314.

SAS INSTITUTE INC. 1982. SAS user's guide: statistics, 1982 edition. SAS Inst. Inc., Cary, N.C. 584pp.

SPIELMAN, A., M. L. WILSON, J. F. LEVINE, AND J. PIESMAN. 1985. Ecology of *Ixodes dammini*-borne human babesiosis and Lyme disease. Annu. Rev. Entomol. 30:439–460.

SPSS, INC. 1983. SPSSX: user's guide. McGraw-Hill Book Co., New York, N.Y. 806pp.

U.S. BUREAU OF THE CENSUS. 1985. Population profile of the United States: 1983–84. Current Popul. Rep. Ser. P-23, No. 145. U.S. Gov. Print. Off., Washington, D.C. 47pp.

Received 21 January 1986.
Accepted 5 August 1986.

Author Index